UNIVERSITY OF WASHINGTON PUBLICATIONS IN BIOLOGY, VOLUME 17

VASCULAR PLANTS OF THE PACIFIC NORTHWEST

By C. Leo Hitchcock, Arthur Cronquist, Marion Ownbey,
and J. W. Thompson

PART 1: *Vascular Cryptogams, Gymnosperms, and Monocotyledons*
By C. Leo Hitchcock, Arthur Cronquist, and Marion Ownbey

PART 2: *Salicaceae to Saxifragaceae*
By C. Leo Hitchcock and Arthur Cronquist

PART 3: *Saxifragaceae to Ericaceae*
By C. Leo Hitchcock and Arthur Cronquist

PART 4: *Ericaceae through Campanulaceae*
By C. Leo Hitchcock, Arthur Cronquist, and Marion Ownbey

PART 5: *Compositae*
By Arthur Cronquist

# VASCULAR PLANTS
# OF THE PACIFIC NORTHWEST

### By
### C. Leo Hitchcock
### Arthur Cronquist
### Marion Ownbey
### J. W. Thompson

***PART 4: ERICACEAE THROUGH CAMPANULACEAE***

**By C. Leo Hitchcock, Arthur Cronquist, and Marion Ownbey**

**Illustrated by Jeanne R. Janish**

*University of Washington Press*

SEATTLE AND LONDON

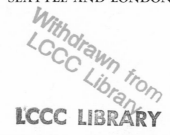

*Vascular Plants of the Pacific Northwest*

PART 4: ERICACEAE THROUGH CAMPANULACEAE

## Introduction*

The primary responsibility for the preparation of Part 4 is divided as follows: Ericaceae through Cuscutaceae, C. Leo Hitchcock, University of Washington; Polemoniaceae through Campanulaceae (except Castilleja), Arthur Cronquist, New York Botanical Garden; Castilleja, Marion Ownbey, State College of Washington. Manuscript prepared by each author has been made available to the other authors, and the first two, in particular, have collaborated in reviewing each other's manuscript. The treatment of the Campanulaceae has been reviewed by Rogers McVaugh, and that of Mimulus by Robert K. Vickery. Lincoln Constance was consulted on a number of problems in the Hydrophyllaceae, and Verne Grant on problems in the Polemoniaceae. The treatment of Penstemon was prepared with the benefit of daily consultation with David D. Keck. Discussion of the horticultural merits of various taxa was written in collaboration with B. O. Mulligan, University of Washington Arboretum, and Carl S. English, Jr., Seattle horticulturist. The final responsibility for all taxonomic decisions rests with the individual author of each part of the volume; in some cases these decisions are contrary to the advice received.

The standard Englerian sequence of families is followed, because we do not think it advisable to introduce a new and unfamiliar system in a regional flora the several volumes of which do not all appear at the same time. A more nearly natural arrangement of the families and orders of dicotyledons in our area will be presented at the beginning of Part 2.

## ERICACEAE. Heath Family

Flowers white to red or reddish-purple, usually regular but sometimes slightly irregular, perianth mostly 4- or 5-merous; sepals distinct or more usually connate, at least at the base, sometimes fleshy; corolla gamopetalous to polypetalous, rotate to funnelform or urn-shaped, rarely lacking; stamens free from and equal to, to twice as many as, the corolla lobes, anthers often inverted, frequently with tubes prolonged above the anther sacs, and sometimes with awnlike protuberances from the back or the tip, 2-celled, opening by terminal or lateral pores or chinks, or by longitudinal slits; ovary superior to inferior, 4- to 12-carpellary, usually 4- to 12-locular, infrequently 1-celled, very rarely 2- to 3-celled, placentation usually axile, rarely parietal; style 1, straight or curved to one side; stigma capitate to distinctly lobed; fruit a capsule, fleshy and truly a berry, drupelike, or a capsule surrounded by the fleshy calyx; perennial, often glandular or hairy trees, shrubs, or herbs with usually alternate, leathery or coriaceous, evergreen or occasionally deciduous leaves, less commonly, the plants fleshy, nonchlorophyllous, white to brownish- or reddish-purple saprophytes with the leaves reduced to bracts.

Perhaps 70 genera and 2000 species of all continents, usually inhabiting acid, rather moist soil; especially common around bogs. The family contains many unusual species, including the saprophytes, and numerous very valuable ornamental flowering shrubs and groundcovers, Rhododendron with about 800 species probably being the most familiar. Often the Ericaceae as here considered is divided into two or more families, including the Vacciniaceae, Pyrolaceae, and Monotropaceae.

1 Plants without green leaves
  2 Styles well developed, exserted from the flowers; flowering stems usually less than 2 dm.
     tall with 1-3 bracts below the flowers               PYROLA
  2 Styles short, usually included in the flowers; stems freely bracteate to the inflorescence, often well over 2 dm. tall

   * This investigation was supported in part by funds provided for biological and medical research by the State of Washington Initiative Measure No. 171, in part by a grant from the Penrose fund of the American Philosophical Society, and also in part by a grant from the National Science Foundation.

3 Corolla lacking; stems longitudinally pink- and white-striped; anthers unawned, opening by basal (falsely terminal) pores                                        ALLOTROPA
3 Corolla present, but calyx sometimes lacking; stems not striped; anthers opening nearly full length by slits
    4 Corolla urn-shaped, gamopetalous; anthers awned on the back        PTEROSPORA
    4 Corolla not urn-shaped, usually polypetalous; anthers unawned
      5 Corolla gamopetalous, fimbriate, hairy within; placentation parietal; anthers ob-
        long, about 2 mm. long, dehiscent full length but not over the tip    HEMITOMES
      5 Corolla polypetalous
        6 Flowers single, waxy-white, drying to black            MONOTROPA
        6 Flowers more than 1, not waxy-white
          7 Anthers 2-3 mm. long, linear, not much broader than the glabrous or min-
            utely puberulent, flattened filaments; corolla not densely hairy within; cap-
            sule nearly globose, glabrous or nearly so; placentation parietal
                                          PLEURICOSPORA
          7 Anthers about 1 mm. long, oval, as broad as long, dehiscent by a continuous
            slit over the tip; filaments hairy; corolla usually densely hairy inside; cap-
            sules oblong-ovoid, hairy
            8 Placentation axile; plants drying to brownish       HYPOPITYS
            8 Placentation parietal; plants drying to blackish    PITYOPUS
1 Plants with well-developed, green leaves
  9 Ovary inferior or apparently inferior because of the fleshiness of the persistent calyx;
    fruit a true berry or a capsule surrounded by the thickened, fleshy calyx which forms a
    pseudo berry
    10 Ovary truly inferior; mature fruit never white; erect shrubs, mostly deciduous
                                        VACCINIUM
    10 Ovary superior but surrounded by the fleshy calyx when ripe and apparently inferior,
       if the ovary at all inferior, the mature fruit white; mostly prostrate to semierect
       (G. shallon sometimes erect) evergreen shrubs or shrublets    GAULTHERIA
  9 Ovary superior, free of the calyx
    11 Fruit fleshy; usually evergreen shrubs or trees with rather large, leathery leaves
      and often with deep reddish to purplish bark; flowers urn-shaped
      12 Plants arborescent; fruit warty, many-seeded; leaves averaging over 6 cm. in
        length                                                          ARBUTUS
      12 Plants shrubby; fruit smooth to viscid but not warty, with 1 seed per carpel;
        leaves averaging less than 6 cm. in length
        13 Leaves evergreen, entire                      ARCTOSTAPHYLOS
        13 Leaves deciduous, sharply serrulate        Arctous alpina (L.) Niedzu.
        (The purplish-berried var. alpina and the red-fruited var. ruber Rehd.
        and Wilson may be found in the region of Glacier Nat. Pk., Mont., but they
        are unknown at present s. of the Canadian Rockies. Both are sometimes in-
        cluded in Arctostaphylos)
    11 Fruit a dry capsule; herbs to large shrubs; corolla often not urn-shaped; leaves often
      scalelike
      14 Leaves scalelike, less than 6 mm. long, usually 4-ranked; anthers awned
                                          CASSIOPE
      14 Leaves not scalelike, usually well over 6 mm. long if at all 4-ranked
        15 Corolla polypetalous or nearly so, flowers usually open and not at all tubular
          16 Plants more herbaceous than woody, seldom over 3 dm. tall; leaves large
            and coriaceous, evergreen
            17 Leaves cauline, scattered or whorled, the plants scapose; styles very
              short, usually not noticeable; filaments enlarged and hairy about mid-
              length; flowers corymbose                   CHIMAPHILA
            17 Leaves mostly basal, the plants scapose; styles elongate, visible in

flower, often strongly curved; filaments not enlarged and hairy; flowers racemose or single     PYROLA
16  Plants woody shrubs, usually over 3 dm. tall
    18  Flowers axillary; petals 8-12 mm. long; leaves not glandular  CLADOTHAMNUS
    18  Flowers in terminal corymbs; petals less than 8 mm. long, leaves often glandular
          LEDUM
15  Corolla gamopetalous; flowers often urceolate
  19  Corolla rotate (saucerlike) with 10 pouches in which the anthers fit in the bud; leaves often strongly revolute     KALMIA
  19  Corolla either not rotate or not saccate
    20  Leaves either deciduous or else large and leathery and over 6 cm. long; plants usually 1-5 m. tall
      21  Corolla urnlike, less than 1 cm. in length and width    MENZIESIA
      21  Corolla not urnlike, well over 1 cm. in length or width  RHODODENDRON
    20  Leaves evergreen, less than 3 cm. long; plants less than 1 m. tall
      22  Anthers unawned; capsule septicidal; leaves less than 1.5 cm. long, 1-2 mm. broad     PHYLLODOCE
      22  Anthers awned; capsule loculicidal; leaves (ours) mostly over 1.5 cm. long, 3-5 mm. broad     ANDROMEDA

### Allotropa T. and G. Candystick; Sugarstick

Flowers in terminal, elongate, spikelike racemes, axillary and exceeded by the subtending bract, often with 1-2 bracteoles below the calyx; sepals 5, distinct; corolla lacking; stamens 10, unappendaged, opening by basal (falsely terminal) pores; pistil 5-carpellary, styles very short, stigma shallowly 5-lobed, ovary superior, 5-celled, with axile placentation; fruit a capsule; fleshy, simple-stemmed, saprophytic herb with lanceolate, pinkish to yellow-brown scalelike leaves.

One species, on the Pacific Coast. (Name from the Greek allos, other, and tropos, turn, direction, way, referring to the inflorescence, to distinguish it from Monotropa.)

Reference:

Copeland, H. F. The structure of Allotropa. Madroño 4:137-68. 1938.

Allotropa virgata T. & G. ex Gray, Proc. Am. Acad. 7:368. 1867. (Wilkes Expedition, Cascade Mts., Wash.)

Plant 1-4 dm. tall, stems white- and pink-striped, 5-10 mm. thick; leaves linear-lanceolate; racemes spikelike, 5-20 cm. long; sepals white or pinkish to brownish, about 5 mm. long; stamens purplish, from about equal to, to twice as long as, the sepals.

At lower elevations of coniferous forests, in deep humus; s. Sierra Nevada and the coastal ranges of Calif., n. to B.C. from the e. slope of the Cascade Range to near the coast. May-Aug.

### Andromeda L.

Flowers in small, terminal umbels; perianth 5-merous; calyx deeply cleft, widely flared; corolla gamopetalous, globose-urceolate, the 5 lobes short, spreading to reflexed; stamens 10, included, filaments flattened, anthers 2-locular, slenderly 2-awned, opening by terminal pores; ovary superior, 5-carpellary, subglobose, stigma capitate; fruit a loculicidal, 5-celled, 5-valved, subglobose capsule; small, glabrous, evergreen shrubs, with narrow, coriaceous, alternate, entire, usually revolute leaves.

Two (or perhaps only 1) boreal to arctic species of the Northern Hemisphere. (Named for the Andromeda of Greek mythology.)

Andromeda polifolia L. Sp. Pl. 1:393. 1753. ("Habitat in Europae frigidiaris paludibus tur-
   fosis")
   Low spreading shrub 1-8 dm. tall; leaves narrowly elliptic to nearly linear in outline, mostly 1.5-4 cm. long, 2-6 mm. broad, apiculate, with entire, revolute margins and whitish-
glaucous lower surface; pedicels 5-12 mm. long, usually somewhat recurved; sepals about 1
mm. long; corolla 5-8 mm. long, pinkish; filaments hairy; capsule glabrous, 4-6 mm. thick.
N=24. Bog rosemary.
   Acid bogs; circumpolar, in N. Am. from Alas. to Labrador s. to B.C. and Alta.; reported
from Wash. and Ida. and to be expected there, but confirming specimens not seen. June-July.
   In e. Can. and n.e. U.S., A. polifolia is replaced by A. glaucophylla Link. (A. polifolia
ssp. glaucophylla [Link.] Hultén) which differs chiefly in having larger leaves that are tomentose beneath. The species is highly regarded as an ornamental, and there are several strains
of the plant grown under such names as var. angustifolia, var. minima, and var. nana.


### Arbutus L. Madrona

   Flowers in large, terminal, compound racemes, perianth 5-merous; corolla gamopetalous,
urn-shaped; stamens 10, included in and borne on the corolla, filaments enlarged and pilose
near the base, anthers with 2 short, reflexed, curved awns, opening by 2 slitlike pores; ovary
superior, 5(4)-celled; fruit a finely granular-tuberculate, many-seeded berry; ours, trees with
large, smooth, shiny, leathery, evergreen, alternate leaves.
   Perhaps 15 species of N. Am. and Eurasia. (Latin name for one of the members, the strawberry tree, Arbutus unedo.)

Arbutus menziesii Pursh, Fl. Am. Sept. 282. 1814. (Menzies, "on the north-west coast of
   America")
   A. procera Dougl. ex Lindl. Bot. Reg. 21: pl. 1753. 1836. (Horticultural specimen from
      seeds collected by Douglas in "mountainous woody parts of N. W. Coast of North America")
   Plant arborescent, or sometimes with many stems from the base and more shrublike, 6-30
m. tall; bark smooth, the older portions dark brownish-red, exfoliating, young bark char-
treuse, aging to deep red; leaves ovate-oblong to elliptic, glabrous, serrate to entire, mostly
7-15 cm. long; flowers white, 6-7 mm. long; berry about 1 cm. in diameter, nearly globose,
orange to red.
   West of the Cascades, chiefly in drier areas, B.C. to Baja Calif. Apr.-May.
   One of our most beautiful native plants and a highly prized ornamental.


### Arctostaphylos Adans. Manzanita

   Flowers mostly in terminal panicles but sometimes reduced to a single raceme, perfect,
regular, (4)5-merous, white or pink, each flower bracteate; sepals nearly distinct; corolla
urn-shaped, gamopetalous, the short lobes usually spreading; stamens 10(8), included, filaments expanded and hairy near the base, anthers opening by falsely terminal pores, each with
2 curved, reflexed, hornlike appendages; ovary superior, 5(4-10)-celled; fruit fleshy, berrylike, containing 5(4-10) stony 1-seeded nutlets; evergreen, prostrate and creeping to erect and
much-branched shrubs; stems usually very crooked, with reddish-purple or brownish, smooth
bark; leaves alternate, leathery, entire.
   About 45 species, nearly all confined to the w. coast of N. Am., one circumpolar. All of
our species are desirable evergreen shrubs, especially A. uva-ursi (ground cover) and the
larger plants, A. patula and A. columbiana. (Name from the Greek arktos, bear, and sta-
phyle, bunch of grapes, the fruits often fed upon by bears.)
   References:
   Adams, J. E. A systematic study of the genus Arctostaphylos. Journ. Elisha Mich. Soc.
      56:1-62. 1940.
   Eastwood, A. A revision of the genera formerly included in Arctostaphylos. Leafl. West.
      Bot. 1:97-100. 1934.
   Jepson, W. L. Fl. Calif. 3:29-51. 1939.

1  Plants prostrate, often rooting along the stems, usually not over 2 dm. tall; leaves usually
    obovate, averaging not over 2.5 cm. in length
    2  Leaves rounded to retuse; berry bright red; plants mainly of lower elevations
                                                                                      A. UVA-URSI
    2  Leaves abruptly mucronate to apiculate; berry brownish-red; plants montane
                                                                                      A. NEVADENSIS
1  Plants erect to spreading shrubs mostly well over 2 dm. tall; leaves mostly ovate to ellip-
    tic, often averaging well over 2.5 cm. in length
        3  Young stems grayish with long, stiff hairs mixed with finer tomentum or pubescence;
            leaves usually grayish-puberulent                                        A. COLUMBIANA
        3  Young stems glabrous to moderately puberulent, hairs uniform; leaves glabrous or
            lightly puberulent
            4  Leaves glossy green; berry red; plants less than 1 m. tall
                                                          see A. media under A. UVA-URSI
            4  Leaves yellow-green; berry brownish; plants usually over 1 m. tall  A. PATULA

Arctostaphylos columbiana Piper in Piper & Beattie, Fl. N.W. Coast 279. 1915. (Piper 898,
    Union City, Mason Co., Wash.)
    A. tomentosa of several authors for our area but not A. tomentosa (Pursh) Lindl. Bot. Reg.
    21: pl. 1791. 1835. (Menzies, "Northwest Coast," but probably from Monterey, Calif.)
    A. virgata Eastw. ex Sarg. Trees & Shrubs 1:203. 1905. A. glandulosa var. virgata Jeps.
    Madroño 1:87. 1922. A. columbiana var. virgata McMinn in Jeps. Fl. Calif. 3:49. 1939.
    (Eastwood, Mt. Tamalpais, Calif.)
    A. setosissima Eastw. Leafl. West. Bot. 1:78. 1933. (Eastwood, Mendocino City, Calif.)
    A. tracyi Eastw. Leafl. West. Bot. 1:79. 1933. A. columbiana var. tracyi Adams ex McMinn, Ill.
    Man. Calif. Shrubs 408. 1939. (Tracy 6141, Big Lagoon, Humboldt Co., Calif.)
    Erect or spreading shrub 1-3(4) m. tall, usually simple at the base; old branches with pur-
plish-red bark, young twigs and petioles grayish-puberulent to tomentose and usually distinctly
bristly with longer, scattered, glandular or eglandular hairs; leaves ovate or lanceolate to elliptic, blades 2-5(6) cm. long, 1-2.5 cm. broad, usually acute or apiculate, finely grayish-
puberulent, especially beneath, petioles about 5 mm. long, inflorescence usually paniculate
(racemose), densely pubescent, bracts lanceolate, 5-15 mm. long, usually exceeding pedicels;
flowers white or pinkish, 6-7 mm. long; ovary copiously pubescent, eglandular; berry depressed, 6-8 mm. broad; nutlets coarsely reticulate-pitted on the back.
    Along the coast on the w. slopes of the Cascades, Mendocino Co., Calif., to s. B.C. May-
July.

Arctostaphylos nevadensis Gray, Syn. Fl. 2[1]:27. 1878.
    Uva-ursi nevadensis Abrams, N. Am. Fl. 29:94. 1914. ("Sierra Nevada, California, com-
    mon at 8-10,000 ft.")
    A. parvifolia Howell, Fl. N.W. Am. 416. 1903. (Howell, Josephine Co., Oreg.)
    Very similar to A. uva-ursi, differing essentially as follows: leaves acute or abruptly mu-
cronate rather than rounded; berry more brownish-red than scarlet; the plants sometimes less
prostrate, more typically montane in distribution.
    Common in the Cascades from Wash. s. to the Sierra Nevada, e. to the Blue Mts., Oreg.
June-July.
    There are many specimens intermediate in nature between this taxon and A. uva-ursi, which
indicates that the two are interfertile, and that they might correctly be treated as varieties of
one species.

Arctosphylos patula Greene, Pitt. 2:171. 1891.
    Uva-ursi patula Abrams, Bull. N.Y. Bot. Gard. 6:433. 1910. ("Sierra Nevada, California,
    from Calaveras Co. southward to Fresno")
    A. pungens var. platyphylla Gray, Syn. Fl. 2[1]:28. 1878. A. platyphylla Kuntze, Rev. Gen.
    2:385. 1891. (Calif., no specimens cited)

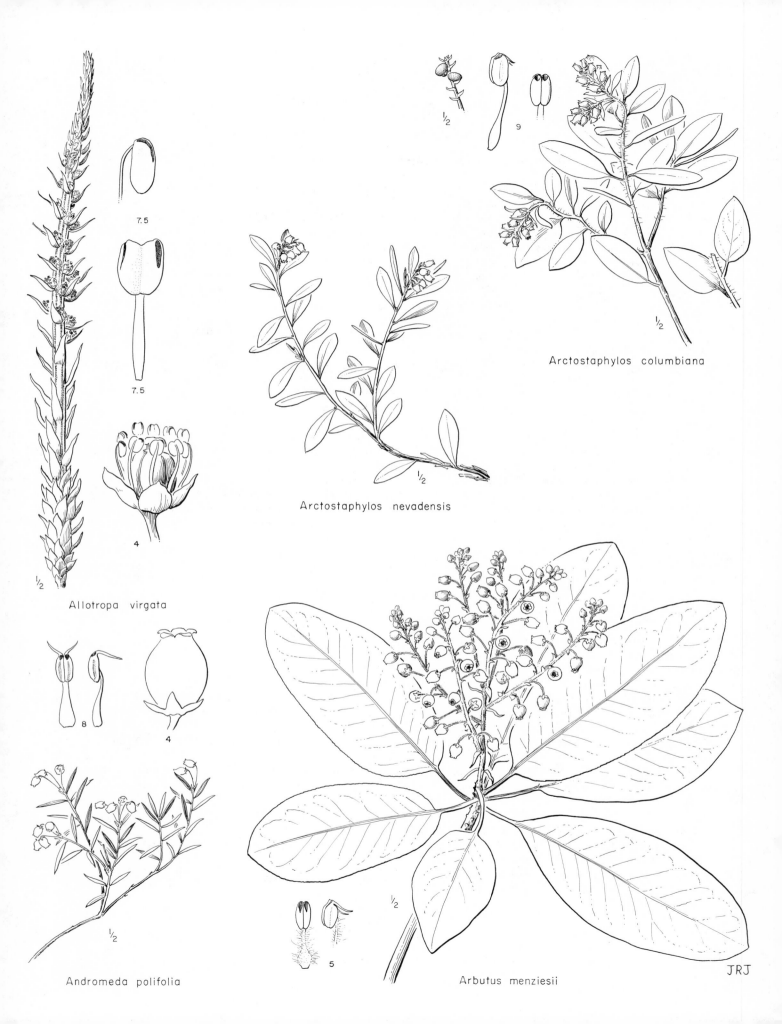

7.5

7.5

4

½

Allotropa virgata

8

4

½

Andromeda polifolia

5

Arctostaphylos nevadensis

½

Arctostaphylos columbiana

½

½

Arbutus menziesii

9

½

JRJ

A. obtusifolia Piper, Bull. Torrey Club 29:642. 1902. (Cusick 2688a, Black Butte, Oreg.,
    July 30, 1901)
A. patula var. incarnata Jeps. Madroño 1:80. 1922. (Harriet P. Kelley, Dunsmuir, Calif.)
    A spreading shrub 1-2 m. tall, glabrous, or the young growth and inflorescence minutely
puberulent, sometimes somewhat glandular, old bark reddish-brown; leaves bright yellow-
green, smooth, ovate-lanceolate or oblong-elliptic to more or less rotund, blades 2-5 cm.
long, usually over half as broad, rounded to acute but nearly always somewhat apiculate; panicles rather large, bracts triangular, 1-2.5 mm. long, glabrous or puberulent; pedicels glabrous; flowers pink, about 6 mm. long; berry 7-10 mm. broad, glabrous, brownish.
    S. Calif. and n. in the coastal ranges and Sierra Nevada to the Cascades of s. Oreg., e. to
Colo. Reported, but not seen from Wash. May-June.

Arctostaphylos uva-ursi (L.) Spreng. Syst. 2:287. 1825.
    Arbutus uva-ursi L. Sp. Pl. 395. 1753. Uva-ursi procumbens Moench. Meth. 470. 1794.
        Mairania uva-ursi Desv. Journ. Bot. II. 1:37, 292. 1813. Uva-ursi buxifolia S. F. Gray,
        Nat. Arr. Brit. Pl. 2:400. 1821. Arctostaphylos officinalis Wimm. & Grab. Siles. 1:391.
        1827. Arctostaphylos procumbens Patze in Meyer & Elkan, Fl. Preuss. 188. 1850. Uva-
        ursi uva-ursi Britt. in Britt. & Brown, Ill. Fl. 2:693. 1913. (Europe)
    Arctostaphylos media Greene, Pitt. 2:171. 1891. (Piper, Mason Co., Wash.)
    A. uva-ursi var. coactilis Fern. & Macbr. Rhodora 16:212. 1914. (E. B. Chamberlain, May
        18, 1899, Brunswick, Me.).
    A. uva-ursi var. adenotricha Fern. & Macbr. Rhodora 16:213. 1914. (C. F. Batchelder,
        May 11, 1888, Golden, B. C.).
    Prostrate shrub with trailing and rooting stems sometimes forming mats several meters
broad, the tips often ascending and 5-15 cm. tall, bark reddish to brownish, tardily exfoliating; stems puberulent, sometimes glandular; leaves oblong to obovate or spatulate, rounded to
obtuse, blades mostly 1.5-3 cm. long, glabrous to puberulent, especially on the margins and
midrib, base acute or narrowed abruptly to petioles 2-5 mm. long; flowers pink, about 5 mm.
long, borne in terminal, few-flowered racemes in the axils of small bracts, usually about
equaling the pedicels; calyx lobes rounded, scarcely 1/4 the length of the urceolate, 5-lobed corolla; anthers pilose near the broadened base; berry bright red, 7-10 mm. broad. N=26. Kin-
nikinnick, bearberry, sandberry.
    Coastal Calif. n. and more widespread from Oreg. to Alas., to N. M., Ill., the Middle At-
lantic states, and Labrador; Eurasia. Apr.-June.
    Very desirable as a ground cover ornamental. In w. Wash. this plant sometimes hybridizes
with A. tomentosa to produce low-growing, more erect shrubs which have been called A. media Greene.

### Cassiope D. Don. White Heather; Mountain Heather

    Flowers 1-several, axillary or terminal; perianth 5(4)-merous; sepals nearly distinct, imbri-
cate, persistent; corolla white or pinkish, campanulate, connate 1/2-2/3 the length but free at
the base, the (usually) 5 lobes much shorter than the tube; stamens 10(8), included, filaments
glabrous, anthers nearly globose, recurved-awned from the back, opening by falsely terminal
pores; ovary superior, fruit a near-globose, 5-celled, many-seeded, loculicidal capsule; low
shrubs often with creeping stems; leaves 4-ranked, overlapping, scalelike to needlelike,
rounded or grooved on the back, sessile or very short-pedicellate.
    Perhaps 12 species of the N. Temp. and Arctic zones; all desirable rock garden plants that
are grown fairly easily but usually too sulky to flower. (Name from Greek mythology.)
    Reference:
    Good, R. D'O. The genera Phyllodoce and Cassiope. Lond. Journ. Bot. 64:1-10. 1926.
1 Leaves alternate, spreading, short-petiolate; flowers usually terminal, solitary, and erect
    (Harrimanella)                                                    C. STELLERIANA
1 Leaves opposite, appressed, sessile; flowers usually several, lateral or subterminal on the
    branches, and pendent

2 Leaves prominently grooved nearly full length on the lower surfaces, pubescent near the
    base at least                                            C. TETRAGONA
2 Leaves not prominently grooved on the lower surfaces, glabrous, or at most ciliolate
                                                      C. MERTENSIANA

**Cassiope mertensiana** (Bong.) G. Don, Gen. Hist. Pl. 3:829. 1834.
   Andromeda mertensiana Bong. Mem. Acad. St. Petersb. VI. 2:152, pl. 5. 1831. (Sitka,
    Alas.)
   A. cupressina Hook. Fl. Bor. Am. 2:38. 1834. (Drummond, Rocky Mts., n. of the Smoking
    R.)
   CASSIOPE MERTENSIANA var. GRACILIS (Piper) C. L. Hitchc. hoc loc. C. mertensiana
    ssp. gracilis Piper, Smith. Misc. Coll. 50:195. 1907. (Piper 2472, Wallowa Mts., Oreg.)
   Plant forming a widespread mat, stems 5-30 cm. tall, nearly completely concealed by the
leaves, glabrous or finely puberulent; leaves opposite, distinctly 4-ranked, sessile, appressed, 2-5 mm. long, ovate-lanceolate, rounded on the back, grooved only at the extreme base,
glabrous or ciliate with minutely glandular tiny hairs or longer chafflike, fugacious, white
hairs; flowers usually several near the branch tips, 5-8 mm. long; pedicels 5-30 mm. long,
glabrous or puberulent; sepals ovate, entire to erose-denticulate, reddish; corolla campanulate, white, the lobes ovate, about 1/3 length of the tube; filaments not enlarged at the base.
   Montane, usually not much below timber line, Alas. to Calif. and Nev., e. to Mont. and the
Canadian Rockies. July-Aug.
   In our area there are two very distinctive phases of the plant: the var. mertensiana, which
has puberulent stems and pedicels, entire calyx lobes, and usually glabrous leaves, ranges in
the Cascades from Alas. to s. Oreg. and in the Rockies nearly to Mont.; the var. gracilis (Piper) C. L. Hitchc., with glabrous stems and peduncles and minutely ciliate leaves, ranges
from n. e. Oreg. through Ida. and Mont. Other equally distinctive variants occur s. of our area.

**Cassiope stelleriana** (Pall.) DC. Prodr. 7:611. 1839.
   Andromeda stelleriana Pall. Fl. Ross. 1:58. 1788. Erica stelleriana Willd. Sp. Pl. 2:387.
    1799. Bryanthus stelleri D. Don, Edinb. New Phil. Journ. 17:160. 1834. Menziesia stelle-
    riana Fisch. ex Hook. Fl. Bor. Am. 2:37. 1834. Harrimanella stelleriana Coville, Proc.
    Wash. Acad. Sci. 3:574. 1901. (Siberia)
   Low spreading shrub 3-10 cm. tall, the branches trailing and matting, minutely puberulent;
leaves alternate, spreading, linear-oblanceolate, 3-5 mm. long, short-petiolate, glabrous or
ciliolate, the margins often somewhat erose; flowers mostly single and terminal, 5-6 mm.
long, on stout, puberulent pedicels 3-10 mm. long, sepals oblong-ovate, reddish, erose-denticulate; corolla broadly campanulate, white or pinkish, the lobes nearly equaling the tube; filaments conspicuously broadened at the base; anthers nearly globose, not much wider than the
filaments. Moss heather.
   Alpine meadows and bogs, Mt. Rainier n. to Alas.; Japan and e. Siberia. July-Aug.

**Cassiope tetragona** (L.) D. Don, Edinb. New Phil. Journ. 17:158. 1834.
   Andromeda tetragona L. Sp. Pl. 393. 1753. (Lapland)
   CASSIOPE TETRAGONA var. SAXIMONTANA (Small) C. L. Hitchc. hoc loc. C. saximonta-
    na Small, N. Am. Fl. 29:59. 1914. C. tetragona ssp. saximontana Porsild, Can. Field Nat.
    54:68. 1940. (McCalla 2161, Sulphur Mt. near Banff, Alta., July 18, 1899)
   Spreading shrub 5-30 cm. tall; branches puberulent; leaves opposite, nearly concealing the
stems, 4-ranked, 3-5 mm. long, thickish, lanceolate, prominently dorsally grooved, sessile,
conspicuously puberulent and ciliolate; flowers subterminal, 4-7 mm. long, white; pedicels
10-25 mm. long; sepals reddish, ovate, entire; corolla campanulate, lobes ovate, about 1/2
the length of the tube; filaments not dilated at the base. N=13.
   Circumboreal, s. as an alpine in the Rockies to Glacier Nat. Pk. and on the Pacific Coast to
Okanogan Co., Wash. June-Aug.
   All our material is referable to the var. saximontana (Small) C. L. Hitchc., having small
flowers (usually not over 5 mm. long) with pedicels seldom over twice the length of the leaves,

whereas the Alaskan plants, var. tetragona, have flowers 5-7 mm. long and proportionately longer pedicels.

### Chimaphila Pursh. Pipsissewa; Prince's Pine

Flowers 2-several in small umbels or corymbs, pendent on short, leafless, terminal peduncles; perianth commonly 5-merous; sepals distinct nearly to the base, persistent; petals distinct, spreading, somewhat reflexed, concave; stamens 10, filaments conspicuously enlarged and usually hairy above the base; anthers awnless, inverted, opening by (falsely) terminal pores on short tubes; ovary superior, 5-lobed; fruit a 5-celled capsule, loculicidally dehiscent from the apex; low semishrubs from slender rootstocks, glabrous or nearly so throughout, the stems but slightly woody; leaves leathery, persistent, mostly whorled, not in basal rosettes.

A small genus of about 6 species of N. Am. and Asia. (Name from the Greek cheima, winter, and philos, loving, referring to the evergreen habit. )
Reference:
Blake, S. F. The varieties of Chimaphila umbellata. Rhodora 19:237-44. 1917.
1 Flowers 1-3; filaments hairy over the entire swollen portion of the base; leaves mostly elliptic                                                                              C. MENZIESII
1 Flowers usually more than 3; filaments ciliate on the swollen base; leaves oblanceolate
                                                                              C. UMBELLATA

Chimaphila menziesii (R. Br.) Spreng. Syst. Veg. 2:317. 1825.
  Pyrola menziesii R. Br. ex D. Don, Mem. Wern. Soc. 5:245. 1824. (Menzies, N.W. coast of N. Am. )
  Plant 5-15 cm. tall; leaves lanceolate or elliptic-lanceolate to elliptic-oblanceolate, blades 2-6 cm. long, sharply serrate to entire, petioles 5-8 mm. long; peduncles 2-5 cm. long; flowers 1-3; sepals lacerate-erose; petals 5-7 mm. long, pinkish; filaments hairy on the swollen, obcordate, lower portion; capsule about 5 mm. broad.
  Coniferous woods, B.C. s. to the Sierra Nevada and s. Calif., e., reputedly, to Ida. and Mont. June-Aug.

Chimaphila umbellata (L.) Bart. Veg. Mat. U.S. 1:17, pl. 1. 1817.
  Pyrola umbellata L. Sp. Pl. 396. 1753. (Europe)
  Chimaphila occidentalis Rydb. N. Am. Fl. 29:30. 1914. C. umbellata var. occidentalis Blake, Rhodora 19:242. 1917. C. umbellata ssp. occidentalis Hultén, Fl. Alas. 8:1203. 1948. (Sandberg, MacDougal, & Heller 519, valley of Pine Creek, near Farmington, Latah Co., Ida., July 28, 1892)
  Stems 1-3 dm. tall; leaves oblanceolate to oblanceolate-obovate, tapered to narrowly acute bases, blades 3-7 cm. long, 0.5-2.5 cm. broad, sharply serrate, petioles 3-7 mm. long; peduncles 5-10 cm. long, with the pedicels usually finely puberulent and often somewhat glandular; flowers 5-15, racemose-corymbose; sepals denticulate-erose; petals 5-7 mm. long, pinkish to somewhat rose; swollen base of the filaments ciliate on the margins but not hairy; capsule 5-7 mm. broad. N=13.
  Woods, especially under conifers, Alas. to s. Calif. e. to the Rockies of Colo. and to e. U.S.; Eurasia. June-Aug.
  All our material is referable to the var. occidentalis (Rydb.) Blake.

### Cladothamnus Bong.

Flowers single and terminal on short, axillary, leafy shoots; calyx connate only at the base, the 5 lobes narrowly linear-lanceolate; corolla more or less rotate, the 5 petals distinct; stamens 10 (sometimes fewer), anthers oval, unawned, dehiscing nearly full length by lateral slits; ovary superior, 5-lobed, style elongate and recurved, enlarged below the discoid stigma; fruit a 5(6)-valved, many-seeded, septicidal capsule; shrub with alternate, thin, pale green, deciduous leaves.
  One species only. (Name from the Greek klados, sprout or slip, and thamnos, bush. )

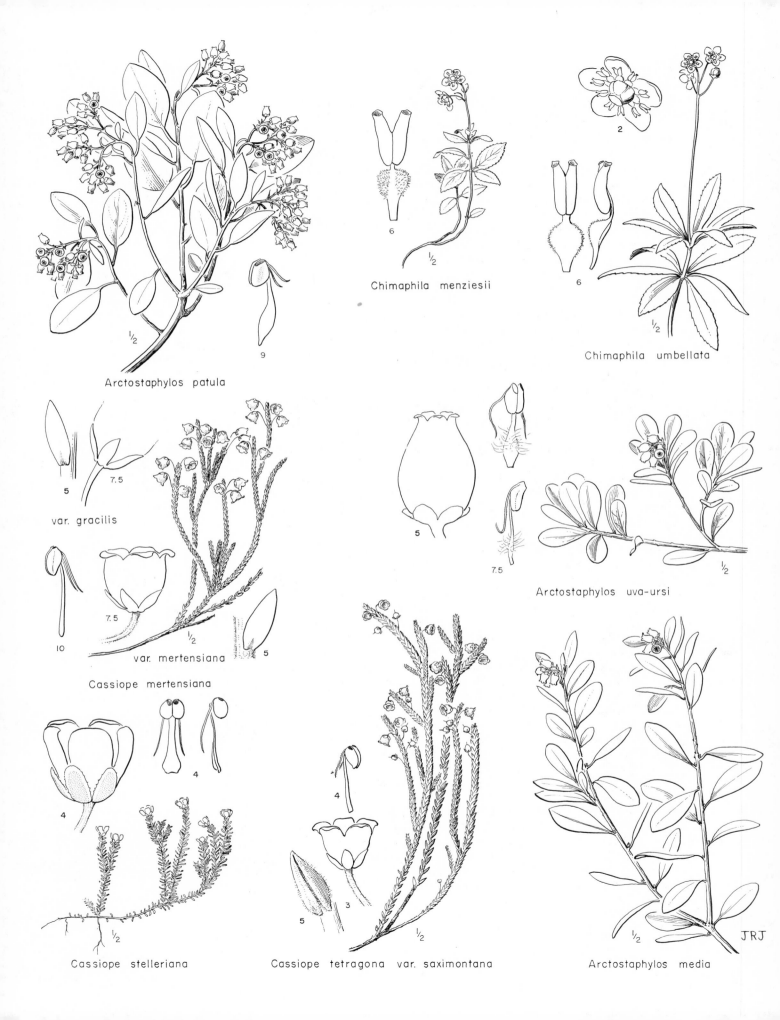

Chimaphila menziesii

Chimaphila umbellata

Arctostaphylos patula

var. gracilis

var. mertensiana

Cassiope mertensiana

Arctostaphylos uva-ursi

Cassiope stelleriana

Cassiope tetragona var. saximontana

Arctostaphylos media

JRJ

Cladothamnus pyrolaeflorus Bong. Mém. Acad. St. Pétersb. VI. 2:155. 1832. (Sitka, Alas.)
  Tolmiea occidentalis Hook. Fl. Bor. Am. 2:45. 1834. (Menzies, N.W. coast of Am.)
    Plant glabrous or very sparingly puberulent on the stems and pedicels, 0.5-3 m. tall; leaves
pale green, somewhat glaucous, elliptic-oblanceolate to oblanceolate, mucronate, entire, 2-5
cm. long, nearly sessile; sepals 7-10 mm. long; petals oblong-elliptic, spreading, 10-15 mm.
long, salmon or coppery; filaments glabrous, 5-8 mm. long; style about 1 cm. long; capsule
4-7 mm. broad, nearly globose.
    Moist forests and stream banks, Alas. s. to B.C. and occasional on the w. side of the Cas-
cades to Saddle Mt., n.w. Oreg. June-July.
    A very distinctive and unusual-flowered shrub, desirable for the acid garden.

## Gaultheria L.

    Flowers white or pinkish, solitary and axillary, or in axillary or terminal racemes, the
perianth 4- to 5-merous; calyx usually deeply lobed; corolla gamopetalous, campanulate to
urceolate-campanulate; stamens 8 or 10, included; filaments expanded near their base, anthers
opening by lateral or terminal pores, with or without awns; pistil 4- to 5-carpellary, stigma
entire or obscurely lobed; fruit a many-seeded capsule, but surrounded by the persistent,
thickened, and pulpy calyx, forming a fleshy pseudo berry; low, depressed to semierect shrubs
with persistent, leathery, shining leaves.
    About 150 species in the cooler regions of N. and S. Am., Asia, and Australia. (Named in
honor of Jean François Gaultier, 1708-1756, physician and botanist of Quebec.)
    The Gaultherias are all highly desirable ground-cover ornamentals in the northwest and have
edible fruit readily eaten by birds and mammals.
1 Leaves less than 1 cm. long; flowers 4-merous, less than 3 mm. long; fruit white
                                                                G. HISPIDULA
1 Leaves usually over 1 cm. long; flowers usually 5-merous and at least 3 mm. long; fruit
    red or bluish-black
    2 Calyx glabrous; leaves oval, rounded or obtuse, averaging less than 2 cm. long and less
        than 1.5 cm. broad, entire to inconspicuously crenate-serrulate  G. HUMIFUSA
    2 Calyx either hairy or the leaves larger and usually noticeably serrulate, often acute
        3 Flowers single in the leaf axils; corolla glabrous; anthers unawned; fruit reddish;
            leaves less than 5 cm. long                                    G. OVATIFOLIA
        3 Flowers in racemes; corolla hairy; anthers awned; fruit purplish; leaves at least 5 cm.
            long                                                          G. SHALLON

Gaultheria hispidula (L.) Muhl. Cat. Pl. 44. 1813.
  Vaccinium hispidulum L. Sp. Pl. 352. 1753. Oxycoccus hispidulus Pers. Syn. 1:419. 1801.
  Gaultheria serpyllifolia Pursh, Fl. Am. Sept. 283. 1814. Chiogenes serpyllifolia Salisb.
  Trans. Hort. Soc. Lond. 2:94. 1817. Phalaerocarpus serpyllifolia G. Don, Gen. Syst. 3:841.
  1834. Lasierpa hispidula Torr. in Geol. Reg. N.Y. 152. 1840. Chiogenes hispidula T. & G.
  ex Torr. Fl. N.Y. 1:450, pl. 68. 1843 (Can. to Pa.)
    A creeping, slender-stemmed shrub, brownish-bristly with somewhat appressed hairs on
the stems, calyxes, and lower surfaces of the leaves; leaves elliptic to obovate, coriaceous,
entire, revolute, 4-10 mm. long, petioles 1.5-2.5 mm. long; flowers mostly axillary and
single, subtended by 2 ovate bracts that are longer than the calyx; corolla campanulate, deeply 4-lobed; stamens 8, filaments flattened, shorter than the anthers, anthers usually with 4
very short terminal points, opening by 2 large lateral pores; berry clear white, 3-5 mm.
thick, surrounded by the calyx, juicy, somewhat spicy, and aromatic. Creeping snowberry,
moxieplum, maidenhairberry.
    Mostly in sphagnum or deep coniferous woods, Labrador w. to B.C., s. into n. Ida. May-
June.
    One of the most beautiful of tiny, creeping ground covers for the wet, acid area of the gar-
den. Perhaps more frequently placed in the genus Chiogenes, but technically not separable
from Gaultheria.

Gaultheria humifusa (Grah.) Rydb. Mem. N.Y. Bot. Gard. 1:330. 1900.

    Vaccinium humifusum Grah. Edinb. New Phil. Journ. 11:193. 1831. (Seeds from the Rocky
      Mts., collected by Drummond)

    Gaultheria myrsinites Hook. Fl. Bor. Am. 2:35, pl. 129. 1834. (Drummond, "Rocky Moun-
      tains, between lat. 52⁰ and 54⁰")

    Low, depressed shrublet scarcely 3 cm. tall, the stems trailing but seldom over 10 cm.
long, glabrous or finely puberulent and sometimes with longer, reddish pilosity; leaves broadly
ovate or oval to nearly elliptic, 1-2(2.5) cm. long, 0.5-1.5 cm. broad, rounded or obtuse
acute), margins thickened, entire to inconspicuously crenulate or serrulate; flowers 3-4 mm.
long, single in the leaf axils on short, bracteate pedicels; calyx glabrous, nearly as long as the
campanulate, 5-lobed, pinkish corolla; stamens included, anthers opening by terminal pores,
without appendages; fruit reddish, 5-7 mm. broad. Alpine wintergreen, matted wintergreen.

    Subalpine to alpine, usually where moist or wet; B.C. s. in the Cascades and Olympic Mts.
to n. Calif., e. to the Rocky Mts. from Alta. to Colo. July-Aug.

    Distinct from G. ovatifolia and not commonly intergrading with it since it usually grows
above the latter in the mountains, but in the Cascades the two occasionally hybridize and produce intermediate plants. (G. N. Jones 4139 and J. W. Thompson 15166b)

Gaultheria ovatifolia Gray, Proc. Am. Acad. 19:85. 1884. (Lyall, Cascade Mts.)

    Low, spreading shrublet seldom over 3 cm. tall, the branches slender, 5-20 cm. long, pu-
berulent and copiously brownish-pilose; leaves ovate, acute, (1.5)2-4 cm. long, (1)1.5-3 cm.
broad, margins thickened and usually conspicuously serrulate; flowers 3.5-5 mm. long, white or
pinkish, 5-merous, single in the axils on short, bracteate pedicels; calyx copiously brownish-
reddish pilose; corolla campanulate, nearly twice the length of the calyx; anthers opening by terminal pores, without appendages; berry bright red, 6-8 mm. broad. Slender or Oregon wintergreen.

    From fairly dry, yellow pine forests to subalpine bogs, B.C. to n. Calif. e. to Ida; report-
ed from Glacier Nat. Pk., Mont. June-Aug.

    The best of our species as a ground cover.

Gaultheria shallon Pursh. Fl. Am. Sept. 283. 1814. (Lewis, Columbia R.)

    Creeping to erect or partially scandent shrub with pilose to hirsute stems 1-12 dm. long;
leaves ovate to ovate-elliptic, 5-9 cm. long, 3-5 cm. broad, sharply serrulate; flowers pinkish, 7-10 mm. long, 5-15 in terminal and subterminal, axillary, bracteate racemes, the rachis
puberulent and glandular-hirsute; calyx glandular-pilose, about half the length of the 5-lobed,
campanulate-urceolate, glandular-puberulent corolla; anthers with 4 slender apical awns, dehiscent by 2 large subterminal pores below the awns; fruit purplish, 6-10 mm. broad. N=44.
Salal.

    From the coast to the w. slope of the Cascades and coastal ranges, B.C. to s. Calif. May-
July.

    So common as to seem undesirable, but really one of the choice native shrubs of our area;
apt to be somewhat invasive. The fruit is a source of food for many animals.

Hemitomes Gray. Gnome Plant

    Flowers bracteate, pinkish-yellow, 1-2 cm. long, the perianth segments usually somewhat
erose-fimbriate; sepals 2-4, essentially distinct; petals usually 4, membranous, connate over
half their length, hairy within and strongly ciliate-pectinate; stamens usually 8, included, filaments hairy; anthers 2-celled, linear, dehiscent full length by slits; ovary superior, 1-celled
with 7-9 parietal placentae; style short, hairy; stigma slightly lobed; fruit baccate-capsular;
fleshy saprophytes from small, fleshy roots with clusters of short, simple stems, and thin,
scalelike, yellowish to brown, nongreen leaves; inflorescence congested, a simple, headlike
spike, or with 1-several short side branches at the base, whitish to pink.

    One species of the Pacific Northwest. (From the Greek hemitomes [hemitomias], half eu-
nuch, one anther cell originally believed to be sterile.)

Reference:
Copeland, H. F. Further studies on Monotropoideae. Madroño 6:109. 1941.

**Hemitomes congestum** Gray, Pac. R.R. Rep. 6:80. 1857.
Newberrya congesta Torr. ex Gray, Bot. Calif. 1:464. 1876. (Williamson Expedition, upper Deschutes Valley, Oreg.)
N. spicata Gray, Proc. Am. Acad. 15:44. 1879. Hemitomes spicatum Heller, Cat. N. Am. Pl. 5. 1898. (Rattan, N. Fork Mad R., Humboldt Co., Calif.)
Hemitomes pumilum Greene, Erythea 2:121. 1894. Newberrya pumila Small, N. Am. Fl. 29: 18. 1914. (Wright, Mendocino Co., Calif.)
Newberrya subterranea Eastw. Proc. Calif. Acad. Sci. III. 1:80. 1897. (Plaskett, Willow Creek Canyon, Monterey Co., Calif.)
N. longiloba Small, N. Am. Fl. 29:18. 1914. (Suksdorf 2168, Skamania Co., Wash., Aug. 19, 1892)
Stems 3-20 cm. tall, fleshy; calyx and corolla similar in color and texture, subequal, pinkish when fresh, drying to brownish; sepals oblong, 12-18 mm. long, corolla tubular-campanulate, 12-20 mm. long, often unequally lobed; anthers about 2 mm. long; top of stigma yellowish; capsule ovoid, hairy.
In deep humus, under conifers, coastal forests from the Olympic Peninsula, Wash., s. to Monterey Co., Calif. June-Aug.

### Hypopitys Hill. Pinesap; American Pinesap

Flowers axillary in terminal racemes, bracteate; perianth 4(3-5)-merous, erose to lacerate-fimbriate; sepals oblanceolate, distinct; petals distinct, saccate at the base, usually hairy on both surfaces; stamens twice the number of petals, filaments hairy, anthers ovoid, opening by a continuous slit full length and across the top; pistil 4(5)-carpellary, ovary superior, 4- to 5-celled and with axile placentation below but 1-celled above; fruit capsular-baccate; fleshy, white to yellowish or pinkish saprophytes with scalelike, nongreen leaves and unbranched stems.
Perhaps 3 species of N. Am. and Eurasia. (Named from the Greek hypo, beneath, and pitys, pine tree, referring to the habitat.)
Reference:
Copeland, H. F. Further studies on Monotropoideae. Madroño 6:99-104. 1941.

**Hypopitys monotropa** Crantz, Inst. Rei. Herb. 2:467. 1766.
Monotropa hypopitys L. Sp. Pl. 1:387. 1753. Hypopitys hypopitys Small, Mem. Torrey Club 4:137. 1894. (Europe)
Monotropa lanuginosa Michx. Fl. Bor. Am. 1:266. 1803. Hypopitys lanuginosa Nutt. Gen. Pl. 1:271. 1818. (N. C.)
Hypopitys europaea Nutt. Gen. Pl. 1:271. 1818. H. multiflora var. americana DC. Prodr. 7:780. 1839. H. americana Small, Fl. S. E. U. S. 880. 1903. Monotropa hypopitys var. americana Domin. Ges. Wiss. Prag. II. 1915:24. 1915. (Can.)
Monotropa fimbriata Gray, Proc. Am. Acad. 8:629. 1873. Hypopitys fimbriata Howell, Fl. N. W. Am. 429. 1901. Monotropa hypopitys var. fimbriata Domin. Ges. Wiss. Prag. II. 1915:24. 1915. (E. Hall 357, Oreg.)
Hypopitys lutea sensu Howell, Fl. N. W. Am. 429. 1901, but not of S. F. Gray. in 1821.
Hypopitys latisquama Rydb. Bull. Torrey Club 40:461. 1913. Monotropa hypopitys var. latisquama Kearney & Peebles, Journ. Wash. Acad. Sci. 29:487. 1939. M. latisquama Hultén, Fl. Alas. 8:1216. 1948. (Flodman 708, Bridger Mts., Mont.)
Hypopitys brevis Small, N. Am. Fl. 29:13. 1914. (Moses Craig 5214, Independence Valley, Cascade Mts., Oreg., Aug., 1892)
Plant 5-25 cm. tall, pinkish to straw-colored, drying to black; leaves entire to fimbriate; racemes usually recurved at anthesis, becoming erect in fruit; pedicels 3-6 mm. long; sepals 5-9 mm. long, petals about twice as long, erect, overlapping one another, usually hairy on one or both surfaces, sometimes glabrous, somewhat saccate at base; stamens shorter than the

corolla, filaments not dilated; style hairy, stigma slightly lobed; capsule subglobose, 5-8 mm. long. N=24.

In humus of chiefly coniferous forests; B. C. to Mendocino Co., Calif., e. to the Atlantic Coast; Europe. May-July.

There is considerable variation evident in the material seen, but it seems that little, if any, basis exists for the recognition of more than one taxon in our area. According to Hultén, Fl. Alas. 8:1216. 1948, our plant is not at all the same thing as the European one; if true, ours should be called Hypopitys fimbriata (Gray) Howell. However, as judged from herbarium material, our plants are not sufficiently different to be separated, even varietally, from those of the Old World.

## Kalmia L. Laurel

Flowers terminal on the shoots, long-pedicellate in leafy-bracted, several-flowered corymbs; perianth 5-merous; calyx lobed nearly to the base; corolla gamopetalous, spreading, shallowly bowl-shaped, with 10 pronounced sacs running nearly to the sinuses in which the anthers are recessed in the bud and from which they spring as the flower opens; stamens 10(8-12), filaments slender, anthers narrowly oblong, dehiscent full length; ovary superior, 5-celled; style slender, straight; stigma scarcely lobed; fruit a small, 5-valved, septicidal capsule; low, evergreen shrubs with alternate to whorled, leathery, entire leaves.

About 6 species of N. Am. (Named for Peter Kalm, 1715-1779, one of Linnaeus' students.)

Kalmia polifolia Wang. Beob. Ges. Nat. Freunde Berl. 2:130. 1787. (Garden specimens from plants introduced by Banks from Newf.)

K. glauca L'Herit. ex Ait. Hort. Kew. 2:64. 1789. (Banks, Newf.) = var. polifolia.

K. glauca var. rosmarinifolia Pursh, Fl. Am. Sept. 296. 1814. (Pursh, Albany, N.Y.) = var. polifolia.

K. glauca var. microphylla Hook. Fl. Bor. Am. 2:41. 1834. K. microphylla Heller, Bull. Torrey Club 25:581. 1898. K. polifolia var. microphylla Rehd. in Bail. Cycl. Am. Hort. 2:854. 1900. (Drummond, swamps in the Rocky Mts.)

K. occidentalis Small, N. Am. Fl. 29:53. 1914. (Mrs. Bailey Willis, foothills of Mt. Rainier, Wash., in 1883)

Plant 1-5 dm. tall, much branched, often matted, spreading by layering and short rhizomes, young stems puberulent but soon becoming glabrous; leaves opposite, dark green and glabrous above, grayish and very finely and densely granular-puberulent beneath, entire, revolute or flat, blades mostly oblong-lanceolate to narrowly oblong or linear-elliptic, 1-4 cm. long; petioles 2-5 mm. long; pedicels 1-4 cm. long, usually glabrous; flowers deep pinkish-rose, 1-2 cm. broad; sepals ovate, 2-3 mm. long, sparsely ciliolate, otherwise glabrous; stamens about equaling the style, slightly exserted, filaments densely hairy just above the base, otherwise glabrous; capsules subglobose, 2-3 mm. long. N=22. Swamp laurel, American laurel.

Lowland bogs to mountain meadows. June-Sept.

A delightful little shrub (especially the var. microphylla) for the wet area in the acid garden. There are 2 phases of the species in our area: (1) var. polifolia, usually over 2 dm. tall; leaves mostly 2-4 cm. long and less than half as broad, strongly revolute; flowers 12-18 mm. broad; widespread in e. N. Am. and extending across the continent in n. Can., ranging down the Pacific Coast, on the w. side of the Cascades, to Calif.; (2) var. microphylla (Hook.) Rehd., usually not over 1 dm. tall; leaves 1-2(3) cm. long, 1/2-1/4 as broad; flowers usually less than 12 mm. broad. An alpine form of the species found from Alas. and the Yukon s. in the Cascades to the Sierra Nevada in Calif., e. in B. C. to the Rocky Mts. and s. to Colo. In the s. part of its range the leaves are more usually revolute, in the Rocky Mts. they are usually plane. Where the two varieties intermingle, as on the w. side of the Cascades, they intergrade; K. occidentalis Small was described from such an intergradient plant.

## Ledum L. Labrador Tea

Flowers many in terminal racemes or corymbs, 5(4-6)-merous; calyx small, divided near-

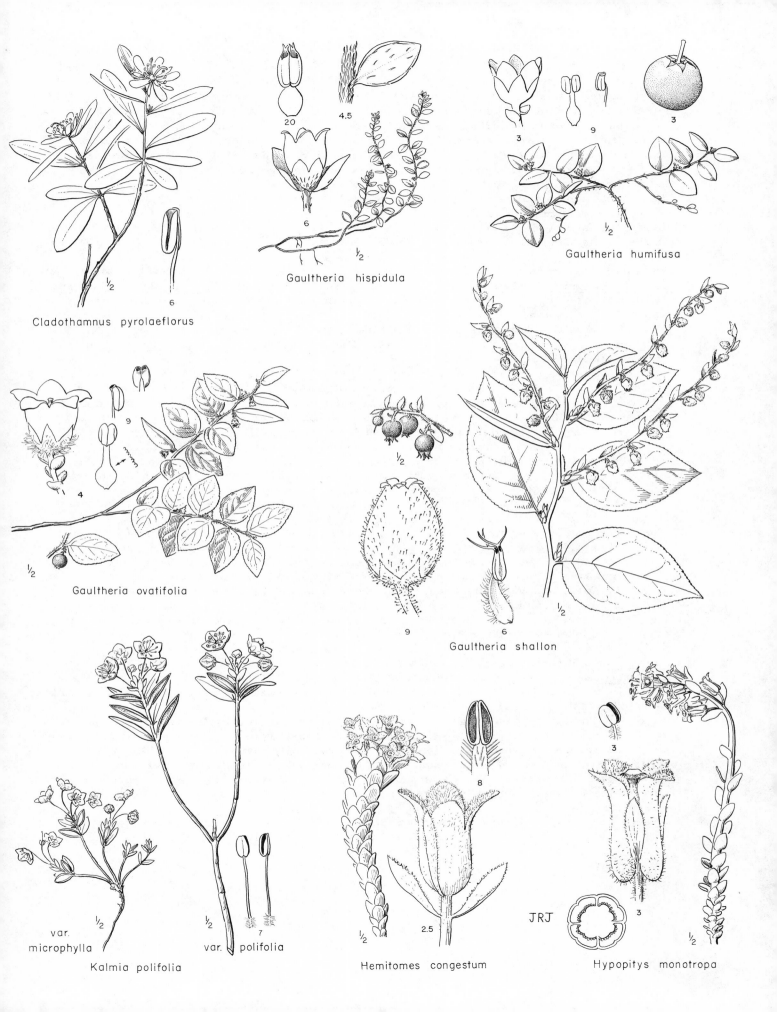

20    4.5

6

Gaultheria hispidula

3    9    3

Gaultheria humifusa

½

6

Cladothamnus pyrolaeflorus

9

4

½

Gaultheria ovatifolia

½

9

6    ½

Gaultheria shallon

var.
microphylla    ½

½    polifolia    7

var. polifolia

Kalmia polifolia

½    2.5

Hemitomes congestum

JRJ

3

3    ½

Hypopitys monotropa

ly to the base; petals distinct to the base, white, spreading to nearly rotate, stamens 10(5-12), filaments slender; anthers oblong-oval, unawned, opening by small round pores near the tip; style slender, essentially straight; stigmas nearly globose; ovary superior; fruit capsular, 5-valved, dehiscing upward septicidally; evergreen shrubs with coriaceous, entire, often revolute-margined, petiolate leaves, usually conspicuously glandular, at least in the inflorescence.

Perhaps 4 species of N. Am. and Eurasia. (Name from the Greek ledon, mastic, a name used by the Greeks for Cistus.)

Hardy, but none too attractive shrubs for the wet to exposed area of the garden. All species more or less poisonous to stock

1 Lower surface of leaves rusty-lanate                                 L. GROENLANDICUM
1 Lower surface of leaves greenish or white, not lanate                L. GLANDULOSUM

Ledum glandulosum Nutt. Trans. Am. Phil. Soc. II. 8:270. 1843. (Nuttall, Rocky Mts.)
   L. californicum Kell. Proc. Calif. Acad. Sci. 2:14. 1863. L. glandulosum ssp. glandulosum
      var. californicum C. L. Hitchc. Leafl. West. Bot. 8:6. 1956. (Hutchings, Sierra Nevada,
      Calif.) = var. glandulosum.
   LEDUM GLANDULOSUM var. COLUMBIANUM (Piper) C. L. Hitchc. hoc loc. L. columbia-
      num Piper, Contr. U. S. Nat. Herb. 11:441. 1906. L. glandulosum ssp. columbianum C. L.
      Hitchc. Leafl. West. Bot. 8:7. 1956. (Piper 6451, Ilwaco, Pacific Co., Wash.)
   L. glandulosum ssp. columbianum var. australe C. L. Hitchc. Leafl. West. Bot. 8:8. 1956.
      (Elmer 4944, Pt. Reyes P.O., Marin Co., Calif., July, 1903) = var. columbianum.
   Plant 0.5-2 m. tall, twigs puberulent and glandular-dotted, leaves ovate and rounded at base to oblong-ovate, elliptic-oblong, or elliptic and acute-based, deep green on upper surfaces, lighter green and often grayish, finely mealy-puberulent, densely glandular, and occasionally puberulent on the lower surfaces, more or less revolute; flowers white; pedicels mostly 1-2 cm. long, puberulent near the base; stamens (5)8-12, usually considerably longer than the style, filaments densely hairy on the lower half; capsules from nearly globose to ovoid, 3-5(6) mm. long, puberulent and glandular. N=13. Trapper's tea.
   B. C. to Calif., mostly e. of the Cascades in Wash., eastward to the Rocky Mts. and southward to n.w. Wyo., also in Ida. and n.e. Oreg. June-Aug.
   There are several geographic races of this species, the two most conspicuous of which are recognized here. The var. columbianum (Piper) C. L. Hitchc. has strongly revolute, narrow leaves, 3-5 cm. long and averaging less than 1 cm. in width, and capsules that tend to be ovoid in form and 4-5.5 mm. long. It ranges from Pacific Co., Wash., southward along the coast to Marin Co., Calif. The var. glandulosum is a more montane plant, occurring in the Cascades of B. C. and Wash., e. in B. C. to the Rocky Mts., to Mont., n. Wyo., c. Ida., and w. to n. e. Oreg. It has plane or slightly revolute leaves that usually average half as broad as long, and more nearly globose capsules 1-3(4.5) mm. long.

Ledum groenlandicum Oeder, Fl. Dan. 4:5. 1777.
   L. palustre ssp. groenlandicum Hultén, Fl. Alas. 8:1219. 1948. (Greenl.)
   L. pacificum Small, N. Am. Fl. 29:37. 1914. L. palustre var. dilatatum Gray, Syn. Fl. 2[1].
      43. 1878. (Tiling, Sitka)
   Plant 0.5-2 m. tall; young twigs strigose-pubescent; leaves linear-elliptic, blades 2-6 cm. long, leathery, deep green, usually glabrous and somewhat rugose to somewhat reddish-lanate above, margins distinctly revolute, lower surfaces copiously rusty-lanate, petioles 2-5 mm. long; flowers white, about 1 cm. broad, the bracts and pedicels finely puberulent and often somewhat glandular; sepals very short; petals spreading; stamens 5-10, slightly exceeding the styles; styles 4-6 mm. long, nearly straight; capsules ovoid, 4-5 mm. long, puberulent. N=13. Labrador tea.
   Greenl. to Alas., s. to New England, in the w. along the coast, chiefly in swamps and bogs to the Olympic Peninsula and n.w. Oreg. and in B. C. to possibly n. Ida. May-July.

## Menziesia Smith

Flowers in terminal clusters, on shoots of the previous year, appearing with the leaves; perianth 4 (ours)- or 5-merous; calyx very short, saucer-shaped, shallowly lobed, glandular-ciliate, persistent; corolla greenish-white or pinkish (ours) to reddish or purplish, gamopetalous,
urceolate-tubular to campanulate, finely puberulent inside; stamens 8 (ours) or 5 or 10, included; filaments flattened, slender, longer than the linear, unawned anthers which are dehiscent
by terminal pores; style about equaling the corolla, stigma slightly enlarged, not lobed; ovary
superior, 4-celled (ours); fruit a many-seeded, 4(5)-celled, septicidal capsule; seeds pointed
or caudate; alternate-leaved, deciduous shrubs.

Six or seven species of N. Am. and Japan, the Japanese species having more showy flowers,
but our species well worth cultivation if only for the autumn coloration. (Named for Archibald
Menzies, 1754-1842, physician and naturalist with the Vancouver Expedition of 1790-95.)

Menziesia ferruginea Smith, Pl. Ic. Ined. pl. 56. 1791.
  M. urceolaris Salisb. Parad. Lond. pl. 44. 1806. (Menzies, "Western North America")
  M. globularis Hook. Fl. Bor. Am. 2:41. 1834, not of Salisb. in 1806. M. glabella Gray, Syn.
    Fl. 2¹:39. 1878. M. ferruginea var. glabella Peck, Man. High. Pl. Oreg. 542. 1941.
    (Drummond, "Alpine woods of the Smoking River, in lat. 56°")
Straggling shrub 0.5-2 m. tall, young branches finely puberulent and somewhat glandular-
pilose, older branches eventually puberulent only, or glabrous; leaves thin, clear green and
usually brownish-pilose, often glandular above, paler beneath and usually glandular-pubescent,
crenulate-serrulate, ovate-elliptic to elliptic-obovate, 4-6 cm. long, apiculate and acute to
rounded; pedicels 1-2 cm. long, short-pilose to conspicuously glandular-bristly and often also
finely puberulent; calyx puberulent to conspicuously long glandular-ciliate; corolla yellowish-
red, 6-8 mm. long; filaments globose to sparsely hairy near the base; ovary glabrous to
sparsely glandular-pilose and often also puberulent; capsules ovoid, 5-7 mm. long. Fool's
huckleberry.

Moist woods to mountain stream banks. May-Aug.

Rather desirable as an ornamental, the foliage being much like that of Enkianthus, a somewhat similar, widely cultivated, ornamental ericaceous shrub. There are two distinct phases
of the species: var. ferruginea, from Alas. s. along the coast and in the Cascades to n. Calif.,
characterized by the glandular hairs of the foliage, usually apiculate or acute leaves, the glandular-ciliate calyx, and the glandular but not puberulent ovary. The var. glabella (Gray) Peck
differs in that the branches are usually finely puberulent and less conspicuously glandular, the
pedicels especially with much shorter pilosity, the glandular hairs but 2 or 3 times as long as
the eglandular; leaves more nearly rounded than acute, often finely puberulent on both surfaces
and less conspicuously pilose above and less glandular beneath; calyx usually puberulent as
well as glandular-ciliate; ovary always finely puberulent (although sometimes very sparsely so)
as well as glandular-pilose; Rocky Mts., Alta. and B.C. to Wyo., westward to e. Wash. and
Oreg. and down the Columbia to Mt. Hood and Mt. Adams, where the two varieties freely inter-
breed.

## Monotropa L. Indian Pipe

Flowers large, single, terminal, nodding but becoming erect in fruit, waxy-white, aging to
black; sepals probably lacking but the upper 1-4 bracts of the stem proximal to the flower and
calyxlike, usually slightly erose-lacerate, deciduous; petals distinct, usually 5(4-6), glabrous
on the outer surface, erect, tardily deciduous; stamens usually 10, included; filaments somewhat hairy, anthers broader than long, unawned, dehiscent by two curving slits extending about
2/3 of the length of the pouchlike chamber; pistil usually 5-carpellary, ovary superior, 5-
celled, with axile placentation; fruit a loculicidal capsule; fleshy-white to pinkish saprophytes
with simple stems and scalelike nongreen leaves.

Three species of the New World and Asia. (Name from the Greek monos, one, and tropos,
direction, the flowers turned to one side and pendulous.)

Reference:
Copeland, H. F. Further studies on Monotropoideae. Madroño 6:104-8. 1941.

Monotropa uniflora L. Sp. Pl. 387. 1753. (E. U. S. )
  M. morisoniana Michx. Fl. Bor. Am. 1:266. 1803. M. morisoni Pers. Syn. 1:469. 1805.
  (E. U. S. )
  Plant with a cluster of flowering stems 5-25 cm. tall, waxy-white, blackening with age; flow-
ers 1. 5-2 cm. long, narrowly campanulate, curved to one side or even drooping; petals sac-
cate at the base, conspicuously broadened above, more or less hairy within; style included;
stigma discoid-lobed; capsule subglobose, about 6 mm. long.
  In deep, shaded woods, Alas. to n. Calif., e. to the Atlantic Coast. July-Aug.

Phyllodoce Salisb. Mountain Heath; Mountain Heather

  Flowers single from the axils, clustered at the stem tips; pedicels glandular-pubescent,
slender, 2-bracteate at the base; perianth 5-merous; calyx divided to near the base; corolla
gamopetalous, campanulate to narrowly urceolate; stamens (7-9)10, filaments slender, included, anthers narrowly oblong, unawned, opening by terminal spreading slits; ovary superior,
glandular-pubescent, style about equaling the corolla; stigma scarcely lobed; capsule septicidally dehiscent from the apex; dwarf evergreen alpine shrubs, often forming extensive mats,
with linear, alternate, closely crowded, revolute leaves borne on short pedicels and leaving a
raised peglike leaf scar.
    About seven species, N. Am. and Eurasia. (From the Greek name for a sea nymph. )
    Nice low shrubs for the moist area, but not easily satisfied in the lowland garden.
    Reference:
    Stoker, Fred. The genus Phyllodoce. New Flora & Silva 12:30-42. 1939.
  1 Corolla campanulate, pink to rose, glabrous; sepals ovate to ovate-lanceolate, obtuse, cil-
      iolate but not pubescent                                                                    P. EMPETRIFORMIS
  1 Corolla narrowly urceolate, dirty yellowish to greenish-white, glandular-puberulent on the
      outside; sepals lanceolate, acute, glandular-puberulent                                     P. GLANDULIFLORA

Phyllodoce empetriformis (Sw. ) D. Don, Edinb. New Phil. Journ. 17:160. 1834.
  Menziesia empetriformis Sw. Trans. Linn. Soc. 10:380. 1811. Bryanthus empetriformis
    Gray, Proc. Am. Acad. 7:367. 1868. (Menzies, "West Coast of North America")
  Menziesia grahamii Hook. Fl. Bor. Am. 2:40. 1834. Phyllodoce grahamii Pursh, Trans. Am.
    Phil. Soc. II. 8:269. 1843. (Drummond, "Alpine woods and open elevated situations of the
    Rocky Mts., lat. 55⁰")
  Matted shrub 1-4 dm. tall, young stems finely puberulent and more or less glandular, soon
glabrous; leaves persistent, 8-16 mm. long, 1-2 mm. broad, deeply grooved beneath and apparently revolute, usually glabrous except for the minutely glandular-serrulate margins; flowers deep pinkish-rose; calyx lobed nearly to the base, the segments ovate-lanceolate, obtuse,
finely ciliolate, otherwise glabrous; corolla campanulate, about 7 mm. long, the lobes recurved; stamens 10, included, filaments glabrous; anthers reddish, dehiscent by apical slits;
ovary shortly yellowish glandular-pubescent; style about equaling the corolla. Pink mountain
heather.
  Alas. s., in the high mountains, to Calif., Ida., and Mont. June-Aug.

Phyllodoce glanduliflora (Hook. ) Coville, Mazama 1:196. 1897.
  Menziesia glanduliflora Hook. Fl. Bor. Am. 2:40. 1834. Bryanthus glanduliflorus Gray,
    Proc. Am. Acad. 7:368. 1868. (Drummond, "mountains north of the Smoking River, lat.
    56⁰")
  Plant 1-4 dm. tall; young stems glandular-puberulent; leaves 6-12 mm. long, 1-1. 5 mm.
broad, finely glandular-puberulent, apparently revolute, pedicels densely glandular-puberulent,
as are the calyx and corolla; calyx lobes narrowly lanceolate, acute; corolla dirty yellowish to
greenish-white, 5-7 mm. long, narrowly urceolate, lobes ovate-lanceolate, spreading; fila-

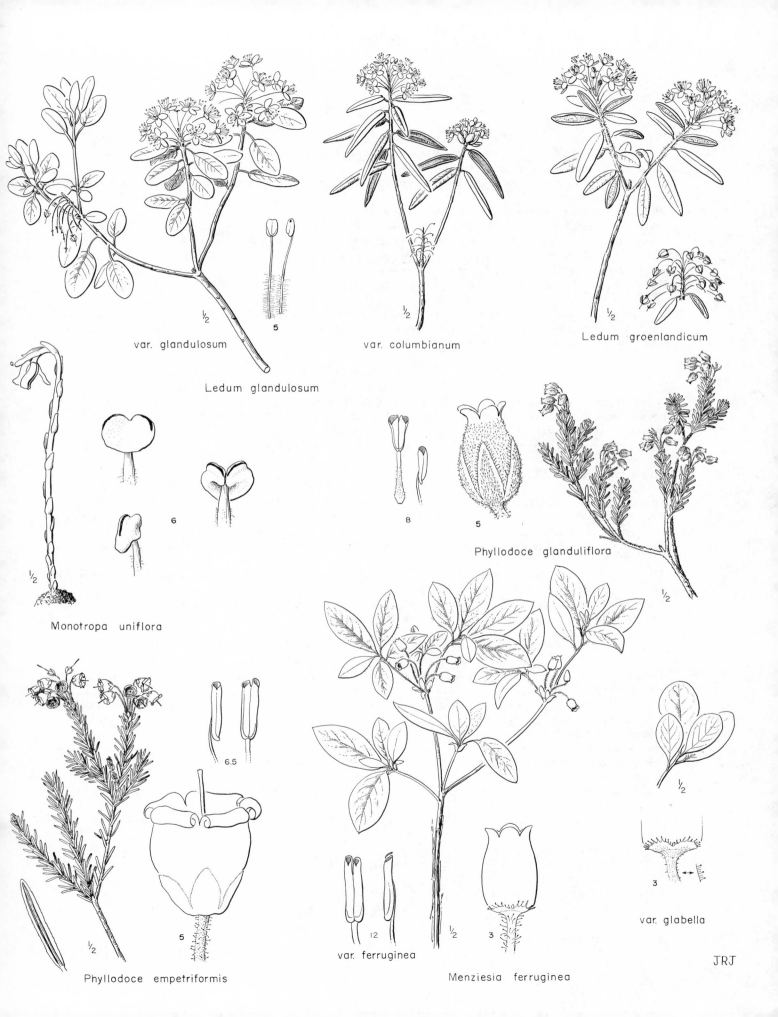

var. glandulosum

5

var. columbianum

Ledum groenlandicum

Ledum glandulosum

6

Monotropa uniflora

8          5

Phyllodoce glanduliflora

½

6.5

var. glabella

3

½

Phyllodoce empetriformis

5

12

var. ferruginea

½

3

Menziesia ferruginea

JRJ

ments pubescent, anthers about 1 mm. long; style included; ovary glandular. Yellow mountain heather.

Yukon and Alas. s. in the Rocky Mts. to Wyo., and in the Cascades to Oreg. July-Aug.

The two species, P. glanduliflora and P. empetriformis, tend to hybridize fairly frequently, producing plants intermediate in nature; these variant hybrids are readily recognizable as such since they are never found except in the presence of the two parents. Their nomenclatural treatment follows:

x Phyllodoce intermedia (Hook.) Camp, New Fl. & Silva 12:210. 1940.

Menziesia intermedia Hook. Fl. Bor. Am. 2:40. 1834. Bryanthus empetriformis var. intermedius Gray, Syn. Fl. 2¹:37. 1876. Phyllodoce intermedia Rydb. Mem. N.Y. Bot. Gard.
1:298. 1900. P. glanduliflora var. intermedia A. Nels. in Coult. & Nels. New Man. Bot.
Rocky Mts. 370. 1909. (Drummond, open places in the Rocky Mts.)

Phyllodoce hybrida Rydb. Mem. N.Y. Bot. Gard. 1:298. 1900. (Rydberg & Bessey 4657,
Old Hollowtop, Pony Mts., Mont.)

x Phyllodoce intermedia hort. clon. "Fred Stoker" of Camp, New Fl. & Silva 12:210. 1940.
P. "pseudoempetriformis" Stoker, New Fl. & Silva 12:40. 1939. (Horticultural form, all
probably propagated from one hybrid plant)

## Pityopus Small. Pine Foot

Similar to Hypopitys in nearly all respects but the plants drying to black; perianth segments also distinct but blotched with chocolate and often less fimbriate; stamens and pistil similar but the placentation parietal.

One species. (Name from the Greek pitys, pine tree, and pus, foot, from the habitat.)
Reference:
Copeland, H. F. On the genus Pityopus. Madroño 3:154-68. 1935.

Pityopus californica (Eastw.) Copeland, Madroño 3:155. 1935.

Monotropa californica Eastw. Bull. Torrey Club 29:75, pl. 7. 1902. M. hypopitys var. californica Domin. Ges. Wiss. Prag. II. 2:24. 1915. (Eastwood, Marin Co., Calif.)

Pityopus oregana Small, N. Am. Fl. 29:16. 1914. (Thomas Howell, Mt. Hood, Oreg., July
3, 1891)

Plant pinkish to yellow, drying to black; occurring with and very similar to Hypopitys monotropa in appearance, differing chiefly in the placentation.

Very rare, or at least very seldom collected; known only from a few stations w. of the Cascades in Oreg. (Mt. Hood and near Corvallis), and from n. Calif. June-July.

## Pleuricospora Gray. Sierra Sap; Fringed Pinesap

Flowers closely crowded into terminal, spikelike racemes, each 1-bracteate; sepals and pet-
als 4 (rarely 5-6), distinct; stamens 8(10), anthers unawned, 2-celled, linear, longitudinally dehiscent; ovary superior, 4-carpellary, 1-celled with 4-6 parietal placentae; stigma slightly 4-lobed; fruit baccate-capsular; fleshy, saprophytic, whitish to yellow-brown herbs with simple stems and scalelike, nongreen, usually fimbriate leaves.

One species of the Pacific Coast. (Name from the Greek pleura, side, and sporos, seed, re-
ferring to the parietal placentation.)
Reference:
Copeland, H. F. The reproductive structure of Pleuricospora. Madroño 4:1-16. 1937.

Pleuricospora fimbriolata Gray, Proc. Am. Acad. 7:369. 1868. (Bolander, Mariposa Grove,
Calif.)

P. longipetala Howell, Fl. N.W. Am. 429. 1901. (Howell, Clackamas Co., Oreg.)

P. densa Small, N. Am. Fl. 29:17. 1914. (C. F. Sonne, in Aug., 1896, Martis Valley,
Placer Co., Calif.)

Plant essentially glabrous throughout; stems 3-12(18) cm. tall; flowers 8-15 mm. long;
bracts slightly shorter than flowers, fimbriate-pectinate like the perianth segments; pedicels

lengthening to as much as 2 cm. after anthesis; sepals oblong-ovate, about half as long as the oblong petals that are a little broadened above; stamens included, slightly exceeding style; filaments broadened and flattened, glabrous or minutely puberulent; anthers linear, 2-3 mm. long, somewhat broader than the filaments.

Plants of deep forests, usually barely pushing above the duff; Sierra Nevada to n. w. Calif. and n. along the w. side of the Cascades to the Olympic Mts. of Wash. June-Aug.

## Pterospora Nutt. Pinedrops; Albany Beechdrops

Flowers many in a much elongated raceme; calyx deeply 5-parted; corolla urn-shaped, connate to near the top, the 5 lobes spreading; stamens 10, anthers ovoid, dehiscent nearly full length, with 2 dorsal, recurved awns; pistil 5-carpellary; ovary superior, 5-celled, the placentation axile; fruit a capsule; tall yellowish to reddish-brown saprophytes without chlorophyll.

One species of N. Am. (Name from the Greek pteron, wing, and sporos, seed, the seeds having a broad, netlike wing attached to one end. )

Reference:

Copeland, H. F. Further studies on Monotropoideae. Madroño 6:97-99. 1941.

Pterospora andromedea Nutt. Gen. Pl. 1:269. 1818. (<u>Whitlow</u>, Niagara Falls)

Stems unbranched, 3-10 dm. tall, reddish-brown, fleshy at anthesis but remaining as fibrous, dried stalks for one or more years, glandular-hairy, narrowly bracteate, especially below; raceme usually equal to the rest of the stem; flowers 5-8 mm. long, pendulous from the axils of much-reduced linear bracts; pedicels recurved, 5-15 mm. long, glandular-hairy, as are the sepals which are about half as long as the glabrous, pale yellow, globose-urceolate corolla; capsule depressed-globose, 8-12 mm. broad.

Deep humus in coniferous forests, especially under yellow pines (in our area); Alas. s. to s. Calif. and in the Rocky Mts. to Mex., e. to the Atlantic Coast. June-Aug.

## Pyrola L. Pyrola; Shinleaf; Wintergreen

Flowers regular to irregular, in terminal, several-flowered racemes on slender pedicels from linear-lanceolate bractlets, sometimes single and terminal; calyx persistent, the 5 lobes shorter than, to much longer than, the united portion; petals 5, distinct, often somewhat unequal, usually concave, deciduous; stamens 10, bent inward, anthers unawned, pendent, inverted, the basal portion (uppermost) dehiscent by 2 pores, these often lateral or terminal on short, tubular extensions; ovary superior, 5-locular; style straight or conspicuously bent to one side, often with small collar just below the inconspicuously lobed to large- and discoid-lobed stigma; fruit a capsule; autophytic to almost completely saprophytic, perennial, glabrous herbs with slender rhizomes, the sterile branches usually with rosettes of leaves; flowering stems leafless but with 1-few bracts, occasionally foliage leaves apparently entirely lacking.

About 15 species of N. Am. and Eurasia, barely reaching the tropics; ours usually in coniferous forests. The genus is sometimes divided, on questionably significant characters, into several genera. (From diminutive of Latin pirus, pyrus, the pear tree, the leaves of some species somewhat pearlike in shape. )

Attractive in the wild where they should be left, since they cannot be cultivated successfully.

References:

Camp, W. H. Aphyllous forms of Pyrola. Bull. Torrey Club 67:453-65. 1940.

Copeland, H. F. Observations on the structure & classification of the Pyroleae. Madroño 9:65-102. 1947.

1 Flowers single and terminal (Moneses)                                    P. UNIFLORA
1 Flowers two or more in racemes
  2 Style straight or nearly so, without a collar or ring below the peltate-lobed stigma; flowers less than 1 cm. broad; anthers opening at the upper end by large pores and without either horns or tubes projecting above the pores

3 Styles usually well over 2 mm. long; racemes secund; petals white, 4-5 mm. long
   (Ramischia)                                                                              P. SECUNDA
3 Styles not over 2 mm. long; racemes not secund; petals flesh color or pinkish, about
   3 mm. long (Erxlebenia)                                                          P. MINOR
2 Style bent to one side, often with a collar or ring below the stigma; flowers usually at least
  1 cm. broad; anthers usually with short horns or tubes and lateral pores (Pyrola)
    4 Plants without green leaves on the flowering branches, occasionally with 1-2 leaves on
      the sterile branches in addition to numerous scales                   P. APHYLLA
    4 Plants usually with 1-many green leaves on all the branches
      5 Leaves deep green but mottled on the upper surface with pale streaks above the main
        veins; blades mostly ovate to ovate-elliptic                     P. PICTA
      5 Leaves not mottled; blades various
        6 Leaves tapered to an acute base, spatulate or oblanceolate to rhombic-elliptic, usu-
          ally at least 3 cm. long and not much over 1/2 as broad, seldom over 2.5 cm. in
          width, pale green or bluish-green, thickish; petals greenish-white; racemes 10-
          to 30-flowered                                                  P. DENTATA
        6 Leaves usually ovate to orbicular and well over half as long as broad, many over 2.5
          cm. in width, often rounded at the base
          7 Flowers pale yellowish or greenish-white
            8 Racemes less than 10-flowered; largest leaf blades usually less than 3 cm.
              long, equaled or exceeded by the petioles; sepals as broad as long, rounded
              to somewhat acute; tubes of the anthers straight      P. VIRENS
            8 Racemes often with 10 or more flowers; most leaf blades over 3 cm. long,
              usually conspicuously exceeding the petioles; sepals acuminate, considerably
              longer than broad; tubes of the anthers curved       P. ELLIPTICA
          7 Flowers pinkish to rose-purple                                 P. ASARIFOLIA

Pyrola aphylla Smith in Rees, Cycl. 29: Pyrola no. 7. 1814.
  Thelaia aphylla Alef. Linnaea 28:39. 1856. Pyrola picta f. aphylla Camp, Bull. Torrey
    Club 67:464. 1940. (Menzies, west coast of N. Am.)
  Pyrola aphylla var. leptosepala Nutt. Trans. Am. Phil. Soc. II. 8:271. 1842. (Presumably
    collected by Nuttall near Ft. Vancouver)
  P. aphylla var. paucifolia Howell, Fl. N. W. Am. 425. 1901. (Howell, Cascade Mts.)
  P. sparsifolia Suksd. Allg. Bot. Zeit. 2:26. 1906. P. picta ssp. picta var. sparsifolia H.
    Andr. Allg. Bot. Zeit. 20:113. 1915. P. picta ssp. picta var. suksdorfii H. Andr. Allg.
    Bot. Zeit. 20:113. 1915. (Suksdorf 2695, Skamania Co., Wash.)
  Plant spreading by rather long, slender rootstocks, more or less reddish-brown; sterile
branches usually with several narrow, chlorophyll-bearing bracts and often with 1-3 elliptic
leaves 1-2 cm. long; flowering stems 5-20 cm. tall, leafless but with several lanceolate bracts
5-10 mm. long; racemes 10- to 20-flowered; pedicels spreading, 3-5 mm. long, from the
axils of lanceolate bracts about as long; flowers about 1 cm. broad; sepals reddish; petals
greenish-white to flesh colored or pinkish-brown, 6-8 mm. long; anthers apiculate at the lower
tips, the pores on the side of very short tubes.
  Wooded areas, chiefly in coniferous forests, B. C. to s. Calif., e. to Ida. and Mont. June-
Aug.
  It has been shown conclusively that plants of the above nature do not constitute a true species, but that they are leafless, or nearly leafless, forms of three or four other species,
most commonly of P. picta Smith, but occasionally of P. dentata, P. asarifolia, or P. virens.
Since the species of Pyrola are separable in considerable part on the basis of leaf peculiarity,
it probably will not be possible to assign these leafless forms to their proper taxa other than
"aphylla," although pinkish-flowered forms most probably will be P. asarifolia. Cf. Camp,
Bull. Torrey Club 67:453-65. 1940.

Pyrola asarifolia Michx. Fl. Bor. Am. 1:251. 1803.

P. rotundifolia var. asarifolia Hook. Fl. Bor. Am. 2:46. 1834. Thelaia asarifolia Alef.
Linnaea 28:54. 1856. (No type designated)

P. bracteata Hook. Fl. Bor. Am. 2:47. 1834. Thelaia bracteosa Alef. Linnaea 28:57. 1856.
Pyrola rotundifolia var. bracteata Gray, Bot. Calif. 1:460. 1876. P. asarifolia var. bracteata Jeps. Fl. Calif. 3:59. 1939. (Scouler, "N.W. Coast") = var. asarifolia.

P. rotundifolia var. purpurea Bunge, Mem. Sav. Acad. St. Pétersb. 2:542. 1835. P. asarifolia var. purpurea Fern. Rhodora 51:103. 1949. (Asia)

P. incarnata Fisch. in DC. Prodr. 7:773. 1839. P. rotundifolia var. incarnata Fisch. in
DC. Prodr. 7:773. 1839. P. asarifolia var. incarnata Fern. Rhodora 6:178. 1904. P.
asarifolia var. ovata Farw. Rep. Mich. Acad. Sci. 19:262. 1917. (Dahuria) = var. asarifolia.

P. uliginosa T. & G. ex Torr. Fl. N.Y. 1:453. 1843. P. rotundifolia var. uliginosa Gray,
Man. No. U.S. 2nd ed., 259. 1856. P. asarifolia var. uliginosa Farw. Rep. Mich. Acad.
Sci. 19:260. 1917. (Dr. Knieskern, Oneida Co., N.Y.) = var. asarifolia

P. elata Nutt. in Trans. Am. Phil. Soc. II. 8:270. 1843. (Nuttall, "shady woods of the
Oregon, near the confluence of the Wahlamet") = var. asarifolia.

P. bracteata var. hillii J.K. Henry, Torreya 14:32. 1914. (A. J. Hill, Mayne I., B.C.) =
var. purpurea.

Plant widely rhizomatous; flowering stems 1.5-4 dm. tall, usually with numerous basal
leaves, their blades rotund to elliptic or obovate, rounded to somewhat acute at the base and
rounded to retuse at the apex, coriaceous, entire to noticeably serrulate with excurrent veins,
(2)3-8 cm. long and nearly as broad, dark green and shining on the upper surfaces, usually
somewhat purplish beneath, not mottled; petioles usually at least as long as the blades;
racemes elongate, (5)10- to 25-flowered; pedicels 3-8 mm. long, nearly equaled by the linear-
lanceolate bractlets; flowers 10-15 mm. broad; calyx lobes acute to acuminate, 2.5-4 mm.
long; petals pinkish to rose or purplish-red, 5-7 mm. long; anthers apiculate at the lower end,
the tubes short and the pores nearly terminal; style strongly curved, 5-8 mm. long, with a
collar below the stigma.

Moist ground almost throughout w. U.S. to Alas., across Can. to n.e. N. Am. June-Sept.

There is a considerable variation in the length of the sepals and in the shape and degree of
serrulation of the leaves, and by these characters two or three species are sometimes recognized. It would seem that this variation is largely fortuitous and that only two poorly differentiated varieties exist: var. asarifolia, the more widespread, has nearly entire, often cordate
leaves, and sepals less than 3.5 mm. long; var. purpurea (Bunge) Fern. has sepals usually
at least 3.5 mm. long and leaf blades usually acute at one or both ends and noticeably serrulate owing to the excurrent veins; it occurs with the var. asarifolia but in our region is more
common w. of the Cascades.


Pyrola dentata Smith in Rees, Cycl. 29: Pyrola no. 6. 1814.

P. picta ssp. dentata Piper, Contr. U.S. Nat. Herb. 11:434. 1906. (Menzies, west coast of
N. Am., probably at Nootka, B.C.)

P. dentata var. integra Gray, Pac. R.R. Rep. 12:54. 1860. P. picta ssp. integra Piper,
Contr. U.S. Nat. Herb. 11:434. 1906. (Cooper, e. of Mt. Adams, Wash.)

P. pallida Greene, Pitt. 4:39. 1899. P. picta var. pallida Parish, Plant World 20:248. 1917.
(Greene 933, Yreka, Calif.)

P. dentata var. apophylla Copeland, Madroño 9:70. 1947. (H. F. Copeland, Jonesville, Butte
Co., Calif.)

Plant with widespread, slender rhizomes, the sterile branches with conspicuous rosettes of
leaves; flowering stems 1-several, 10-25 cm. tall, usually with several (rarely 1 or 2) basal
leaves; leaves mostly oblanceolate or obovate to rhombic or elliptic, the blades 2-6 cm. long,
often with petioles as long, acute at the base, glaucous and pale green or bluish-green, not
mottled, rather leathery, entire to serrulate; racemes 10- to 20-flowered; flowers about 1 cm.
broad; sepals greenish; petals cream to greenish-white, about 6 mm. long; anthers apiculate
at the lower tip, pores on the sides of short tubes; style strongly curved.

Coniferous forests, especially yellow pine, B.C. to the Sierra Nevada of Calif., e. to Mont., Ida., and Wyo. June-Aug.

Often very similar to P. picta and possibly only a phase of that taxon; entire- and serrulate-leaved forms often occur together throughout the range. Occasional leafless forms may be called P. dentata var. apophylla Copeland.

Pyrola elliptica Nutt. Gen. Pl. 1:273. 1818.
  Thelaia elliptica Alef. Linnaea 28:47. 1856. (Nuttall, Phila.)
  Plant spreading from slender rhizomes; flowering stems 15-25 cm. tall, with many basal leaves; leaves rather coriaceous, the blades broadly elliptic to oblong or obovate, mostly 3.5-7 cm. long and about 3/4 as broad, crenulate-serrulate, petioles very rarely as long; racemes (6)9- to 20-flowered; pedicels 3-8 mm. long, nearly equaled by the linear-lanceolate bractlets; flowers 10-12 mm. broad, white or greenish-white; calyx lobes longer than broad, usually acuminate, the tips somewhat reflexed; petals 6-7 mm. long; tubes of anthers usually somewhat recurved; style curved, 5-7 mm. long, with a distinct collar below the stigma.

In woods, n.e. N. Am. s. to Md., w. in Can. to the Rocky Mts. and B.C.; reported from Ida. June-July.

Pyrola minor L. Sp. Pl. 396. 1753.
  Amelia minor Alef. Linnaea 28:25. 1856. Erxlebenia minor Rydb. N. Am. Fl. 29:28. 1914.
    (Europe)
  Pyrola minor var. conferta C. & S. Linnaea 1:514. 1826. P. conferta Fisch. ex Ledeb. Fl.
    Ross. 2:930. 1846. (Chamisso, Unalaschka)
  Plant spreading from slender rhizomes, the flowering stems usually single, 1-2 dm. tall, with several basal leaves; leaves broadly elliptic to subrotund, rounded to subcordate at base, crenulate-serrulate, 1.5-3.5 cm. long, bright green; racemes 5- to 20-flowered, flowers rather crowded, 5-7 mm. broad; sepals pinkish; petals pale pinkish to rose, about 3 mm. long; anthers about 1 mm. long, ovate, apiculate, the pores terminal; filaments thin, flattened; style 1-2 mm. long, included, straight, without a collar; stigma 5-lobed.

In coniferous forests in our area, Alas. to s. Calif. e. to the Atlantic Coast, s. in the Rocky Mts. to Colo.; Eurasia. June-Aug.

Pyrola picta Smith in Rees, Cycl. 29: Pyrola no. 8. 1819. (Menzies, "N.W. Coast")
  P. septentrionalis H. Andr. Öst. Bot. Zeits. 63:71. 1913. (H. D. Langdille, Mt. Hood, Oreg.)
  P. conardiana H. Andr. Öst. Bot. Zeits. 63:73. 1913. (Conard 273, in part, Mt. Baldy,
    Quinault R., Wash.)
  P. paradoxa H. Andr. Verh. Bot. Ver. Brandenb. 54:220. 1913. (Conard 273, in part, Mt.
    Baldy, Quinault R., Wash.)
  P. picta f. aphylla (Smith) Camp, Bull. Torrey Club 67:464. 1940.
  Plant spreading by slender rhizomes, the sterile shoots with several leaves; flowering stems usually single, frequently leafless at base but with several lanceolate bracts (then usually classified as P. aphylla); leaves ovate to elliptic-rotund, 2-7 cm. long, coriaceous, deep green but grayish-mottled along the upper surfaces of the main veins, usually somewhat purplish on the lower surfaces, the margins thickened, entire to serrulate; stems reddish-brown, 10-25 cm. tall; racemes mostly 10- to 25-flowered; pedicels spreading, 4-8 mm. long, from lanceolate bracts often as long; flowers about 1 cm. broad; sepals reddish; petals 6-8 mm. long, yellowish or greenish-white to purplish; anthers apiculate on the lower tips, pores on the sides of short tubes.

Coniferous forests, B.C. to s. Calif., e. to the Rocky Mts. from Mont. to Colo. June-Aug.

Pyrola secunda L. Sp. Pl. 396. 1753.
  Ramischia secundiflora Opiz, Seznam 82. 1852. Actinocyclus secundus Klotzsch, Monats.
    Akad. Berl. 1857:14. 1857. Ramischia secunda Garcke, Fl. Deuts. 222. 1858. Orthilia
    secunda House, Am. Midl. Nat. 7:134. 1921. (Europe)
  Pyrola secunda var. obtusata Turcz. Bull. Soc. Nat. Mosc. 213-14:507. 1848. Orthilia se-
    cunda var. obtusata House, Am. Midl. Nat. 7:134. 1921. (Ircut R., near Tunka, s. Siberia)

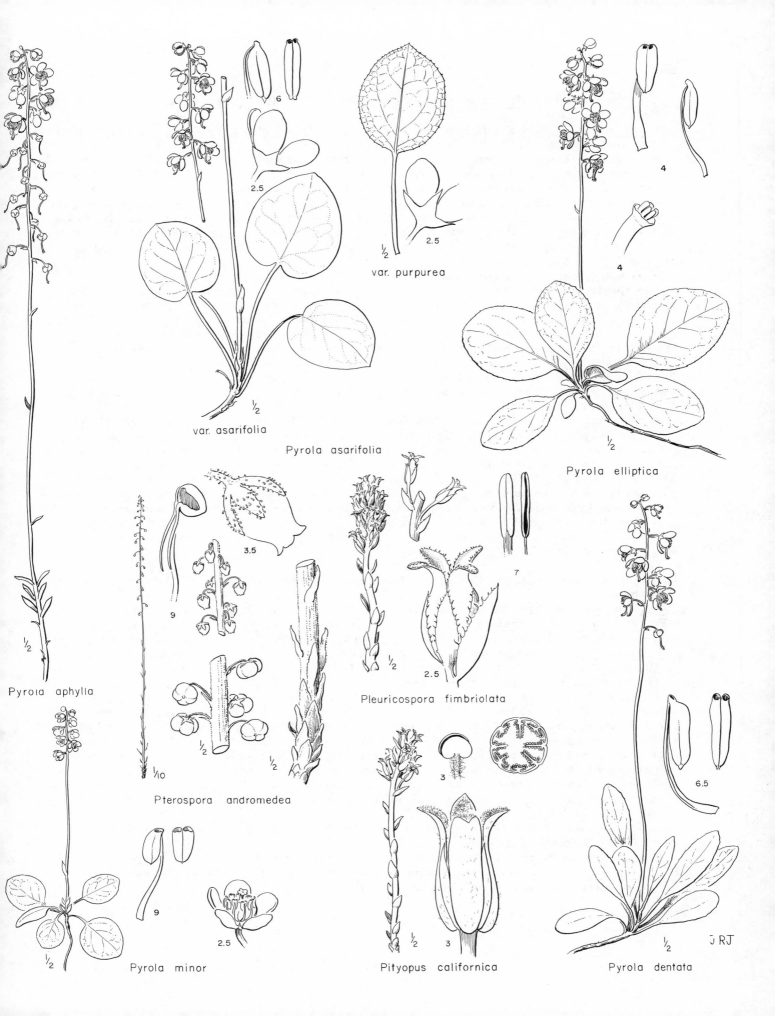

var. purpurea

var. asarifolia

Pyrola asarifolia

Pyrola elliptica

Pyrola aphylla

Pterospora andromedea

Pleuricospora fimbriolata

Pyrola minor

Pityopus californica

Pyrola dentata

J R J

P. secunda var. pumila Paine, Cat. Pl. Oneida Co. 135. 1865. (Paine, Summit Lake, Onei-
    da Co., N.Y.)
P. secunda f. eucycla Fern. Rhodora 28:223. 1926. (Fernald & Long 28801, n. of Doctor Hill,
    St. John Bay, Newf.)

Plant widespreading by slender rhizomes, the flowering stems usually single, 5-15 cm. tall;
leaves numerous, ovate to ovate-elliptic, 1.5-6 cm. long, crenulate to serrulate, rounded at
the base, clear green; racemes closely 6- to 20-flowered, secund; pedicels 3-8 mm. long,
from linear-lanceolate bracts nearly as long; flowers 5-6 mm. broad; calyx lobes erose; petals
4-5 mm. long, white, each with 2 small tubercles at the base on the inner surface; filaments
slender, about equaling the petals; anthers about 1.5 mm. long, not apiculate, the pores terminal, tubes lacking; style 3-4 mm. long, exserted, straight or nearly so, collar lacking;
stigma peltate, 5-lobed.

In woods, usually under conifers. June-Aug.

The species is separable into two rather poorly defined varieties: (1) var. secunda, a taller
plant with mostly elliptic to ovate, acute leaves, and racemes mostly more than 10-flowered;
Alas. to s. Calif. and Mex., e. to the Atlantic Coast and in Eurasia; (2) var. obtusata Turcz.,
a smaller plant, with smallish, more nearly rotund, obtuse leaves and often with racemes of
fewer than 10 flowers. Chiefly in the Rocky Mts., to Alas. and e. to Newf.; Siberia.

Pyrola uniflora L. Sp. Pl. 397. 1753.
Moneses uniflora Gray, Man. 273. 1848. M. grandiflora S. F. Gray, Nat. Arr. Brit. Pl.
    2:403. 1821. (N. Europe)
M. reticulata Nutt. Trans. Am. Phil. Soc. II. 8:271. 1843. M. uniflora var. reticulata
    Blake, Rhodora 17:28. 1915. (Nuttall, "firwoods and the Columbia not far from the sea")

Leaf blades ovate-elliptic to obovate, 1-2.5 cm. long, with petioles from 1/2 as long to
nearly equal in length, lightly to prominently serrate-crenate; scapes 3-15 cm. tall, usually
with 1(2) bracts about midlength; flowers single and terminal, 1.5-2.5 cm. broad, white; sepals about 1/4 the length of the petals, usually reflexed, erose-denticulate; petals ovate-lanceolate, spreading; style 2-4 mm. long, straight; capsule nearly globose, 6-7 mm. thick.

Woods, from sea level to well up in the mountains, usually on rotting wood, e. N. Am. s.
as far as Pa., w. to Alas., on both sides of the Cascades and in the Rocky Mts., to Calif.,
Ida., and N.M. June-July.

Pyrola virens Schweigg. in Schweigg. & Koerte, Fl. Erlang. 1:154. 1804. Date uncertain.
    (Europe)
P. chlorantha Sw. Svensk. Vet. Akad. Handl. 31:190. 1810. (Stockholm, Sweden)
P. chlorantha var. saximontana Fern. Rhodora 22:51. 1920. P. virens var. saximontana
    Fern. Rhodora 43:167. 1941. (Mrs. J. Clemens, in 1908, Yellow Bay, Flathead Lake, Mont.)
P. chlorantha var. paucifolia Fern. Rhodora 22:51. 1920. P. virens f. paucifolia Fern.
    Rhodora 43:167. 1941. P. chlorantha f. paucifolia Camp, Bull Torrey Club 67:464. 1940.
    (E. F. Williams, July 26, 1910, Atwell Hill, Piermont, N.H.)

Plant from widespread slender rhizomes; flowering stems usually single, 1-2 dm. tall,
naked or with 1-several leaves at the base, sterile shoots leafy; leaves rather coriaceous, pale
green on the upper surfaces, deeper green beneath, the blades from broadly elliptic or oblong-
elliptic to oblong-obovate or rotund, 1-2.4(3.5) cm. long, crenulate to crenulate-serrulate,
usually exceeded by the petioles; racemes usually (1)2- to 8(10)-flowered, not secund; pedicels
3-8 mm. long, the bracts linear-lanceolate, often as long; flowers 9-13 mm. broad, pale yellowish; calyx lobes rounded, broader than long; petals 5-6 mm. long; anthers with well developed straight tubes, the pores nearly terminal; style curved, 3-6 mm. long, with a distinct
collar below the lobed stigma. N=23.

Chiefly in coniferous forests, usually where moist, Alas. to Calif., e. in Canada to the Atlantic Coast; Eurasia. June-Aug.

Those plants of P. aphylla that can be seen to belong to this species might properly be called
P. virens f. paucifolia Fern. The species is sometimes called P. chlorantha Sm. because of
the uncertainty of the date of publication of P. virens.

## Rhododendron L.

Flowers showy, somewhat irregular; perianth 5-merous; calyx foliaceous to very much reduced and shallowly lobed; corolla gamopetalous, funnelform to shallowly campanulate, deeply lobed; stamens 5 or 10 (ours) to 25, anthers unawned, oblong, opening by terminal pores; ovary superior, pubescent, 5-celled; style straight or nearly so, stigma capitate or very shallowly lobed; fruit a capsule; deciduous or evergreen shrubs.

Several hundred species, mostly of s. e. Asia, a few in N. Am. and Europe. (Name from the Greek rhodon, rose, and dendron, tree.)

The genus includes some of the most highly prized ornamental shrubs in cultivation.

1 Stamens 5; leaves deciduous                                    R. OCCIDENTALE
1 Stamens usually 10
  2 Leaves deciduous; flowers in axillary clusters of 1-4      R. ALBIFLORUM
  2 Leaves persistent, leathery; flowers many in terminal corymbs  R. MACROPHYLLUM

Rhododendron albiflorum Hook. Fl. Bor. Am. 2:43. 1834.

  Azalea albiflora Kuntze, Rev. Gen. 2:387. 1891. Azaleastrum albiflorum Rydb. Mem. N.Y.
    Bot. Gard. 1:297. 1900 (Drummond, "Alpine woods of the Rocky Mountains")

  Cladothamnus campanulatus Greene, Erythea 3:65. 1895. (High mountains of Wash. and B.C.)

Plant 1-2 m. tall, young twigs finely puberulent and rather copiously strigose-hirsute with coarse reddish hairs, this same hairiness present on young leaves, bracts, and calyx; leaves deciduous, thin, deep clear green on the upper surfaces, lighter beneath, the blades elliptic-oblanceolate, entire to undulate, 4-9 cm. long, petioles 5-10 mm. long; flowers in axillary clusters of 1-4 along the stems, white or ochroleucous; pedicels 1-1.5 cm. long, glandular-pubescent as well as coarsely reddish-hirsute; calyx divided to the base, the lobes foliaceous, 8-10 mm.. long,. usually pubescent with a mixture of fine puberulence, coarser gland-tipped hairs, and (at least near tips) long, reddish, appressed hairs; corolla 1.5-2 cm. broad, lobed at least half the length, the lobes spreading, nearly equal, entire; stamens 10, exserted, filaments densely hairy on the lower half; anthers ovoid; ovary long-hairy, 4- to 5-celled; capsules short, heavy-walled. White rhododendron.

Wet places, usually along streams, chiefly montane, B.C. s. to Oreg., e. to w. Mont. June-Aug.

Rhododendron macrophyllum G. Don, Gen. Hist. Pl. 3:843. 1834.

  Hymenanthes macrophyllum Copeland, Leafl. West. Bot. 5:140. 1948. (Menzies, "N.W.
    Coast of America")

  Rhododendron californicum Hook. Curtis' Bot. Mag. 81: pl. 4863. 1855 (Lobb, mountains of
    Calif.)

  R. macrophyllum f. album Rehd. Journ. Arn. Arb. 28:254. 1947. Hymenanthes macrophyl-
    lum f. album Copeland, Leafl. West. Bot. 5:140. 1948. (J. E. Barto, May 3, 1930, Junc-
    tion City, Lane Co., Oreg.) An albino form.

Evergreen shrub 1-5 m. tall, branches glabrate, puberulent when young; leaves leathery, glabrous or nearly so, oblong-elliptic, 8-20 cm. long, entire; flowers many in terminal corymbs; pedicels glabrous; calyx minute, very shallowly lobed; corolla tubular-campanulate, 2.5-4 cm. long, pale pink to deep rose-purplish, deeply 5-lobed, lobes spreading, crisped-undulate; stamens 10, unequal, the longer ones well exserted; filaments sparsely short-hairy on the lower half; anthers oblong, 2-3 mm. long; ovary hairy; capsules woody, 1.5-2 cm. long. Rhododendron.

Coastal from B.C. through Wash. and into the Cascades through Oreg. s. to n. Calif. May-July.

Frequently cultivated west of the Cascades; probably hardier and in other respects superior to many of the Asiatic species. This, the state flower of Washington, has perhaps more commonly been called R. californicum, because Don described the Menzies plant as being white-flowered. Since this is the only Rhododendron on the west coast to which the two names could apply, there is no good reason to reject the earlier name R. macrophyllum.

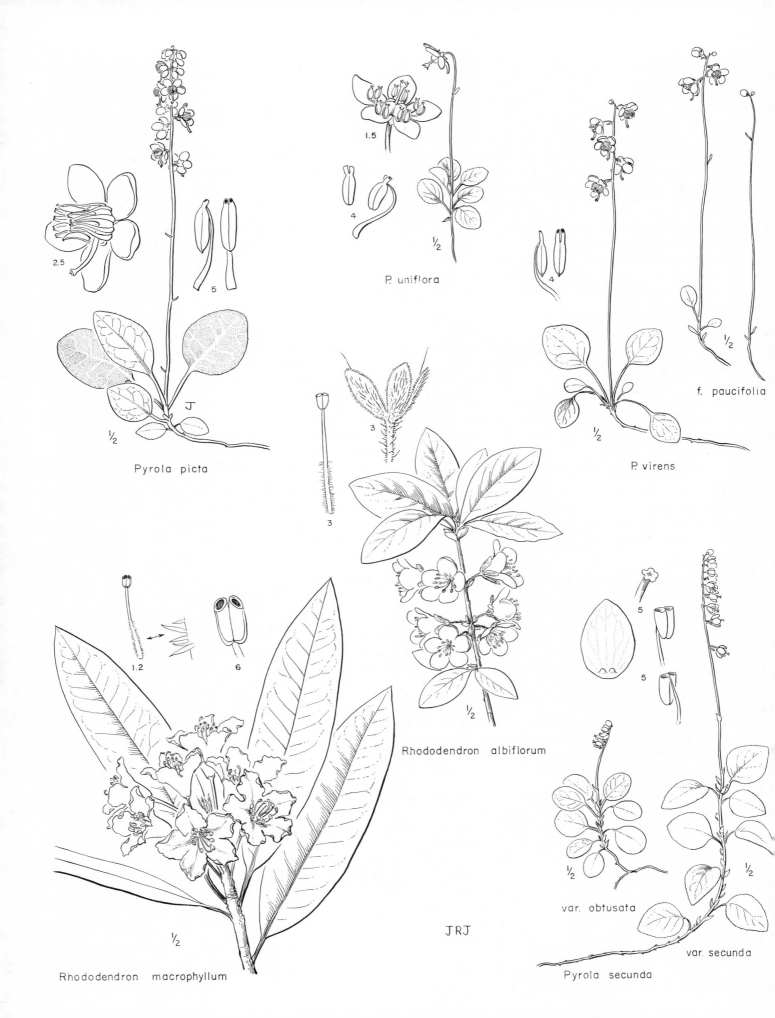

2.5

5

Pyrola picta

½

1.5

4

½

P. uniflora

3

3

4

½

f. paucifolia

½

½

P. virens

Rhododendron albiflorum

½

1.2

6

Rhododendron macrophyllum

½

5

5

5

½

var. obtusata

½

var. secunda

Pyrola secunda

JRJ

Rhododendron occidentale (T. & G.) Gray, Bot. Calif. 1:458. 1876.

Azalea occidentalis T. & G. Pac. R.R. Rep. 4:116. 1857. (Bigelow, Laguna de Santa Rosa, Calif.)

A. californica T. & G. ex. Dur. Journ. Acad. Phila. II. 3:94. 1855, (Pratten, Deer Creek, Nevada Co., Calif.) not R. californicum Hook.

Spreading shrub 1-5 m. tall, young twigs more or less hirsute; leaves deciduous, thin, bright yellow-green, entire, ciliate, elliptic to oblanceolate or narrowly obovate, 3-9 cm. long; flowers 5-20 in terminal corymbs, usually very fragrant; pedicels glandular, 1-3 cm. long; calyx lobes 2-6 mm. long, ovate-oblong, glandular; corolla irregular, glandular-pubescent on the outside, white to deep pink, the tube narrowly funnelform, 2-3 cm. long, about equal to the widely spreading undulate lobes, lower lobe with a deep yellowish, elongate blotch; stamens 5, filaments copiously hairy 2/3 of their length; ovary densely glandular-hairy. N=13. Western azalea.

Mountains of s. Calif. through much of the Sierra Nevada and coastal ranges to s. w. Oreg. as far as the Umpqua Valley, probably not quite reaching our range; very widely cultivated and one of the choicest of the Azaleas. May-July.

## Vaccinium L. Bilberry; Blueberry; Huckleberry; Cranberry

Flowers 1-several, axillary to terminal, and single or racemose; calyx obscurely to conspicuously 4- to 6-lobed, persistent or deciduous; corolla gamopetalous and globose or urceolate, to essentially polypetalous and the 4-5(6) segments reflexed; filaments glabrous or hairy; anthers dorsally awned or awnless, dehiscent by pores at the top of short to prominent tubes; ovary inferior; fruit a reddish to bluish berry; creeping to erect, deciduous to evergreen shrubs or shrublets with thin to leathery, entire to serrulate leaves.

Perhaps 150 species of colder or mountainous areas of all continents except Australia. (Latin name for blueberry.)

Most species are good ornamental shrubs as well as sources of edible fruit.

References:

Camp, W. H. A survey of the American species of Vaccinium, subgenus Euvaccinium. Britt. 4:205-47. 1942.

-------. A preliminary consideration of the biosystematy of Oxycoccus. Bull. Torrey Club 71: 426-37. 1944.

1 Leaves glossy-green, serrulate, evergreen; erect shrubs 0.3-4 m. tall; berry deep blue; filaments hairy, anthers unawned                                                   V. OVATUM

1 Leaves usually deciduous or entire or plants otherwise not as above

  2 Berry red; plants evergreen, 5-20 cm. tall; corolla gamopetalous; common in the Canadian Rockies and in n. B. C. but probably not native within our area where commonly grown as a ground cover                                      V. vitis-idaea L.

  2 Berry usually blue, but if red the leaves deciduous or the plants creeping and prostrate or the corolla polypetalous

    3 Corolla polypetalous, the petals reflexed; stamens well exserted, anthers unawned; stems slender, prostrate; leaves persistent (Oxycoccus)

      4 Bracts lanceolate to oblong, often foliaceous, 2-5 mm. long, borne well above midlength of the pubescent pedicel; flowers lateral along the stems
                                                                 V. MACROCARPON

      4 Bracts linear or linear-lanceolate, not foliaceous, usually less than 2 mm. long, borne at or below midlength of the pedicel; flowers often terminal on the stems
                                                                V. OXYCOCCOS

    3 Corolla gamopetalous, usually somewhat urn-shaped; stamens usually included, anthers awned; stems usually more or less erect; leaves deciduous

      5 Berry red; branchlets green, sharply angled; plants usually well over 1 m. tall; leaves usually entire                                                 V. PARVIFOLIUM

5 Berry blue or if red then either the plants less than 0.5 m. tall or the leaves serrulate
  6 Branches bright green or yellow-green, sharply angled, very numerous; leaves sharply
    serrulate, usually less than 15 mm. long; berry bright, but not deep, red; pedicels less
    than 3 mm. long; flowers about 4 mm. long; plants usually less than 2.5 dm. tall
                                             V. SCOPARIUM
  6 Branches usually neither bright green nor sharply angled, but if angled then the leaves
    mostly over 15 mm. long, or the berry not red, or plants over 2.5 dm. tall
    7 Plants usually 2-3 dm. tall; berry dark red to bluish; branches green and angled;
      leaves sharply serrulate, 1-3 cm. long; flowers single in the axils
                              V. MYRTILLUS
    7 Plants either over 4 dm. tall or branches not angled, or leaves entire or but slightly
      serrulate; flowers often more than one per axil; berry mostly deep bluish-black
      8 Calyx deeply lobed, the lobes triangular, persistent in fruit; flowers 1-4 per axil;
        buds with 4-7 scales, these conspicuous at the base of the pedicels; branches not
        angled; leaves entire
        9 Leaves usually at least half as broad as long, strongly veiny on the lower sur-
          faces; berry 6-8 mm. broad                    V. ULIGINOSUM
        9 Leaves usually much less than half as broad as long, not strongly veiny beneath;
          berry 4-5 mm. broad                      V. OCCIDENTALE
      8 Calyx shallowly lobed, the rounded lobes deciduous; tube of calyx forming a crowning
        ring on the fruit; branches often angled; flowers single in the leaf axils; bud scales
        two
      10 Plants usually 2-3 (less than 5) dm. tall; leaves more or less serrate above the
        middle, mostly no more than indistinctly serrulate on the lower half; twigs in-
        conspicuously angled to terete
        11 Corolla narrowly urn-shaped, twice as long as thick; filaments longer than
          the anthers (including tubes); leaves oblanceolate, strongly reticulate be-
          neath, not glaucous                    V. CAESPITOSUM
        11 Corolla more nearly globular, not twice as long as thick; filaments shorter
          than the anthers; leaves more obovate than oblanceolate, glaucous on the
          lower surfaces                     V. DELICIOSUM
      10 Plants either well over 5 dm. tall, or the leaves either entire or serrulate be-
        low midlength
        12 Leaves sharply serrulate nearly full length; twigs somewhat angled
        13 Leaves "globular," i. e. oblong-obovate, rounded to abruptly acute; co-
          rolla as broad as long                V. GLOBULARE
        13 Leaves ovate or ovate-oblong, less commonly oblong-obovate, but usual-
          ly long-pointed to somewhat acuminate; corolla longer than broad
                            V. MEMBRANACEUM
        12 Leaves entire or very inconspicuously serrulate and the serrulations most-
        ly or all below midlength
        14 Fruiting pedicels often 1 cm. long or longer, straight or nearly so,
          somewhat enlarged immediately below the ovary; leaves sparsely glan-
          dular along the midnerve on the lower surfaces, often puberulent, the
          veins not prominent; plants mostly over 6 dm. tall; corolla usually
          broader than long, bronzy-pink; style exserted  V. ALASKAENSE
        14 Fruiting pedicels mostly much less than 1 cm. long, considerably curved
          but not enlarged immediately below the fruit; leaves glabrous, not glan-
          dular along the midrib, the nerves prominent; plants often less than 6
          dm. tall; corolla usually longer than broad, pink; style usually included
                        V. OVALIFOLIUM

Vaccinium alaskaense Howell, Fl. N. W. Am. 412. 1901. (Cascade Mts. of Oreg. to Alas.,
collector not mentioned)

V. ovalifolium Bong. Mém. Acad. St. Pétersb. VI. 2:150. 1833, not of Smith in 1819.
V. oblatum Henry, Fl. So. B. C. 228. 1915. (Henry, Vancouver, B. C. )

Deciduous shrub 5-12 dm. tall, flowering when the leaves are considerably expanded; old bark grayish, young twigs somewhat angled, yellow-green, glabrous or very finely puberulent; leaves ovate-elliptic to elliptic-obovate, 2.5-6 cm. long, entire or very inconspicuously serrulate, the teeth mostly below midlength, the lower surfaces distinctly lighter, more or less glaucous, sparsely glandular-hirsute on the midnerve, and usually puberulent; flowers single in the axils, about 7 mm. long; pedicels 5-15 mm. long, usually at least 10 mm. long in fruit, straight or nearly straight, somewhat enlarged immediately below the ovary; calyx obscurely lobed; corolla bronzy-pink, gamopetalous, globose-urceolate, as broad as, or broader than, long, widest just above the base; style usually slightly exserted; filaments glabrous, shorter than the dorsally awned anthers which dehisce through slender, terminal pores; berry bluish-black (glaucous) to purplish-black and nonglaucous, 7-10 mm. broad, of good flavor. Blueberry.

Along the coast and in the Cascades from n. w. Oreg. to Alas. May-June.

Vaccinium caespitosum Michx. Fl. Bor. Am. 1:234. 1803. (Hudson Bay, Can. )
V. caespitosum var. cuneifolium Nutt. Trans. Am. Phil. Soc. II. 8:263. 1843. (Nuttall, plains of the Oregon, near the Wahlamet)
V. caespitosum var. angustifolium Gray, Proc. Am. Acad. 8:393. 1872. (Hall 340, Silver Cr., Oreg. )
V. caespitosum var. arbuscula Gray, Syn. Fl. 2¹:24. 1878. V. arbuscula Merriam, N. Am. Fauna 16:159. 1899. (Austin, Plumas Co., Calif. )
V. nivictum Camp, Britt. 4:211. 1942. (Abrams 12742, Desolation Valley, Eldorado Co., Calif. )

Plant spreading widely by rootstocks and forming mats 1.5-3(5) dm. tall; twigs usually somewhat angled, with yellowish-green to reddish bark, usually finely puberulent, sometimes glabrous; leaves usually oblanceolate, 1-3(5) cm. long, 1/4-1/2 as broad, acute to obtuse at the apex, cuneate-based, light green and glabrous (puberulent) on the upper surface, paler and usually somewhat glandular on the lower surface, obscurely to plainly serrulate from the tips to near midlength or below, the teeth usually gland-tipped; flowers single in the axils, whitish to pink, 5-6 mm. long; calyx obscurely lobed, the upper portion deciduous from the fruit; corolla gamopetalous, narrowly tubular-urceolate, twice as long as thick; filaments glabrous, exceeding the anthers; anthers with dorsal awns, the pollen sacs about equaling the slender apical pore-bearing tubes; berry glaucous-blue, nearly globose, 5-8 mm. broad, sweet and palatable. Dwarf bilberry, dwarf huckleberry.

Meadows and mountain slopes, Alas. to Calif., e. through Ida. to the Rocky Mts., and, further north, to the Atlantic Coast. May-July.

Rather variable and sometimes separated, by seemingly intangible characters, into two or three additional taxa.

Vaccinium deliciosum Piper, Mazama 2:103. 1901. (Allen 217, Mt. Rainier, Wash. )

Low, often matted shrub 1.5-3(6) dm. tall; branches inconspicuously angled, greenish-brown, glabrous or minutely puberulent; leaves obovate to obovate-oblanceolate, 1.5-5(6) cm. long, 1/3-1/2 as broad, rounded to obtuse (acute), cuneate-based, pale greenish, glaucous but not glandular on the lower surfaces, obscurely to plainly serrulate for the upper 1/2-2/3 of their length; flowers single in the axils, pinkish, 6-7 mm. long; calyx very obscurely lobed, the upper portion deciduous from the fruit; corolla gamopetalous, globular-urceolate, considerably less than twice as long as broad; filaments glabrous, shorter than the dorsally awned anthers which bear slender, terminal tubes; berry glaucous-blue, nearly globular, 6-8 mm. broad, very palatable. Blue huckleberry.

Montane in the Olympics Mts. and in the Cascades from s. B. C. to n. Oreg. May-June.

Vaccinium globulare Rydb. Mem. N. Y. Bot. Gard. 1:300. 1900. (Tweedy 1170, Spanish Cr., Mont., in 1886, first of several specimens cited)

Shrub (3)5-12 dm. tall, young twigs angled, greenish-yellow, glabrous, the older bark grayish and shredding; leaves globular, that is, more or less oblong-obovate but nearly cuneate at the base, rounded to abruptly acute at the apex, 2-4(5) cm. long, usually about half as broad, finely serrulate most of the length, distinctly glaucous, paler, and glandular on the lower surfaces but otherwise usually glabrous; flowers single in the axils, on pedicels 5-10 mm. long, pale pinkish-yellow, 6-7 mm. long; calyx obscurely lobed, the basal portion persistent on the fruits; corolla gamopetalous, more or less depressed-globular-urceolate, nearly as broad as long; filaments glabrous, shorter than the dorsally awned anthers that dehisce through slender terminal tubes; berry bluish-purple, 6-8 mm. broad. Blue huckleberry.

E. Wash. and Oreg. through Ida. to Mont. and Wyo. at lower and middle elevation in the mountains. May-June.

Vaccinium macrocarpon Ait. Hort. Kew. 2:13, pl. 7. 1789.
  Oxycoccus palustris var. macrocarpus Pers. Syn. 1:419. 1805. Schollera macrocarpa Britt.
    Mem. Torrey Club 5:253. 1894. (e. N. Am.)
  Creeping evergreen shrublet with stems 1-4 dm. long, finely puberulent-lanate when young; leaves narrowly elliptic-oblong to oblong, glossy on the upper surfaces, very pale beneath, 7-15 mm. long, margins slightly revolute, with petioles about 1 mm. long; flowers lateral on leafy branches from the axils of much-reduced bracts, pedicels pubescent, 2-4 cm. long, bearing, well above midlength, more or less foliaceous bractlets 3-10 mm. long; petals 4, distinct, deep pink, spreading or recurved, 6-10 mm. long, very narrow; filaments broad, sparsely ciliolate, about 1/3 the length of the anther sacs which are about equal to the terminal tubes, the anthers unawned; berry 1-2 cm. broad, deep red, more or less globose. N=12. Common cranberry.

Native over much of e. N. Am., occasionally escaped from cultivation and established near the coast in Wash. May-June.

Vaccinium membranaceum Dougl. ex Hook. Fl. Bor. Am. 2:32. 1834.
  V. myrtilloides var. macrophyllum Hook. Fl. Bor. Am. 2:32. 1834. V. macrophyllum Piper,
    Contr. U.S. Nat. Herb. 11:443. 1906. (Menzies, "N.W. Coast")
  Spreading shrub 0.5-2 m. tall, young twigs somewhat angled, yellow-green, glabrous or slightly puberulent, the old bark grayish and shredding; leaves ovate or oblong-ovate to elliptic-obovate but conspicuously attenuate and somewhat acuminate, rounded-acute at the base, 2-5 cm. long, finely serrulate-crenate nearly full length, sparsely glandular and somewhat paler on the lower surfaces; flowers about 6 mm. long, single in the axils on pedicels 5-10 mm. long, pale yellowish-pink; calyx obscurely lobed; corolla gamopetalous, ovoid-urceolate, about 1/3 again as long as broad; filaments glabrous, shorter than the dorsally awned anthers; berries purple or dark purplish-reddish, not glaucous, 7-9 mm. broad. Mountain huckleberry.

Mountain slopes, B.C. to n. Calif. on both sides of the Cascades and Olympic Mts., to e. Mont. and Ida. Apr.-June.

Occasional red-fruited plants are to be found (e.g. Stevens Pass, J. H. Schultz 4422) which otherwise are not separable from those with bluish-black fruits.

Vaccinium myrtillus L. Sp. Pl. 1:349. 1753. (Europe)
  V. oreophilum Rydb. Bull. Torrey Club 33:148. 1906. (Watson 735, Uintah Mts.)
  Plant (1.5)2-3(4) dm. tall, branches many but thicker and less numerous than in V. scoparium, strongly angled, usually puberulent, greenish; leaves ovate to elliptic-lanceolate, 1-3 cm. long, light green, sharply serrulate, strongly veiny on the lower surfaces; flowers single in the axils, about 5 mm. long, pedicels 2-3 mm. long; calyx very shallowly lobed; corolla pinkish, gamopetalous, urceolate-campanulate; filaments glabrous; anthers with dorsal awns, the pore-bearing terminal tubes about equaling the pollen sacs; berry dark red to bluish, 5-8 mm. broad. N=12. Dwarf bilberry.

Rocky Mts. from Alta. to N.M. w. to B.C. and s., on the e. side of the Cascades, to the Wenatchee Mts. of Wash; Eurasia. May-Aug.

Very similar to V. scoparium but more often pubescent, and with somewhat larger flowers and leaves, less numerous branches, and a darker red to bluish berry.

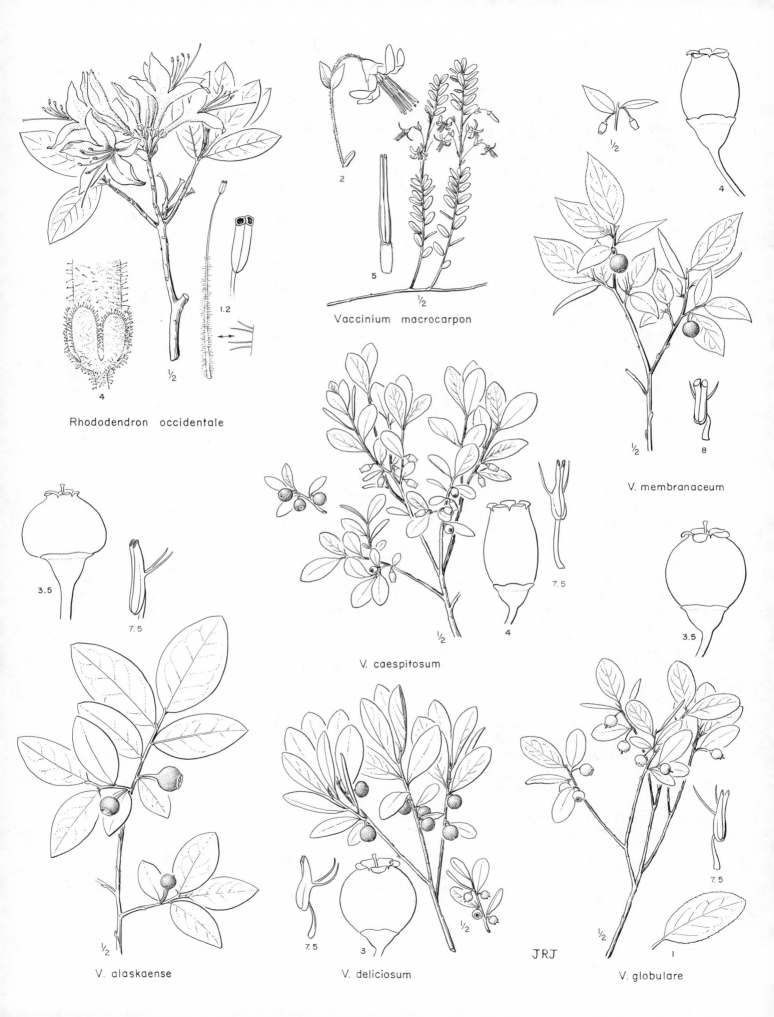

Rhododendron occidentale

Vaccinium macrocarpon

V. membranaceum

V. caespitosum

V. alaskaense

V. deliciosum

V. globulare

JRJ

Vaccinium occidentale Gray, Bot. Calif. 1:451. 1880. ("Sierra Nevada at 6,000 or 7,000 feet, from Mariposa to Sierra Co.," Bolander, the first of three collections cited)

Very similar to V. uliginosum; plant 2-6 dm. tall, branches not angled, yellowish-green, usually glabrous; leaves mostly oblanceolate, 1-3 cm. long, usually less than 1/2 as broad, entire, glaucous and not strongly reticulate on the lower surfaces; buds with 4-7 scales; flowers 1-4 per axil, about 6 mm. long; corolla pinkish, gamopetalous, urceolate-campanulate; filaments glabrous; anthers awned, the pore-bearing terminal tubes about equaling the pollen sacs; berries blue, glaucous, 4-5 mm. broad. Western huckleberry.

B. C. s., mostly on the e. side of the Cascades, to the Sierra Nevada and n.w. Calif., e. to Mont. and n. Utah. June-July.

Vaccinium ovalifolium Smith in Rees, Cycl. 36: Vaccinium no. 2. 1817. (Menzies, "West Coast of North America")

V. chamissonis Bong. Mém. Acad. St. Pétersb. VI. 2:151. 1833. (Chamisso, Unalaska)

Spreading deciduous shrub 4-10 dm. tall, flowering before the leaves have reached half their mature size; twigs yellowish-green, conspicuously angled, glabrous, old branches grayish; leaves ovate-elliptic, 2-4(5) cm. long, glabrous, more or less glaucous on the lower surfaces, entire or very slightly serrulate; flowers borne singly, about 7 mm. long; pedicels 1-5 mm. long, strongly recurved in fruit, not enlarged immediately under the berry; corolla pinkish, gamopetalous, tubular-urceolate, usually longer than broad, broadest just below midlength; style usually included; filaments glabrous, shorter than the dorsally awned, terminal-pored anthers; berry purplish-black to bluish-black, 6-9 mm. long, of good flavor but drier and not so acid as that of V. alaskaense. Blueberry.

Alas. s. to Oreg. in the Cascades and w., e. to Mont. and in e. U.S. and e. Asia. May-July.

Vaccinium ovatum Pursh, Fl. Am. Sept. 290. 1814.

Metagonia ovata Nutt. Trans. Am. Phil. Soc. II. 8:264. 1843. (Lewis, Columbia R.)

Vaccinium lanceolatum Dunal in DC. Prodr. 7:570. 1839. (Douglas, "Nova California")

V. ovatum var. saporosum Jeps. Man. Fl. Pl. Calif. 751. 1925. (Robt. Brandt, Gualala, Mendocino Co., Calif.)

Spreading to upright evergreen shrub, 0.5-4 m. tall; young stems puberulent; leaves disposed in horizontal rows, the branches with the leaves thus noticeably dorsiventral, the blades ovate to ovate-lanceolate, 2-5 cm. long, acute, leathery, bright glossy green on the upper surfaces, paler beneath, the margins thickened and sharply serrulate; flowers axillary in small 3- to 10-flowered racemes or corymbs; corolla bright pinkish, gamopetalous, narrowly campanulate, about 7.5 mm. long, the lobes short and spreading; stamens included; filaments densely pubescent, anthers with long, straight tubes above the pollen sacs but not awned; berry deep purplish-black, 4-7 mm. broad, glaucous or blackish and shining, sweet and edible but with a fairly strong musky flavor. N=12. Evergreen huckleberry.

From the w. side of the Cascades to the coast, B. C. to the redwood area of Calif., sporadic southward to s. Calif. Apr.-Aug.

A very ornamental shrub especially when used as ground cover or with Rhododendrons, tending to become spindly and more or less clambering with extremes of moisture and shade; gathered extensively by "brush-pickers" for florists' use. There is much variation in the color and quality of the berries, which are gathered extensively for table use. The glaucous phase has been designated var. saporosum Jeps. In our area the berries of some plants turn reddish-brown and are not nearly so juicy as the blackish ones.

Vaccinium oxycoccos L. Sp. Pl. 351. 1753.

Oxycoccus oxycoccos MacM. Bull. Torrey Club 19:15. 1892. O. palustris Pers. Syn. 1:419. 1805. (Europe)

Vaccinium oxycoccus var. intermedium Gray, Syn. Fl. 2nd ed. $2^1$:396. 1886. Oxyxoccus palustris var. intermedium Howell, Fl. N.W. Am. 413. 1901. O. oxycoccus ssp. intermedius Piper, Contr. U.S. Nat. Herb. 11:444. 1906. O. intermedius Rydb. Fl. Rocky Mts. 646, 1065. 1917. (E. U.S.)

Oxycoccus quadripetalus Gilib. Fl. Lithuan. 1:5. 1781. (Europe) = var. intermedium.
   Creeping shrub with very slender, glabrous to finely pubescent stems; leaves persistent, ovate to lanceolate, acute, 5-15 mm. long, deep green and shining on the upper surface, grayish beneath, the margins strongly revolute; flowers 1-several, terminal or lateral on the stems; pedicels very slender, 2-4 cm. long, glabrous or finely pubescent, with 2 tiny bractlets 1-2.5 mm. long, borne usually well below midlength but sometimes slightly above; flowers deep pinkish; perianth 4-merous, petals distinct, 5-8 mm. long, usually recurved; filaments ciliate, usually at least half as long as the unawned anthers (including the pore-bearing apical tubes); berry 5-10 mm. broad, deep red. N=36. Wild cranberry.
   Across the continent in Can., s., usually in sphagnum bogs, to Oreg. and Ida. in the west; Eurasia. May-July.
   There are two phases of the species in much of its range: one, which has been called var. intermedium Gray, is a larger leaved, coarser plant with leaves usually over 7 mm. long, petals usually over 6 mm. long, and pedicels usually puberulent; the other, var. oxycoccos, usually has leaves not over 7 mm. long, petals not over 6 mm. long, and glabrous pedicels.

Vaccinium parvifolium Smith in Rees, Cycl. 36: Vaccinium no. 3. 1817. (Menzies, "West
   Coast of North America")
   Erect shrub 1-4 m. tall; branches green, very prominently angled, glabrous or minutely puberulent when young; leaves tardily deciduous, often a few persistent, oval to oblong-elliptic, usually rounded, 1-2.5 cm. long, glabrous or somewhat puberulent, usually entire but on juvenile growth sometimes serrulate; flowers axillary on short pedicels; corolla pale, waxy, yellowish-pink, gamopetalous, rather broadly campanulate-urceolate, about 4 mm. long; filaments glabrous, anthers with prominent spreading-erect awns, the apical pore-bearing tubes short; berry globose, bright red, 6-9 mm. broad. N=12. Red huckleberry.
   From s.e. Alas. to c. Calif. on the w. side of Cascades in our area. Apr.-June.
   The berries are rather sour but have a good flavor; the plants are attractive and desirable as ornamentals, some of the reddish-colored leaves tending to hang on the bright green branches until ruined by heavy freezes.

Vaccinium scoparium Leiberg, Mazama 1:196. 1897.
   V. myrtillus var. microphyllum Hook. Fl. Bor. Am. 2:33. 1834. V. microphyllum Rydb.
      Bull. Torrey Club 24:251. 1897, not of Rein. in 1826. V. erythrococcum Rydb. Mem. N.Y.
      Bot. Gard. 1:301. 1900. (Drummond, "Alpine woods near the Height of Land and Columbia
      Portage")
   Plant usually more or less matted, 1-2.5 dm. tall, the branches many, slender, broomlike, strongly angled, greenish or yellowish-green, glabrous or occasionally sparsely finely puberulent; leaves lanceolate to ovate-lanceolate, 8-15 mm. long, finely serrulate, light-green, glabrous or very minutely puberulent, conspicuously veiny on the lower surfaces; pedicels 2-2.5 mm. long; flowers about 4 mm. long; corolla gamopetalous, pinkish, campanulate-urceolate; calyx very shallowly lobed; filaments glabrous; anthers awned, the pore-bearing terminal tubes about equal to the pollen sacs; berry globose, bright red, 3-5 mm. broad, sweetish. Grouseberry, whortleberry.
   B.C. to n. Calif., e. through Ida. to Alta. and S.D., s. in the Rocky Mts. to Colo., usually rather well up in the mountains. May-Aug.
   The berries are readily eaten by birds and other animals.

Vaccinium uliginosum L. Sp. Pl. 350. 1753. (Europe)
   V. uliginosum var. mucronatum Herder, Acta. Hort. Petrop. 1:321. 1872, nom. nud.; ex
      Gray, Syn. Fl. $2^1$:23. 1878. (Unalaska and the Alaska panhandle)
   Plant 2-5 dm. tall, freely branched; stems not angled, young branches yellowish-green, finely puberulent, old bark grayish-red; buds with 4-7 scales; leaves entire, oblanceolate to obovate, 1-3 cm. long, from not quite half so broad to nearly equally wide, firm in texture, strongly reticulate on the lower surfaces; flowers 1-4 per axil, the several bud scales usually evident; flowers 5-6 mm. long, pink; sepals ovate-deltoid, persistent on the mature fruit; co-

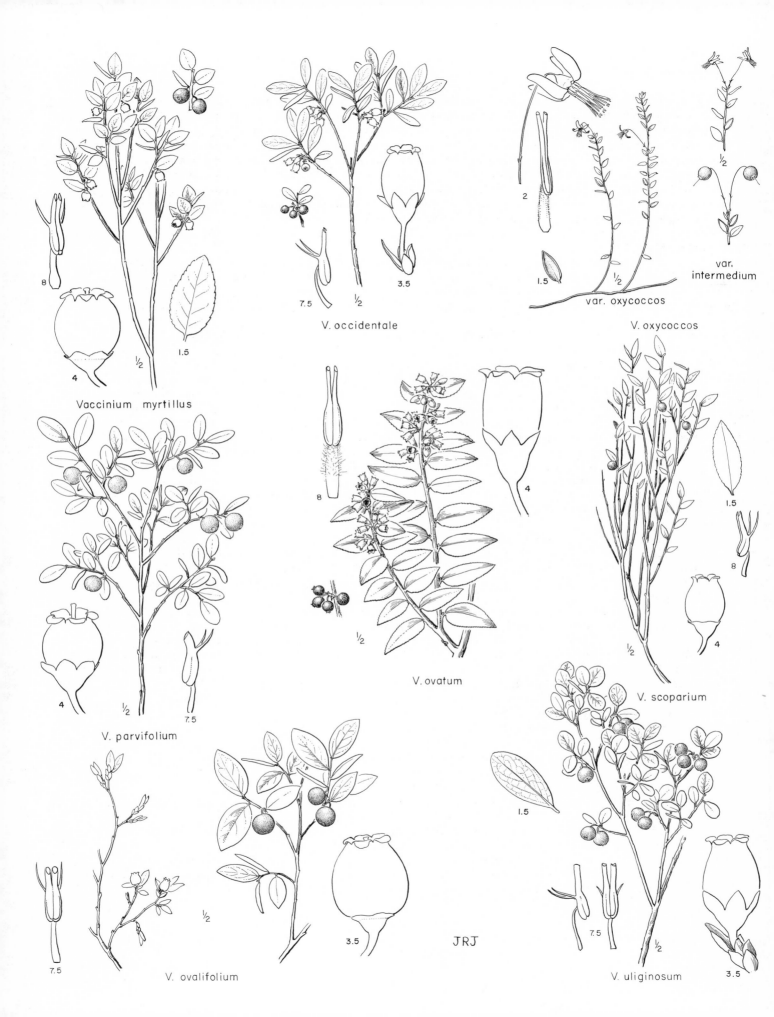

8

4

½        1.5

Vaccinium myrtillus

7.5        ½        3.5

V. occidentale

2

1.5        ½

var. oxycoccos

½

var.
intermedium

V. oxycoccos

4

½        7.5

V. parvifolium

8

½

V. ovatum

4

1.5

8

½        4

V. scoparium

7.5

V. ovalifolium        ½        3.5

JRJ

1.5

7.5        ½

V. uliginosum        3.5

rolla awned, the pore-bearing terminal tubes about equaling the pollen sacs; berry blue, glaucous, 6-8 mm. broad. N=12, 24. Bog blueberry.

Bogs along the coast, Alas. to n. Calif., e. in Can. to the Atlantic Coast: Eurasia. May-July. The berries have a good flavor.

## PRIMULACEAE. Primrose Family

Flowers perfect, regular, hypogynous (all ours), 5(4-9)-merous, axillary and single, or in terminal or axillary racemes sometimes on leafless scapes; calyx shallowly to deeply lobed; corolla (wanting in Glaux) gamopetalous, from rotate and often very deeply lobed to tubular or salverform and shallowly lobed; stamens equal to and opposite the petals, inserted on the tube, occasionally alternating with small staminodia, included to much exserted; ovary superior (ours), rarely partially inferior, 1-celled with free-central placentation; style single, stigma usually capitate; fruit a valvate or circumscissile, few- to many- seeded capsule; annual or perennial herbs, often scapose, with simple, mostly entire, alternate or opposite (whorled), exstipulate leaves.

About 25 genera and 700 species of wide distribution, chiefly in the N. Temp. and Arctic zones. Cyclamen, Dodecatheon, Primula, and Douglasia are well-known cultivated members of the family.

1 Plant a caespitose perennial with cushions of persistent, small, narrow leaves; flowering
    stems leafless, flowers 1-several, showy, pink to violet      DOUGLASIA
1 Plants not caespitose, or if so, the corollas white or the stems leafy
  2 Leaves in basal rosettes; flowering stems leafless
    3 Lobes of the corolla several times as long as the tube, sharply reflexed; stamens pro-
      truding their full length      DODECATHEON
    3 Lobes of the corolla less than twice as long as the tube, not sharply reflexed; stamens
      usually included
      4 Flowers white, usually less than 5 mm. long, if larger, then the basal leaves dense-
        ly silky-hairy      ANDROSACE
      4 Flowers usually pink to lavender, well over 5 mm. long; leaves never silky-hairy
        PRIMULA
  2 Leaves not confined to basal rosettes; flowering stems leaf-bearing
    5 Flowers sessile in the leaf axils
      6 Corolla absent, sepals petaloid; perennials with (mostly) opposite leaves; capsules
        valvate      GLAUX
      6 Corolla present; tiny annuals with (mostly) alternate leaves; capsules circumscis-
        sile      CENTUNCULUS
    5 Flowers pedicellate, often terminal
      7 Leaves less than 2 cm. long; prostrate annual; capsules circumscissile
        ANAGALLIS
      7 Leaves usually well over 2 cm. long; perennial; capsules valvate
        8 Stems less than 3 dm. long; leaves alternate, usually clustered at the top of the
          stems; perianth 5- to 9-merous      TRIENTALIS
        8 Stems usually over 3 dm. long; leaves opposite or whorled, not clustered at the
          stem tips; perianth 5(6)-merous
          9 Flowers solitary in the axils; leaves ovate; plants not dark-maculate; stami-
            nodia present      STEIRONEMA
          9 Flowers usually racemose, if solitary and axillary the leaves usually oval;
            plants reddish- or purplish-black-maculate; staminodia usually absent
            LYSIMACHIA

### Anagallis L. Pimpernel

Flowers axillary, pedicellate, 5-merous; calyx deeply cleft; corolla rotate, divided nearly to the base; capsule circumscissile; low prostrate annual with opposite (ternate) leaves.

About 20 species, widespread. (From the Greek name for pimpernel.)

Anagallis arvensis L. Sp. Pl. 148. 1753. (Europe)

Stems prostrate or ascending, 1-4 dm. long, glabrous; leaves ovate, 5-15 mm. long, entire, sessile and somewhat clasping-based; pedicels filiform, 1-4 cm. long, gracefully curved, usually recurved in fruit; sepals linear-lanceolate, acuminate, 2-4 mm. long, nearly distinct; corolla 5-8 mm. broad, salmon, the lobes very finely ciliolate; filaments hairy; capsules spherical, 3-5 mm. broad. Scarlet pimpernel, poor man's weatherglass.

Native of Eurasia, now widely introduced as a weed in temperate regions including the Pacific Coast; fairly abundant in Calif., less common in Oreg., and only occasional in w. Wash. and s. Vancouver I.; reported from Ida. May-June.

## Androsace L.

Flowers in involucrate umbels of 2-25, small, white or cream-colored, 5-merous; calyx turbinate to hemispheric, connate about half the length; corolla tubular-campanulate, contracted and sometimes fornicate in the throat, lobes spreading, shorter than the tube; stamens included; capsule membranous, valvular-dehiscent from the top; seeds pitted, brownish; low annual or perennial herbs with basal rosettes of simple leaves and 1-several naked scapes.

About 100 species of the N. Temp. and Arctic zones, especially common in Asia. (Name from the Greek androsakes, an uncertain sea plant.)

Reference:

Robbins, G. Thomas. North American species of Androsace. Am. Midl. Nat. 32:137-63. 1944.

1 Perennial, grayish-pilose; scapes single from each rosette; corolla lobes over 2 mm. long
                                                                                    A. LEHMANNIANA
1 Annuals or biennials, not pilose; scapes usually several from the leaf rosette
  2 Flowers less than 3 mm. long; calyx hemispheric; plant glabrous or sparsely glandular-puberulent above; leaves abruptly narrowed to distinct petioles; seeds light yellow, less than 0.5 mm. long                                                      A. FILIFORMIS
  2 Flowers often more than 3 mm. long; calyx more nearly turbinate; plants usually puberulent; leaves narrowed gradually to the base; seeds dark brown, over 0.5 mm. long
    3 Involucral bracts oblong to oblong-obovate, less than 4 times as long as broad; corolla scarcely exceeding the calyx tube; calyx lobes ovate-lanceolate, about equaling the tube                                                                    A. OCCIDENTALIS
    3 Involucral bracts lanceolate to linear, at least 4 times as long as broad; corolla exserted from the calyx tube; calyx lobes shorter than the tube, usually nearly deltoid
                                                                                    A. SEPTENTRIONALIS

Androsace filiformis Retz. Obs. Bot. 2:10. 1781. (Siberia)

A. capillaris Greene, Pitt. 4:148. 1900. (Rocky Mts.)

Delicate many-stemmed annual; leaves glabrous, 1-3 cm. long, the blades ovate to deltoid, denticulate-serrulate, narrowed abruptly to broad petioles about as long; scapes 3-12 cm. tall, glabrous or very sparsely glandular-puberulent above; umbels many-flowered; bracts several, 2-4 mm. long; pedicels very slender, 1-4 cm. long, glabrous or sparsely glandular-puberulent; calyx hemispheric, about 2 mm. long, lobes triangular, 3-nerved; corolla about equaling the sepals, lobes oblong-ovate, 1-1.5 mm. long; capsules about twice the length of the calyx, walls thin and papery; seeds light yellowish-brown, 0.2-0.3 mm. long.

Rocky Mts. from Mont. to Colo., w. along the Columbia R. to Klickitat Co., Wash., and n. Willamette Valley, Oreg.; Asia. June-July.

Androsace lehmanniana Spreng. Isis 1:1289, pl. 9. 1817. ("Oriente")

A. chamaejasne of many American authors, not of Host.

A. albertina Rydb. Bull. Torrey Club 40:462. 1913. Drosace albertina Rydb. Fl. Rocky Mts. 649. 1917 (Mr. & Mrs. C. Van Brunt 77, "Lake Agnes, National Park, Banff, Alta.," Aug., 1897)

Perennial with a branched caudex and prostrate stems each with a terminal rosette, forming mats as much as 10 cm. broad, copiously long-pilose; leaves oblanceolate, 5-15 mm. long; scapes single from each rosette, 3-10 cm. tall; umbels 2- to 8-flowered, compact, the pedicels usually shorter than the flowers; calyx about 3 mm. long; corolla 5-7 mm. broad, white, with yellow eye.

Arctic America; Alas. s. to the Big Snowy Mts. on the e. side of the Rockies in Mont.; Asia. June-July.

Potentially this is a beautiful rock garden plant, but it cannot be grown on the w. side of the Cascades and perhaps it is not amenable to cultivation elsewhere.

Androsace occidentalis Pursh, Fl. Am. Sept. 137. 1814.

Amadea occidentalis Lunell, Am. Midl. Nat. 4:504. 1916. (Nuttall, Banks of the Missouri)

Androsace platysepala Woot. & Standl. Bull. Torrey Club 34:519. 1907. (Metcalfe 1547, Kingston, Sierra Co., N.M., March 30, 1905)

A. simplex Rydb. Bull. Torrey Club 40:462. 1913. A. occidentalis var. simplex St. John, Vict. Mem. Mus. Mem. 126:53. 1922. A. occidentalis f. simplex Robbins, Am. Midl. Nat. 32:153. 1944. (Elrod et al. 33, Missoula, Mont., May, 1897)

Pubescent annual, leaves in a single rosette, lanceolate to oblanceolate or spatulate, 1-3 cm. long, entire to denticulate, grayish-puberulent with simple hairs; scapes 1-many, 3-10 cm. tall, pubescent with branched hairs; umbels 3- to 10-flowered, bracts oblong to oblanceolate or oblong-obovate, 4-10 mm. long, about 1/3 as broad; pedicels rather slender, 1-3 cm. long; calyx turbinate-campanulate, 4-5 mm. long, pubescent, carinate below each lobe, tube about equaled by the expanded and more greenish lanceolate-ovate lobes; corolla white, nearly included in the calyx tube; capsule globose, about equaling the calyx tube; seeds dark brown, nearly 1 mm. long.

Mississippi Valley w. to the w. slopes of the Rocky Mts., from B.C. to N.M., also into Ariz. and the Sierra Nevada of Calif. Apr. -June.

Androsace septentrionalis L. Sp. Pl. 142. 1753. (Europe)

A. septentrionalis var. subulifera Gray, Syn. Fl. 2¹:60. 1878. Primula septentrionalis var. subulifera Derganc, Allg. Bot. Zeit. 10:110. 1904. Androsace subulifera Rydb. Bull. Torrey Club 33:148. 1906. A. septentrionalis ssp. subulifera Robbins, Am. Midl. Nat. 32:158. 1944. (H. G. French, near Boulder, Colo.)

A. septentrionalis var. subumbellata A. Nels. Bull. Wyo. Exp. Sta. 28:149. 1896. A. subumbellata Small, Bull. Torrey Club 25:319. 1898. A. septentrionalis ssp. subumbellata Robbins, Am. Midl. Nat. 32:160. 1944. (A. Nelson 998, Union Pass, Wyo.)

A. diffusa Small, Bull. Torrey Club 25:318. 1898. A. septentrionalis var. diffusa Knuth, Engl. Pflanzenr. IV. 237:215. 1905. Amadea diffusa Lunell, Am. Midl. Nat. 4:504. 1916. (No type specified)

Androsace arguta Greene, Pitt. 4:148. 1900. (W. G. Hay, Bering Strait)

A. gormani Greene, Pitt. 4:149. 1900. (Gorman 981, Yukon Valley)

A. puberulenta Rydb. Bull. Torrey Club 30:260. 1903. A. septentrionalis var. puberulenta Knuth, Engl. Pflanzenr. IV. 237:216. 1905. Amadea puberulenta Lunell, Am. Midl. Nat. 4:504. 1916. Androsace septentrionalis ssp. puberulenta Robbins, Am. Midl. Nat. 32:161. 1944. (Rydberg & Vreeland 5772, Veta Pass, Colo.)

Pubescent annual 3-25 cm. tall; leaves in a single rosette, oblanceolate, 1-3 cm. long, entire to denticulate, sparsely to densely hairy with simple or forked hairs; scapes usually many, 1-10(13) cm. long, from nearly glabrous to sparsely glandular-pubescent or more usually densely hairy with fine branched hairs; umbels 3- to 25-flowered, bracts linear to lanceolate, 3-6 mm. long, usually not much over 1 mm. broad; pedicels rather slender, 1-5(12) cm. long, the outer ones strongly curved, from nearly glabrous to fairly densely pubescent and more or less glandular; calyx 2.5-4 mm. long, turbinate, strongly carinate, the tube considerably longer than the deltoid to lanceolate-deltoid lobes; corolla white, slightly longer than the calyx; capsule turbinate-globose, about equaling the calyx tube; seeds dark brown, about 1 cm. long. N=10.

Circumpolar in the Arctic, s. in the mountains of w. U.S. to Calif., Ariz., and N.M. May-Aug.

A variable species, both as to pubescence and stature, but not satisfactorily divisible in our area into natural infraspecific taxa, our plants belonging mostly either to var. subumbellata

Nels. , if sparsely hairy, or var. puberulenta (Rydb. ) Knuth if generally pubescent.

## Centunculus L. Chaffweed

Flowers axillary, subsessile, inconspicuous, 4(5)-merous; sepals greenish; corolla white or pinkish, papery, nearly globular; stamens borne at the throat of the corolla; capsule circumscissile; seeds pitted; delicate annual with alternate (or the lowermost opposite), entire leaves.
Probably only the one species. (A diminutive of the Latin cento, a patchwork. )

Centunculus minimus L. Sp. Pl. 116. 1753. (Europe)
Plant glabrous; stems decumbent (and rooting at the nodes) to erect, 2-10 cm. long; leaves obovate to spatulate or elliptic, 5-10 mm. long, decurrent on the stems; flowers solitary in the axils, pedicels about 1 mm. long; calyx 2-3 mm. long, lobed nearly to the base, the segments linear, minutely serrulate; corolla scarcely half the length of the calyx, drying and calyptrate on the capsule, tube about twice the length of the acute, deltoid lobes; capsule globose, about 2 mm. long; seeds brown. N=11.
Moist ground or around vernal pools from the coast to the interior valleys, more or less cosmopolitan. May-June.

## Dodecatheon L. Shooting Star

Flowers showy, 4- to 5-merous, borne in terminal, involucrate umbels (sometimes single), on slender, recurved pedicels; calyx short-tubular, the lobes lanceolate; corolla showy, short-tubular, the lobes long and strongly reflexed, white to purple, the tube very short; stamens connivent around the style; filaments short, free or connected by a membrane; anthers long and slender, basally attached, dehiscent on the inner surface, connective prominent, highly colored, smooth to transversely rugose; style slightly exceeding the stamens; stigma capitate, sometimes rather conspicuously enlarged; fruit a 1-celled capsule, valvate to the tip, or the tip operculate with the style and the walls valvate below; seeds many; scapose, herbaceous perennial from slender to thick rhizomes or very short caudices, often with small, ricelike bulblets among the roots, glabrous to conspicuously glandular-pubescent, with petioled, entire to dentate leaves.
About 13 species of N. Am. (Name from the Greek dodeka, twelve, and theos, god, the plant protected by the Greek gods. )
The species are all desirable garden subjects and are easily grown, D. dentatum and D. jeffreyi adjusting readily to wet areas and stream banks and readily seeding themselves.
References:
Beamish, Katherine I. Studies in the genus Dodecatheon of Northwestern America. Bull. Torrey Club 82:357-66. 1955.
Thompson, H. J. The biosystematics of Dodecatheon. Contr. Dudley Herb. 4:73-154. 1953.
1 Leaves abruptly contracted to petioles about as long as the rounded to cordate-based, sinuate-dentate blades; petals white, drying and persistent with the stamens, the capsules protruding beyond them                                                               D. DENTATUM
1 Leaves usually gradually attenuate to winged petioles, blades seldom conspicuously toothed; petals usually colored, forced off by the growing capsules and deciduous with the stamens
  2 Stigma conspicuously capitate, usually at least twice as broad as the style at midlength; filaments usually not much more than 1 mm. in length
    3 Flowers 4-merous; leaves linear-oblanceolate, usually less than 1 (to 1.5) cm. broad, glabrous; inflorescence usually glabrous; capsule not operculate, dehiscent from the tip by valves                                                                        D. ALPINUM
    3 Flowers 4- or 5-merous; leaves mostly over 1 cm. broad, frequently glandular-hairy; inflorescence often conspicuously glandular; tip of capsule usually operculate
                                                                                      D. JEFFREYI
  2 Stigma not conspicuously capitate, less than twice as thick as the style; filaments usually united to form a tube over 1 mm. in length
    4 Leaves contracted abruptly to the petiole, blade ovate to deltoid, usually over 2 cm.

broad and less than 2.5 times as long as broad; bulblets present among the roots at flowering time; inflorescence glandular; stamen tube over 1.5 mm. long; capsules operculate
                                                            D. HENDERSONII
4   Leaves usually narrowed gradually to the petiole, blade usually over 3 times as long as
    broad; bulblets lacking; inflorescence sometimes eglandular; stamen tube often less than
    1.5 mm. long; capsules often not operculate
    5   Filaments usually less than 1 (to 1.5) mm. long, free or united into a tube, the connec-
        tives cross-rugulose; capsule operculate                     D. CONJUGENS
    5   Filaments usually over 1.5 mm. long, connective often smooth or longitudinally wrinkled
        rather than transversely wrinkled; capsule valvate to the tip
        6   Filament tube 1.5-2 mm. long, purplish, the connectives more or less transversely
            rugose; plant densely and finely glandular-puberulent throughout including the cap-
            sules; leaves often conspicuously denticulate              D. POETICUM
        6   Filament tube often over 2 mm. long, usually yellow or orange, if purplish then the
            connectives smooth or longitudinally wrinkled rather than transversely rugose, or
            plants glabrous, or the leaves entire                      D. PAUCIFLORUM

Dodecatheon alpinum (Gray) Greene, Erythea 3:39. 1895.
    D. meadia var. alpinum Gray, Bot. Calif. 1:467. 1876. D. jeffreyi var. alpinum Gray, Bot. Gaz.
    11:232. 1886. D. alpinum ssp. alpinum Thomps. Contr. Dudley Herb. 4:141. 1953. (Brewer
    s. n., Mt. Dana and Mono Pass summit, Calif.; lectotype by Thompson)
    D. alpinum f. nanum Hall, U. Calif. Pub. Bot. 4:205. 1912. (Hall & Babcock 3618, Mt. Dana,
    Calif.)
    D. alpinum ssp. majus Thomps. Contr. Dudley Herb. 4:142. 1953. (H. J. Thompson 1055,
    Big Bear Lake, San Bernardino Co., Calif.)
    Plant tufted, often bulblet-bearing at the base, usually glabrous but sometimes sparsely
glandular-pubescent in the inflorescence; leaves linear-oblanceolate, obtuse or rounded, 3-10
cm. long, 3-15 mm. broad, entire to somewhat sinuate, gradually contracted to winged petioles; scapes 1-3 dm. long; flowers 1-9, 4-merous; calyx usually glabrous, finely purple-
flecked, sepals narrowly lanceolate, 4-7 mm. long; corolla 10-18 mm. long, lobes purplish,
the tube yellowish but purplish-red-ringed at base; filaments not over 1 mm. long, free or
united into a tube; connectives deep purplish, cross-rugose; anthers 5-8 mm. long, purplish; stigma capitate, usually about twice the thickness of the style at midlength; capsules oblong-ovoid, 5-8 mm. long, dehiscent by valves to the tip, not operculate. N=22.
    Mountain meadows or along streams, Wallowa Mts. of e. Oreg. s.w. to the Cascades of s.
Oreg. and through the Sierra Nevada to s. Calif., e. to Ariz. and Utah. June-July.
    The size of the leaves and flowers and the length of the scape are influenced greatly by
conditions of the local habitat, especially elevation.

Dodecatheon conjugens Greene, Erythea 3:40. 1895 (Kelsey, near Helena, Mont.)
    D. glastifolium Greene, Erythea 3:71. 1895. (Mrs. Austin, in 1894, Lava Beds, Modoc Co.,
    Calif.) = var. conjugens.
    D. acuminatum Rydb. Mem N.Y. Bot. Gard. 1:304. 1900. (F. W. Anderson, in 1885, Mis-
    souri R., above mouth of Sand Coulee, Mont.) = var. conjugens.
    D. pulchrum Rydb. Mem. N.Y. Bot. Gard. 1:304. 1900. (Tweedy 432, Indian Creek, Yellow-
    stone Nat. Pk., in 1885) = var. conjugens.
    D. cylindrocarpum Rydb. Mem. N.Y. Bot. Gard. 1:305. 1900. (Rydberg & Bessey 4674,
    Bridger Mts., Mont., June 17, 1897; lectotype, according to Thompson) = var. conjugens.
    D. pubescens Rydb. Mem. N.Y. Bot. Gard. 1:306. 1900. (Tweedy, in 1883, Missoula, Mont.)
    = var. viscidum.
    D. hendersonii var. leptophyllum Suksd. Deuts. Bot. Monats. 18:132. 1900. D. conjugens
    ssp. leptophyllum Piper, Contr. U.S. Nat. Herb. 11:446. 1906. (Suksdorf, Falcon Valley,
    Wash.) = var. conjugens.
    D. viscidum Piper, Bull. Torrey Club 28:43. 1901. D. conjugens var. viscidum Mason ex
    St. John, Fl. S.E. Wash. 311. 1937. (Piper 2832, 10 miles w. of Spangle, Wash., May 24,
    1898)

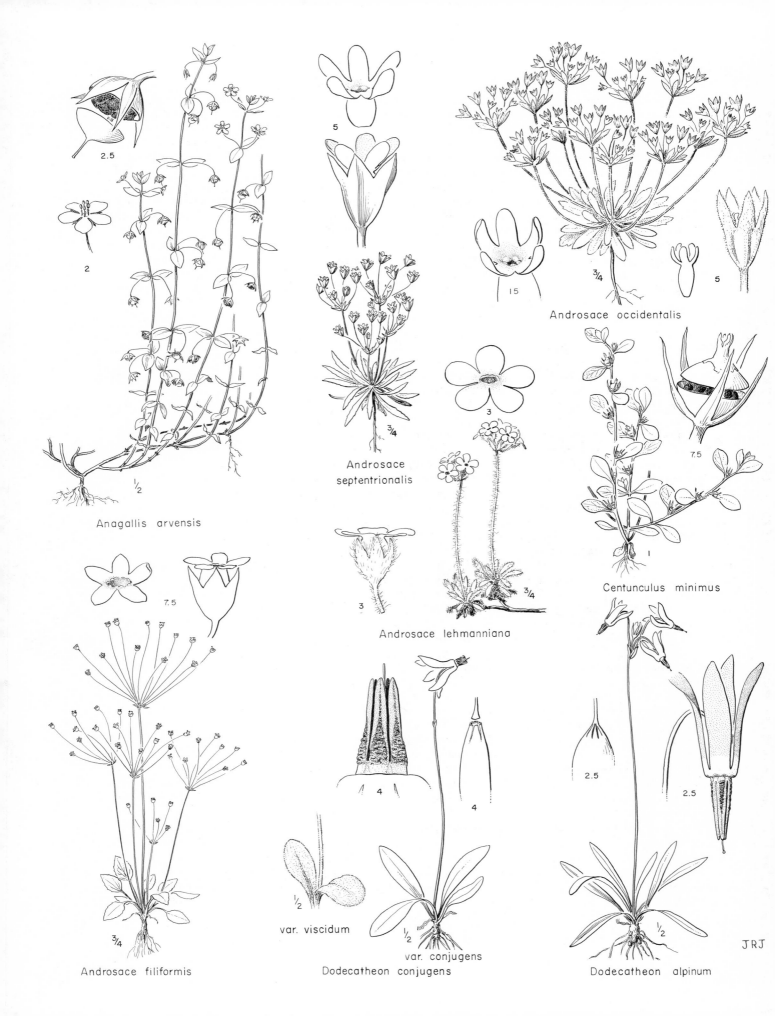

2.5

2

1/2

Anagallis arvensis

5

15

3/4

Androsace septentrionalis

3

3/4

Androsace lehmanniana

7.5

Androsace occidentalis

5

7.5

1

Centunculus minimus

7.5

3/4

Androsace filiformis

4

4

var. viscidum

1/2

var. conjugens

1/2

Dodecatheon conjugens

2.5

2.5

Dodecatheon alpinum

1/2

JRJ

D. campestrum Howell, Fl. N. W. Am. 432. 1901. (Suksdorf, Klickitat Hills, Wash. ) = var.
    conjugens.
D. albidum Greene in Fedde, Rep. Sp. Nov. 13:323. 1914. (Wyo. and Mont. , 3 collections
    cited) = var. conjugens.
    Plants without rootstocks or bulblets; leaves glabrous to densely granular-puberulent, 3-20
cm. long, blades lanceolate to oblanceolate, spatulate, or even obovate, usually several times
as long as broad, entire, narrowed gradually (usually) or somewhat abruptly to winged petioles
somewhat shorter; scapes usually glabrous, 1-3 dm. long; flowers 1-10, usually 5-merous;
calyx finely purple-maculate, glabrous, lobes lanceolate, 3-6 mm. long; corolla 1-3 cm. long,
lobes often white but usually rose-pink to deep orchid, the tube yellowish, undulately purplish-
red ringed at base; filaments mostly not over 1(1. 5) mm. long, free or united, usually yellowish (purplish), smooth; connectives deep red to purple and transversely rugose; anthers 6-8
mm. long, yellow (purple-maculate) to light or deep purple; stigma scarcely at all enlarged;
capsule cylindric-ovoid, 8-12 mm. long, walls thin, tip operculate with the style. N=22.
    In seepages in sagebrush to montane meadows, e. slopes of the Cascades from B. C. to n.
Calif. , e. to Alta. and Wyo. Apr. -June.
    There are two poorly marked varieties of this species in our range; var. conjugens, con-
sisting of plants that are essentially glabrous, is the more widespread, whereas var. viscidum
(Piper) Mason ex St. John, which has pubescent leaves and (often) pubescent scapes, occurs
sporadically with var. conjugens in the more e. part of its range, sometimes to the exclusion
of the latter.

Dodedatheon dentatum Hook. Fl. Bor. Am. 2:119. 1838.
    D. meadia var. dentatum Gray, Bot. Gaz. 11:234. 1886. D. meadia var. latilobum Gray,
    Syn. Fl. 2nd. ed. , 2¹:58. 1886. (Douglas, "N. W. interior," last journey)
    Plant from short, thick, ascending rhizomes, without bulblets; leaves glabrous, blade ovate
to oblong or oblong-lanceolate, 3-10 cm. long, 2-6 cm. broad, conspicuously sinuate to
sharply dentate or undulate-dentate, the base from cordate to abruptly rounded, narrowed abruptly to slender petiole from about equal to, to nearly twice as long as, the blade; scapes
1. 5-4 dm. long, glabrous or very sparsely glandular-hairy; flowers 2-12, 5-merous; pedicels
as much as 7 cm. long; sepals lanceolate, 3-4 mm. long; corolla 12-20 mm. long, the lobes
creamy white, the tube yellowish, with an undulate reddish-purple ring at base; filaments free,
less than 1 mm. long, deep reddish-purple, smooth (as are the connectives) except for a longitudinal groove; anthers 6-7 mm. long, deep red; stigma very slightly enlarged; capsules
cylindric-ovoid, thin-walled, 6-10 mm. long, dehiscent to the tip by valves, growing past the
persistent, dried corolla and stamens. N=22.
    Around waterfalls, stream banks, and shaded, moist slopes, e. side of the Cascades from
s. B. C. and Wash. to n. Oreg. and in c. Ida. ; not known from n. e. Oreg. May-July.

Dodecatheon hendersonii Gray, Bot. Gaz. 11:233. 1886.
    D. meadia var. hendersoni Brandg. Zoë 1:20. 1890. Meadia hendersoni Kuntze, Rev. Gen.
    2:398. 1891. (Henderson 81, Tualatin Plains, Oreg.)
    Dodecatheon integrifolium var. latifolium Hook. Fl. Bor. Am. 2:119. 1838. D. latifolium
    Piper, Contr. U. S. Nat. Herb. 11:446. 1906. ("Dry banks about Ft. Vancouver on the Co-
    lumbia," Dr. Scouler, Dr. Gairdner, Tolmie)
    D. cruciatum Greene, Pitt. 1:213. 1888. D. hendersoni var. cruciatum Greene, Pitt. 2:75.
    1890. Meadia hendersonii var. cruciata Greene, Man. Bay Reg. 237. 1894. Dodecatheon
    hendersoni ssp. cruciatum Thomps. Contr. Dudley Herb. 4:131. 1953. (Greene, Mt. Ta-
    malpais, Calif. )
    D. atratum Greene, Fedde Rep. Sp. Nov. 13:323. 1914. (Foster, Gate, Wash. , in 1912)
    Plant lacking rootstocks but with numerous bulblets at flowering time; leaves usually gla-
brous, 3-14 cm. long, blades ovate to deltoid-elliptic or more or less spatulate, entire to
shallowly sinuate or remotely denticulate, narrowed abruptly to petioles 1/4 as long to nearly
as long; scapes 1-3 dm. long, glabrous to glandular-puberulent as is the inflorescence; flowers 2-15, 4- or 5-merous on the same plant; calyx finely purple-flecked, usually sparsely
glandular-puberulent, with lanceolate lobes 6-8 mm. long; corolla 15-25 mm. long, the lobes

deep magenta to light orchid, grading to yellow at the base and on the tube, base of tube deep reddish-purple; filaments united to form a tube 2-4 mm. long, deep reddish-purple, usually cross-rugose, sometimes nearly smooth; anthers 4-6 mm. long, deep red to purple, the connective more deeply colored, usually cross-rugose; stigma not noticeably enlarged; capsules cylindric-ovoid, 7-12 mm. long, the tip operculate. N=about 44.

Woods and prairies, Vancouver I., s. on the w. side of the Cascades to s. Oreg. and in both the Coast Range and Sierra Nevada to s. Calif. March-June.

All our material is referable to the var. hendersonii. Other varieties occur in California.

Dodecatheon jeffreyi van Houtte, Fl. Serres 16:99, pl. 1662. 1865.
    D. meadia var. lancifolium Gray, Bot. Calif. 1:467. 1876. D. meadia var. jeffreyi Brandg.
    Zoë 1:20. 1890. Meadia jeffreyi Kuntze, Rev. Gen. 2:398. 1891. (Described from a plant
    grown in Europe from California seed)
    Dodecatheon crenatum Greene, Pitt. 2:74. 1890, but not of Raf. in 1883. Meadia crenata
    Kuntze, Rev. Gen. 2:398. 1891. Dodecatheon viviparum Greene, Erythea 3:38. 1895. D.
    jeffreyi var. viviparum Abrams, Ill. Fl. Pac. St. 3:340. 1951. (Greene, Aug. 20, 1889,
    Mt. Rainier, Wash.)
    D. tetrandrum Suksd. ex Greene, Erythea 3:40. 1895. D. jeffreyi var. tetrandrum Jeps.
    Man. Fl. Pl. Calif. 754. 1925. (Cusick, e. Oreg.)
    D. dispar A. Nels. Bot. Gaz. 52:269. 1911. (Macbride 672, Trinity Lakes, Elmore Co.,
    Ida., Aug. 29, 1910)
    D. exilifolium Macbr. & Pays. Contr. Gray Herb. n. s. 49:63. 1917. (Macbride & Payson
    3744, Smoky Mts., Custer Co., Ida.)
    Plants in large clumps connected by slender rootstocks, glabrous or (especially the inflorescence) sparsely to densely glandular-pubescent, without bulblets; leaves oblanceolate, acute to rounded, gradually attenuate to long petioles, entire to callus-crenulate or serrulate, 0.5-4 dm. long, 1-6 cm. broad; scapes 1-6 dm. long, 3- to 20-flowered; flowers 4- or 5-merous; calyx tube 2-5 mm. long, the lobes acute, 5-10 mm. long; corolla lobes 1-2.5 cm. long, purplish, reddish, mauve, or lavender-magenta to light yellow or whitish, corolla tube lighter than the lobes, cream or yellow, usually purplish or reddish-banded just below the sinuses of the lobes; filaments scarcely 1 mm. long, free to the base or shallowly united, deep reddish to blackish-purple, anthers 6-10 mm. long, yellow to reddish but connectives usually more deeply colored and conspicuously cross-rugose; stigmas capitate, approximately twice the width of the style at midlength; capsules ovoid, 7-11 mm. long, with a circumscissile and operculate tip, the remaining ovary wall valvate from the top. N=21, 22, 43.

On wet ground, usually in meadowland or along streams, Alas. s. in the mountains through the Cascades and Olympics to the s. Sierra Nevada of Calif., e. to Ida. and Mont. June-Aug.

The leaves of this species vary greatly in size, sometimes being only 4-5 cm. in length, but this variation apparently is due to local ecological conditions, such as moisture, shade, and elevation.

Dodecatheon pauciflorum (Durand) Greene, Pitt. 2:72. 1890.
    D. meadia var. pauciflora Dur. Journ. Acad. Phila. II. 3:95. 1855. (No collections cited)
    D. meadia var. puberula Nutt. Journ. Acad. Phila. 7:48. 1834. D. puberulum Piper, Contr.
    U.S. Nat. Herb. 11:445. 1906. (Wyeth, Flathead R., Mont.) = var. pauciflorum.
    D. integrifolium var. vulgare Hook. Fl. Bor. Am. 2:118. 1838. D. vulgare Piper, Contr.
    U.S. Nat. Herb. 11:445. 1906. (Dr. Richardson, Drummond, Hudson Bay territories to
    Carlton House Ft., and prairies of the Rocky Mts.) = var. pauciflorum.
    D. cusickii Greene, Pitt. 2:73. 1890. Meadia cusickii Kuntze, Rev. Gen. 2:398. 1891.
    Dodecatheon pauciflorum var. cusickii Mason ex St. John, Fl. S. E. Wash. 312. 1937. (Cu-
    sick 1527, e. Oreg.)
    D. pauciflorum var. monanthum Greene, Pitt. 2:73. 1890. D. pauciflorum ssp. monanthum
    Knuth, Pflanzenr. IV. 237:243. 1905. D. radicatum ssp. monanthum Thomps. Contr. Dud-
    ley Herb. 4:146. 1953. (Cusick 1528, e. Oreg.)
    ?D. radicatum Greene, Erythea 3:37. 1895. (Fendler 549, N. M.)

D. puberulentum Heller, Bull. Torrey Club 24:311. 1897. (Heller 2985, Clearwater R.,
near the Upper Ferry, Ida., Apr. 29, 1896) = var. cusickii.

D. salinum var. pauciflorum A. Nels. Bull. Torrey Club 26:131. 1899. D. pauciflorum ssp.
salinum Knuth, Pflanzenr. IV. 237:243. 1905. Meadia salina Lunell, Am. Midl. Nat. 5:239.
1919. (A. Nelson 3012, Evanston, Wyo., May 29, 1897) = var. pauciflorum.

Dodecatheon uniflorum Rydb. Mem. N. Y. Bot. Gard. 1:307. 1900. (Rydberg & Bessey 4668,
Old Hollowtop, Mont.) = var. watsonii.

D. philoscia A. Nels. Bull. Torrey Club 28:227. 1901. (A. Nelson 8063, Laramie R. at
Jelm, Wyo., Aug. 11, 1900) = var. pauciflorum.

D. multiflorum Rydb. Bull. Torrey Club 31:631. 1905. (Rydberg & Vreeland 5781, Sangre
de Cristo Creek, Colo.) = var. pauciflorum.

D. radicatum var. sinuatum Rydb. Bull. Torrey Club 31:631. 1905. D. sinuatum Rydb. Bull.
Torrey Club 33:148. 1906. (Crandall, in 1890, Larimer Co., Colo.) = var. pauciflorum.

D. pauciflorum var. shoshonensis A. Nels. Bot. Gaz. 54:143. 1912. (Nelson & Macbride
1362, Shoshone Falls, Ida., July 26, 1911) = var. pauciflorum.

D. pauciflorum var. exquisitum Macbr. & Pays. Contr. Gray Herb. n. s. 49:63. 1917. (Macbride & Payson 3747, Smoky Mts., Ida.) = var. pauciflorum.

DODECATHEON PAUCIFLORUM var. WATSONII (Tidestr.) C. L. Hitchc. hoc loc. D. watsoni Tidestr. Proc. Biol. Soc. Wash. 36:183. 1923. D. radicatum ssp. watsonii Thomps.
Contr. Dudley Herb. 4:147. 1953. (Watson 756, Humboldt Mts., Nev.)

D. cusickii var. album Suksd. Werdenda 1:30. 1927. D. pauciflorum var. cusickii f. album
St. John, Fl. S. E. Wash. 312. 1937. D. cusickii f. album St. John, Fl. S. E. Wash. 2nd
ed. 550. 1956. (Suksdorf 8601, Spangle, Wash.) = var. cusickii.

D. superbum Pennell & Stair, Bartonia 24:20. 1947. (Anderson 6962, Alas.) = var. alaskanum.

DODECATHEON PAUCIFLORUM var. ALASKANUM (Hultén) C. L. Hitchc. hoc loc. D. macrocarpum var. alaskanum Hultén. Fl. Alas. 8:1289. 1948. (Eyerdam 47, Pt. Hobron,
Alas.)

D. radicatum ssp. macrocarpum (Gray) Beamish, Bull. Torrey Club 82:363. 1955, as to
material considered but not as to the type, which is referable to D. clevelandii, a California species. (Dall s. n., June 2, 1874, Middleton I., Alas.; neotype by Beamish) = var.
alaskanum.

Plant with very short erect rootstocks, without bulblets, glabrous to conspicuously glandular-pubescent throughout including the capsules; leaves 2-15(20) cm. long, blades oblong-lanceolate to spatulate or oblanceolate, entire to somewhat denticulate, narrowed gradually to winged petioles nearly as long; scapes (2)5-40 cm. long; flowers 1-25, 5-merous; calyx usually purple-
flecked, lobes 3-5 mm. long; corolla 10-20 mm. long, the lobes (white) purplish-lavender,
tube yellowish, usually with a purplish undulate line at the base; filaments united into a yellowish, orange, or (less commonly) purple tube 1.5-3 mm. long, smooth or only slightly wrinkled,
connectives purple, smooth or somewhat longitudinally wrinkled when dried; anthers usually
yellowish to reddish-purple, 4-7 mm. long; stigma slightly larger than the style; capsule
ovoid-cylindric, glabrous to glandular-hairy, membranous to firm-walled, 5-15 mm. long,
dehiscent to the tip by valves.

Coastal prairies to inland saline swamps and mountain meadows and streams, from near
sea level to above timber line, Alas. to Mex., e. to Pa. Apr.-Aug.

An exceedingly variable taxon which has been subdivided by various workers into many
species, most of which appear to be extremes of a continuous, variant series. Aside from a
coastal, polyploid (N=44, 66), glabrous or very slightly pubescent, large-leaved form, var.
alaskanum (Hultén) C. L. Hitchc., which is scarcely recognizable from similar diploid plants
of the interior, four races doubtfully merit recognition. (1) Var. watsonii (Tidestr.) C. L.
Hitchc. is merely the extremely reduced plant of alpine situations, found chiefly in the Rocky
Mts. Such plants often are dwarfed to 2-5 cm. in height and have only 1 flower per scape; they
merge by all degrees to the taller, pluriflowered plants of the lowlands. (2) Var. monanthum
Greene is characterized by a purplish, rather than yellow, staminal tube; it occurs through
the desert ranges from e. Oreg. to Utah, largely to the exclusion of var. pauciflorum; usually such plants are glabrous, but not uncommonly they are fairly densely pubescent (Hitchcock

& Muhlick 13093 and 12976) and therefore would be referable to var. cusickii as well as to var. monanthum. (3) Var. cusickii (Greene) Mason ex St. John (2N=44, 45) is the pubescent phase of the plant, and is often recognized as a distinct species, alleged to differ from var. pauciflorum also in having firmer capsules. It is sympatric throughout its range with var. pauciflorum, but apparently isolated therefrom ecologically since it tends to blossom earlier because it occurs on more open grassy areas, whereas var. pauciflorum prefers moister, cooler areas. As judged by intermediates, the two are freely interfertile. (4) The greater part of the material (staminal tube yellow, plants 2- to many-flowered, glabrous or nearly so) constitutes the var. pauciflorum.

1 Staminal tube purplish; e. Oreg. to Utah               var. monanthum Greene
1 Staminal tube yellow
  2 Plants dwarfed, usually but 2-5 cm. tall and 1- to 2-flowered; alpine to subalpine in the
    Rockies and n. Oreg.                     var. watsonii (Tidestr.) C. L. Hitchc.
  2 Plants taller and usually with more flowers
    3 Plants glandular-pubescent throughout; s. B. C. to Oreg., e. to Mont.
                                   var. cusickii (Greene) Mason ex St. John
    3 Plants glabrous to sparsely pubescent
      4 Leaves ovate to ovate-lanceolate, rather abruptly narrowed to petioles; near the
        coast from Alas. to Oreg.              var. alaskanum (Hultén) C. L. Hitchc.
      4 Leaves more commonly spatulate to oblanceolate, narrowed gradually to petioles;
        Alas. to n. Calif., e. chiefly n. of the Great Basin to the Central and Middle At-
        lantic states, s. in the Rockies to Mex.           var. pauciflorum

There is some doubt that D. pauciflorum is the proper name for this taxon, but there seems to be as good reason for using that binomial as for accepting D. radicatum Greene as the specific epithet, since the latter name was based on a type (apparently lost) which was described by Greene as having "staminal tube very short," a character that very definitely does not fit these plants.


Dodecatheon poeticum Henderson, Rhodora 32: 27. 1930. (Henderson 503, Columbia R. near Hood River Co., Oreg.)

Plant finely glandular-puberulent throughout, without bulblets; leaves 4-15 cm. long, blades oblong-lanceolate to spatulate or oblanceolate, from rather coarsely serrate-dentate to nearly entire, narrowed rather gradually to slender winged petioles sometimes longer; scapes 1-4 dm. long; flowers 2-10, 5-merous; calyx greenish, usually not purple-maculate, glandular-pubescent, its lobes 3-5 mm. long; corolla 12-20 mm. long, the lobes broad, bright pink to orchid, tube with an undulate carmine-purplish ring at base, yellow above; filaments 1.5-2.5 mm. long, connate into a tube usually 1.5-2 mm. long, deep purplish, smooth or, like the purplish connectives, more or less transversely rugose; anthers 5-7 mm. long, purple; stigma but slightly enlarged; capsules 6-9 mm. long, ovoid, firm-walled, finely glandular-pubescent on the upper half, dehiscent to the tip by valves.

Wet soil on n. e. side of the Cascade Range from n. of Satus Pass, Yakima Co., Wash., to Wasco and Hood River cos., Oreg. Mar.-May.


Douglasia Lindl. Nom. Conserv.


Flowers single to several in involucrate umbels on leafless peduncles, sessile to long-pedicellate, commonly 5-merous; calyx united half its length, carinate below the lobes; corolla tubular-funnelform, tube usually about equaling the calyx, constricted and 5-fornicate at the throat within, the lobes spreading; stamens attached above midlength of the corolla, filaments shorter than the anthers; capsule 5-valvular from the apex; seeds 1-3, brown, finely pitted; caespitose, matted to cushion-forming perennials, usually finely stellate, at least on the peduncles; stems prostrate or ascending, dichotomously branched, ending in rosettes of small, entire or dentate leaves.

Six species, n. w. and Arctic N. Am.; Eurasia. (Named in honor of David Douglas, 1798-1834, famous early plant explorer in the northwest.)

Our species are all prized rock garden subjects, D. laevigata being the most suitable for

areas w. of the Cascades. Propagated by cuttings and layers as well as by seed.
  Reference:
  Constance, Lincoln. A revision of the genus Douglasia. Am. Midl. Nat. 19:249-59. 1938.
1 Flowers single or sometimes in pairs, not involucrate          D. MONTANA
1 Flowers 2-10, pedicellate in involucrate umbels
  2 Leaves grayish-pubescent with tiny stellae; involucral bracts usually several times as
     long as broad; flowers usually conspicuously pedicellate          D. NIVALIS
  2 Leaves glabrous or barely ciliolate; involucral bracts usually not more than twice as
     long as broad; flowers usually subsessile          D. LAEVIGATA

Douglasia laevigata Gray, Proc. Am. Acad. 16:105. 1880.
  Primula laevigata Derganc, Allg. Bot. Zeit. 10:111. 1904. Gregoria laevigata House, Bull.
    N.Y. State Mus. nos. 233-34:69. 1921. (J. & T. J. Howell, mountains near Mt. Hood,
    Oreg.)
  Douglasia laevigata var. ciliolata Const. Am. Midl. Nat. 19:254. 1938. (Henderson 3878,
    Mt. Henderson, 8 miles n. of Lake Cushman, Olympic Mts., Mason Co., Wash.)
  Laxly spreading and forming rather extensive mats; leaves oblong-oblanceolate to oblanceolate, 5-20 mm. long, 2-5 mm. broad, entire to, more commonly, few-toothed, glabrous or
ciliolate; peduncles 2-7 cm. long, finely stellate; umbels tightly 2- to 10-flowered, pedicels
2-15 mm. long, finely stellate; involucres of 4-8 ovate-lanceolate to broadly ovate, acute, 3-
8 mm. long bracts; calyx 6-7 mm. long, glabrous to sparsely finely stellate, the lobes lanceolate, acuminate, subequal to the tube; corolla deep pinkish-rose, fading to lavender, tube 6-7
mm. long, lobes oblong-obovate, 4-5 mm. long, rounded or erose but not retuse.
  Talus slopes and rocky alpine ledges to moist coastal bluffs, w. side of the Cascade Range,
from Snohomish Co. to Mt. Rainier, the Olympic Mts., the mountains of s.w. Wash., and adjacent Oreg. (Mt. Hood and Saddle Mt.), and in the Columbia gorge. Mar.-Aug.
  There are two intergradient forms of the plant; var. laevigata, from the Columbia gorge,
has leaves that are entirely glabrous or very inconspicuously ciliolate and more often entire,
the umbels usually tending to be loose, with pedicels 5-10 mm. long; var. ciliolata Const.,
the more widespread variety, is characterized by more conspicuously ciliolate leaves and
more compact umbels.
  In the Christ Herbarium of the New York Botanical Garden there is a fragmentary collection
which was supposedly made at the Red River Ranger Station, Nez Perce Nat. Forest, Idaho
Co.,, Ida. The specimen is more nearly referable to D. laevigata than to any other of our species though smaller-flowered and pubescent with simple, rather than branched, hairs. It has
not been possible to re-collect the plant in Idaho, and there is some doubt that it actually came
from there. For this reason it is not here proposed as a new variety of D. laevigata although
that is what it is believed to represent.

Douglasia montana Gray, Proc. Am. Acad. 7:371. 1868.
  Primula montana Derganc, Allg. Bot. Zeit. 10:111. 1904. Gregoria montana House, Bull.
    N.Y. State Mus. nos. 233-34:69. 1921. (Winslow J. Harvard, Rocky Mts.)
  Androsace uniflora Haussk. Mitt. Bot. Ver. Ges. Thür. 9:23. 1890. Douglasia montana var.
    uniflora Knuth, Pflanzenr. IV. 237:169. 1905. (Röll, near Garrison, Mont., Aug., 1888)
  D. biflora A. Nels. Bull. Torrey Club 25:277. 1898. D. montana var. biflora Knuth, Pflan-
    zenr. IV. 237:169. 1905. (A. Nelson 2450, Dome Lake, Big Horn Mts., Wyo., July 18,
    1896)
  Leaves linear-lanceolate, 4-8 mm. long, minutely scabrous-serrulate; scapes 1-4 from the
rosettes, 5-25 mm. long, pubescent with tiny branched hairs, with a single (2) terminal flower, bractless or with 1 or 2 bracts at the base of the flower, or with 2 or 3 bracts if the flowers are geminate; flowers 6-8 mm. long, pedicels lacking or (if more than 1 per scape) as
much as 5 mm. long; calyx reddish, strongly carinate, pubescent like the scapes or glabrous,
lobes acuminate, equaling the tube; corolla bright pink to "rose-violet," lobes oblong, 4-5 mm.
long, retuse, tube usually exceeded by the calyx lobes; capsules slightly longer than the calyx
tube.

4

4

Dodecatheon jeffreyi

3.5

2.5

4

Dodecatheon poeticum

4

4

Dodecatheon hendersonii

var. laevigata

3

var. ciliolata

Douglasia laevigata

3

2

½

Dodecatheon dentatum

JRJ

3

var. watsonii

3

var. pauciflorum

Dodecatheon pauciflorum

Foothills and open ridges to scree slopes in the mountains, n. Wyo. to Waterton Lakes, B. C. , westward to Ida. , on noncalcareous soil as well as limestone. May-July.

Douglasia nivalis Lindl. in Quart. Journ. Sci. 383. 1827.
  Primula douglasii Kuntze, Rev. Gen. 1:400. 1891. Gregoria nivalis House, Bull. N. Y. State
    Mus. nos. 233-34:68. 1921. (Douglas, Canadian Rocky Mts. , Alta. )
  Douglasia dentata Wats. Proc. Am. Acad. 17:375. 1882. D. nivalis var. dentata Gray, Syn.
    Fl. 2nd ed. 2[1]:399. 1886. Androsace dieckeana Haussk. Mitt. Bot. Ver. Ges. Thür. 9:22.
    1890. (Watson 264, Peshastin Canyon, Wash. )
  Plant matted, more or less grayish overall with dense, fine stellae; leaves linear to oblanceolate, 1-3 cm. long, 1.5-4 mm. broad, entire to conspicuously serrate; scapes 1-several,
peduncles 1-7 cm. long, umbels 2- to 8-flowered, involucrate with several (4-10) narrowly
lanceolate to ovate-acutish bracts 3-8 mm. long, pedicels 3-40 mm. long; calyx 6-7 mm. long,
prominently sulcate, the lobes deltoid, acuminate, subequal to the tube; corolla bright red to
magenta-purple, the tube about equaling or slightly exceeding the calyx, lobes ovate-cuneate,
4-5 mm. long, rounded or erose but not retuse.
  Sagebrush slopes to alpine ridges and talus slides. Apr. -Aug.
  The var. nivalis, (plants with entire leaves) ranges from the Wenatchee Mts. of Kittitas Co.
n. to Chelan and Douglas cos. , and in n. e. Wash. and into the Rocky Mts. of B. C. and Alta.
(according to Constance) but apparently has not been collected outside of c. Wash. in recent
years; the var. dentata Gray, with serrulate leaves, is found only in the Wenatchee Mts. area
of Wash. , where it sometimes occurs with var. nivalis, but more usually the plants of any one
locality are all of one variety or the other.

## Glaux L. Saltwort; Sea Milkwort

  Flowers small, single and sessile in the axils; calyx campanulate, the 5 lobes more or less
petaloid; corolla lacking; stamens 5, free of the calyx, alternate with its lobes; capsule valvate the full length; seeds several, flattened, coherent to the placenta and shed with it as a
unit; glabrous, fleshy, perennial herbs; leaves opposite below, alternate above.
  A single species. (Name from the Greek glaucos, bluish green. )

Glaux maritima L. Sp. Pl. 207. 1753.
  Glaucoides maritima Lunell, Am. Midl. Nat. 4:505. 1916. (Europe)
  Glaux maritima var. obtusifolia Fern. Rhodora 4:215. 1902. Glaucoides maritima var. ob-
    tusifolia Lunell, Am. Midl. Nat. 5:97. 1917. (Pursh, Que. , is the first specimen cited)
  Plant spreading by shallow rhizomes; stems leafy to the tip, 3-30 cm. tall; leaves crowded,
oval to oblong or oblanceolate, 5-25 mm. long, sessile, 1.5-10 mm. broad, articulate at the
base, obtuse to subacute; calyx 4-5 mm. long, white or pinkish, petal-like, lobes ovate, about
equaling the tube; capsule about 2.5 mm. long. N=15.
  Moist saline soil, inland marshes and meadowland, and coastal tidelands, over much of
Arctic and temperate N. Am.; Eurasia. May-July.
  The var. obtusifolia Fern. is often recognized as distinct from var. maritima on the basis
of its erect habit and oval to oblong, rounded leaves, as compared with the prostrate or ascending habit and narrower, pointed leaves of the latter. The two forms grow together and are
of little significance since there are all degrees of intermediacy.

## Lysimachia L. Loosestrife

  Flowers pedicellate, single and axillary or in terminal or axillary racemes, yellow, often
purple-dotted or -streaked, mostly 5(6-7)-merous; calyx lobed nearly full length; corolla
usually rotate, divided nearly to the base, convolute in the bud; filaments attached at the base
of the corolla, and often more or less basally connate, exserted; ovary more or less glabrous;
capsule valvate, few-seeded; seeds pitted; usually glabrous perennial with opposite or whorled
leaves, finely spotted or streaked with red.

At least 100 species, world wide in distribution but mostly in the N. Temp. Zone. (Derivation from Greek name, Lysimachos, from lysis, a loosing, and mache, strife.)
Lysimachia nummularia is a vigorous ground cover.
Reference:
Ray, James Davis, Jr. The Genus Lysimachia in the New World. Ill. Biol. Monog. 24[3-4]:1-
    160. 1956.
1 Flowers single or in pairs in the leaf axils of creeping stems; petioles not ciliate
                                                                              L. NUMMULARIA

1 Flowers in racemes
    2 Racemes terminal; petals not clawed                                      L. TERRESTRIS
    2 Racemes axillary; petals clawed                                         L. THYRSIFLORA

Lysimachia nummularia L. Sp. Pl. 148. 1753 ("Habitat in Europa")
   Prostrate herb with creeping stems, rooting at the nodes, glabrous or nearly so, finely
punctate with tiny red (blackish) dots; leaves short-petiolate, the blades suborbicular to oblong-oval, 1-3 cm. long, nearly as broad; flowers single in the axils on pedicels 1-3 cm. long;
calyx 6-9 mm. long, the lobes cordate-lanceolate; corolla rotate, yellow, the lobes 8-12 mm.
long, glandular-ciliolate and glandular-pubescent near the base on the upper surfaces; stamens
exserted, filaments glandular-pubescent, unequal, several times the length of the anthers, shortly
connate at the base, capsule included in the calyx. N=18. Moneywort or creeping loosestrife.
   Naturalized from Europe and sparingly escaped along roadsides and railroads, where moist;
Thurston Co., Wash., Sauvies I., the Columbia R. lowlands, and the Willamette Valley of
Oreg., and in many other parts of the U.S.; much more common in c. and e. U.S. June-Aug.

Lysimachia terrestris (L.) B.S.P. Prelim. Cat. N.Y. Pl. 34. 1888.
   Viscum terrestre L. Sp. Pl. 1023. 1753. (Phila.)
   Lysimachia stricta Ait. Hort. Kew. 1:199. 1789. (N. Am.)
   Erect, glabrous herb 2-9 dm. tall, from rootstocks; lowermost leaves sessile and clasping,
5-15 mm. long, the upper ones with distinct petioles and linear-lanceolate to narrowly elliptic
or lanceolate blades 3-8 cm. long; sterile shoots often bulblet-bearing in the axils; flowers in
terminal, leafy, bracteate, simple (compound) racemes, the lower flowers often whorled, pedicels slender, 1-2 cm. long; perianth maculate and streaked with purplish-black; calyx segments
lanceolate, 2-4 mm. long; corolla yellow, rotate, lobes oblong-lanceolate, 6-8 mm. long; filaments about 3 mm. long; seeds few, chocolate brown, very finely pitted. Bog loosestrife.
   Bogs and swamps, abundant and native in e. U.S., introduced in cranberry bogs on Vancouver I. and in Pacific and Kitsap cos., Wash., and perhaps elsewhere. June-Aug.

Lysimachia thyrsiflora L. Sp. Pl. 147. 1753.
   Naumburgia thyrsiflora Reichb. Fl. Germ. Excurs. 410. 1830. N. guttata Moench. Meth.
     Suppl. 23. 1802. (Europe)
   Erect, glabrous herb 2-8 dm. tall, from creeping rhizomes, finely dark purplish or black-
ish-maculate almost throughout; lowermost leaves sessile and greatly reduced, the upper ones
linear-lanceolate to elliptic-lanceolate, long-acuminate, as much as 15 cm. long, narrowed
gradually to the base, essentially sessile; flowers pale yellow, crowded and subsessile in
short, dense, slender, pedunculate racemes in the 2 or 3 pairs of larger leaves near midstem;
calyx lobes maculate, glabrous, lanceolate, 2-3 mm. long; corolla lobes somewhat clawed at
the base, linear-lanceolate, about twice the length of the calyx, sparingly maculate near the
tip, sometimes bearing a tiny toothlike scale just below each sinus; filaments not connate, exceeding the corolla, glabrous; style about equaling the stamens; ovary and style maculate; capsules about 2.5 mm. long. Tufted loosestrife.
   Swamps, lakes, and ditches throughout much of N. Am.; Eurasia. May-July.


Primula L. Primrose


   Flowers showy, in involucrate umbels, 5-merous; calyx persistent; corolla salverform, the
lobes usually emarginate, fornices absent or inconspicuous; stamens attached in the upper

third of the corolla tube, included, filaments very short; ovary superior; style usually included; capsules valvate; scapose, herbaceous or rarely suffrutescent perennials.

Perhaps 200 species, chiefly boreal to alpine in the n. Temp. Zone, especially abundant in s. c. Asia. (Diminutive of the Latin primus, early or first, since many species flower early in spring.)

Reference:

Williams, L. O. Revision of the western Primulas. Am. Midl. Nat. 17:741-48. 1936.

1 Leaves erect, 10-25 cm. long, prominently denticulate; corolla limb usually at least 1.5 cm.
  broad                                                                 P. PARRYI
1 Leaves mostly in flat rosettes, less than 10 cm. long; corolla limb less than 1.5 cm. broad
  2 Scapes seldom over 10 cm. tall; corolla lobes about 4 mm. long, rounded
                                                             P. CUSICKIANA
  2 Scapes usually over 10 cm. tall; segments of the corolla 2-3 mm. long, deeply bilobed
                                                             P. INCANA

Primula cusickiana Gray, Syn. Fl. 2nd. ed. 2$^1$:399. 1886.

P. angustifolia var. cusickiana Gray, Syn. Fl. 2:393. 1878. (Cusick, Union Co., Oreg.)
P. brodheadae M. E. Jones, Zöe 3:306. 1893. (Brodhead, Ketchum, Ida.)
P. brodheadae var. minor M. E. Jones, Zöe 3:306. 1893. (M. E. Jones, Bayhorse, Ida.)

Plant glabrous or sparsely glandular-puberulent; leaves oblanceolate to oblong-obovate, 1.5-5 cm. long, entire to inconspicuously few-toothed; scapes 2-9 cm. tall, 1- to 3-flowered; involucral bracts 2-3, lanceolate, 2-9 mm. long; pedicels 1-5 mm. long; calyx 5-8 mm. long, the lanceolate, acute lobes about equaling the tube; corolla about 1 cm. long, bluish-violet to purplish, fornices lacking in the throat, lobes about 4 mm. long, obovate, shallowly retuse; capsule ovoid, about 5 mm. long.

Foothills to subalpine slopes, often on talus or spring-fed ground, c. Ida. to the Wallowa and Blue mts. of Oreg. Mar.-June.

A beautiful little perennial desirable for the moist rock garden, but very difficult to establish.

Primula incana M. E. Jones, Proc. Calif. Acad. Sci. II. 5:706. 1895.

P. farinosa var. incana Fern. Rhodora 9:16. 1907. (Jones 5312av, Tropic, Utah)
P. americana Rydb. Bull. Torrey Club 28:500. 1901. (Rydberg 2746, Deer Lodge, Mont., in 1895)

Plant farinose; leaves oblanceolate to spatulate, 3-6 cm. long, denticulate; scapes 6-40 cm. tall; flowers 3-12, bracts several, linear-lanceolate, 5-10 mm. long, gibbous-based; calyx slightly shorter than the corolla, lobed about 1/3 of its length, the lobes ovate-oblong; corolla 8-11 mm. long, lilac, lobes 2-3 mm. long, deeply bifid, fornices inconspicuous; capsule about equaling the calyx.

Stream banks and moist meadows, Rocky Mts. from Colo. and Utah to n. Can.; known from s. e. Ida., but doubtful if in our range. May-July.

Primula parryi Gray, Am. Journ. Sci. 84:257. 1862. (Parry 311, Middle Park, Colo.)

P. mucronata Greene, Pitt. 3:251. 1897. P. parryi f. mucronata Cockerell, Torreya 15:204. 1915. (Greene, Rocky Mts., Nev.)
P. parryi var. brachyantha Rydb. Mem. N.Y. Bot. Gard. 1:302. 1900. (Tweedy 82, Sheep Mt., Park Co., Mont., in 1887)

Plant rather viscid, somewhat mephitic, fleshy; leaves erect, denticulate to dentate, oblanceolate, as much as 30 cm. long and 5 cm. broad, narrowed to very broad petioles; scapes usually about equaling the leaves at anthesis; flowers 3-15; bracts lanceolate, 5-15 mm. long; calyx purplish, campanulate, 8-11 mm. long, lobed over half the length; corolla reddish-purple, the eye yellow, the tube 8-11 mm. long, lobes oval, usually retuse, 5-8 mm. long, fornices lacking but callosities present just below the sinuses of the lobes; capsule ovoid, about equaling the calyx.

Alpine rock crevices, talus, meadows, and stream banks; Rocky Mts., from c. Mont. and Ida. to N.M., into Utah and Ariz. June-Aug.

5

Primula cusickiana

2

3/4

Glaux maritima

4

Lysimachia thyrsiflora

1/2

var. dentata

1/2

2.5

Douglasia nivalis

var. nivalis

1/2

1.5

Douglasia montana

1/2                2

Lysimachia nummularia

1.5

1.5

Lysimachia terrestris

1/2

JRJ

A most beautiful garden subject for the moist, well-drained area, but not easily grown.

## Steironema Raf. Fringed Loosestrife

Flowers single in the axils, pedicellate, 5(6)-merous; calyx divided to near the base; corolla rotate, yellow, lobed to near the base, each segment enveloping its opposed stamen in bud; stamens alternating with simple (bifurcate) staminodia, the filaments distinct or nearly so; ovary superior, globose; capsule valvular; erect perennial herbs with opposite or whorled leaves.

Six species of temperate N. Am., often included in Lysimachia. (Name from the Greek steiros, sterile, and nema, thread, referring to the staminodia.)

1 Medial cauline leaves ovate to ovate-lanceolate, usually over 3 cm. broad, generally ciliate; petioles usually abundantly ciliate; corolla mostly 9-13 mm. long     S. CILIATUM

1 Medial cauline leaves narrower, the blades linear to lanceolate, mostly 1-3 cm. broad, rarely at all ciliate; petioles sparsely ciliate; corolla mostly 7-9 mm. long
                                                                    S. LANCEOLATUM

Steironema ciliatum (L.) Raf. Ann. Gén. Phys. 7:193. 1820.

  Lysimachia ciliata L. Sp. Pl. 147. 1753. Nummularia ciliata Kuntze, Rev. Gen. 1:398. 1891.
    ("Habitat in Virginia, Canada")
  Steironema laevigatum Howell, Fl. N. W. Am. 436. 1901. (Howell, Rogue R., Oreg.)
  S. ciliatum var. occidentale Suksd. Allg. Bot. Zeit. 12:26. 1906. (Suksdorf 1530, Bingen,
    Wash.)
  S. pumilum Greene, Leafl. 2:111. 1910. (Dr. Lunell, Leeds, N. D.)
  Plant 3-12 dm. tall, with creeping rhizomes, not maculate; blades of the medial cauline leaves 5-15 cm. long, (2.5)3-6 cm. broad, ovate to broadly lanceolate, attenuate-acuminate, finely ciliate-serrulate, rounded to cordate at the base; petioles 5-20 mm. long, usually conspicuously long-ciliate; pedicels slender, arched, 3-8 cm. long; sepals lanceolate-acuminate, 5-7 mm. long; corolla lobes more or less obovate, about 1 cm. long, erose and usually abruptly mucronate to subcaudate, densely granular-puberulent near the base within; filaments shorter than the anthers, not connate, puberulent like the corolla; staminodia simple (to bifurcate); anthers about 3 mm. long; capsule ovoid, subequal to the sepals.

  Damp meadows, ponds, and along streams, rather general in temperate N. Am.; in our area e. of the Cascades but extending down the Columbia gorge to Multnomah Co., Oreg. June-Aug.

Steironema lanceolatum (Walt.) Gray, Proc. Am. Acad. 12:63. 1877.

  Lysimachia lanceolata Walt. Fl. Carol. 92. 1788. (S. C.)
  L. hybrida Michx. Fl. Bor. Am. 1:126. 1803. L. lanceolata var. hybrida Gray, Man. 283.
    1848. L. ciliata var. hybrida Chapman, Fl. S. U. S. 280. 1860. Steironema lanceolatum
    var. hybridum Gray, Proc. Am. Acad. 12:63. 1877. S. hybridum Raf. ex Jackson, Ind.
    Kew. 2:985. 1895. S. ciliatum var. hybridum Chapman, Fl. S. U. S. 3rd ed. 298. 1897.
    Nummularia hybrida Farw. Am. Midl. Nat. 11:67. 1928. Lysimachia lanceolata ssp. hy-
    brida Ray, Ill. Biol. Monog. 24:39. 1936. ("Hab. in Carolina")
  Very similar to S. ciliatum, differing chiefly in having less extensive rhizomes and narrower medial cauline leaves, the blades linear to lanceolate, 5-10 cm. long, 1-3 cm. broad, and rarely at all ciliate but tapered gradually to the 1-3 cm. long, much less abundantly ciliate petioles; also the corolla is smaller, the lobes being mostly about 8(6-9) mm. long; staminodia simple.

  Swamps, damp meadows, and margins of streams and ponds, abundant and common in c. and and e. U. S., much less frequent in w. U. S. July-Aug.

  Our material, as described above, has been referred to the var. hybridum (Michx.) Gray. It differs from the var. lanceolatum in several minor ways and is slightly more western in distribution. Although known in our area only from collections from the Yakima Valley, Wash., it is more widespread in Ariz. and N. M.

## Trientalis L. Starflower

Flowers axillary on slender, curved pedicels from 1 or more leaf axils, usually 6 or 7(5-9)-merous; calyx parted to near the base, segments linear-lanceolate; corolla white to pinkish-rose, rotate, divided nearly to the base, the lobes convolute; stamens exserted, filaments shortly connate at the base, glabrous, slender and considerably longer than the anthers; ovary globose, style slender; capsule valvate; seeds numerous; perennial herbs from slender root-stocks and short, thickened tubers; stems erect, simple, rather inconspicuously glandular-puberulent with reddish to purplish hairs; leaves entire, reduced and often scalelike on the lower part of the stem, the larger ones mainly crowded or whorled at or near the tip of the stem.

Three species, N. Am. and n. Eurasia. (Name from the Latin, meaning containing one-third of a foot, referring to the height.)

1 Flowers usually white; reduced foliage leaves scattered along the stem below the crowded terminal cluster, largest leaves usually less than 5 cm. long; tubers small, horizontal
                                                                    T. ARCTICA
1 Flowers pinkish; leaves of terminal whorl often over 5 cm. long, all others reduced to tiny bracts; tubers enlarged, usually erect or ascending                T. LATIFOLIA

Trientalis arctica Fisch. ex Hook. Fl. Bor. Am. 2:121. 1838.
   T. europaea var. arctica Lehm. Fl. Ross. 3:25. 1847. T. europaea ssp. arctica Hultén, Svensk. Vet. Akad. Handl. Ill. 8[2]:56. 1930. (Tolmie, Clarence Strait)

Tubers short, horizontal, not conspicuously enlarged; aerial stems erect, 5-20 cm. tall; leaves reduced below, sessile, oval to obovate, enlarged upward, the main leaves 3-8, elliptic to obovate, 1.5-5 cm. long, petiolate; corolla usually white, 12-16 mm. broad.

Alas. s. to Alta., B.C. and in n. Ida., in the Cascades to n. Oreg. and along the coast to extreme s. Oreg., in bogs and swamps. May-Aug.

Our plant is often considered to be only a geographic race (variety or subspecies arctica) of a circumpolar species, T. europaea L.

Trientalis latifolia Hook. Fl. Bor. Am. 2:121. 1838.
   T. europaea var. latifolia Torr. Pac. R.R. Rep. 4:118. 1857. Alsinanthemum europaeum var. latifolium Greene, Man. Bay Reg. 238. 1894. (Tolmie, Ft. Vancouver)

Tubers usually erect, 1-2 cm. long, as much as 6 mm. thick; stems 1-2.5 dm. tall; the (3)4-8 leaves of the terminal whorl broadly ovate-elliptic to obovate, 3-10 cm. long, 1.5-4 cm. broad; corolla 8-12 mm. broad, pinkish to rose or somewhat pinkish-lavender. Indian potato.

Woods and prairies, s. B.C. to Alta. s. to n. Ida., and (chiefly on the w. side of the Cascades) s. to the Sierra Nevada and along the coast to some distance s. of San Francisco. Apr.-July.

## PLUMBAGINACEAE. Plumbago Family

Flowers 5-merous, perfect, regular, hypogynous, usually cymose or spicate to racemose or paniculate, sometimes involucrate; calyx lobed, prominently nerved, often papery and conspicuously plicate; corolla gamopetalous to nearly polypetalous, scarious in ours; stamens borne on the corolla tube, opposite its lobes; pistil 1, 5-carpellary; styles 5, distinct or basally united; ovary 1-locular with a single, basally attached ovule; fruit achenelike, often enclosed in the calyx; scapose (ours) to leafy-stemmed, herbaceous (ours) to shrubby perennials with alternate or (ours) basal leaves.

About 10 genera and 300 species, widely distributed, but most common in drier, warmer regions of the Old World.

## Armeria Willd. Nom. Conserv. Thrift

Flowers subsessile in involucrate clusters, the outer bracts reflexed and sheathing the scape;

calyx slightly oblique at the base, dry and scarious, funnelform; corolla long-clawed, connate only at the base, delicate and papery, usually pinkish; styles distinct, glandular-hairy at the base; scapose perennials with dense tufts of narrowly linear, tough, persistent, basal leaves.

Approximately 35 species of the N. Temp. Zone and s. S. Am. (Derivation uncertain.)

Reference:

Lawrence, G. H. M. The genus Armeria in North America. Am. Midl. Nat. 37:757-79. 1947.

Armeria maritima (Mill.) Willd. Enum. Pl. Hort. Berol. 1:333. 1809.

Statice maritima Mill. Gard. Dict. 8th ed. no. 3. 1768. Armeria vulgaris var. maritima Rosenv. in Meddel. om Grönl. 3:683. 1891. A. elongata var. maritima Skottsb. Svensk. Vet. Akad. Handl. 56:286. 1916. (Europe)

Statice armeria L. Sp. Pl. 274. 1753, in part only.

Armeria vulgaris Willd. Enum. Pl. Hort. Berol. 1:333. 1809. (Europe)

A. purpurea Koch, Flora 6:710. 1823. A. vulgaris var. purpurea Mert. & Koch in Roehling, Deutschl. Fl. 2:488. 1826. A. maritima var. purpurea Lawrence, Gentes Herb. 4:405. 1940. (Germany)

A. vulgaris f. arctica Cham. Linnaea 6:566. 1831. A. arctica Wallr. Beit. Heft 1:207. 1844. Statice arctica Blake var. genuina Blake, Rhodora 19:8. 1917. Armeria vulgaris ssp. arctica Hultén, Fl. Aleut. Isl. 275. 1937. A. scabra ssp. arctica Iversen, Dansk. Vidensk. Selsk. Biol. Meddel. 15:18. 1940. (Chamisso, Unalaska, is the first of several localities cited)

A. andina var. californica Boiss. in DC. Prodr. 12:682. 1848. Statice arctica var. californica Blake, Rhodora 19:9. 1917. Armeria macloviana ssp. californica Iversen, Dansk. Vidensk. Selsk. Biol. Meddel. 15:18. 1940. A. maritima var. californica Lawrence, Gentes Herb. 4:406. 1940. (Coulter, in Calif.)

Leaves 5-10 cm. long, 1-3 mm. broad, glabrous to ciliate or puberulent, drying and persistent; scapes 1-several, 1-5 dm. tall, glabrous; headlike flower clusters 1.5-3 cm. broad; outer involucral bracts usually purplish, ovate to lanceolate, the margins papery; flowers short-pedicellate, in clusters of 3 subtended by 2 transparent bracts; pedicels 1-5 mm. long; calyx 5-6 mm. long, 10-nerved and soft-hairy at the base and along the 5 ribs above, ending in short teeth; corolla pinkish to lavender. N=9.

Along beaches and coastal bluffs and sometimes somewhat inland as on Tacoma "prairies"; Arctic N. Am. to Newf.; on the Pacific Coast to s. Calif.; Eurasia. Mar.-July.

Our plants are considered to belong to two varieties of the species. The var. purpurea (Mert. & Koch) Lawrence, is poorly differentiated and doubtfully distinct from var. californica (Boiss.) Lawrence, the two being rather arbitrarily separated as follows: var. californica, outer involucral bracts triangular to lanceolate, often exceeding the head, leaves glabrous, Vancouver I. to Calif.; var. purpurea, involucral bracts ovate to obovate, leaves ciliate to pubescent, Wash. to Alas.; Europe. The ranges of the two overlap in the Puget Sound area where they intergrade completely.

## OLEACEAE. Olive or Ash Family

Flowers usually racemose to paniculate, hypogynous, regular, perfect, or unisexual in a few dioecious or polygamo-dioecious species; calyx usually 4-merous, valvate, rarely absent; corolla gamopetalous (rarely polypetalous), usually 4-merous, or (as in ours) lacking; stamens mostly 2(4), anthers apiculate, the cells back to back, dehiscing lengthwise; disc lacking; pistil 1, 2-carpellary, 2-celled; fruit various but in ours a samara; shrubs or trees with opposite (very rarely alternate), simple to pinnately compound, usually exstipulate leaves.

About 20 genera and at least 400 species, found on all continents, especially common in warmer, temperate and subtropical regions. A family of considerable interest in our region because of the many ornamentals, such as Forsythia (golden bells), Syringa (lilac), Ligustrum (privet), Chionanthus (fringe tree), Jasminum (jasmine), and Osmanthus; other genera are cultivated in warmer regions, including Olea, the olive. Occasionally any of these, but especially the first two, may persist in waste areas and appear to be members of the native flora.

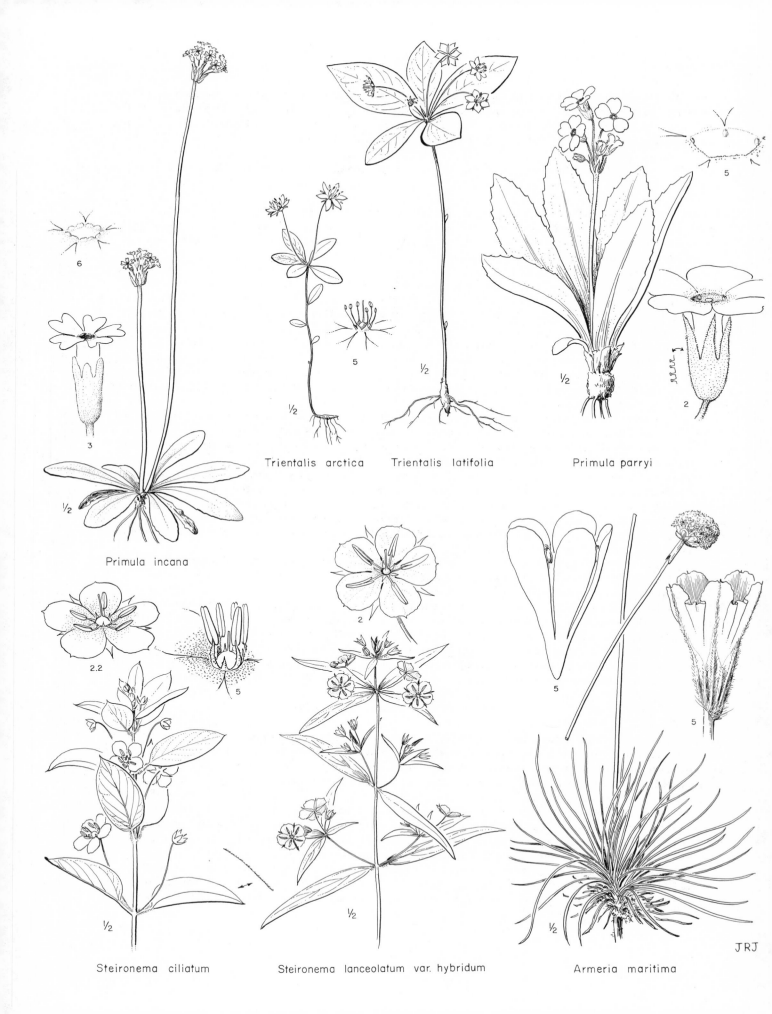

6

3

½

Primula incana

Trientalis arctica    Trientalis latifolia    Primula parryi

5

½

5

½

2

2.2

5

2

5

5

½

½

½

Steironema ciliatum    Steironema lanceolatum var. hybridum    Armeria maritima

JRJ

Ligustrum vulgare L., the common privet, is sometimes reported from wooded areas, undoubtedly due to bird dissemination.

## Fraxinus L. Ash

Flowers usually preceding or developing with the leaves, small, thyrsoid to glomerate; calyx much reduced, usually 4-lobed; corolla (2)4-merous or very often lacking; stamens 2(3 or 4), filaments slender, anthers 2-celled; ovary 1- to 2-celled, usually 1-seeded; fruit a winged, elongate samara; monoecious or polygamous to dioecious trees (ours) with odd-pinnate leaves.
About 40 species of temperate N. Am. and Eurasia. (The Latin name for the ash tree.)

Fraxinus latifolia Benth. Bot. Sulph. 33. 1844.
  F. oregona var. latifolia Lingels. Engl. Bot. Jahrb. 40:220. 1907. (Hinds, San Francisco)
  F. oregona Nutt. N. Am. Sylva 3:59, pl. 99. 1849. F. americana ssp. oregana Wesmael.
    Bull. Soc. Bot. Belg. 31:110. 1892. (Nuttall, Oreg. Terr.)
Dioecious trees with trunks 10-20 m. tall and as much as 1 m. thick, bark rough, grayish-brown, young twigs and herbage usually puberulent to somewhat tomentose, but glabrate; leaflets mostly 5-7(3-9), ovate or ovate-oblong to oblong-obovate, as much as 15 cm. long, entire to crenate-serrate, abruptly short-acute to acuminate, light green above, paler beneath, sessile or subsessile or the terminal ones short-petiolulate; flowers in crowded panicles, appearing with the leaves, the staminate with minute bractlike calyx, stamens usually 2, the anthers much longer than the filaments, without vestiges of pistils, the pistillate with larger calyx, the 4 lobes more or less laciniate, about 1/4 the length of the ovary; style longer than the ovary, stigmatic most of its length, bilobed; samaras 3-5 cm. long, 3-9 mm. broad, wing extending halfway to the base of the main seed-containing body. N=23.
In rather deep, fertile, usually moist soil along the coast from B.C. to the Sierra Nevada and Coast Ranges of Calif. to well s. of San Francisco. Mar.-May.
One of our few deciduous trees of commercial importance, the wood used for furniture, tool handles, etc.

## GENTIANACEAE. Gentian Family

Flowers usually showy, perfect (usually), regular, gamopetalous, hypogynous, 4-5(6-7)-merous (except the pistil), terminal and single or in simple to compound cymes with numerous axillary flowers; calyx persistent, regularly lobed to more deeply cleft, or the lobes sometimes lacking; corolla rotate to salverform or tubular, usually marcescent, convolute in the bud, frequently plicate at the sinuses, often glandular at the base or at the base of the lobes; stamens inserted on the corolla tube and alternate with the lobes, anthers 2-celled, usually versatile, dehiscent full length; ovary sessile to long-stipitate, 1-celled with 2 parietal placentae, often on a basal, annular disc; style usually rather short and thick, cleft above and stigmatic toward the tip or more obviously ending in enlarged stigmas; fruit capsular, usually valvate; seeds numerous, small, usually reticulate, striate, or pitted, often winged; annual or perennial, usually glabrous herbs with simple, mostly entire, opposite to whorled, exstipulate leaves.
Sixty to 70 genera and nearly 800 species, cosmopolitan, often in moist, cool habitats.
1 Corolla salverform to tubular, the lobes usually no longer than the tube
  2 Anthers spirally twisted after flowering (sometimes only slightly so); flowers usually
      pinkish or red (white or yellowish); calyx narrowly tubular-funnelform   CENTAURIUM
  2 Anthers not twisting after dehiscence; flowers usually bluish or bluish-tinged or -streaked,
      if yellow the calyx not narrowly funnelform
    3 Corolla yellowish, the tube about equaling the calyx; small annual      MICROCALA
    3 Corolla usually blue, if yellow the plants perennial                    GENTIANA
1 Corolla rotate, the lobes usually longer than the tube
  4 Plant annual or biennial; stigmas decurrent for half the length of the ovary
                                                                      LOMATOGONIUM

4  Plant perennial; stigmas not decurrent on the ovary
   5  Styles thick, scarcely 1 mm. long; flowers 5-merous                  SWERTIA
   5  Styles slender, at least 2 mm. long; flowers 4-merous                FRASERA

## Centaurium Hill.  Centaury

Flowers often single, but more usually subsessile to long-pedunculate in few- to many-flow-ered cymes, white to yellowish-pink or deep salmon, 5(4)-merous except for the pistil; calyx slender, deeply lobed, the segments linear and often membranous margined; corolla salver-form, contorted-convolute in the bud, the lobes narrow, mostly about half the length of the tube, usually slightly emarginate; stamens inserted about midlength of the corolla tube, well exserted, the filaments slender and glabrous, the anthers oblong, spirally coiling after de-hiscence, the anther sacs sometimes slightly unequal; ovary 1-celled, the 2 parietal placentae intruded; style simple or shortly branched at the tip; stigma flattened, oval to flabelliform or triangular; capsule slender, 2-valved; seeds minute; annual (ours), strict glabrous herbs with opposite, sessile or clasping, entire leaves.

About 20 species, very widespread. (From the old Latin name, centaureum, long used by the early herbalists, referring to the centaur, Chiron, who supposedly discovered medicinal prop-erties for the plant.)

1  Basal leaves several, often forming a tuft or rosette, rather conspicuously 2(5)-veined from the base; flowers usually many and crowded in compact clusters, nearly sessile; anthers usually at least 1.5 mm. long                                      C. UMBELLATUM
1  Basal leaves usually well spaced, not forming rosettes, inconspicuously nerved with usually only 1 main vein; flowers usually few, often conspicuously pedicellate; anthers seldom as much as 1.5 mm. long
   2  Pedicels longer than the central flowers, usually over 2 cm.       C. EXALTATUM
   2  Pedicels all less than 2 cm. long, usually much shorter than the flowers
                                                                          C. MUHLENBERGII

Centaurium exaltatum (Griseb.) Wight ex Piper, Contr. U.S. Nat. Herb. 11:449. 1906.
   Cicendia exaltata Griseb. in Hook. Fl. Bor. Am. 2:60, pl. 157. 1838. Erythraea douglasii
   Gray in Bot. Calif. 1:480. 1876. E. exaltata Coville, Contr. U.S. Nat. Herb. 4:150. 1893.
   (Douglas, "not uncommon between the Kettle Falls, 'and Narrows' of the Columbia River,
   N.W.C.")
   Plant glabrous; stems 1 to several, simple (usually) to branched, 5-25 cm. tall; basal leaves
few to several but not forming a definite rosette, elliptic-lanceolate to oblanceolate, 5-25 mm.
long, usually acute, with one main vein; cauline leaves longer and narrower, acute; flowers
few, often single; pedicels strict, slender, usually exceeding the subtending bracts, that of
the central flower often 2-4 cm. long; calyx slender, 6-9 mm. long, lobed nearly to the base,
not membranous in the sinuses between the lobes; corolla light salmon to nearly white, the
tube up to 1.5 times the length of the calyx, the lobes about 4 mm. long, lanceolate, usually
slightly retuse; stamens exserted, the anthers usually about 1 (to 1.7) mm. long, slightly twist-ed; style shortly branched at the tip, the stigmatic lobes obovate-deltoid; capsule nearly twice
the length of the calyx.

Moist places, especially around hot springs and alkaline lakes, e. of the Cascade Range,
Wash. to Neb., s. to Colo. and e. Calif. June-July.

Centaurium muhlenbergii (Griseb.) Wight ex Piper, Contr. U.S. Nat. Herb. 11:450. 1906.
   Erythraea muhlenbergii Griseb. Gen. & Sp. Gent. 146. 1839. Centaurodes muhlenbergii
   Kuntze, Rev. Gen. 2:426. 1891 (Douglas, Calif.)
   Erythraea curvistaminea Wittr. Bot. Zentralb. 26:317. 1886. Centaurium curvistamineum
   Abrams, Ill. Fl. Pac. St. 3:352. 1951. (Suksdorf, Lincoln Co., Wash.)
   Erythraea minima Howell, Fl. N.W. Am. 443. 1901. Centaurium minimum Piper in Piper &
   Beattie, Fl. N.W. Coast 288. 1915. (Howell, w. Oreg.)
   C. muhlenbergii var. albiflorum Suksd. Werdenda 1:30. 1927. C. muhlenbergii f. albiflorum
   St. John, Fl. S.E. Wash. 314. 1937 (Suksdorf 8903, Latah Creek, s.e. of Spangle, Wash.)
   Plant slender, usually single-stemmed, 3-30 cm. tall; basal leaves few to several but not in

a conspicuous rosette, ovate or oblong to obovate, 5-15 mm. long, rounded, the single indis-
tinct vein sometimes branched near the base; cauline leaves narrower, mostly linear-oblanceolate and acute; flowers few; pedicels usually considerably shorter than the calyx; calyx 6-8
mm. long, not membranous in the sinuses between the slender, linear lobes; corolla white to
deep pink, usually about half again as long as the calyx, the lobes narrow, somewhat retuse,
3-4 mm. long; stamens exserted, the anthers about 1 mm. long, coiling slightly after dehiscence, often the pollen sacs somewhat unequal at the base; stigma lobes ovate-deltoid; capsules
about twice the length of the calyx.

Moist soil, e. Wash. from Kittitas to Spokane cos., s. to Nev., w. through the Columbia
gorge to the Willamette Valley and s.; rather general in coastal and c. Calif. June-Aug.

Centaurium umbellatum Gilib. Fl. Lithuan. 35. 1781.(Europe)
    Gentiana centaurium L. Sp. Pl. 229. 1753. Centaurium erythraea Raf. Dan. Holst. Fl. 2:75.
    1796. Erythraea centaurium Pers. Syn. 1:283. 1805. Centaurium centaurium Wight ex
    Piper, Contr. U. S. Nat. Herb. 11:449. 1906.(Europe)
    Plant 1-5 dm. tall, 1-many stemmed, simple to branched from near the base, glabrous; bas-
al leaves several, often forming a distinct tuft or rosette, sessile, obovate or oblong-oblanceolate, 1.5-4 cm. long, rounded, rather prominently 3- to 5-veined, the nerves nearly parallel;
lower cauline leaves more or less similar, gradually reduced, narrower and more acute above;
flowers many in crowded cymes, essentially sessile, each of the 2 subtending bracts usually
with a rudimentary flower in its axil; calyx 4-6 mm. long, the lobes slender, acute, membranous in the sinuses; corolla yellowish to salmon red, the tube nearly twice the length of the
calyx, the lobes 4-6 mm. long, oblong-lanceolate; stamens exserted slightly, unequal; anthers
1.5-2 mm. long, conspicuously twisted; lobes of the stigma oval; capsules slender, about 10
mm. long, the parietal placentae intruded and nearly touching. N=21.

Usually where moist, on wasteland, meadows, and prairies. Native of Europe, established
from n.w. Wash. to n. Calif., chiefly w. of the Cascades, less common e. to Ida. June-Aug.


Frasera Walt.


Flowers 4-merous except for the 2-carpellary pistil, rotate to shallowly campanulate, numerous in a large, compact to loose, often interrupted thyrse, withering-persistent; calyx
cleft nearly to the base, the lobes lanceolate to linear-subulate; corolla white or yellowish-
green with purplish maculations to bluish or purplish, deeply divided, the tube almost none,
often with short processes (squamellae) or (in F. speciosa) with a broader, scalelike, lacerate or fimbriate process (corona) fused to the base of the corolla lobe and with a single or
(in F. speciosa) paired pit (fovea) near the base of each segment, the fovea more or less completely surrounded by a usually fimbriate membrane (the hood); stamens distinct or very conspicuously united at the base, alternate with scalelike processes (crown scales) borne just
within the stamens, or the scales reduced to squamellae or even lacking; ovary subsessile,
1-locular, with 2 parietal placentae; style short to elongate; capsule septicidal from the apex,
often compressed; seeds compressed, often narrowly wing-margined, minutely alveolate-
puncticulate; glabrous to puberulent perennial herbs from a simple to branched caudex, with
1-several flowering stems and opposite or whorled, often basally connate-perfoliate and sheathing, prominently nerved, usually entire leaves.

About 12 N. Am. species, mostly in w. U.S. (Named in honor of John Fraser, 1750-1811,
an English nurseryman who collected plants in N. Am.)
    References:
    Card, Hamilton H. A revision of the genus Frasera. Ann. Mo. Bot. Gard. 18:245-82. 1931.
    Post, Douglas Manner. A revision of the genus Frasera. U. of Wash. Thesis. 1950.
    St. John, Harold. Revision of the genus Swertia (Gentianaceae) of the Americas and the re-
        duction of Frasera. Am. Midl. Nat. 26:1-29. 1941.
    The following treatment is adopted from the Post manuscript, except for certain name
changes necessitated by treating several infraspecific taxa in the varietal rather than subspecific category.
    Morphologically, Frasera approaches Swertia rather closely and frequently has been merged

with it. Recent cytological study of the two taxa suggest, however, that Frasera, with a basic chromosome number of 13, is perhaps more closely related to Gentiana (which has a similar number) than to Swertia, where haploid numbers of 9, 12, and 14 have been reported.

1 Cauline leaves in whorls of 3-5(6); flowering stems nearly always single; styles shorter than the ovary
  2 Corolla greenish-yellow with purplish maculations, the lobes with paired, more or less elliptic foveae        F. SPECIOSA
  2 Corolla more nearly clear blue or bluish-purple, the lobes with a single, nearly orbicular fovea        F. FASTIGIATA
1 Cauline leaves usually opposite; flowering stems usually more than 1; styles longer than the ovary
  3 Flowers white or cream; crown scales seldom over 2 mm. long, entire to setiform; foveae obovate, completely surrounded by the uniformly fringed hood    F. MONTANA
  3 Flowers blue (except for the occasional albino); crown scales 1-6 mm. long, lacerate to entire; foveae oblong, the hood smaller and scarcely fringed toward the outer edge        F. ALBICAULIS

Frasera albicaulis Dougl. ex Griseb. in Hook. Fl. Bor. Am. 2:67. 1838.
  Swertia albicaulis Kuntze, Rev. Gen. 2:430. 1891. Leucocraspedum albicaule Rydb. Fl. Rocky Mts. 665. 1917. Frasera nitida var. albicaulis Card, Ann. Mo. Bot. Gard. 18:269. 1931. (Douglas, "in the mountain vallies between Spokan and Kettle Falls")
  FRASERA ALBICAULIS var. NITIDA (Benth.) C. L. Hitchc. hoc loc. F. nitida Benth. Pl. Hartw. 322. 1849. Swertia albicaulis var. nitida Jeps. Fl. Calif. 3:95. 1939. (Hartweg 1833, mountains near Sacramento, Calif.)
  FRASERA ALBICAULIS var. CUSICKII C. L. Hitchc. hoc loc. F. cusickii Gray, Proc. Am. Acad. 22:310. 1887. F. nitida var. cusickii Nels. & Macbr. Bot. Gaz. 61:33. 1916. (Cusick 1427, Grande Ronde Valley, Oreg.)
  F. caerulea Mulford, Bot. Gaz. 19:118. 1894. Leucocraspedum coeruleum Rydb. Fl. Rocky Mts. 666. 1917. (Mulford, near Wagonville, Owyhee Mts., Ida., July 8, 1892) = var. albicaulis.
  Frasera nitida var. albida Suksd. Werdenda 1:30. 1927. Swertia columbiana f. albida St. John, Am. Midl. Nat. 26:24. 1941. (Suksdorf 11458, e. of Husum, Klickitat Co., Wash.) = var. columbiana.
  Frasera albicaulis f. alba St John, Proc. Biol. Soc. Wash. 41:196. 1928. Swertia albicaulis f. alba St. John, Am. Midl. Nat. 26:17. 1941. (C. S. Parker 446, dry bluff above highway, Pullman to Moscow, in Whitman Co., Wash.) = var. albicaulis.
  Swertia watsonii St. John, Am. Midl. Nat. 26:21. 1941. (Watson 269, Battleground, Big Hole Valley, Beaverhead Co., Mont., July 24, 1880) = var. albicaulis.
  FRASERA ALBICAULIS var. IDAHOENSIS (St. John) C. L. Hitchc. hoc loc. Swertia idahoensis St. John, Am. Midl. Nat. 26:24. 1941. (Cusick 2226, near Cuprum, Ida., July 10, 1899)
  FRASERA ALBICAULIS var. COLUMBIANA (St. John) C. L. Hitchc. hoc loc. Swertia columbiana St. John, Am. Midl. Nat. 26:22. 1941. (Suksdorf 40, Klickitat Co., Wash., May 27, Aug., 1881)

Plant densely puberulent to glabrous, with (usually) several flowering stems 1-7 dm. tall from a branched caudex; leaves prominently 3-nerved, white-margined, the basal 5-30 cm. long, 5-20 mm. broad, linear-oblanceolate to narrowly spatulate; cauline leaves opposite, considerably reduced upward; thyrse narrow, congested to open, mostly interrupted; calyx lobes narrowly ovate to linear-lanceolate and subulate, 4-9 mm. long; corolla rotate, pale to fairly dark bluish or purplish, often darker mottled, occasionally white, the lobes 5-11 mm. long, ovate to oblong, acute to acuminate, often squamellate on the inner surface and usually with several long linear processes at the base; foveae 1 per lobe, narrowly oblong, the hood saccate at the base, narrowly fimbriate-margined above; filaments usually united at the base by conspicuous, linearly lacerate to entire crown scales 1-6 mm. long (scales rarely lacking); style slender, considerably longer than the ovary; capsule 10-15 mm. long, somewhat compressed.

A wide-ranging species of the plains and lower mountains from s. B. C. s. on the e. side of

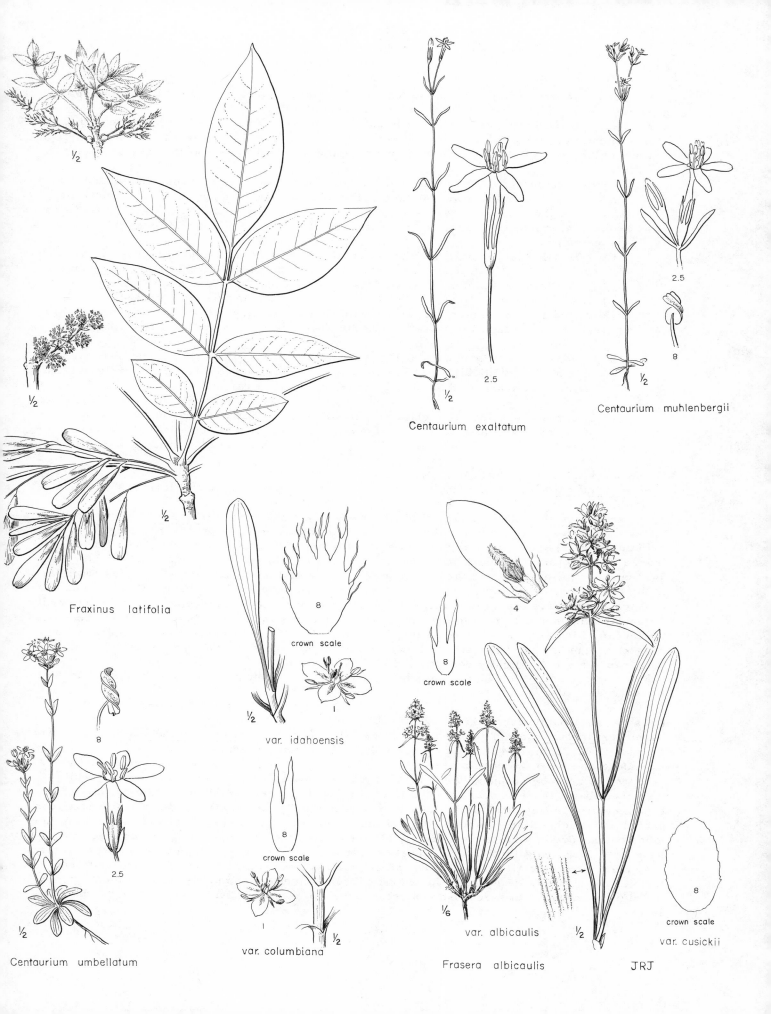

½

½

½

Fraxinus latifolia

Centaurium exaltatum

2.5

½

2.5

8

½

Centaurium muhlenbergii

8

crown scale

var. idahoensis

1

8

crown scale

var. columbiana

8

½

1

½

4

8

crown scale

½

2.5

Centaurium umbellatum

⅙

var. albicaulis

½

Frasera albicaulis

8

crown scale

var. cusickii

JRJ

the Cascades to Nev. and Calif., e. to Ida. and w. Mont., and represented by several intergradient races, most of which occur in our area. May-July.

1 Stems and leaves both rather uniformly puberulent; crown scales mostly lacerate into linear
    segments; c. Wash. through Ida. to Mont.                                        var. albicaulis
1 Stems glabrous, the leaves also glabrous or sometimes puberulent at the base and along the
    midrib
    2 Crown scales subentire, semipetaloid, 2.5-4.5 mm. long; leaves puberulent at the base
        and sometimes on the lower surface of the midrib; Blue Mts. of Grant and Union cos.,
        Oreg., to Owyhee Co., Ida.                                  var. cusickii (Gray) C. L. Hitchc.
    2 Crown scales lobed to lacerate, not petaloid, often less than 2.5 mm. long; plants some-
        times glabrous
        3 Corolla pale blue, usually not darker-mottled; crown scales 2-6 mm. long, mostly
            deeply lacerate into linear segments; sheath of the basal leaves usually not over 1.5
            (2) cm. long, not bluish; Wallowa Mts., Baker Co., Oreg., and adj. Adams Co.,
            Ida.                                         var. idahoensis (St. John) C. L. Hitchc.
        3 Corolla pale to dark blue but usually with darker mottling; crown scales 1-4 mm. long,
            from erose to lacerate; sheath of the basal leaves usually over 1.5 cm. long, often
            bluish-tinged
            4 Lower leaf sheaths usually puberulent; crown scales (1)3-4 mm. long; near the Co-
               lumbia R. in Klickitat and Yakima cos., Wash., and adj. n. Oreg., well isolated
               from, but scarcely distinguishable from, the next variety
                        var. columbiana (St. John) C. L. Hitchc.
            4 Lower leaf sheaths often glabrous; crown scales usually less than 3 mm. long; s. e.
               Oreg. to c. Calif., on the w. side of the Sierra Nevada, but often attributed to our
               area                                         var. nitida (Benth.) C. L. Hitchc.

**Frasera fastigiata** (Pursh) Heller, Bull. Torrey Club 24:312. 1897.
   Swertia fastigiata Pursh, Fl. Am. Sept. 101. 1814. (Lewis, Quamash Flats, Ida., June 4,
     1806)
   Frasera thyrsiflora Hook. Journ. Bot. & Kew Misc. 3:288. 1851. (Douglas, "Mountain Val-
     leys, Spokan and Kettle Falls, Valley of the Columbia"; locality probably in error)
   Plant glabrous and more or less glaucous; flowering stems usually single, 5-15 dm. tall,
from a usually simple caudex; basal leaves obovate to spatulate-oblanceolate, 20-45 cm. long,
5-12 cm. broad, acute, thin; cauline leaves in whorls of 3 or sometimes 4, the lower similar
to the basal, reduced and more nearly ovate-lanceolate or lanceolate above; inflorescence narrow, congested, the branches erect; calyx lobes 5-13 mm. long, lanceolate, acuminate; corolla tubular-campanulate, pale to dark blue, with numerous squamellae between the stamens and
the foveae which are about 1.5 mm. long, suborbicular, slightly sunken, and completely surrounded by the hoods; crown scales or fimbriae lacking, the stamens distinct; style about 1 mm.
long; capsule ovoid, slightly compressed, 12-18 mm. long.
   Moist woods and meadowland, Blue Mts. of Oreg. and Wash., e. to n. e. Ida., from Idaho to
Kootenai cos. May-July.

**Frasera montana** Mulford, Bot. Gaz. 19:119. 1894.
   Leucocraspedum montanum Rydb. Fl. Rocky Mts. 665. 1917. Swertia montana St. John, Am.
     Midl. Nat. 26:16. 1941. (Mulford, July 26, 1892, Pioneer, Ida.)
   Plant glabrous or very minutely puberulent near the base, the flowering stems usually several from a branched caudex, 3-8 dm. tall; leaves conspicuously 3-nerved, narrowly white-
margined; basal leaves linear-lanceolate to narrowly spatulate, 7-20(30) cm. long, 0.5-1.5 cm.
broad; cauline leaves opposite, reduced, mostly narrowly lanceolate, acute; thyrse short, congested; calyx lobes linear-lanceolate, 3-7 mm. long; corolla rotate, clear white to cream, the
lobes oblong-ovate, rounded and usually abruptly apiculate, 5-9 mm. long; corona lacking; foveae 1 per lobe, obovate, the hood saccate at the base for 1/3-1/4 the length of the fovea, otherwise deeply lacerate all around; crown scales usually present, variable, from entire to setiform; style slender, longer than the ovary; capsule somewhat compressed, 7-10 mm. long.
   Drier meadowland to sagebrush and yellow pine valleys and hillsides, c. and w. c. Ida. May-
July.

Frasera speciosa Dougl. ex Griseb. in Hook. Fl. Bor. Am. 2:66. 1838.
  Tessaranthium speciosum Rydb. Fl. Rocky Mts. 666. 1917. (Douglas, "on the low hills near
    Spokan and Salmon Rivers and subalpine parts of the Blue Mountains, near the Kooskooska
    River, N. W. C. ")
  T. radiatum Kell. Proc. Calif. Acad. Sci. I. 2:144. 1862. Swertia radiata Kuntze, Rev. Gen.
    2:430. 1891. (Gibbs, headwaters of the Carson R. , Calif. , in 1861)
  Frasera macrophylla Greene, Pitt. 4:186. 1900. Tessaranthium macrophyllum Rydb. Fl.
    Rocky Mts. 666. 1917. Swertia radiata var. macrophylla St. John, Am. Midl. Nat. 26:11.
    1941. (C. F. Baker, about Pagosa Springs, Colo. , July 25, 1899)
  Frasera speciosa var. angustifolia Rydb. Bull. Torrey Club 31:632. 1904. F. angustifolia
    Rydb. Bull. Torrey Club 33:149. 1906. Tessaranthium angustifolium Rydb. Fl. Rocky Mts.
    666. 1917. (Shear 3369, Lima, Mont. , in 1895)
  Plant glabrous to scabrous-puberulent throughout; flowering stems single, 1-2 m. tall, from
a (usually) simple caudex; basal leaves 25-50 cm. long, 3-15 cm. broad, elliptic-oblong to
oblanceolate or spatulate; cauline leaves in whorls of 3-5, much reduced; inflorescence elongate, open or congested; calyx lobes 15-30(35) mm. long, lanceolate and more or less subulate; corolla rotate-campanulate, yellowish-greenish, purplish-blue maculate, the lobes 10-
25 mm. long, oval to broadly obovate, rounded to acute or acuminate; corona nearly as broad
as the base of the corolla lobes, deeply lacerate into several linear segments, nearly equaling
the paired, oblong-elliptic foveae which are 3.5-9 mm. long and completely encircled by conspicuous lacerate-fimbriate hoods; crown scales lacking; style about 1/2 as long as the ovary;
capsule 2-2.6 cm. long, slightly obcompressed.
  Open (to wooded) hills and valleys to well up in the mountains, e. Wash. to the Dakotas, s.
to Calif. , N.M. , and n. Mex. June-Aug.


Gentiana L. Gentian


  Flowers solitary to numerous in flat-topped to much elongate cymose clusters, 4- to 5(6)-
merous (except the pistil), white or yellowish to bluish or purplish, often with considerable
green mottling; calyx fused to near the tip and tubular, to lobate nearly to the base, sometimes
lined with an inner membrane that projects above the base of the often unequal lobes; corolla
narrowly funnelform to salverform but usually closing quickly, persistent, 4- to 5(6)-lobed
1/5-1/3 of the length, often either plicate in the sinuses (the plaits rounded to acute or lobed
or toothed) or with setaceous scales at the base of the lobes on the inner surface; stamens adnate to the corolla tube for 1/3-3/5 of its length, the adnate portion often with free, winglike
margins, the free filaments often conspicuously flattened; anthers versatile, erect or recurved; ovary stipitate to sessile; style usually short and rather stout, ending in 2 stigmatose
lobes or with enlarged, crenate-margined stigmas; capsules 1-celled, 2-valved, many-seeded;
annual, biennial, or perennial herbs from fleshy roots or slender rhizomes, ours usually glabrous; leaves opposite, petiolate to sessile, and sometimes clasping.
  A large genus, mostly of temperate and arctic regions usually on wet soil, many of the perhaps 300 species very widely distributed. (Named for King Gentius of Illyria, who supposedly
discovered medicinal properties for the plant. )
  The plants usually flower in late summer or fall and many are regarded as choice ornamentals, especially for the rock garden.
  Reference:
  Gillet, John M. A revision of the North American species of Gentianella Moench. Ann. Mo.
    Bot. Gard. 44:195-269, 1957.
1 Perennials; corollas usually 5-merous and well over 2.5 cm. long, if shorter then the plants
    either somewhat prostrate and matted and spreading by very slender rhizomes, or the
    leaves with long, sheathing, connate bases and the blades then less than 1 cm. long; stig-
    mas never flabellate-erose
  2 Corollas usually well over 2.5 cm. long; calyx with an inner membranous lining extending
      above the base of the lobes; plants not producing slender rhizomes
    3 Flowers white or pale yellowish, blotched and striped with purple; leaves 4-12 cm.
      long, usually less than 1/6 as broad                              G. ALGIDA

    3 Flowers bluish or purple, if yellowish then the leaves less than 6 times as long as broad
        4 Plaits or appendages of the corolla sinuses truncate or rounded, entire; flowers of the lower axils with bractless pedicels; seeds fusiform, not winged
                                                                  G. SCEPTRUM
        4 Plaits or appendages of the corolla sinuses usually lacerate into narrow segments; flowers bracteate at the base of the calyx
            5 Leaves ovate to obovate, rarely as much as twice as long as broad; plants glabrous; flowers usually single; seeds elongate, more or less terete, not flattened or wing-margined                                  G. CALYCOSA
            5 Leaves usually at least twice as long as broad, if at all ovate or obovate then the stems very finely pubescent in lines below the leaves; leaves finely ciliolate; calyx lobes somewhat glandular-ciliolate; flowers usually several; seeds flattened, broadly wing-margined                G. AFFINIS
  2 Corollas less than 2.5 cm. long; calyx not membranous-lined above the base of the lobes; plants usually spreading by very slender rhizomes, often more or less matted
      6 Flowers usually 5-merous; cauline leaves 2-4 pairs, very shortly connate
                                        G. GLAUCA
      6 Flowers 4-merous; cauline leaves many, long-sheathing      G. PROSTRATA
1 Annuals; corollas mostly less than 2 cm. long, if longer, then usually 4-merous or with lacerate-fimbriate lobes, or the stigmas broadly flabellate-erose
  7 Corollas deep blue, usually over 2 cm. long, 4-merous; stigmas broadly flabellate-erose
    8 Stems simple; flowers nearly always single; basal leaves usually not more than 2; anthers 1-1.5 mm. long; style thick, 1-2 mm. long; seeds fusiform, longitudinally striate                                      G. SIMPLEX
    8 Stems mostly branched; flowers and basal leaves usually several; anthers 3-4 mm. long; style usually slender and over 2 mm. long
      9 Upper leaves mostly acute; base and some of the keels of the calyx whitish- or hyaline-papillate; approaching our range on the n. and e. G. macounii Holm, referred by Gillett to Gentianella crinita as a subspecies
      9 Upper leaves mostly obtuse; calyx wrinkled or smooth but not papillate
                                      G. DETONSA
  7 Corollas less than 2 cm. long or not deep blue, sometimes 5-merous; stigmas not flabellate-erose
    10 Sinuses of the corolla plicate, often 2-toothed, the lobes not fringed within at the base; calyx lobed not over half the length
      11 Flowers 5-merous; leaves not long-sheathing or overlapping below
                                      G. DOUGLASIANA
      11 Flowers 4(5?)-merous; leaves long-sheathing, overlapping near the base
                                      G. PROSTRATA
    10 Sinuses of corolla not plicate, lobes sometimes fringed within at the base; calyx often lobed over half its length
      12 Corolla lobes not fringed within at the base            G. PROPINQUA
      12 Corolla lobes fringed within at the base, acute to rounded
        13 Calyx lobed to the base; fringe of the corolla composed of 2 distinct scales per lobe; flowers on long, naked pedicels            G. TENELLA
        13 Calyx lobed 2/3-4/5 of its length; fringe composed of 1-10 filiform setae per lobe; flowers mostly subtended by bractlets            G. AMARELLA

Gentiana affinis Griseb. ex Hook. Fl. Bor. Am. 2:56. 1838.
  Pneumonanthe affinis Greene, Leafl. 1:71. 1904. Dasystephana affinis Rydb. Bull. Torrey
    Club 33:149. 1906. (Several collections by Drummond and by Douglas are cited)
  Gentiana parryi Engelm. Trans. Acad. Sci. St. Louis 2:218, pl. 10. 1863. Pneumonanthe

parryi Greene, Leafl. 1:71. 1904. <u>Dasystephana</u> <u>parryi</u> Rydb. Bull. Torrey Club 33:149.
1906. (<u>Parry</u> <u>304</u>, Snowy Range, Colo. )

Gentiana <u>affinis</u> var. <u>ovata</u> Gray, Bot. Calif. 1:483. 1876. (<u>Bolander</u>, near San Francisco)

G. <u>oregana</u> Engelm. ex Gray, Syn. Fl. 2[1]:122. 1878. <u>Dasystephana</u> <u>oregana</u> Rydb. Bull.
Torrey Club 40:464. 1913. (Specimens from B. C., Ida., Oreg., and Calif. are cited)

Gentiana <u>forwoodii</u> Gray, Proc. Am. Acad. 19:86. 1883. G. <u>affinis</u> var. <u>forwoodii</u> Kusnezow,
Acta Hort. Petrop. 15:202. 1898. <u>Pneumonanthe</u> <u>forwoodii</u> Greene, Leafl. 1:71. 1904.
<u>Dasystephana</u> <u>forwoodii</u> Rydb. Bull. Torrey Club 33:149. 1906. (<u>Dr</u>. <u>W</u>. <u>H</u>. <u>Forwood</u>, Wind
River Mts., Wyo.

Gentiana <u>affinis</u> var. <u>parvidentata</u> Kusnezow, Acta Hort. Petrop. 15:201. 1898. (Collections
from Wash., Ida., and Colo. are cited)

G. <u>bracteosa</u> Greene, Pitt. 4:180. 1900. <u>Pneumonanthe</u> <u>bracteosa</u> Greene, Leafl. 1:71. 1904.
Gentiana <u>parryi</u> var. <u>bracteosa</u> Nels. in Coult. & Nels. New Man. Bot. Rocky Mts. 382.
1909. (<u>Greene</u>, Marshall Pass, Colo., in 1896)

Gentiana <u>remota</u> Greene, Pitt. 4:182. 1900. (<u>Greene</u>, Humboldt R. at Deeth, Nev., Aug. 5,
1895)

G. <u>interrupta</u> Greene, Pitt. 4:182. 1900. <u>Dasystephana</u> <u>interrupta</u> Rydb. Bull. Torrey Club
33:149. 1906. (<u>C</u>. <u>F</u>. <u>Baker</u>, Pagosa Sprs., Colo., Aug. 30, 1899)

Gentiana <u>menziesii</u> sensu Peck, Man. High. Pl. Oreg. 557. 1941, but not of Griseb.

Caespitose perennial from rather thick fleshy roots, without rhizomes; stems 1-several,
erect to decumbent at the base, 1.5-8 dm. tall, minutely puberulent in lines below the slightly decurrent leaf bases; leaves 8-15 pairs, the lowermost usually reduced to bladeless, connate, sheathing bases, middle cauline blades narrowly lanceolate, oblong, or oblong-oblanceolate
to broadly lanceolate or elliptic-ovate, 2-5 cm. long, 5-20(25) mm. broad, usually very finely glandular-ciliolate, at least near their bases, floral leaves similar or shorter and broader; flowers
5-merous, few, closely crowded, often arising from the top 2 or 3 nodes only or the upper 3-5
nodes all floriferous; peduncles 3-25 mm. long, bracteate at the summit; bracteoles foliaceous to
somewhat scarious, linear to ovate; calyx tube 3-9 mm. long, tubular-funnelform, greenish-
to bluish- or purplish-tinged, with an inner membranous lining projecting above the bases of
the lobes and toothed inside them, lobes usually unequal, from ovate to linear and from longer
than the tube to reduced to mere teeth, or lacking entirely, the tube then nearly entire to
erose, and from nearly truncate to oblique or deeply parted once or twice; corolla tubular-
funnelform, (2)2.5-4(4.5) cm. long, deep blue but usually variously mottled or streaked with
green, lobes (3)4-6(7) mm. long, oblong-ovate to ovate, rounded to acute or abruptly pointed,
the plaits of the sinuses 1/2-3/4 the length of the lobes and usually laciniately deeply cleft into 2-5 narrow segments (entire); stamens slightly shorter than the corolla tube, filaments adnate to near midlength of the corolla, the adnate portion broadly wing-margined; anthers 2.5-
4 mm. long; ovary long-stipitate; style short, cleft above, the stigmatic surfaces oblong-oval;
seeds flattened and wing-margined, very finely reticulate.

Meadowland or damp soil, valleys and foothills to well up in the mountains; B. C. s. on the
e. side of the Cascades through Wash. to s. c. and s. w. Oreg., and Calif. and Ariz., e. to
Alta. and the Rocky Mt. states to n. Mex. July-Sept.

As here treated, <u>G</u>. <u>affinis</u> is a very wide-ranging species, showing considerable variation
in height, leaf shape, number of flowers per stem, and size of the calyx lobes; since all variants are so much more similar to one another than to any other species, and since most of
them are usually found throughout the entire range of the species, rather than localized in definite geographic areas, there seems to be very good reason for their inclusion in the one specific complex. Lengthy and detailed study might reveal several distinctive geographic races
within the species as a whole, but it is not believed that such exist in our particular range.
The several species that have been recognized by other workers might be separated by the key
which follows. It should be stressed that such separation is purely mechanical, and that the
resulting segmentation will not represent geographic races, regardless of the nomenclatural
rank assigned to them.

1 Calyx lobes nearly or quite lacking, or the tube usually deeply cleft once or twice; found
   throughout the range of <u>G</u>. <u>affinis</u>                                              G. forwoodii Gray

1 Calyx lobes present but often small, the tube usually not deeply cleft
  2 Subtending floral leaves usually broader than the other leaves, often somewhat scarious,
    as is the calyx tube; limited largely to the Rocky Mt. area, occasional throughout the
    range                                                                    G. parryi Engelm.
  2 Subtending leaves mostly not conspicuously broader than, or different in texture from,
    the other leaves; calyx tube not scarious, or scarious only between the bases of the
    lobes
    3 Stems low, nearly prostrate; leaves fleshy; plants markedly puberulent on the calyx
      and leaves; calyx not scarious; along the coast, s. Oreg. to c. Calif.
          G. affinis var. ovata Gray, the same as G. menziesii in Peck's Manual
    3 Stems usually erect or no more than decumbent; plants not fleshy; calyx usually scar-
      ious between the lobes
      4 Stems very elongate and flowers well spaced in interrupted inflorescences; through-
        out the entire range although allegedly peculiar to the Rocky Mts.
                        G. interrupta Greene
      4 Stems not elongate; leaves and flowers more closely approximate, the latter usual-
        ly closely clustered
        5 Leaves ovate, scarcely 3 times as long as broad; general but mainly in w. Ida.
          and e. Oreg.                                               G. oregana Engelm. ex Gray
        5 Leaves lanceolate to oblong, at least 3 times as long as broad; general
                           G. affinis

Gentiana algida Pall. Fl. Ross. 1:2, pl. 95. 1788. (Siberia)
  G. romanzovii Ledeb. ex Bunge in Nouv. Mém. Soc. Nat. Mosc. 1:215. 1829. G. algida var.
    romanzovii Kusnezow, Acta Hort. Petrop. 15:252. 1898. Dasystephana romanzovii Rydb.
    Fl. Rocky Mts. 662. 1917. (Siberia)
  Caespitose perennial with 1-several stems 5-20 cm. tall; basal leaves linear-oblanceolate,
4-12 cm. long, cauline leaves 3-5 pairs, linear-oblong to oblong-lanceolate, 3-5 cm. long,
mostly about 5(10) mm. broad, their bases connate for 5-8 mm., the subtending leaves of the
usually 5-merous, closely crowded, subsessile flowers often considerably broader, sometimes
ovate-lanceolate; calyx narrowly funnelform, mostly about 2 cm. long, usually purplish-
blotched, tube truncate between the lobes which are subequal, linear to lanceolate, from about
half as long as, to subequal to, the tube, and somewhat carinate and more or less transrugose;
corolla usually about twice the length of the calyx (3.5-5 cm. long), white or pale yellowish,
purple-blotched and purplish-streaked from the back of the lobes nearly to the calyx, strongly
plicate between the acuminate short lobes; filaments freed slightly below midlength of the corolla, the adnate portion broadly wing-margined, free filaments about twice the length of the
3-4 mm. anthers; ovary long-stipitate; style deeply 2-cleft, the stigmatic portion elliptic-oblong. N=13.
  Alpine bogs and meadows, Rocky Mts. from Colo. to Alas.; e. Siberia. June-Aug.
  One of our most beautiful gentians, more usually called G. frigida Haenke or G. romanzovii
Ledeb. by American authors, but according to Hultén (Fl. Alas. 8:1301. 1947) the former name
belongs to a closely related European species and the latter applies to a small nondistinctive
Siberian variant of this taxon.

Gentiana amarella L. Sp. Pl. 230. 1753.
  Amarella amarella Cockerell, Am. Nat. 40:871. 1906. Gentianella amarella Börner, Fl.
    Deut. Volk. 543. 1912. (Europe)
  Gentiana acuta Michx. Fl. Bor. Am. 1:177. 1803. Amarella acuta Raf. Fl. Tellur. 3:21.1836.
    Ericala acuta G. Don, Gen. Syst. 4:190. 1838. Gentiana amarella var. acuta Herder, Acta
    Hort. Petrop. 1:428. 1872. Gentiana amarella f. michauxiana Fern. Rhodora 19:151. 1917.
    Gentianella acuta Hiit. in Mem. Soc. Faun. Fl. Fenn. No. 25:76. 1950. (Michaux, Tadous-
    sac, Can.)
  Gentiana plebeja Cham. ex Bunge, Nouv. Mém. Soc. Nat. Mosc. 1:250, pl. 9, fig. 5. 1824.
    Gentiana acuta ssp. plebeja Wettst. in Öst. Bot. Zeits. 50:194. 1900. Amarella plebeia
    Greene, Leafl. 1:53. 1904. Gentiana acuta var. plebeja Hultén, Fl. Aleut. Isl. 276. 1937.

Gentiana acuta var. stricta Griseb. in Hook. Fl. Bor. Am. 2:63. 1838. Gentiana amarella
var. stricta Wats. Bot. King Exp. 277. 1871. Gentiana acuta var. strictiflora Rydb. Mem.
N.Y. Bot. Gard. 1:309. 1900. Gentiana stricta Howell Fl. N.W. Am. 445. 1901. Gentiana strictiflora A. Nels. Bot. Gaz. 34:26. 1902. Amarella strictiflora Greene, Leafl. 1:53.
1904. (Can. to the Rocky Mts. and Slave Lake, 4 collectors mentioned)

Gentiana acuta var. nana Engelm. Trans. Acad. Sci. St. Louis 2:214, pl. 9, figs. 6-9. 1863.
(Parry 309, Colo.)

Gentiana heterosepala Engelm. Trans. Acad. Sci. St Louis 2:215. 1863. Amarella heterosepala Greene, Leafl. 1:53. 1904. Gentianella amarella ssp. heterosepala Gillett, Ann. Mo.
Bot. Gard. 44:256. 1957. (Engelmann, Uinta Mts. in 1859) - a variant of questionable
significance, with very unequal calyx lobes, recognized by Gillett as confined mostly to Utah
and Wyo., s. to N.M., but with one specimen cited from Nez Perce Co., Ida.

G. tortuosa M. E. Jones, Proc. Calif. Acad. Sci. II. 5:707. 1895. Amarella tortuosa Rydb.
Bull. Torrey Club 40:463. 1913. (M. E. Jones 6008, Panguitch Lake, Utah)

Gentiana anisosepala Greene, Pitt. 3:309. 1898. Amarella anisosepala Greene, Leafl. 1:53.
1904. (A. A. Heller 3440, Nez Perce Co., Ida.)

Gentiana distegia Greene, Pitt. 4:182. 1900. Amarella distegia Greene, Leafl. 1:53. 1904.
(C. F. Baker, high mountains about Pagosa Peak, Colo., Aug. 1899)

Amarella copelandi Greene, Leafl. 1:53. 1904. (E. B. Copeland, Mt. Eddy, Siskiyou Co.,
Calif., Sept., 1903)

A. californica Greene, Leafl. 1:54. 1904. (Mrs. Austin, Sierra Nevada in Plumas and Butte
cos., Calif.)

A. lemberti Greene, Leafl. 1:54. 1904. (J. B. Lembert, in 1893, Yosemite Valley, Calif.)

A. macounii Greene, Leafl. 1:54. 1904. (Macoun, July 21, 1893, Vancouver I.)

A. conferta Greene, Leafl. 1:55. 1904. (Spreadborough, Chaplin, Assiniboia, Can., Aug.
28, 1895)

A. revoluta Greene, Leafl. 1:55. 1904. (Wooton 552, White Mts., N.M., in 1897)

A. scopulorum Greene, Leafl. 1:55. 1904. Gentiana scopulorum Tidestr. Contr. U.S. Nat.
Herb. 25:415. 1925. (Rocky Mts. region from Colo. to Mont.)

A. cobrensis Greene, Leafl. 1:56. 1904. (Greene, Santa Rita del Cobre, N.M., Oct. 11,
1880)

Gentianella clementis Rydb. Bull. Torrey Club 31:631. 1905. (Clements 253, Minnehaha,
Colo., in 1901)

Gentiana polyantha A. Nels. Bot. Gaz. 56:68. 1913. (E. P. Walker, Iron Springs Mesa,
Colo., Aug. 21, 1912)

Annual or biennial 5-40 cm. tall, entirely glabrous or sometimes with scaberulous-ciliolate
leaves and calyx lobes; stems simple to freely branched, lightly angled; basal leaves usually
several, mostly oblanceolate, 5-40 mm. long, cauline leaves mostly 5-8 pairs, from ovate-
lanceolate and clasping-based to lanceolate, oblong, or oblong-oblanceolate, as much as 6 cm.
long and 3 cm. broad, not connate at the base; flowers immediately subtended by bractlets or
with pedicels 3-20 mm. long, 4- or 5-merous (even on the same plant), varying greatly in
size according to their position and the time of blossoming but mostly 1-2 cm. long, sometimes few or even solitary but usually numerous, the plants often floriferous from near the
base, the axillary cymes sometimes very elongate; calyx 1/3-1/2 the length of the corolla,
lobed 2/3-4/5 of its length, often more deeply cleft on one side, the lobes unequal, from linear to lanceolate; corolla salverform but usually closed and apparently tubular, violet, dark
purplish-blue, pale bluish-purple, lavender, or clear blue, to pale yellowish and lightly bluish-tinged, lobes about 1/2 the length of the tube, oblong to lanceolate, rounded to obtuse, appendaged at the base with slender fimbriae 1/2-3/4 as long, sinuses not plaited; stamens well
included in, and attached to the lower third of the corolla tube, the adnate portion not wing-
margined; ovary sessile; style essentially lacking; stigmatic lobes oblong, rounded; capsule
slightly exceeding the persistent corolla; seeds ovoid to spherical, yellow, nearly smooth.
N=18.

Meadows and moist areas in general, over much of N. Am.; in the west, from Alas. to
Mex., from the coast (in Wash.) or coastal mountains to the Rocky Mts.; Eurasia. June-Sept.

Although all workers have realized that our plants are very similar to those of the Old World, many species have been proposed for N. Am. on the basis of variations in habit, leaf shape, calyx-lobing, and other extremely plastic characters. In our range none of these variants is peculiar to any one geographic area as would be expected of taxa significant enough to name either at the specific or infraspecific level.

When all our plants are included in one species, there still is no agreement as to the proper name for the group; some authors consider them so nearly identical with European plants as to call them G. amarella L.; others claim that they are different enough to merit status as the separate species, G. acuta Michx., whereas others believe them to be similar to, but not identical with, Old World plants, and therefore treat them at the infraspecific level as G. amarella var. acuta (Michx.) Herder or spp. acuta (Michx.) Gillett.

Gentiana calycosa Griseb. in Hook. Fl. Bor. Am. 2:58, pl. 146. 1838.
  Pneumonanthe calycosa Greene, Leafl. 1:71. 1904. Dasystephana calycosa Rydb. Bull. Torrey Club 40:464. 1913. (Tolmie, Mt. Rainier)
  Gentiana calycosa var. stricta Griseb. in Hook. Fl. Bor. Am. 2:58. 1838. G. calycosa var. monticola Rydb. Bull. Torrey Club 24:252. 1897. Dasystephana monticola Rydb. Bull. Torrey Club 40:464. 1913. (Tolmie, Mt. Rainier) = var. calycosa.
  Gentiana gormani Howell, Fl. N. W. Am. 446. 1901. (N. E. Wash. to Alas.) = var. calycosa.
  G. calycosa var. xantha A. Nels. Bot. Gaz. 34:26. 1902. Dasystephana calycosa var. xantha Rydb. Fl. Rocky Mts. 663. 1917. (Merrill & Wilcox 1108, above Leigh's Lake, Teton Mts., Wyo., July 26, 1901) = var. calycosa.
  GENTIANA CALYCOSA var. OBTUSILOBA (Rydb.) C. L. Hitchc. hoc loc. Dasystephana obtusiloba Rydb. Bull. Torrey Club 40:464. 1913. (Vreeland 1162, Mary Baker Lake and Sperry Glacier, Mont., Aug. 21, 1901)
  Gentiana cusickii Gand. Bull. Soc. Bot. France 65:60. 1918. (Cusick 1772, Oreg.) = var. calycosa.
  G. idahoensis Gand. Bull. Soc. Bot. France 65:60. 1918. (Evermann, Petit Lake, Ida.) = var. asepala.
  G. myrsinites Gand. Bull. Soc. Bot. France 65:60. 1918. (Suksdorf 6057, Mt. Paddo, Wash.) = var. calycosa.
  G. saxicola English, Proc. Biol. Soc. Wash. 47:192. 1934, not of Griseb. (English 1650, Little Rockies, Lewis Co., Wash.)
  GENTIANA CALYCOSA var. ASEPALA (Maguire) C. L. Hitchc. hoc loc. G. calycosa ssp. asepala Maguire, Madroño 6:151. 1942. (Maguire et al. 4225, saddle w. of Mt. Agassiz, Uintah Mts., Utah)
  G. (Dasystephana) parryi sensu auct. in large part. Probably not of Engelm. (cf. G. affinis)
  Caespitose perennial from fleshy thick roots, without rhizomes; stems usually several, erect to decumbent-based, usually simple, 5-30 cm. long, glabrous; leaves mostly 7-9 pairs, the lowest connate-sheathing but usually with blades at least as long as the sheath; main cauline leaves 1-2.5(3) cm. long, broadly lanceolate to ovate-rotund, mostly semicordate and sheathing at the base, glabrous; flowers mostly single and terminal or 3 in a terminal cyme, less commonly also terminal on peduncles 5-30 mm. long from the next lower one or two nodes, 5-merous, (2)2.5-4 cm. long, subtended just below the calyx by lanceolate to broadly ovate bracteoles; calyx tube 4-10 mm. long, from scarious and bluish- or purplish-tinged to greenish overall and not at all scarious, the membranous lining toothed inside of, and above the bases of, the lobes, lobes glabrous or minutely granular-ciliolate, from broadly cordate-oval and often spreading, to lanceolate and erect and from twice as long as the tube to greatly reduced and linear to oblong, or lacking entirely, the calyx tube then usually deeply cleft to unequally 2-lipped; corolla tubular-funnelform, deep blue and variously streaked or mottled with green, to (rarely) yellowish, the lobes oblong-ovate to ovate-oval, rounded or more usually abruptly pointed, 5-10 mm. long, plaits of the sinuses from nearly equal to the lobes to only half as long, usually laciniately cleft into 2-4 narrow segments; stamens slightly shorter than, and freed slightly below midlength of, the corolla tube, the adnate portion of the filaments broadly wing-margined, anthers 3-4 mm. long; ovary long-stipitate; style short, cleft above, the stigmatic lobes flattened, oblong, rounded; seeds oblong-fusiform, flattened but little, not at all winged, finely longitudinally reticulate.

Meadows, swamps, and stream banks, montane, usually alpine or subalpine, B.C. to Calif., e. to the Rocky Mts. July-Oct.

One of the best native species, but difficult to get into cultivation successfully. Like G. affinis, a very wide-ranging species showing a great deal of variation in the habit, and in leaf shape and floral characters, especially the texture and lobing of the calyx. Much of this variation is fortuitous, but certain fairly clearly marked tendencies are evident; most of the plants that lack calyx lobes are found, in our area, in c. Ida. and adj. Mont. They constitute the var. asepala (Maguire) C. L. Hitchc. Plants of the Rocky Mts. from Alta. to Mont. more commonly have green calyces with very large lobes that are green and similar to the tube in texture; they were called Dasystephana obtusiloba by Rydberg, but might much more appropriately be recognized at the infraspecific level as var. obtusiloba (Rydb.) C. L. Hitchc. Most plants from the Cascades and Olympic Mts. have somewhat fleshier, thicker, often more spreading calyx lobes than the plants of the Rocky Mts. and are somewhat taller and more frequently branched-floriferous below the stem tips; since the species was described from such plants, they constitute the var. calycosa; G. saxicola English appears to be an ecological variant of this variety, with widely flaring calyx lobes; it is apparently restricted to rock slides and drier areas, rather than moist meadows. Its status at present is uncertain.

Occasional sporadic plants with more yellowish than bluish corollas are known to occur; although called var. xantha by Nelson, they seem not to constitute a natural population in any locality.

Gentiana detonsa Rottb. Acta Hafn. 10:435. 1770.
  Anthopogon detonsa Raf. Fl. Tellur. 3:25. 1837. Ge..tianella detonsa G. Don, Gen. Syst. 4:179. 1838. (Iceland)
Gentiana barbata Froel. Gent. Diss. 114. 1796. Anthopogon barbata Raf. Fl. Tellur. 3:25. 1837. Gentiana detonsa var. barbata Griseb. in Hook. Fl. Bor. Am. 2:64. 1838. Gentianopsis barbata Ma, Acta Phytotax. 1:8. 1951. (Beyond the Tom R. in Siberia) = var. detonsa.
Gentiana thermalis Kuntze, Rev. Gen. 2:427. 1891. Anthopogon thermalis Rydb. Fl. Rocky Mts. 659. 1917. (Yellowstone Park) = var. unicaulis.
Gentiana elegans A. Nels. Bull Torrey Club 25:276. 1898. Anthopogon elegans Rydb. Bull. Torrey Club 33:148. 1906. Gentianella detonsa ssp. elegans Gillett, Ann. Mo. Bot. Gard. 44:217. 1957. (A. Nelson 1539, Cummins, Wyo., July, 1895) = var. unicaulis.
GENTIANA DETONSA var. UNICAULIS (A. Nels.) C. L. Hitchc. hoc loc. Gentiana elegans var. unicaulis A. Nels. Bull Torrey Club 25:277. 1898. (A. Nelson 4173, Battle Lake, Carbon Co., Wyo., Aug. 6, 1897)
Gentiana macounii Holm. Ott. Nat. 15:110, 179, pl. 11, figs. 1-2. 1901. Anthopogon macounii Rydb. Fl. Rocky Mts. 659. 1917. (Indicated to have been collected by Macoun at several stations in Alta. and Sask.) = var. unicaulis.

Glabrous annual 1-3.5(4) dm. tall; stems usually several from the base, simple or branched; leaves numerous in basal tuft, oblanceolate to spatulate, 1.5-4 cm. long; cauline leaves 2-4 pairs, narrowly lanceolate to oblong or oblanceolate, 1.5-5 cm. long; flowers 4(3)-merous, terminal on long, naked peduncles and usually also axillary on elongate peduncles with a pair of scarcely reduced, leaflike bracts at or below midlength; calyx 15-25 mm. long, lobes acuminate, subequal, alternately narrower and broader at the base, about equaling the tube; corolla 3.5-5.5 cm. long, deep blue or purplish, glandular at the base between the bases of the filaments, lobes subequal to the tube, oblong-obovate, erose-lacerate and rounded at the tip, more deeply lacerate on the sides, sinuses neither appendaged nor plaited; stamens slightly shorter than, and inserted about midlength of, the corolla tube, the adnate portion broadly wing-margined, free portion thin and flat; anthers oblong, 3-4 mm. long; ovary stipitate; style comparatively slender, 4-6 mm. long; stigma lobed, broad and flattened, pectinately fringed-margined; seeds prismatic, terete, about 0.5 mm. long, finely alveolate, dark-brown. N=22. Fringed gentian.

Meadows, bogs, and moist ground, circumboreal; in N. Am. from Newf. to Alas., extending s. and e. to Calif., Ariz., Mex., S.D., Ind., and N.Y., but found in our range only in Mont. and c. Ida. July-Aug.

Differing from G. simplex in seed character and the longer anthers, broader filaments,

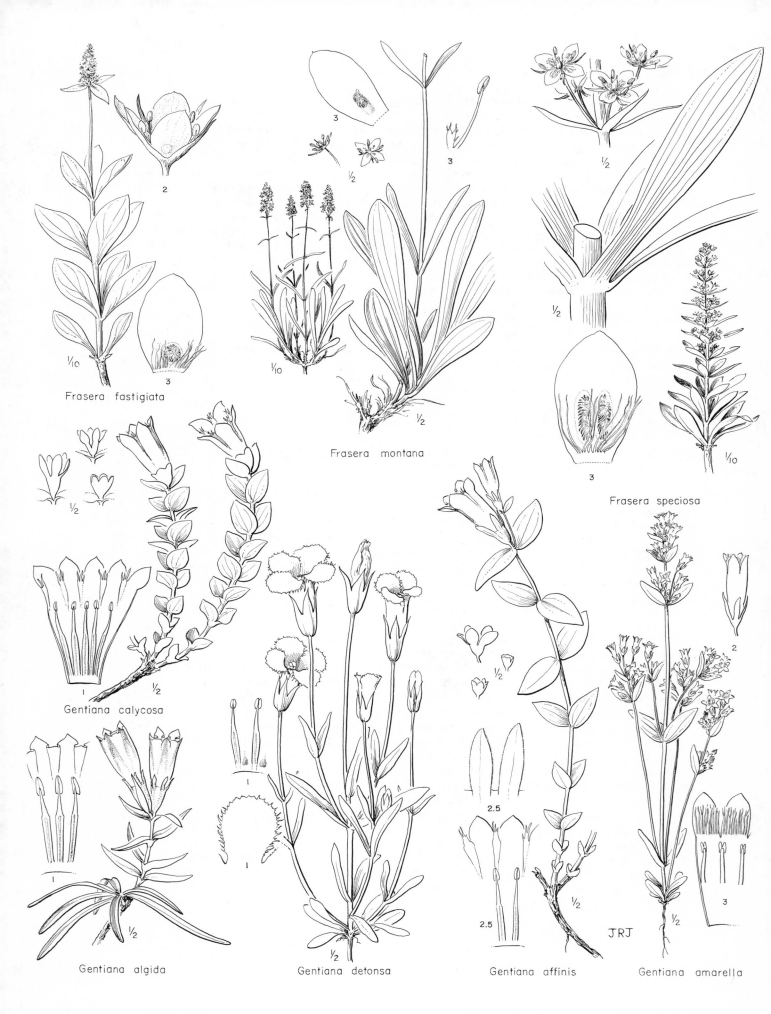

Frasera fastigiata

Frasera montana

Frasera speciosa

Gentiana calycosa

Gentiana algida

Gentiana detonsa

Gentiana affinis

Gentiana amarella

JRJ

longer, thinner style, lacerate petals, and branched stems. A highly polymorphic species with numerous geographic races, our material, as described above, all referable to var. unicaulis (A. Nels.) C. L. Hitchc., ranging from Ida. and Mont. to Nev., Utah, Colo., and N. Mex. A second variety, originally described as Gentiana serrata var. holopetala Gray, occurs to the s. of our region, chiefly in e. Oreg. and Calif.

Gentiana douglasiana Bong. Mém. Acad. St. Pétersb. VI. 2:156, pl. 6. 1833. (Douglas, mountains of w. N. Am.)
   G. douglasiana var. patens Hook. Fl. Bor. Am. 2:60. 1838. (Dr. Scouler, Menzies, "in the marshes of the N. W. C.")
   Glabrous, usually freely branching annual 5-20 cm. tall; basal leaves 5-15 mm. long, ovate to deltoid-obovate; cauline leaves few, ovate, 5-10 mm. long, connate-based and slightly decurrent, the stems distinctly angled, usually floriferous from most of the nodes, the cymose inflorescence rather rounded to nearly flat-topped; flowers 5-merous, all closely subtended by ovate, acute, keeled bracteoles; calyx 4-6 mm. long, the lobes acuminate, carinate, about half the length of the tube which is membranous within above the base of the lobes; corolla slightly more than twice the length of the calyx, narrowly salverform but quickly closing and apparently funnelform, plicate, the plaits of the sinuses 2/3 the length of the lobes, bifid, tube whitish, drying sordid yellow, lobes blue on the back, about 1/2 the length of the tube, oblong-lanceolate; stamens attached about midlength of the corolla tube, filaments slender, anthers linear, about 1 mm. long; ovary sessile; style very short, stigmatic lobes oblong-rounded; capsule shorter than the corolla, flattened and wing-margined and crested above; seeds dark brown, fusiform, about 1.5 mm. long, very finely and indistinctly longitudinally striate-reticulate.
   Bogs and tundra from Alas. s. to Vancouver I. and known from two localities, Lake Ozette and Snoqualmie Pass, in Wash. July-Sept.

Gentiana glauca Pall. Fl. Ross. 2:209, pl. 93. 1790.
   Dasystephana glauca Rydb. Fl. Rocky Mts. 662. 1917. (Kamchatka)
   Perennial with stems 4-15 cm. long from creeping rootstocks, forming small rosettes of obovate-oblanceolate leaves 1-2 cm. long; cauline leaves 2-4 pairs, obovate or oblong-obovate, usually smaller than the basal, very shortly connate; flowers 3-5(7), closely crowded, subsessile or the lowest one with pedicels as much as 8 mm. long, subtending bracts similar to the cauline leaves or sometimes broader; flowers usually 5-merous, dark blue, 1-2 cm. long; calyx usually bluish-tinged, about 1/3 the length of the corolla, inconspicuously carinate, lobes unequal, obtuse, about 1/2 the length of the tube; corollas plicate, narrowed at the limb, lobes broadly triangular, obtuse, sinuses with small, triangular-ovate, plicate lobes; filaments adnate about 1/3 the length of the corolla, the adnate portion narrowly wing-margined, slightly shorter than the free portion; anthers linear-oblong, 1-1.5 mm. long; seeds flattened, lamellate-winged, reticulate-alveolate.
   Tundra and alpine meadows, Alas. and Yukon s. to B. C. and in the Rocky Mts. to Mont.; Asia. July-Sept.

Gentiana propinqua Richards in App. Frankl. Journ. 734. 1823.
   Amarella propinqua Greene, Leafl. 1:53. 1904. Gentianella propinqua Gillett, Ann. Mo. Bot. Gard. 44:236. 1957. (Drummond, alpine swamps in the Rocky Mts.)
   Annual, glabrous throughout or the upper leaves and calyx lobes very minutely scabrous-ciliolate; stems angled, freely branching to simple, usually with many smaller branches and very much reduced flowers from near the base; basal leaves oblanceolate, 5-20 mm. long; cauline leaves rather numerous, ovate-lanceolate, usually cordate-based, 1-4 cm. long; flowers loosely cymose, 4-merous, closely to distantly subtended by foliaceous bractlets, upper flowers 15-22 mm. long, lower ones often less than half that size; calyx 1/3-1/2 the length of the corolla, not membranous-lined within, usually lobed over half the length, the lobes unequal, lanceolate-acuminate; corolla light purple, salverform but quickly closing, the lobes ovate, decidedly acuminate, sinuses not at all plicate; stamens about equaling, and attached above midlength of, the corolla tube, the adnate portion very slightly winged below, free filaments

flattened; anthers oblong, 1-1.5 mm. long, bluish-purple; ovary nearly sessile; style very short; stigmatic lobes flattened, oblong-rounded; capsule about equaling the corolla; seeds ovoid, pale yellow, nearly smooth, about 0.7 mm. long.

Meadowland and stream banks, Alas. s. in the Rocky Mts. to B. C. and Alta. and to Beaver-head Co., Mont., eastward to Que. and Newf. June-July.

Gentiana prostrata Haenke in Jacq. Coll. Bot. 2:66, pl. 17, fig. 2. 1788.
  Chondrophylla prostrata Anderson, Proc. Iowa Acad. Sci. 25:445. 1918. (Mountains of Salz-
    burg, Austria)
  Gentiana fremonti Torr. in Frem. Rep. 94. 1843. Chondrophylla fremonti A. Nels. Bull.
    Torrey Club 31:245. 1904. (Fremont, Wind River Mts., Wyo.)
  Gentiana prostrata var. americana Engelm. Trans. Acad. Sci. St. Louis 2:217, pl. 9, figs.
    10-15. 1863. Chondrophylla americana A. Nels. Bull. Torrey Club 31:245. 1904. (Alpine
    peaks of the Snowy Range, Colo.)
  Gentiana humilis sensu Gray, Syn. Fl. 2¹:120. 1878, perhaps not of Ster. in 1812; not of
    Salisb. in 1796.

Low glabrous annual or biennial 2-15 cm. tall; stems usually several from a slender root (or rhizome?); leaves mosslike, light green, conspicuously lighter-margined, the basal close-ly crowded and overlapping, the upper cauline more distant, all connate-sheathing, with orbic-ular or ovate to obovate blades 3-10 mm. long; flowers single and terminal, 4(5)-merous; ca-lyx very narrowly funnelform, (4)6-14 mm. long, greenish to bluish- or purplish-tinged, lobes ovate-lanceolate, about 1/3 the length of the tube, subequal, tube not membranous-lined above the bases of the lobes; corolla (8)12-22 mm. long, usually narrowly funnelform, quick-ly closing and contorting after anthesis, from clear blue to purplish and often somewhat green-ish-mottled, lobes ovate-lanceolate, acute, about 1/5 the length of the tube, plaits of the si-nuses conspicuous, somewhat shorter than the lobes, entire and acute to minutely erose-toothed; stamens slightly shorten than, and inserted slightly above midlength of, the corolla tube; filaments very slender; anthers oblong, about 0.7 mm. long; ovary stipitate, the stipe elon-gating considerably in fruit and exceeding the withered corolla by as much as 2 cm., but often the corolla tightly shut and retaining the capsule within, the stipe then twisting and coiling and sometimes bulging into a loop, rupturing the calyx; style branched, stigmatic lobes flattened, oval; capsule papery-walled, finely cross-reticulate, linear-ellipsoid, about 1 cm. long; seeds brown, narrowly wing-margined on one side, inconspicuously longitudinally striate. N=about 18.

Alpine bogs and meadows of high mountains, Rocky Mts., Colo. to Alas., w. to Utah, Nev., and c. Ida., and in n. Calif.; Eurasia and S. Am. July-Aug.

Most workers have considered that two species of this nature exist in our area, distinguished chiefly on the basis that the capsules of one, G. fremontii (Alta. to Colo.), eventually are exserted from the calyx and corolla, whereas the capsules of the other, G. americana or G. prostrata (supposedly the more widespread), are retained in the calyx.

In our material, as well as in plants from Europe, both exserted and included capsules are found throughout the range of the plant, and often on the same specimen, proving that no basis exists for the belief that more than one species is involved.

Gentiana sceptrum Griseb. in Hook. Fl. Bor. Am. 2:57, pl. 145. 1838.
  Pneumonanthe sceptrum Greene, Leafl. 1:71. 1904. (Douglas, near Ft. Vancouver)
  Gentiana menziesii Griseb. in Hook. Fl. Bor. Am. 2:59. 1838. (Menzies, "North-West
    Coast")
  G. orfordii Howell, Fl. N.W. Am. 446. 1901. (Joseph Howell, Port Orford, Oreg.)
Caespitose perennial with 1-many stems from thick, fleshy roots (and short rhizomes?), 2-12 dm. tall; leaves 10-15 pairs, the lowest 2-5 reduced to short bracts that are connate most of their length, the upper leaves enlarging gradually to about the seventh to tenth pairs, these mostly oblong-lanceolate, 3-6 cm. long, 7-15(25) mm. broad; flowers usually several, 5-merous, 3-4.5 cm. long, some at least (usually the lower ones) with bractless pedicels 3-20 mm. long; calyx from less than 1/2 to as much as 2/3 as long as the corolla, usually bluish-tinged, the tube usually 8-14 mm. long, membranous above the point of separation of the lobes,

which are from narrowly lanceolate and shorter than the tube, to broadly lanceolate and nearly twice as long; corolla bluish, often greenish-streaked or -mottled, the lobes oval, 6-9 mm. long, plaits of the sinuses low, truncate or rounded, nearly or quite entire; filaments inserted slightly below midlength of the corolla, the adnate protion broadly wing-margined, free portion 10-15 mm. long; anthers about 4 mm. long; ovary stipitate; style short, branched near the tip; stigmas flattened, oblong-rounded; capsule about equaling the dried corolla; seeds fusiform, nearly terete, 1.5-2 mm. long, finely reticulate.

Bogs and wet places, B.C. s., on the w. side of the Cascades, especially near the coast, to n.w. Calif. July-Sept.

Gentiana simplex Gray, Pac. R.R. Rep. 6:87, pl. 16. 1857.
 Anthopogon simplex Rydb. Fl. Rocky Mts. 659. 1917. Gentianella simplex Gillett, Ann. Mo.
  Bot. Gard. 44:232. 1957. (Newberry, Upper Klamath Lake, Oreg.)
 Erect, usually single-stemmed, simple annual 1-2 dm. tall; leaves 2-5(7) pairs, the lowermost ovate to ovate-lanceolate, 3-6 mm. long, the upper ones longer, oblong, 1-2.5 cm. long, rounded to acute, the uppermost pair considerably below the single, terminal, 4(3)-merous flower; calyx 15-20 mm. long, lobed about half the length, the segments slightly unequal, more or less rugose-carinate; corolla 2-4 cm. long, deep blue, lobed nearly half the length, not appendaged or plaited in the sinuses, with small oblong glands within at the base of, and alternate with, the stamens, lobes oblong, slightly dentate or laciniate on the sides, rounded; stamens included, inserted about midlength of the corolla tube, the adnate portion inconspicuously wing-margined, free filaments broad and flattened; anthers cordate or oblong, 1-1.5 mm. long; ovary distinctly stipitate; style short, 1-2 mm. long; stigma discoid-lobed, the margins fimbriate; seeds tiny, fusiform, longitudinally reticulate-pitted and more or less wing-margined.
 Mountain bogs and meadows; Cascades of Oreg. from Deschutes Co. s. to the Sierra Nevada of c. Calif., and e. through Oreg. to c. Ida. July-Aug.

Gentiana tenella Rottb. Acta Hafn. 10:436. 1770.
 Amarella tenella Cockerell, Am. Nat. 40:871. 1900. Gentianella tenella Börner, Fl. Deut.
  Volk. 542. 1912. Lomatogonium tenellum Löve & Löve, Acta Hort. Gotoburg. 20⁴:117.
  1956. (Eurasia)
 Gentiana monantha A. Nels. Bull. Torrey Club 31:244. 1904. Amarella monantha Rydb. Bull.
  Torrey Club 33:148. 1906. Gentiana tenella var. monantha Rouss. & Raym. in Nat. Can.
  79:77. 1952. (Dr. Clements 456, Mirror Lake, Colo., Sept. 6, 1901)
 Glabrous annual 4-15 cm. tall; stems simple to freely branched near the base, very slender, 4-angled; basal leaves usually several, oblanceolate, 3-10 mm. long, cauline leaves few, oblanceolate, 5-15 mm. long; flowers 8-15 mm. long, 4(5)-merous, single and terminal on long naked peduncles, or few to many from the lower leaves on elongate pedicels as much as 10 cm. long; calyx half the length of the corolla or longer, lobes nearly distinct, slightly gibbous-based, the outer 2 broader and shorter than the inner, usually more rugose and with membranous margins; corolla white to bluish purple, tubular, not plicate in the sinuses, lobes about 1/3 the length of the tube, oblong-lanceolate, obtuse to acute, each fringed within by 2 basal, erect, lacerate scales about 1/3 as long; stamens shorter than, and freed about midlength of, the corolla tube, filaments rather broad, not wing-margined below the point of insertion; anthers sagittate-cordate; ovary sessile; style very short, stigmatic lobes oblong; capsule slightly exceeding the corolla; seeds yellow, ovoid, nearly smooth. N=5.
 Widely separated mountain ranges of Eurasia, Greenl., Alas. and Can., s. in the Rocky Mts. from Wyo. to N.Mex., and Ariz., disjunct in the s. Sierra Nevada and White Mts. of Calif. and Nev., known from Yellowstone Nat. Pk., reported from Ida., and to be expected there as well as in Mont. July-Aug.

## Lomatogonium A. Br.

Flowers 4- or 5-merous; calyx deeply divided; corolla convolute, nearly rotate, lobed nearly to the base, each segment with 2 small, scaly, basal appendages; stamens attached at the base of the corolla; ovary sessile, style lacking, the stigmatic surfaces decurrent over half the length of the ovary; capsule 1-celled; glabrous opposite-leaved annual or biennial.

About 10 species, mainly in n. Eurasia. (Name from the Greek loma, fringe, and gonos, referring to the ovary, because of the hemlike stigma on top of the ovary.)

Reference:

Fernald, M. L. Lomatogonium the correct name for Pleurogyne. Rhodora 21:193. 1919.

Lomatogonium rotatum (L.) Fries ex Nyman, Consp. Fl. Eur. 3:500. 1881.

Swertia rotata L. Sp. Pl. 1:226. 1753. Gentiana rotata Froel. Gent. 105. 1796. Narketis rotata Raf. Fl. Tellur. 2:26. 1836. Pleurogyne rotata Griseb. Gen. & Sp. Gent. 309. 1839. (Siberia)

Swertia pusilla Pursh, Fl. Am. Sept. 101. 1814. Pleurogyne carinthiaca var. pusilla Gray, Syn. Fl. 2¹:124. 1878. (N.H.)

Pleurogyne rotata var. tenuifolia Griseb. Gen. & Sp. Gent. 309. 1839. Lomatogonium rotatum f. tenuifolium Fern. Rhodora 21:197. 1919. (Altai Mts. and Hudson's Bay)

Pleurogyne rotata var. americana Griseb. Gen. & Sp. Gent. 309. 1839. Lomatogonium rotatum f. americanum Fern. Rhodora 21:197. 1919. (Labrador)

Pleurogyne fontana A. Nels. Proc. Biol. Soc. Wash. 17:177. 1904. (A. Nelson, Crow Creek, Wyo., in 1903)

Lomatogonium rotatum f. ovalifolium Fern. Rhodora 21:197. 1919. (St. John 1970, Amherst I., Magdalen Islands, Que., Aug. 25, 1914)

Stems simple to freely branched, 10-25 cm. tall; lower leaves oblanceolate to spatulate, the upper narrowly lanceolate, 1-3 cm. long; flowers mostly axillary on long, slender pedicels from the scattered upper leaves; sepals (2)4-5, linear or linear-lanceolate, usually exceeding the corolla; corolla (white to) blue, the lobes elliptic-ovate, 6-15 mm. long; capsule oblong-ovoid, about equaling the persistent corolla, stigmatic in 2 lines on the upper 1/2-3/4 of the length; seeds brown, prismatic, deeply few-pitted. N=5.

Wet, often saline soil, in our area strictly montane; Greenl. to Alas. s. to Alta., in the Rocky Mts. from Ida. and Wyo. to N.M., to be expected in Mont.; Siberia. July-Sept.

## Microcala Hoffm. & Link. Timwort

Flowers 4-merous, yellowish; corolla salverform; stamens freed above midlength of the corolla tube, the adnate filaments not winged; anthers cordate-ovate; ovary sessile; style slender, stigmas flabellate-lobed; capsule 1-celled; seeds flattened, ovoid, coarsely reticulate; small, glabrous, opposite-leaved annuals.

Two species of N. and S. Am. and Eurasia. The flowers are usually closed, opening only in bright sunlight. (Name from the Greek mikros, small, and kalos, beautiful.)

Microcala quadrangularis (Lam.) Griseb. in DC. Prodr. 9:63. 1845.

Gentiana quadrangularis Lam. Encyc. Meth. 2:645. 1788. Exacum quadrangulare Willd. Sp. Pl. 1:636. 1797. (Dombey, near Lima, Peru)

Stems very slender, simple or sparingly branched, angled, 3-8 cm. tall; leaves lanceolate to oblanceolate, 4-9 mm. long, mostly near-basal; flowers 5-8 mm. long, terminal on the main stem or on elongate, axillary, usually naked pedicels; calyx broadly campanulate, distinctly carinate on the back of the lobes, plicate-carinate between, the teeth acute, scarcely 1/2 the length of the tube; corolla salverform but usually closed, the tube about equaling the calyx, the lobes ovate-rounded, about 2/5 of the length of the tube, not plicate; stamens subequal to the corolla tube; capsule about equaling the calyx.

Prairies and open, usually moist flats, rarely collected, Willamette and Umpqua valleys of Oreg., s. to much of n.e. Calif.; S. Am. Mar.-June.

## Swertia L.

Flowers cymose to thyrsoid, 5- (or 4)-merous except for the pistil; corolla bluish-purple to white or greenish, deeply 5-lobed, the segments each with a basal pair of glands (foveae),

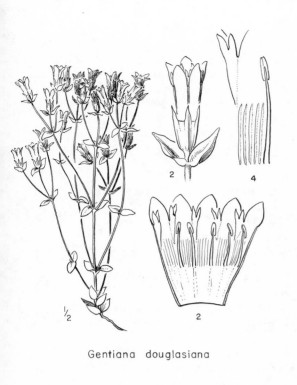

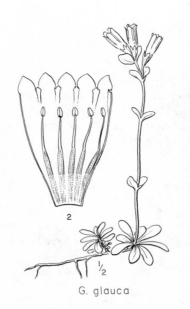

G. glauca

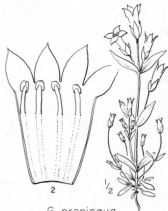

G. propinqua

Gentiana douglasiana

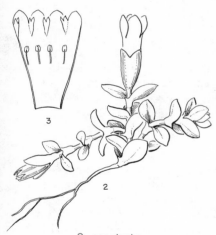

G. prostrata

Microcala quadrangularis

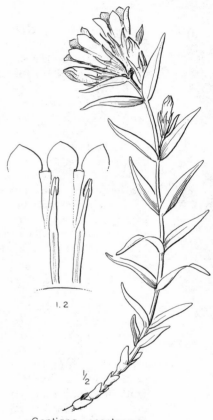

Gentiana sceptrum

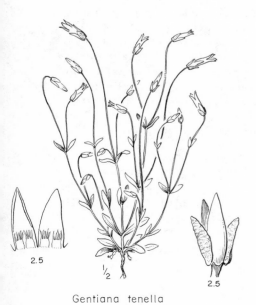

Gentiana tenella

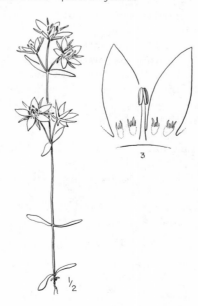

Lomatogonium rotatum

JRJ

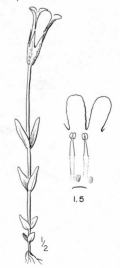

Gentiana simplex

which are surrounded by fringed hoods; crown scales lacking in ours; stamens inserted near the base of the corolla tube; style very short; fruit a 1-celled capsule with 2 parietal placentae; glabrous annual or perennial herbs with opposite entire leaves.

Over 50 species, nearly cosmopolitan but primarily in e. Asia. (Named in honor of Emanuel Sweert, a Dutch gardener and author, 1552-   .)

Swertia perennis L. Sp. Pl. 1:226. 1753. ("Helvetia, Bavaria")
  S. obtusa Ledeb. Mem. Acad. St. Pétersb. 5:526. 1814. S. perennis var. obtusa Griseb. in Hook. Fl. Bor. Am. 2:66. 1838.
  S. occidentalis Greene, Pitt. 4:184. 1900. (Cusick 2100, Hurricane Creek, e. Oreg.)
  S. ovalifolia Greene, Pitt. 4:185. 1900. (Cusick, Blue Mts., Oreg.)
  S. palustris A. Nels. Bull. Torrey Club 28:227. 1901. (A. Nelson 7774, Nash's Fork, s. Wyo.)
  S. congesta A. Nels. Bull. Torrey Club 28:228. 1901. (A. Nelson 7852, Medicine Bow Range, Wyo.)
  S. parallela Greene, Leafl. 1:92. 1904. (Rydberg & Bessey 4699, Jack Creek Canyon, Mont.)
  S. fritillaria Rydb. Bull. Torrey Club 40:465. 1913. (Garrett 1566, Big Cottonwood Canyon, Utah)
  Plant with rather thick, short rhizomes, the flowering stems usually single at the ends of the rootstocks, 5-50 cm. tall; leaves mostly basal, (2)5-12(20) cm. long, obovate to oblong-elliptic, narrowed to rather slender petioles sometimes as long as the blades; cauline leaves opposite (or sometimes one or more alternate), considerably reduced except at the first of the 2-3(5) nodes; inflorescence from several-flowered and loosely thyrsoid to 1- to 3-flowered and cymose; corolla bluish-purple and variously maculate with greenish or white (occasionally albino), the lobes narrowly oblong, entire to erose (6)8-11 mm. long, the tube with a few small squamellae below the foveae; foveae orbicular, completely surrounded by fringed hoods; style less than 1 mm. long; capsule lance-ellipsoid, compressed, 7-12 mm. long.· N=9, 12, 14.

Montane to subalpine, on meadowland, streambanks, and other moist places, Alas. to Calif., e. through the Rockies; Eurasia. July-Aug.

## MENYANTHACEAE. Buck Bean Family

Flowers in simple to compound racemes or cymes on elongate, naked, scapelike peduncles, 5(4-6)-merous except the pistil, regular, perfect, gamopetalous; calyx shallowly to deeply lobed, adnate to the lower 1/5-1/2 of the ovary; corolla marcescent or deciduous, valvate to valvate-imbricate in bud, salverform to rotate and with a very short tube, lobes either with elongate fringed scales or copiously bearded with very slender bristlelike scales; stamens 5, borne on the corolla and alternate with the lobes; anthers versatile and often sagittate, dehiscent full length; pistil 1, 2-carpellary; ovary partially inferior, 1-celled with 2 parietal placentae; styles dimorphic, usually either short and thick or considerably elongate and more slender; capsule firm-walled, valvate, or tardily dehiscent to indehiscent; seed coat smooth and shining; glabrous perennial herbs with thick rhizomes that are covered with the old leaf bases; leaves alternate, sheathing-based, long-petioled, the blades simple to trifoliate.

A small family often combined with the Gentianaceae, in which case the two genera here recognized are usually combined into one, Menyanthes, differing from the Gentianaceae chiefly in the alternate stipulate leaves, nonconvolute fringed or bearded corolla, partially inferior ovary, and smooth seeds. Five genera and about 35 species, very widely distributed.
  Reference:
  Lindsay, Alton A. Anatomical evidence for the Menyanthaceae. Am. Journ. Bot. 25:480-85. 1938.
1 Leaves trifoliate, not crenate; corolla bearded with narrow hairlike scales
                                                               MENYANTHES
1 Leaves simple, crenate; corolla crested with fringed scales running the length of the lobes
                                                               NEPHROPHYLLIDIUM

## Menyanthes L.  Buck Bean

Flowers on long naked scapes in simple to compound racemes, 5(4-6)-merous, usually dimorphic, some with well-exserted stamens but included style, others with exserted style and shorter stamens; corolla salverform, lobes and throat densely scaly-hairy; filaments freed in the throat of the corolla; ovary about 1/3 inferior, 1-celled; capsule usually indehiscent; perennial, glabrous, fleshy herbs with shallow rhizomes; leaves alternate, all basal, the long petioles with very conspicuous decurrent sheathing stipules, the blades trifoliate.

A single species. (Name from the Greek meniaios, monthly, and anthos, flower.)

Menyanthes trifoliata L. Sp. Pl. 145. 1753. (Europe)

Rhizomes thick, covered with old leaf bases; stems prostrate or the floral branches ascending; petioles 10-30 cm. long, the leaflets short-stalked, elliptic-ovate to elliptic-obovate, 4-12 cm. long, entire to distinctly coarsely undulate-dentate; peduncles 5-30 cm. long; pedicels 5-20 mm. long, bracteate-based; calyx 3-5 mm. long, 5- to 6-lobed nearly to the base, lobes obtuse; corolla tube usually about twice the length of the calyx, whitish, the lobes 5 or 6, usually purplish-tinged, ovate-lanceolate, acutish, 5-7 mm. long, covered with short scales that extend down into the tube; anthers purplish, versatile, about 2 mm. long; ovary globose; style slender; stigmas oval; capsule thick-walled, many-seeded; seeds brownish-yellow, smooth and shining. N=27.

Bogs and lakes, Greenl. to Alas. s. to Pa. and Ind., in the Rockies to Colo., and on the Pacific Coast mostly w. of the Cascades to s. Oreg. and into the Sierras of Calif., also in n.e. Oreg.; Europe. May-Aug.

## Nephrophyllidium Gilg.  Deer Cabbage

Flowers 5-30 in loose cymes on long naked scapes, 5(4-6)-merous except for the pistil; corolla rotate, the lobes fringed; ovary at least half inferior; style shorter than, to much longer than, the stamens; capsule ovoid-conic, 1-celled, with 2 parietal placentae, valvate; perennial glabrous herbs from shallow rhizomes; leaves simple, the stipules sheathing.

One species. (Name from the Greek, meaning like Nephrophyllum, another genus which has kidney-shaped leaves, from nephros, kidney, and phyllon, leaf.)

Nephrophyllidium crista-galli (Menzies) Gilg in Eng. & Prantl, Nat. Pflanzenf. IV. 2:106. 1895.

Menyanthes crista-galli Menzies ex Hook. Bot. Misc. 1:45, pl. 24. 1830. (Prince William's Sound)

Rhizomes thick and fleshy, covered with old leaf bases; petioles 20-30 cm. long, the decurrent stipules 5-30 cm. long; blades cordate-ovate, broadly reniform or reniform-obcordate, 3-12 cm. broad, the length considerably less, finely to coarsely crenate; peduncles 1-5 dm. long; cymes usually open and loosely flowered; floral tube turbinate; sepals lanceolate, spreading, 3-5 mm. long; corolla white, rotate, the tube but 2-4 mm. long, the lobes ovate-lanceolate, 4-6 mm. long, the midnerve and usually the margins with erect, erose-undulate membranes running lengthwise; filaments broad, adnate to the lower part of the corolla tube; anthers about 3 mm. long; styles 1-5 mm. long; ovary over half inferior at anthesis, but the capsule elongated, the superior portion 10-18 mm. long.

Bogs, swamps, and wet "prairies" in our area, Olympic Peninsula, Wash., to Alas.; Japan. June-Aug.

The plants are attractive and desirable for cultivation on wet soil, looking considerably like some of the large-leaved saxifragas.

## APOCYNACEAE.  Dogbane Family

Flowers solitary or cymose (ours), perfect, hypogynous, gamopetalous, regular (slightly irregular), 5-merous except the pistil; calyx usually divided nearly to the base; corolla campanulate to salverform, convolute in the bud; stamens adnate to the corolla and alternate with its lobes, the anthers 2-locular, connivent around and sometimes adherent to the style, often sterile in the basal portion; pollen mostly granular; pistil 1, 2-carpellary, the ovaries distinct and 1-locular (in ours), usually surrounded by 5 nectaries, style 1, the stigma enlarged; fruit follicular (ours) or capsular, baccate, or drupaceous, many-seeded; seeds with or without a terminal tuft of soft hairs; perennial herbs (ours) or woody vines, shrubs, or trees with opposite (alternate or whorled), simple, entire leaves and milky juice.

A large family of perhaps 250 genera and over 1000 species, widely distributed but chiefly tropical; many species cultivated as ornamentals.

1  Flowers blue (white), over 1.5 cm. long                                VINCA
1  Flowers greenish-white to pink, less than 1.5 cm. long                 APOCYNUM

### Apocynum L.  Dogbane

Inflorescence cymose, often considerably compounded; calyx lobed 1/3-2/3 its length; corolla tubular to campanulate, the lobes oblong-lanceolate to ovate, from 1/2 as long as the tube to subequal thereto, erect to spreading or reflexed, the tube with triangular subulate appendages alternate with the bases of the stamens; filaments freed at the base of the corolla tube, short, incurved, usually sparsely puberulent; anthers versatile, introrse, narrowly lanceolate, sagittate at the base, their tips convergent and slightly adnate to the stigma; style very short, enlarged and broadly clavate-ovoid above, stigmatic at the top; pistils surrounded at the base by 5 peglike tiny nectaries; follicles very elongate; seeds terete, apically long-comose; rhizomatous, perennial herbs with opposite, subsessile to petiolate leaves and milky juice.

About 5 species, all North American. (Name from the Greek apo, away from, and kyon, dog.)

The fiber and bark of the taller species especially have been widely used by the American Indians for making cordage and rope.

Reference:

Woodson, R. E. Studies in the Apocynaceae I. Ann. Mo. Bot. Gard. 17:1-212. 1930. (For more complete synonymy see this work.)

There are few floral variations in Apocynum, species delimitation being dependent almost entirely on leaf and fruit position, size of flower, relative length of calyx and corolla, pubescence, and the amount of compounding of the inflorescence. These several characters vary independently, and it is not surprising therefore that some workers have recognized many more species than others; Greene, for instance, proposed 43 new species at one time (Leafl. 2:164-89. 1912). Most recent treatments for the species of our area have followed Woodson's monograph, with the recognition of 5 or 6 species. However, as Woodson pointed out, there are few sharp lines of demarcation in the genus and much of the pattern of intergradation is believed to be due to hybridization. The following species are all that are readily distinguishable in our region, and two of these, A. medium and A. sibiricum, are of dubious validity.

1  Flowers less than 5(2-4.5) mm. long, corolla greenish-white to white, usually less than
     twice as long as the calyx, lobes erect or but slightly spreading; leaves ascending
  2  Follicles usually more than 12 cm. long; coma of seeds 2-3 cm. long; leaves, except
       sometimes the lower cauline, distinctly petioled and not cordate-based
                                                                          A. CANNABINUM
  2  Follicles less than 12 cm. long; coma of seeds 1-2 cm. long; leaves of main stem usually
       all sessile or subsessile and cordate-based                       A. SIBIRICUM
1  Flowers usually at least 5 mm. long; corolla pinkish, often more than twice the length of
     the calyx, the lobes spreading to reflexed; leaves mostly drooping to spreading
  3  Calyx usually at least half as long as the corolla, the lobes narrowly lanceolate, acute to
       acuminate; leaves often ascending                                 A. MEDIUM
  3  Calyx usually less than half as long as the corolla, the lobes lanceolate to deltoid or ovate
       often obtuse; leaves mostly spreading or drooping                 A. ANDROSAEMIFOLIUM

Apocynum androsaemifolium L. Sp. Pl. 2nd. ed. 311. 1762. ("Habitat in Virginia, Canada")
  A. androsaemifolium var. incanum DC. Prodr. 8:439. 1844. A. incanum Miller, Proc. Biol.
     Soc. Wash. 13:81. 1899. (No type specified) = var. androsaemifolium.

A. androsaemifolium var. pumilum Gray, Syn. Fl. 2nd. ed. 2[1]:83. 1886. A. pumilum
  Greene, Man. Bay Reg. 240. 1894. (Calif. to B.C.)
A. rhomboideum Greene, Pitt. 5:66. 1902. A. pumilum var. rhomboideum Bég. & Bel.
  Monog. Apocynum 98. 1913. (Jepson, Napa Valley, Calif.) = var. pumilum.
A. scopulorum Greene ex Rydb. Fl. Colo. 269. 1906. (Colo., no type designated) = var. an-
  drosaemifolium.
A. androsaemifolium ssp. detonsum Piper, Contr. U.S. Nat. Herb. 11:453. 1906. (Vasey
  429, e. Wash., in 1889) = var. pumilum.
A. xylosteaceum Greene, Leafl. 2:185. 1912. (Dr. H. M. Cronkhite, Klamath Valley,
  Oreg.) = var. pumilum.
A. macranthum Rydb. Fl. Rocky Mts. 669, 1065. 1917. (Sandberg, MacDougal & Heller 372, Big
  Potlatch R., Ida., July 9, 1892) = var. androsaemifolium.
  Plant freely branched, glabrous to tomentose, 2-5 dm. tall; leaves subsessile to petiolate,
drooping, oblong-elliptic to ovate, oval, or cordate-ovate, 2.5-7 cm. long, usually mucronate;
cymes from chiefly terminal to more diffuse, some axillary and subterminal; flowers pinkish,
mostly 5-7(4-10) mm. long; calyx 1/4-1/2 the length of the corolla tube, the lobes ovate and rounded to lanceolate and somewhat acuminate; corolla from tubular and of nearly uniform diameter
from the base to the limb, to distinctly campanulate and nearly twice as broad at the limb as at
the base, the lobes oblong-lanceolate, about 1/2 the length of the tube, erect to spreading or
slightly reflexed; follicles 5-14 cm. long, erect to reflexed; hairs of the seeds white to tawny,
1-2 cm. long. N=8?
  General throughout much of Can. and most of the U.S. except in the s.e.; usually on rather
dry soil of the valleys and foothills to medium levels in the mountains. June-Sept.
  In many places a serious orchard weed. Woodson and most other workers recognize at least
two species instead of this one. (1) A. androsaemifolium, with a campanulate corolla 5-10 mm.
long that is twice as broad at the top as at the base, a predominantly terminal inflorescence,
pendulous follicles, and the coma of the seeds tawny and 12-20 mm. long. (2) A. pumilum
with a cylindrical corolla 4-6 mm. long that is as broad at the base as above, axillary as well
as terminal cymes, erect follicles, and the coma of the seeds mainly white and 10-20 mm.
long. Each of these two species is usually subdivided into varieties on the basis of the amount
of pubescence present, the more hairy phases being called A. androsaemifolium var. incanum
DC. and A. pumilum var. rhomboideum (Greene) Bég. & Bel. These extreme phases are
readily recognizable, but the vast majority of the material is of an intermediate nature. For
this reason it is believed that the complex can be treated more consistently (in this work, at
least) as one species, with two poorly differentiated varieties.
1 Corolla campanulate, 5-10 mm. long; follicles usually pendulous; cyme terminal, simple;
    with the range of the species                                          var. androsaemifolium
1 Corolla more tubular, 4-7 mm. long; follicles usually erect, inflorescence frequently
    larger and partially axillary as well as terminal; with var. androsaemifolium in most of
    w. U.S. and Can.                                                        var. pumilum Gray


Apocynum cannabinum L. Sp. Pl. 213. 1753.
  Cynopaema cannabinum Lunell, Am. Midl. Nat. 4:509. 1916. ("Hab. in Virginia, Canada")
  Apocynum pubescens Mitchell ex R. Br. Mem. Wern. Soc. 1:68. 1809. A. cannabinum var.
    pubescens DC. Prodr. 8:440. 1844. (No type designated) = var. glaberrimum.
  A. cannabinum var. glaberrimum DC. Prodr. 8:439. 1844. (No type designated)
  A. suksdorfii Greene, Pitt. 5:65. 1902. A. cannabinum var. suksdorfii Bég. & Bel. Atti R.
    Accad. Lincei II. 9:104. 1913. (Suksdorf 1522, "sandy banks of the Columbia R.," Oreg.)
  Plant glabrous (ours) to pubescent, usually erect, simple to freely branched, 3-10 dm. tall;
leaves yellowish-green, erect or ascending, oblong-ovate or oblong-lanceolate to lanceolate,
usually rounded (acute) and abruptly mucronate, 5-11 cm. long, short-petiolate and rounded to
acute at the base, or the lower cauline somewhat cordate; cymes usually both terminal and
lateral, the terminal considerably surpassed by the leafy branches of the lateral cymes; flowers 2-4(4.5) mm. long, whitish to greenish-white; calyx from less than half as long as, to
nearly equal to, the corolla, the lobes mostly lanceolate and acute to oblong-lanceolate, usual-

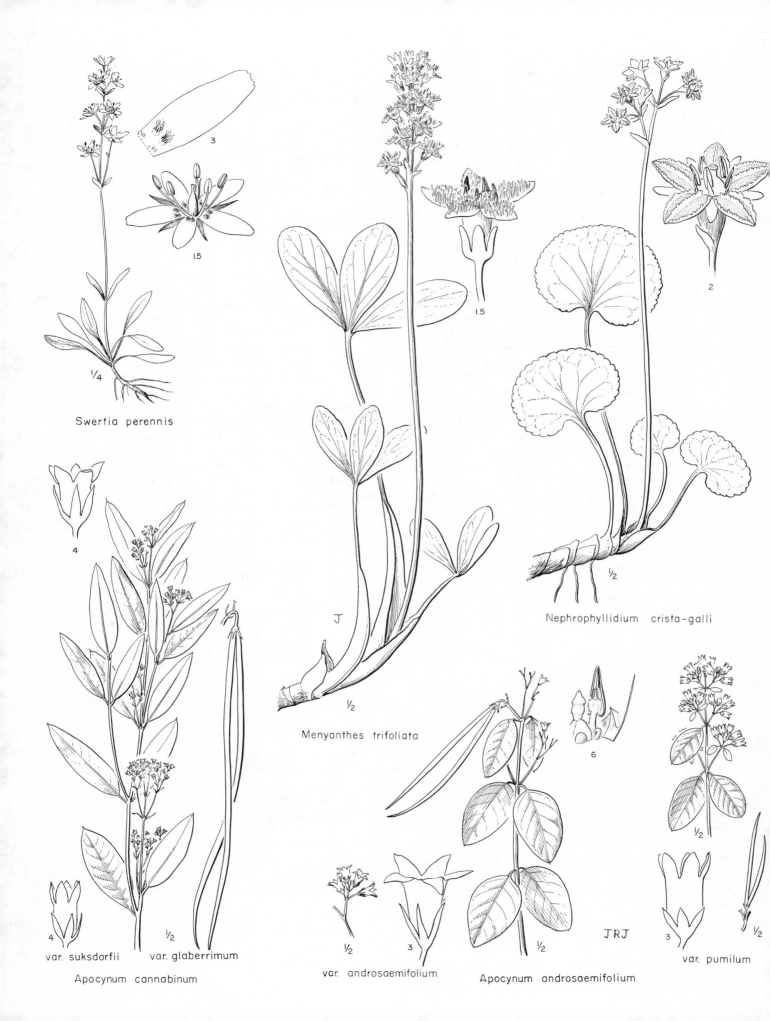

3

1.5

Swertia perennis

4

var. suksdorfii    var. glaberrimum

Apocynum cannabinum

J

1/2

Menyanthes trifoliata

1.5

2

Nephrophyllidium crista-galli

6

1/2    3

var. androsaemifolium    Apocynum androsaemifolium

JRJ    3    1/2

var. pumilum

ly erect or very slightly spreading; follicles 12-18 cm. long, more or less sickle-shaped; coma of the seeds 2-3 cm. long. N=8? 11.

Rather general throughout nearly all of the U.S. and much of Can. June-Sept.

Often a very serious weed on wasteland or areas that are infrequently plowed, such as some orchards. The plant intergrades freely with A. sibiricum, A. androsaemifolium, and A. medium. Although nearly all recent workers have followed Woodson and described both A. cannabinum and A. sibiricum as having corollas only slightly, if at all, longer than the calyx, such distinction is largely artificial since the plants vary to the condition where the corolla is at least twice as long as the calyx.

1 Calyx lobes about half the length of the corolla; with the var. glaberrimum in w. U.S.
                                                         var. suksdorfii (Greene) Bég. & Bel.

1 Calyx lobes more than half as long as the corolla; general and more common in our area,
   usually referred to as                                var. glaberrimum DC.


X Apocynum medium Greene, Pitt. 3:230. 1897. (No type designated)

  A. floribundum Greene, Erythea 1:151. 1893. A. lividum var. floribundum Bég. & Bel. Atti
    R. Accad. Lincei V. 9:128. 1913. A. medium var. floribundum Woodson, Ann. Mo. Bot.
    Gard. 17:113. 1930. A. cannabinum var. floribundum Jeps. Fl. Calif. 3:103. 1939. (Greene,
    in 1889, "Dry ground bordering pine woods, in the higher mts. of the Mohave Desert, in
    Kern Co., Calif.")
  A. vestitum Greene, Man. Bay Reg. 240. 1894. A. medium var. vestitum Woodson, Ann.
    Mo. Bot. Gard. 17:116. 1930. (Greene, hills west of Napa Valley, Calif.)
  A. lividum Greene, Pl. Baker. 3:17. 1901. A. medium var. lividum Woodson, Ann. Mo. Bot.
    Gard. 17:115. 1930. (C. F. Baker, July 8, 1901. Black Canyon, Colo.)
  A. ciliolatum Piper, Contr. U.S. Nat. Herb. 11:453. 1906. (Lake & Hull 549, Wawawai,
    Wash., July 17, 1892)
  A. sarniense Greene, Leafl. 2:167. 1912. A. medium var. sarniense Woodson, Ann. Mo.
    Bot. Gard. 17:111. 1930. (C. K. Dodge, Sarnia, Ont.)
  A. convallarium Greene, Leafl. 2:174. 1912 (Leiberg, near Meeksville, in the Clark's Fork
    Valley, Mont., Aug. 22, 1895)
  A. denticulatum Suksd. Werdenda 1:31. 1927. (Suksdorf 4049, "einige km. nordlich von Rock-
    land," Klickitat Co., Wash., June 8, 1904)

Plant 2-6 dm. tall, glabrous to pubescent overall, rather freely branched; leaves subsessile to distinctly petiolate, usually spreading, ovate-lanceolate to ovate, obtuse to acute and mucronulate, cordate to rounded at the base; cymes usually axillary as well as terminal; flower (3.5)4.5-6 mm. long; calyx usually nearly equaling the corolla tube, the lobes oblong-lanceolate to narrowly lanceolate, acute to acuminate; corolla campanulate, pinkish, the lobes spreading; follicles 7-14 cm. long, pendulous.

General in the valleys and lower levels of the mountains of w. N. Am., e. through the central states to the Atlantic Coast. June-Sept.

As pointed out by Greene, these plants are almost exactly intermediate between A. androsaemifolium and A. cannabinum, and tend to blend with both species, the corolla varying from about 3.5 to 6 mm. in length. Anderson (Ann. Mo. Bot. Gard. 23:159-68. 1936) has shown that this is not a natural species as that taxon is recognized in most genera; but that it consists of hybrids and hybrid segregates involving the two species it resembles. Where the intergradation approaches closest to cannabinum, at least some plants that have been called A. medium var. lividum might as well be called pubescent forms of var. suksdorfii (E. R. Lake, Wawawai, Wash.)

Various species (or varieties) have been proposed on the bases of the degree of hairiness of the plants, e.g. var. sarniense (Greene) Woodson with hirtellous corolla and soft pubescent leaves; var. floribundum (Greene) Woodson, completely glabrous; var. lividum (Greene) Woodson, glabrous except the under surface of the leaves and with erose-ciliate calyx lobes; var. vestitum (Greene) Woodson, the entire plant pubescent. Since there are all degrees of intermediacy and none of these forms is limited to a distinctive geographic range, there seems little reason to subdivide the "species" mechanically, especially in view of its hybrid nature.

Apocynum sibiricum Jacq. Hort. Vindob. 3:37, pl. 66. 1770.
  A. hypericifolium Ait. Hort. Kew. 1:304. 1789. A. cannabinum var. hypericifolium Gray,
    Man. 365. 1848. Cynopaema hypericifolium Lunell, Am. Midl. Nat. 4:509. 1916. (N. Am.)
  Apocynum salignum Greene, Pitt. 5:64. 1902. A. hypericifolium var. salignum Bég. & Bel.
    Atti R. Accad. Lincei V. 9:115. 1913. A. sibiricum var. salignum Fern. Rhodora 37:328.
    1935. (Chestnut & Drew, Humboldt Co., Calif.)
  Very similar to A. cannabinum, differing in that most of the lower leaves at least are sessile or subsessile, cordate-based, and often clasping, and the bracts of the inflorescence are
often slightly larger and broader; follicles significantly different, 4-11 cm. long, and nearly
straight; coma of the seeds 1-2 cm. long.
  General over much of the U.S. and Can. June-Sept.
  Occurring with A. cannabinum throughout much of its range and intergrading (therefore interfertile?) with it. The separation of the two ordinarily will be an arbitrary one on the basis
of sessile and cordate versus petioled and rounded to acute-based leaves, since only occasional specimens bear follicles. Our form of the plant is glabrous and is the var. salignum
(Greene) Fernald.

### Vinca L. Periwinkle

  Flowers axillary, long-pedicellate; corolla salverform, crested-fornicate in the throat; stamens inserted below the throat, filaments short; style elongate; stigma ovoid, annular-bor-
dered at center, tapering to a truncate apex; seeds not comose; trailing, perennial herbs.
  Perhaps 7 or 8 species of the Old World. (From the ancient Latin name, vincapervinca,
from which comes the name periwinkle.)

Vinca major L. Sp. Pl. 209. 1753. (Europe)
  Branches 1-3 dm. long, often trailing and freely rooting; leaves ovate to ovate-lanceolate,
ciliolate, 3-9 cm. long; calyx lobes linear; corolla blue, 3-5 cm. broad, hairy in the throat;
tube about 2 cm. long, lobes more or less truncate; follicles 3-5 cm. long. N=46.
  Ours an infrequent garden escape w. of the Cascades, considered to be an excellent ornamental ground cover; native of Europe. June-Aug.

### ASCLEPIADACEAE. Milkweed Family

  Flowers perfect, regular, gamopetalous, hypogynous, 5-merous except for the pistil; sepals
free or connate at the base; corolla usually lobed over half its length and with the lobes reflexed,
valvate in the bud; stamens adnate to the base of the corolla tube, monadelphous (rarely
distinct) and forming a "column" to which are attached, alternate with the corolla lobes, 5 sac-
like "hoods," each usually with an elongate, inwardly curved appendage or "horn" fastened
near the base of the hood and either included within it or exserted from it; anthers 2-locular,
basally attached, introrse, usually appendaged at their tips with a scarious membrane or "corona," their hardened winged margins forming elongate, elevated, liplike "commissural
grooves," the pollen of each anther sac coalescent into a single suspended mass or "pollinium,"
the pollinia of pairs of anther sacs of adjacent anthers joined by an appendage, the "translator
arm," the two arms attached at the top of, and over, the commissural groove, to a single cleft
"gland" which is also at the edge of a lobe of the stigma (these glands fasten to the legs of visit-
ing insects and are transported with their attached pollinia to other flowers in effecting cross-
pollination); pistil 1, 2-carpellary, the ovaries superior (slightly inferior) and distinct, styles
2, stigma 1, peltate-discoid, 5-lobed, adnate to the stamens; fruit follicular, many-seeded;
seeds flattened, apically comose; mostly perennial herbs (ours) that are often vinelike or
shrubby to arborescent and often very fleshy, with opposite or whorled (alternate), simple,
entire leaves, and milky juice, without stipules or these very small and usually fugacious.
  A large family of over 200 genera and about 1500 species, widespread but most abundant in
the tropics and subtropics.

## Asclepias L. Milkweed

Flowers umbellate; sepals and petals (in ours) reflexed; one follicle usually abortive; plants from rhizomes or thickened, fleshy roots; leaves fleshy, usually with tiny, caducous stipules.

Perhaps 100 species of Africa and the Americas. (Named for the legendary Greek physician and god of medicine, Asklepios.)

Reference:

Woodson, Robert E., Jr. The North American species of Asclepias L. Ann. Mo. Bot. Gard. 41:1-211. 1954.

1 Leaves narrowly linear to linear-lanceolate, verticillate to scattered; hoods attached above the base of the staminal column     A. FASCICULARIS
1 Leaves usually lanceolate or broader, opposite; hoods attached at the base of the column
   2 Horns considerably exceeding the stamens, divergent; follicles warty-spiny; plants usually grayish-lanate; petals pinkish to purple     A. SPECIOSA
   2 Horns included in the cleft hoods; follicles smooth; plants greenish, glabrate; petals pale greenish-yellow     A. CRYPTOCERAS

Asclepias cryptoceras Wats. Bot. King Exp. 283, pl. 28. 1879. (Watson, W. Humboldt Mts., Nev.)

   Acerates latifolia Torr. & Frem. in Frem. Rep. 317. 1845, not of Raf. (Fremont, locality uncertain)

   Asclepias davisii Woodson, Ann. Mo. Bot. Gard. 26:261. 1939. A. cryptoceras ssp. davisii Woodson, Ann. Mo. Bot. Gard. 41:180. 1954. (R. J. Davis 85, Glenn's Ferry, Elmore Co., Ida., May 15, 1938)

Plant from enlarged, woody, often fusiform root, rather fleshy; stems 1-3 dm. long, usually prostrate or decumbent; leaves opposite, from obovate and rounded but usually abruptly apiculate, to oval or oblong or (the upper ones) ovate-lanceolate, 2-6 cm. long, mostly nearly as broad, ciliolate, otherwise usually glabrous; umbels usually terminal, the central one 5- to 10-flowered, with a peduncle 2-4 cm. long and pedicels about equaling the peduncle, the lateral terminal umbels sessile and fewer-flowered, occasionally also with axillary few-flowered umbels from the next lower node; sepals linear to lanceolate, 5-9 mm. long, green or somewhat reddish-tinged, more or less floccose; corolla pale greenish-yellow, the lobes lanceolate, 8-12 mm. long, often reddish-tinged on the back; hoods pinkish, 5-6 mm. long, attached at the base of the staminal column, adnate about 2/3 of their length, the body saccate and bilobed above, the lobes projecting into short, erect to slightly recurved teeth that slightly exceed the anthers and completely enclose the short horn; follicles ovoid, 3-5 cm. long, smooth. N=11.

Gravelly to heavy clay soil in the hills and lower mountains of Asotin Co., Wash., and Grant Co., Oreg., to Payette Co., Ida., s. to Calif., e. to Colo., Wyo., and Ariz. Apr.-June.

Asclepias fascicularis Dcne. in DC. Prodr. 8:569. 1844. (Douglas, Calif.)

   A. macrophylla Nutt. Journ. Acad. Phila. II. 1:180. 1847. (Nuttall, near Monterey, Calif.)

   A. macrophylla var. comosa Dur. & Hilg. Pac. R.R. Rep. 5³:10, pl. 8. 1855. (Heermann, Posé Creek, Kern Co., Calif.)

   A. mexicana of Rydb. and other authors, but not of Cav. in 1791.

Plant widely rhizomatous, 3-8 dm. tall, glabrous to sparsely puberulent; leaves mostly whorled and 3-6 per node, the upper often opposite, linear-lanceolate to narrowly oblanceolate, 5-15 cm. long, (2)3-12(16) mm. broad, stipules minute; umbels many-flowered, commonly 2 or more from the upper axils, usually crisp-puberulent; peduncles 2-6 cm. long, pedicels 2-15 mm. long; sepals greenish to pinkish- or purplish-tinged, about 2 mm. long, puberulent; corolla lobes 2-4.5 mm. long, pale to dark pinkish-purple; hoods erect, 1-2 mm. long, considerably exceeded by the horns, subequal to the stamens; follicles erect, 6-12 cm. long, narrowly lanceolate-fusiform, smooth; seeds about 6 mm. long, flattened and wing-margined.

Dry to moist soil, especially along water courses, n.e. Wash. to Utah and Ariz., w. to the Cascades in the sagebrush and yellow pine region and to the coast in Oreg., s. to Baja Calif. June-Aug.

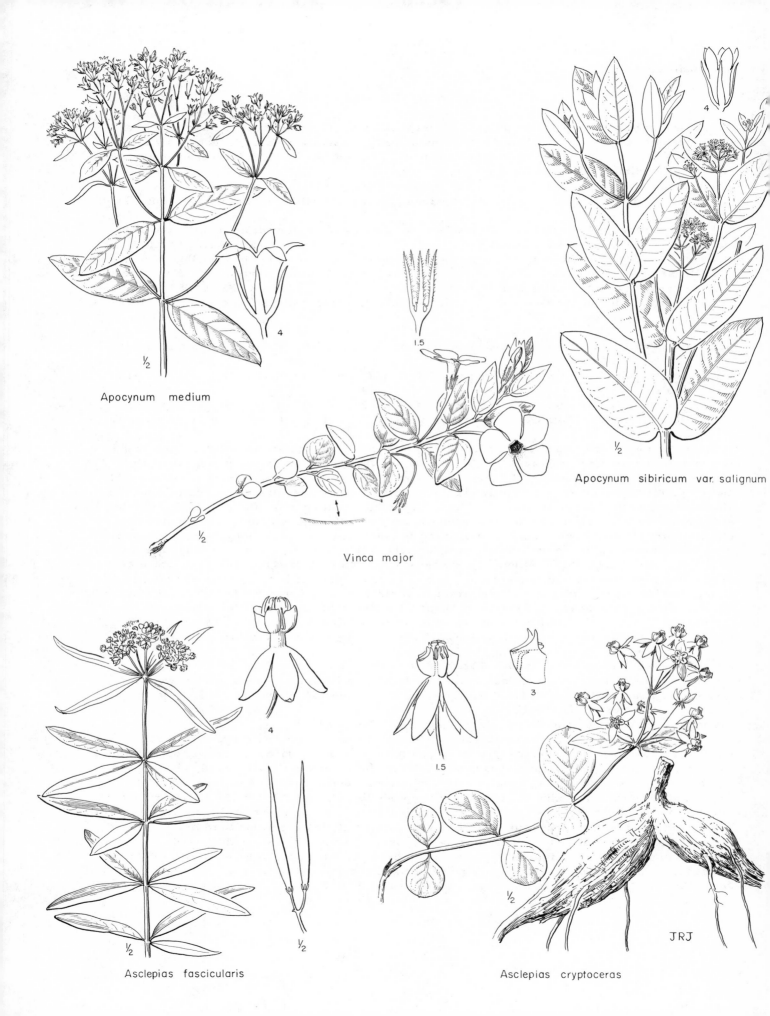

Apocynum medium

4

1.5

Apocynum sibiricum var. salignum

½

Vinca major

½

4

1.5

3

Asclepias fascicularis

½

½

Asclepias cryptoceras

½

JRJ

Often a noxious weed; somewhat poisonous to livestock, although usually not eaten in sufficient quantity to cause loss.

Asclepias speciosa Torr. Ann. Lyc. N.Y. 2:218. 1828. (James, Canadian R.)
A. douglasii Hook. Fl. Bor. Am. 2:53, pl. 142. 1838. (Douglas, Rocky Mts.)
Plant with widespread rhizomes, 4-12 dm. tall, soft grayish-tomentose throughout or the stems glabrous below; leaves sometimes glabrate, opposite, petiolate, oblong-lanceolate to ovate-oblong, 10-20 cm. long, as much as 10 cm. broad, conspicuously transversely veined; umbels usually several with peduncles 3-8 cm. long, the pedicels 1-3 cm. long; sepals greenish, tinged with red; petals about 1 cm. long, pinkish to reddish-purple; hoods pink, considerably longer than the petals, somewhat divergent; horns incurved; follicles narrowly ovoid, verrucose-exasperate, 7-11 cm. long; seeds flattened, rugose, about 8 mm. long. N=11.
Sandy to loamy, often moist soil, especially along waterways; e. side of the Cascades from B.C. s. to much of nonmontane Calif., e. to the central states. June-Aug.

### CONVOLVULACEAE. Morning-glory Family

Flowers complete, gamopetalous, hypogynous, regular, usually 5-merous except for the pistil, single and axillary to terminal and cymose; calyx often subtended by involucral bracts, the sepals distinct or basally connate, imbricate; corolla often very showy, campanulate or more usually funnelform to salverform, from entire-limbed to deeply lobed, usually twisted in the bud; stamens borne on the corolla alternate with the lobes, anthers 2-celled; pistil 1, ovary 2(1)-celled, usually 4-ovuled, often surrounded by a small disc; styles 1 or 2; stigmas capitate to elongate-attenuate; fruit usually capsular, 1- to 4(6)-seeded, loculicidal (circumscissile); mostly erect to procumbent or scandent herbs (ours), or shrubs to small trees, with alternate simple exstipulate leaves and often milky juice.
Generally distributed, but especially common in the tropics; about 50 genera and perhaps as many as 1200 species, including Ipomoea to which genus the common sweet potato and cultivated morning-glory belong.
1 Flowers well over 1 cm. long, funnelform
　2 Annual; stigmas capitate, often 3; occasionally escaping and at least temporarily becoming established, but probably not persistent
　　　　　　　　　　　　　　Ipomoea purpurea (L.) Lam. Common morning-glory
　2 Perennial; stigmas 2, elongate　　　　　　　　CONVOLVULUS
1 Flowers less than 1 cm. long, often campanulate; alkaline valleys of s.e. Oreg., s. and along the coast to Mex., probably not getting into our area　　　Cressa cretica L.

### Convolvulus L. Bindweed; Morning-glory

Flowers funnelform, large and showy, single, or sometimes in pairs on axillary peduncles, 2-bracteate, the bracts either slender and much smaller than the calyx lobes and usually borne somewhat below the calyx, or broader and borne contiguous to the calyx and usually equaling or exceeding it and largely concealing it; sepals subequal or the outer considerably the broadest; corolla usually plaited, white to pinkish or somewhat purplish, conspicuously contorted in the bud; stamens freed near the base of the corolla tube, included; style slender, usually exceeding the stamens, stigmatic lobes flattened, narrowly lanceolate to oval; ovary 2-celled, 4-ovuled; capsule ovoid to globose, tardily 2- to 4-valvate, 4-seeded or with 1 or 2 of the seeds abortive; perennial (ours) herbs from deep-seated, rather slender rhizomes, usually with traling to twining stems and simple, alternate, usually more or less hastate leaves.
About 200 species, found in most temperate and tropical parts of the world, many as very noxious weeds. (Name from the Latin convolvere, meaning to twine.)
Reference:
House, H.D. Synopsis of the Californian species of Convolvulus. Muhl. 4:49-56. 1908.
1 Leaves thick and fleshy, reniform, glabrous; confined to coastal dunes　C. SOLDANELLA
1 Leaves mostly not fleshy, sometimes hairy, seldom reniform; seldom, if ever, on coastal dunes

2  Flowers double, bright pink, single in the axils; plants pubescent; stems twining; escaped
      from cultivation; not seen from our area although occasionally reported from Idaho
      (California rose)                                                                      C. japonicus Thunb.
2  Flowers not double, usually white, sometimes paired; plants often glabrous, not always
      twining
   3  Bracts linear, often borne considerably below the calyx
      4  Plant pubescent; flowers 1 per peduncle; bracts commonly close to the calyx, usually
            well over 3 mm. long                                                      C. POLYMORPHUS
      4  Plant often glabrous; flowers often 2 per peduncle; bracts minute, often rather distant
            from the calyx                                                                C. ARVENSIS
   3  Bracts broader, often cordate, usually borne adjacent to the calyx and largely concealing
         it
      5  Plants of dry, rocky, open slopes, glabrous; stems erect to trailing but not twining;
            flowers white                                                              C. NYCTAGINEUS
      5  Plants usually of moist lowland areas, often hairy; stems twining; flowers white to
            deep pink                                                                        C. SEPIUM

Convolvulus arvensis L. Sp. Pl. 153. 1753. (Europe)
   C. ambigens House, Bull. Torrey Club 32:139. 1905. (C. S. Crandall 4218, near Ft. Col-
      lins, Colo., June 22, 1896)
   Perennial with widespread and deeply descending rhizomes, glabrous or sparsely to densely
pubescent; stems trailing to somewhat twining, 2-20 dm. long; leaf blades from ovate-lanceolate to truncate-based or more commonly sagittate or hastate, obtuse or abruptly apiculate to
acute, 2-6 cm. long; petioles 5-30 mm. long; peduncles usually exceeding the leaves, mostly
2-bracteate about midlength, some, at least, usually 2-flowered and the longer pedicel mostly 2-bracteolate, the other ebracteate; calyx campanulate-cupuliform, lobed full length, the
segments oblong, ovate, 4-5 mm. long, hyaline-margined, usually slightly gibbous at base;
corolla white or pinkish-purple, at least on the outside, 1.5-2.5 cm. long; stigmatic lobes narrow, slightly flattened; capsule ovoid-obconic, 5-7 mm. long; seeds about 4 mm. long, smooth.
N=25. Small or field bindweed; orchard morning-glory.
   Introduced from Europe and only too well established in much of N. Am. Midspring to late
fall.
   The plant is very difficult to eradicate because of its low growth and deep, widespread rhizomes, and in many areas it is considered to be the worst weed.

Convolvulus nyctagineus Greene, Pitt. 3:327. 1898. (Th. Howell, in 1882, Oreg.)
   C. macounii Greene, Pitt. 3:331. (Macoun, August, 1895, Milk R., Assiniboia, Sask., Can.)
   Calystegia atriplicifolia Hallier f., Bull. Herb. Boiss. 5:385, pl. 13, fig. 2. 1897. Convol-
      vulus atriplicifolius House, Muhl. 4:54. 1908. Not Convolvulus atriplicifolius Poir. (Th.
      Howell, Oreg.)
   Plant from slender rootstocks, glabrous; stems from semierect and only 1-3 dm. long to
trailing and 2-6 dm. long; leaves with petioles from shorter than, to 2-3 times as long as, the
blades, the blades rather thick, fleshy, deltoid to hastate or cordate-hastate, 3-5(7) cm. long,
acuminate to rounded and mucronulate; flowers single on peduncles that are from much shorter
than, to as long as, the leaves, usually with near-basal, ovate to ovate-lanceolate bracts with rounded to semi-cordate bases that usually exceed and largely conceal the pedicel; sepals rather thin
in texture, 12-15 mm. long; corolla creamy-white or pinkish-tinged, 3-5 cm. long; stigmas elliptic-lanceolate, flattened; capsule about 1 cm. long; seeds 4-6 mm. long, essentially smooth.
   Rocky open ground, mostly in yellow pine woodland in our area, from near Mt. Adams,
Wash., s. in the Cascades and through the Willamette Valley to the n. Sierras and Lake Co.,
Calif.; possibly to the Canadian Rockies, if, as believed, C. macounii Greene, is referable
here. June-Aug.

Convolvulus polymorphus Greene, Pitt. 3:331. 1898. (N. Calif. and s. Oreg.; Howell 1948,
   from near Roseburg, Oreg., the only specimen cited, was said to differ from the "more typical Californian plant" in several ways)

Perennial from deep rootstocks, pale green and finely puberulent throughout; stems more trailing than twining, 2-10 dm. long; blades mostly 2-4 cm. long, usually longer than the petioles, markedly variable, but mostly hastate to sagittate-hastate, from broader than long to the reverse, terminal lobe from less than 1/2 to over 3/4 of the total length of the blade, lanceolate-acuminate to oval and abruptly mucronulate, the lateral lobes from rounded to distinctly squarish-angled; flowers single on peduncles shorter than the leaves, the bracts linear to linear-lanceolate, 3-10(12) mm. long, borne 1-6 mm. below the calyx; calyx lobes very unequal, oblong-ovate, 10-13 mm. long, obscurely to prominently mucronate; corolla 3-4.5 cm. long, creamy-white or occasionally somewhat pinkish-streaked; stigmatic lobes narrowly oblong-lanceolate, slightly flattened.

Dry valleys and hillsides, often on serpentine outcrops, chiefly from Klamath and Josephine cos., Oreg., s. into the Sierra Nevada and Coast Ranges to Lake Co., Calif., but also in Maupin Valley, Wasco Co., Oreg. June-July.

Convolvulus sepium L. Sp. Pl. 153. 1753.
  Calystegia sepium R. Br. Prodr. 483. 1810. (Europe)
  Convolvulus repens L. Sp. Pl. 158. 1753. C. sepium var. pubescens Gray, Man. 376. 1867.
    C. sepium var. repens Gray, Syn. Fl. 2[1]:215. 1878. C. sepium var. pubescens Fern.
    Rhodora 10:55. 1908. Volvulus sepium var. pubescens Farw. Am. Midl. Nat. 12:130. 1930.
    (E. N. Am.)
  Convolvulus sepium var. americanus Sims in Curtis' Bot. Mag. 19:pl. 732. 1804, at least as
    to usage. C. americanus Greene, Pitt. 3:328. 1898. Calystegia americana Daniels, U. Mo.
    Stud. Sci. I. 2:195. 1907. (Cultivated plants, presumably of American origin)
  Convolvulus sepium var. fraterniflorus Mack. & Bush, Man. Jackson Co. Mo. 153. 1902. C.
    fraterniflorus Mack. & Bush, Rep. Mo. Bot. Gard. 16:104. 1905. (Martin City, Mo.)
  C. sepium var. communis Tryon, Rhodora 41:419. 1939. (Wilkinson 7662, Mansfield, Ohio
    June 17, 1895)
  Plant widespread from slender, elongate rhizomes, glabrous to softly pubescent; stems trailing or climbing, 2-3 m. long; petioles 2-8 cm. long; leaf blades sagittate to hastate, 5-12 cm. long, from less than half, to more than half as broad as long, basal lobes rounded to distinctly angled, the tips acuminate; flowers usually single in the axils on peduncles subequal to the leaves; bracts broadly cordate-ovate, obtuse to acute, enclosing and often exceeding the calyx; calyx lobed nearly to the base; corolla white to deep pink, 4-7 cm. long; stigmas ovoid, much flattened; capsules ovoid-globose, about 1 cm. long; seeds about 5 mm. long, finely papillate-verrucose. N=11, 12. Hedge bindweed.

Moist soil, especially along river bottoms and coastal marshes, introduced on the west coast in scattered localities and especially common in the Puget Sound area; native in e. U.S. A difficult weed. May-Sept.

The species, originally described from European plants, shows much variation in the lobing of the leaves, flower color, and amount of pubescence, and has been divided into several very poorly defined varieties with our white-flowered plants usually called the var. repens (L.) Gray. In the latest comprehensive treatment of the group the following two taxa are reported for our region: var. communis Tryon, plants glabrous, corolla pink; and var. fraterniflorus Mack. & Bush, plants glabrous or pubescent, corollas white.

Convolvulus soldanella L. Sp. Pl. 159. 1753.
  Calystegia soldanella R. Br. Prodr. 483. 1810. (England)
  C. reniformis R. Br. Prodr. 483. 1810. (Brown, Pt. Jackson, near Sydney, Australia)
  Plant glabrous, fleshy, from deep rootstocks; stems creeping but not twining, usually several dm. long; leaves with petioles 1-3 times as long as the blades; blades mostly 1.5-4 cm. broad, reniform, rounded to conspicuously retuse, mucronulate, with basal, often somewhat angled lobes; calyx immediately subtended and nearly equaled by broad, usually cordate bracts; sepals oblong-ovate, commonly mucronulate, about 1 cm. long, accrescent and about 2 cm. in fruit; corolla broadly funnelform, 3-5(6) cm. long, pinkish-purple; stigmas greatly flattened, oblong-oval; capsule subglobose, about 1.5 cm. long; seeds 6-8 mm. long, essentially smooth. N=11. Beach morning-glory.

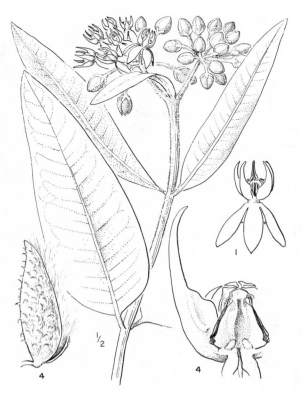

Asclepias speciosa

Convolvulus sepium

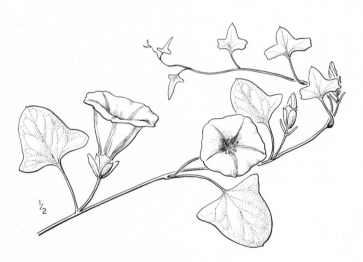

Convolvulus polymorphus

Convolvulus nyctagineus

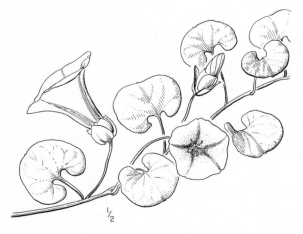

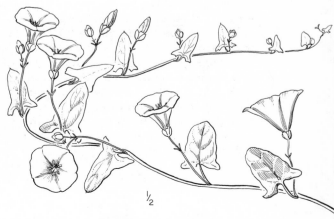

Convolvulus soldanella

JRJ

Convolvulus arvensis

Coastal beaches and sand dunes, often extending down to the high tide level, B. C. to San Diego Co. , Calif. ; islands of the Pacific; Europe. Apr. -Sept.

## CUSCUTACEAE. Dodder Family

Flowers small, complete, gamopetalous, hypogynous, often fleshy at base, usually pinkish, 5(4 or 3)-merous except for the 2-carpellary ovary; parasitic herbs closely related to, and perhaps more usually included in, the Convolvulaceae, but differing in the parasitic habit and the constantly distinct styles and the small, usually scale-appendaged corollas. The seeds of the plants germinate to produce rooted seedlings which soon die (since they lack chlorophyll) unless they come into contact with, and are able to parasitize, other flowering plants. This is accomplished by means of special rootlike branches or haustoria that are sunk into the host and through which all absorption by the parasite occurs.

Only the 1 genus.

### Cuscuta L. Dodder; Love Tangle; Coral Vine

Flowers small, sessile to short-pedicellate in small cymules or aggregated into larger, often globular masses, whitish to yellow, (3)4- or more commonly 5-merous excepting the pistil; calyx deeply lobed, often somewhat fleshy at the base; corolla narrowly tubular to campanulate, the lobes imbricate in the buds, reflexed to erect in flower; stamens inserted just below the sinuses of the lobes, the free filaments from much shorter to longer than the anthers, the portion fused with the corolla often prominent and usually covered at the base by mostly fimbriate-margined scales which are united somewhat above the base of the corolla (scales sometimes lacking); ovary 2-celled, 4-ovuled; styles 2, distinct; stigmas capitate to linear-elongate; capsule membranous, circumscissile near the base or indehiscent, 1- to 4-seeded, often rupturing the corolla and pushing it upward as a calyptralike crown; parasitic, twining, leafless, perennial herbs (ours), with very slender, pinkish-yellow to white, glabrous stems, the leaves reduced to tiny scales.

About 150 species, world-wide in distribution, but largely American. (Derivation uncertain, commonly supposed to be Arabic. )

Parasitic on many flowering plants and sometimes causing considerable damage, especially to clover and alfalfa.

References:

Yuncker, T. G. Revision of the North American and West Indian species of Cuscuta. Ill. Biol. Monog. 6:91-231. 1921.

-------. The genus Cuscuta. Mem. Torrey Club 18:113-331. 1932.

1 Stigmas attenuate, not capitate; capsules irregularly circumscissile near the base
  2 Calyx more fleshy than membranous, the lobes as broad as long, the tips turgid and
      slightly recurved; corolla usually not ruptured or pushed upward by the developing cap-
      sule, the tube seldom any longer than the calyx          C. APPROXIMATA
  2 Calyx membranous, scarcely at all fleshy, the lobes longer than broad, acute or acumi-
      nate, the tips not turgid or recurved; corolla often pushed upward by the developing cap-
      sule, the tube often exceeding the calyx          C. EPITHYMUM
1 Stigmas capitate; capsules not circumscissile
  3 Scales lacking entirely or greatly reduced and neither free-margined nor erose or pec-
      tinate
    4 Corolla lobes reflexed; anthers oblong, over 0. 5 mm. long     C. CALIFORNICA
    4 Corolla lobes spreading to erect; anthers less than 0. 5 mm. long
      5 Flowers sessile; corolla tube equaling or exceeding the calyx     C. OCCIDENTALIS
      5 Flowers with pedicels 1-3 mm. long; corolla tube shorter than the calyx
                                                                          C. SUKSDORFII
  3 Scales present, their free margins usually erose to pectinate
    6 Flowers about 2 mm. long, usually 4(3 or 5)-merous, nearly or quite sessile; sepals
        free nearly to the base, very unequal, the outer greatly overlapping the inner; co-
        rolla withering and calyptrate on the capsule          C. CEPHALANTHI

6  Flowers either at least 2.5 mm. long, or 5-merous and pedicellate; sepals usually nearly
    equal and not conspicuously overlapping; corollas persistent or deciduous
   7  Flowers usually 4(5)-merous; capsules globose or nearly so; corolla calyptrate on the
       capsule; scales very inconspicuously fringed or more usually entire-margined
                                                                            C. SUKSDORFII
   7  Flowers usually 5-merous; capsules often ovoid; corolla often persistent as the capsule
       matures; scales conspicuously erose to fimbriate-margined
      8  Capsules usually somewhat beaked or crested at the base of the styles, longer than
          broad, usually globose-ovoid to ovoid-obconic; corolla not at all papillose
         9  Calyx lobes conspicuously erose-denticulate, oval to orbicular; flowers about 2 mm.
             long; corolla tube only slightly exceeding the calyx          C. DENTICULATA
         9  Calyx lobes usually acute, but if rounded not at all erose-denticulate; flowers usual-
             ly well over 2 mm. long; corolla tube frequently greatly surpassing the calyx
            10  Sepals nearly or quite as broad as long, rounded; flowers 2-3 mm. long; corolla
                 lobes obtuse; seeds 3-4                                       C. UMBROSA
            10  Sepals considerably longer than broad, acute to acuminate; flowers often well
                 over 3 mm. long; corolla lobes acute; seeds usually one
               11  Flowers 5-6 mm. long, often sessile; calyx about half the length of the co-
                    rolla tube, the lobes conspicuously overlapping at the base; corolla tube
                    about twice the length of the lobes, cross-wrinkled-corrugate
                                                                             C. SUBINCLUSA
               11  Flowers 2-4.5 mm. long, short-pedicellate; corolla tube not cross-wrinkled,
                    about equaling the corolla lobes and little if any longer than the calyx; ca-
                    lyx lobes scarcely overlapped at the base               C. SALINA
      8  Capsules globose, usually not at all beaked, if somewhat pointed then the corolla in-
          conspicuously papillose
         12  Corolla very inconspicuously papillose on the margins and the outer surfaces of the
              often erect lobes; calyx lobed nearly to the base, the segments very conspicuous-
              ly overlapping; capsule slightly crested-pointed                C. INDECORA
         12  Corolla not papillose, the lobes spreading or reflexed; calyx lobed about 1/2 of its
              length; capsule depressed at the tip                           C. PENTAGONA

Cuscuta approximata Babington, Ann. & Mag. Nat. Hist. 13:253, pl. 4, fig. 3. 1844.
   C. planiflora var. approximata Engelm. Trans. Acad. Sci. St. Louis 1:465. 1859. (Described
      from plants growing in England, brought in as seed, supposedly from Asia)
   C. urceolata Kuntze, Flora 4:651. 1846. C. approximata var. urceolata Yuncker, Mem. Tor-
      rey Club 18:297. 1932. (Kuntze 263, Sierra Nevada, Spain)
   C. cupulata Engelm. Bot. Zeit. 4:276. 1846. (Caucasus and Alta., by Godet and by Ledebour)
   C. gracilis Rydb. Bull. Torrey Club 28:501. 1901. (Tweedy 3292, between Sheridan and
      Buffalo, Wyo., in 1900)
   C. anthemi A. Nels. Bot. Gaz. 37:277. 1904. (E. Nelson 4936, Seminole Mts., Wyo., in 1898)
   Flowers 2-3 mm. long, sessile, several to many in tight clusters; calyx yellowish, subequal to the corolla, rather fleshy, the lobes as broad as long, triangular-ovate, the tips turgid and slightly recurved; corolla campanulate-globose, not calyptrate on the developing capsule, the lobes slightly shorter than the tube, oblong-ovate; stamens visible in the corolla but
not exserted therefrom; anthers subequal to the filaments; scales nearly equaling the corolla
tube, oblong, erose-fringed, rounded to slightly bifid even in the same flower, united for
about 1/4 of their length; ovary globose; stigmas about equaling the slender styles, tapered to
the tip; capsule circumscissile near the base. N=14.
   Parasitic on leguminous crops, especially alfalfa, and doing considerable damage; probably
in all our western states; Eurasia and Africa. June-Aug.
   The species is thought to be native to the Old World, but several varieties are now well established in N. Am. although our plants are all referable to the var. urceolata (Kuntze)
Yuncker.

Cuscuta californica Choisy, Mem. Soc. Phys. Nat. Genèv. 9:279. 1841. (Douglas, Calif.)
  C. californica var. graciliflora Engelm. Trans. Acad. Sci. St. Louis 1:499. 1859. (Calif.,
    by Douglas, Fremont, & Bigelow)
  C. californica var. longiloba Engelm. Trans. Acad. Sci. St. Louis 1:499. 1859. (S. Calif.,
    several specimens cited)

  Flowers scattered, occasionally in few-flowered cymes, exceeding the pedicels; calyx about 2 (to 3) mm. long, the lower half fleshy, the lobes triangular to lanceolate, acute, scarcely 1 mm. long; corolla tube 2-3 mm. long, tubular or slightly campanulate, the lobes from slightly more than half as long, to as long, as the tube, lanceolate, acute, entire, usually sharply reflexed; stamens well exserted; anthers oblong, about 0.7 mm. long, slightly exceeding the free portion of the filaments; adnate portion of the filaments fairly prominent, the scales completely adnate and so inconspicuous as to seem entirely lacking, but their outline barely discernible; styles 2-3 mm. long, exserted; stigmas capitate; capsule globose, included in the corolla and not greatly distending it, not circumscissile; seeds usually single (2-4).

  Wash. to Baja Calif., on both sides of the Cascades, on many hosts. June-Aug.

Cuscuta cephalanthi Engelm. Am. Journ. Sci. 43:336, pl. 6, figs. 1-6. 1842.
  Epithymum cephalanthi Nieuwl. & Lunell, Am. Midl. Nat. 4:511. 1916. (Engelmann, near St.
    Louis, Mo.)
  Cuscuta tenuiflora Engelm. Lond. Journ. Bot. 2:197. 1843.

  Flowers about 2 mm. long, 4(3 or 5)-merous, nearly or quite sessile, aggregated into large globular to irregular masses; calyx not fleshy, often partially concealed by a sepal-like bract, the segments obtuse, 2/3-3/4 the length of the corolla tube, free nearly to the base of the calyx, the outer ones markedly overlapping the inner; corolla tubular-campanulate, much distended and sometimes ruptured by the growing capsule on which it is calyptrate, the lobes ovate, obtuse to rounded, about 1/3 the length of the tube, erect to spreading; stamens usually slightly exserted, the anthers oval, about equaling the filaments; scales oblong, from very inconspicuously short-fringed to conspicuously fringed at the apex, about 3/4 of the length of the tube, united for about 1/4 of their length; styles about 1 mm. long; stigmas capitate, globose; capsule globose, not circumscissile, about 2.5 mm. in diameter, not at all crested, the interstylar opening small; seeds 1-4. N=30.

  Wash. and Oreg. e. to the Atlantic Coast, s. to Tenn. and Tex., on many hosts. Summer.

Cuscuta denticulata Engelm. Am. Nat. 9:348. 1875. (Parry 205, St. George, Utah)
  Flowers sessile or subsessile in small clusters, 5-merous, about 2 mm. long; calyx not fleshy, deeply divided, the lobes oval to orbicular, considerably overlapped, the margins irregularly erose-denticulate; corolla not calyptrate on the developing capsule, the tube campanulate, slightly exceeding the calyx, the lobes ovate-lanceolate to ovate, overlapping at base, spreading to ascending, about equaling the tube; stamens slightly exserted; anthers 0.3 mm. long, about equaling the filaments, oval; scales about reaching the base of the free portion of the filaments, obovate-oblong, subentire to irregularly erose-dentate, united to midlength; styles about 0.5 mm. long, stigmas capitate; capsule narrowly ovoid, not circumscissile, crested-thickened around the base of the style, the interstylar opening small or none; seeds usually single.

  Mostly of the desert regions of Calif., Nev., and Utah, on many desert shrubs, reported for Ida., but possibly not getting into our area. May-July.

Cuscuta epithymum Murr. Syst. Veg. 13th ed. 140. 1774. (Europe)
  Flowers often somewhat purplish-tinged, even on the calyx, but more usually on the anthers and stigmas, 2.5-3 mm. long, subsessile in few-flowered glomerules or aggregated into dense compact clusters, 5-merous; calyx membranous, not fleshy, about equal to the corolla tube or somewhat shorter, the lobes ovate-lanceolate, somewhat acuminate, the tips sometimes slightly spreading; corolla often calyptrate on the developing capsule, the tube cylindric-campanulate, slightly exceeding the ovate-lanceolate to lanceolate, acute to acuminate, spreading lobes; stamens usually conspicuously exserted, the filaments freed just below the sinuses of the corolla, usually slightly exceeding the oblong-oval anthers, the adnate portion

not at all prominent; scales obovate-spatulate, about 2/3 the length of the corolla tube, free above for nearly half their length, fringed at least above; styles longer than the ovary, tapered to purplish, attenuate, stigmatic tips; capsule spherical, circumscissile near the base, mostly 4-seeded. N=7.

Introduced in many parts of w. U.S., on both sides of the Cascades in Wash. and Oreg., chiefly on leguminous plants but also on <u>Collomia</u>, <u>Achillea</u>, <u>Spiraea</u>, <u>Stellaria</u>, etc. July-Sept.

Cuscuta indecora Choisy, Mem. Soc. Phys. Nat. Genèv. 9:278, pl. 3, fig. 3. 1841.
    <u>Epithymum</u> <u>indecorum</u> Nieuwl. & Lunell, Am. Midl. Nat. 4:511. 1916. (<u>Berlandier</u> <u>2285</u>,
       Matamoros, Mex.)
    <u>Cuscuta</u> <u>neuropetala</u> Engelm. Am. Journ. Sci. 45:75. 1843. <u>C</u>. <u>indecora</u> var. <u>neuropetala</u>
       Hitchc. Contr. U.S. Nat. Herb. 3:549. 1896. (<u>Lindheimer</u> <u>124</u>, near Houston, Tex.)
    <u>C</u>. <u>pulcherrima</u> Scheele, Linnaea 21:750. 1848. <u>C</u>. <u>decora</u> var. <u>pulcherrima</u> Engelm. Trans.
       Acad. Sci. St. Louis 1:502. 1859. (<u>Lindheimer</u>, New Braunfels, Tex.)
    <u>C</u>. <u>decora</u> var. <u>subnuda</u> Engelm. Trans. Acad. Sci. St. Louis 1:502. 1859. <u>C</u>. <u>indecora</u> var.
       <u>subnuda</u> Yuncker, Am. Journ. Bot. 10:9. 1923. (<u>Tweedie</u>, islands of the Parana, Brazil)
    Flowers 5-merous, 2-5 mm. long, on pedicels at least as long, forming small, loose clusters, more or less cellular-papillose, especially on the back and margins of the corolla lobes; calyx not at all or but slightly fleshy, inconspicuously pouched at the base, lobed 3/4 of the length, the lobes ovate, strongly overlapping, obtuse; corolla tube campanulate, slightly exceeding the calyx, not forced off by the developing capsule, the lobes lanceolate, about 2/3 the length of the tube, usually erect or slightly inflexed at the tip; anthers nearly 1 mm. long, oval, about equaling the stout filaments; scales obovate, deeply fringed, usually extending past the base of the stamens, united for slightly less than half their length; styles about 1 mm. long; stigmas capitate; capsule not circumscissile, globular but slightly crested-thickened at the tip, interstylar opening essentially lacking; seeds 2-4.

One phase of this wide-ranging species of N. Am. and the W.I., the var. <u>neuropetala</u> (Engelm.) Hitchc., may get into our range in Ida., but is more common s., from Calif. to Ill. and to Mex. and S. Am.; on a wide variety of hosts, but usually in fairly low moist or irrigated areas. July-Sept.

Cuscuta occidentalis Millspaugh, Field Mus. Pub. Bot. 5:204. 1923. (<u>Hartweg</u> <u>1863</u>, Monterey,
    Calif.)
    <u>C</u>. <u>californica</u> var. <u>breviflora</u> Engelm. Trans. Acad. Sci. St. Louis 1:499. 1859. (<u>Hartweg</u>,
       Monterey, Calif.)
    Flowers sessile or subsessile in small clusters; calyx about 2 mm. long, the lower half fleshy, the lobes triangular, nearly 1 mm. long, acuminate, their tips spreading slightly; corolla tube slightly longer than the calyx, broadly campanulate, usually conspicuously pouched between the sepals, the lobes lanceolate-acuminate, spreading-ascending, usually slightly longer than the tube; stamens mostly slightly exserted; anthers oval, less than 0.5 mm. long, usually slightly exceeded by the filaments; scales lacking or completely adnate, their outline scarcely discernible; styles 1-1.5 mm. long, usually exserted, stigmas very small, capitate; capsule globose, not circumscissile, closely enveloped by the corolla; seeds usually 4.

In most western states, from Wash. to Calif. and inland to Colo., on many different hosts. June-Aug.

Cuscuta pentagona Engelm. Am. Journ. Sci. 43:340, pl. 6, figs. 22-24. 1842.
    <u>C</u>. <u>arvensis</u> var. <u>pentagona</u> Engelm. Trans. Acad. Sci. St. Louis 1:494. 1859. (<u>Rugel</u>, Nor-
       folk, Va., on <u>Euphorbia</u> <u>tragia</u>)
    <u>C</u>. <u>pentagona</u> var. <u>calycina</u> Engelm. Am. Journ. Sci. 45:76. 1843. <u>C</u>. <u>arvensis</u> var. <u>calyci</u>-
       <u>na</u> Engelm. Trans. Acad. Sci. St. Louis 1:495. 1859. <u>C</u>. <u>campestris</u> Yuncker, Mem.
       Torrey Club 18:138. 1932. (<u>Lindheimer</u> <u>126</u>, Tex.)
    <u>C</u>. <u>pentagona</u> var. <u>microcalyx</u> Engelm. Am. Journ. Sci. 45:76. 1843. (Ill.)
    <u>C</u>. <u>arvensis</u> Beyrich ex Engelm. in Gray, Man. 2nd ed. 336. 1856. <u>Epithymum</u> <u>arvense</u>
       Nieuwl. & Lunell, Am. Midl. Nat. 4:511. 1916. (e. and s. U.S., no type cited)
    Flowers 1.5-3 mm. long, 5-merous, in small clusters on pedicels usually no longer than

the calyx; calyx not fleshy, shallowly cup-shaped, the 5 lobes broadly ovate-triangular, about equaling the corolla, overlapping slightly at the base; corolla persistent, pushed off by the developing capsule, the tube broadly campanulate, usually but slightly exceeding the 5 spreading or reflexed, lanceolate, slightly acute lobes; stamens exserted, the filaments exceeding the small oval anthers; scales obovate-oblong, conspicuously fringed above, equaling or extending above the bases of the free portion of the filaments, joined for about 1/3 their length; styles about 1 mm. long; stigmas capitate; capsule depressed-globose, about 2.5-3 mm. broad; interstylar opening conspicuous; seeds usually 2. N=about 28.

On a wide variety of hosts, especially legumes, over most of the U.S. and in many other parts of the world. July-Sept.

There are two forms of the species in our area, the var. pentagona having flowers usually less than 2 mm. long, and the calyx lobes conspicuously overlapping at the sinuses, whereas the var. calycina Engelm. (C. campestris Yuncker) has flowers 2 mm. long or longer, and less conspicuously overlapping calyx lobes; the two plants intergrade and would appear not to represent separate species.

Cuscuta salina Engelm. in Gray, Bot. Calif. 1:536. 1876.
  C. subinclusa var. abbreviata Engelm. Trans. Acad. Sci. St. Louis 1:500. 1859. (Wright, Mare I., San Francisco Bay)
  C. californica var. squamigera Engelm. Trans. Acad. Sci. St. Louis 1:499. 1859. C. squamigera Piper, Contr. U.S. Nat. Herb. 11:455. 1906. C. salina var. squamigera Yuncker, Ill. Biol. Monog. 6:161, fig. 126. 1921. (Remy, Rio Virgen, Utah)
  C. salina var. major Yuncker, Ill. Biol. Monog. 6:161, figs. 32a-e, 121, 140. 1921. (S. F. Baker 41, Palo Alto, Calif.)

Flowers (2)2.5-4.5 mm. long, 5(4)-merous, in few-flowered loose clusters on pedicels not quite so long; calyx rather fleshy at the base, subequal to the corolla tube, its lobes ovate-lanceolate, acute to slightly shortly acuminate; corolla tube narrowly campanulate, about equaling the broadly lanceolate, acuminate, entire to erose, erect to slightly spreading lobes; stamens freed just below the sinuses of the corolla, the free filaments prominent; scales about half the length of the corolla tube, poorly developed, narrowly oblong, their free margins very narrow (or even lacking), entire to few-toothed, usually the free tips with a few somewhat longer teeth; styles about 1 mm. long, stigmas capitate; capsule globose-ovoid, crested just below the base of the styles, the interstylar opening small or none; seeds usually one.

Mostly on chenopodiaceous plants (but also on others, especially the Compositae) of saline soils, along the coast from B.C. to Mex., inland to Utah and Ariz. June-Aug.

There is considerable variation in the flower size and in the degree of development of the scales, which although more prominent than in C. suksdorfii, are not very conspicuous. The larger flowered plants are usually referred to the variety major Yuncker, but there seems to be little reason for according them particular nomenclatural status.

Cuscuta subinclusa Dur. & Hilg. Journ. Acad. Phila. II. 3:42. 1855. (Heermann, on a willow, Tejon Pass, Calif.)

Flowers 5-6 mm. long, 5-merous, sessile or short-pedicellate in small glomerules or aggregated into large globular masses; calyx about half the length of the corolla tube, the lobes oblong-lanceolate, acute to acuminate or slightly cuspidate; corolla tubular-cylindric, usually inconspicuously transversely corrugate-wrinkled above the calyx, the lobes 1/2-1/3 the length of the tube, ovate, acute to acuminate, slightly erose-crenulate; filaments adnate to the sinuses of the corolla lobes, the anthers essentially sessile, oblong, about 1 mm. long; scales narrowly oblong, slightly more than half the length of corolla tube, united for 1/4-1/3 their length, their free margins irregularly short-fimbriate; styles considerably exceeding the ovary; stigmas peltate-capitate; capsule ovoid, thickened at the apex and forming a collarlike base to the styles; seeds usually one.

Western Oreg., s. to Mex., on a wide range of hosts, such as Salix, Prunus, Ceanothus, Rhus, Sambucus, and many Compositae; rather late flowering.

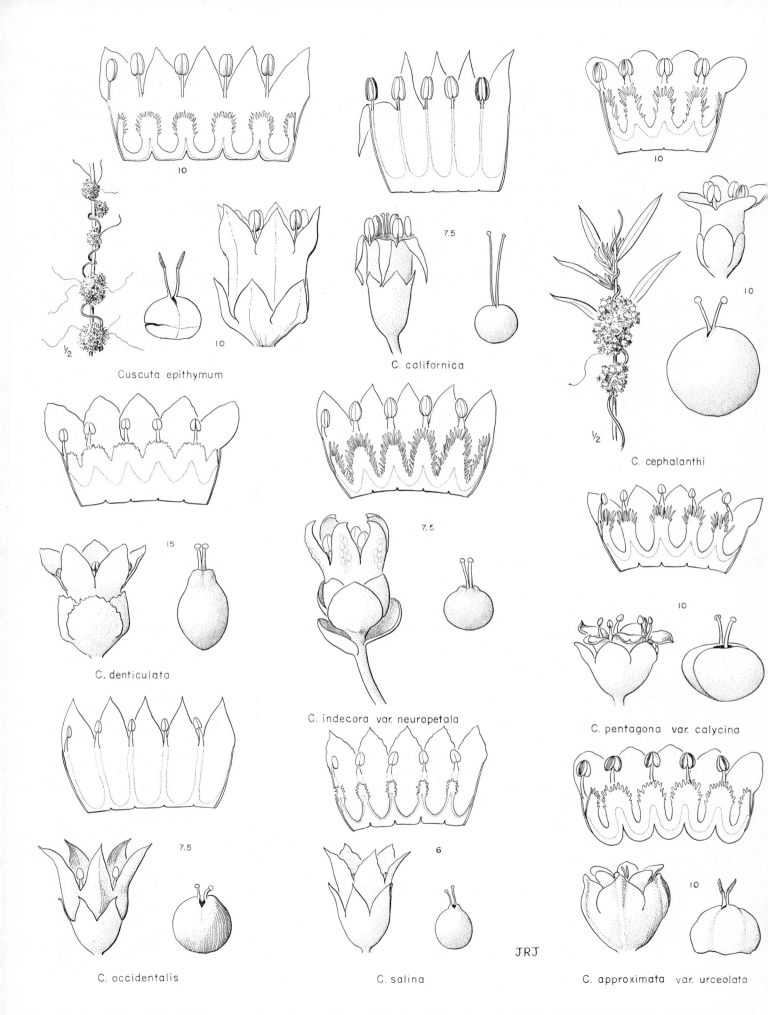

Cuscuta epithymum

C. californica

C. cephalanthi

C. denticulata

C. indecora var. neuropetala

C. pentagona var. calycina

C. occidentalis

C. salina

C. approximata var. urceolata

JRJ

Cuscuta suksdorfii Yuncker, Mem. Torrey Club 18:167. 1932.

  C. salina var. acuminata Yuncker, Ill. Biol. Monog. 6:162, figs. 32f-g, 89. 1921. (Suksdorf
    1487, Skamania Co., Wash.)

  Flowers 2.5-3 mm. long, 4- to 5-merous, borne in few-flowered clusters on pedicels 1-3
mm. long; calyx broad, nearly equaling the corolla, its lobes ovate-lanceolate, the tips long-
acuminate, somewhat spreading; corolla campanulate, calyptrate on the developing capsule,
lobed slightly over half the length, the lobes ovate- lanceolate, somewhat acuminate, erect or
somewhat spreading; stamens inserted just below the sinuses of the corolla; anthers oval, less
than 0.5 mm. long, from subequal to, to only half as long as, the filaments; scales variable,
sometimes almost completely lacking, but more usually very narrowly free-margined, entire
or with 2 divergent teeth at the tip or the margins slightly erose, occasionally the scales not
joined at their bases; styles usually considerably shorter than the ovoid-conic ovary; stigmas
capitate; capsule ovoid, very inconspicuously glandular, thin-walled, with a large interstylar
opening; seeds usually 2-4, but not infrequently one.

  On Aster, Spraguea, and other hosts, s. Wash. to the s. Sierra Nevada, Calif., at medium
to higher levels in the mountains. July-Sept.

Cuscuta umbrosa Beyrich ex Hook. Fl. Bor. Am. 2:78. 1838. (Several collections cited)

  C. gronovii var. (?) curta Engelm. Trans. Acad. Sci. St. Louis 1:508. 1859. C. curta Rydb.
    Bull. Torrey Club 40:466. 1913. (Douglas, n.w. Am.; lectotype by Yuncker)

  C. megalocarpa Rydb. Bull. Torrey Club 28:501. 1901. (Vreeland 670, Cucharas Creek,
    near La Veta, Colo., in 1900, first of two collections cited)

  Flowers 2-3 mm. long, 5-merous, borne in large aggregates of small clusters, pedicels
somewhat shorter than the flowers; calyx not fleshy-based, its lobes ovate, rounded, about as
broad as long, extending to about midlength of the corolla tube; corolla tubular-campanulate,
its lobes ovate-triangular, obtuse, spreading to reflexed; stamens usually slightly exserted,
the anthers oval, about equaling the filaments; scales about 2/3 the length of the tube, fringed-
margined and truncate to lobate at the tip, joined to about midlength; styles scarcely 1/3 the
length of the ovary, stigmas capitate; capsule globose, usually slightly crested at the base of
the styles, interstylar opening fairly conspicuous; seeds 3-4.

  Mostly N. M. to s. e. Mont. and e., probably not in our range (and therefore not illustrated),
occurring on many hosts.

POLEMONIACEAE. Phlox Family.

  Flowers gamopetalous, perfect, generally 5-merous as to the calyx, corolla, and androecium; calyx lobes equal or unequal, the tube often with alternating costae and hyaline intervals;
corolla regular or occasionally somewhat bilabiate, contorted in bud; filaments alternate with
the corolla lobes, attached near the base or more often near the sinuses; ovary superior, with
ordinarily 3(2-5) carpels; style simple, usually with separate stigmas; placentation axile; capsule usually loculicidal, sometimes irregularly or scarcely dehiscent; ovules and seeds few to
many; embryo straight; endosperm present; annual to perennial herbs, less commonly shrubs
or vines, with opposite or alternate, entire to pinnately compound or variously dissected
leaves, the flowers solitary or more often in open or compact (often headlike), variously modified cymes.

  About 17 genera and 250 species, native to N. and S. Am. and Eurasia, best developed in w.
N. Am., especially Calif. The genera are notoriously ill-defined, and the genus Gilia has often been considered to include all the genera here treated except Phlox and Polemonium. As
thus expanded, Gilia is not only heterogeneous, but still is not sharply limited from either Phlox
or Polemonium. There is little doubt that the segregate genera here accepted represent natural, recognizable groups of species, and it is convenient to maintain them for the present.

  The family contains several genera of considerable horticultural value, and some of our native species of Collomia, Gilia, Phlox, and Polemonium are worthy of cultivation.

  Reference:

  Brand, A. Polemoniaceae. Pflanzenr. IV. 250 (Heft 27). 1907

1 Leaves represented only by the persistent cotyledons and by a whorl of entire, often basally

connate bracts just beneath the compact inflorescence; small annuals    GYMNOSTERIS
1  Leaves more or less well developed, either clustered at the base, or distributed along the
   stem, or both; annual to perennial
   2  Calyx tube of essentially uniform texture throughout, somewhat accrescent, not ruptured
      by the developing capsule; leaves not at once palmatifid and sessile
      3  Leaves pinnately compound, with definite leaflets (these broader than linear except in
         one species); calyx tube herbaceous at anthesis; perennials, one species annual
                                                                          POLEMONIUM
      3  Leaves entire to variously dissected, but without definite leaflets; calyx tube charta-
         ceous at anthesis; mostly annual, one of our species perennial  COLLOMIA
   2  Calyx tube with green costae separated by hyaline intervals, or, if greenish essentially
      throughout, then the leaves sessile and palmatifid into linear segments
      4  Filaments very unequally inserted; leaves entire, partly or wholly opposite
         5  Perennial; leaves all, or nearly all, opposite; seeds not becoming mucilaginous
            when moistened                                               PHLOX
         5  Annual; upper leaves alternate; seeds becoming mucilaginous when moistened
                                                                          MICROSTERIS
      4  Filaments about equally, or occasionally somewhat unequally, inserted; leaves seldom
         at once opposite and entire
         6  Calyx lobes generally more or less unequal; leaves typically pinnatifid to bipinnati-
            fid with narrow rachis and narrow, more or less spinulose-tipped segments, vary-
            ing to linear and entire; annuals with the flowers in leafy-bracteate heads
            7  Heads distinctly tomentose with very fine, interwoven hairs; anthers generally
               (including our species) well over 0.5 mm. long, nearly linear and deeply sagit-
               tate; leaves relatively soft                             ERIASTRUM
            7  Heads glabrous or glandular to rather coarsely villous, not tomentose; anthers
               small, up to about 0.5 mm. long, not evidently sagittate; leaves mostly rather
               firm and prickly                                         NAVARRETIA
         6  Calyx lobes equal or nearly so; our species either perennial, or with the flowers
            not in heads, or with the heads essentially bractless, or with the leaves neither
            deeply pinnatifid nor entire
            8  Leaves sessile, palmatifid into linear segments
               9  Annual; leaves soft, mostly or all opposite; seeds often becoming mucilaginous
                  when moistened                                        LINANTHUS
               9  Perennial; seeds remaining unchanged when moistened
                  10  Hyaline intervals of the calyx very narrow and inconspicuous, not reaching
                      the base; leaves rather soft, opposite; plants woody only at the base
                                                                        LINANTHASTRUM
                  10  Hyaline intervals of the calyx well developed and conspicuous; leaves firm
                      and prickly, ours with only the lower ones opposite; plants shrubby
                                                                        LEPTODACTYLON
            8  Leaves diverse, but not at once sessile and palmatifid; herbs or subshrubs with
               alternate (or all basal), entire to dissected leaves
               11  Leaf segments or teeth very conspicuously bristle-tipped; corolla tending to
                   be bilabiate (only obscurely so in our species); annual    LANGLOISIA
               11  Leaf segments or teeth sometimes shortly and inconspicuously spinulose, but
                   not evidently bristle-tipped; corolla regular; annual, biennial, or perennial
                                                                        GILIA


## Collomia Nutt.

Flowers borne in terminal headlike clusters (these reduced to a single flower in one species)
which may be overtopped by their subtending branches so that the flowers appear to be lateral
or in the forks of the branches; calyx tube chartaceous, of nearly uniform texture throughout,
not ruptured by the developing capsule; calyx lobes greenish and commonly more herbaceous;
corolla tubular-funnelform to nearly salverform, bluish or pinkish to white, or sometimes

salmon or yellow; stamens equally or unequally inserted below the sinuses of the corolla, of equal or unequal length; seeds 1-3 per locule, those of the annual species becoming mucilaginous when moistened; annual or perennial herbs, mostly taprooted, slightly or scarcely mephitic, with alternate (or the lower opposite), entire to variously dissected leaves, but without well-defined leaflets.

About 13 species, native to temperate N. and S. Am., chiefly in w. U.S., ours all sharply defined. The three perennial species form a well-marked group within the genus. (Name from the Greek kolla, glue, referring to the mucilaginous quality of the moistened seeds.)

1  Plants perennial, with numerous sprawling stems                                   C. DEBILIS
1  Plants annual, often single-stemmed, commonly more or less erect
  2  Lower leaves more or less pinnatifid or subbipinnatifid; locules 2- to 3-seeded
                                                  C. HETEROPHYLLA
  2  Lower leaves, like the others, entire; locules 1-seeded
    3  Corolla 2-3 cm. long, the lobes 5-10 mm.                          C. GRANDIFLORA
    3  Corolla 4-15 mm. long, the lobes 1-3 mm.
      4  Stamens unequally inserted; plants simple or branched, the compact, leafy-bracteate flower clusters usually many-flowered and always appearing terminal to the main stem or to the short or elongate branch on which they are borne; leaves often over 5 mm. wide
        5  Calyx lobes merely acute, mostly 3-4 mm. long in fruit; plants often well over 1 dm. tall; widespread species                          C. LINEARIS
        5  Calyx lobes elongate, aristate-attenuate, the longer ones 5-11 mm. long in fruit; plants 1 dm. tall or less; local in Gilliam Co., Oreg.    C. MACROCALYX
      4  Stamens equally inserted; plants tending to branch just beneath the individual flowers or flower clusters, so that many or all of the flowers in well-developed specimens appear to be borne in the forks or scattered along the branches; leaves up to 5 mm. wide
        6  Flowers, or most of them, borne in small clusters (2-5 together); corolla 8-14 mm. long; calyx lobes 2-3 mm. long at anthesis, sometimes 4 mm. in fruit
                                            C. TINCTORIA
        6  Flowers borne singly, or a few of them in pairs; corolla 4-6 mm. long; calyx lobes about 1 mm. long at anthesis, up to 2 mm. in fruit    C. TENELLA

Collomia debilis (Wats.) Greene, Pitt. 1:127. 1887.
  Gilia debilis Wats. Am. Nat. 7:302. 1873. Navarretia debilis Kuntze, Rev. Gen. 2:433. 1891. (Wheeler, Utah)
  Gilia larseni Gray, Proc. Am. Acad. 11:84. 1876. Collomia debilis var. larsenii Brand, Pflanzenr. IV. 250:52. 1907. Gilia debilis var. larseni Macbr. Contr. Gray Herb. n. s. 56:57. 1918. Collomia larsenii Pays. U. Wyo. Pub. Bot. 1:85. 1924. (Lemmon & Larsen, Lassen's Peak, Calif.)
  Gilia howardii M. E. Jones, Zoë 2:250. 1891. Collomia debilis f. howardii Wherry, Am. Midl. Nat. 32:224. 1944. (Howard, mountains of s. Mont. in 1884)
  Collomia debilis var. ipomoea Pays. U. Wyo. Pub. Bot. 1:82. 1924. C. debilis ssp. ipomoea Wherry, Am. Midl. Nat. 31:222. 1944. (Payson 3013, Gros Ventre Mts., 15 miles n. of Bondurant, Wyo.)
  C. debilis var. camporum Pays. U. Wyo. Pub. Bot. 1:83. 1924. C. debilis f. camporum Wherry, Am. Midl. Nat. 31:222. 1944. (Kittredge s.n., Missoula, Mont., June 6, 1915, at 3250 ft.)
  C. debilis var. integra Pays. U. Wyo. Pub. Bot. 1:83. 1924. C. hurdlei var. integra A. Nels. Am. Journ. Bot. 18:435. 1931. C. debilis ssp. trifida f. integra Wherry, Am. Midl. Nat. 31:222. 1944. (Nelson & Macbride 1456, Bear Canyon, near Mackay, Custer Co., Ida.) = var. debilis.
  C. debilis var. trifida Pays. U. Wyo. Pub. Bot. 1:85. 1924. C. debilis ssp. trifida Wherry, Am. Midl. Nat. 31:222. 1944. C. hurdlei var. trifida A. Nels. Am. Journ. Bot. 18:435. 1931. (Macbride & Payson 3435, Bonanza, Custer Co., Ida.) = var. debilis.
  C. hurdlei A. Nels. Am. Journ. Bot. 18:435. 1931. C. debilis ssp. trifida f. hurdlei Wher-

ry, Am. Midl. Nat. 31:224. 1944. (Hurdle, Twin Peaks, Challis Nat. Forest, Ida., Aug.,
   1928) = var. debilis.
C. debilis var. dentata C. L. Hitchc. Leafl. West. Bot. 4:203. 1945. (Hitchcock & Muhlick
   9058, 10 miles s. of Gibbonsville, Lemhi Co., Ida.) = var. camporum.
   Perennial from a usually deep-seated taproot and crown, with numerous sprawling, simple
or branched stems, these commonly becoming very slender and rhizomelike toward the base,
the whole plant forming a loose mat often several dm. across; herbage puberulent or glandular-puberulent to subglabrous; leaves wholly alternate, tending to be crowded distally, short-
petiolate, the blade up to about 3 cm. long and 13 mm. wide, entire to dissected; flowers sessile or short-pedicellate in small, leafy-bracteate clusters at the ends of the stems; corolla
showy, tubular-funnelform, 12-35 mm. long, blue or lavender to pink, white, or even ochroleucous, the tube (including the gradually expanded throat) much longer than the lobes; stamens
equally inserted well below the sinuses, from shortly included to shortly exserted, the filaments equal or unequal; locules 1-seeded.
   Shifting talus slopes at high elevations in the mountains (var. camporum at lower elevations);
Wash. to Mont., southward to n. Calif., c. Utah, and w. Wyo. June-Aug.
   Collomia debilis is variable in pubescence, form of the leaves, size and color of the flowers,
and length of stamens with relation to the corolla, but much of the variation is haphazard and
without taxonomic significance. I believe that only four infraspecific taxa can properly be maintained. One of these, the var. larsenii, is well marked and has often been treated as a separate species, but it intergrades with typical C. debilis in the Cascade region of Wash. It might
reasonably be treated as a distinct subspecies, but no new combination is here proposed. Two
other local populations may be distinguished, with some difficulty, from the widespread and
highly variable var. debilis. The several varieties, as I understand them, may be characterized as follows:
1 Principal leaves entire or few-toothed to rather deeply 3- to 5-lobed, the lobes entire; co-
      rolla 1.5-3.5 cm. long
   2 Leaves relatively narrow and elongate, all entire or merely with a few sharp, small
         teeth, more or less strongly acute (or the lower more obtuse)
      3 Corolla mostly 1.5-2.5 cm. long; low elevations, vicinity of the n. fork of the Salmon
            R., Lemhi Co., Ida., northward to Missoula, Mont.            var. camporum Pays.
      3 Corolla mostly 2.5-3.5 cm. long; high elevations in the mountains of w. Wyo.
                                                                    var. ipomoea Pays.
   2 Leaves mostly shorter and broader, varying from all entire to often some or many of
         them more or less deeply 3- to 5-lobed or -cleft, the blade (when unlobed) rounded to
         acutish; corolla mostly 1.5-2.5 cm. long; high elevations in the mountains from c. Utah
         and n.e. Nev. to w. Wyo., w. Mont., c. Ida., c. and n.e. Oreg., and the Cascades of
         Wash.                                                          var. debilis
1 Principal leaves deeply and sometimes irregularly 3- to 7-cleft, with some or all of the
      segments again cleft; corolla mostly 12-15 mm. long; Cascade (and Olympic) region from
      Wash. to n. Calif.                                          var. larsenii (Gray) Brand

Collomia grandiflora Dougl. ex. Lindl. Bot. Reg. 14: sub pl. 1166. 1828.
   Gilia grandiflora Gray, Proc. Am. Acad. 17:223. 1882. Navarretia grandiflora Kuntze, Rev.
      Gen. 2:433. 1891. (Cultivated specimens, derived eventually from seeds taken by Douglas
      in the "northwest of North America, in all the country bordering on the river Columbia, as
      far to the eastward as the valleys of the Rocky Mountains, but not beyond that great dividing
      ridge")
   Gilia grandiflora var. diffusa Mulford, Bot. Gaz. 19:120. 1894. Collomia grandiflora ssp.
      diffusa Piper, Contr. U.S. Nat. Herb. 11:465. 1906. C. grandiflora f. diffusa Wherry,
      Am. Midl. Nat. 31:223. 1944. (Mulford, foothills about Boise, Ida.) A branching form.
   Collomia scabra Greene, Leafl. 2:88. 1910. C. grandiflora f. scabra Wherry, Am. Midl.
      Nat. 31:226. 1944. (Cotton 702, Rattlesnake Mts., Wash.)
   C. grandiflora var. axillaris A. Nels. Bot. Gaz. 52:270. 1911. Gilia grandiflora var. axil-
      laris Nels. & Macbr. Contr. Gray Herb. n.s. 56:57. 1918. Collomia grandiflora f. axil-

laris Wherry, Am. Midl. Nat. 31:223. 1944. (Macbride 580, Trinity, Elmore Co. , Ida.)
A form with numerous short branches.

Similar to C. linearis, but averaging more robust and a little broader-leaved, and with
larger, differently colored flowers; plants up to 1 m. tall; calyx lobes up to 4 mm. long at anthesis; corolla mostly salmon or yellowish, 2-3 cm. long, the lobes 5-10 mm. long; filaments
unequally or subequally inserted well below the sinuses, slightly to strongly unequal, the longest one commonly 5-7 mm. long, the shortest sometimes only 1-2 mm. N=8.

In rather dry, open or lightly wooded places, from the lowlands to moderate elevations in
the mountains; both sides of the Cascades, from Wash. and adj. B. C. to w. Mont. and w.
Wyo. , southward to Calif. and Ariz. , wholly w. of the continental divide. May-Aug.

An attractive annual that readily maintains itself in the wild garden.

Collomia heterophylla Hook. Curtis' Bot. Mag. 56: pl. 2895. 1829.
  Courtoisia bipinnatifida Reichenb. Icon. Exot. pl. 208. 1830. Navarretia heterophylla Benth.
  in DC. Prodr. 9:309. 1845. Gilia heterophylla Dougl. ex. Gray, Syn. Fl. 2nd ed. 2¹:408.
  1886. (Cultivated specimens, from seeds collected by Douglas near Ft. Vancouver, Wash.)
  Annual (sometimes biennial?) 0. 5-4 dm. tall, viscid-villous at least above, simple or more
often bushy-branched, often with several stems from the base; leaves nearly all alternate, the
lower more or less pinnatifid or subbipinnatifid, with the blade up to 3. 5 cm. long and 2 cm.
wide; middle and upper leaves progressively less cleft, those subtending the dense, terminal
clusters of flowers entire or merely toothed, shortly or scarcely petiolate, more or less elliptic, smaller than the others; corolla pink or lavender to white, 8-17 mm. long, much longer than the calyx, with slender, only slightly flaring tube and short (3-4 mm. ), somewhat
spreading lobes; filaments unequally inserted well below the sinuses, the lower very short,
the upper sometimes more elongate, up to 1 mm. long or more; locules 2- to 3-seeded. N=8.

Woods, forest openings, and loose stream banks; at lower elevations w. of the Cascade summits from Wash. and adj. s. B. C. to Calif. ; also in n. Ida. June-Aug.

Collomia linearis Nutt. Gen. Pl. 1:126. 1818.
  Collomia parviflora Hook. Curtis' Bot. Mag. 56: pl. 2893. 1829. Gilia linearis Gray, Proc.
  Am. Acad. 17:223. 1882. Navarretia linearis Kuntze, Rev. Gen. 2:433. 1891. (Nuttall,
  "near the banks of the Missouri, about the confluence of the Shian River, and in the vicinity
  of the Arikaree village") Nuttall cited the South American Phlox linearis Cav. (Collomia
  cavanillesii H. & A.) as a questionable synonym of his own Collomia linearis. I concur in
  the customary interpretation that Nuttall's treatment should be regarded as the publication
  of a new species, based on his own specimen, rather than as a transfer of Cavanilles' name.
  Annual, mostly 1-6 dm. tall, finely puberulent or in part subglabrous below, the pubescence
becoming longer and glandular or viscid above; plants simple and unbranched, especially when
smaller, or often with several or many short or elongate axillary branches when more robust,
the main stem and each of the branches terminating in a dense, leafy-bracteate cluster of essentially sessile flowers, that of the main stem the largest; leaves numerous, nearly all alternate, lanceolate or linear, sessile or nearly so, entire, 1-7 cm. long, 1-13 mm. wide, those
subtending the flower clusters often relatively broader than the others; calyx lobes narrowly
triangular, acute, herbaceous, 1. 5-3 mm. long at anthesis, commonly 3-4 mm. in fruit; corolla pink or bluish to white, much longer than the calyx, 8-15 mm. long, with slender tube
and short (1. 5-3 mm. ) lobes; filaments unequally inserted, about 1 mm. long or less; locules
uniovulate. N=8.

In dry or moderately moist, open or lightly shaded places from the lowlands to moderate
elevations in the mountains; the most widespread species of the genus, with a natural range
apparently from B. C. to Calif. , eastward to Ont. , Wis. , Neb. , and N. M. , and also in the
Gaspé region of Que. ; introduced elsewhere in the U. S. and in Alas. ; in our range occurring
chiefly but not wholly e. of the Cascade summits. May-Aug.

Collomia macrocalyx Leiberg ex Brand, Fedde Rep. Sp. Nov. 17:317. 1921. (Leiberg 113,
  near Lone Rock, Gilliam Co. , Oreg. )
  Annual, up to 1 dm. tall, finely puberulent below, more scaberulous and slightly viscid

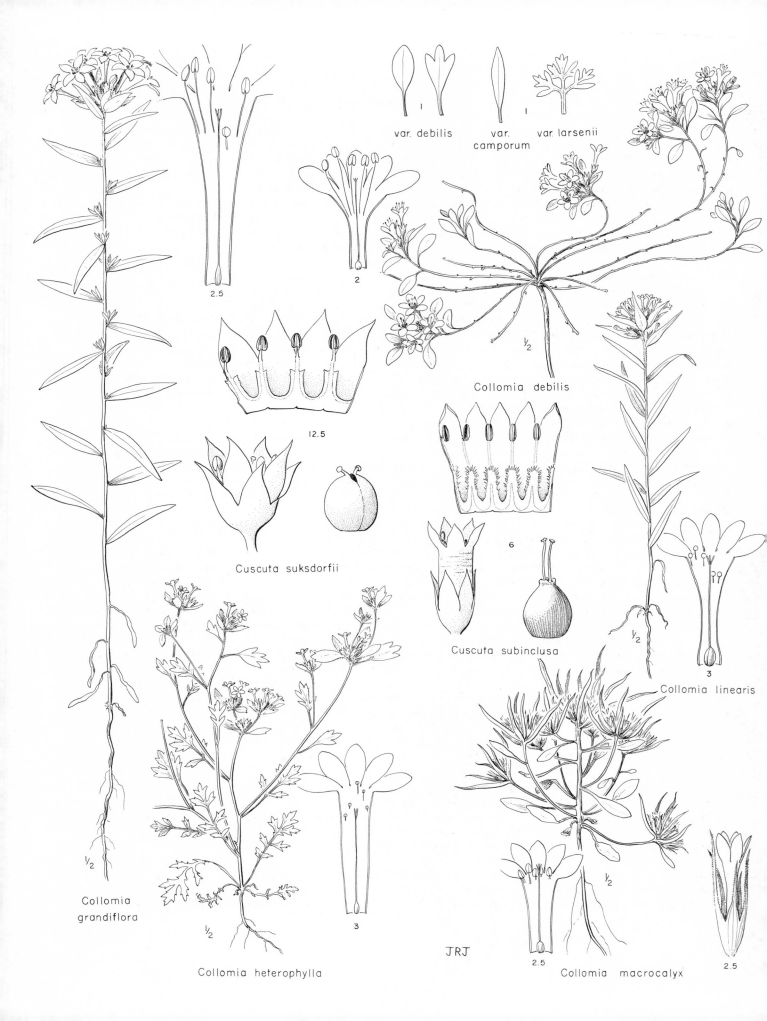

var. debilis

var. camporum

var. larsenii

2.5

2

12.5

Collomia debilis

6

Cuscuta suksdorfii

Cuscuta subinclusa

3

Collomia linearis

Collomia grandiflora

1/2

Collomia heterophylla

3

JRJ

2.5

Collomia macrocalyx

2.5

above, branched when well developed, the main stem and branches terminating in dense, leafy-bracteate flower clusters; leaves entire, those below the inflorescence few and sometimes de-ciduous, the lower petiolate, with small, elliptic blade, sometimes opposite; leaves subtending the flower clusters linear, 1-3 cm. long, 1-2 mm. wide, tapering to a slender point; calyx teeth narrow, firm, aristate-attenuate, unequal, the longer ones 5-11 mm. long in fruit; co-rolla blue, about 1 cm. long, the lobes 2-2.5 mm. long; filaments short, less than 1 mm. long, somewhat unequally inserted a little below the sinuses; locules uniovulate.

Dry, open places at low elevations; local in Gilliam Co., Oreg. May-June.

Collomia tenella Gray, Proc. Am. Acad. 8:259. 1870.
  Gilia leptotes Gray, Proc. Am. Acad. 17:223. 1882. Navarretia leptotes Kuntze, Rev. Gen.
    2:433. 1891. Gilia tenella Nels. & Macbr. Bot. Gaz. 61:34. 1916, not of Benth. in 1849.
    (Watson 900, "Wahsatch Mts. about Parley's Park")
  Slender annual up to 1.5(2) dm. tall, stipitate-glandular at least above, floriferous to near the base, freely branched in such a way that the solitary (or in small part paired) subsessile flowers appear to be borne in the forks or to be axillary and scattered along the branches; leaves linear or nearly so, 1-5 cm. long, 1-5 mm. wide; corolla pale lavender or pinkish to white, 4-6 mm. long, seldom as much as twice as long as the calyx, which is angular beneath the squarrosely subappendiculate sinuses; calyx teeth triangular, acute, about 1 mm. long at anthesis, up to 2 mm. in fruit; filaments equally inserted about 0.5 mm. below the sinuses of the corolla, about 1 mm. long, or some shorter and only 0.5 mm. long; locules uniovulate.

Dry, open places from the plains to moderate elevations in the mountains, often with sage-brush; c. Wash. to c. Ida. and n.w. Wyo., southward to Utah and Nev. June-July.

Collomia tinctoria Kell. Proc. Calif. Acad. Sci. 3:17. 1863.
  Gilia tinctoria Kell. ex Curran, Bull. Calif. Acad. Sci. 1:142. 1885. (Dorr, w. slope of the
    Sierra Nevada)
  Collomia linearis var. subulata Gray, Proc. Am. Acad. 8:259. 1870. Gilia linearis var.
    subulata Gray, Syn. Fl. 2nd ed. 2$^1$:408. 1886. Collomia tinctoria f. subulata Wherry, Am.
    Midl. Nat. 31:225. 1944. (Typification obscure; perhaps nomenclaturally identical with Col-
    lomia tinctoria Kell.)
  Gilia aristella Gray, Syn. Fl. 2nd. ed. 2$^1$:408. 1886. Navarretia aristella Kuntze, Rev. Gen.
    2:433. 1891. Collomia aristella Rydb. Mem. N.Y. Bot. Gard. 1:318. 1900. (Greene, n.
    Calif., is the first of several specimens cited)
  Slender annual up to 1.5(2) dm. tall, strongly stipitate-glandular or glandular-puberulent, floriferous to near the base, nearly simple when depauperate, otherwise freely branched in such a way that many of the (1)2- to 5-flowered clusters of short-pedicellate flowers appear to be borne in the forks; leaves linear or nearly so, 1-5 cm. long, 1-5 mm. wide, the lower of-ten opposite; calyx tube bowed out at the sinuses; calyx teeth narrowly triangular-attenuate, with firm tip, 2-3 mm. long at anthesis, often 4 mm. in fruit; corolla mostly pinkish or laven-der, about twice as long as the calyx, or a little longer, 8-14 mm. long, with slender tube and short (1.5-2.5 mm.) lobes; filaments equally inserted about 0.5 mm. below the sinuses of the corolla, 2 or 3 of them very short, 0.5 mm. long or less, the others 1-2.5 mm. long; locules uniovulate.

Dry, open places from the foothills to moderate or rather high elevations in the mountains, often in talus or on gravelly, unstable slopes; c. Wash. (Kittitas Co.) to Calif., eastward to c. Ida., s.e. Oreg., and w. Nev. June-Aug.

## Eriastrum Woot. & Standl.

Flowers borne in leafy-bracteate, often rather small heads which are more or less strongly tomentose with very fine, tangled hairs; calyx with broad, hyaline intervals between the more herbaceous costae, not much accrescent, commonly ruptured by the developing capsule; calyx lobes entire, scarcely to strongly unequal; corolla funnelform to subsalverform, the lobes of-ten relatively longer than in Navarretia; filaments short or elongate, equally or subequally in-serted at or shortly below the sinuses; anthers mostly narrow and strongly sagittate, generally

over 0. 5 mm. long; ovary 3-carpellary, with 1-several ovules in each locule; seeds usually
becoming mucilaginous when wet; annual (1 species perennial) herbs with alternate, remotely
pinnatifid to linear and entire leaves which (including the segments) are spinulose-tipped, but
less rigid than in Navarretia.

About a dozen species of w. N. Am., especially Calif. Eriastrum is closely allied to Navarretia, and with some justification might be subordinated to that genus, as done by Brand.
(Name from the Greek erion, wool, and aster, star, referring to the woolly plants with star-
like flowers)

Reference:

Mason, H. L. The genus Eriastrum and the influence of Bentham and Gray upon the problem
of generic confusion in the Polemoniaceae. Madroño 8:65-91. 1945.


Eriastrum sparsiflorum (Eastw.) Mason, Madroño 8:86. 1945.

Gilia sparsiflora Eastw. Proc. Calif. Acad. Sci. III. 2:291. June 3, 1902. Navarretia fili-
folia ssp. sparsiflora Brand, Pflanzenr. IV. 250:167. 1907. Gilia filifolia var. sparsiflora
Macbr. Contr. Gray Herb. n. s. 49:57. 1917. Hugelia filifolia var. sparsiflora Jeps. Man.
Fl. Pl. Calif. 792. 1925. (Eastwood, Kings River Canyon, Calif., July, 1899)
ERIASTRUM SPARSIFLORUM var. WILCOXII (A. Nels.) Cronq. hoc loc. Gilia wilcoxii A.
Nels. Bot. Gaz. 34:27. July 16, 1902. Navarretia wilcoxii Brand, Pflanzenr. IV. 250:165.
1907. Welwitschia wilcoxii Rydb. Fl. Rocky Mts. 688. 1917. Eriastrum wilcoxii Mason,
Madroño 8:85. 1945. (Merrill & Wilcox 822, 862, and 952, all in the vicinity of St. Anthony,
Ida.; syntypes)
Gilia, Welwitschia, and Hugelia floccosa, misapplied by authors, incl. Gray; not Gilia floc-
cosa Gray sens. strict., which is technically founded on the Californian Eriastrum luteum
(Benth.) Mason.

Annual up to 3 dm. tall, freely branched when well developed, thinly floccose-tomentose or
(except the more evidently tomentose heads) subglabrous; leaves 1-3 cm. long, narrowly linear and entire, or with 1-3 pairs of lateral segments near or below the middle; heads few-flowered, compact, their bracts generally cleft; calyx lobes more or less strongly unequal; corolla
blue to white, 7-13 mm. long, the lobes 2.5-5 mm.; filaments 0.7-2.0 mm. long, attached
less than 1.5 mm. from the sinuses, often somewhat unequal and not quite equally inserted;
anthers 0.8-1.3 mm. long, narrow, deeply sagittate; ovules 2-4 in each locule, sometimes
not all maturing.

Dry, open, often sandy places in the plains, foothills, and valleys, wholly e. of the Cascades
in our range; c. Wash. to s. Calif., eastward to c. Ida. and w. Utah. May-Aug.

Eriastrum sparsiflorum has sometimes been subordinated to the closely allied E. filifolium
of coastal s. Calif., but the geographic hiatus reinforces the rather weak morphological differences sufficiently to permit the segregation suggested by Mason. The varieties sparsiflorum and wilcoxii, although differing in relatively minor features, have more than once been
reported to occur together without intergradation, and Mason has alleged several differences,
in addition to those here admitted, which he believes warrant specific distinction of these two
taxa. My own observations indicate random rather than correlated variation in the suggested
characters other than flower size and leaf form. Since both of the latter features are also subject to some failure, even in areas well outside the known range of var. sparsiflorum (e. g.,
c. Wash. and c. Ida.), it seems likely that the apparent distinctness of the two taxa at some
points is due to one or another of the phenomena (e. g., polyploidy, self-pollination, apomixis)
which are now well known to inhibit interbreeding between local infraspecific populations in
many other groups. Our plants, as I understand them, may be treated as follows:

1 Leaves all entire, or occasionally some of them with a single pair of lateral lobes; corolla
(boiled) mostly 7-9 mm. long; c. Oreg. to w. Nev. and s. Calif., eastward in the Snake R.
plains to Twin Falls Co., Ida.                                    var. sparsiflorum
1 Leaves, or many of them, with 1-3 pairs of lateral lobes; corolla (boiled) (8)9-12(13) mm.
long; Grant Co., Wash., southward through e. Oreg. (where rarely collected) to s. Calif.,
eastward throughout the Snake R. plains of Ida. (and in the valley of the Salmon R. from
near Challis to near Salmon) and to w. Utah                var. wilcoxii (A. Nels.) Cronq.

## Gilia R. & P.

Flowers borne in various sorts of basically cymose (determinate) inflorescences; calyx slightly or scarcely accrescent, with prominent scarious or hyaline intervals between the more herbaceous costae or segments, commonly ruptured by the developing capsule; corolla campanulate to more often funnelform or salverform, variously colored; filaments generally equal and equally (occasionally unequally) inserted in the corolla tube or at the sinuses; seeds 1-many per locule (or some locules empty), in most species becoming mucilaginous when moistened; taprooted annual to perennial herbs with alternate (or all basal, or the lower opposite), entire or toothed to more often pinnatifid or ternate (or sometimes palmatifid) leaves, without well-defined leaflets.

About 40 or 50 species, as here defined, native to N. and S. Am., chiefly in w. U.S., especially in Calif. The segregate genus Ipomopsis, as recently delimited by Grant, seems to be scarcely definable on a morphological basis, and is here retained in Gilia. Within our range, the species of Gilia are all sharply delimited. (Named for Felipe Luis Gil, Spanish botanist)

References:

Constance, L., and Reed Rollins. A revision of Gilia congesta and its allies. Am. Journ. Bot. 23:433-40. 1936.

Grant, V. Genetic and taxonomic studies in Gilia. I. Gilia capitata. El Aliso 2:239-316. 1950.

-------. VI. Interspecific relationships in the leafy-stemmed Gilias. El Aliso 3:35-49. 1954.

Grant, V., and A. Grant. VII. The woodland Gilias. El Aliso 3:59-91. 1954.

-------. VIII. The cobwebby Gilias. El Aliso 3:203-87. 1956.

Grant, V. A synopsis of Ipomopsis. El Aliso 3:351-62. 1956.

Wherry, E. T. The Gilia aggregata group. Bull Torrey Club 73:194-202. 1946.

1 Corolla relatively very large (the undivided portion 1.5-3.5 cm. long), often partly or wholly bright red; biennial or short-lived perennial     G. AGGREGATA
1 Corolla smaller (the undivided portion not over 1 cm. long), never bright red
   2 Flowers borne in one or more very dense, capitate or spicate-capitate clusters
    3 Plants distinctly perennial; corolla white
     4 Filaments shorter than the anthers; style distinctly shorter than the corolla tube; stems simple and herbaceous throughout     G. SPICATA
     4 Filaments longer than the anthers (commonly 2-4 times as long); style equaling or exserted from the corolla tube; stems basally branched and somewhat woody
                                        G. CONGESTA
    3 Plants annual; corolla bluish             G. CAPITATA
   2 Flowers scattered in a more or less open inflorescence; plants annual
    5 Leaves mostly crowded at or near the base of the stem, the principal ones more or less strongly toothed to pinnatifid
     6 Lower part of the plant with some loose, cottony, more or less deciduous pubescence; seeds well over 1 mm. long, becoming mucilaginous when wet; leaves often more or less strongly pinnatifid; corolla 3-11 mm. long    G. SINUATA
     6 Plant without cottony pubescence; seeds under 1 mm. long, remaining unchanged when wet; leaves toothed or lobed, but scarcely pinnatifid; corolla (2)3-7 mm. long             G. LEPTOMERIA
    5 Leaves well distributed along the stem, mostly or all entire, mostly linear or nearly so
     7 Corolla 1.5-3 mm. long; seeds solitary (reputedly sometimes 2) in each locule, 1-1.5 mm. long             G. TENERRIMA
     7 Corolla (3.5)4-8(9) mm. long
      8 Filaments inserted just beneath the sinuses; seeds mostly 2-5 per locule, 1-1.5 mm. long; foothills and mountains    G. CAPILLARIS
      8 Filaments inserted well below the sinuses; seeds 1 per locule, 2.5-3.5 mm. long; dry lowlands             G. MINUTIFLORA

Gilia aggregata (Pursh) Spreng. Syst. 1:626. 1825.

Cantua aggregata Pursh, Fl. Am. Sept. 147. 1814. Ipomeria aggregata Nutt. Gen. Pl. 1:124.

1818. Batanthes aggregata Raf. Atl. Journ. 1:145. 1832. Collomia aggregata Porter ex
Rothr. Rep. U.S. Geogr. Surv. 6:198. 1878. Ipomopsis aggregata V. Grant, El Aliso 3:
360. 1956. (Lewis, Hungry Creek; this is in Ida., along the Lolo Trail, fide Wherry)
Gilia aggregata var. attenuata Gray, Syn. Fl. 2¹:145. 1878. G. attenuata A. Nels. Bull. Torrey Club 25:278. 1898. Callisteris attenuata Greene, Leafl. 1:160. 1905. Batanthes attenuata Greene, Leafl. 1:224. 1906. Ipomopsis aggregata ssp. attenuata V. Grant, El Aliso 3:
360. 1956. (Parry, Middle Park, Colo.)
Callisteris formosissima Greene, Leafl. 1:160. 1905. Batanthes formosissima Greene,
Leafl. 1:224. 1906. Gilia formosissima Woot. & Standl. Contr. U.S. Nat. Herb. 16:161.
1913. G. aggregata ssp. formosissima Wherry, Bull. Torrey Club 73:198. 1946. (Metcalfe
1318, Black Range, N.M.) = var. aggregata.
Gilia aggregata subvar. helleri Brand, Pflanzenr. IV. 250:115. 1907. G. attenuata f. helleri
Wherry, Bull. Torrey Club 73:198. 1946. (Heller 3253, Lake Waha, Nez Perce Co., Ida.)
= var. attenuata.
G. aggregata f. aurea Macbr. & Pays. Contr. Gray Herb. n. s. 49:64. 1917. (Macbride &
Payson 3082, Martin, Ida.)
G. pulchella Dougl. ex Hook. Fl. Bor. Am. 2:74. 1838. Callisteris pulchella Greene, Leafl.
1:160. 1905. Batanthes pulchella Greene, Leafl. 1:224. 1906. Gilia aggregata f. pulchella
Wherry, Bull. Torrey Club 73:199. 1946. (Douglas, "banks of the Spokan R., near its junction with the Columbia") A white-hairy form of var. aggregata.
Mephitic biennial or short-lived perennial, generally (always?) blooming only once, mostly
2-10 dm. tall, with 1 or several stems, stipitate-glandular at least upward, or the glands
sometimes in large part replaced by long, spreading, crisped, white hairs; leaves seldom approaching 1 dm. long, pinnatifid or subbipinnatifid with narrow rachis and segments, the cauline ones well developed, gradually reduced upward, the basal ones forming a persistent or
deciduous rosette; flowers more or less numerous and rather short-pedicellate in a loosely
thyrsoid to elongate-paniculiform, irregularly (leafy-) bracteate inflorescence of basically cymose nature; corolla showy, with an elongate, gradually or scarcely flaring tube 1.5-3.5 cm.
long, the spreading lobes 6-13 mm. long, acute to attenuate; filaments equally or a little unequally inserted well above the middle of the corolla tube, shortly included to more often shortly exserted; style elongate; locules each with several ovules, but sometimes maturing only 1
or 2 seeds, these 2.5-4 mm. long, becoming mucilaginous when wet. N=7. Scarlet gilia, sky-
rocket.
    Open or lightly wooded, often rocky slopes and banks, and drier meadows, from the low-
lands to rather high elevations in the mountains; s. B.C. to Mont., southward to Calif. and
Mex., wholly e. of the Cascade summits in our range. May-Aug.
    A showy species, suitable for sunny sites in the wild garden, especially e. of the Cascades.
It is monocarpic, maintaining itself in the garden by seed.
    This is a variable but sharply limited and readily recognizable taxon, divided by some bot-
anists into as many as 7 scarcely separable species. Most of our plants have the corolla largely or wholly bright red (typically speckled with whitish) with gradually flaring tube, and represent the widespread var. aggregata. The var. attenuata Gray, with typically white or yellowish
corollas that may be speckled or tinged with red, and with notably slender, scarcely flaring
corolla tube, is sometimes found in our range also. Even in Colo., the region of its greatest
abundance, the var. attenuata hybridizes freely and frequently with var. aggregata, and there
are many mixed colonies which range from one extreme to the other, with numerous and var-
ied intermediate types. Most of the colonies of var. attenuata within our area appear to have
been influenced by interbreeding with the ubiquitous var. aggregata.

Gilia capillaris Kell. Proc. Calif. Acad. 5:46. 1873.
    Navarretia capillaris Kuntze, Rev. Gen. 2:433. 1891. G. leptalea ssp. capillaris Brand,
    Pflanzenr. IV. 250:98. 1907. (Kellogg, Cisco, Placer Co., Calif.)
    Gilia sinister M. E. Jones, Contr. West. Bot. 10:57. 1902. Collomia sinistra Brand, Pflan-
    zenr. IV. 250:54. 1907. (Jones 6458, Middle Valley, Washington Co., Ida., July 7, 1899)
    Openly branched annual up to 3(4) dm. tall, the stem, and sometimes to a lesser extent the
leaves, copiously stipitate-glandular, especially above, the glands often blackish; leaves whol-

ly cauline, all linear and entire, up to 3. 5(5) cm. long and 3(5) mm. wide, some of the lower
often opposite; flowers mostly terminal and solitary, but appearing scattered and leaf-opposed
because of the sympodial branching of the stem; corolla bright pink-lavender to pale bluish or
nearly white, 4-9 mm. long, the undivided portion at least twice as long as the lobes; filaments
short, equal, inserted just beneath the sinuses; seeds 2-5 in each locule, mostly 1-1. 5 mm.
long, becoming mucilaginous when wet.

Meadows and moist to moderately dry, open or lightly wooded slopes in the foothills and at
moderate elevations in the mountains; Cascade-Sierra region from c. Wash. southward, and
in the Ochoco-Wallowa-Blue Mt. region of c. and n. e. Oreg. and s. e. Wash., extending also into
Washington Co., Ida., and reputedly to Bonner and Clark cos.; wholly e. of the Cascade summits in our range. June-July.

The allied, larger-flowered species G. leptalea (Gray) Greene, of Calif. and s. Oreg., has
been reported (Grant, El Aliso 3:69, 88. 1954) from n. Oreg., on the basis of a specimen in
the Marcus E. Jones Herbarium at Pomona College, supposedly collected by C. Davidson at
The Dalles, but it is believed that the data with the specimen are incorrect (a conclusion now
concurred in by Grant).

Gilia capitata Sims, Curtis' Bot. Mag. 53: pl. 2698. 1826. (Cultivated plants, from seeds collected by Douglas near Ft. Vancouver, Wash.)

Erect, mostly rather slender, simple or sparingly branched annual 1. 5-10 dm. tall, the
stem glabrous to stipitate-glandular or obscurely floccose; leaves basal and cauline, gradually
reduced upward, the lower mostly bipinnatifid with slender rachis and narrow ultimate segments which are 1-10(15) mm. long and up to 2 mm. wide; flowers essentially sessile in dense,
naked-pedunculate, essentially bractless, 50- to 100-flowered cymose heads terminating the
stem and branches; corolla light bluish, 6-10 mm. long, the slender lobes commonly about
1(2) mm. wide or less, about equaling the tube; filaments inserted in the sinuses of the corolla,
shorter or longer than the lobes; capsule only tardily dehiscent; seeds about 1. 5(2) mm. long,
becoming mucilaginous when wet, (1)2-3 in each locule, or 1 or even 2 locules sometimes
empty. N=9.

Dry slopes and other open places at lower elevations; s. Vancouver I., Wash., and n. Ida.,
southward to Oreg. and Calif., commoner w. of the Cascade summits. June-July.

A common component of wildflower seed mixtures, maintaining itself in dry, sunny areas.
Only the var. capitata, as described above, occurs in our range. Several other varieties, all
marked by the capitate inflorescence and annual habit, occur in Calif. and s. Oreg.

Gilia congesta Hook. Fl. Bor. Am. 2:75. 1838.
  Ipomopsis congesta V. Grant, El Aliso 3:361. 1956. (Douglas, "sandy plains of Columbia";
  actual locality doubtful)
  Gilia crebrifolia Nutt. Journ. Acad. Phila. II. 1:156. 1848. G. congesta var. crebrifolia
  Gray, Proc. Am. Acad. 8:274. 1870. (Nuttall, Big Sandy Creek of the Colorado of the West)
  G. iberidifolia Benth. Journ. Bot. & Kew Misc. 3:290. 1851. Navarretia iberidifolia Kuntze,
  Rev. Gen. 2:433. 1891. Gilia congesta ssp. iberidifolia Brand, Pflanzenr. IV. 250:121.
  1907. (Geyer 46 in part, presumably from the Blackwater of the Platte, Neb.) Apparently =
  var. congesta.
  G. spergulifolia Rydb. Bull. Torrey Club 31:633. 1904. (Tweedy 576, headwaters of the
  Tongue R., Big Horn Mts., Wyo.) = var. crebrifolia, but the name principally misapplied
  by Rydberg to what is here treated as var. frutescens.
  G. montana Nels. & Kennedy, Proc. Biol. Soc. Wash. 19:37. 1906. G. congesta var. mon-
  tana Const. & Roll. Am. Journ. Bot. 23:439. 1936. Ipomopsis congesta ssp. montana V.
  Grant, El Aliso 3:361. 1956. (Kennedy 1170, summit of Mt. Rose, Nev.)
  GILIA CONGESTA var. PALMIFRONS (Brand) Cronq. hoc loc. G. congesta ssp. palmifrons
  Brand, Pflanzenr. IV. 250:122. 1907. G. palmifrons Rydb. Bull. Torrey Club 40:470. 1913.
  (Specimens are cited from Oreg., Calif., Nev., and Utah; the illustration may have been
  drawn from Heller 7100, Castle Peak, Nevada Co., Calif., at 9000 ft.; a duplicate of the
  Heller specimen at New York closely resembles the illustration)
  G. burleyana A. Nels. Bot. Gaz. 54:144. 1912. G. congesta var. burleyana Const. & Roll.

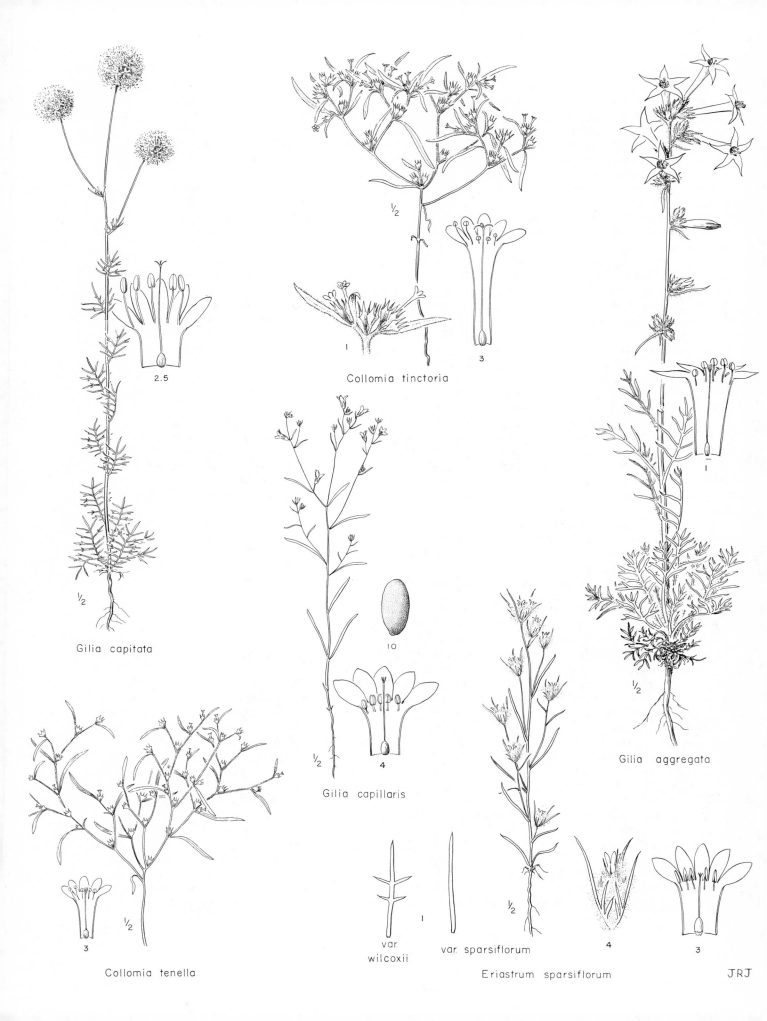

Collomia tinctoria

2.5

Gilia capitata

1/2

Gilia capillaris

10

1/2      4

Collomia tenella

3      1/2

var.
wilcoxii

var. sparsiflorum

Eriastrum sparsiflorum

Gilia aggregata

1/2

4      3

JRJ

Am. Journ. Bot. 23:440. 1936. (Nelson & Macbride 1126, King Hill, Elmore Co., Ida.) =
var. congesta, but misapplied by Constance & Rollins to what is here treated as var. frutescens

GILIA CONGESTA var. FRUTESCENS (Rydb.) Cronq. hoc loc. G. frutescens Rydb. Bull.
Torrey Club 40:471. 1913. Ipomopsis frutescens V. Grant, El Aliso 3:361. 1956. (Jones
5247, Springdale, Washington Co., Utah)

GILIA CONGESTA var. VIRIDIS Cronq. hoc loc. Caulibus supra basin simplicibus, ramis
basalibus sterilibus foliosis plus minusve evolutis, foliis viridibus glabris vel subglabris
plerumque trifidis nonnumquam partim pinnatifidis vel integris. (Hitchcock & Muhlick 8620,
6 miles s. of Lowman, Boise Co., Ida., at 6000 ft., June 1, 1944; N.Y. Bot. Gard.)

Perennial, branched and woody at the base, with several or numerous ascending or erect
stems up to 2(3) dm. tall; herbage more or less arachnoid-tomentose, or in large part glabrous; leaves mostly less than 2 cm. long (in our varieties), trifid, palmatifid, pinnatifid, or
entire, the narrow segments occasionally again cleft; flowers essentially sessile in dense, inconspicuously bracteate, cymose heads terminating the stem and branches; corolla white, the
tube 3-4 mm. long, shortly or scarcely surpassing the calyx, the spreading lobes 2 mm. long;
filaments attached at the sinuses, longer than the 0.4-0.6 mm. (dry) anthers, commonly (1.5)
2-4 times as long; style equaling or commonly a little exserted from the corolla tube; locules
1- to 2-ovulate, 1-seeded, or some empty; seeds 2 mm. long, becoming mucilaginous when
wet.

Dry, open slopes, from the lowlands to high elevations in the mountains; e. and s. Oreg. to
Calif., eastward to w. N.D., w. Neb., and Colo. June-Aug.

A complex species consisting of some 7 varieties, 4 of which occur in our range. It is possible that these varieties might profitably be organized into two or more subspecies, but no
such organization is here attempted. Many of the specimens of var. palmifrons from Calif.
and Nev. approach the dwarf alpine var. montana in having fairly prominent, sterile, leafy
shoots at the base, but within our range it (var. palmifrons) appears properly placed in the
key. Of the 3 extra-limital varieties, var. montana (Nels. & Kennedy) Const. & Roll. occurs
in the Sierra Nevada and s. Cascades (to s. Oreg.); var. frutescens (Rydb.) Cronq., differing
from var. congesta in its mostly entire and often more elongate cauline leaves, and often taller
stems, is largely confined to the Colorado Plateau region of w. Colo. and e. Utah; and var.
pseudotypica Const. & Roll., with the basal and often also the cauline leaves entire and elongate, and approaching smaller forms of G. spicata in habit, occurs in Neb., S.D., and adj.
Wyo. The 4 varieties occurring in our range may be characterized as follows:

1 Leaves green and glabrous or nearly so, generally many of them crowded on short, sterile
     shoots near the base; stems generally simple above the branching base; moderate to high
     altitudes in the mountains
   2 Leaves all, or nearly all, entire; s.w. Mont. to n. and w. Wyo.
                                                           var. crebrifolia (Nutt.) Gray
   2 Leaves mostly trifid, often some pinnatifid or entire; c. Ida. to n.e. Oreg. (Wallowa
     Mts.), and in n.e. Nev.                             var. viridis Cronq.
1 Leaves sparsely to copiously arachnoid-woolly, rarely glabrous; sterile shoots seldom
     much developed; stems often branched above; foothills, valleys, and plains
   3 Many of the leaves palmatifid or subpalmatifid; valley of the Salmon R. in Custer and
     Lemhi cos., Ida.; sand dunes about St. Anthony, Ida.; Great Basin region of Nev., w.
     Utah, and s.e. Oreg., extending into the Sierra Nevada and s. Cascades
                                                           var. palmifrons (Brand) Cronq.
   3 Leaves trifid to pinnatisect, or some of them entire, rarely any of them subpalmatifid;
     widespread to the s. and e. of our range (though rare or absent from most of the range
     of var. palmifrons), extending n. to Baker Co., Oreg., the n. margin of the Snake R.
     plains in Ida., and Beaverhead Co., Mont.          var. congesta


Gilia leptomeria Gray, Proc. Am. Acad. 8:278. 1870. (Watson, mountain valleys of Nev. and
Utah)

GILIA LEPTOMERIA var. MICROMERIA (Gray) Cronq. hoc loc. G. micromeria Gray, Proc.

Am. Acad. 8:279. 1870. G. leptomeria ssp. micromeria Mason & Grant, Madroño 9:214.
1948. (Watson, mountain valleys of Nev. and Utah)

Annual, seldom over 2.5 dm. tall, freely branched above or nearly throughout, the stem
and branches stipitate-glandular above or more often throughout; leaves tufted at the base,
glabrous or sometimes glandular, oblanceolate, up to 6 cm. long and 1.5 cm. wide, rather
shallowly pinnatilobate or remotely and coarsely dentate, the teeth or lobes evidently spinulose-tipped; cauline leaves all, or nearly all, reduced to mere bracts; flowers terminal and
solitary on short, stout, ascending pedicels mostly 0.5-3 mm. long, but appearing scattered
because of the sympodial branching of the stem; corolla commonly pink-lavender with a yellow
throat, or nearly white, mostly 3-7 mm. long, the undivided portion 2-4 times as long as the
often tridentate or mucronate-acuminate lobes; filaments equal, inserted at the sinuses of the
corolla, scarcely longer than the 0.5 mm. anthers; each locule with mostly 8-12 seeds in 2
rows; seeds ovoid, 0.5-0.7 mm. long, remaining unchanged when moistened.

Dry, open places at low elevations, especially in sand or sandy soil; Grant Co., Wash., to
s. Calif., eastward throughout the Snake R. plains of Ida. (and along the Salmon R. from Challis to Salmon), and to Colo. and N.M. May-July.

Most or all of our plants belong to the var. leptomeria, as described above. The var. micromeria (Gray) Cronq., often occurring in more alkaline soil, barely or scarcely enters our
range from the south, but is otherwise found throughout most of the range of the species. It
averages smaller (to 15 cm. high) than var. leptomeria, with more spreading, often longer
pedicels, and has smaller flowers (corolla 2-3.5 mm. long) with fewer seeds (these commonly 6-8 per locule).

Gilia minutiflora Benth. in DC. Prodr. 9:315. 1845.
    Ipomopsis minutiflora V. Grant, El Aliso 3:361. 1956. (Douglas, n.w. Am.)
Subglabrous to densely and finely stipitate-glandular or glandular-scaberulous annual 1-6
dm. tall, commonly with a more or less well-defined central axis and numerous elongate, ascending, monopodial or sympodial, floriferous but otherwise often simple branches; leaves
wholly cauline, gradually reduced upward, some of the lower often trifid with narrow segments,
otherwise all linear and entire, up to 4 cm. long, scarcely over 1 mm. wide; flowers leaf-
opposed, or axillary, or terminating short or more elongate axillary branches; corolla white
or pale bluish, 4-7 mm. long, the undivided portion mostly about 2(1.5) times as long as the
lobes; filaments inserted well below the sinuses, near, or a little above, the middle of the
tube; seeds solitary in each locule, 2.5-3.5 mm. long, becoming mucilaginous when wet.

Dry, sandy places in the plains and valleys, often with sagebrush; c. and e. Wash. and adj.
B.C. and n. Ida., southward to n. Oreg., and eastward throughout the Snake R. plains of Ida.
June-Aug.

Gilia sinuata Dougl. ex Benth. in DC. Prodr. 9:313. 1845.
    G. inconspicua var. sinuata Gray, Bot. Calif. 1:498. 1876. G. inconspicua ssp. sinuata
      Brand, Pflanzenr. IV. 250:105. 1907. G. tenuiflora var. sinuata Jeps. Fl. Calif. 3:179.
      1943. (Douglas, near Okanogan on the Columbia R.)
    G. inconspicua (Smith) Sweet, often misapplied to our plants. The identity of the original ma-
      terial of G. inconspicua is still doubtfull, but it is probably not the same as G. sinuata. See
      Jeps. Fl. Calif. 3:188-90. 1943; Mason, Madroño 6:200-2. 1942; and Grant and Grant, El
      Aliso 3:249-53. 1956.
    G. arenaria ssp. leptantha var. borealis Brand, Pflanzenr. IV. 250:103. 1907. (Howell s.n.,
      Columbia R. near Umatilla, Oreg., May 1, 1882; lectotype by Grant and Grant) = var. si-
      nuata.
    GILIA SINUATA var. TWEEDYI (Rydb.) Cronq. hoc loc. G. tweedyi Rydb. Bull. Torrey Club
      Club 31:634. 1904. G. minutiflora var. tweedyi Brand, Pflanzenr. IV. 250:92. 1907.
      (Tweedy 4422, Encampment, Carbon Co., Wyo.)
    G. modocensis Eastw. Leafl. West. Bot. 2:283. 1940. (Eastwood & Howell 8073, between
      Likely and Jess Valley, Modoc Co., Calif.) A form of var. sinuata with irregularly cleft
      leaves

G. ochroleuca M. E. Jones, sometimes misapplied to our plants, the name properly belonging to a closely allied taxon of s. Calif.

G. ochroleuca ssp. transmontana Mason & Grant, Madroño 9:215. 1948, sometimes misapplied to our plants, the name properly belonging to a closely allied taxon of s. Calif.

Openly branched annual up to 4 dm. tall, stipitate-glandular above or nearly to the base, the
lower part of the plant rather sparingly provided with some loose, cottony, eventually more or
less deciduous pubescence; leaves borne chiefly at or near the base, up to 7 cm. long and 2
cm. wide, typically rather deeply pinnatilobate with entire or few-toothed lobes, varying to
deeply pinnatifid, or scarcely more cleft than in G. leptomeria, but with less strongly spinulose-tipped segments; cauline leaves strongly and progressively reduced, but often better developed than in G. leptomeria, pinnatifid to entire; flowers borne in an open, sparsely (and often irregularly) bracteate, terminal cymose inflorescence which is generally less clearly or
less regularly sympodial than in G. leptomeria; corolla blue or blue-lavender with yellowish
throat, 3-11 mm. long, the undivided portion 2-4 times as long as the lobes; filaments inserted at the sinuses of the corolla, equaling to several times as long as the 0.5 mm. anthers;
capsule in our varieties oblong-ovoid, each locule with 2-8 angular seeds in 2 rows; seeds
about 1.5 mm. or sometimes 2 mm. long, becoming mucilaginous when wet. N=9, 18.

Dry, open, often sandy places in the plains and foothills, e. Wash. to s. e. Calif., e.
through the Snake R. plains (and thence northward to the valley of the Salmon R. in Ida.) and
to Wyo. and N. M.; not yet recorded from Mont. May-July.

A variable species of uncertain limits, but within and near our range sharply defined. The
widespread var. sinuata has the corolla mostly (6)7-11 mm. long, and has mostly 4-8 seeds in
each locule. The var. tweedyi (Rydb.) Cronq., also widespread, but in our range known only
along the n. margin of the Snake R. plains, and n. through the Lost R. region to the valley of
the Salmon R., has smaller flowers, the corolla mostly 3-6 mm. long, the seeds mostly 2-4
in each locule. The var. tweedyi averages more slender and smaller-leaved than the var. sinuata, and often has relatively better developed cauline leaves.

Grant and Grant (El Aliso 3:203-87. 1956) maintain G. tweedyi as a distinct species, and divide what is here treated as G. sinuata var. sinuata into two species: G. sinuata proper, with
less dissected leaves, the cauline ones entire or with the lobes shorter than the width of the
rachis; and G. inconspicua, with more dissected leaves, the cauline ones with the lobes longer than the width of the rachis. G. inconspicua and G. sinuata, as defined by Grant and Grant,
are similar in habitat and geographic distribution, and the morphological distinction is not always clear. Aside from the taxonomic question, I am not convinced by Grant and Grant's ar-
gument for applying the name G. inconspicua to the plant under consideration, especially since
the seeds from which the type of G. inconspicua was grown must have been collected outside
the presently known range of the species to which they attach the name. No other binomial is
available, however, for G. inconspicua in the restricted sense as defined by them. G. inconspicua sensu Grant and Grant is also not sharply separable from G. tweedyi, although here
it is obvious that two different taxa, with quite different range outlines, do exist.

In their published paper Grant and Grant report G. sinuata and G. tweedyi to be tetraploids,
while they report G. inconspicua to be a diploid, and they consider these differences in ploidy
level to be taxonomically important. A. Grant now reports (personal communication) that true
G. inconspicua (sensu Grant and Grant) is actually a tetraploid like G. sinuata and G. tweedyi,
and that some southern diploids previously referred to G. inconspicua should be excluded from
that taxon. The pattern of relationship between chromosome number and morphology which is
beginning to be exposed in this group is suggestive of the pattern which has been demonstrated
in the Artemisia tridentata complex. In this latter group each of several related taxa consists
of morphologically similar diploids and tetraploids. Where two taxa occur together at the
same ploidy level they may hybridize freely, but where they occur together at different ploidy
levels hybrids are much less frequently observed. In such complexes ploidy level has little direct taxonomic significance, and a proper taxonomic treatment must continue to be founded
largely on classical morphologic-geographic methods. The cytogenetic information, of course,
continues to be useful in helping to explain how the existing dynamic equilibrium is maintained.

Gilia spicata Nutt. Journ. Acad. Phila. II. 1:156. 1848.

Ipomopsis spicata V. Grant, El Aliso 3:361. 1956. (Nuttall, Scotts Bluffs, on the Platte R.,
Neb.)

Gilia cephaloidea Rydb. Bull. Torrey Club 24:293. 1897. G. spicata var. cephaloidea Const.
& Roll. Am. Journ. Bot. 23:436. 1936. (Rydberg 2764, Lima, Beaverhead Co., Mont.,
one of the several specimens cited in the original description, is marked in Rydberg's hand
as the type)

GILIA SPICATA var. ORCHIDACEA (Brand) Cronq. hoc loc. G. congesta var. orchidacea
Brand, Ann. Cons. Jard. Bot. Genèv. 15-16:333. 1913. (Blankinship 782, Mt. Baldy, Anaconda, Deerlodge Co., Mont.)

Perennial, the taproot surmounted by a branched caudex or simple crown; stems several or
solitary, herbaceous, erect, simple, 0.5-3 dm. tall; inflorescence and herbage loosely and
often rather thinly arachnoid-tomentose; leaves mostly trifid or pinnatifid with narrow rachis
and segments, or in depauperate individuals sometimes all entire; basal leaves tufted and persistent, 1-6 cm. long; cauline leaves often smaller but still generally well developed; flowers
sessile in a terminal, dense, capitate or more often spicate-capitate, bracteate inflorescence
of eventually cymose origin, the lower glomerules often more or less removed from the main
mass of flowers; corolla persistent, white, the tube 5-9 mm. long, well surpassing the calyx,
the spreading lobes 2.5-5 mm. long; filaments attached at the sinuses of the corolla, shorter
than the 0.6-1.0 mm. (dry) anthers; style short, not reaching much, if at all, beyond the middle of
the corolla tube; locules bearing several ovules but only one (or no) seeds; seeds 2.5-3 mm.
long, becoming mucilaginous when wet.

Dry, open places, from the plains to high elevations in the mountains; c. Ida. to S. D.,
Kans., Colo., and Utah. June-Aug.

Our plants, as described above, belong to a regional phase of the species, occurring in c.
Ida., Mont., and n. Wyo., that has often been treated under a separate binomial, G. cephaloidea Rydb. When this taxon is reduced to a variety of G. spicata, as done by Constance &
Rollins, it must for technical, nomenclatural reasons bear the name G. spicata var. orchidacea (Brand) Cronq., rather than G. spicata var. cephaloidea (Rydb.) Const. & Roll. The var.
orchidacea as thus defined is weakly divisible into two more or less altitudinally segregated
phases; a dwarf, mostly alpine plant up to 1 dm. tall, with mostly capitate inflorescence (represented by the type of var. orchidacea), and a taller plant, of more moderate elevations,
with the inflorescence generally a little more elongate (represented by the type of G. cephaloidea). It is possible that these two phases, together with a third form from the mountains of
Colo., should be recognized as varieties of a single subspecies, which for nomenclatural reasons would have to be founded on the Colorado plant, but the necessary new combinations are
not here proposed. With or without the interpolation of a subspecies, the name G. spicata var.
orchidacea (Brand) Cronq. must be used for at least the smaller plants from our area, and the
name G. spicata var. cephaloidea (Rydb.) Const. & Roll. is available for use by those who
wish to distinguish the more robust element or our regional population.

Gilia tenerrima Gray, Proc. Am. Acad. 8:277. 1870. (Watson, hills above the Bear R., near
Evanston, "Utah")

Diffusely branched annual up to 2 dm. tall, the stem and branches (and sometimes to a lesser extent the leaves) copiously stipitate-glandular nearly or quite to the base, the glands often
dark; leaves all, or nearly all, entire, the lowermost more or less oblanceolate and often
forming a small rosette, the others more linear or linear-oblong, up to 2 cm. long and 4 mm.
wide, progressively reduced upward, becoming small bracts; flowers terminal and solitary,
but appearing scattered and leaf-opposed because of the sympodial branching of the stem, the
pedicels well developed, slender, divergent or retrorse; corolla ochroleucous or faintly lavender, 1.5-3 mm. long, the lobes from about equaling to only half as long as the tube; filaments attached well below the sinuses, a little surpassed by the lobes; seeds solitary (reputedly sometimes 2) in each locule, broadly ellipsoid, 1-1.5 mm. long, becoming mucilaginous
when wet.

Open or lightly wooded slopes, from the foothills to moderate elevations in the mountains;

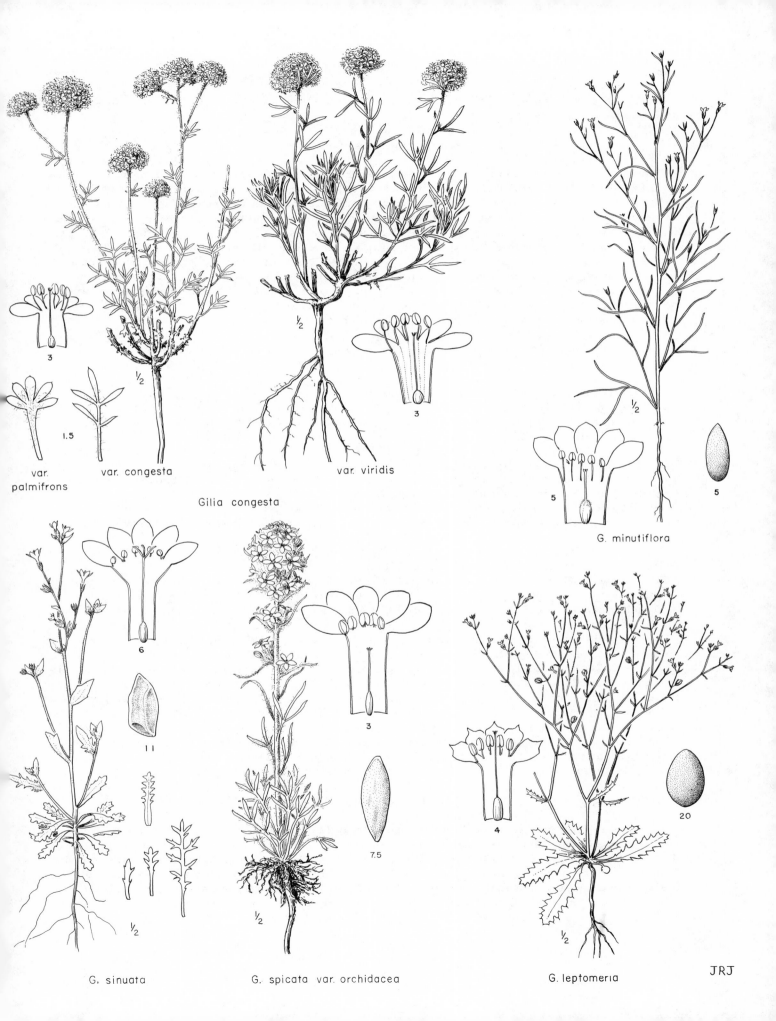

3

var.
palmifrons

1.5

var. congesta

½

var. viridis

3

Gilia congesta

½

5

5

G. minutiflora

6

11

½

G. sinuata

3

7.5

½

G. spicata var. orchidacea

4

20

½

G. leptomeria

JRJ

c. Ida. and s. w. Mont. to Wallowa and Harney cos., Oreg., and to Nev., Utah, and Wyo. June-Aug.

## Gymnosteris Greene

Flowers sessile in a compact, terminal cluster which may be reduced to a single flower; calyx almost wholly scarious except for the commonly more herbaceous teeth, a little inflated at anthesis, generally not ruptured by the growing capsule; corolla white to yellow or pink, salverform or nearly so, with slender tube, very short throat, and more or less spreading lobes; filaments very short, inserted at the corolla throat; seeds several or rather numerous in each locule, becoming mucilaginous when wet; diminutive, essentially glabrous annuals with persistent, connate cotyledons, short naked stem, and a whorl of entire, often basally connate leaves forming an involucre just beneath the flower cluster.

The genus consists of the two following sharply distinct species. (Name from the Greek gymnos, naked, and sterizo, to support, referring to the leafless stem.)

1 Corolla showy, the tube 6-10 mm. long, the lobes mostly 3-6 mm. long
                                                                    G. NUDICAULIS
1 Corolla inconspicuous, the tube 2.5-5 mm. long, the lobes 0.7-1.5 mm. long
                                                                    G. PARVULA

Gymnosteris nudicaulis (H. & A.) Greene, Pitt. 3:304. 1898.
  Collomia nudicaulis H. & A. Bot. Beechey Voy. 368. 1838. Gilia nudicaulis Gray, Proc. Am.
    Acad. 8:266. 1870. Navarretia nudicaulis Kuntze, Rev. Gen. 2:433. 1891. Linanthus nudi-
    caule Howell, Fl. N. W. Am. 456. 1901. (Tolmie, Green R., Snake Country)
  Plants 2-10 (or reputedly 15) cm. tall, simple or with a pair of basal branches; bracteal
leaves linear or lanceolate, 4-15 mm. long; calyx 2.5-5 mm. long; corolla white or yellow,
showy, the tube 6-10 mm. long, the lobes mostly 3-6 mm. long, cuneate, broadest at the mucronate, otherwise mostly subtruncate tip. N=6.
  Dry, open, often sandy places in the plains and foothills; Baker Co., Oreg., westward along
the n. border of the Snake R. plains to Blaine Co., Ida., and southward to Ormsby Co., Nev.
Apr.-May.

Gymnosteris parvula Heller, Muhl. 1:3. 1900.
  Gilia parvula Rydb. Mem. N. Y. Bot. Gard. 1:320. 1900, not of Greene in 1887. Gymnosteris
    nudicaulis var. parvula Jeps. Man. Fl. Pl. Calif. 809. 1925. Gymnosteris rydbergii
    Tidestr. Contr. U.S. Nat. Herb. 25:437. 1925. (Tweedy 823, Swan Lake, Yellowstone Nat.
    Pk.)
  Gymnosteris leibergii Brand, Fedde Rep. Sp. Nov. 17:318. 1921. G. parvula f. leibergii
    Wherry, Am. Midl. Nat. 31:231. 1944. (Leiberg 2144, Dry Creek, Malheur Co., Oreg.)
  Plants 0.7-4 (reputedly 8) cm. tall, simple or occasionally branched at the base, often uniflorous in small individuals; bracteal leaves narrowly lanceolate to ovate, 3-13 mm. long; calyx 2.5-4 mm. long; corolla inconspicuous, white or pinkish, sometimes (? regularly) with a
yellow eye, the tube 2.5-5 mm. long, shortly or not at all exserted, the lobes 0.7-1.5 mm.
long, varying from oblong and rounded or acutish to nearly the shape of those of G. nudicaulis.
  Open, dry to moderately moist slopes, flats, and drier meadows, from the foothills and adjacent plains to moderate or even high elevations in the mountains; Deschutes and Grant cos.,
Oreg., to Yellowstone Nat. Pk., southward to Calif. and Colo. May-July.

## Langloisia Greene

Flowers sessile or nearly so in small, terminal, leafy-bracteate clusters; calyx with equal, bristle-tipped segments, the hyaline intervals narrow and readily rupturing with the growth of the capsule; corolla white to bluish or pink, tubular-funnelform, evidently or (in ours) obscurely bilabiate, the upper lip 3-lobed, the lower 2-lobed; filaments inserted at, or shortly below, the sinuses, equal or unequal, surpassed by to equaling or shortly surpassing the corolla lobes; capsule triangular, with mostly 2-10 seeds per locule, these becoming mucilaginous

when moistened; low, compact, much-branched annuals with alternate, deeply cleft or pinnatifid leaves; leaf segments prominently bristle-tipped, the lower segments sometimes reduced
to a bristle or cluster of bristles.

A technically weak but habitally well-marked group of about 5 species, native to desert regions of w. N. Am. (Named for Rev. Father Auguste Barthélemy Langlois, 1832-1900, ama-
teur botanist of Louisiana. )

Langloisia setosissima (T. & G.) Greene, Pitt. 3:30. 1896.
  Navarretia setosissima T. & G. ex Torr. Ives' Rep. 22. 1860. Gilia setosissima Gray,
    Proc. Am. Acad. 8:271. 1870. Loeselia setosissima Peter in Engl. & Prantl, Nat. Pflan-
    zenf. IV. 3a:54. 1891. (Fremont 414, "most westerly part of N.M., near Virgen R.," ac-
    cording to the label with the apparent type)
  Plants forming dense, low mats or cushions up to 1 or even 2 dm. wide, or simple when
depauperate, evidently villous-puberulent when young, later subglabrate; leaves 1-2.5 cm.
long, up to 1 cm. wide, subcuneate, one or two distal pairs of lateral lobes well developed,
the proximal ones reduced and often represented merely by a prominent bristle or cluster of
bristles; corolla commonly light blue or lavender, nearly regular, 13-20 mm. long, the lobes
shorter than the tube, the limb 7-12 mm. wide; stamens surpassed by the corolla lobes.

Dry, open places in desert and semidesert regions, often in sandy soil; along the Salmon R.
from near Challis to near Salmon, Ida.; e. Malheur Co., Oreg., and Snake R. plains of Ida.,
southward to Baja Calif. and Sonora, Mex. May-July.

Leptodactylon H. & A.

Flowers sessile, solitary in the axils (terminal internodes sometimes foreshortened) or at
the ends of short, leafy branches; calyx scarcely accrescent, the firm costae joined by prominent hyaline intervals, and each shortly exserted into a spinulose tip; corolla white to pink or
salmon, with long, slender tube and well-developed, basally narrowed lobes; filaments about
equally inserted above the middle of the corolla tube, slightly, if at all, longer than the rather
small anthers; seeds several or many in each of the (2)3(4) locules, remaining unchanged when
moistened; taprooted shrubs or subshrubs with small, alternate or opposite, sessile, rigid,
palmatifid or sometimes pinnatifid leaves, these with narrow, spinulose-tipped segments.

About half a dozen species, native to w. U.S. and adjacent Can. and Mex. (Name from the
Greek leptos, thin, fine, and daktylos, finger, referring to the segments of the leaves.)

Leptodactylon pungens (Torr.) Nutt. Journ. Acad. Phila. II. 1:157. 1848.
  Cantua pungens Torr. Ann. Lyc. N.Y. 2:221. 1827. Batanthes pungens Raf. Atl. Journ. 145.
    1832. Aegochloa torreyi G. Don, Gen. Syst. 4:246. 1837. Gilia pungens Benth. in DC.
    Prodr. 9:316. 1845. Navarretia pungens Kuntze, Rev. Gen. 2:433. 1891. (James, "Valley
    of the Loup Fork," Neb. )
  Phlox hookeri Dougl. ex Hook. Fl. Bor. Am. 2:73. 1838. Gilia hookeri Benth. in DC. Prodr.
    9:316. 1845. Leptodactylon hookeri Nutt. Journ. Acad. Phila II. 1:157. 1848. Gilia pungens
    var. hookeri Gray, Proc. Am. Acad. 8:268. 1870. Cantua pungens var. hookeri Howell,
    Fl. N.W. Am. 453. 1901. Leptodactylon pungens var. hookeri Jeps. Man. Fl. Pl. Calif.
    807. 1925. L. pungens ssp. hookeri Wherry, Am. Midl. Nat. 34:383. 1945. (Douglas,
    "near the narrows of the Okanogan and Priest's Rapid of the Columbia")
  Gilia pungens var. squarrosa Gray, Proc. Am. Acad. 8:268. 1870. Cantua pungens var.
    squarrosa Howell, Fl. N.W. Am. 453. 1901. Leptodactylon patens Heller, Muhl. 1:146.
    1906. L. pungens squarrosa Tidestr. Proc. Biol. Soc. Wash. 48:42. 1935. L. pungens
    ssp. squarrosa Wherry, Am. Midl. Nat. 34:385. 1945. (Arid districts of Nev. and Utah,
    collectors Anderson, Watson et al. )
  Gilia lilacina Brand, Pflanzenr. IV. 250:128. 1907. (Baker 1307, Clear Creek Canyon,
    Ormsby Co., Nev. )
  G. pungens ssp. pulchriflora Brand, Ann. Cons. Jard. Bot. Genèv. 15-16:333. 1913. Lep-
    todactylon lilacinum f. pulchriflorum Wherry, Am. Midl. Nat. 34:384. 1945. L. pungens

ssp. pulchriflorum Mason in Abrams, Ill. Fl. Pac. St. 3:455. 1951. (Culbertson, Farewell
  Gap., Tulare Co., Calif.)
Leptodactylon brevifolium Rydb. Bull. Torrey Club 40:474. 1913. L. pungens ssp. brevifo-
  lium Wherry, Am. Midl. Nat. 34:383. 1945. (Purpus 6306, Juniper Range, Utah)
L. hazelae Peck, Proc. Biol. Soc. Wash. 49:111. 1936. L. pungens ssp. hookeri f. hazelae
  Wherry, Am. Midl. Nat. 34:383. 1945. (Barton, Snake River Canyon near the mouth of
  Battle Creek, Wallowa Co., Oreg., May, 1934)

Dense or openly branched shrub 1-6 dm. tall, sweetly aromatic, glabrous to puberulent or
glandular; leaves numerous and often crowded, alternate or the lower opposite, mostly 5-12
mm. long, palmatifid into 3-7 rigid, linear-subulate, spinulose-tipped segments (the central
one the largest) and bearing axillary fascicles; dead leaves usually persistent through one or
more seasons; flowers solitary in the axils (the uppermost internodes sometimes foreshortened); calyx 7-10 mm. long; corolla 1.5-2.5 cm. long, characteristically white, washed or
marked with lavender outside, varying to yellowish or almost salmon, apparently nocturnal,
the rather slender lobes 6-10 mm. long, loosely contorted-closed during the day.

Dry, open, often sandy or rocky places, from the deserts and plains to moderate elevations
in the drier mountains; Wash. (e. of the Cascades) and adjacent s. B.C. to Baja Calif., eastward to Mont., w. Neb., and N.M. May-July.

This species is worth a trial in the arid rock garden e. of the Cascades, but is wholly un-
suited to the coastal climate. Some authors recognize several infraspecific or specific segregates, based on vesture, size of flowers and leaves, and other minor details. At least as
regards our range, such taxa are morphologically confluent and not evidently correlated with
either ecology or geographic distribution, so that they appear essentially artificial and arbitrary.

## Linanthastrum Ewan

Flowers borne in leafy-bracteate, small, cymose, terminal inflorescences which may be
compact and headlike; calyx only slightly or scarcely accrescent, rather firm and subherbaceous nearly throughout, the hyaline intervals narrow and inconspicuous, not reaching the
base, or nearly wanting; corolla salverform, white or creamy; stamens about equal and equally inserted at the base of the short throat; locules with mostly 2-4 ovules, but often maturing
only one seed; valves of the capsule persistent after dehiscence; seeds remaining unchanged
when moistened; taprooted, perennial, woody-based herbs with sessile, opposite, palmatifid
leaves, the segments linear and elongate.

One or two closely allied species of w. U.S. and n. Mex. (Name from Linanthus and the
Latin suffix aster, expressing incomplete resemblance.)

Linanthastrum is intermediate between Linanthus and Leptodactylon, but forms a discordant
element in either genus, and its inclusion in Linanthus breaks down what is otherwise the only
absolute key distinction (duration) between the two larger groups. Recent students of the family are in accord that Linanthastrum is more closely allied to Linanthus than to Leptodactylon,
but the suggestion of affinity provided by the identical odor of Linanthastrum nuttallii and Leptodactylon pungens should not be overlooked. The maintenance of Linanthastrum as a distinct
genus seems less disturbing to the system of genera of the family than its inclusion in either
Linanthus or Leptodactylon.

Linanthastrum nuttallii (Gray) Ewan, Journ. Wash. Acad. Sci. 32:139. 1942.
  Gilia nuttallii Gray, Proc. Am. Acad. 8:267. 1870. Navarretia nuttallii Kuntze, Rev. Gen.
    2:433. 1891. Linanthus nuttallii Greene ex Milliken, U. Calif. Pub. Bot. 2:54. 1904. Lep-
    todactylon nuttallii Rydb. Bull. Torrey Club 33:149. 1906. Siphonella nuttallii Heller, Muhl.
    8:57. 1912. (Nuttall, Rocky Mts.) Siphonella Heller is a later homonym of Siphonella Small.
  Siphonella montana Nutt. ex Gray, Proc. Am. Acad. 8:267. 1870, as a synonym. Gilia nut-
    tallii var. montana Brand, Pflanzenr. IV. 250:125. 1907. (Part of the original material of
    Gilia nuttallii Gray)
  Siphonella parviflora Nutt. ex Gray, Proc. Am. Acad. 8:267. 1870, as a synonym. Gilia nut-
    tallii var. parviflora Brand, Pflanzenr. IV. 250:125. 1907. (Part of the original material of
    Gilia nuttallii Gray)

Sweetly aromatic perennial from a stout, woody taproot and branching caudex, the numerous slender, simple or sparingly branched stems up to 3 dm. tall, puberulent at least above, woody at the base; principal leaves 5- to 9-cleft into spinulose-tipped linear segments up to 2 cm. long, firmer than in Linanthus, softer than in Leptodactylon, each commonly with an axillary fascicle of smaller leaves; flowers subsessile in compact, leafy-bracteate clusters at the ends of the stems; calyx 6-9 mm. long; corolla tube woolly-puberulent, about equaling the calyx; corolla limb about 1 cm. wide; anthers just reaching the orifice of the corolla.

Open or sparsely wooded, often rocky slopes well up in the mountains, sometimes above timber line; Cascade-Sierra region from Wash. to s. Calif., eastward to c. Ida., w. Wyo., Colo., and N.M. June-Aug.

A depressed form occurring near the upper altitudinal limits of the species, resembling some of the compact perennial species of Phlox in habit, may be worthy of varietal segregation, but is thus far nameless. The closely allied, more southern plant, with more open inflorescence, which Ewan has treated as Linanthastrum nuttallii ssp. floribundum (Gray) Ewan, is variously considered to be a poorly defined species or a geographical race of L. nuttallii.

## Linanthus Benth.

Flowers borne in open, commonly dichasial inflorescences, or in compact, headlike, terminal clusters; calyx only slightly or scarcely accrescent, usually with more or less evident hyaline intervals between the more herbaceous costae or segments, the membrane in a few species inconspicuous and not reaching the base of the calyx, or almost wanting; corolla campanulate to salverform, variously colored; stamens mostly equal or nearly so and about equally inserted in the throat or tube; valves of the capsule persistent after dehiscence; seeds 1-several per locule, sometimes becoming mucilaginous when moistened, or emitting spiracles; taprooted annual herbs with opposite (or the uppermost alternate), sessile, palmatifid leaves with mostly linear and elongate segments, or the uppermost (rarely all) leaves simple and undivided.

About three dozen species, native to w. N. Am. and to Chile, best developed in Calif. (Name from the Greek linon, flax, and anthos, flower.)

1 Flowers clustered, subsessile; corolla 1-3 cm. long, salverform, with very narrow, exserted tube and abruptly spreading limb; chiefly w. of the Cascades     L. BICOLOR
1 Flowers evidently pedicellate in an open inflorescence; corolla up to about 1 cm. long, and (except generally in L. bakeri) more funnelform or campanulate, with the tube scarcely or not at all exserted from the calyx; Cascades and eastward
  2 Seeds solitary in each locule; corolla 1.5-2.5 mm. long, less than 1.5 times as long as the calyx, glabrous, as also the filaments; habitat similar to L. septentrionalis; Cascades, eastward occasionally to c. Ida.     L. HARKNESSII
  2 Seeds 2-8 in each locule; corolla 2.5-10 mm. long, about 1.5 times as long as the calyx, or longer, and usually either with an internal ring of hairs or with the filaments hairy at the base
    3 Corolla lobes only about half as long as the tube; filaments only 1-2 times as long as the anthers; ring of hairs in the corolla tube borne well below the level of insertion of the glabrous filaments; rare species     L. BAKERI
    3 Corolla lobes about equaling or more often distinctly longer than the tube; filaments several or many times as long as the anthers; common species
      4 Corolla 2.5-5(6) mm. long, mostly less than 2 (commonly about 1.5) times as long as the calyx; corolla tube hairy within about at the level of insertion of the glabrous or subglabrous filaments, or occasionally subglabrous; widespread e. of the Cascades, mostly in the foothills, mountains, and high plains

                                                             L. SEPTENTRIONALIS
      4 Corolla (5)6-10 mm. long, mostly 2-3.5 times as long as the calyx; filaments basally hairy, the corolla tube generally glabrous or nearly so; Wash. (e. of the Cascades) and adjacent n. Ida., s. through e. Oreg. to Calif., mostly in the dry lowlands, occasionally extending into the ponderosa pine zone

                                                             L. PHARNACEOIDES

Linanthus bakeri Mason, Madroño 9:250. 1948. (Mason 7015, Pilot Hill, Eldorado Co., Calif.)

Slender annual up to 2.5 dm. tall, stipitate-glandular on the pedicels and sometimes also beneath the nodes, otherwise inconspicuously strigose-puberulent or subglabrous; leaves 3- to 7-parted, less than 1 cm. long; pedicels slender and elongate; calyx 3.5-5 mm. long, the tube longer than the teeth, the herbaceous portions wider than the connecting membranes and somewhat 3-nerved; corolla white to pink or violet, sometimes bicolored, the slender tube distinctly (rarely scarcely) exserted from the calyx, and usually with a ring of hairs near or below the middle; corolla lobes only about half as long as the tube; filaments attached at or just below the sinuses, only 1-2 times as long as the anthers; seeds several in each locule.

Dry, open places at lower elevations, often on serpentine; Sierra Nevada and n. Coast Ranges of Calif., and n. (according to Mason) near the e. base of the Cascades to Klickitat Co., Wash. Spring.

Linanthus bicolor (Nutt.) Greene, Pitt. 2:260. 1892.

Leptosiphon bicolor Nutt. Journ. Acad. Phila. II. 1:156. 1848. Gilia bicolor Piper, Contr. U.S. Nat. Herb. 11:460. 1906. (Nuttall, on the Oregon near the outlet of the Wahlamet) LINANTHUS BICOLOR var. MINIMUS (Mason) Cronq. hoc loc. L. bicolor ssp. minimus Mason, Madroño 9:250. 1948. (Roush, Gages Point, Skagit Co., Wash., May 8, 1927)

Slender annual up to 1.5 dm. tall; stem puberulent, often branched at the base, otherwise usually simple or nearly so; leaves rather firm, 3- to 7-cleft, up to nearly 2 cm. long, the segments harshly ciliate, linear, or the middle one sometimes wider; flowers subsessile in a dense, terminal, leafy cluster, the leaves of the inflorescence longer than those below and more prominently ciliate; flowers fragrant; calyx firm, 5-10 mm. long, with narrow, acerose, somewhat 3-nerved segments, the intervening membranes small and inconspicuous, not extending to the base; corolla usually prominently bicolored, with very slender, long-exserted tube up to 3 cm. long, short, abruptly flaring, yellow throat, and spreading, deep pink to purplish or white lobes 2-4 mm. long; filaments attached in the corolla throat, shortly exserted from the orifice; seeds several in each locule.

Open places at lower altitudes w. of the Cascades, and extending through the Columbia gorge to Klickitat Co., Wash.; s. Vancouver I. to s. Calif. Apr.-June.

The var. minimus (Mason) Cronq., a coastal ecotype extending from s. Vancouver I. (and about Puget Sound) to n. Calif., has relatively small, often scarcely bicolored corollas with the tube only 8-14 mm. long. The more common var. bicolor occurs chiefly inland (and about Puget Sound), and has larger, characteristically bicolored corollas with the tube 13-30 mm. long.

Linanthus harknessii (Curran) Greene, Pitt. 2:255. 1892.

Gilia harknessii Curran, Bull. Calif. Acad. Sci. 1:12. 1884. Navarretia harknessii Kuntze, Rev. Gen. 2:433. 1891. Gilia pharnaceoides var. harknessii M. E. Jones, Contr. West. Bot. 12:55. 1908. (Harkness, Donner Pass, Sierra Nevada, Calif. Aug., 1883)

Slender annual up to 2.5 dm. tall, freely branched above (or throughout) when well developed, glabrous or partly puberulent; leaves 3- to 7-parted into soft, linear segments up to 1.5 cm. long, or the uppermost entire; flowers long-pedicellate, borne in the forks and terminating the ultimate branches; calyx 2-3 mm. long at maturity, nearly or quite equaling the capsule, the prominent hyaline intervals extending nearly to the base between the faintly or scarcely 3-nerved, greener segments; corolla white or reputedly sometimes pale blue, glabrous, 1.5-2.5 mm. long, equaling or slightly surpassing the calyx, cleft about halfway to the base, the limb up to 2.5 mm. wide; filaments glabrous; seeds elongate, each filling a locule.

Forest openings, dry meadows, and open slopes and ridges, mostly in the foothills or at moderate elevations in the mountains; Cascade-Sierra region from Kittitas Co., Wash., to Calif. (more common southward), eastward occasionally (chiefly in the less arid areas) to c. Ida. and n.e. Nev. June-Aug.

Linanthus pharnaceoides (Benth.) Greene, Pitt. 2:254. 1892.

Gilia pharnaceoides Benth. Bot. Reg. 19: sub pl. 1622. 1833. G. liniflora var. pharnaceoides

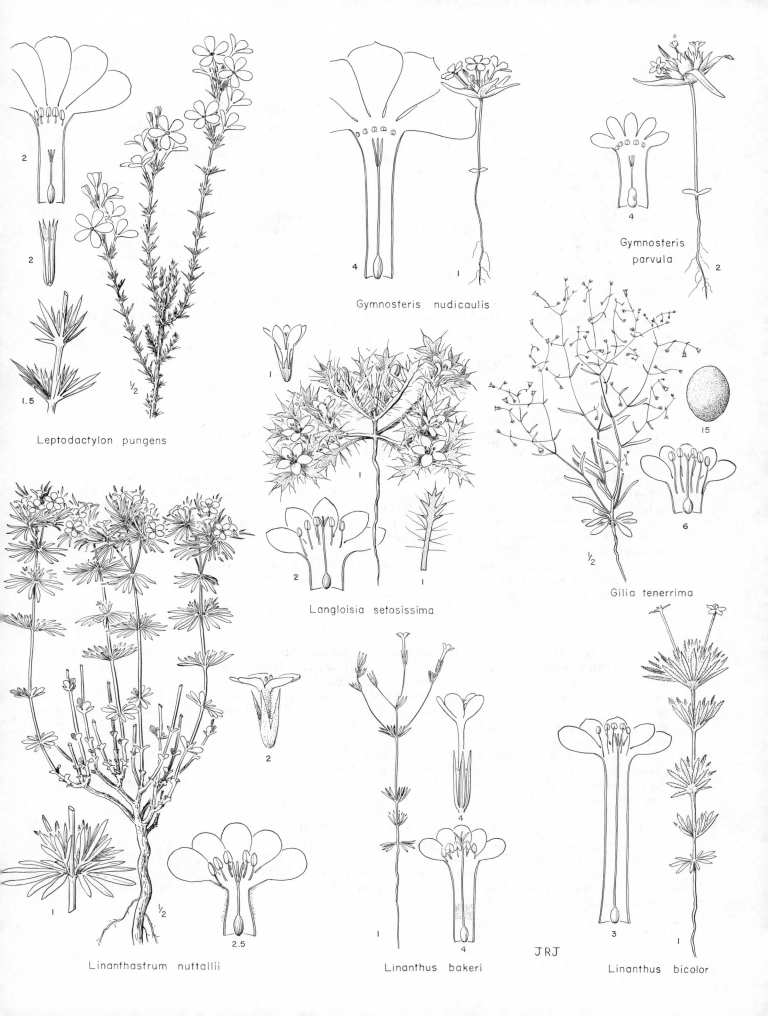

Leptodactylon pungens

Gymnosteris nudicaulis

Gymnosteris parvula

Langloisia setosissima

Gilia tenerrima

Linanthastrum nuttallii

Linanthus bakeri

JRJ

Linanthus bicolor

Gray, Proc. Am. Acad. 8:263. 1870. Linanthus liniflorus ssp. pharnaceoides Mason in
Abrams, Ill. Fl. Pac. St. 3:418. 1951. (Douglas, Calif.)

Glabrous to scabrous-puberulent annual, 1-3 dm. tall, freely (sometimes diffusely) branched
when well developed; leaves 3- to 9-cleft into linear segments (0.5)1-2 cm. long; or the
uppermost entire; flowers numerous and long-pedicellate, but the branching often less regular
than in L. septentrionalis and L. harknessii; calyx 2-3 mm. long in flower, up to 5 mm. in
fruit, the prominent hyaline intervals extending essentially to the base between the more or
less 3-nerved, more herbaceous segments; corolla commonly pale blue or bluish with a yellow
throat, 2-3.5 times as long as the calyx, (5)6-10 mm. long, funnelform, often broadly so, the
limb 5-12 mm. wide, the lobes mostly 2-3 times as long as the short tube; filaments generally
hairy at the base, the corolla glabrous or nearly so; seeds 2-several in each locule.

Dry, open places in the lowlands, often among sagebrush, sometimes extending into the ponderosa pine zone in e. Wash. and adj. Ida.; Wash. (e. of the Cascades) and adj. n. Ida.,
southward through e. Oreg. to Calif. May-July.

Linanthus pharnaceoides is not always sharply distinct, at least in the herbarium, from the
usually much larger-flowered L. liniflorus (Benth.) Greene, of the California coast ranges
from San Francisco southward, and monographic study may well force the union of the two under the binomial L. liniflorus. Meanwhile, it may be pointed out that the now generally accepted California species L. filipes (Benth.) Greene is scarcely more sharply limited from L.
pharnaceoides, at least in the herbarium, than the latter is from L. liniflorus. Each of the
three taxa has its own characteristic aspect.

Linanthus septentrionalis Mason, Madroño 4:159. 1938.
   L. harknessii var. septentrionalis Jeps. & Bailey in Jeps. Fl. Calif. 3:210. 1943. Gilia sep-
      tentrionalis St. John, Fl. S.E. Wash. 2nd. ed. 550. 1956. (Mason 3497, Tower Jct., Yel-
      lowstone Nat Pk., Wyo.)

Similar to L. harknessii, and often confused with it, but apparently sharply distinct; corolla
2.5-5(6) mm. long, distinctly surpassing the calyx, commonly about 1.5 (and fairly consistently less than 2) times as long, white to light blue or lavender, the limb up to 5 mm. wide, the
lobes equaling or more often distinctly surpassing the tube, which usually bears a ring of short hairs
within at about the level of insertion of the glabrous or subglabrous filaments; seeds 2-8 in
each locule.

Forest openings, dry meadows, and other open places, mostly in the foothills, high plains,
or at moderate elevations in the mountains, occasionally in the drier lowlands, wholly e. of
the Cascades; s. B.C. to Calif., e. to Alta., Wyo. and Colo. May-July.


Microsteris Greene


Flowers terminal and paired (or solitary), but seemingly scattered or loosely aggregated because of the branching of the stems; calyx somewhat accrescent, with prominent hyaline intervals between the more herbaceous costae or segments, ruptured by the developing capsule; corolla salverform, with slender tube and short, spreading lobes; filaments short, very unequally inserted; valves of the capsule disarticulating completely on dehiscence; seeds one per locule, becoming mucilaginous when moistened; annual herbs with entire leaves, the lower opposite, the upper alternate.

A single, highly variable species, evidently allied to Phlox, and sometimes forced into that
genus as a discordant element. Resembling some species of Collomia in aspect, but readily
distinguished by the calyx. (Name from the Greek mikros, small, and sterizo, to support, presumably referring to the small size of the plant.)

Microsteris gracilis (Hook.) Greene, Pitt. 3:300. 1898.
   Gilia gracilis Hook. Curtis' Bot. Mag. 56: pl. 2924. 1829. Collomia gracilis Dougl. ex Hook. loc.
      cit. as a synonym; ex Benth. Bot. Reg. 19: sub. pl. 1622. 1833. Phlox gracilis Greene, Pitt.
      1:141. 1887. Navarretia gracilis Kuntze, Rev. Gen. 2:433. 1891. Polemonium morenonis
      Kuntze, Rev. Gen. 3³:203. 1898. (Cultivated plants, from seeds collected by Douglas near
      "the banks of the Spoken river, and on high grounds near Flathead river")

MICROSTERIS GRACILIS var. HUMILIOR (Hook.) Cronq. hoc loc. Collomia gracilis var. humilior Hook. Fl. Bor. Am. 2:76. 1838. C. humilis Dougl. ex Hook. loc. cit. as a synonym. Gilia gracilis ssp. humilis Brand, Pflanzenr. IV. 250:91. 1907. (Douglas, "Fort Vancouver, and at the Multnomack") Here accepted in the customary sense, but the original description is inadequate, and the plant to which the name has been applied is not otherwise known w. of the Cascades in Wash. and Oreg.

Collomia micrantha Kell. Proc. Calif. Acad. 3:18. 1863. Microsteris micrantha Greene, Pitt. 3:303. 1898. Gilia gracilis ssp. humilis var. micrantha Brand, Pflanzenr. IV. 250: 91. 1907. (Dunn, near Silver City, Nev.) Name customarily misapplied to what is here treated as M. gracilis var. humilior, but the original description and accompanying illustration are of var. gracilis.

Microsteris humilis Greene, Pitt. 3:301. 1898. Gilia microsteris Piper in Piper & Beattie, Fl. Palouse Region 142. 1901. Phlox gracilis ssp. humilis Mason in Abrams, Ill. Fl. Pac. St. 3:413. 1951. (Typification obscure; Collomia humilis Dougl. is cited as a probable synonym; name clearly intended to apply to what is here treated as M. gracilis var. humilior)

Microsteris glabella Greene, Pitt. 3:301. 1898. Gilia gracilis var. glabella Suksd. Deuts. Bot. Monats. 18:132. 1900. (Suksdorf 2206, Falcon Valley, Klickitat Co., Wash.) = var. humilior.

Microsteris stricta Greene, Pitt. 3:302. 1898. Gilia gracilis var. stricta Brand, Pflanzenr. IV. 250:89. 1907. (Baker & Nutting s.n., Burney Valley, Shasta Co., Calif., May 25, 1894; lectotype by Wherry) = var. gracilis.

Microsteris diffusa Heller, Bull. Torrey Club 26:313. 1899. (Heller 3098, mouth of the Potlatch R., Nez Perce Co., Ida.) = var. humilior.

Gilia gracilis var. elatior Suksd. Deuts. Bot. Monats. 18:132. 1900. (Suksdorf 2114, w. Klickitat Co., Wash.) = var. gracilis.

G. gracilis var. pratensis Suksd. Deuts. Bot. Monats. 18:132. 1900. (Suksdorf 1508, Falcon Valley, Klickitat Co., Wash.) = var. gracilis.

G. gracilis ssp. humilis var. micrantha subvar. angustifolia Brand, Pflanzenr. IV. 250:91. 1907. (Numerous specimens from N. Am. are cited) = var. humilior.

G. gracilis ssp. humilis var. micrantha subvar. bellidifolia Brand, Pflanzenr. IV. 250:91. 1907. Collomia bellidifolia Dougl. ex Hook. Fl. Bor. Am. 2:76. 1838, as a synonym. (Douglas, locality not clearly given) = var. humilior.

Gilia gracilis ssp. spirillifera Brand, Pflanzenr. IV. 250:92. 1907. (Synthetic; no type indicated)

G. gracilis ssp. spirillifera var. euphorbioides Brand, Pflanzenr. IV. 250:92. 1907. (Elmer 143, Whitman Co., Wash.) = var. humilior.

G. longisepala Gand. Bull. Soc. Bot. France 65:59. 1918. (Suksdorf 2314, Skamania Co., and 5144, Bingen, Wash., are cited in that order) = var. humilior.

Subsimple or much-branched plants up to 3 dm. tall, puberulent or glandular-puberulent at least above; leaves linear or lance-linear to elliptic, or the lower obovate, up to about 5 cm. long and 8 mm. wide; flowers mostly in pairs at the ends of the stem and branches, one subsessile, the other evidently pedicellate, or sometimes borne singly; corolla 5-15 mm. long, with white or yellowish tube and pink to lavender limb. N=7.

Dry to moderately moist, open places, mostly in the foothills and lowlands, the var. gracilis sometimes in meadows or along streams; s. B.C. to Mont., southward to Baja Calif. and N.M.; Bolivia, Argentina, and Chile. Mar.-June.

The species is divisible into two strongly marked but intergradient varieties:

1 Primary stem (up to the first flower) (5)8-25 cm. high; plants branched only above the middle, or sometimes essentially throughout, but in any case higher than broad; corolla (8)9-15 mm. long, the lobes (1.5)2-4 mm.; chiefly w. of the Cascades, from s. B.C. to Calif., also eastward, in less arid regions or habitats than var. humilior, to c. and n.e. Oreg., n. Ida., w. Mont., and Yellowstone Nat. Pk., Wyo. var. gracilis

1 Primary stem 1-5(8) cm. high; plants at maturity much branched, commonly as broad as high, or broader; corolla 5-8(10) mm. long, the lobes 1-2(2.5) mm.; arid regions from e. Wash. and s. B.C. to Baja Calif., eastward to w. Mont., Wyo., w. Neb., and N.M.,

wholly e. of the Cascade summits in our range; also in S. Am.

var. humilior (Hook.) Cronq.

## Navarretia R. & P.

Flowers borne in dense, leafy-bracteate, glandular or glabrous to coarsely villous heads terminating the stem and branches; calyx with broad hyaline intervals between the more herbaceous costae, not much accrescent, sometimes ruptured by the developing capsule; calyx lobes more or less unequal, rarely essentially equal, commonly spine-tipped (like the bracts and their segments), the larger ones often trifurcate; corolla salverform or funnelform, with slender tube and mostly short lobes; filaments elongate or very short, equally, or a little unequally, inserted, usually not far below the sinuses of the corolla; anthers small, elliptic, commonly less than 0.5 mm. long; stigmas 2 or 3, sometimes almost wholly connate; capsule (1)2- to 3-locular, the partitions sometimes imperfect, dehiscent or indehiscent; seeds solitary to rather numerous in each locule, becoming mucilaginous when moistened; prickly, annual herbs with alternate (or the lower opposite), mostly pinnatifid or more often irregularly bipinnatifid leaves (or leaves all entire in depauperate individuals), the segments narrow and spine-tipped.

About 30 species, mostly of w. N. Am., especially Calif.; one species in Chile and Argentina. (Named for Fr. Ferdinand Navarrete, Spanish physician.)

Reference:

Crampton, Beecher. Morphological and ecological considerations in the classification of Navarretia (Polemoniaceae). Madroño 12:225-38. 1954.

The floral measurements given hereunder are taken from boiled flowers, in which the veins of the corolla, here used (following Crampton) as a major key character, are readily visible at 10x magnification against a black background.

1 Petal traces trifurcate well down in the corolla tube (commonly below the middle), so that each corolla lobe receives 3 veins; stigmas 3 (2 of them often partly united in N. divaricata); capsules 3-locular (or 2-locular by failure of one partition in N. divaricata, but still with 3 valves), and (except in N. tagetina) regularly dehiscent by 3 valves
  2 Filaments inserted at least 2.5 mm. below the sinuses; terminal segment of the leaves not elongate; capsule generally with (6)8-9 or more seeds per carpel (unique among our species in all of the foregoing characters); corolla 9-12 mm. long, deep to pale blue; w. of the Cascades                                                 N. SQUARROSA
  2 Filaments inserted within 1.5 mm. of the sinuses; terminal segment of the leaves notably elongate; capsule with mostly 1-4 seeds per carpel
    3 Corolla yellow (unique among our species in this respect), 5-8 mm. long; bracts distinctly pinnatifid; widespread e. of the Cascades                        N. BREWERI
    3 Corolla pinkish or bluish to white
      4 Corolla 3.5-5 mm. long; filaments shorter than the anthers, up to about 0.3 mm. long; branches tending to arise just beneath or even within the heads (unique among our species in the latter two characters); bracts more or less palmatifid; widespread e. of the Cascades                                                  N. DIVARICATA
      4 Corolla 8-11 mm. long; filaments much longer than the anthers, commonly 1.5-2 mm. long; branches usually not closely associated with the heads; bracts distinctly pinnatifid; Klickitat Co., Wash., southward to Calif.    N. TAGETINA
1 Petal traces simple up to about the level of the sinuses, or beyond, so that each corolla lobe receives only one vein; stigmas 2, often almost wholly united, and capsules more or less 2-locular (rarely stigmas 3 and capsules perhaps 3-locular); capsules indehiscent, or tardily and irregularly dehiscent by disintegration of the lower part of the lateral walls
  5 Corolla lobes only slightly or scarcely (to 1.5 times) longer than wide, with branched midvein; ovules (2)3-6 per carpel; leaves and bracts firm; corolla 4-11 mm. long; widespread                                                                        N. INTERTEXTA
  5 Corolla lobes narrow, about twice as long as wide, with unbranched midvein; ovules 1 or less often 2 per carpel; leaves and bracts relatively soft; corolla 4-6 mm. long; e. of the Cascades                                                                    N. MINIMA

Navarretia breweri (Gray) Greene, Pitt. 1:137. 1887.

Gilia breweri Gray, Proc. Am. Acad. 8:269. 1870. (Brewer, "Sierra Nevada at Ebbett's and Amador Pass")

Annual up to about 1 dm. tall, freely and widely branched when well developed, finely and rather densely glandular-puberulent or stipitate-glandular; leaves up to 2.5 cm. long, firm and prickly, pinnatifid, the terminal segment distinctly longer than any of the lateral ones; bracts pinnatifid, scarcely different from the proper leaves; calyx 6-9.5 mm. long, the segments entire, somewhat unequal; corolla yellow, from a little shorter to a little longer than the calyx, 5-8 mm. long, the lobes 0.8-1.3 mm.; petal traces trifurcate well down in the corolla tube, so that 3 veins enter each corolla lobe; filaments about equal, 1.5-2 mm. long, subequally inserted 1-1.3 mm. below the sinuses; stigmas 3; capsule 3-locular, dehiscent, usually from the base upward; ovules and seeds 1-2 (reputedly sometimes 3) in each locule.

Dry meadows and other dry, open places, often in shallow clay soil, from the foothills and high plains to moderate altitudes in the mountains; Douglas Co., Wash., to c. Ida. and w. Wyo., southward to Calif. and Ariz. May-July.

Navarretia divaricata (Torr.) Greene, Pitt. 1:136. 1887.

Gilia divaricata Torr. ex Gray, Proc. Am. Acad. 8:270. 1870. (Shelton, Sierra Nevada, Calif.; the first specimen cited by Gray, and the only one annotated by Torrey as "n. sp.")

Low, slender annual, viscidly villous-puberulent at least in the inflorescences, often also finely glandular; primary head seldom more than 5 cm. above the ground; divaricate, subnaked, lateral branches commonly arising just beneath, or even within, the head and again producing terminal heads which may have associated branches, so that mature plants tend to be broader than high; proper leaves few, seldom 2.5 cm. long, pinnatifid with narrow rachis and segments, the lateral segments few and near the base (or wholly wanting in depauperate plants), the terminal one elongate; bracts foliaceous with chartaceous base, palmatifid or subpalmatifid; calyx 5-10 mm. long, generally surpassing the corolla, 2 or 3 of the lobes evidently longer than the others; corolla white to pale pink or lavender, 3.5-5 mm. long, the lobes less than 1 mm. long; petal traces trifurcate well down in the corolla tube, so that each lobe receives 3 veins; filaments very short, up to about 0.3 mm. long, subequally inserted less than 1 mm. below the sinuses; stigmas 3, often 2 of them partly connate; ovules and seeds mostly 6-12(2-4 per carpel); capsule regularly dehiscent (usually from the base upward) by 3 valves, but commonly only 2-locular by complete failure of one of the partitions, the odd valve narrower than the others.

Dry, open slopes, from the foothills to moderate elevations in the mountains; Klickitat Co., Wash., southward through e. Oreg. to Calif., eastward to n.w. Mont. and c. Ida. June-Aug.

Navarretia intertexta (Benth.) Hook. Fl. Bor. Am. 2:75. 1838.

Aegochloa intertexta Benth. Bot. Reg. 19: sub pl. 1622. 1833. Gilia intertexta Steud. Nom. 2nd ed. 1:683. 1840. (Douglas, Calif. and N.W. Am.)

Navarretia propinqua Suksd. Allg. Bot. Zeit. 12:26. 1906. N. intertexta var. propinqua Brand, Pflanzenr. IV. 250:163. 1907. (Suksdorf cites several of his own collections, the first from Spokane Co., Wash., in 1889)

N. pilosifaucis St. John & Weitman, Proc. Biol. Soc. Wash. 41:196. 1928. (Weitman 312, Rock Lake, Whitman Co., Wash.) = var. propinqua.

Simple to freely branched annual up to 2.5 dm. tall, puberulent (at least the stem), becoming more or less villous in the inflorescences; leaves up to 3 cm. long, firm, pinnatifid or bipinnatifid with narrow rachis and segments, the terminal segment generally elongate, or the lower leaves sometimes entire; bracts with broader, firmer rachis than the leaves; calyx white-hairy at the mouth within, the lobes more or less unequal, the larger ones often trifid; corolla white to pale lavender or pale bluish, 4-11 mm. long, the lobes mostly less than 2 mm. long, 1-1.5 times as long as wide; petal traces with a pair (rarely only one) of lateral branches near the level of the sinuses; filaments inserted within 1.5(2) mm. of the sinuses; stigmas 2 (very rarely 3), evident; ovary 2-locular, the partition sometimes imperfect, with (2)3-5(6) ovules and seeds per locule; capsule indehiscent, or tardily dehiscent by disintegration of the lower lateral walls.

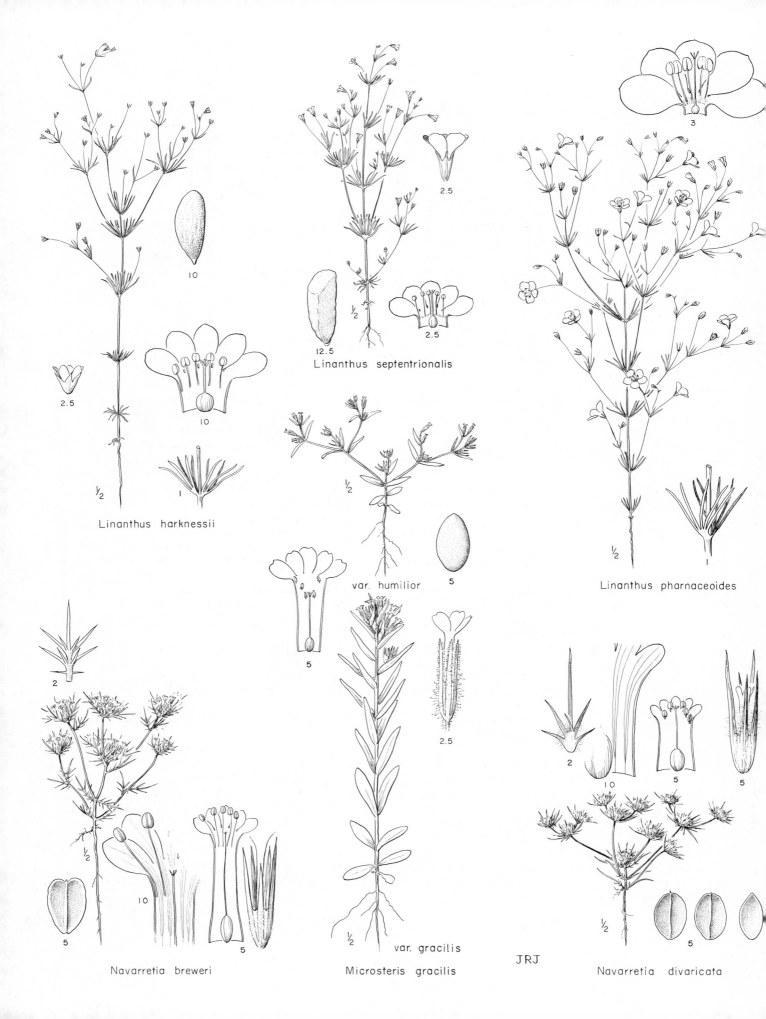

10

2.5

10

2.5

½

**Linanthus harknessii**

2.5

½

12.5

2.5

**Linanthus septentrionalis**

½

var. humilior

5

5

2.5

½

var. gracilis

**Microsteris gracilis**

3

½

1

**Linanthus pharnaceoides**

2

2

10

5

5

½

5

**Navarretia divaricata**

2

½

10

5

**Navarretia breweri**

JRJ

Open slopes, meadows, and margins of pools, from the lowlands to moderate elevations in the mountains; Wash. to Calif., eastward to Sask., Colo., and Ariz. June-Aug.

The species is divisible into two technically well-marked but intergradient varieties, as follows:

1 Corolla 7-11 mm. long, generally surpassing the calyx; lateral branches from the midvein of the corolla lobes generally again branched; filaments 1.5-3.5 mm. long; inflorescence conspicuously hairy; plants relatively robust, often well over 1 dm. tall; w. of the Cascades, and in n.e. Oreg., s.e. Wash., and adj. Ida.                          var. intertexta

1 Corolla 4-7 mm. long, equaling or more often surpassed by the calyx; lateral branches from the midvein of the corolla lobes generally simple; filaments 0.5-2 mm. long; inflorescences usually less strongly hairy; plants mostly less robust, seldom over 1 dm. tall; e. of the Cascades                                    var. propinqua (Suksd.) Brand

Navarretia minima Nutt. Journ. Acad. Phila. II. 1:160. 1848.
  Gilia minima Gray, Proc. Am. Acad. 8:269. 1870. (Nuttall, plains of the Oregon, near Walla Walla, Wash.)
  Navarretia suksdorfii Howell, Fl. N.W. Am. 457. 1901. N. minima var. suksdorfii Brand, Pflanzenr. IV. 250:164. 1907. (Suksdorf, Falcon Valley, Klickitat Co., Wash.)

Low annual up to 1 dm. tall, commonly branched from near the base, puberulent or subglabrous, the leaves and bracts relatively soft and often rather pale green; leaves few, up to 3.5 cm. long and with the rachis up to 2 mm. wide, the lateral segments few and small (or none) but often trifid, the terminal one elongate, entire or with spinulose teeth; inflorescence subglabrous to occasionally evidently villous-puberulent; bracts shorter and more dissected than the leaves; calyx mostly 5-6 mm. long, shortly white-hairy at the mouth within, the lobes more or less unequal, the larger ones often trifid; corolla white, 4-6 mm. long, the lobes oblong, about 1 mm. long and half as wide, each with a single, unbranched midvein; filaments 0.7-1.2(2.2) mm. long, inserted 0.3-0.5 mm. below the sinuses of the corolla; stigmas 2, very small, sometimes almost wholly connate; ovary 2-locular, sometimes imperfectly so, with 1 or less often 2 (very rarely 3) ovules per locule; capsule indehiscent, sometimes eventually opening by irregular disintegration of the lower part of the lateral walls.

Moist meadows, sites of vernal pools, and other moist, open places in the foothills, plains, and lowlands; e. of the Cascades in Oreg. (and Klickitat Co., Wash.) southward to n. Calif. and Nev., and eastward along the n. margin of the Snake R. plains to Fremont Co., Ida. May-July.

Much of the material which has passed as N. minima is actually N. intertexta var. propinqua.

Navarretia squarrosa (Esch.) H. & A. Bot. Beechey Voy. 368. 1838.
  Hoitzia squarrosa Esch. Mém. Acad. St. Pétersb. 10:283. 1826. Gilia squarrosa H. & A. Bot. Beechey Voy. 151. 1833. (Eschscholtz, Calif.)
  Gilia pungens Dougl. ex Hook. Curtis' Bot. Mag. 57: pl. 2977. 1830. Aegochloa pungens Benth. Bot. Reg. 19: sub pl. 1622. 1833. Navarretia pungens Hook. Fl. Bor. Am. 2:75. 1838. (Cultivated plants, from seeds collected by Douglas in "mountains valleys, near the sources of the Multnomack")

Erect, simple or moderately branched, glandular-hairy, and mephitic annual up to 4 dm. tall; leaves firm and strongly spiny, pinnatifid or subbipinnatifid, up to 6 cm. long, the terminal segment not elongate; heads relatively large, often 3 cm. thick, the bracts pinnatifid to subpalmatifid; calyx 8-14 mm. long, the lobes slightly or scarcely unequal, entire or with a few, sharp, slender, lateral teeth; corolla pale to deep blue, from a little shorter to a little longer than the calyx, 9-12 mm. long, the lobes 2-3 mm. long; petal traces branched well down in the tube, so that each corolla lobe receives 3 veins; filaments equal or unequal, mostly 1-4 mm. long, equally or unequally inserted near or above the middle of the tube, at least 2.5 mm. below the sinuses; stigmas 3; capsule 3-locular, dehiscent from the top downward; seeds mostly (6)8-9 per locule, the ovules sometimes as many as 15, but probably not all maturing.

Open places w. of the Cascades, from s. Vancouver I. to Calif. June-Sept.

Navarretia tagetina Greene, Pitt. 1:137. 1887. (Greene, Siskiyou Co., Calif.)

Gilia klickitatensis Suksd. Deuts. Bot. Monats. 18: 133. 1900. Navarretia klickitatensis
Suksd. loc. cit., nom. altern. (Suksdorf 991, near the mouth of the Klickitat R., Klickitat
Co., Wash.)

Superficially resembling N. intertexta var. intertexta, but sharply distinguished by a series
of well-correlated, technical characters; calyx less strongly hairy at the mouth within, the
lobes very strongly unequal and at least the larger ones generally spiny-cleft; corolla 8-11 mm.
long, the lobes 2-2.5 mm.; petal traces trifurcate well down in the tube so that 3 veins enter
each corolla lobe; stigmas 3; ovary 3-locular; ovules and seeds 1-2 (reputedly sometimes 3)
per locule.

Dry, open places at low elevations; Klickitat Co., Wash., southward to Calif.; rare in our
range. May-June.

### Phlox L.  Phlox; Wild Sweet William

Flowers borne in terminal cymes which are often reduced to a single flower; calyx scarcely
accrescent, with prominent scarious or hyaline intervals between the more herbaceous costae,
ruptured by the developing capsule; corolla salverform, with slender tube and abruptly spreading lobes, white to pink, purple, or blue; filaments short and unequally placed in the corolla
tube; anthers included, or often some of them partly exserted; seeds 1(2-4) per locule, remaining unchanged when moistened; annual or more often (including all our species) perennial
herbs, taprooted or less often fibrous-rooted, with opposite (or the uppermost alternate), entire, often narrow and needlelike, frequently basally connate leaves.

About 50 or 60 species, native to N. Am. and n. Asia, best developed in w. N. Am. (Name
a direct transliteration of the Greek word for flame, referring to the brightly colored flowers.)

Most of the species of Phlox are desirable rock garden subjects, but our native ones do not
do well in the climate w. of the Cascades. They are propagated rather easily by late summer
cuttings, by which they may be introduced into the garden.

Reference:

Wherry, E. T. The genus Phlox. Morris Arb. Monog. 3. 1955.

The tall species of Phlox are all reasonably well marked, at least in our area, and should
cause little trouble once it is understood that the conspicuous variations in vesture of the commoner species are haphazard and without taxonomic significance. The low species are quite
otherwise. The taxa are real, but they lack sharp boundaries. Short of considering all of the
low Phloxes to be varieties of a single species (the treatment adopted by Jepson in dealing
with the ones occurring in Calif.), it does not seem possible to arrive at sharply limited taxa,
and even such a desperate and unsatisfactory expedient would not provide for the fact that P.
aculeata, one of the low species, is uncomfortably close to P. longifolia, one of the tall species. It has therefore been necessary to use binomials for a series of taxa which in their degree of distinctness from each other are more nearly comparable to varieties. Under such
circumstances it is perhaps not surprising that the construction of a key should present extra-
ordinary difficulties. The key here presented, while admittedly no model of excellence,
should serve to identify properly the large majority of specimens. Measurements of floral
parts in the key and text are based on dry material, measurements of styles do not include the
branches, and measurements of leaves do not include the connate portion at the base.

1 Plants more or less erect, or loosely repent, with well-developed internodes, and with the
   larger leaves often more than 3.5 cm. long or more than 5 mm. wide; flowers mostly
   slender-pedicellate in a 3- to many-flowered terminal cyme (P. viscida sometimes approaches the following group in habit)
  2 Style short, 0.5-2 mm. long, from barely longer to more often distinctly shorter than
     the stigmas
    3 Corolla lobes nearly always notched at the tip; intercostal membranes of the calyx flat
       or nearly so; plants seldom wholly glabrous; leaves linear to broadly lanceolate,
       some of them usually over 2 mm. (to 11 mm.) wide; widespread, but not in the can-
       yons of the Snake and Salmon rivers                              P. SPECIOSA

      3 Corolla lobes not notched at the tip; intercostal membranes of the calyx evidently car-
inate toward the base; plants wholly glabrous (except within the calyx); leaves linear,
up to about 2 mm. wide; region of the canyons of the Snake and Salmon rivers of Oreg.
and Ida., barely reaching Wash.                                                    P. COLUBRINA

  2 Style elongate, 6-15 mm. long, several times as long as the stigmas

    4 Leaves linear or nearly so, not over 5 mm. wide

      5 Intercostal membranes of the calyx strongly and permanently carinate toward the
base; widespread                                                            P. LONGIFOLIA

      5 Intercostal membranes of the calyx flat or nearly so; local in s. e. Wash., n. e.
Oreg., and adj. Ida.                                                        P. VISCIDA

    4 Leaves distinctly broader than linear, at least the larger ones (7)10 mm. wide or more

      6 Plants erect, 5-10 dm. tall; larger leaves more than 5 cm. long; Clearwater Co.,
Ida.                                                                        P. IDAHONIS

      6 Plants repent or decumbent, up to 3 dm. tall; larger leaves 1-3 cm. long; Linn Co.,
Oreg., and southward                                                        P. ADSURGENS

1 Plants compact and tending to form mats, occasionally looser and up to 1 or even 1.5 dm.
tall; leaves short and crowded, rarely more than 3 cm. long (to 3.5 cm. in P. aculeata),
never more than 5 mm. wide; flowers mostly solitary (3), short-pedicellate or sessile at
the ends of the numerous stems, scarcely forming distinct inflorescences

  7 Intercostal membranes of the calyx generally more or less carinate

    8 Keel of the intercostal membrane conspicuous, strongly bulged; style 5-10 mm. long;
Snake R. plains of Ida., barely extending into adj. Oreg.                    P. ACULEATA

    8 Keel of the intercostal membrane inconspicuous, low, linear; style 2-5 mm. long;
Great Basin region, extending northward to c. and n. e. Oreg. and c. Ida., but apparently not on the Snake R. plains                                          P. AUSTROMONTANA

  7 Intercostal membranes of the calyx essentially flat

    9 Leaves only 2-4(5) mm. long, evidently arachnoid-woolly, narrowly to subdeltoidally
triangular, closely crowded and appressed, tending to form quadrangular shoots;
widespread, but in our range mostly from c. Ida. eastward    P. MUSCOIDES

    9 Leaves, if not over 5 mm. long, either narrower and essentially linear, or not at all
arachnoid-woolly, or both, never forming compact quadrangular shoots

      10 Leaves narrowly linear, mostly (2.5)4-10(13) mm. long and about 0.5(1) mm. wide
near the middle, firm and pungent, very often arachnoid; plants forming dense
cushions; style 2-5(6) mm. long; e. of the Cascades, chiefly in the foothills, valleys, and plains                                                           P. HOODII

      10 Leaves never arachnoid, and usually longer, or wider, or distinctly softer than
those of P. hoodii, but if similar to those of P. hoodii in size and texture, then
the plants looser and suberect

        11 Plants tending to be loosely erect, 5-15 cm. tall, often resembling a small
Leptodactylon in habit; leaves narrow, firm, commonly 5-13 mm. long and
0.5(1) mm. wide near the middle; calyx and pedicels (often also the herbage)
usually glandular; style 3-7(8) mm. long; chiefly in and near the ponderosa
pine zone, from n. w. Mont. to e. and (occasionally) c. Wash.
P. CAESPITOSA

        11 Plants either more compact and mat-forming, or, if looser and suberect, then
with the leaves distinctly longer, or wider, or softer than those of P. caespitosa; pubescence and styles diverse

          12 Styles 1-5 mm. long; leaves fairly firm, rarely as much as 1.5 cm. long,
the margins somewhat thickened, but not strongly whitish; calyx usually
glandular-hairy; moderate to high elevations in the mountains

            13 Style about 1-2 mm. long; Cascade Range of c. Wash. to n. Oreg.
P. HENDERSONII

            13 Style mostly 2-5 mm. long; wholly e. of the Cascades
P. PULVINATA

          12 Styles mostly 5-12 mm. long, or occasionally only 4 mm. in plants which
differ in other respects from the foregoing group

14  Leaves 2-5(7) mm. long, prominently white-margined; style 5-8 mm. long; moderate to
        high elevations in the mountains of Mont. and e. Ida.              P. ALBOMARGINATA
14  Leaves (or at least the larger ones)(7)10-25 mm. long, with or without white margins
        15  Leaves mostly 2-5 mm. wide near the middle, firm, tending to be white-margined;
                style 6-12 mm. long; high plains, intermontane valleys, and foothills, wholly e. of
                the continental divide                                          P. ALYSSIFOLIA
        15  Leaves (0.5)1-2.5 mm. wide near the middle, scarcely white-margined; style 4-8(10)
                mm. long
                16  Leaves very finely and sparsely scaberulous, otherwise essentially glabrous, the
                        better developed ones mostly (10)12-30 mm. long and 1-2 mm. wide; plants never
                        evidently glandular; mountains from s. w. Mont. and adj. Ida. southward
                                                                                    P. MULTIFLORA
                16  Leaves not at all scaberulous, but often otherwise hairy or glandular, and common-
                        ly more or less ciliate-margined near the base
                        17  Leaves either succulent, or evidently hairy (and often glandular) or both, the
                                larger ones 10-25 mm. long; foothills, intermontane valleys, and high plains,
                                on both sides of the continental divide, in Mont., e. Ida., and Wyo.
                                                                                    P. KELSEYI
                        17  Leaves neither succulent nor hairy (except for the basally usually arachnoid-
                                ciliate margins), often less than 1 cm. long, generally not at all glandular;
                                wholly w. of the continental divide, at moderate to high elevations in the
                                mountains from s. B. C. to Calif., eastward to c. Ida. and n. w. Mont.
                                                                                    P. DIFFUSA

Phlox aculeata A. Nels. Bot. Gaz. 52:270. 1911. (Macbride 73, New Plymouth, Payette Co.,
  Ida.)
    Caespitose, taprooted perennial, glandular-hairy or occasionally merely hirsute above,
commonly subglabrous below, the numerous, crowded, branching stems less than 1 dm. tall;
leaves firm, narrowly linear, mostly 1-3.5 cm. long and 0.5-1.5 mm. wide; flowers short-
pedicellate, 1-3 at the ends of the stems; hyaline intervals of the calyx strongly and permanently bulged-carinate toward the base; calyx lobes narrow, acerose; corolla pink to white,
the tube 10-15 mm. long, less than twice as long as the calyx, the lobes 7-12 mm. long;
style 5-10 mm. long.
    Dry, open, sometimes alkaline places, often with sagebrush; Snake R. plains of Ida., chief-
ly from Gooding Co. westward, extending into adj. n. Malheur Co., Oreg., only barely or
scarcely entering our range. Apr. -May.
    Phlox aculeata combines the habit of the depressed phloxes with the flowers (especially the
calyx) of P. longifolia, the latter species belonging to the elate group. Further study may possibly warrant the subordination of P. aculeata to P. longifolia.

Phlox adsurgens Torr. ex Gray, Proc. Am. Acad. 8:256. 1870.
    Armeria adsurgens Kuntze, Rev. Gen. 2:432. 1891. (Wood, Canyon Pass, Oreg.)
    Perennial from an eventual taproot, often more or less rhizomatous as well, the several
stems loosely curved-ascending or more or less repent, up to about 3 dm. tall, glabrous below, becoming glandular in the inflorescence; leaves firm, glabrous, ovate or lance-ovate to
elliptic or almost obovate, the better developed ones mostly 1-3 cm. long and 7-18 mm. wide,
the internodes well developed; flowers slender-pedicellate in open, terminal cymes; intercostal membranes of the calyx sometimes loosely carinate, but readily distended; corolla pink,
the tube 14-20 mm. long, nearly or fully twice as long as the calyx, the obovate lobes 9-15
mm. long; style elongate, nearly or fully reaching the orifice of the corolla, several times as
long as the linear stigmas. N=7. Also 2N=21.
    Wooded slopes at moderate elevations in the mountains; Klamath region of s. w. Oreg. and
adj. Calif., extending northward along the w. side of the Cascades occasionally as far as Linn
Co., Oreg. May-July.

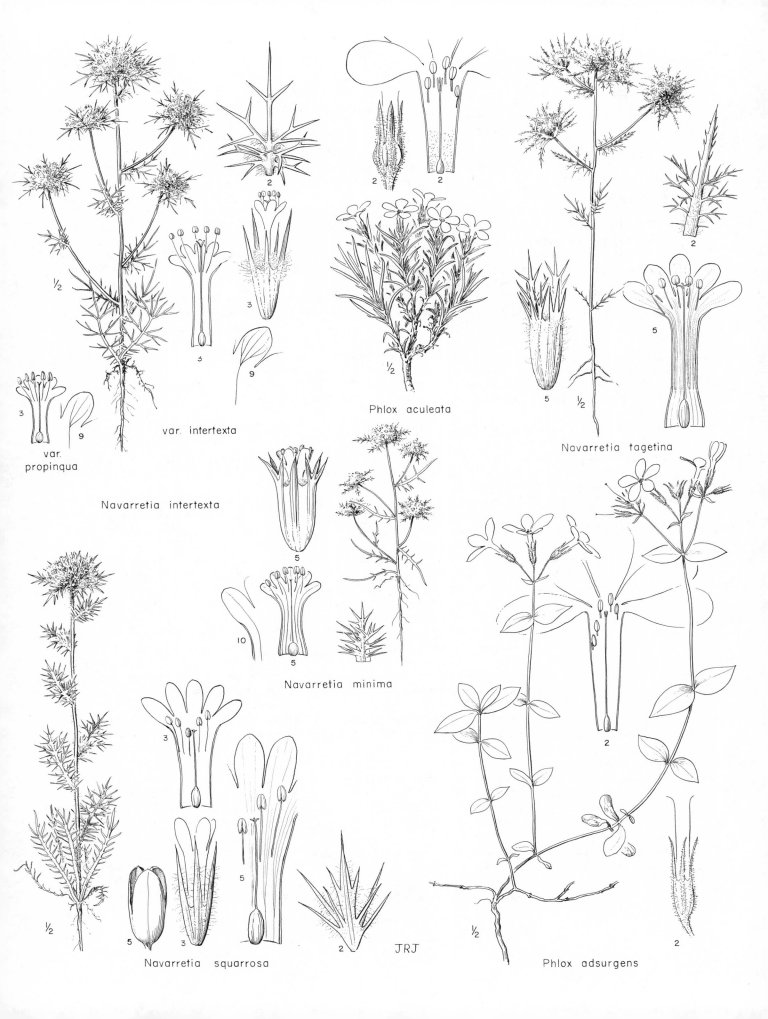

var. intertexta

var.
propinqua

Navarretia intertexta

Phlox aculeata

Navarretia tagetina

Navarretia minima

Navarretia squarrosa

JRJ

Phlox adsurgens

Phlox albomarginata M. E. Jones, Zoë 4:367. 1894.

  P. kelseyi var. albomarginata Brand, Pflanzenr. IV. 250:79. 1907. (Kelsey, e. face of Mt.
    Helena, Mont., May, 1891)

  P. albomarginata var. minor M. E. Jones, Zoë 4:368. 1894. P. kelseyi var. albomarginata
    subvar. minor Brand, Pflanzenr. IV. 250:80. 1907. P. kelseyi var. minor A. Nels. U.
    Wyo. Pub. Bot. 1:53. 1924. P. albomarginata ssp. minor Wherry, Not. Nat. Acad. Phila.
    87:11. 1941. (Presumably the same station and collector as P. albomarginata)

  P. diapensioides Rydb. Mem. N.Y. Bot. Gard. 1:317. 1900. P. kelseyi var. diapensioides
    Brand, Pflanzenr. IV. 250:80. 1907. P. albomarginata ssp. diapensioides Wherry, Not.
    Nat. Acad. Phila. 87:11. 1941. (Tweedy 282, Madison Co., Mont.)

  Taprooted perennial, caespitose, often pulvinate, not over about 5 cm. tall, the numerous
stems (when not hidden by the leaves) spreading-hairy and becoming glandular upward; leaves
firm, mostly 2-5(7) mm. long and 1-2.5 mm. wide, the surfaces glabrous or hairy, the thickened, prominently whitish-cartilaginous margins ciliate toward the base; flowers solitary at
the ends of the stems, sessile or short-pedicellate; intercostal membranes of the calyx flat;
calyx lobes firm, but the midrib not much thickened; corolla pink or purplish to white, the
tube 9-12 mm. long, often twice as long as the calyx, the lobes 6-9 mm. long; style 5-8 mm.
long.

  Open, rocky places at moderate to high elevations in the mountains; s. w. Mont. (as far n. e.
as Helena) and adj. Ida. (Lost River Mts., Custer Co.) May-July.

  Phlox albomarginata is obviously allied to the larger P. alyssifolia of lower elevations, and
the two are not always sharply distinct.


Phlox alyssifolia Greene, Pitt. 3:27. 1896. (Macoun 11813, Twelve-Mile Lake, Wood Mt.,
  Sask.)

  P. collina Rydb. Mem. N.Y. Bot. Gard. 1:316. 1900. P. kelseyi var. collina Brand, Pflan-
    zenr. IV. 250:80. 1907. P. alyssifolia ssp. collina Wherry, Not. Nat. Acad. Phila. 146:8.
    1944. (Tweedy 154, Madison Co., Mont.)

  P. variabilis Brand, Pflanzenr. IV. 250:87. 1907. P. kelseyi ssp. variabilis Wherry, Not.
    Nat. Acad. Phila. 87:8. 1941. (Hall & Harbour 454, Rocky Mts. of Colo., lat. 39-41 degrees)

  P. abdita A. Nels. U. Wyo. Pub. Bot. 1:47. 1924. P. alyssifolia ssp. abdita Wherry, Not.
    Nat. Acad. Phila. 146:8. 1944. (Alice Pratt s. n., Piedmont, S. D., June, 1895)

  P. sevorsa A. Nels. U. Wyo. Pub. Bot. 1:128. 1926. P. alyssifolia f. sevorsa Wherry, Not.
    Nat. Acad. Phila. 146:8. 1944. (A. Nelson 9116, Lusk, Converse Co., Wyo.)

  Taprooted perennial, compactly caespitose, or looser and up to 1 dm. tall, the numerous
stems spreading-hairy, especially above, and commonly becoming glandular near the mostly
glandular-hairy calyx; leaves firm, linear-oblong to lanceolate or elliptic, the larger ones
mostly (7)10-25 mm. long, 2-5 mm. wide, the surfaces glabrous or sometimes hirsute, the
margins thickened, cartilaginous, and tending to be whitish, evidently ciliate toward the base;
flowers 1-2 at the end of the stems, short-pedicellate; hyaline intervals of the calyx flat; calyx lobes firm but flattened, the midrib not very conspicuous; corolla white to lilac or pink,
the tube 10-15 mm. long, less than twice as long as the calyx, the broad lobes 7-13 mm. long;
style 6-12 mm. long.

  Dry, open places in the high plains, intermontane valleys, and lower foothills; s. w. Mont.
to s. Sask., w. N. D. and S. D., and n. c. Colo., wholly e. of the continental divide. May-July.


Phlox austromontana Coville, Contr. U.S. Nat. Herb. 4:151. 1893.

  P. douglasii var. austromontana Jeps. & Mason in Jeps. Man. Fl. Pl. Calif. 786. 1925.
    (V. Bailey 1944, Beaverdam Mts., Utah)

  P. densa Brand, Pflanzenr. IV. 250:83. 1907. P. austromontana ssp. densa Wherry, Journ.
    Wash. Acad. Sci. 29:519. 1939. (M. E. Jones 2021, Frisco, Utah, is the first of 5 speci-
    mens cited)

  P. diffusa ssp. subcarinata Wherry, Journ. Wash. Acad. Sci. 29:517. 1939. P. diffusa var.
    subcarinata Peck, Man. High. Pl. Oreg. 572. 1941. (Heller 9910a, Mt. Rose, Washoe Co.,
    Nev.)

  Caespitose perennial from a taproot, forming loose to rather dense mats, the numerous

stems occasionally more erect and nearly 1 dm. tall; plants subglabrous, or loosely hairy upward, not at all glandular; leaves firm, mostly 8-15(20) mm. long and 0.5-1.5 mm. wide;
flowers solitary and sessile or nearly so at the ends of the stems; intercostal membranes of
the calyx with a low, median linear keel; corolla white or less often bluish or pink, the tube
8-15 mm. long, less than twice as long as the calyx, the lobes 5-8 mm. long; style 2-5 mm. long.

Dry to moderately moist, open or sparsely wooded, often stony slopes, from the foothills to
moderate or rather high elevations in the mountains, mostly avoiding the open lower plains; Great
Basin and adj. Colorado Plateau region of Utah, Nev., and n. Ariz., extending to the mountains of
s. Calif. and to the crest of the Sierra Nevada, and northward to c. and n.e. Oreg. (Grant and Union cos.) and to the Seven Devils Mts. (Idaho Co.) and the valley of the Salmon R. (Lemhi Co.) in
Ida., but apparently not on the Snake R. plains. June-July.

At the northern and western margins of its range, P. austromontana intergrades with P.
diffusa. Here, as elsewhere among the low phloxes, a reasonable system requires the maintenance at specific level of taxa which are scarcely more distinct from each other than are
varieties in many other groups.

Phlox caespitosa Nutt. Journ. Acad. Phila. 7:41. 1834.
  Armeria caespitosa Kuntze, Rev. Gen. 2:432. 1891. Phlox douglasii var. caespitosa Mason
    ex Jeps. Man. Fl. Pl. Calif. 786. 1925. (Wyeth, dry hills along the Flathead R. in Mont.)
  P. douglasii Hook. Fl. Bor. Am. 2:73. 1838. Armeria douglasii Kuntze, Rev. Gen. 2:432.
    1891. (Douglas, limestone range of the Blue Mts., and on the Rocky Mts.)
  P. rigida Benth. in DC. Prodr. 9:306. 1845. P. caespitosa var. rigida Gray, Proc. Am.
    Acad. 8:254. 1870. P. douglasii ssp. rigida Wherry, Proc. Acad. Phila. 90:137. 1938. P.
    douglasii var. rigida Peck, Man. High. Pl. Oreg. 571. 1941. (Douglas, Blue Mts.) A
    small form.
  P. piperi E. Nels. Rev. W. N. Am. Phl. 18. 1899. P. caespitosa var. rigida subvar.
    piperi Brand, Pflanzenr. IV. 250:83. 1907. (Piper 2286, Spokane, Wash.)
  ?P. kelseyi var. ciliata Brand, Pflanzenr. IV. 250:79. 1907. (Kelsey, Coeur d'Alene, Ida.,
    in 1891) No original material seen.

Taprooted, subshrubby perennial, resembling a small Leptodactylon in aspect, the numerous slender stems tending to be ascending or loosely erect, commonly 5-15 cm. tall; plants
commonly more or less glandular or glandular-hairy, at least upward, less often merely
hairy or even subglabrous; leaves firm and pungent, narrowly linear, mostly 5-13 mm. long
and about 0.5(1) mm. wide near the middle, often shortly and inconspicuously ciliate-margined
toward the base; flowers solitary (3) and short-pedicellate or sessile at the ends of the
branches; intercostal membranes of the calyx flat; calyx lobes narrow and commonly thickened; corolla tube 8-14 mm. long, less than twice as long as the calyx, the lobes 7-10 mm. long;
style mostly 3-7(8) mm. long.

Open or partly shaded, rather dry places, commonly in the ponderosa pine zone, less often
descending into the sagebrush or ascending to the higher forest zones; n. Ida. (rarely as far
as Adams Co.) and adj. n.w. Mont. (rarely as far s.e. as the Belt Mts.), s. B.C., and e.
Wash., and occasionally westward to Chelan Co., Wash.; apparently also in n.e. Oreg. Apr.-
May(June).

The name P. caespitosa has usually been misapplied to P. pulvinata, q.v.

Phlox colubrina Wherry & Const. Am. Midl. Nat. 19:433. 1938. (Constance et al. 1823, 1
  mile above the mouth of Sheep Creek, a tributary of the Snake R., Idaho Co., Ida.)
  Perennial, 1.5-5 dm. tall, tending to become shrubby toward the base, essentially glabrous
throughout, except that the calyx tube and lobes are puberulent within; leaves linear, 2-7 cm.
long, 0.5-2 mm. wide, the internodes well developed; inflorescence loosely cymose, leafy-
bracteate at least below; intercostal membranes of the calyx more or less strongly carinate,
sometimes eventually distended by the developing capsule; corolla pink or occasionally white,
the tube 8-13 mm. long, shortly or scarcely surpassing the calyx, the lobes 10-20 mm. long,
oblanceolate, entire; anthers included well down in the corolla tube; style 0.5-2 mm. long,
from distinctly shorter to barely longer than the linear stigmas.

Dry, open slopes and cliffs in and near the canyon of the Snake R. (and below Riggins on the

Salmon R. ), at elevations up to 4200 ft. , in Oreg. and Ida. , barely or scarcely reaching Wash.
Apr. -June.

Phlox colubrina is a well-marked local species which combines some of the characters of P.
speciosa and P. longifolia; both of these latter species approach but scarcely encroach upon
its area.

Phlox diffusa Benth. Pl. Hartw. 325. 1849.
   P. douglasii var. diffusa Gray, Proc. Am. Acad. 8:254. 1870. (Hartweg 380, near Bear Val-
      ley in the Sacramento Mts. )
   P. diffusa ssp. longistylis Wherry, Proc. Acad. Phila. 90:139. 1938. P. diffusa var. longi-
      stylis Peck, Man. High. Pl. Oreg. 572. 1941. (Wherry, s. slope of Mt. Adams, Yakima
      Co. , Wash. , July 30, 1931)
   Taprooted perennial, caespitose and commonly mat-forming, occasionally looser and up to
1 dm. tall; leaves softer than those of P. hoodii and P. caespitosa, green and glabrous except
for the often basally arachnoid-ciliate margins, 5-20 mm. long, (0.5)1-2 mm. wide; flowers
solitary and sessile or nearly so at the ends of the stems; calyx commonly arachnoid-villous,
the intercostal membranes flat, the lobes narrow and often thickened; corolla white to pinkish
or light bluish, the tube 9-17 mm. long, less than twice as long as the calyx, the lobes 5-9 mm.
long; style (4)4.5-8(10) mm. long.
   Forests and open rocky slopes at moderate to high elevations in the mountains, wholly w. of
the continental divide; mountains of s. Vancouver I. and n. w. Wash. ; Saddle Mt. , Clatsop Co. ,
Oreg. ; Cascade region from s. B. C. through Oreg. and into the Sierra Nevada of Calif. , eastward through n. Wash. to n. Ida. and n. w. Mont. , sometimes extending as far s. as Valley and
Custer cos. , Ida. ; occasional in the mountains of n. e. Oreg. May-Aug.
   Our plants, as described above, constitute the well-marked var. longistylis (Wherry) Peck,
which might perhaps with almost equal propriety be treated at the specific level. South of our
range (in the Cascade Range from n. Klamath Co. southward, and on the w. slopes of the Sierra Nevada) the var. longistylis gives way to the var. diffusa, characterized by its shorter
styles, these 2-4(4.5) mm. long. The Klamath species P. cyanea Eastw. , with bright purple
or blue flowers, has sometimes been mistaken for P. diffusa.

PHLOX HENDERSONII (E. Nels.) Cronq. hoc loc.
   P. condensata var. hendersoni E. Nels. Rev. W. N. Am. Phl. 14. 1899. P. caespitosa ssp. mus-
      coides var. hendersonii Brand, Pflanzenr. IV. 250:84. 1907. P. douglasii ssp. hendersonii
      Wherry, Proc. Acad. Phila. 90:137. 1938. P. douglasii var. hendersonii Peck, Man. High.
      Pl. Oreg. 571. 1941. (Henderson, Mt. Adams, Wash. , Aug. 10, 1902)
   Taprooted, pulvinate-caespitose perennial; leaves crowded, often ternate, nearly linear,
mostly 5-10 mm. long and scarcely 1 mm. wide near the middle, copiously glandular-hairy,
the margins strongly thickened but not prominently whitish; flowers solitary and sessile or
nearly so at the ends of the stems, often 6-merous, calyx glandular-hairy like the leaves, the
intercostal membranes flat, the lobes firm and commonly thick-margined; corolla tube about
1 cm. long, not much longer than the calyx, the lobes about 5 mm. long; style only about 1-2
mm. long.
   Exposed, rocky places; Chelan Co. , Wash. , to Mt. Hood, Oreg. July-Aug.
   Although known from only a few collections, P. hendersonii appears to be a relatively well-
marked species.

Phlox hoodii Rich. App. Frankl. Journ. 733. 1823.
   Armeria hoodii Kuntze, Rev. Gen. 2:432. 1891. Fonna hoodii Nieuwl. & Lunell, Am. Midl.
      Nat. 4:512. 1916. (Richardson, Carlton House, Sask. )
   Phlox canescens T. & G. Pac. R. R. Rep. 2:122. 1857. Armeria canescens Kuntze, Rev.
      Gen. 2:432. 1891. Phlox douglasii var. canescens Mason ex Jeps. Man. Fl. Pl. Calif. 786.
      1925. P. hoodii ssp. canescens Wherry, Proc. Acad. Phila. 90:139. 1938. P. hoodii var.
      canescens Peck, Man. High. Pl. Oreg. 572. 1941. (Beckwith, Cedar Mts. , s. of Great
      Salt lake)
   P. hoodii var. glabrata E. Nels. Rev. W. N. Am. Phl. 11. 1899. P. glabrata Brand, Pflan-

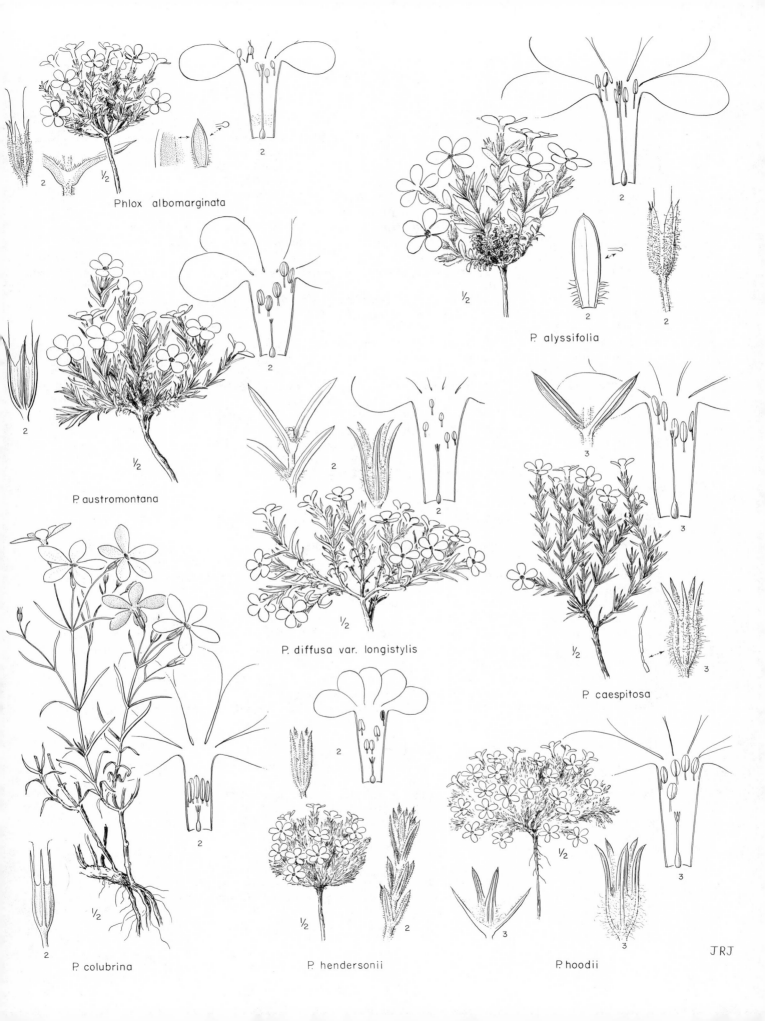

Phlox albomarginata

P. alyssifolia

P. austromontana

P. diffusa var. longistylis

P. caespitosa

P. colubrina

P. hendersonii

P. hoodii

JRJ

zenr. IV. 250:86. 1907. P. hoodii ssp. glabrata Wherry, Not. Nat. Acad. Phila. 87:14.
1941. (A. Nelson 8, Laramie, Wyo.)
P. scleranthifolia Rydb. Mem. N.Y. Bot. Gard. 1:313. 1900. P. douglasii var. scleranthifolia Brand, Pflanzenr. IV. 250:85. 1907. P. diffusa ssp. scleranthifolia Wherry, Not. Nat.
Acad. Phila. 87:13. 1941. (Rydberg 880, Hot Springs, S.D.) A subglabrate form.
P. lanata Piper, Bull. Torrey Club 29:643. 1902. (Cusick 2557, Steens Mts., Oreg., June
10, 1901) A strongly woolly form.
P. hoodii ssp. viscidula Wherry, Not. Nat. Acad. Phila. 146:11. 1944. (Rose 10, 5 miles w.
of Drummond, Granite Co., Mont.) A glandular form.

Taprooted perennial, compact and forming a mat or low cushion, more or less strongly
arachnoid (especially on the calyx and leaf margins) to occasionally glandular or subglabrous;
leaves firm and pungent, narrowly linear, mostly (2.5)4-10(13) mm. long and about 0.5(1) mm.
wide near the middle; flowers mostly solitary and sessile or nearly so at the ends of the stems;
intercostal membranes of the calyx flat, often obscured by the pubescence; calyx lobes narrow
and firm, the midrib commonly thickened; corolla bluish or pink to white, the tube mostly 4-
10(12) mm. long (the shorter lengths occurring chiefly to the e. of our range), the lobes 4-7
mm. long; style 2-5(6) mm. long. N=14.

Dry, open places in the foothills, valleys, and plains, commonly with sagebrush, occasionally ascending to moderate elevations in the mountains; e. of the continental divide from n. Colo.
and w. Neb. to Alta. and Sask., and northward to Yukon and e. Alas.; passing westward
through Wyo. to s. Ida. and the n. half of Utah, thence to n.e. Calif., e. Oreg. (as far as the
foot of the Cascade Range), and c. Wash., and occasionally to c. Ida. Apr.-June.

There is some tendency for the plants of our area (and southward) to have the corolla tube
longer than in typical P. hoodii, of the high plains and adj. foothills and valleys e. of the continental divide, but there is so much morphological and geographical overlapping that a taxonomic distinction scarcely appears profitable. The name P. hoodii var. canescens (T. & G.)
Peck is available for the larger-flowered, more western segment of the species, should the
need be demonstrated.

Phlox idahonis Wherry, Not. Nat. Acad. Phila. 87:6. 1941. (Wherry, near Headquarters,
Clearwater Co., Ida., July 17, 1940)

Perennial, 5-10 dm. tall, the stem solitary from the upturned end of a slender rhizome,
evidently glandular in the inflorescence, otherwise loosely spreading-hairy or partly glabrate;
leaves glabrous or subglabrous above, loosely spreading-hairy beneath, lanceolate or lance-
ovate, acuminate, the lower several pairs more or less reduced and soon withering, the
others up to 9 cm. long and 3 cm. wide, the broadest ones not far below the openly branched
inflorescence; intercostal membranes of the calyx tending to be carinate at anthesis, but distended by the growing capsule; corolla pink, the tube 17-20 mm. long, about twice as long as
the calyx, the obovate lobes 8-9 mm. long; style elongate, nearly or quite reaching the orifice
of the corolla.

Moist meadows and stream banks; known only from a few collections in the immediate vicinity of the type locality in Clearwater Co., Ida. June-July.

Phlox idahonis is without close allies in w. U.S., being related instead to P. carolina L.
and P. maculata L., of e. N. Am.

Phlox kelseyi Britt. Bull. Torrey Club 19:225. 1892.
Fonna kelseyi Nieuwl. & Lunell, Am. Midl. Nat. 4:512. 1916. (Kelsey s.n., Helena, Mont.,
May, 1891)
Phlox kelseyi ssp. glandulosa Wherry, Not. Nat. Acad. Phila. 87:8. 1941. (Davis 827, 2
miles w. of Soda Springs, Caribou Co., Ida., May 30, 1939)
PHLOX KELSEYI var. MISSOULENSIS (Wherry) Cronq. hoc loc. P. missoulensis Wherry,
Not. Nat. Acad. Phila. 146:7. 1944. (Reed 8, Missoula, Mont.)

Taprooted perennial, caespitose, the numerous stems up to 1 dm. long, closely crowded
and suberect, or looser and more prostrate, glabrous to spreading-hirsute and sometimes
glandular; leaves mostly 1-2.5 cm. long, or some of them a little shorter, 1-2.5 mm. wide
near the middle, the surfaces glabrous to hairy or glandular, the margins thickened but not

evidently whitish, ciliate at least toward the base; flowers short-pedicellate or sessile, solitary at the ends of the stems; intercostal membranes of the calyx flat; calyx lobes flattened, with prominent or inconspicuous midrib; corolla light blue to white, the tube 10-13 mm. long, equaling or well surpassing the calyx, the lobes 6-9 mm. long; styles 4-7.5 mm. long.

Foothills and valleys; Meagher and Missoula cos., Mont., to c. Colo. and Custer and Caribou cos., Ida.; White Pine Co., Nev. May-July.

The var. kelseyi, with essentially the range of the species (except the Nevada record), occurs in alkaline meadows and is more or less succulent. The var. missoulensis (Wherry) Cronq., occurring chiefly about Missoula, Mont., but known also from Meagher Co., grows on open slopes and is more rigid. A third variety, differing from var. kelseyi in its shorter leaves and disjunct distribution, occurs in White Pine Co., Nev. The smaller-leaved specimens of P. kelseyi var. missoulensis may be difficult to distinguish from the larger-leaved plants of P. pulvinata.

Phlox longifolia Nutt. Journ. Acad. Phila. 7:41. 1834.
Armeria longifolia Kuntze, Rev. Gen. 2:432. 1891. (Wyeth, valleys of the Rocky Mts.)
Phlox humilis Dougl. ex Hook. Fl. Bor. Am. 2:72. 1838, pro syn.; ex Benth. in DC. Prodr. 9:306. 1845. P. longifolia ssp. marginata var. humilis Brand, Pflanzenr. IV. 250:66. 1907. P. longifolia ssp. humilis Wherry, Proc. Acad. Phila. 90:135. 1938. P. longifolia var. humilis Peck, Man. High. Pl. Oreg. 571. 1941. (Douglas, subalpine range of the Rocky Mts., and on the Blue Mts., fide Hooker; dry sands along the Columbia R., fide Bentham)
P. speciosa var. linearifolia Hook. Kew Journ. Bot. 3:289. 1851. P. linearifolia Gray, Proc. Am. Acad. 8:255. 1870. Armeria linearifolia Kuntze, Rev. Gen. 2:432. 1891. Phlox longifolia var. linearifolia Brand, Helios 22:80. 1905. P. longifolia ssp. linearifolia Brand, Pflanzenr. IV. 250:66. 1907. (Geyer 340, valley of the Kooskooskie R. and the adjoining plains)
P. viridis E. Nels. Rev. W. N. Am. Phl. 24. 1899. P. stansburyi ssp. compacta var. puberula subvar. viridis Brand, Pflanzenr. IV. 250: 67. 1907. P. longifolia ssp. viridis Wherry, Not. Nat. Acad. Phila. 87:5. 1941. (Piper 2689, Ellensburg, Wash., May 20, 1897)
P. longifolia var. puberula E. Nels. Rev. W. N. Am. Phl. 26. 1899. P. stansburyi ssp. compacta Brand, cum var. puberula Brand, Pflanzenr. IV. 250:67. 1907. P. puberula Nels in Coult. & Nels New Man. Bot. Rocky Mts. 397. 1909. P. longifolia ssp. compacta Wherry, Proc. Acad. Phila. 90:135. 1938. P. longifolia var. compacta Peck, Man. High. Pl. Oreg. 571. 1941. P. viridis ssp. compacta Wherry, Morris Arb. Monog. 3:87. 1955. (A. Nelson 4544, Evanston, Wyo.) P. longifolia ssp. compacta Brand was synthetic, composed of varieties viscida and puberula. By his exclusion of var. viscida, and his citation of P. longifolia var. puberula as a synonym of P. longifolia ssp. compacta, Wherry in 1938 effectively typified Brand's subspecies on P. longifolia var. puberula. His attempted retypification in 1955, on a specimen collected by himself in 1948, is illegitimate.
P. longifolia ssp. marginata Brand, Pflanzenr. IV. 250:65. 1907. P. marginata A. Nels. Am. Journ. Bot. 18:442. 1931, where incorrectly attributed to Brand. (Cusick 2517, Snake R., Oreg.)
P. pinifolia Brand, Pflanzenr. IV. 250:80. 1907. (Cusick 183, Union Co., Oreg.)
P. linearifolia var. longipes M. E. Jones, Contr. West Bot. 12:53. 1908. P. longifolia ssp. longipes Wherry, Proc. Acad. Phila. 90:135. 1938. P. longifolia var. longipes Peck, Man. High. Pl. Oreg. 571. 1941. P. viridis ssp. longipes Wherry, Morris Arb. Monog. 3:88. 1955. (Jones 6450, Weiser, Ida.)
P. longifolia var. filifolia A. Nels. Bot. Gaz. 54:143. 1912. (Nelson & Macbride 1192, Ketchum, Ida.)
P. cortezana A. Nels. Am. Journ. Bot. 18:434. 1931. P. longifolia ssp. cortezana Wherry, Not. Nat. Acad. Phila. 87:5. 1941. (Nelson 10436, between Cortez, Colo., and Mesa Verde Nat. Pk.)
P. longifolia ssp. calva Wherry, Proc. Acad. Phila. 90:136. 1938. P. longifolia var. calva Peck, Man. High. Pl. Oreg. 571. 1941. (Wherry, 13 miles s.w. of Darlington, Custer Co., Ida., June 21, 1931)

Perennial from an eventual taproot, but often branched and creeping below the ground level, (0.5)1-4 dm. tall, often woody at the base, but generally less so than in P. speciosa, glabrous to strongly glandular (especially in the inflorescence) or hairy; leaves linear, mostly 1.5-8 cm. long and 1-2.5(4) mm. wide, the internodes well developed; inflorescence leafy-bracteate at least below, rather loosely cymose, the sweet-scented flowers with well-developed, slender pedicels; intercostal membranes of the calyx strongly and permanently carinate toward the base; corolla pink to white, the tube 10-18 mm. long, 1-2 times as long as the calyx, the lobes 7-15 mm. long, more or less obovate, entire or merely erose, half as long to fully as long as the tube; filaments mostly attached above the middle of the corolla tube, often some of the anthers partly exserted; style elongate, 6-15 mm. long, several times as long as the linear stigmas.

Dry, open rocky places, from the lowlands to moderate or occasionally high elevations in the mountains; Wash. and adj. s. B. C. to s. Calif., eastward to w. Mont., w. Wyo., Colo., and N.M. Apr.-July.

The several specific and infraspecific segregates, based largely on pubescence and size of the plant, have similar or identical ranges, do not show any ecologic differentation, and sometimes occur together, with intermediates. It thus seems clear that they are mere formae, rather than proper varieties or subspecies. No new combinations are here proposed.

The related P. stansburyi (Torr.) Heller, differing in the longer tube and relatively shorter lobes of the corolla, may eventually prove to be better subordinated to P. longifolia, but is in any case wholly to the s. of our range.

Phlox multiflora A. Nels. Bull. Torrey Club 25:278. 1898. (A. Nelson 3175, Laramie Hills, Albany Co., Wyo.)
  P. multiflora var. depressa E. Nels. Rev. W. N. Am. Phl. 20. 1899. P. depressa Rydb. Bull. Torrey Club 33:149. 1906. P. multiflora ssp. depressa Wherry, Not. Nat. Acad. Phila. 87:12. 1941. (A. Nelson 3084, Point of Rocks, s. w. Wyo.) A small form.
  P. costata Rydb. Mem. N.Y. Bot. Gard. 1:315. 1900. P. kelseyi var. costata Brand, Pflanzenr. IV. 250:80. 1907. (Rydberg & Bessey 4807, Cedar Mts., Mont.) A small form.
  P. patula A. Nels. U. Wyo. Pub. Bot. 1:48. 1924. P. multiflora ssp. patula Wherry, Morris Arb. Monog. 3:148. 1955. (Platte Canyon, Colo., May 19, 1894.)
  Taprooted, more or less mat-forming perennial, the numerous stems occasionally loosely suberect, but rarely as much as 1 dm. tall; leaves inconspicuously scaberulous, much less rigid than in P. hoodii, linear, the better developed ones mostly (10)12-30 mm. long and 1-2 mm. wide; flowers 1-3 at the ends of the stems, sessile or short-pedicellate; calyx glabrous or sometimes arachnoid-puberulent, the intercostal membranes flat, the lobes rather narrow and often thickened, but not very firm; corolla white or occasionally bluish, the tube 10-14 mm. long, the lobes 6-11 mm. long; style 5.5-8 mm. long.

Open or wooded, often rocky places, from the higher foothills to above timber line in the mountains; s. w. Mont. and adj. Ida. to Colo. and n. e. Utah; Elko Co., Nev. May-Aug.

Phlox muscoides Nutt. Journ. Acad. Phila. 7:42. 1834.
  Armeria muscoides Kuntze, Rev. Gen. 2:432. 1891. Phlox caespitosa ssp. muscoides Brand, Pflanzenr. IV. 250:84. 1907. P. hoodii ssp. muscoides Wherry, Morris Arb. Monog. 3:164. 1955. (Wyeth, in alpine situations at the sources of the Missouri)
  Phlox bryoides Nutt. Journ. Acad. Phila II. 1:153. 1848. Armeria bryoides Kuntze, Rev. Gen. 2:432. 1891. (Nuttall, dividing range of the Rocky Mts.)
  Pulvinate-caespitose, taprooted perennial, more or less strongly arachnoid-woolly; leaves numerous, 2-4(5) mm. long, narrowly to subdeltoidally triangular, firm, tending to be whitish-margined, closely appressed, commonly (except in vigorous young shoots) concealing the stem and tending to form quadrangular shoots; flowers solitary and sessile or short-pedicellate at the ends of the stems; hyaline intervals of the calyx narrow, flat, obscured by the wool; calyx lobes somewhat resembling the leaves; corolla white to bluish or purple, the tube mostly 5-10 mm. long, equaling or nearly twice as long as the calyx, the lobes 3-5 mm. long; style 2-5 mm. long.

Dry, open or sparsely wooded, sometimes alkaline slopes, mostly in the foothills and dry valleys and plains, sometimes ascending to moderate elevations in the mountains, often occurring with sagebrush; w. Neb. and adj. Colo. and Wyo. to s. w. Mont. , e. Ida. (Lemhi, Custer, and Teton cos. ), and irregularly to Lake Co. , Oreg. , and Nye Co. , Nev. May-July.

Phlox muscoides is not always sharply distinct from P. hoodii, at least in the herbarium, and Wherry (Not. Nat. Acad. Phila. 87:15. 1941) has concluded that the type of P. muscoides is actually a depressed form of P. hoodii. Although the leaves of Wyeth's specimen are not so broad as those of some others, I believe the plant falls well within the normal range of variability of the species to which the name is here attached.

PHLOX PULVINATA (Wherry) Cronq. hoc loc.
  P. caespitosa ssp. pulvinata Wherry, Not. Nat. Acad. Phila. 87:10. 1941. (Wherry s. n. ,
    Brainard Lake, Boulder Co. , Colo. , June 21, 1937)
  P. caespitosa ssp. platyphylla Wherry, Not. Nat. Acad. Phila 87:9. 1941. (Davis 1673, head
    of Wild Horse Creek, e. side of Mt. Hyndman, Custer Co. , Ida. )
  P. variabilis ssp. nudata Wherry, Morris Arb. Monog. 3:131. 1955. (C. L. Hitchcock 16961,
    Brandon Lakes, Madison Co. , Mont. )
  P. caespitosa sensu auct. , misapplied, not of Nutt.
  Taprooted, caespitose perennial, always depressed and mat-forming; leaves crowded, fairly firm, mostly 5-12(15) mm. long, the surfaces glabrous to glandular or hairy, the margins only slightly thickened and not markedly whitish, usually more or less ciliate toward the base; flowers solitary (3) at the ends of the stems, sessile or short-pedicellate; calyx usually glandular-hairy, occasionally eglandular or even glabrous, the intercostal membranes flat, the lobes flattened, with inconspicuous midvein; corolla white or light bluish, the tube 9-13 mm. long, up to twice as long as the calyx, the lobes 5-7 mm. long; style 2-5 mm. long.

  Open, often rocky places at moderate to high elevations in the mountains; s. w. Mont. (n. e. to Judith Basin Co. ) to n. N. M. , westward to Union Co. , Oreg. , Elko Co. , Nev. , and Juab and Iron cos. , Utah. June-Aug.

  This is the species which has regularly been known as P. caespitosa Nutt. The type of P. caespitosa, however, from dry hills on the Flathead R. in Mont. , blooming about April 20, proves to be a compact plant of the taxon usually known as P. douglasii. The latter name, being four years later than P. caespitosa, must subside.

  Phlox condensata (Gray) E. Nels. (incl. P. covillei E. Nels. and P. tumulosa Wherry), a closely allied species ranging from Colo. to Calif. , usually but not always well up in the mountains, has the style mostly only 1-2 mm. long, and is more condensed than P. pulvinata, resembling P. muscoides in habit.

Phlox speciosa Pursh, Fl. Am. Sept. 149. 1814.
  Armeria speciosa Kuntze, Rev. Gen. 2:432. 1891. (Lewis, plains of the Columbia; probably
    actually on the Clearwater R. below Kamiah, Ida. , fide Piper)
  Phlox speciosa var. elatior Hook. Fl. Bor. Am. 2:72. 1838. P. sabinii Dougl. ex Hook. loc.
    cit. as a synonym; ex Benth. in DC. Prodr. 9:305. 1845. P. speciosa var. sabinii Gray,
    Proc. Am. Acad. 8:256. 1870. (Douglas, limestone rocks of the Blue Mts. )
  P. speciosa var. latifolia Hook. Kew Journ. Bot. 3:289. 1851. (Geyer 375, "plateaus of the
    Coeur d'Aleine and Nez Perce")
  P. occidentalis Dur. ex Torr. Pac. R. R. Rep. 4:125. 1857. P. speciosa var. latifolia f. oc-
    cidentalis Brand, Pflanzenr. IV. 250:74. 1907. P. speciosa ssp. occidentalis Wherry,
    Proc. Acad. Phila. 90:133. 1938. P. speciosa var. occidentalis Peck, Man. High. Pl.
    Oreg. 570. 1941. (Durand, near Nevada City, Nevada Co. , Calif. )
  P. lanceolata E. Nels. Rev. W. N. Am. Phl. 29. 1899. P. speciosa var. latifolia f. lanceo-
    lata Brand, Pflanzenr. IV. 250:74. 1907. P. speciosa ssp. lanceolata Wherry, Proc. Acad.
    Phila. 90:133. 1938. P. speciosa var. lanceolata Peck, Man. High. Pl. Oreg. 570. 1941.
    (Piper, Ellensburg, Wash. , in 1897)
  P. whitedii E. Nels. Erythea 7:167. 1899. P. speciosa ssp. lignosa var. whitedii Brand,
    Pflanzenr. IV. 250:74. 1907. (Whited 1036, Wenatchee, Wash. )
  P. speciosa var. nitida Suksd. Deuts. Bot. Monats. 18:132. 1900. P. speciosa var. latifolia
    subvar. nitida Brand, Pflanzenr. IV. 250:74. 1907. P. speciosa ssp. nitida Wherry, Proc.

Phlox idahonis

P. kelseyi

P. multiflora

P. longifolia

P. pulvinata

P. muscoides

JRJ

Acad. Phila. 90:134. 1938. (Suksdorf 2208, w. Klickitat Co., Wash., June 2, 1893) A rare, glabrate form.

P. speciosa ssp. lignosa Brand, Pflanzenr. IV. 250:74. 1907. (Synthetic, composed of varieties suksdorfii and whitedii)

P. speciosa ssp. lignosa var. suksdorfii Brand, Pflanzenr. IV. 250:74. 1907. P. suksdorfii St. John, Res. Stud. State Coll. Wash. 1:104. 1929. (Suksdorf 883, Columbus, Klickitat Co., Wash.)

P. imminens St. John, Res. Stud. State Coll. Wash. 1:102. 1929. (Gabby 53, Columbia R. Valley, Stevens Co., Wash.)

Perennial from a woody taproot, 1.5-4 dm. tall, tending to become shrubby toward the base, generally glandular or glandular-hairy above, at least on the pedicels and calyx, often also more or less hairy below, rarely glabrous throughout; leaves linear to broadly lanceolate, up to 7 cm. long and 1 cm. wide, the internodes evident; inflorescence loosely cymose, leafy-bracteate at least below; intercostal membranes of the calyx flat or slightly carinate; corolla pink to white, the tube 1-1.5 cm. long, shortly or scarcely surpassing the calyx, the lobes 1-1.5 cm. long, nearly always evidently notched at the tip; filaments inserted above or below the middle of the corolla lobes; style very short, 0.5-2 mm. long, mostly shorter than the linear stigmas.

Sagebrush and ponderosa pine areas; Okanagan Valley of s. B.C., southward to Wasco Co., Oreg., and irregularly eastward to n. Ida. (Bonner Co. to Lewis Co.), and w. Mont. (Lewis & Clark Co., Gallatin Co.); Klamath region of s.w. Oreg., southward into the Sierra Nevada of Calif. Apr.-June.

Phlox viscida E. Nels. Rev. W. N. Am. Phl. 24. 1899.

P. stansburyi ssp. compacta var. viscida Brand, Pflanzenr. IV. 250:67. 1907. (Piper 2397, Blue Mts., Columbia Co., Wash.)

P. speciosa f. alba St. John, Fl. S. E. Wash. 329. 1937. (St. John & Brown 3879, opposite Zindel, Asotin Co., Wash.)

P. mollis Wherry in Davis, Fl. Ida. 569. 1952., without Latin diagnosis; Wherry, Morris Arb. Monog. 3:128. 1955. (J. H. Christ 16557, Zaza, Nez Perce Co., Ida.)

Perennial from a taproot, often somewhat woody at the base, and tending to creep below ground as in P. longifolia, 0.5-2 dm. tall; leaves linear (often broadly so) or lance-linear, 1-4 cm. long, 1-5 mm. wide, rather crowded, but the internodes evident; flowers ill-scented, slender-pedicellate in small and compact, leafy-bracteate, terminal cymes; intercostal membranes of the calyx flat or nearly so; corolla pink or purple to sometimes white, the tube 11-15 mm. long, slightly shorter to somewhat longer than the calyx, the lobes 8-14 mm. long, obovate, entire or merely erose; filaments attached above the middle of the corolla tube, often some of the anthers partly exserted; style elongate, mostly 6-9 mm. long.

Open, gravelly or rocky places at moderate elevations in the mountains of Wash. (s. Garfield, Columbia, and Asotin cos.), Oreg. (e. Wallowa Co.), and Ida. (Nez Perce and Idaho cos.); reputedly irregularly southward to Utah and Nev. May-June.

The typical form of the species is evidently glandular or glandular-hairy above. P. mollis Wherry is a loosely woolly, scarcely viscid or glandular form so far known only from several collections about Zaza, Nez Perce Co., Ida. Further exploration may warrant the distinction of P. mollis as a local variety, but the failure of vesture to mark proper varieties in the allied P. longifolia and P. speciosa suggests the need for caution here, especially since P. viscida is still known from relatively few collections.

## Polemonium L.

Flowers borne in diverse sorts of basically cymose (determinate) inflorescences; calyx essentially herbaceous, somewhat accrescent and becoming chartaceous; corolla tubular-funnelform or subsalverform to nearly rotate, mostly blue or white, less often purple, yellow, or flesh-colored to salmon; stamens about equally inserted; seeds 1-10 per locule, sometimes becoming mucilaginous when moistened; perennial (1 species annual) herbs with alternate, pinnately compound or very deeply pinnatifid leaves, ordinarily with well-defined leaflets, these either entire or so deeply 2- to 5-cleft as to appear verticillate; plants generally more or less glandular (at least in the inflorescence) and often strongly mephitic.

About 20 species of N. Am. and S. Am. and Eurasia, best developed in the cordillera of w. N. Am. The larger species, such as P. occidentale, are sometimes called Jacob's-ladder, the dwarf alpine ones, such as P. viscosum, sky pilot. (Said to be named for Polemon, a Greek philosopher; also said to be derived from the Greek polemos, strife. According to Hooker, Curtis' Bot. Mag. 57: pl. 2979. 1830, "It is said the discovery of its supposed properties, occasioned a war between two kings.")

Polemonium takes well to cultivation. P. occidentale and P. carneum especially, among our species, seed prolifically and are most desirable additions to the wild garden.

References:

Davidson, J. F. The genus Polemonium. U. Calif. Pub. Bot. 23: 209-82. 1950. (19 species recognized.)

Wherry, E. T. The genus Polemonium in North America. Am. Midl. Nat. 27:741-60. 1942. (42 species recognized.)

1 Annual; corolla shorter than or merely equaling the calyx            P. MICRANTHUM
1 Perennial; corolla much longer than the calyx
  2 Corolla more or less funnelform or tubular-funnelform to subsalverform, longer than wide, the lobes shorter than the tube; dwarf (rarely to 4 dm.) alpine and subalpine, petrophilous, very strongly glandular and mephitic plants with numerous stems from a taproot and much-branched (sometimes elongate) caudex
    3 Leaflets opposite or offset, undivided; corolla mostly 12-15 mm. long; Cascades
                                                      P. ELEGANS
    3 Leaflets (or most of them) so deeply 2- to 5-cleft as to appear verticillate; corolla (13) 17-25(30) mm. long; e. of the Cascades, and also in the Coast Mts. of s. B. C.
                                                   P. VISCOSUM
  2 Corolla campanulate or broader, about as wide as long, or wider, the lobes about as long as the tube, or longer; plants of various habit and habitat, usually less strongly glandular and mephitic
    4 Flowers large, the calyx 7.5-14 mm. long at anthesis, the corolla (dry) (15)18-28 mm. long; corolla salmon or flesh-colored to yellow, white, or purple, only rarely blue
                                                   P. CARNEUM
    4 Flowers smaller, the calyx up to 8 mm. long at anthesis, the corolla up to 17 mm. long; corolla mostly blue (often with a yellow eye) or white
      5 Leaflets narrowly linear; flowers white or creamy; tall, erect plants with clustered stems; low altitudes in e. Wash.            P. PECTINATUM
      5 Leaflets broader; flowers usually blue; habit otherwise; widespread montane plants
        6 Stems solitary from the upturned end of a mostly rather short and simple horizontal rhizome, erect, (1.5)4-10 dm. tall; very wet places at moderate elevations in the mountains            P. OCCIDENTALE
        6 Stems more or less clustered from a branched (sometimes elongate and rhizome-like) caudex which usually surmounts a taproot; plants lax, up to 3(5) dm. tall, of moist to dry, often rocky places, at moderate to high elevations in the mountains            P. PULCHERRIMUM

Polemonium carneum Gray, Syn. Fl. 2¹:151. 1878.

  P. incarnatum attr. to Gray by J. D. Hook. Curtis' Bot. Mag. 115: pl. 6965. 1887. (Greene 796, near Yreka, Siskiyou Co., Calif.; type of both P. carneum and P. incarnatum, the latter apparently merely a clerical error by Greene on labels distributed with the type collection)

  P. amoenum Piper, Erythea 7:174. 1899. P. carneum ssp. amoenum Brand, Pflanzenr. IV. 250:41. 1907. P. carneum f. amoenum G. N. Jones, U. Wash. Pub. Biol. 5:215. 1936. (Lamb 1178, Humptulips, Grays Harbor Co., Wash.) A form with blue (or purple?) flowers, otherwise not especially unusual.

  P. carneum var. luteum Gray, Syn. Fl. 2nd ed. 2¹:412. 1886. P. luteum Howell, Fl. N. W. Am. 463. 1901. P. carneum ssp. luteum Brand, Pflanzenr. IV. 250:41. 1907. P. carneum

f. luteum Wherry, Am. Midl. Nat. 27:748. 1942. (Howell, Cascade Range, Oreg., June, 1885)

Perennial with loosely clustered (sometimes solitary?) stems from a woody rhizome or caudex, loosely erect, 3-10 dm. tall, viscid-villous in the inflorescence, otherwise glabrous or nearly so except for the villous-ciliate margins of the petioles or the lower portion thereof, or sometimes the stem viscid-villous throughout; leaflets mostly 11-19, lanceolate to ovate or elliptic, generally acute, thin, mostly 1.5-4.5 cm. long and 6-23 mm. wide, the 3 terminal ones sometimes partly confluent; basal leaves long-petiolate, cauline progressively less so, but still fairly well developed, the stem appearing amply leafy; flowers rather few and often long-pedicellate, in an open, terminal, generally leafy inflorescence; calyx 7.5-14 mm. long at anthesis, the lobes shorter or longer than the tube; corolla campanulate, (15)18-28 mm. long, the lobes longer than the tube, variable in color, often flesh-colored, salmon, or yellow, sometimes lavender to dark purple, rarely blue or white. N=9.

Thickets, woodlands, and forest openings, from near sea level to moderate elevations in the mountains; Klamath region of s.w. Oreg. and adj. Calif., n. through the Cascades to Skamania Co., Wash.; Olympic and Grays Harbor region of w. Wash., and reported from Tillamook Co., Oreg.; about San Francisco Bay, Calif. Seldom collected. May-July.

The color form from the Olympic Peninsula is especially suitable for the wild garden, maintaining itself easily without being weedy. It transplants readily and seeds freely.

Polemonium elegans Greene, Pitt. 3: 305. Apr. 8, 1898.
   P. viscosum ssp. elegans Brand, Pflanzenr. IV. 250:44. 1907. (Piper, Mt. Rainier, Wash., Aug., 1895)
   P. bicolor Greenm. Bot. Gaz. 25:262. Apr. 15, 1898. P. elegans f. bicolor Wherry, Am. Midl. Nat. 27:743. 1942. (Allen 294, Mt. Rainier, Wash.)

Low perennial from a stout taproot and much-branched (sometimes elongate) caudex, up to 1.5 dm. tall, densely stipitate-glandular or glandular-villosulous, and strongly mephitic; leaves chiefly but not wholly basal, rather short-petiolate, with conspicuously expanded and persistent chartaceous base; leaflets mostly 13-27, opposite or offset, crowded, 2.5-6 mm. long and 1-3.5 mm. wide, entire; inflorescence persistently capitate-cymose; calyx 5.5-8.5 mm. long at anthesis, the acute or acutish lobes about equaling or a little shorter than the tube; corolla blue, funnelform or tubular-funnelform, 12-15 mm. long, often only half as wide, the lobes shorter than the tube; stamens scarcely equaling the corolla, but exserted from the tube and often exposed by the somewhat spreading limb.

Open, rocky places at high altitudes in the mountains, commonly above timber line; Cascades of Wash. and adj. B.C. July-Aug.

Polemonium micranthum Benth. in DC. Prodr. 9:318. 1845.
   Polemoniella micrantha Heller, Muhl. 1:57. 1904. (Douglas, Columbia R.)

Slender, taprooted annual, up to 3 dm. tall, simple and erect (especially when small) to more often freely branched and loosely ascending (or nearly prostrate), glandular-puberulent throughout, or sometimes partly glabrate; leaves well distributed along the stems, with mostly 7-15 opposite or offset leaflets 2-9 mm. long and 1-4 mm. wide; flowers terminal and solitary, but soon appearing leaf-opposed because of the sympodial development of the stem, the peduncle at first short, later sometimes elongate to 1-2 cm.; corolla inconspicuous, ordinarily white, 2-5 mm. long, equaling or commonly a little shorter than the calyx, the corolla lobes about equaling the tube; stamens included.

Open, dry or vernally moist places in the plains and foothills, often with sagebrush, sometimes as a weed in waste places; Wash. and s. B.C. to Calif., Utah, and w. Mont., chiefly e. of the Cascades; Argentina and Chile. Mar.-May.

This species is unique in Polemonium in its annual habit, inconspicuous corolla, and sympodial stems with solitary, leaf-opposed flowers. Its segregation as a distinct, monotypic genus is therefore not wholly unwarranted.

Polemonium occidentale Greene, Pitt. 2:75. 1890.
   P. acutiflorum ssp. occidentale Hultén, Fl. Alas. 8:1323. 1948. P. caeruleum ssp. occi-

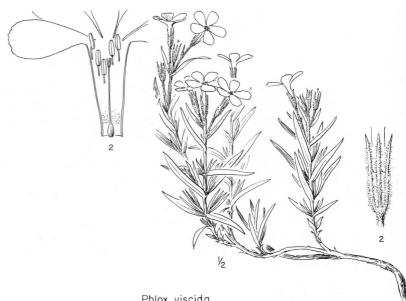

Phlox viscida

Phlox speciosa

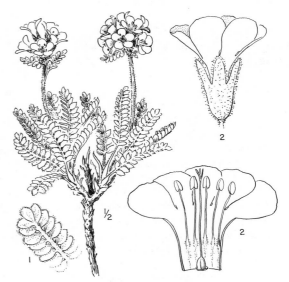

Polemonium elegans

Polemonium carneum

JRJ

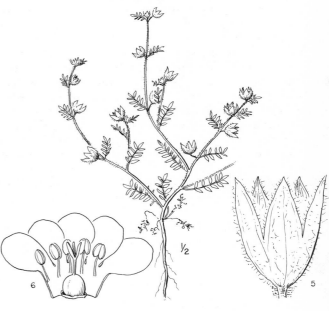

Polemonium micranthum

dentale Davids. U. Calif. Pub. Bot. 23:225. 1950. P. coeruleum var. occidentale St. John, Fl. S. E. Wash. 2nd ed. 551. 1956. (Baker 365, Gunnison, Colo.; lectotype by Davidson)

P. caeruleum var. pterosperma Benth. in DC. Prodr. 9:317. 1845. P. pterospermum Brand, Helios 22:77. 1905, not of Nels. & Cockerell in 1903. (Drummond, Rocky Mts.) This appears to furnish the proper epithet should the taxon here treated as P. occidentale be reduced to varietal rank under any other species.

P. occidentale var. intermedium Brand, Pflanzenr. IV. 250:33. 1907. P. intermedium Rydb. Bull. Torrey Club 40:478. 1913. (Macoun 66622, mouth of the Salmon R., s. B. C., is the first of two specimens cited) A form with relatively broad and openly branched inflorescence.

P. occidentale ssp. amygdalinum Wherry, Am. Midl. Nat. 27:750. 1942. (Suksdorf, Mt. Adams, Wash., Aug. 30, 1881) A depauperate, subscapose extreme. This appears to furnish the nomenclaturally proper epithet should the taxon here treated as P. occidentale be reduced to subspecific rank under any other species.

Stems solitary from the upturned end of a mostly fairly short and simple, horizontal rhizome, (1.5)4-10 dm. tall; basal and lower cauline leaves well developed and rather long-petiolate, with commonly 11-27 more or less lanceolate and acute leaflets (0.5)1-4 cm. long and (0.2)2-13 mm. wide, the three terminal ones sometimes confluent; middle and upper leaves more or less reduced and less petiolate, more conspicuously so in smaller plants; plants more or less strongly glandular-villous or -villosulous in the inflorescence or sometimes nearly throughout; flowers crowded in a typically elongate and somewhat thyrsiform to sometimes more openly branched and corymbiform (or in smaller plants even subcapitate) inflorescence; calyx 5-8 mm. long at anthesis, the lobes equaling or commonly a little shorter than the tube; corolla mostly sky blue or nearly so (white in rare individuals), mostly 1-1.6 cm. long and generally at least as wide, the lobes longer than the tube; stamens usually about equaling or a little shorter than the corolla; style generally exserted and conspicuously surpassing the stamens; seeds winged or wingless.

Swamps, stream edges, and similar very wet places, mostly at moderate elevations in the mountains, occasionally in the foothills or near timber line; cordilleran region from Alta. and B. C. to Calif. and Colo., almost entirely e. of the Cascades in Wash., and mostly so in Oreg.; also rarely in n. Minn. and probably n. Mich. June-Aug.

Polemonium occidentale, P. vanbruntiae Britt., P. acutiflorum Willd., and P. caeruleum L. are geographically and morphologically characterized taxa which are not always clearly distinguishable in the herbarium. The Eurasian P. caeruleum proper, unlike the American P. occidentale and P. vanbruntiae, is a plant of open, often rocky slopes, which tends to have a descending rootstock or root, and which at least frequently has clustered stems. It would seem to be connected to P. occidentale, if at all, only through P. acutiflorum [P. caeruleum ssp. villosum (Rud.) Brand], which to judge from the herbarium specimens may vary both in habitat and root-habit, but which usually differs from P. occidentale, at least, in its larger flowers and relatively shorter styles, as well as in some other not wholly stable features. P. vanbruntiae, of the Appalachian region, is geographically isolated from the rest of its group. It remains to be determined whether the other three taxa actually fade into one another in a geographical replacement series, or whether the populations remain essentially distinct and the herbarium evidence of intergradation is illusory. The reduction of P. occidentale to P. caeruleum raises serious difficulties to the continued maintenance of such universally recognized species as P. reptans L., P. foliosissimum Gray, and P. pulcherrimum Hook., since the nature of the basal and underground parts and the characteristic habitat furnish some of the best characters by which these species are distinguished from P. occidentale and P. vanbruntiae. Furthermore, the retention of P. occidentale in infraspecific status under either P. caeruleum or P. acutiflorum will require a new combination (and a different epithet, according to whether one prefers subspecific or varietal status) if one wishes to abide by the rules of nomenclature. It therefore seems preferable to retain P. occidentale as a species until the necessity to subordinate it to P. caeruleum or P. acutiflorum is more clearly demonstrated.

Polemonium pectinatum Greene, Bull. Calif. Acad. Sci. 1:10. 1884. (Hilgard, e. Wash. in 1882)

Stems clustered on an apparent taproot, 3-8 dm. tall; plants glandular-hairy in the inflorescence, otherwise essentially glabrous; leaves wholly cauline, short-petiolate, with mostly 11-17 narrowly linear, rather distant leaflets that are mostly 1.5-5 cm. long and up to 2 mm. wide; inflorescence branched and corymbiform, the individual flowers rather short-pedicellate; calyx 3-6 mm. long at anthesis, the lobes equaling or a little shorter than the tube; corolla white or slightly creamy, campanulate or broader, 9-14 mm. long, the lobes longer than the tube; stamens equaling or a little shorter than the corolla.

Moist bottom lands in n. Whitman and s. Spokane cos., Wash. May-June.

Polemonium pulcherrimum Hook. Curtis' Bot. Mag. 57: pl. 2979. 1830.
  P. caeruleum var. pulcherrimum Hook. Fl. Bor. Am. 2:71. 1838. (Drummond, highest
  Rocky Mts.)
  P. mexicanum Nutt. Journ. Acad. Phila. 7:41. 1834, not of Cerv. in 1816. P. parvifolium
  Nutt. ex Rydb. Bull. Torrey Club 24:253. 1897. P. pulcherrimum ssp. parvifolium Brand,
  Pflanzenr. IV. 250:35. 1907. P. pulcherrimum var. parvifolium Nels. in Coult. & Nels.
  New Man. Bot. Rocky Mts. 404. 1909. P. pulcherrimum f. parvifolium Wherry, Am. Midl.
  Nat. 27:751. 1942. (Wyeth, Flathead R., Mont.) = var. pulcherrimum.
  P. viscosum var. pilosum Greenm. Bot. Gaz. 25:263. 1898. P. viscosum ssp. pilosum
  Piper, Contr. U.S. Nat. Herb. 11:467. 1906. P. pulcherrimum ssp. parvifolium var. pi-
  losum Brand, Pflanzenr. IV. 250:36. 1907. P. pilosum G. N. Jones, U. Wash. Pub. Biol.
  5:215. 1936. P. shastense f. pilosum Wherry, Am. Midl. Nat. 27:752. 1942. (Allen 261,
  Goat Mts., Cascades, Wash.) = var. pulcherrimum.
  P. haydeni A. Nels. Bull. Torrey Club 26:353. 1899. P. pulcherrimum ssp. parvifolium
  var. haydeni Brand, Pflanzenr. IV. 250:35. 1907. (Hayden, Jackson's Hole, Wyo., June
  15, 1860) = var. pulcherrimum.
  POLEMONIUM PULCHERRIMUM var. DELICATUM (Rydb.) Cronq. hoc loc. P. delicatum
  Rydb. Bull. Torrey Club 28:29. 1901. P. pulcherrimum ssp. delicatum Brand, Pflanzenr.
  IV. 250:35. 1907. (Rydberg & Vreeland 5720, West Spanish Peak, Colo.)
  P. californicum Eastw. Bot. Gaz. 37:437. 1904. (Eastwood, Yosemite Nat. Pk., July, 1902)
  = var. calycinum.
  P. calycinum Eastw. Bot. Gaz. 37:438. 1904. P. pulcherrimum ssp. delicatum var. calycinum
  Brand, Pflanzenr. IV. 250:35. 1907. (Bruce 1212, Mt. Lassen, Calif.)
  P. tricolor Eastw. Bot. Gaz. 37:439. 1904. P. pulcherrimum ssp. tricolor Brand, Pflanzenr.
  IV. 250:35. 1907. (Chandler 1671, Marble Mt., Siskiyou Co., Calif.) The oldest available
  epithet in subspecific status for the taxon here treated as P. pulcherrimum var. calycinum.
  P. rotatum Eastw. Bot. Gaz. 37:441. 1904. (Maclean, Klondike, Yukon Terr.) = var. pul-
  cherrimum.
  P. shastense Eastw. Bull. Torrey Club 32:205. 1905. P. pulcherrimum ssp. parvifolium var.
  berryi subvar. shastense Brand, Pflanzenr. IV. 250:36. 1907. (Copeland 3515, Mt. Shasta,
  Calif.) = var. pulcherrimum.
  P. columbianum Rydb. Bull. Torrey Club 40:477. 1913. (Leiberg 1205, divide between St.
  Joe and Clearwater rivers, Ida.) = var. calycinum.
  P. orbiculare Gand. Bull. Soc. Bot. France 65:58. 1918. P. haydeni f. orbiculare Wherry,
  Am. Midl. Nat. 27:751. 1942. (Scheuber, Bridger Mts., Mont., June 9, 1901) = var. pul-
  cherrimum.
  P. oreades Gand. Bull. Soc. Bot. France 65:58. 1918. P. californicum f. oreades Wherry,
  Am. Midl. Nat. 27:751. 1942. (Cusick 2750, mountains of Cycan Valley, s. Oreg.) = var.
  pulcherrimum.
  P. oregonense Gand. Bull. Soc. Bot. France 65:58. 1918. P. californicum f. oregonense
  Wherry, Am. Midl. Nat. 27:751. 1942. (Cusick 1717, Blue Mts., Oreg.) = var. calycinum.
  P. paddoense Gand. Bull. Soc. Bot. France 65:58. 1918. P. californicum f. paddoense Wherry, Am. Midl. Nat. 27:751. 1942. (Suksdorf 2766, Mt. Paddo (Adams) Wash.) = var. calycinum, passing to var. pulcherrimum.
Perennial, 0.5-3(5) dm. tall, generally with several loosely erect, more or less clustered
stems from a branched (sometimes elongate and rhizomelike) caudex that usually surmounts a
taproot; plants more or less strongly glandular or glandular-villous in the inflorescence, else-

where mostly less so or subglabrate; basal leaves well developed and often tufted, the cauline ones often more or less reduced; leaflets mostly 11-25, opposite or offset; inflorescence a congested but scarcely capitate cyme; calyx 4-6 mm. long at anthesis, with acute or merely acutish lobes; corolla blue, often with a yellow eye, more or less campanulate, or broader, 7-13 mm. long and commonly about as wide, or wider, the lobes about equaling or nearly twice as long as the tube. N=9, 18.

At moderate to high elevations in the mountains, often in moist or shaded places, or in exposed sites above timber line; Alas. to Calif. and Colo. May-Aug.

Polemonium pulcherrimum as here defined consists of three varieties. The varieties pulcherrimum and calycinum, with broadly overlapping range, constitute well-marked ecotypes which, in the absence of other considerations, might be treated as closely allied and poorly defined species. In the s. Rocky Mts. both of these varieties give way to the var. delicatum (Rydb.) Cronq., which lies squarely between the two both morphologically and ecologically, and which overlaps so much with both that it could scarcely be separated from either were it not geographically segregated. Thus, many of the specimens of var. delicatum could pass readily for var. calycinum, and many others for var. pulcherrimum, if the labels were covered, while more northern specimens intermediate between varieties pulcherrimum and calycinum may so exactly simulate var. delicatum that some of them have been so labeled (as a species) by a well-known student of the family. The var. delicatum typically has leaflets near the upper size limit for var. pulcherrimum and near the lower limit for var. calycinum, with the three terminal ones weakly or scarcely confluent, and has the calyx lobes a little longer than the tube, but it is variable in all these features. The extensive synonymy reflects the variability of the species, but the numerous local races are scarcely definable when viewed against the background of the whole population. The two varieties of the species which occur in our range may be characterized as follows:

1 Relatively small and compact plants, mostly of high altitudes in the mountains, often above timber line, regularly taprooted, seldom over 2(3) dm. tall; leaflets rarely as much as 1 cm. long and 5 mm. wide, generally all essentially distinct; calyx lobes usually shorter than or about equaling the tube; Alas. and Yukon to Wyo., w. Utah, c. Nev., and the Sierra Nevada of Calif., mostly at progressively higher altitudes southward

var. pulcherrimum

1 Larger, laxer, and more robust plants of moderate elevations in the mountains, often 3 dm. and sometimes as much as 5 dm. tall, the taproot often poorly or scarcely developed; leaflets fewer and larger than in var. pulcherrimum, the larger ones seldom less than 1 cm. long and 5 mm. wide, sometimes up to 3.5 cm by 15 mm., the 3 terminal ones more or less confluent; calyx lobes mostly longer than the tube, sometimes twice as long; Cascade-Sierra region of Wash., Oreg., and Calif. (and adj. Nev.), e. to c. Ida. and rarely w. Mont.

var. calycinum (Eastw.) Brand

Polemonium viscosum Nutt. Journ. Acad. Phila. II. 1:154. 1848. (Nuttall, toward the headwaters of the Platte)

?P. confertum Gray, Proc. Acad. Phila. 1863:73. 1864. P. confertum var. elatius Brand, Pflanzenr. IV. 250:44. 1907. (Hall & Harbour 450 in part, Colo.; lectotype by Rydberg) This name, which has sometimes been misapplied to typical P. viscosum, properly applies to a group of specimens, chiefly from Colo., which approach P. pulcherrimum var. delicatum in their larger (often 1 cm. long) and less strongly or only partly (or not at all) verticillate leaflets. The proper taxonomic status of these specimens is dubious. Some of the more extreme among them have been regarded as hybrids between the two species they resemble. See Am. Midl. Nat. 27:745. 1942 for nomenclature.

Low perennial from a stout taproot and much-branched (sometimes elongate) caudex, up to 2(4) dm. tall, densely stipitate-glandular or glandular-villosulous, and strongly mephitic; leaves chiefly but not wholly basal, up to 1.5(2) dm. long including the rather short petiole, this with conspicuously expanded and chartaceous, persistent base; leaflets numerous, crowded, mostly or all 2- to 5-cleft to the base or nearly so, thus apparently verticillate, the individual segments 1.5-6 mm. long and 1-3 mm. wide; inflorescence densely cymose-capitate, usually elongating a little in fruit; calyx 7-12 mm. long at anthesis, the narrow, acute lobes shorter

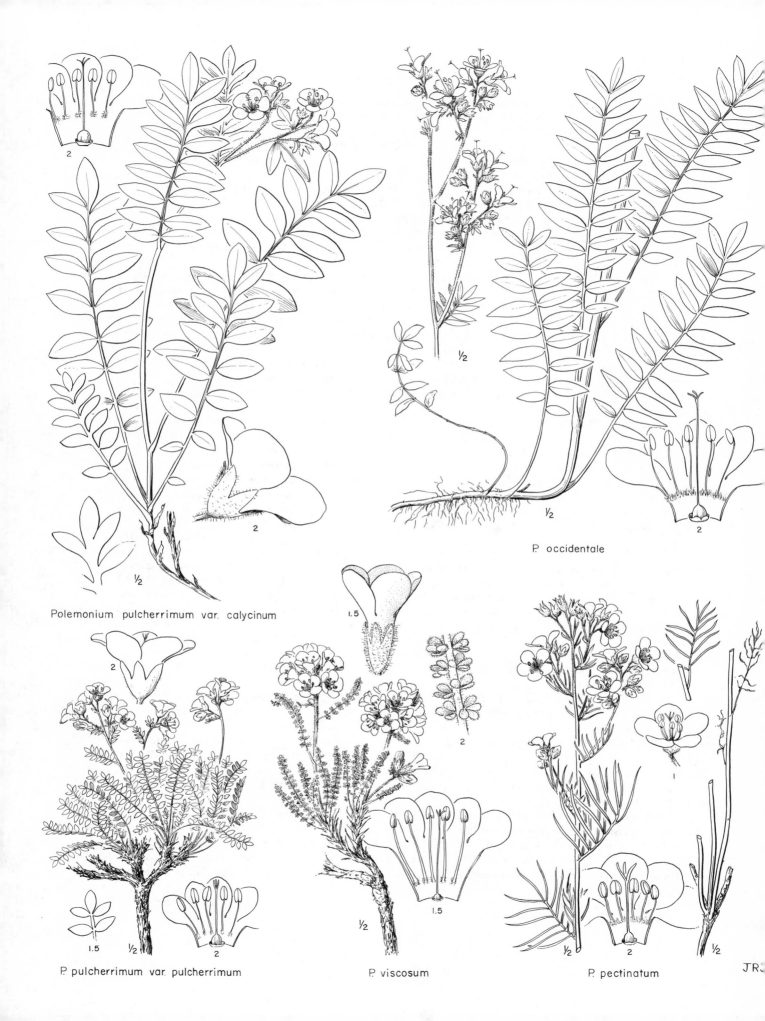

2

Polemonium pulcherrimum var. calycinum

P. occidentale

2

P. pulcherrimum var. pulcherrimum

P. viscosum

P. pectinatum

JR

than the tube; corolla blue, more or less funnelform or tubular-funnelform, (13)17-25(30) mm. long, longer than wide, often very conspicuously so, the lobes shorter than the tube; stamens distinctly shorter than to sometimes nearly equaling the corolla.

Open, rocky places at high altitudes in the mountains, commonly above timber line; Rocky Mt. region, from s.w. Alta. to n. N.M., w. to Okanogan Co., Wash., the Wallowa Mts. of n.e. Oreg., and the mountains of c. Nev.; reported by Davidson from the Coast Ranges of s. B.C. Apparently wholly absent from the Cascade-Sierra axis, being replaced in the Sierra Nevada, White Mts., and extreme s. Cascades by the closely allied but perhaps specifically distinct P. eximium Greene (with two varieties), and in the n. Cascades by the wholly distinct P. elegans. July-Aug.

## HYDROPHYLLACEAE. Waterleaf Family

Flowers gamopetalous, perfect, regular or nearly so, generally 5-merous as to the calyx, corolla, and androecium; calyx cleft to the middle or more commonly to the base or nearly so; corolla imbricate (rarely contorted) in bud; filaments alternate with the corolla lobes, attached near the base or well up in the corolla tube, and very often flanked by a pair of small scales; ovary superior, ordinarily 2-carpellary, with distinct or partly (rarely wholly) united styles and mostly capitate stigmas; placentae 2, usually parietal, but often more or less intruded, and sometimes meeting and joined, the ovary then 2-locular and the placentation axile; capsule loculicidal (sometimes also septicidal), or irregularly dehiscent; ovules few-many; seeds 1-many; embryo straight; endosperm present; annual to perennial herbs, rarely shrubs, with alternate or sometimes partly or wholly opposite, entire to cleft or compound leaves, the flowers solitary or in variously modified (often helicoid) cymes.

About 25 genera and 250-300 species (fide Constance), of wide distribution, best developed in w. U.S. Some of our species of Hesperochiron, Hydrophyllum, Nemophila, and Phacelia are good garden prospects.

Reference:

Brand, A. Hydrophyllaceae. Pflanzenr. IV. 251 (Heft 59). 1913.

1 Flowers (at least in our species) borne singly, or sometimes a few of them in a lax, few-flowered, terminal inflorescence
  2 Leaves entire or subentire; placentae more or less strongly intruded and partitionlike
    3 Plants acaulescent, perennial, the flowers borne on basal peduncles (habit unique in the family); mesophytes         HESPEROCHIRON
    3 Plants caulescent, ours dichotomously branched annual xerophytes with the flowers sessile in the forks         NAMA
  2 Leaves coarsely toothed to pinnatifid; placentae enlarged but not at all partitionlike, nearly filling the young ovary, and forming a lining for the mature capsule
    4 Calyx provided with spreading or reflexed auricles at the sinuses; seeds with a partial or complete, persistent or deciduous outer covering (cucullus) over the seed coat (unique among our genera in both of these regards)         NEMOPHILA
    4 Calyx exauriculate; cucullus wanting         ELLISIA
1 Flowers in definite inflorescences, not solitary
  5 Style entire or nearly so; more or less fibrous-rooted, mesophytic perennials with reniform-orbicular, chiefly basal leaves; placentae strongly intruded and partitionlike         ROMANZOFFIA
  5 Style evidently though sometimes shortly lobed; habit otherwise
    6 Inflorescence of subdichotomously branched (sometimes capitate) cymes, without an evident central axis; placentae enlarged but not at all partitionlike, nearly filling the young ovary, and forming a lining for the mature capsule; fibrous-rooted, perennial mesophytes with cleft leaves and exserted stamens         HYDROPHYLLUM
    6 Inflorescence of one or more sympodial, more or less helicoid cymes which may be aggregated into a compound, often thyrsoid inflorescence; placentae evidently intruded and partitionlike; taprooted annuals or perennials, mesophytic or xerohytic, with entire to variously dissected leaves and included or exserted stamens         PHACELIA

## Ellisia L. Nom. Conserv.

Flowers solitary in or opposite the axils, the stem sometimes also terminating in a lax, few-flowered cyme; calyx cleft to near the base, exappendiculate, accrescent and subrotate at maturity; corolla campanulate, white to partly or wholly lavender, about equaling or slightly surpassing the calyx; filaments included, attached to the base of the obscurely appendiculate corolla tube; style surpassed by the corolla, cleft up to half its length; ovules 2 on each of the 2 expanded parietal placentae; capsule 1-locular, dehiscent by two valves; seeds finely reticulate, without a cucullus; small, branching annual with pinnatifid leaves, at least the lower ones opposite.

A single species, as defined by Constance. (Named for John Ellis, 1710-76, English botanist)

Reference:

Constance, Lincoln. The genera of the tribe Hydrophylleae of the Hydrophyllaceae. Madroño 5:28-33. 1939.

Ellisia nyctelea L. Sp. Pl. 2nd ed. 1662. 1763.

Ipomoea nyctelea L. Sp. Pl. 160. 1753. Polemonium nyctelea L. Sp. Pl. 2nd ed. 231. 1762. Macrocalyx nyctelea Kuntze, Rev. Gen. 2:434. 1891. Nyctelea nyctelea Britt. in Britt. & Brown, Ill. Fl. 2nd ed. 3:67. 1913. N. americana Moldenke, Bull. Torrey Club 59:155. 1932. (Gronovius, Va.)

Plants 0.5-4 dm. tall, usually freely branched, rather sparsely strigose or in part hirsute, the petioles coarsely ciliate toward the base; leaf blades up to 6 cm. long and 5 cm. wide, pinnatifid, with wing-margined rachis and mostly 3-6 pairs of rather narrow, entire or few-toothed lateral segments; pedicels mostly less than 1 cm. long at anthesis, sometimes up to 5 cm. in fruit; calyx foliaceous-accrescent, the ovate or lance-ovate lobes veiny and commonly 1 cm. long at maturity; seeds mostly 4, 2.5-3 mm. long. N=10.

In moist, shaded bottoms; Mich. to Sask., Alta., and Mont., s. to N.M., Tex., and La.; and along the Atlantic seaboard; barely entering our range from the east. June-July.

## Hesperochiron Wats. Nom. Conserv.

Flowers borne singly on basal peduncles; calyx cleft to near the base; corolla campanulate to rotate, white, often tinged or marked with lavender or purple, the tube generally evidently hairy within; filaments shorter than the corolla, attached to the base of the tube, dilated at the base, often unequal; style shortly 2-cleft; placentae shortly intruded; capsule loculicidal, 1-celled, many-seeded; small, scapose, perennial herbs from an often caudexlike taproot, with petiolate, entire or subentire, oblanceolate to elliptic or ovate leaves. N=8.

The genus consists of the two following species. (Name from the Greek hesperos, of the evening, and Chiron, the centaur; of uncertain significance.)

The species are tempting to native-garden enthusiasts, and they might be coaxed into doing well, e. of the Cascades.

1 Corolla saucer-shaped or rotate, the lobes generally much longer than the tube; leaves glabrous on the lower (or both) surfaces; flowers mostly 1-5(8)          H. PUMILUS
1 Corolla campanulate or funnelform, the lobes seldom much, if at all, longer than the tube; leaves often short-hairy on both sides; flowers more or less numerous, seldom as few as five                                                                          H. CALIFORNICUS

Hesperochiron californicus (Benth.) Wats. Bot. King Exp. 281. 1871.

Ourisia californica Benth. Pl. Hartw. 327. 1839. (Hartweg, "in montibus Sacramento")

Hesperochiron latifolius Kell. Proc. Calif. Acad. 5:44. 1873. H. californicus var. latifolius Brand, U. Calif. Pub. Bot. 4:226. 1912. (Kellogg, Cisco, Calif., June 19, 1870)

Capnorea watsoniana Greene, Pitt. 5:44. 1902. Hesperochiron californicus var. watsonianus Brand, Pflanzenr. IV. 251:166. 1913. (Watson 956, near Carson City, Nev.)

Capnorea lasiantha Greene, Pitt. 5:47. 1902. Hesperochiron californicus var. watsonianus subvar. lasianthus Brand, Pflanzenr. IV. 251:166. 1913. H. lasianthus St. John, Fl. S.E.

Wash. 332. 1937. (Several specimens from Wash., Oreg. and Ida. are cited)
Capnorea macilenta Greene, Pitt. 5:48. 1902. (Henderson 2747, Moscow, Ida.)
Capnorea incana Greene, Pitt. 5:49. 1902. Hesperochiron californicus var. incanus Brand,
    Pflanzenr. IV. 251:167. 1913. H. incanus Garrett, Spring Fl. Wasatch Reg. 4th ed. 126.
    1927. (Nelson & Nelson 5409, Monida, Mont.)
Perennial from a taproot which is sometimes surmounted by a branching caudex, without
rhizomes; leaf blades oblanceolate to elliptic or ovate, up to 7.5 cm. long and 2.5 cm. wide,
equaling or longer than the petiole, commonly shortly spreading-hairy on both sides, varying
to occasionally glabrous or marginally strigose-ciliate; peduncles more or less numerous,
seldom as few as 5, mostly shorter than the leaves; corolla white or tinged with lavender or
purple, sometimes lavender-veined, campanulate or funnelform, 1-2(2.5) cm. long, as long
as, or a little longer than, wide, often slightly oblique and unequally lobed; corolla lobes mostly 0.8-1.5 times as long as the internally densely hairy tube, often with scattered long hairs
on the inner surface. N=8.
Mostly in more or less alkaline meadows and flats in the plains, foothills, and intermontane
valleys; Wash. to n. Baja Calif., e. to w. Mont., w. Wyo., and n.e. Utah; wholly e. of the
Cascades in our range. Apr.-June.


Hesperochiron pumilus (Griseb.) Porter, Hayden Geol. Rep. 1872: 778. 1873.
    Villarsia pumila Griseb. in Hook. Fl. Bor. Am. 2:70. 1838. Capnorea pumila Greene, Ery-
    thea 2:193. 1894. (Douglas, "vallies of the Rocky Mountains, between Kettle Falls and Spo-
    kan")
Hesperochiron ciliatus Greene, Pitt. 1:282. 1889. Capnorea ciliata Greene, Erythea 2:193.
    1894. Hesperochiron pumilus var. ciliatus Brand, U. Calif. Pub. Bot. 4:227. 1912. (Shock-
    ley, Soda Springs, Esmeralda Co., Nev., April, 1888)
Capnorea fulcrata Greene, Pitt. 5:51. 1902. Hesperochiron pumilus f. fulcratus Brand,
    Pflanzenr. IV. 251:167. 1913. (Vasey 473, Wash.)
Capnorea nervosa Greene, Pitt. 5:51. 1902. Hesperochiron pumilus f. nervosus Brand,
    Pflanzenr. IV. 251:167. 1913. (Henderson 695, Moscow, Ida., May 13, 1894)
Capnorea hirtella Greene, Pitt. 5:51. 1902. Hesperochiron pumilus var. vestitus f. hirtellus
    Brand, U. Calif. Pub. Bot. 4:227. 1912. (Howell s.n., prairies of e. Wash., April, 1880)
Capnorea villosula Greene, Pitt. 5:52. 1902. Hesperochiron pumilus var. vestitus cum f.
    villosulus Brand, U. Calif. Pub. Bot. 4:227. 1912. H. villosulus Suksd. Wash. Fl. Fasc.
    13:3. 1928. (Elmer 1001, Pullman, Wash.)
Similar in general appearance to H. californicus, averaging smaller and less robust; taproot
more caudexlike, often premorse, and producing slender rhizomes from near the summit;
leaves commonly glabrous except for the strigose or ciliate margins, occasionally strigose or
even shortly spreading-hairy on the upper surface only; peduncles few, mostly 1-5(8); corolla
white or whitish, purplish-penicillate, saucer-shaped or rotate, 1.5-3 cm. wide, distinctly
wider than high even as pressed, the glabrous lobes (1.5)2-4 times as long as the internally
densely hairy tube.
In meadows, swales, and moist, open slopes (seldom if ever in alkaline soil) from the valleys and plains to more commonly in the foothills or well up in the mountains; Wash. to Calif.,
e. barely to Mont. (Big Hole Pass), and to Utah, Colo., and Ariz., wholly e. of the Cascade
summits in our range; most common in Ida. Apr.-June.


## Hydrophyllum L. Waterleaf

Flowers borne in compact (often capitate), mostly subdichotomously branched cymes that
lack a well-developed main axis; calyx divided to below the middle or near the base, usually
exappendiculate; corolla campanulate or a little broader, white to purple; filaments exserted,
attached to the corolla tube at or near its base, each flanked by a pair of ciliate, linear corolla appendages; style exserted, rather shortly but distinctly 2-cleft; ovules 4; seeds 1-3; capsule strictly 1-locular, dehiscent by two valves; herbs, mostly perennial and with fleshy-fibrous roots attached to a very short or well-developed rhizome; leaves alternate, best developed toward the base of the plant, variously cleft or pinnatifid.

Eight species of the U.S. and s. Can. Our species form a morphological series with only small gaps between each taxon and its neighbors in the key. (Name from the Greek hydor, water, and phyllon, leaf; without obvious significance)

Reference:

Constance, L. The genus Hydrophyllum L. Am. Midl. Nat. 27:710-31. 1942.

1 Leaflets evidently toothed along the margins (teeth relatively few and large in H. occidentale, which thus approaches or nearly matches the following group); rhizome generally more or less evident, though often short; anthers 1.0-2.4 mm. long

  2 ˙Leaf blades scarcely longer than wide, the 5(7-9) segments or leaflets mostly approximate and pinnipalmately disposed, all but the lowermost pair commonly confluent; pedicels mostly 5-12 mm. long; w. of the Cascade summits      H. TENUIPES

  2 Leaf blades evidently longer than wide, the 7-15 leaflets or segments pinnately disposed, the lower ones remote, only the upper ones confluent; pedicels mostly 2-7 mm. long

    3 Leaflets acuminate and sharply toothed, the teeth usually 4-8 to a side; Wash. and s. B.C. to n. Calif., eastward to n.e. Oreg. and c. Ida., and in the s. Rocky Mts.      H. FENDLERI

    3 Leaflets obtuse to abruptly acute; teeth obtuse or acute, usually 2-4 to a side; Marion and Polk cos., Oreg., to Calif., thence eastward into Ariz. and Utah; also in Elmore Co., Ida.      H. OCCIDENTALE

1 Leaflets entire or more commonly some of them with 1 or 2 deep incisions (the segments directed forward) or distal teeth; rhizome scarcely evident; anthers 0.6-1.3 mm. long; widespread e. of the Cascade summits      H. CAPITATUM

Hydrophyllum capitatum Dougl. ex Benth. Trans. Linn. Soc. 17:273. 1835. (Douglas, "fissures in moist rocks in the interior of the Columbia in Northwest America")

H. capitatum var. pumilum Hook Kew Journ. Bot. 3:292. 1851. (Geyer 326, Oreg.) = var. capitatum.

H. capitatum var. alpinum Wats. Bot. King Exp. 249. 1871. H. alpestre Nels. & Kennedy, Muhl. 3:142. 1908. (Watson 868, East Humboldt Mts., Nev.)

H. capitatum var. pumilum subvar. densum Brand, Pflanzenr. IV. 251:33. 1913. (Macoun 66617, between the Kettle and Columbia rivers, B.C., is the first of 7 specimens cited) = var. capitatum.

H. capitatum var. alpinum subvar. laxum Brand, Pflanzenr. IV. 251:33. 1913. (Elmer 1002 and Piper 1893, both from near Pullman, Wash., are cited in that order) = var. capitatum.

H. thompsoni Peck, Torreya 28:55. 1928. H. capitatum var. thompsoni Const. Am. Midl. Nat. 27:726. 1942. (Peck 7782, near Multnomah Falls, Oreg.)

Plants 1-4 dm. tall, loosely hirsute or partly strigose; stems solitary or few, delicately attached to the rather deep-seated and very short rhizome from which the cluster of fleshy-fibrous roots descends; leaves few and relatively large, long-petiolate, some of them attached below the ground, the blade sometimes 1 dm. wide and nearly 1.5 dm. long, pinnatifid into 7-11 sessile leaflets, the upper confluent, the lower approximate or remote; leaflets acute to more often obtuse or rounded and mucronate, with rounded, entire margins, or commonly some of them with one or two large, obliquely forward-pointing, entire-margined lateral lobes, the central lobe sometimes also with one or two distal teeth; peduncles (except in var. thompsonii) well surpassed by the leaves, short, seldom any of them over 5 cm. long, often reflexed in fruit; corolla 5-9 mm. long, white to more often lavender or purplish blue; anthers 0.6-1.3 mm. long. N=9.

In thickets, woodlands, and moist, open slopes, from the foothills and valleys to well up in the mountains; s. B.C. and Alta. to c. Calif. and Colo., almost wholly e. of the Cascade summits in our range (except in the Columbia gorge). Mar.-July.

A desirable early-flowering perennial, adapted to the region e. of the Cascades, and worth trying w. of the Cascades as well. It can be transplanted in late summer.

The species consists of three well-marked varieties. The var. thompsonii resembles H. occidentale in the nature of its peduncles, but is otherwise scarcely different from var. capitatum. Except for the local var. thompsonii, the length of the peduncles furnishes the most convenient and tangible character to distinguish H. capitatum from H. occidentale, even though occasional specimens of H. capitatum var. capitatum approach H. occidentale in that

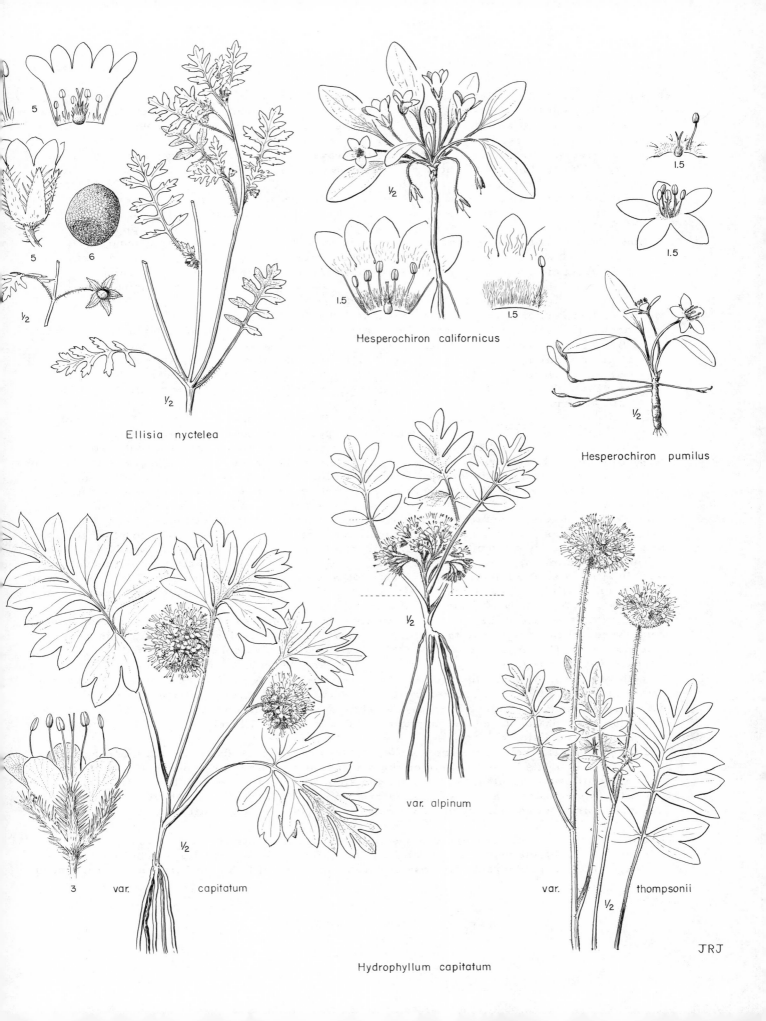

5

5      6

½

Ellisia  nyctelea

1.5

Hesperochiron  californicus

1.5

1.5

1.5

½

Hesperochiron  pumilus

½

var. alpinum

3      var.      capitatum

var.      thompsonii

½

JRJ

Hydrophyllum  capitatum

regard. It is possible that careful field study will show the var. thompsonii to constitute a distinct species, but for the present I defer to Constance's judgment in the matter.
1 Stem aerial as well as subterranean, the cymes somewhat elevated above the ground level;
    cymes generally capitate even in fruit
    2 Inflorescences equaling or surpassing the leaves, the peduncles elongate and erect, mostly 5-20 cm. long; in and near the Columbia gorge in Oreg. and Wash., and extending n. as far as s. Yakima Co. in Wash.       var. thompsonii (Peck) Const.
    2 Inflorescences well surpassed by the leaves, the peduncles short, seldom any of them over 5 cm. long, often reflexed in fruit; Wash. (e. of the Cascade summits) and adj. s. B.C. to s.w. Alta., s. to n.e. Oreg., Utah, and Colo.      var. capitatum
1 Stem almost wholly subterranean, the cymes scarcely elevated above the ground level; cymes tending to be lax at least in fruit, the pedicels often eventually recurved; Oreg. e. of the Cascade summits (and exclusive of the Wallowa-Blue Mt. region) to Calif., Nev., s.w. Ida., and w. Utah      var. alpinum Wats.

Hydrophyllum fendleri (Gray) Heller, Plant World 1:23. 1897.
  H. occidentale var. fendleri Gray, Proc. Am. Acad. 10:314. 1875. H. albifrons var. fendleri Brand, Pflanzenr. IV. 251:34. 1913. (Fendler 841, Santa Fe Creek, N.M.)
  H. albifrons Heller, Bull. Torrey Club 25:267. 1898. H. fendleri var. albifrons Macbr. Contr. Gray Herb. n. s. 49:23. 1917. (Heller 3269, Lake Waha, Nez Perce, Co., Ida.)
  H. congestum Wieg. Bull. Torrey Club 26:136. 1899. (Flett, Mt. Tacoma, Wash., in 1896) = var. albifrons.
  H. albifrons var. eualbifrons subvar. pendulum Brand, Pflanzenr. IV. 251:34. 1913. (Macoun 76745, Skagit R., B.C., is the first of 3 specimens cited) = var. albifrons.

    Perennial from a short, stout rhizome, with somewhat thickened, fibrous roots; stems solitary, 2-8 dm. tall, retrorsely hairy; leaves rather few but large, the blade sometimes 2.5 dm. long and 1.5 dm. wide, mostly long-petiolate, pinnatifid with mostly 7-11(15) sessile or subsessile leaflets, the lower remote, the upper confluent; leaflets more or less acuminate and sharply toothed, with commonly 4-8 teeth to a side, hairy (or scabrous) on both sides, often thinly so; cymes only moderately compact, the pedicels mostly 2-7 mm. long; calyx lobes mostly 3-6 mm. long, long-ciliate on the margins and sometimes more shortly hairy on the back; corolla 6-10 mm. long, white to partly or wholly lavender or purplish; anthers 1.3-2.2 mm. long. N=9.

    Thickets and moist, open places, from the valleys and foothills to well up in the mountains; Marble Mts. of s. B.C. to the Olympic Mts. of Wash., and through the Cascades of Wash. and Oreg. to n. Calif.; Palouse and Blue Mt. regions of s.e. Wash. and n.e. Oreg., thence s.e. to c. Ida.; s. Wyo. to N.M. and s.e. Utah. May-Aug.

    The fernlike aspect of the leaves is attractive, but the plants are aggressive and tend to become weedy when brought into the garden. The range of the species is divided into three wholly discrete parts. Of these, the s. Rocky Mt. region is occupied solely by var. fendleri, the Cascade-Olympic region solely by var. albifrons, and the Palouse-Blue Mt. -c. Ida. region by both varieties. Although the two varieties are superficially similar and sometimes occur together, they are so sharply marked by the pubescence that I have not seen a single really doubtful specimen. Careful field study may show them to be distinct species, but I defer here to the judgment of Constance and maintain them at the varietal level instead. The two may be characterized as follows:
1 Pubescence relatively harsh and sparse, the leaves scabrous above and not white-hairy beneath, the sepals bristly-ciliate on the margins and glabrous to inconspicuously strigillose or strigillose-puberulent on the back, the stem generally retrorsely hispid or hispid-hirsute; inflorescence shortly surpassed by to more often equaling or shortly surpassing the leaves; leaflets sometimes up to 15; calyx lobes 4-6 mm. long; corolla mostly 6-8 mm. long      var. fendleri
1 Pubescence relatively soft and copious, the leaves merely strigose above and softly white-hairy beneath, the sepals softly long-ciliate on the margins and villous-puberulent on the back, the stem retrorsely strigose or hirsute-puberulent to hispid-hirsute; inflorescences

usually shortly surpassed by the leaves; leaflets rarely more than 11; calyx lobes 3-5 mm.
long; corolla 7-10 mm. long                                     var. albifrons (Heller) Macbr.


Hydrophyllum occidentale (Wats.) Gray, Proc. Am. Acad. 10:314. 1875.
  H. macrophyllum var. occidentale Wats. Bot. King Exp. 248. 1871. (Bigelow, Duffield
    Ranch, Sierra Nevada; this is near Sonora, in Tuolumne Co., according to Jepson)
  H. occidentale var. watsoni Gray, Proc. Am. Acad. 10:314. 1875. H. watsonii Rydb. Bull.
    Torrey Club 40:478. 1913. (Watson 865, West Humboldt Mts., Pershing Co., Nev.)
  Single-stemmed perennial (often branched near the ground level and thus seemingly tufted),
1.5-5 dm. tall, with a short, stout rhizome from which the fleshy-fibrous roots descend; herbage more or less hispid-hirsute, or the leaves more strigose, the stem often also with some
shorter and finer hairs; leaves rather large (the blade sometimes 2 dm. long and 1 dm. wide),
long-petiolate, pinnatifid, with 7-15 leaflets, the lower leaflets remote, the upper confluent;
leaflets more or less acutish and coarsely few-toothed or -lobed, commonly with 2-4 teeth to
a side, rarely many of them subentire; peduncles more or less elongate, mostly 0.5-2 dm.
long, seldom much, if at all, surpassed by the leaves; cymes subcapitate, the pedicels seldom
as much as 5 mm. long; corolla 6-10 mm. long, white to blue-violet; anthers 1.2-2.4 mm.
long. N=9.
  Thickets, dense or open woods, and moist open places; from the Coast Ranges to the base of
the Cascades, in Polk and Marion cos., Oreg., s. to c. Calif., and thence e. to Utah and
Ariz.; an outlying station in Elmore Co., Ida. (Hitchcock & Muhlick 8682 and 8716, both near
Dog Creek). Apr.-July.
  The nomenclaturally typical phase of the species, as described above, occurs mostly w. of
the Sierra Nevada and Cascade crests; plants of the Great Basin and of Ida. tend to be more
softly pubescent, and may be distinguished, with some difficulty, as var. watsonii Gray.


Hydrophyllum tenuipes Heller, Bull. Torrey Club 25:582. 1898. (Heller 3853, Chehalis R. at
  Montesano, Wash.) The type represents the phase with cream-colored flowers.
  H. tenuipes var. viride Jeps. Man. Pl. Calif. 811. 1925. (Matthews, Ft. Bragg, Calif.)
    Corolla greenish-white.
  H. viridulum G. N. Jones, U. Wash. Pub. Biol. 7:175. 1938. (Piper 260, Seattle, Wash.)
    Corolla greenish-white.
  Perennial 2-8 dm. tall from a short or elongate rhizome; stem solitary, retrorsely hispid-
hirsute, or rarely subglabrous; leaves few and large, long-petiolate, rather thinly strigose on
both sides, or often scabrous above, the blade scarcely longer than wide, sometimes 15 cm.
long and wide, pinnipalmately divided into 5(7-9) acute or acuminate, more or less sharply
toothed leaflets, the lower pair approximate but discrete, the upper confluent; peduncles 0.5-
1.5 dm. long, sometimes forked above the middle, commonly about equaling or shortly surpassing the leaves; inflorescence less compact than in our other species, the pedicels mostly
5-12 mm. long; calyx lobes 4-6 mm. long, bristly-ciliate on the margins, glabrous or sparsely strigulose on the back; corolla 5-7 mm. long, green or greenish-white to sometimes (especially in Wash. and in the Columbia gorge) cream-colored or (mouth of the Columbia; Pt. Angeles; Hoh R.) blue or purple; anthers 1.0-1.7 mm. long. N=9.
  Moist woods at lower elevations, from the w. flank of the Cascades to the coast; Wash. and
s. Vancouver I. to n. Calif. May-July.


                                     Nama L. Nom. Conserv.


  Flowers solitary or in small cymes; calyx cleft to near the base; corolla tubular to funnelform or narrowly campanulate, purple to white; filaments included, unequally inserted, the
adnate basal portion commonly (but not in our species) with somewhat dilated, free margins or
marginal appendages; style cleft only at the tip (ours) or more commonly to the base; placentae more or less intruded (and the ovary thus appearing 2-celled) but at least in our species
apparently not joined; ovules and seeds more or less numerous; capsule loculicidal; annual
(ours) or perennial herbs or subshrubs with alternate, entire leaves.
  About 40 species, chiefly of s.w. U.S. and n. Mex., a few in s. S. Am., and one in Hawaii.

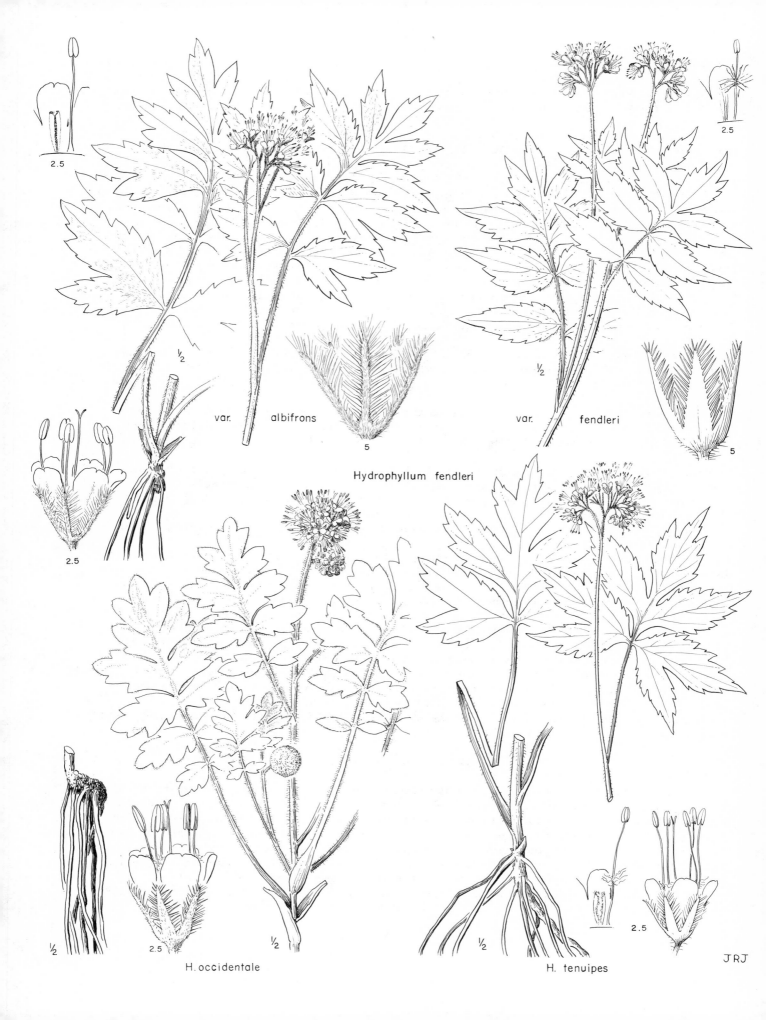

var. albifrons

var. fendleri

Hydrophyllum fendleri

H. occidentale

H. tenuipes

JRJ

Our two species are sharply defined and constitute the well-marked section Conanthus. (Name a transliteration of the Greek word meaning a spring; of doubtful significance.)

Reference:

Hitchcock, C. L. A taxonomic study of the genus Nama. Am. Journ. Bot. 20:415-30; 518-34. 1933.

1 Corolla 2.5-5 mm. long             N. DENSUM
1 Corolla 8-14 mm. long              N. ARETIOIDES

Nama aretioides (H. & A.) Brand, U. Calif. Pub. Bot. 4:224. 1912.

  Eutoca aretioides H. & A. Bot. Beechey Voy. 374. 1838. Conanthus aretioides Wats. Bot. King Exp. 256. 1871. Marilaunidium aretioides Coville, Contr. U.S. Nat. Herb. 4:161. 1893. (Tolmie, between the Burnt and Malheur rivers, Oreg.)

  Eutoca aretioides var. perpusilla H. & A. Bot. Beechey Voy. 374. 1838. (Tolmie, Burnt R., Oreg.)

Dwarf, prostrate, dichotomously branched, taprooted annual, sparsely or moderately spreading-hispid throughout; leaves oblanceolate, often approximate, tending to be crowded toward the tips of the branches, up to 4 cm. long and 6 mm. wide; flowers solitary and terminal, immediately subtended by a leaf, soon appearing to be in the forks of the branches, because of the development of a pair of axillary buds; corolla 8-14 mm. long, with bright pink to deep rose limb 5-10 mm. wide, the tube mostly yellow; style 2-6 mm. long, cleft only at the tip. N=7.

Dry, sandy plains and hills, often with sagebrush; Yakima Co., Wash., to Inyo Co., Calif., and e. in the Snake R. plains to Gooding Co., Ida. May-June.

Some of the plants occurring to the s. of our range have a somewhat larger corolla, with broader limb, and may be varietally separable.

Nama densum Lemmon, Bull. Torrey Club 16:222. 1889.

  Conanthus densus Lemmon ex Heller, Cat. N. Am. Pl. 6. 1898. (Lemmon, Edgewood, Siskiyou Co., Calif., in 1889)

  Gilia hispida Piper, Erythea 6:30. 1898. (Henderson 2402, Hindshaw s.n., May 25, 1896, and Piper s.n., July 11, 1897, all from Pasco, Wash., are cited in that order) = var. parviflorum.

  Conanthus parviflorus Greenm. Erythea 7:117. 1899. Nama densum var. parviflorum C. L. Hitchc. Am. Journ. Bot. 20:420. 1933. N. parviflorum Const., U. Calif. Pub. Bot. 23:370. 1950. (Cusick 1957, Malheur region, e. Oreg.; lectotype by Hitchcock)

Similar to N. aretioides, often more densely and less coarsely hairy; corolla white to lavender, 2.5-5 mm. long, the limb 1-3 mm. wide; style 0.3-1.5 mm. long. N=7, 14.

Dry, sandy places in the deserts and foothills, from Adams Co., Wash., to Inyo Co., Calif., e. to Gooding Co., Ida., and Nye Co., Nev., and with outlying stations in Grand Co., Utah, and Carbon Co., Wyo. May-July.

The species consists of two well-marked varieties, as follows:

1 Plants densely spreading-hirsute throughout, with oblanceolate leaves seldom over 1.5(2) cm. long; corollas mostly 2.5-4 mm. long; style seldom over 1 mm. long; desert and ponderosa pine zones, or at higher elevations southward; Deschutes and Harney cos., Oreg., to Inyo Co., Calif., and along the w. edge of Nev.       var. densum

1 Plants less densely but more coarsely hairy, with longer, often relatively narrower (frequently linear-oblanceolate) leaves, the larger ones generally well over 1.5 (to 4) cm. long; corollas 3.5-5 mm. long; style up to 1.5 mm. long; deserts; Adams Co., Wash., to Modoc Co., Calif., e. to Gooding Co., Ida., and Lander Co., Nev., with outlying stations in Wyo. and Utah       var. parviflorum (Greenm.) C. L. Hitchc.

## Nemophila Nutt. Nom. Conserv.

Flowers solitary at the axils, or some of them in a loose, few-flowered, racemelike, terminal cyme; calyx divided to near the base, and ordinarily bearing a spreading or deflexed appendage (auricle) at each sinus; corolla campanulate to rotate, white to blue or purple; filaments surpassed by the corolla, attached to the corolla tube near or well above the base, each generally flanked by a pair of small corolla appendages; style surpassed by the corolla, shallowly to deeply cleft; ovules 2-several on each of the 2 large parietal placentae; capsule 1-locular, dehiscent by 2 valves; seeds with an obscure to evident, partial or complete covering (cucullus) external to the regular seed coat; delicate, taprooted annuals with opposite or less often alternate, mostly more or less strongly pinnatifid or pinnatilobate leaves, the entire-margined cotyledons often enlarged and foliaceous.

Eleven species, of the U. S. and adj. parts of Can. and Mex. (Name from the Greek nemos, a grove, and philein, to love, referring to the habitat of some species.)

Reference:

Constance, L. The genus Nemophila Nutt. U. Calif. Pub. Bot. 19:341-98. 1941. (The treatment presented below is wholly in accord with that of Constance.)

1 Leaves all alternate; corolla shorter than the calyx; seeds mostly solitary N. BREVIFLORA
1 Leaves all opposite, or some of the upper ones alternate; corolla equaling or surpassing the calyx; seeds mostly 2-22
  2 Corolla mostly 7-25 mm. wide, broadly campanulate to saucer-shaped; style 2-5 mm. long
    3 Corolla mostly 15-25 mm. wide, white or whitish, and conspicuously flecked with blackish-purple, capsule well surpassing the calyx; seeds mostly 8-22; Coast Ranges and Willamette Valley of Oreg., and southward          N. MENZIESII
    3 Corolla mostly 7-15 mm. wide, commonly blue-lavender toward the periphery and nearly white toward the center, not flecked with dark purple; capsule surpassed by the strongly accrescent fruiting calyx; seeds mostly 2-4; near the Snake R. in e. Oreg., s.e. Wash., and adj. Ida., and extending up the Salmon R. to Shoup, Ida.
                                                N. KIRTLEYI
  2 Corolla 2-6 mm. wide, mostly campanulate; style 0.6-1.5 mm. long; widespread
    4 Appendages of the calyx well developed, mostly 1/2-3/4 (or fully) as long as the proper lobes, 1-3 mm. long at maturity; leaves rather deeply pinnatifid, the rachis mostly 1-4 mm. wide; usually in moist, open places, especially in meadows and bottom lands          N. PEDUNCULATA
    4 Appendages of the calyx relatively small and inconspicuous, seldom half as long as the proper lobes, about 1 mm. long or less at maturity; leaves often less deeply cleft; usually on wooded slopes or in other more or less shady places   N. PARVIFLORA

Nemophila breviflora Gray, Proc. Am. Acad. 10:315. 1875.

  Viticella breviflora Macbr. Contr. Gray Herb. n.s. 59:32. 1919. (Watson 869, Parley's Park, Salt Lake Co., Utah)

Plants subprostrate to more often ascending or loosely erect, generally branched, 1-3 dm. tall; stem weak, angled, sparsely beset with fine, short, retrorse prickles; leaves alternate, thin, coarsely and rather sparsely strigose or hirsute, the margins and especially the petiole often conspicuously ciliate, the blade ordinarily with narrow rachis and 2 pairs of spreading lateral lobes, these up to 2 cm. long and 7 mm. wide, more or less acute and often mucronate; pedicels opposite the leaves, short at anthesis, in fruit 5-15 mm. long and more or less spreading or deflexed; flowers tiny, the lavender corolla about 2 mm. long and more or less evidently surpassed by the calyx; calyx accrescent, the stiffly ciliate lobes 3-5 mm. long at maturity and equaling or surpassing the capsule; auricles well developed, 1-2 mm. long at maturity; style 0.5-1 mm. long, cleft only at the tip; seeds mostly solitary and filling the capsule, the cucullus thin and persistent. N=9.

Wooded slopes, thickets, or less often in open places, from the foothills to moderate elevations in the mountains; Marble Mts., B.C., and e. of the Cascades in Wash., Oreg., and n. Calif., eastward to Mont. and Colo. Apr.-July.

Nemophila kirtleyi Henderson, Bull. Torrey Club 27:350. 1900.

  Viticella kirtleyi Macbr. Contr. Gray Herb. n.s. 59:31. 1919. (Henderson 3082, Salmon River Hill, beyond Florence, Idaho Co., Ida.)

Plants more or less erect, 1-3 dm. tall, tending to be freely branched, so that many of the pedicels eventually seem to be borne in the forks of the branches; herbage evidently spreading-hairy throughout, or the leaves more strigose; cotyledons conspicuous and often persistent, the elliptic blade sometimes 4 cm. long; leaves all opposite, or some of the upper alternate, the blade mostly 1.5-5 cm. long and 1-4 cm. wide, commonly with 2 pairs of coarse lateral lobes that are usually broadest toward the base, some of the lobes sometimes with 1 or 2 teeth; pedicels axillary, or the upper opposite the axils, mostly solitary at the nodes, 1-2.5(5) cm. long in flower, elongating to 2-8 cm. in fruit; calyx lobes 3-5 mm. long at anthesis, up to 1 cm. or more (and appendages to 5 mm.) in fruit, more or less strongly surpassing the capsule; corolla broadly campanulate or nearly saucer-shaped, mostly 7-15 mm. wide, blue-lavender toward the periphery, nearly white toward the center; anthers 0.7-1.5 mm. long; style 2-5 mm. long, cleft only at the tip; seeds mostly 2-4, 3-4 mm. long, the prominent pits more or less aligned in rows, the cucullus poorly developed and incomplete, commonly covering only one end.

Rocky slopes, usually in partial shade, at lower elevations near the Snake R. where it forms the Ida.-Oreg. boundary, and shortly northward into Wash., and also extending up the Salmon R. to Shoup, Ida. Apr.-May.

Nemophila menziesii H. & A. Bot. Beechey Voy. 152. 1833.
Viticella menziesii Macbr. Contr. Gray Herb. n.s. 59:30. 1919. (Menzies, Calif., possibly at Monterey)
Nemophila atomaria Fisch. & Mey. Ind. Sem. Hort. Petrop. 2:42. 1835. N. insignis var. atomaria Jeps. Fl. W. Middle Calif. 434. 1901. N. menziesii var. atomaria Chandler, Bot. Gaz. 34:204. 1902. N. menziesii ssp. atomaria Brand, Pflanzenr. IV. 251:49. 1913. Viticella menziesii var. atomaria Macbr. Contr. Gray Herb. n.s. 59:30. 1919. (Near the present Ft. Ross, Sonoma Co., Calif.)

Stems subprostrate to loosely erect, usually branched, at least at the base, up to 3 dm. long, sparsely strigose or retrorse-bristly; herbage somewhat succulent; leaves opposite, thinly strigose or hirsute, pinnatifid, with rather narrow rachis (seldom 3 mm. wide) and mostly (2)3-4 pairs of convex-margined lobes up to about 8 mm. long, the upper lobes larger than the lower and sometimes with 1 or 2 teeth; pedicels solitary at each node, axillary, 2-6 cm. long, or longer in fruit; calyx lobes at anthesis 3-4.5 mm. long and auricles 1-2 mm. long, not much accrescent; corolla saucer-shaped, mostly 1.5-2.5 cm. wide, white or faintly suffused or veined with lavender, and conspicuously flecked with blackish-purple; lobes longer than the tube; anthers blackish-purple, 1-2 mm. long; style 2.5-5 mm. long, cleft less than halfway; seeds mostly 8-22, about 2 mm. long, with a prominent deciduous cucullus at one end. N=9.

Moist, open or shaded places at lower elevations, sometimes in cultivated fields; Willamette Valley and Coast Ranges of Oreg., s. to n. Baja Calif. Apr.

This beautiful little annual maintains itself rather well in the wild garden and is really always a component of "native wild flower seed" mixtures.

Our plants, as described above, are var. atomaria (Fisch. & Mey.) Chandler. Two other varieties, differing in flower color, pubescence, and leaf form, occur to the s. of our range. The species is noteworthy because of the demonstrated existence, at least in Calif., of numerous microraces, many or all of which are interfertile with some and intersterile with others in the group.

Nemophila parviflora Dougl. ex Benth. Trans. Linn. Soc. 17:275. 1835.
Viticella parviflora Macbr. Contr. Gray Herb. n.s. 59:32. 1919. (Douglas, "from the Columbia"; Ft. Vancouver, Wash., fide Constance)
Nemophila inconspicua Henderson, Bull. Torrey Club 27:349. 1900. N. parviflora var. typica subvar. inconspicua Brand, Pflanzenr. IV. 251:55. 1913. (Henderson 3289, Soldier Mt., Blaine Co., Ida.) = var. austiniae.
N. austinae Eastw. Bull. Torrey Club 28:143. 1901. N. parviflora var. austinae Brand, Pflanzenr. IV. 251:55. 1913. Viticella parviflora var. austinae Macbr. Contr. Gray Herb. n.s. 59:32. 1919. (Austin, Davis Creek, Calif.; Modoc Co., fide Constance)
Nemophila pustulata Eastw. Bull. Torrey Club 28:145. 1901. (Macoun 667, Vancouver, B.C.) = var. parviflora.

Stems loosely erect to subprostrate, 0.5-3(5) dm. long, often with widely divaricate branch-

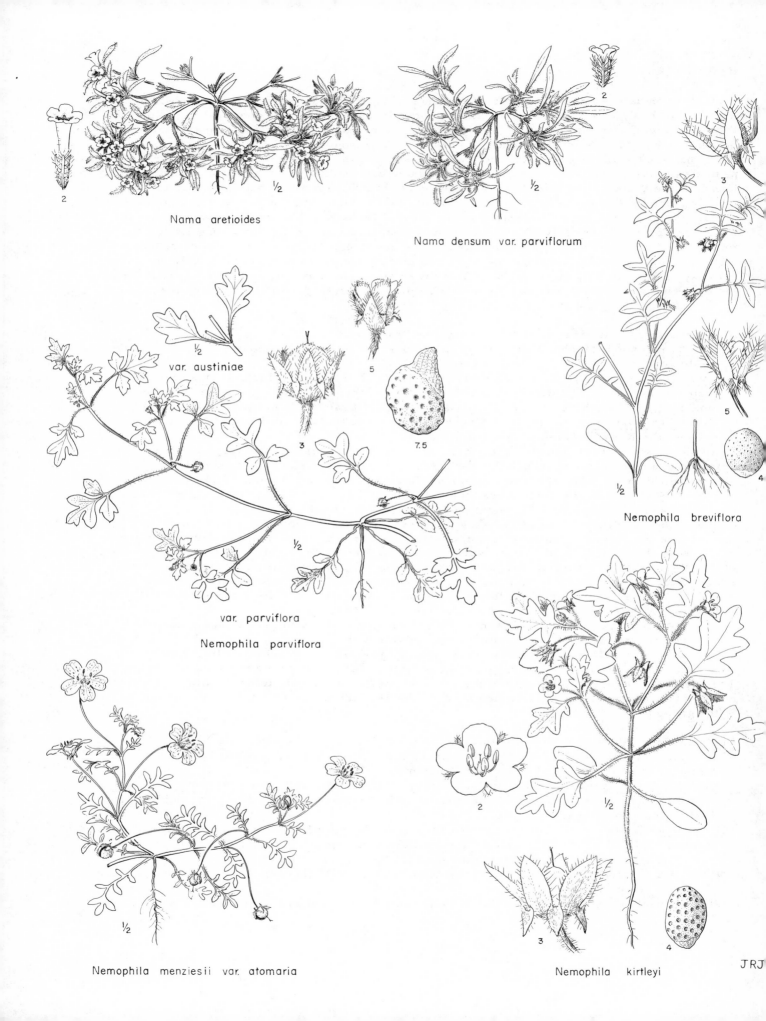

Nama aretioides

Nama densum var. parviflorum

var. austiniae

var. parviflora

Nemophila parviflora

Nemophila breviflora

Nemophila menziesii var. atomaria

Nemophila kirtleyi

JRJ

es, retrorsely and rather sparsely stiff-hairy, or merely thinly strigose; leaves opposite, or some of the upper alternate, strigose on both sides, or spreading-hirsute beneath, the blade mostly 1-3.5 cm. long and 0.8-2.5 cm. wide and commonly either broader or less deeply cleft (or both) than in N. pedunculata, mostly with 2 pairs of lateral lobes or coarse teeth; lower lobes commonly as large as or larger than the upper; pedicels axillary, 2-15 mm. long, or up to 3 cm. in fruit; corolla lavender, about equaling or shortly surpassing the calyx, 1-4.5 mm. wide; calyx lobes 2-3 mm. long in fruit, the appendage up to about 1 mm.; anthers 0.3-0.4 mm. long; style 0.6-1.5 mm. long, commonly cleft about halfway to the base; seeds mostly 2-4, 2-2.5 mm. long, capped at one end by the deciduous cucullus. N=9.

Usually on wooded slopes or in other more or less shady places, from the lowlands to moderate elevations in the mountains; s. B.C. to Calif., e. to c. Ida. and n. Utah. Apr.-July.

The species consists of three well-marked varieties, the two which occur in our range being characterized as follows:

1 Leaves relatively thin and deeply cleft, with prominently mucronate, often toothed lobes, the lower pair of sinuses in particular tending to approach the midrib, and the lower (or both) pair of lobes tending to be narrowed at the base; some of the upper leaves often alternate; flowers averaging a little larger than in var. austiniae; at lower elevations w. of the Cascades                                                                        var. parviflora

1 Leaves relatively firm, averaging smaller, less deeply cleft, the lobes (or coarse teeth) only obscurely or scarcely mucronate and seldom evidently narrowed at the base; leaves all opposite; e. of the Cascades, in the foothills and at moderate elevations in the mountains                                                                var. austiniae (Eastw.) Brand

Nemophila pedunculata Dougl. ex Benth. Trans. Linn. Soc. 17:275. 1835.

  Viticella pedunculata Macbr. Contr. Gray Herb. n.s. 59:32. 1919. (Douglas, on the Columbia; near Ft. Vancouver, fide Constance)

  Nemophila sepulta Parish, Erythea 7:93. 1899. N. pedunculata var. sepulta Nels. & Macbr. Bot. Gaz. 65:65. 1918. Viticella pedunculata var. sepulta Macbr. Contr. Gray Herb. n.s. 59:32. 1919. (Parish 3782, Bear Valley, San Bernardino Mts., Calif.)

  Nemophila menziesii var. minutiflora Suksd. Deuts. Bot. Monats. 18:133. 1900. N. minutiflora Suksd. W. Am. Sci. 14:32. 1903. N. sepulta var. minutiflora Brand, Pflanzenr. IV. 251:52. 1913. (Suksdorf 2198, Falcon Valley, Klickitat Co., Wash.)

  N. densa Howell, Fl. N.W. Am. 466. 1901. N. sepulta var. densa Brand, U. Calif. Pub. Bot. 4:211. 1912. N. pedunculata var. densa Nels. & Macbr. Bot. Gaz. 65:66. 1918. Viticella pedunculata var. densa Macbr. Contr. Gray Herb. n.s. 59:32. 1919. (Wash. to Calif.; no type cited)

  Nemophila reticulata Suksd. W. Am. Sci. 14:32. 1903. (Suksdorf 684, Falcon Valley, Klickitat Co., Wash.)

  N. erosa Suksd. W. Am. Sci. 14:33. 1903. N. pedunculata f. erosa Brand, Pflanzenr. IV. 251:53. 1913. (Suksdorf 2315, Columbia R., Clarke Co., Wash.)

  N. mucronata Eastw. ex Sheld. Bull. Torrey Club 30:309. 1903. (Sheldon 10204, Chenowith Creek, Wasco Co., Oreg.)

  N. pedunculata f. chandleri Brand, Pflanzenr. IV. 251:53. 1913. (Several specimens are cited, including Chandler 6073 and 929)

  N. eriocarpa Gand. Bull. Soc. Bot. France 65:64. 1918. (Suksdorf 2638, Rockland, Wash.)

Prostrate or nearly so, usually with several stems from the base, these up to 3 dm. long, simple or branched, glabrous or with scattered hairs; leaves opposite, thinly strigose or hirsute-strigose on both sides, the mostly somewhat wing-margined petioles often prominently ciliate especially toward the base; leaf blades mostly 6-25 mm. long and 5-17 mm. wide, rather deeply pinnatilobate, with 2-3(4) pairs of convex-margined lateral lobes (the lower ones often as large as the upper) and narrow but foliaceous rachis mostly 1-4 mm. wide; pedicels 4-12 mm. long at anthesis, sometimes up to 3.5 cm. in fruit; calyx lobes 1.5-4 mm. long at maturity, the auricles 1-3 mm. long and 1/2-3/4 (or fully) as long as the lobes; corolla campanulate, pale lavender, or whitish and marked with lavender, 2-6 mm. wide; anthers 0.4-0.8 mm. long; style 0.7-1.0 mm. long, cleft about half; seeds mostly 2-8, 1-4 mm. long, with a prominent, deciduous cucullus at one end. N=9.

Mostly in moist, open places, especially in meadows and bottom lands, in the foothills and lowlands, occasionally at moderate elevations in the mountains; nearly throughout Calif. and Oreg., extending into adj. parts of Baja Calif., Nev., and Ida. (to Adams, Elmore, and Owyhee cos.), and n. through w. Wash. to s. Vancouver I., B.C. Apr.-June.

## Phacelia Juss.

Flowers borne in helicoid cymes, these often aggregated into a compound inflorescence; calyx divided nearly to the base, with equal or unequal, often accrescent lobes; corolla variously blue or purple to pink, white, or yellow, usually deciduous, tubular to rotate; filaments equal or less often unequal, attached to the corolla tube near or at its base; style shortly (but evidently) to very deeply cleft; ovules 2-many on each of the two placentae, these slightly to strongly intruded; capsule loculicidal; annual to perennial, taprooted herbs, with entire to pinnately dissected, chiefly or wholly alternate leaves.

A large and polymorphic genus, of probably at least 150 species, native to the New World, best developed in w. U.S. and n. Mex. (Name from the Greek phakelos, a fascicle, referring to the congested inflorescence.)

Most of the species of Phacelia, especially the annuals, are easily grown from seeds, but among our plants only the perennial species P. sericea and P. lyallii are very attractive. These might be desirable in rock gardens e. of the Cascades, although they will probably require winter protection.

The corolla scales have frequently been used as a source of taxonomic characters, but they are so subject to sporadic and taxonomically insignificant variation as to be better ignored until many flowers of each species have been carefully examined. All of our species are distinguishable without reference to the scales, which are here omitted from the descriptions. Our species, with the exception of P. nemoralis, P. hastata, and P. heterophylla, appear to be well characterized, at least in our region.

Phacelia lenta Piper, Bull. Torrey Club 28:44. 1901 (Brandegee 976, bare hills of the Columbia R., Washington Terr., Sept. 1883) resembles P. sericea in some respects (including the marcescent corollas), but differs in its broader, less thyrsoid inflorescence and in being glandular throughout. It is known only from the type specimen (Gray Herb.), a postmature plant with opened capsules most of which contain only about 2 small seeds that far from fill the locules. The specimen is perhaps a hybrid between P. sericea and some other species, possibly P. glandulosa. Neither P. sericea nor any other species which might reasonably be involved in the origin of P. lenta is known from the bare hills of the Columbia in Washington.

References:

Dundas, F. W. A revision of the Phacelia californica group (Hydrophyllaceae) for North America. Bull. So. Calif. Acad. Sci. 33:152-68. 1934.

Howell, J. T. Studies in Phacelia - A revision of the species related to P. pulchella and P. rotundifolia. Am. Midl. Nat. 29:1-26. 1943.

-------. A revision of Phacelia section Miltitzia. Proc. Calif. Acad. Sci. IV. 25:356-76. 1944.

-------. Studies in Phacelia--Revision of species related to P. douglasii, P. linearis, and P. pringlei. Am. Midl. Nat. 33:460-94. 1945.

-------. A revision of Phacelia sect. Euglypta. Am. Midl. Nat. 36:381-411. 1946.

Voss, John W. A revision of the Phacelia crenulata group for North America. Bull. Torrey Club 64:81-96; 133-44. 1937.

1 Flowers mostly 4-merous; corolla not over 2 mm. long; small annual with entire to shallowly lobed leaves                                                                         P. TETRAMERA
1 Flowers mostly 5-merous; corolla 2.5 mm. long or more; habit diverse
  2 Corolla yellow (becoming whitish or even partly purplish in age), persistent and surrounding the fruit; annual                                                          P. LUTEA
  2 Corolla never wholly yellow (though sometimes ochroleucous, or with yellow tube and bluish limb), ordinarily deciduous except in 2 perennial species
    3 Leaves all entire, or in some species some of them with a large, entire terminal segment and one or several smaller, entire lobes or leaflets below the middle
      4 Plants biennial or perennial; filaments strongly exserted; ovules 4; seeds 1-4

5 Plants robust (5-20 dm. tall), usually perennial and several-stemmed; strongly spreading-bristly, with relatively large leaf-blades that generally bear 1 or more pairs of basal leaflets or lobes; mesic, commonly shady habitats from the coast to the lower slopes of the Cascades P. NEMORALIS

5 Plants not presenting the foregoing combination of morphological characters, and generally either more eastern, or occurring in drier habitats, or both

  6 Plants biennial or short-lived perennial from a taproot, typically with a single erect stem that is often well over 5 dm. tall, or this surrounded by several ascending lesser stems, some of the middle and lower leaves usually with one or two (four) pairs of lateral lobes or leaflets at the base of the blade; herbage often somewhat griseous, but scarcely silvery, often markedly spreading-bristly; plants (except var. pseudohispida) mostly of the foothills, valleys, and plains P. HETEROPHYLLA

  6 Plants perennial from a taproot which is usually surmounted by a branched caudex, usually with several more or less equal, suberect to prostrate stems that are seldom over 5 dm. tall; leaves all entire, or sometimes some of them with a pair of small lateral lobes near the base; herbage, especially in plants from lower altitudes, more or less silvery with a short, dense pubescence, the longer bristles, if present, seldom very conspicuous; plants of all elevations P. HASTATA

4 Plants annual; filaments only shortly or not at all exserted; ovules (and usually seeds) more than 4 except in P. humilis

  7 Corolla somewhat showy, the limb 4-18 mm. wide; style (branches included) 3 mm. long or more; filaments about equaling or shortly exserted from the corolla

    8 Ovules 4; seeds 1-4; corolla 4-7 mm. wide; style cleft to the middle or below; leaves all entire P. HUMILIS

    8 Ovules and seeds more than 4 (seeds 6 or more); corolla 8-18 mm. wide; style not cleft to the middle; larger leaves often with 1-4 lateral segments below the middle P. LINEARIS

  7 Corolla inconspicuous, the limb 1.5-3 mm. wide; style about 1.5 mm. long or less; filaments included; leaves all entire

    9 Principal leaves more or less oblanceolate, 2-4 mm. wide, the blade much longer than the short (to 4 mm.) petiole; calyx segments conspicuously unequal in fruit P. MINUTISSIMA

    9 Principal leaves more or less elliptic or oblong-ovate, 4-10 mm. wide, the blade equaling or shorter than the well-developed (to 2 cm.) petiole; calyx segments slightly or scarcely unequal P. INCANA

3 Leaves coarsely toothed or pinnatilobate to more often pinnatifid or bipinnatifid, mostly without a large, entire, terminal segment

10 Filaments included; desert or lowland annuals, generally glandular, seldom (except P. rattanii) as much as 3 dm. tall, often without a well-defined central axis

  11 Ovules and seeds more or less numerous, the seeds cross-corrugated; style cleft 1/4 of its length, or less; calyx segments linear or linear-oblanceolate, not conspicuously accrescent; leaves subbipinnatifid, with narrow rachis and segments

    12 Corolla mostly 6.5-12 mm. long (dry), the limb 5-10 mm. wide; style (4)5-8 mm. long P. BICOLOR

    12 Corolla mostly 4.5-6.5 mm. long, the limb 2-3.5 mm. wide; style 2-3 mm. long P. GLANDULIFERA

  11 Ovules and seeds 4, or the seeds fewer, these pitted-reticulate, not cross-corrugate; style cleft nearly half its length, or more; calyx segments broader and leaves less dissected than in the foregoing group

    13 Plants rather thinly bristly-hispid throughout, as well as usually glandular; calyx segments unequally accrescent, the larger ones spatulate, not markedly firm and veiny P. RATTANII

    13 Plants merely glandular-hairy, not at all bristly-hispid; calyx segments lance-elliptic or somewhat oblong, conspicuously accrescent, firm and veiny in fruit P. THERMALIS

10 Filaments about equaling to obviously exceeding the corolla; habit and habitat diverse

14  Ovules and seeds 4 (or the seeds fewer); style cleft to well below the middle; plants
    glandular, occurring in the deserts and dry valleys and plains e. of the Cascades
    15  Perennial; stems prostrate or weakly ascending, 5-15 dm. long; corolla white to laven-
        der or dull cream                      P. RAMOSISSIMA
    15  Annual or biennial; stems ascending or erect, 1-3.5 dm. tall; corolla blue-violet
                                            P. GLANDULOSA
14  Ovules and seeds more or less numerous; style not cleft beyond about the middle; plants
    glandular or eglandular, occurring in more mesic habitats, or at higher altitudes, or
    both
    16  Plants annual or biennial, single-stemmed or with the central stem surrounded by as-
        cending lesser stems; corolla glabrous within; filaments only shortly or scarcely ex-
        serted                          P. FRANKLINII
    16  Plants perennial, usually several-stemmed (P. idahoensis sometimes rather short-
        lived and single-stemmed); corolla (except P. bolanderi) hairy within
        17  Plants with basally disposed leaves, the basal and lower cauline ones well devel-
            oped and generally persistent, the middle and upper ones generally more or less
            reduced and less petiolate; helicoid cymes thyrsoidally or even capitately aggre-
            gated; inflorescence not evidently glandular
            18  Leaves more or less pinnatifid or bipinnatifid; plants scarcely glandular, some-
                times silvery-hairy (almost always so when only 1-2.5 dm. tall); corolla per-
                sistent and surrounding the fruit
                19  Filaments barely exserted, less than 1.5 times as long as the corolla
                                    P. IDAHOENSIS
                19  Filaments long-exserted, (1.5)2-3 times as long as the corolla
                                    P. SERICEA
            18  Leaves pinnatilobate or merely coarsely toothed; plants 0.5-2.5 dm. tall, with
                green and generally glandular herbage; corolla deciduous; filaments well ex-
                serted                      P. LYALLII
        17  Plants leafy-stemmed, without tufts of basal leaves; helicoid cymes more or less
            corymbosely aggregated; inflorescence conspicuously glandular
            20  Corolla mostly 5-9 mm. wide; filaments well exserted; plants erect, 5-20 dm.
                tall, merely strigillose below the inflorescence; seeds commonly 12-16
                                    P. PROCERA
            20  Corolla mostly 10-20 mm. wide; filaments about equaling or slightly surpassing
                the corolla; plants sprawling to sometimes apparently erect, seldom as much
                as 1 m. tall, the herbage evidently spreading-hirsute at least in part; seeds
                commonly 30-60                    P. BOLANDERI

Phacelia bicolor Torr. ex Wats. Bot. King Exp. 255. 1871. (Torrey 345, Empire City, Nev.)
  P. leibergii Brand, Pflanzenr. IV. 251:128. 1913. P. bicolor var. leibergii Nels. & Macbr.
    Contr. Gray Herb. n.s. 49:40. 1917. (Leiberg 321, between P[r]ineville and Bear Buttes, Oreg.)
  P. adspersa Brand, Fedde Rep. Sp. Nov. 17:319. 1921. (Leiberg 2220, near Harper Ranch,
  Malheur Valley, Oreg.)
  Similar to P. glandulifera, but larger-flowered; corolla 6.5-12 mm. long, narrowly campa-
nulate, the limb 5-10 mm. wide; longer filaments nearly or quite reaching the sinuses; style
(4)5-8 mm. long; ovules probably averaging more numerous than in P. glandulifera (24-28 fide
Howell). N=13.
  Dry, sandy plains, commonly with sagebrush; Deschutes, Crook, and n. Malheur cos.,
Oreg., s. (chiefly e. of the Sierra crest) to s. Calif. and w. Nev. June.
  Our plants, as described above, represent the technically well-defined var. leibergii (Brand)
Nels. & Macbr., which is confined to Oreg. The more southern var. bicolor (c. Lake Co.,
southward) has shorter filaments, the longer ones falling short of the sinuses of the corolla by
2-5 mm., and the corollas average slightly longer while the styles average shorter.

Phacelia bolanderi Gray, Proc. Am. Acad. 10:322. 1875. (<u>Bolander</u>, Cottonaby Creek, Mendocino Co., Calif.)

Stems sprawling to apparently sometimes erect, stout, 2-8(12) dm. long, arising from a deep-seated, slender, strongly branched taproot, the subterranean portion of the stem resembling a woody rhizome; plants (the aerial portion) loosely hirsute (or in part coarsely strigose) throughout, and generally also stipitate-glandular, at least in the inflorescence, the glands commonly blackish at least in the herbarium; leaves all cauline, gradually reduced upward, the blade ovate to broadly elliptic, mostly 3-11 cm. long and 2-7 cm. wide, shallowly pinnatilobate or very coarsely few-toothed, often with a pair of small, detached segments at the base; lower petioles nearly or quite equaling the blades, the upper shorter; helicoid cymes corymbosely loosely aggregated; corolla lavender to bluish or purplish, open-campanulate or subrotate, 1-2 cm. wide, glabrous or nearly so within; filaments about equaling or slightly surpassing the corolla; style elongate, cleft to about the middle or less; seeds numerous, commonly 30-60, pitted-reticulate. N=11.

Mostly on open, often unstable slopes and banks; near the coast from Sonoma Co., Calif., to Coos Bay, Oreg., mostly at low elevations, but ascending to 4500 ft. in Curry Co., Oreg.; an apparently isolated station in Wahkiakum Co., Wash. June-July.

Phacelia franklinii (R. Br.) Gray, Man. 2nd ed. 329. 1856.

<u>Eutoca</u> <u>franklinii</u> R. Br. App. Frankl. Journ. 764. 1823. (<u>Richardson</u>, Great Bear Lake)

Annual or biennial, 1-7 dm. tall, with a single, usually erect stem, or this surrounded by several lesser stems; herbage hirsute-puberulent (or the stem more spreading-hirsute) and somewhat viscid, many of the hairs swollen and glandular at the base; leaves basal and cauline, only gradually or scarcely reduced upward, petiolate (less so upward, with pinnatifid or subbipinnatifid blade 1.5-9 cm. long and 0.5-5 cm. wide; inflorescences scattered or aggregated but generally not forming a dense, terminal thyrse; corolla purplish, deciduous or sometimes persistent, broadly campanulate, 6-9 mm. long and 8-12 mm. wide, hairy outside, glabrous inside; filaments surpassing the corolla lobes by about 2 mm. or less; style elongate, cleft less than halfway; ovules 26-46 (fide Howell); seeds pitted-reticulate.

Stream banks, meadows, and open slopes, especially in gravelly soil, at moderate elevations in the mountains, sometimes in burns or other disturbed sites; w. Mont., n. Wyo., and c. Ida., to Mack. and Yukon, e. to Lake Superior. June-July.

Phacelia glandulifera Piper, Contr. U.S. Nat. Herb. 11:472. 1906.

P. <u>ivesiana</u> f. <u>glandulifera</u> Brand, Pflanzenr. IV. 251:126. 1913. <u>P.</u> <u>ivesiana</u> var. <u>glandulif-era</u> Nels. & Macbr. Contr. Gray Herb. n. s. 49:40. 1917. (<u>Piper 2954</u>, Pasco, Wash.)

P. <u>luteopurpurea</u> A. Nels. Bot. Gaz. 52:271. 1911. (<u>Macbride 84</u>, New Plymouth, Canyon Co., Ida.)

Mephitic annual, 0.5-2(3) dm. tall, usually branched from the base and often without a well-defined central axis; herbage shortly spreading-hairy and (especially the upper part of the stem) stipitate-glandular; leaves 2-10 cm. long (petiole included) and 0.5-3 cm. wide, subbipinnatifid, with mostly narrowly foliaceous rachis and narrow segments, many of which are again toothed or cleft; inflorescences few-flowered, compact; calyx segments linear or linear-oblanceolate, often unequal at maturity; corolla 4.5-6.5 mm. long (dry), the tube yellowish, the limb lavender or light blue, 2-3.5 mm. wide; filaments not reaching the sinuses; style 2-3 mm. long, the branches 0.5 mm. long or less; ovules 14-23 (fide Howell); seeds commonly somewhat fewer, angular, cellular-reticulate and strongly cross-corrugate, about 1 mm. long. N=13.

Dry, sandy plains and hills, commonly with sagebrush; c. Wash. (Grant Co.) to Mono Co., Calif., e. through the Snake R. plains to s. w. Wyo., and to c. and n. e. Nev. June.

I concur in Howell's conclusion (Am. Midl. Nat. 36:398-404. 1946) that <u>P. glandulifera</u> is properly to be distinguished from the chiefly more southern and eastern <u>P. ivesiana</u> Torr. by several minor but well-correlated characters.

Phacelia glandulosa Nutt. Journ. Acad. Phila. II. 1:160. 1848.

<u>Eutoca</u> <u>glandulosa</u> Hook. Kew Journ. Bot. 3:293. 1851. (<u>Nuttall</u>, "about Ham's Fork, Colorado of the West," Wyo.)

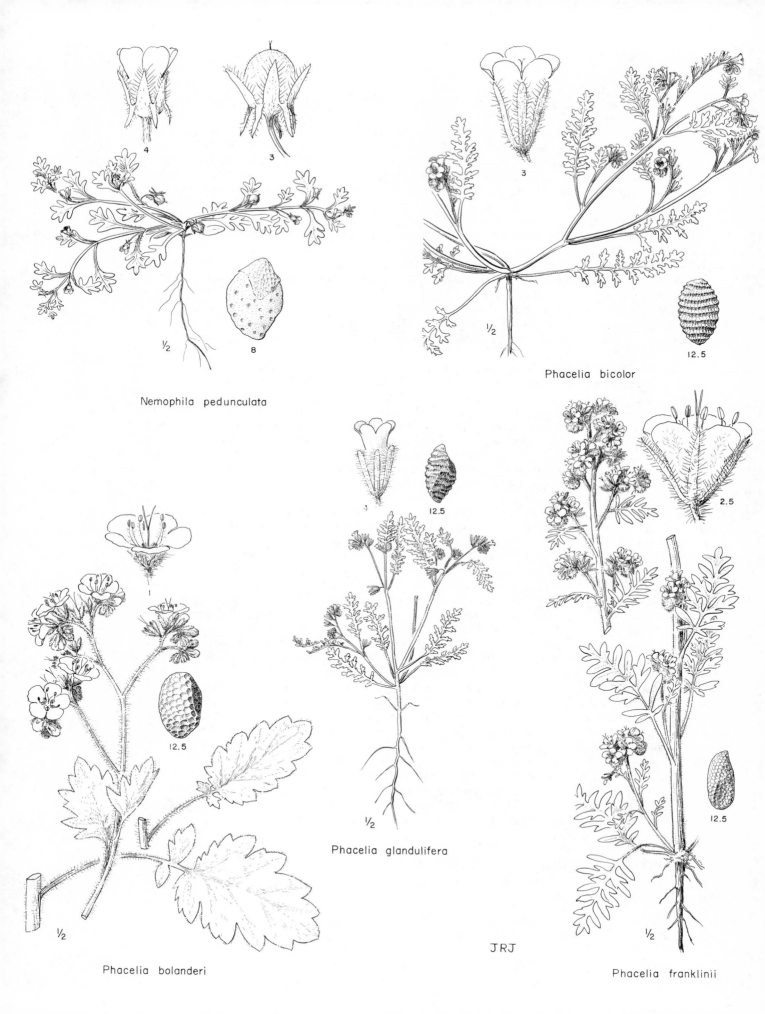

4

3

½

8

Nemophila pedunculata

3

½

12.5

Phacelia bicolor

1

12.5

Phacelia bolanderi

12.5

½

Phacelia glandulifera

2.5

12.5

½

JRJ

Phacelia franklinii

Strongly glandular-hairy and odoriferous annual or biennial with 1-several erect or ascending stems 1-3.5 dm. tall; leaves cauline and basal, mostly 2-11 cm. long and 0.5-3 cm. wide, pinnatifid or pinnately compound with sessile or short-petiolulate, coarsely few-toothed or cleft segments, the upper confluent; helicoid cymes tending to be aggregated toward the summit of a short, common peduncle; sepals oblanceolate, 2.5-4 mm. long; corolla blue-violet, campanulate, often broadly so, 6-9 mm. wide; filaments long-exserted, glabrous; style cleft to near the base; seeds 2-4, 2.5-3 mm. long, reticulate-pitted, excavated on each side of the prominent, ventral ridge.

Loose banks and talus slopes; valleys of the Salmon R. and its major tributaries in the Challis-Salmon region of c. Ida., to s.w. Mont., e. Utah, and w. Colo. June-Aug.

Our plants are relatively uniform; several varieties or closely related species may be distinguished in the Colorado Plateau and s. Rocky Mt. region and southward.

Phacelia hastata Dougl. ex Lehm. Stirp. Pug. 2:20. 1830. (Douglas, barren sandy plains of the Columbia, and on the Blue Mts.)
PHACELIA HASTATA var. LEUCOPHYLLA (Torr.) Cronq. hoc loc. P. leucophylla Torr. in Frem. Rep. 89. 1843. P. magellanica f. leucophylla Brand, Pflanzenr. IV. 251:98. 1913. (Fremont, Goat I., upper n. fork of the Platte)
P. canescens Nutt. Journ. Acad. Phila. II. 1:159. 1848. (Nuttall, Rocky Mts. and Blue Mts.) = var. leucophylla.
P. frigida Greene, Pitt. 4:39. 1899. P. magellanica f. frigida Brand, U. Calif. Pub. Bot. 4:218. 1912. P. heterophylla f. frigida Macbr. Contr. Gray Herb. n.s. 49:35. 1917. P. heterophylla var. frigida Jeps. Man. Fl. Pl. Calif. 819. 1925. P. mutabilis var. frigida G. N. Jones, U. Wash. Pub. Bot. 7:175. 1938. P. magellanica var. frigida Jeps. Fl. Calif. 3:248. 1943. (Merriam, Mt. Shasta, Calif., Aug. 3, 1898) = var. compacta.
PHACELIA HASTATA var. ALPINA (Rydb.) Cronq. hoc loc. P. alpina Rydb. Mem. N. Y. Bot. Gard. 1:324. 1900. P. heterophylla var. alpina Nels. in Coult. & Nels. New Man. Bot. Rocky Mts. 408. 1909. P. magellanica f. alpina Brand, U. Calif. Pub. Bot. 4:217. 1912. P. leucophylla f. alpina Macbr. Contr. Gray Herb. n.s. 49:34. 1917. P. leucophylla var. alpina Dundas, Bull. So. Calif. Acad. Sci. 33:164. 1934. (Rydberg & Bessey 4855, Cedar Mt., Mont.)
PHACELIA HASTATA var. COMPACTA (Brand) Cronq. hoc loc. P. compacta Greene ex Baker, W. Am. Pl. 18. 1902. nom. nud. P. magellanica f. compacta Brand, U. Calif. Pub. Bot. 4:217. 1912. P. leucophylla var. compacta Macbr. Contr. Gray Herb. n.s. 49:34. 1917. P. heterophylla var. compacta Jeps. Man. Fl. Pl. Calif. 819. 1925. (Baker 1142, Spooner, Douglas Co., Nev.)
P. burkei Rydb. Bull. Torrey Club 36:675. 1909. (Burke, Snake country, Ida.) = var. leucophylla.
PHACELIA HASTATA var. LEPTOSEPALA (Rydb.) Cronq. hoc loc. P. leptosepala Rydb. Bull. Torrey Club 36:676. 1909. (Farr 1013, Vermilion Lake, B.C.)
P. magellanica ssp. barbata Brand, U. Calif. Pub. Bot. 4:217. 1912, in part. (Synthetic; no type given)
P. magellanica f. angustifolia Brand, Pflanzenr. IV. 251:98. 1913. (Specimens cited from B.C., Wash., Ida., Wyo., Colo., and Calif.) = var. leucophylla.
P. leucophylla var. suksdorfii Macbr. Contr. Gray Herb. n.s. 49:34. 1917. (Suksdorf 3647, near Bingen, Klickitat Co., Wash.) = var. hastata.
Perennial, the taproot usually surmounted by a branched caudex; stems usually several and more or less similar, prostrate to suberect, up to 5(10) dm. tall; herbage generally more or less silvery with a fine, short, loose pubescence (often less so in forms from higher altitudes), less bristly than in P. heterophylla, the bristles when present mostly ascending or appressed except in the inflorescence; leaves prominently veined, all entire or sometimes some of them with a pair of small lateral lobes or leaflets at the base of the blade; basal leaves tufted and persistent, narrowly to broadly elliptic, petiolate, the cauline ones progressively reduced and becoming sessile; inflorescence usually rather short and compact, sometimes more elongate and narrow; corolla dull whitish to lavender or dull purple, 4-7 mm. long and broad; filaments conspicuously exserted, usually hairy near the middle; ovules 4, commonly only 1 or 2 maturing. X=11.

In dry, open places at all elevations, often in sand; s. B.C. and Alta. to Calif., Colo., and w. Neb. May-Aug.

Phacelia hastata and P. heterophylla belong to a polyploid complex which might conceivably be treated as a single, sharply limited species, P. magellanica (Lam.) Coville, with numerous infraspecific taxa. Such a treatment, which has been approached if not wholly realized by several botanists (notably Brand and Jepson), has the merit of permitting the ready use of a binomial, but is unattractive because of the excessive variability encompassed within a single species. The opposite treatment, in which each of the ultimate taxa is dignified with a binomial (as approached though not wholly realized by Constance for Abrams' flora), misrepresents the relationships within the group, as well as ignoring the numerous intermediate forms. The treatment here suggested is considered to be a reasonable compromise between the two extremes. It has the virtue of holding the species to reasonable limits, as well as preserving some well-known binomials, but is admittedly faulty in that the species are less sharply defined than might be desired. P. heterophylla var. pseudohispida, and P. hastata var. leptosepala, in particular, are closely allied and often difficult to distinguish, since each of these varieties tends to vary in the direction of the other species. The second (and perhaps ultimately more desirable) choice, in my opinion, would be to recognize P. hastata, P. heterophylla, P. nemoralis, and a few (perhaps 4) other major groups from outside our range as subspecies of P. magellanica. Some of these subspecies would then be further subdivided into varieties. It may be noted that P. argentea Nels. & Macbr. (P. heterophylla var. rotundata Dundas), of coastal dunes in s. Oreg. and n. Calif., and P. corymbosa Jeps. (P. dasyphylla var. ophitidis Macbr.) are so closely allied to P. hastata that they might perhaps properly be subordinated to it, but the necessary new combinations are not here proposed. The four varieties of P. hastata which occur in our range may usually be distinguished as follows:

1 Plants of the foothills, valleys and plains; stems ascending to suberect, mostly over 15 cm. tall; herbage ordinarily silvery
  2 Leaves usually all entire, occasionally some of them with a pair of small lateral lobes; widespread e. of the Cascades and Sierra Nevada, but absent from the range of var. hastata                                 var. leucophylla (Torr.) Cronq.
  2 Some of the leaves generally with a pair of small lateral lobes (this feature, sporadic and uncommon in var. leucophylla, is more stabilized here); calyx often very stiffly long-hispid; Kittitas Co., Wash., to Deschutes and Grant cos., Oreg., characteristically developed especially in the sands along both sides of the Columbia for some miles eastward from the gorge                                 var. hastata
1 Plants of moderate to high elevations in the mountains and either with distinctly greener herbage than in the foregoing varieties, or with the stem either more or less prostrate, or less than 15 cm. long
  3 Flowers mostly whitish; stems often more or less erect; leaves often with a pair of lateral lobes; herbage often more bristly than in the following variety
    4 Dwarf, alpine and subalpine plants, generally not over 15(20) cm. tall, often prostrate; Cascade-Sierra region from e. Wash. southward, extending also to c. and s. Nev.                            var. compacta (Brand) Cronq.
    4 Taller plants, mostly 15-50 cm. tall, usually of moderate elevations in the mountains, habitally much like var. leucophylla, but greener and more bristly, and the larger leaves often with a pair of lateral lobes; often (? always) hexaploid, perhaps with 2 genomes of P. heterophylla added to 4 of P. hastata var. leucophylla; mountains of Wash., n.e. Oreg., n. Ida., n.w. Mont., and s. B.C.
                                var. leptosepala (Rydb.) Cronq.
  3 Flowers light lavender to dull purplish (unique among the varieties of the species in this regard); stems prostrate or merely ascending at the tip; leaves usually all entire; mountains of the Great Basin, extending northward to Mont., c. Ida., and n.e. Oreg.
                            var. alpina (Rydb.) Cronq.


Phacelia heterophylla Pursh, Fl. Am. Sept. 140. 1814.
  P. magellanica f. heterophylla Brand, U. Calif. Pub. Bot. 4:218. 1912. P. magellanica var.

heterophylla Jeps. Fl. Calif. 3:246. 1943. (Lewis, "on dry hills and on the banks of the Kooskoosky" [Clearwater])

P. virgata Greene, Erythea 4:54. 1896. P. magellanica f. virgata Brand, U. Calif. Pub. Bot. 4:219. 1912. P. californica var. virgata Jeps. Man. Fl. Pl. Calif. 820. 1925. (Greene 832, Yreka, Calif.) = var. heterophylla.

P. mutabilis Greene, Erythea 4:55. 1896. P. nemoralis var. mutabilis Macbr. Contr. Gray Herb. n. s. 49:37. 1917. (Greene, summit station, Sierra Nevada in 1895; lectotype by Howell) = var. pseudohispida.

P. biennis A. Nels. Bull. Torrey Club 26:132. 1899. P. sericea var. biennis Brand, Pflanzenr. IV. 251:107. 1913. (Nelson 1323, Pole Creek, Wyo.) A small, purple-flowered form of var. heterophylla, perhaps due to influence from P. hastata var. alpina.

PHACELIA HETEROPHYLLA var. PSEUDOHISPIDA (Brand.) Cronq. hoc loc. P. nemoralis var. pseudohispida Brand, U. Calif. Pub. Bot. 4:219. 1912. P. pinnata var. pseudohispida Dundas, Bull. So. Calif. Acad. Sci. 33:167. 1935. (Baker 344, Stalkers, Shasta Co., Calif.)

P. magellanica ssp. barbata Brand, U. Calif. Pub. Bot. 4:217. 1912, in part. (Synthetic; no type given)

Biennial or short-lived perennial from a taproot, with a single erect, often stout stem 2-12 dm. tall, or this often surrounded at the base by several ascending, lesser stems (or more perennial and without an erect central stem in forms of var. pseudohispida that approach P. hastata var. compacta); herbage green or grayish with pubescence, but scarcely silvery, often evidently spreading-bristly, the stem generally covered with fine, short, loose or spreading, often glandular hairs in addition to the longer bristles; leaves prominently veined, the lower petiolate, the upper progressively less so; basal leaves larger or smaller than the well-developed cauline ones; some of the better developed leaves ordinarily with one or two (four) pairs of lobes or leaflets at the base of the blade, these sometimes fairly large, but still much smaller than the terminal segment, rarely all of the leaves entire; inflorescence densely bristly and short-hairy, characteristically elongate and narrow, somewhat virgate, but sometimes more openly branched, or shorter and more compact; corolla dull whitish to occasionally purplish, 3-6 mm. long and broad; filaments conspicuously exserted, hairy near the middle; ovules 4, often only 1 or 2 maturing. X=11.

In dry, open places at lower elevations (var. pseudohispida extending to moderate elevations in the mountains); Wash. and adj. s. B.C. to Mont., s. to c. Calif., Ariz., and N.M. May-July.

See comment under P. hastata.

The var. heterophylla, as principally described above, is widespread e. of the Cascades, and is occasionally found in drier habitats w. of the Cascades as well. At middle altitudes in the Cascade-Sierra region (also rarely in the Oregon Coast Ranges?), it is replaced by the poorly defined var. pseudohispida (Brand) Cronq., a relatively small and weak phase, seldom as much as 5 dm. tall, with the stem(s) frequently curved at the base, or ascending rather than erect, and seldom virgate.

Phacelia humilis T. & G. Pac. R.R. Rep. 2:122. 1855. (Snyder, near the summit of the Sierra Nevada)

P. violacea Brand, Rep. Sp. Nov. 17:319. 1921. (Sandberg & Leiberg 350, Egbert Spring, Douglas Co., Wash., and Heller 10946, Mt. Rose, Nev., are cited in that order)

Annual, 0.5-3 dm. tall, simple or openly branched, the stem spreading-puberulent and thinly spreading-hirsute, the leaves often more hirsute-strigose; leaves cauline, entire, short-petiolate, with oblanceolate to elliptic, lanceolate or ovate blade 0.5-4 cm. long and 2-25 mm. wide; inflorescence small and compact; sepals narrow, bristly-hispid; corolla lavender to blue-violet, campanulate, 4-7 mm. long and wide; filaments shortly (or scarcely) exserted, generally somewhat hirsute; style cleft to the middle or below; seeds 1-4, finely pitted, 2-2.5 mm. long. N=11.

In moist to moderately dry soil at elevations of 1500-5500 ft. in c. Wash. (Chelan and Douglas cos. to Yakima Co.); reputedly in s.e. Oreg.; and throughout the length of the Sierra Nevada and adj. mountains in Calif. and Nev., chiefly e. of the Sierra crest. May-July.

Phacelia humilis is vaguely similar to P. linearis, from which it is readily distinguished by the smaller, relatively as well as actually narrower corollas, mostly more distinctly exserted stamens, more deeply cleft styles, spreading-hirsute as well as puberulent stems, and often broader leaves, as well as by the technical character of the number of ovules and seeds.

**Phacelia idahoensis** Henderson, Bull. Torrey Club 22:48. 1895.
  P. sericea ssp. idahoensis Brand, Pflanzenr. IV. 251:107. 1913. (Henderson 2770, Craig
  Mts., Nez Perce Co., Ida.)
  Perennial (sometimes short-lived) with one, or more often several, stout, erect stems 2-8 dm. tall; herbage rather thinly strigillose, becoming more hirsute-puberulent in the inflorescence; leaves deeply pinnatilobate to more often pinnatifid, often irregularly so, with entire or cleft, often rather broad segments; basal and lower cauline leaves well developed and generally persistent, up to 15 or even 30 cm. long (ciliate-margined petiole included) and 5 cm. wide; middle and upper leaves gradually reduced and less petiolate; inflorescence an elongate terminal thyrse composed of numerous short, helicoid cymes; corolla lavender or bluish, persistent, campanulate, 4-6 mm. long and wide, short-hairy within the tube as well as outside; filaments slightly or scarcely longer than the corolla, but evidently exserted from the broad tube; style short or elongate, cleft less than halfway; seeds more or less numerous, angular, pitted-reticulate.
  Meadows, stream banks, and slopes, sometimes in disturbed sites in the foothills and lower mountains; Shoshone and Benewah cos. to Boise and Custer cos., Ida. May-July.
  The characters of P. idahoensis are such as to promote the suspicion that it may have originated as a hybrid or hybrid segregate from P. franklinii and P. sericea var. ciliosa. It now appears to be a stable population.

**Phacelia incana** Brand, Beil. Jahresb. Kgl. Gymnas. Sorau 8.1911. (M. E. Jones, Dugway,
  Utah [presumably Tooele Co.], May 28, 1891)
  Branching, more or less erect annual 0.5-1.5 dm. tall; herbage softly spreading-hairy throughout, many of the hairs gland-tipped; leaf blades elliptic to nearly ovate, 3.5-15 mm. long, 2-10 mm. wide; petioles well developed, at least the lower longer than the blade; inflorescences mostly relatively lax and elongate, rather few-flowered; calyx 3-4.5 mm. long at anthesis, up to 6 or 7 mm. in fruit, the segments narrow, linear or linear-oblanceolate, slightly or scarcely dissimilar; corolla about equaling the calyx, 3.5-4.5 mm. long, 2-3 mm. wide, the tube white or yellowish, the scarcely spreading limb white or bluish; stamens included; style 1.5 mm. long, the branches very short; ovules 22-38 (fide Howell); seeds angular, reticulate-pitted, nearly 1 mm. long.
  Calcareous, dry slopes in the foothills or at moderate elevations in the mountains; Custer Co., Ida., and (s.w.?) Wyo. to Utah and Nev.; seldom collected, and known from only one station in our range (pass 7 miles n. of Dickey, Ida.) July.

**Phacelia linearis** (Pursh) Holz. Contr. U.S. Nat. Herb. 3:242. 1895.
  Hydrophyllum lineare Pursh, Fl. Am. Sept. 134. 1814. (Lewis, "on the banks of the Missouri"; probably actually near The Dalles, Oreg.)
  Eutoca menziesii R. Br. App. Frankl. Journ. 764. 1823. Phacelia menziesii Torr. ex Wats.
  Bot. King Exp. 252. 1871. (Menzies, Calif.)
  Eutoca multiflora Dougl. ex Lindl. Bot. Reg. 14: pl. 1180. 1828. E. menziesii var. multiflora A. DC. Prodr. 9:294. 1845. (Douglas, N. W. Am.)
  E. congesta Dougl. ex Lehm. Stirp. Pug. 2:18. 1830. E. menziesii var. congesta A. DC.
  Prodr. 9:294. 1845. (Douglas, N. W. Am.)
  Erect annual, 1-5 dm. tall, simple or freely branched, the stem densely puberulent, the leaves often more strigose; leaves cauline, sessile or short-petiolate, narrow, mostly 1.5-11 cm. long, 1.5-12 mm. wide (exclusive of the 1-4 divergent narrow segments which some of them may bear below the middle); flowers crowded; sepals narrow, elongate, bristly-ciliate; corolla blue-lavender, very broadly campanulate, the limb 8-18 mm. wide; filaments sparsely bearded, about equaling the corollas as pressed; style entire to well above the middle; ovules mostly 10-20; seeds mostly 6-15, coarsely pitted, 1.5 mm. long. N=11.

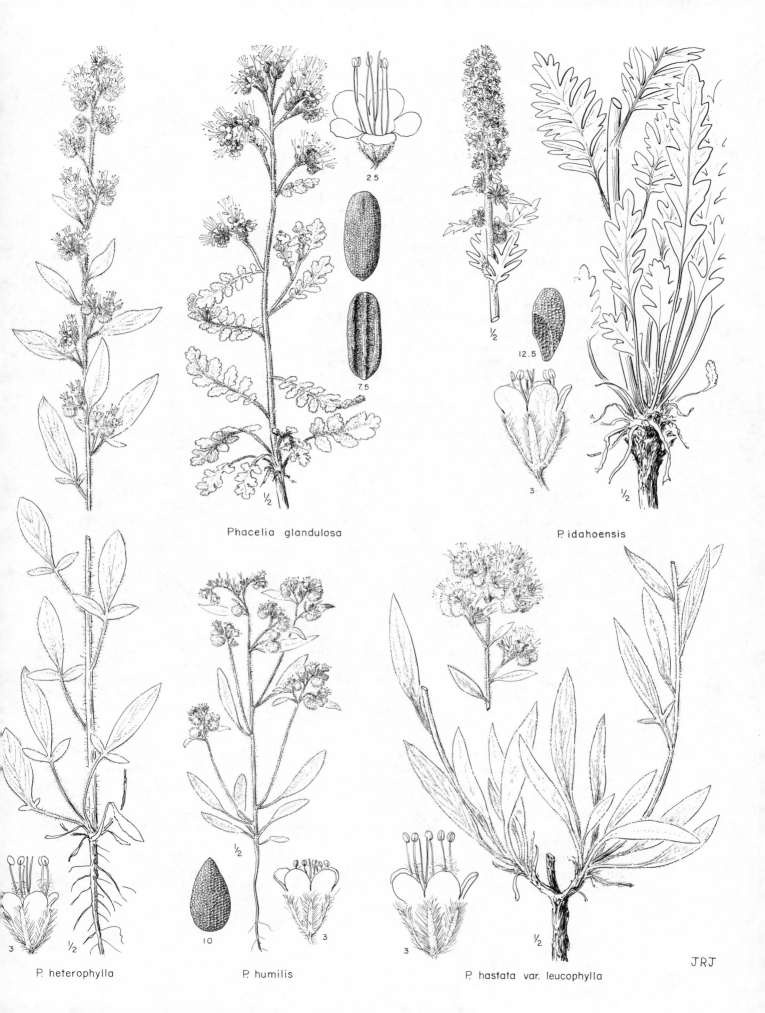

2.5

7.5

Phacelia glandulosa

1/2

12.5

3

1/2

P. idahoensis

3    1/2
P. heterophylla

1/2
10    3
P. humilis

3    1/2
P. hastata var. leucophylla

JRJ

Dry, open places in the foothills and plains; s. B.C. and Alta. to n. Calif., Utah, and Wyo., chiefly e. of the Cascades. Apr.-June.

A very common and showy plant.

**Phacelia lutea** (H. & A.) J. T. Howell, Leafl. West. Bot. 4:15. 1944.

Eutoca lutea H. & A. Bot. Beechey Voy. 373. 1838. Miltitzia lutea A. DC. Prodr. 9:296. 1845. Emmenanthe lutea Gray, Proc. Am. Acad. 10:328. 1875. (Tolmie, Snake Fort, Snake Country; near the mouth of the Boise R., Ida.)

PHACELIA LUTEA var. SCOPULINA (A. Nels.) Cronq. hoc loc. Emmenanthe scopulina A. Nels. Bull. Torrey Club 25:380. 1898. Miltitzia lutea var. scopulina Brand, Pflanzenr. IV. 251:131. 1913. Miltitzia scopulina Rydb. Bull. Torrey Club 40:479. 1913. Phacelia scopulina J. T. Howell, Leafl. West. Bot. 4:16. 1944. (Nelson 3056, Green R., Sweetwater Co., Wyo.)

Emmenanthe salina A. Nels. Bull. Torrey Club 25:381. 1898. Miltitzia salina Rydb. Bull. Torrey Club 40:479. 1913. Phacelia salina J. T. Howell, Leafl. West. Bot. 4:16. 1944. (Nelson 3105, Bitter Creek, Sweetwater Co., Wyo.) A form of var. scopulina with the style only about 1 mm. long.

Phacelia lutea var. purpurascens J. T. Howell, Proc. Calif. Acad. Sci. IV. 25:365. 1944. (Henderson 5092, Humphrey's, Grant Co., Oreg.)

PHACELIA LUTEA var. CALVA Cronq. hoc loc. A var. lutea differt planta glabra vel supra parce glandulose, filamentis paulo longioribus (Type: Maguire & Holmgren 26386, roadside bank along highway U.S. 95, 4 miles n.e. of the Idaho-Oregon state line, Owyhee Co., Ida.; N.Y. Bot. Gard.)

Prostrate or decumbent annual, freely branched from the base, often forming mats as much as 4 dm. across; leaves petiolate, with oblanceolate to elliptic or ovate blade up to 2.5 cm. long and 1.5 cm. wide, these all entire, or some of them coarsely crenate to pinnately lobed; inflorescences numerous and small; corolla persistent and surrounding the fruit, campanulate or tubular-campanulate, the tube 2-3 times as long as the lobes; corolla wholly yellow, or sometimes with some fine red lines, becoming whitish and papery in age and then sometimes acquiring a purplish cast in part; filaments included, or equaling the corolla; ovules and seeds mostly 8-20; seeds reticulate and more or less prominently cross-corrugated. N=12.

Alkaline, usually clay (rarely sandy) flats and banks in the deserts and foothills; c. Oreg. to Mono Co., Calif., e. irregularly to s.w. Mont., s.w. Wyo., w. Colo., and e. Ariz. May-July.

The several varieties of P. lutea here recognized appear to represent real taxa, but are nonetheless wholly confluent. The var. scopulina might perhaps profitably be elevated to subspecific rank, with the inclusion of P. submutica J. T. Howell as a variety. The latter, in addition to its occurrence at the type locality in Mesa Co., Colo., was collected a century ago by Newberry on the Little Colorado R., near the present site of Winslow, Ariz. Newberry's specimen (deposited at New York) forms the basis for the reported occurrence in Ariz. of the related P. glaberrima (Torr.) J. T. Howell. P. adenophora J. T. Howell, of w. Nev., n.e. Calif., and adj. s.e. Oreg., is closely allied to P. lutea, but is perhaps adequately characterized by the presence of some hairs on the filaments and within the corolla tube, and by its consistently pinnatilobate or pinnatifid leaves. The varieties of P. lutea known to occur in our range may usually be recognized as follows:

1 Corolla 2.5-4 mm. long at anthesis, about equaling the calyx, slightly accrescent (to 5 mm.) in fruit but then distinctly surpassed by the calyx; style (0.5)1-2.5 mm. long, usually hairy to near the middle; herbage shortly spreading-hairy, not glandular; longer filaments usually surpassing the sinuses; Silver Bow Co., Mont.; s.w. Wyo.; Tooele Co., Utah, and across Nev. to Mono Co., Calif., and occasionally extending into s.e. Oreg. in forms mostly transitional to var. lutea                                        var. scopulina (A. Nels.) Cronq.

1 Corolla (4)4.5-8 mm. long at anthesis, usually surpassing the calyx, seldom definitely surpassed by the calyx even in fruit

  2 Herbage glandular-villous; style 1.7-2.5 mm. long, hairy only at the base; longer filaments often surpassing the sinuses; corolla averaging a little shorter than in the two following varieties; Ochoco-Blue Mt. region of c. Oreg. in Jefferson, Wheeler, Crook,

Grant, and Baker cos., notably on the alkaline, red clays derived from volcanic tuff
                                                                var. purpurascens J. T. Howell

2 Herbage mostly not glandular, or glandular chiefly in the inflorescence; style mostly 2.5-3.8
   mm. long

   3 Herbage wholly glabrous, or slightly glandular in the inflorescence; longer filaments
      commonly surpassing the sinuses; local in n. Owyhee Co., Ida., where known from sev-
      eral collections                                                 var. calva Cronq.

   3 Herbage more or less densely short-hairy, only occasionally also glandular; filaments
      not reaching the sinuses; Great Basin region of s. e. Oreg., chiefly in Malheur and Har-
      ney (and Deschutes?) cos., occasionally extending to Humboldt Co., Nev.    var. lutea

**Phacelia lyallii** (Gray) Rydb. Mem. N. Y. Bot. Gard. 1:325. 1900.
   P. sericea var. lyallii Gray, Proc. Am. Acad. 10:323. 1875. P. sericea ssp. lyallii Brand,
   Pflanzenr. IV. 251:107. 1913. (Lyall, Rocky Mts. at lat. 49 degrees)

Dwarf perennial with numerous stems up to 2.5 dm. tall from a taproot and branched caudex; plants long-hairy in the short, compact, terminal inflorescence, otherwise green, sparsely hirsute or strigose, and conspicuously to obscurely or scarcely glandular; leaves pinnatilobate or very coarsely few-toothed, the persistent basal or lower ones up to 10 cm. long
(short petiole included) and 3 cm. wide, the others fairly well developed, but smaller and often sessile, sometimes clasping; corolla blue-purple, deciduous, campanulate, 5-9 mm. long
and wide, shortly (sometimes obscurely) hairy inside as well as out; filaments long-exserted,
1.5-2.5 times as long as the corolla; style elongate, cleft less than halfway; seeds presumably
as in P. sericea.

Talus slopes and rock crevices at high elevations in the mountains, often above timber line;
w. Mont. and adj. Alta. July-Aug.

**Phacelia minutissima** Henderson Bull. Torrey Club 27:351. 1900. (Henderson 3386, Soldier
   Mts., Blaine [now Camas] Co., Ida.)

Dwarf, branching annual up to 1 dm. tall; herbage shortly spreading-hairy and stipitate-
glandular throughout; leaves mostly oblanceolate or linear-oblong, the blade up to about 1 cm.
long and 4 mm. wide, tapering to the short petiole or subpetiolar base up to 4 mm. long; inflorescence short, few-flowered; calyx 2.5-3 mm. long at anthesis, the narrow, linear or oblanceolate segments markedly accrescent in fruit and becoming distinctly unequal in length and
width, one sometimes foliaceous and 1 cm. or more long; corolla lavender, 2.5-4 mm. long;
stamens included; style 1 mm. long or less, cleft up to half its length; ovules about a dozen,
the reticulate-pitted seeds of similar number and scarcely 1 mm. long, or fewer and up to 1.5
mm.

At moderate elevations in the mountains; known from the type locality in Ida., from the Wallowa Mts. of Oreg., and from the mountains of Elko Co., Nev.; rarely collected. July.

**Phacelia nemoralis** Greene, Pitt. 1:141. 1887. (Greene, hills behind Oakland, Calif., July 21,
   1887)

Resembling a very robust (to 2 m. tall), greener, and strongly bristly form of P. heterophylla, but tending to have a more open inflorescence, and commonly becoming a several-
stemmed perennial with the taproot surmounted by a branching caudex; cauline leaves usually
better developed than the basal ones.

Thickets and woodlands, commonly in distinctly more mesic (and shady) habitats than P.
heterophylla; Wash. to c. Calif., from the coast to the lower slopes of the Cascades. Apr.-
July.

Phacelia nemoralis has sometimes been referred to the South American P. pinnata (R. & P.)
Macbr. (= P. peruviana Spreng., an illegitimate nomenclatural synonym), but the latter species was originally described and figured as an annual, about a foot high, with large, tufted
basal leaves and more reduced cauline leaves. It therefore seems unwise to reduce P. nemoralis to P. pinnata, even though some of the plants from S. Am. which have been distributed
as P. peruviana or P. pinnata are so similar to P. nemoralis as to be perhaps conspecific.

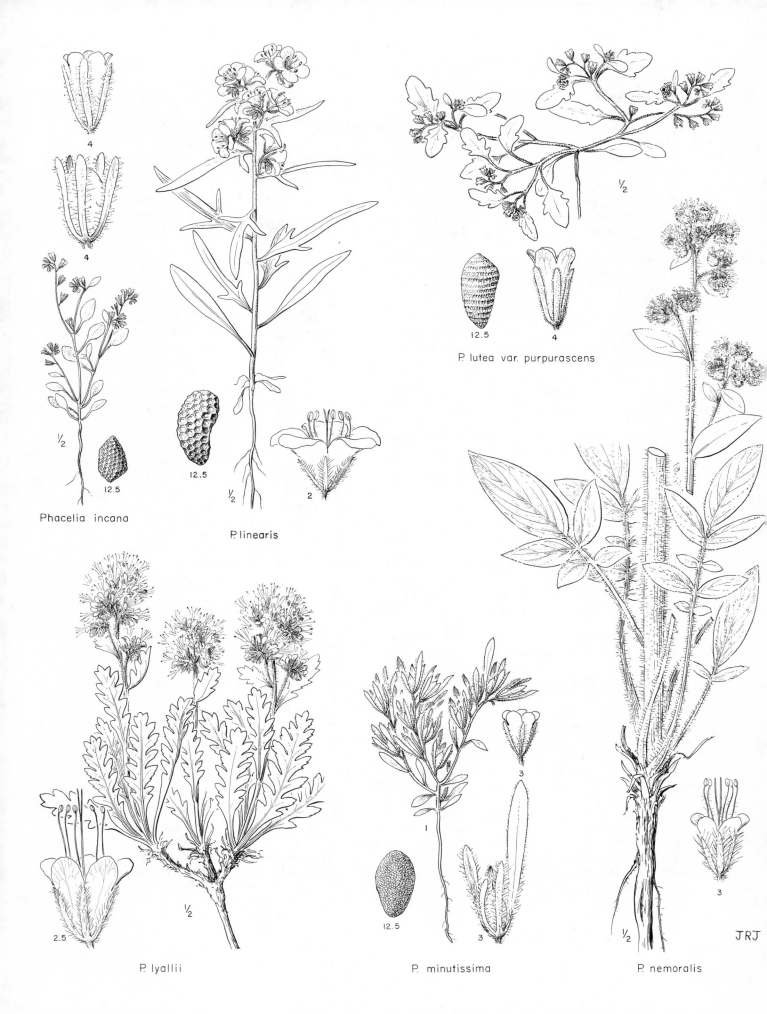

4

4

1/2

Phacelia incana

12.5

12.5

1/2

2

P. linearis

1/2

12.5

4

P. lutea var. purpurascens

2.5

1/2

P. lyallii

1

12.5

3

3

P. minutissima

1/2

3

P. nemoralis

JRJ

Phacelia procera Gray, Proc. Am. Acad. 10:323. 1875. ("Mountain meadows of the Sierra
   Nevada, in Nevada and Sierra counties, Bolander, Lemmon, etc.")
   Robust perennial, 5-20 dm. tall, conspicuously black-glandular in and near the inflorescence,
otherwise green and merely strigillose; leaves numerous, petiolate, the blade up to 12 cm.
long and 7 cm. wide, pinnately few-lobed or cleft (or sometimes merely very coarsely toothed),
the best developed ones borne 1-4 dm. above the base of the stem, the lower cauline ones reduced and the basal ones apparently wanting; inflorescence of several helicoid cymes loosely
or fairly closely aggregated into a terminal, corymbiform cluster; corolla pale, commonly
greenish-white, deciduous, campanulate, sometimes broadly so, 5-7 mm. long, 5-9 mm.
wide; filaments strongly exserted; style elongate, cleft to the middle or less; seeds fairly numerous (commonly 12-16), angular, pitted-reticulate. N=11.
   Meadows and open or lightly wooded slopes at moderate elevations in the mountains; Cascade region of Wash., s. to the n. Sierra Nevada and Coast Ranges of Calif.; also in the Blue
Mt. region of n.e. Oreg. and in adj. Ida. June-Aug..

Phacelia ramosissima Dougl. ex Lehm. Stirp. Pug. 2:21. 1830. (Douglas, "plains of the Columbia, near Priest's Rapid, and at the Stony Islands")
   Strongly glandular-hairy and odoriferous perennial from a taproot and branched caudex,
with numerous coarse but weak and brittle stems 5-15 dm. long, these simple or branched,
prostrate or weakly ascending; leaves all or principally cauline, up to 20 cm. long (short petiole included) and 10 cm. wide, pinnately compound with sessile or subsessile, coarsely toothed,
or cleft and again toothed leaflets, the lower leaflets remote, the upper confluent; peduncles short, commonly once or twice forked before giving rise to the short, dense, helicoid
cymes; sepals oblanceolate or spatulate, 4-6 mm. long; corolla pale, sometimes white, sometimes lavender, or marked and washed with light lavender and dull cream, funnelform-cam-
panulate, often broadly so, the limb 6-12 mm. wide; filaments glabrous, evidently exserted;
style cleft to well below the middle; seeds 2-4, 2.5-3 mm. long, strongly pitted, with an evident ventral ridge, but scarcely excavated. N=11.
   Dry, open places in the plains and foothills, commonly on basaltic talus or about ledges and
cliffs; c. Wash. to Calif., e. to s.w. Ida., Nev., and Ariz., wholly e. of the Cascades in our
range. May-Aug.
   Our plants are fairly uniform; several varieties or closely related species may be distinguished to the s. of our range.

Phacelia rattanii Gray, Proc. Am. Acad. 20:302. 1885. (Rattan, Lake Co., Calif.)
   Slender annual, 1-6 dm. tall, rather thinly bristly-hispid throughout, also with some shorter, softer hairs, at least on the leaves, and often glandular; leaves chiefly wholly cauline,
petiolate, the blade thin, 2-6(10) cm. long, 1-5(8.5) cm. wide, coarsely toothed to more often
shallowly pinnatilobate with the lobes again few-toothed; helicoid cymes few, scattered; calyx
segments oblanceolate or spatulate, unequally accrescent, up to 7 mm. long in fruit; corolla
white to pinkish or bluish, 3-4.5 mm. long, the limb 2-4 mm. wide; stamens and style included; style 1.5-3 mm. long, cleft to near or below the middle; ovules 4; seeds generally 2,
coarsely pitted-reticulate.
   In thickets and under shelving rocks in the foothills, valleys, and plains; Wasco Co., Oreg.
to s.w. Ida. and c. Calif. June-July.

Phacelia sericea (Grah.) Gray, Am. Journ. Sci. II. 34:254. 1862.
   Eutoca sericea Grah. Curtis' Bot. Mag. 56: pl. 3003. 1830. E. pulchella Lehm. Stirp. Pug.
      2:18. 1830. (Drummond, sandy debris of the Rocky Mts.)
   Phacelia sericea var. ciliosa Rydb. Bull. Torrey Club 31:636. 1904. P. ciliosa Rydb. Bull.
      Torrey Club 33:149. 1906. (Osterhout 2619, n. of Merker, Colo.)
   P. sericea var. caespitosa Brand, Pflanzenr. IV. 251:107. 1913. (Suksdorf 886, Mt. Adams,
      Wash.) A densely and rather loosely hairy phase, perhaps properly to be recognized as a
      rather weak variety, which largely or wholly replaces var. sericea in the Cascades of
      Wash. & adj. B.C., the Olympic Mts. of Wash., and the Marble Mts. of B.C.
   Perennial, usually with several or many erect or ascending stems from a taproot and

2.5

½

10

Phacelia procera

3

10

½

P. ramosissima

3

3

½

var. ciliosa

½

var. sericea

12.5

P. sericea

½

12.5

P. rattanii

JRJ

branched caudex, spreading-hairy in the inflorescence, otherwise thinly strigose to densely
sericeous or loosely woolly, not evidently glandular; leaves pinnatifid, with entire or sometimes cleft segments, basally disposed, the basal and lower cauline ones well developed and
persistent, the middle and upper ones more or less reduced and less petiolate; inflorescence a
dense, terminal, usually elongate thyrse composed of many short, compact, helicoid cymes;
corolla purple or dark blue, persistent, campanulate, mostly 5-6 mm. long and wide, hairy
inside and out; filaments long-exserted, (1.5)2-3 times as long as the corolla; style 6-13 mm.
long, cleft to the middle or less; seeds mostly 8-18, pitted-reticulate. N=11.

Open or wooded, often rocky places at moderate to high elevations in the mountains; s. B. C.
and Alta. to Wash., e. Oreg., n. e. Calif., Nev., and Colo. June-Aug.

Two varieties may be recognized:

1 Small, usually densely hairy (more or less sericeous) plants of high elevations in the mountains, often near or above timberline, 1-3 dm. tall; leaf segments usually relatively narrow and blunt; petioles seldom very strongly ciliate; apparently absent from Oreg., Calif.,
Nev., and most of Utah                                                    var. sericea
1 Larger, less densely hairy (more strigose, often thinly so) plants of moderate elevations in
the mountains, (2)3-6(9) dm. tall; leaf segments averaging broader, and sometimes more
acute; petioles tending to be evidently ciliate; absent from Wash.    var. ciliosa Rydb.

Phacelia tetramera J. T. Howell, Leafl. West. Bot. 4:16. 1944.
   Emmenanthe pusilla Gray, Proc. Am. Acad. 11:87. 1876. Miltitzia pusilla Brand, Pflanzenr.
      IV. 251:132. 1913. Not Phacelia pusilla Torr. (Watson 878 in part, Steamboat Springs,
      Washoe Co., Nev.)
   Miltitzia pusilla var. flagellaris Brand, Pflanzenr. IV. 251:132. 1913. (Cusick 758, Union
      Co., Oreg., and Cusick 1946, Malheur R., Oreg., are cited in that order)

Slender annual, freely branched from the base, apparently prostrate and matforming, shortly spreading-hairy throughout; leaves petiolate, oblanceolate to lance-ovate, the blade up to
1.5 cm. long and 7 mm. wide, entire or with a few coarse teeth or shallow lobes; inflorescences
numerous and small; flowers mostly 4-merous, occasionally some of them 5-merous; corolla
persistent and surrounding the fruit, white or reputedly pale yellow, minute, about 1-1.5 mm.
long at anthesis, up to 2 mm. in fruit, the calyx nearly or fully twice as long; filaments
sometimes slightly surpassing the sinuses; style very short, 0.3-0.4 mm. long; seeds commonly 6-10, reticulate and more or less evidently cross-rugulose.

Alkaline flats and washes; Nev. and adj. Calif., n. to s. e. Oreg., sometimes extending as
far n. as Union Co. May-June.

Phacelia thermalis Greene, Erythea 3:66. 1895. (Baker & Nutting, Little Hot Spring Valley,
   Modoc Co., Calif., June 4, 1894)
   P. firmomarginata A. Nels. Bot. Gaz. 54:143. 1912. (Macbride 979, Twilight Gulch, Owyhee
      Co., Ida.)

Glandular-hairy annual, 1-3 dm. tall, branched from the base and commonly without a well-
defined central axis, the several stems prostrate to ascending; leaves 1-9 cm. long (short
petiole included) and 0.5-2.5 cm. wide, pinnatilobate to more often pinnatifid, with mostly
narrow rachis and sessile, generally few-toothed segments or leaflets, these broader than in
P. glandulifera, the upper ones confluent; inflorescences densely flowered even at maturity, up
to 10 cm. long in fruit; corolla about equaling the calyx, 3-4 mm. long and 2-3 mm. wide,
lavender or whitish; calyx segments lance-elliptic, strongly accrescent in fruit, becoming 6-8
mm. long, 2-3 mm. wide, firm and prominently veiny; style 1.5-2 mm. long, cleft to the middle or beyond; ovules 4; seeds 4 or fewer, 2-2.5 mm. long, lanceolate, with ventral keel and
rounded back, prominently pitted-reticulate. N=11.

Heavy clay soil in desert regions; n. e. Calif. to s. w. Ida. and s. e. Oreg., entering our
range in n. Malheur Co. May-June.

## Romanzoffia Cham.

Flowers borne in loose or condensed, pedunculate, naked, racemelike, sympodial cymes;

calyx divided nearly to the base; corolla campanulate, white, commonly with a yellow eye; filaments about equal, attached to the corolla tube at or near its base; style simple, with a capitate, entire or obscurely lobed stigma; capsule loculicidal, wholly or in large part 2-celled; ovules numerous; low, perennial herbs with petiolate, reniform-orbicular, toothed or lobed, chiefly basal leaves, the cauline ones alternate and mostly borne near the base.

Four closely allied species, of w. N. Am., resembling some species of Saxifraga in habit. Two of the species are boreal or montane and lack tubers; the other two occur farther south at lower elevations and bear prominent woolly tubers. Each of these two groups consists of a condensed, evidently hairy species with long calyx and short pedicels and styles, and a laxer, more glabrate species with shorter calyx and long pedicels and styles. (Named for Count Nikolai von Romanzoff, sponsor of Kotzebue's voyage to Calif.)

1 Plants bearing well-developed, brown-woolly tubers at the base
   2 Plants condensed, the inflorescence barely if at all surpassing the leaves, and the pedicels less than 1 cm. long even in fruit; bluffs along the coast   R. TRACYI
   2 Plants laxer, the inflorescence well surpassing the leaves, and the pedicels (or at least the lower ones) commonly 1-4 cm. long in fruit; Coast Ranges, Oregon Cascades, and Columbia gorge, not at high elevations   R. SUKSDORFII
1 Plants without tubers, but the petioles strongly dilated and overlapping to form a bulbous base; well up in the mountains, or boreal   R. SITCHENSIS

Romanzoffia sitchensis Bong. Mém. Acad. St. Pétersb. VI. 2:158. 1833. (Eschscholz, Sitka)
  R. sitchensis var. parviflora Haage & Schmidt, Gartenfl. 22:33, pl. 748a. 1873. (Garden specimens, from seeds originating at Sitka.)
  ?R. sitchensis var. grandiflora Haage & Schmidt, Gartenfl. 22:33, pl. 748b. 1873. (Garden specimens, from seeds presumably originating at Sitka.) The figure is suggestive of the more northern R. unalaschkensis Chamisso.
  R. macouni Greene, Pitt. 5:37. 1902. R. sitchensis var. grandiflora subvar. macouni Brand, Pflanzenr. IV. 251:171. 1913. (Macoun 34921, Chilliwack Valley, B.C.)
  R. rubella Greene, Pitt. 5:37. 1902. (Macoun 34922, Chilliwack Valley, B.C.)
  R. glauca Greene, Pitt. 5:38. 1902. (Macoun 34923, Chilliwack Valley, B.C.)
  R. leibergii Greene, Pitt. 5:38. 1902. (Leiberg 1461, near Stevens Peak, Coeur d'Alene Mts., Ida)
  R. sitchensis var. grandiflora subvar. greenei cum f. vulgaris Brand, Pflanzenr. IV. 251:171. 1913. (Elmer 690, Bridge Creek, Okanogan Co., Wash., is the first of four specimens cited)
  R. minima Brand, Pflanzenr. IV. 251:169. 1913. (Shaw 1017, Selkirk Mts., B.C.)

Plants slender, lax, (0.5)1-2(3) dm. tall; basal leaves with reniform-orbicular blade (0.5)1-4 cm. wide, palmately veined and rather shallowly lobed or coarsely toothed, the petioles well-developed (sometimes 15 cm. long), their bases overlapping, conspicuously expanded, often thickened, and commonly villous-ciliate; plants otherwise generally subglabrous, or finely stipitate-glandular in the inflorescence, rarely somewhat glandular-villous; cauline leaves few and mostly borne near the base, or none; inflorescence loose and elongate, well surpassing the leaves, the pedicels (or at least the lower ones) mostly 1-4 cm. long in fruit; calyx seldom reaching the sinuses of the corolla; corolla 6-11 mm. long and wide; style 2-5 mm. long. N=11.

Wet cliffs and ledges in the mountains, sometimes above timber line, descending to sea level to the n. of our range; s. Alas. (beyond the panhandle) to Alta., n. w. Mont., n. Ida., and n. Oreg. (at 3000 ft. in the Columbia gorge), and rarely to n. Calif. June-Aug.

Romanzoffia suksdorfii Greene, Pitt. 5:38. 1902.
  R. sitchensis var. grandiflora subvar. greenei f. suksdorfii Brand, Pflanzenr. IV. 251:171. 1913. (Suksdorf, Mitchell's Point, Wasco Co., Oreg., April and May, 1884)
  R. californica Greene, Pitt. 5:39. 1902. (San Mateo Co., Calif., n. perhaps to Oreg.; several collectors mentioned, but no type cited)

Almost exactly similar to R. sitchensis, except that the petioles are less expanded and not aggregated to form a bulbous base, and that the plant bears 1 or more well-developed brown-woolly tubers (these sometimes 1.5 cm. long and 1 cm. thick) at the base; plants averaging a little more robust than R. sitchensis, often 3(4) dm. tall, the leaf blades up to 9 cm. wide; cauline leaves often with axillary bulbils. N=11.

Wet cliffs and ledges at lower elevations in the Coast Ranges from San Francisco to n. Oreg.; also on both sides of the river in the Columbia gorge, and rarely in the Cascade Range of Oreg. Apr. -May.

Romanzoffia tracyi Jeps. Fl. Calif. 3:296. 1943. (Parks & Tracy 11008, Trinidad, Humboldt Co., Calif.)

Resembling R. sitchensis and R. suksdorfii, and tuberiferous like the latter; plants condensed, 1 dm. tall or less, and more or less evidently glandular-villous; inflorescence compact, scarcely surpassing the leaves, the pedicels less than 1 cm. long even in fruit; calyx relatively large, generally reaching or surpassing the sinuses of the corolla; style 2-3 mm. long. N=11.

Bluffs along the coast; Humboldt Co., Calif., and n. at widely scattered stations to the Olympic Peninsula of Wash. Mar. -Apr.

Romanzoffia tracyi stands in the same morphological relation to R. suksdorfii as the chiefly Aleutian species R. unalaschkensis Cham. does to R. sitchensis.

## BORAGINACEAE. Borage Family

Flowers gamopetalous, perfect, ordinarily 5-merous as to calyx, corolla, and androecium; sepals more or less united or essentially distinct; corolla ordinarily regular, usually bearing more or less evident and often hairy appendages (the fornices) opposite the lobes at the summit of the tube; stamens epipetalous, alternating with the lobes of the corolla, included or sometimes exserted; ovary superior, basically 2-carpellary, each carpel 2-ovulate and with a secondary partition; fruit of 4 nutlets which are typically attached individually to the short or elongate gynobase; style simple or sometimes cleft into two or even four segments (rarely wanting), typically attached directly to the gynobase and arising between the essentially distinct lobes of the ovary, less commonly attached to the summit of the entire or merely 4-lobed ovary which then separates only tardily into individual nutlets; seeds nearly or quite without endosperm; mostly herbs (tropical members often woody), often rough-hairy, with exstipulate, alternate or sometimes partly (rarely wholly) opposite, mostly entire leaves, and diverse sorts of basically cymose (determinate) inflorescences, these most commonly sympodial, helicoid cymes (false racemes or spikes) which tend to elongate and straighten with maturity.

A large, cosmopolitan family, of nearly 100 genera and well over 1500 species, well developed in w. U.S., especially in drier habitats. Cynoglossum, Hackelia, and Mertensia include species of considerable potential or proven horticultural value.

References:

Brand, A. Borraginaceae--Borraginoideae. Cynoglosseae & Cryptantheae. Pflanzenr. IV. 252. Heft 78. 1921, and Heft 97, 1932.

Johnston, I. M. Studies in the Boraginaceae. II. 1. A synopsis of the American native and immigrant borages of the subfamily Boraginoideae. Contr. Gray Herb. n. s. 70:3-55. 1924.

-------. Studies in the Boraginaceae. III. 1. The Old World genera of the Boraginoideae. Contr. Gray Herb. n. s. 73:42-73. 1924.

1 Style more or less deeply 2-cleft; ours a dichotomously branched, prostrate annual with minute flowers and small, petiolate leaves with the blade nearly or fully as broad as long (Ehretioideae)                                                                     COLDENIA
1 Style entire or minutely lobed at the tip; plants not at once prostrate and with the leaf blades about as broad as long
  2 Ovary merely shallowly lobed, with the style wanting (ours) or borne on its summit, the stigma expanded and often as broad as the ovary; ours wholly glabrous, succulent plants with the white or faintly bluish flowers crowded in naked, secund, false spikes (Heliotropioideae)                                                                HELIOTROPIUM
  2 Ovary deeply 4-parted, the style borne on the gynobase and arising between the essentially distinct lobes of the ovary; stigmas small; plants not both wholly glabrous and with the flowers crowded in naked, secund, false spikes (Boraginoideae)

3 Base of the nutlet produced into a prominent, thickened rim which fits closely to the broad, low gynobase and surrounds the stipelike basal attachment, which latter fits into a distinct pit in the gynobase (body of the nutlet sometimes incurved so that the rim and attachment appear basilateral); ours introduced, weedy species with blue or occasionally ochroleucous flowers (Anchuseae)

    4 Corolla rotate; stamens exserted; anthers evidently appendiculate dorsally, connivent around the style     BORAGO

    4 Corolla tubular-campanulate to funnelform or salverform, the stamens included, exappendiculate, not connivent

        5 Corolla tubular-campanulate, the well-defined throat much longer than the short, erect or apically spreading lobes; style shortly exserted; basal rim of the nutlets evidently toothed     SYMPHYTUM

        5 Corolla funnelform or salverform, the throat often poorly defined, not much if at all longer than the well-developed, more or less spreading lobes; style included; basal rim of the nutlets entire or nearly so     ANCHUSA

3 Base of the nutlet not produced into a thickened rim; attachment various, but rarely approaching that of the foregoing group

    6 Nutlets broadly attached at the base to the broad low gynobase (Lithospermeae)

        7 Corolla evidently irregular (the upper side distinctly the longer), in ours generally blue; filaments (or some of them) strongly exserted     ECHIUM

        7 Corolla regular, white or bluish-white to yellowish, yellow, or greenish; filaments inconspicuous, scarcely or not at all exserted

            8 Corolla lobes rounded and more or less spreading; style in our species included     LITHOSPERMUM

            8 Corolla lobes acute or acuminate, erect; style conspicuously exserted     ONOSMODIUM

    6 Nutlets apically or medially to basilaterally or nearly basally attached, but the attachment, if nearly basal, always small

        9 Nutlets either evidently armed with glochidiate prickles or uncinate bristles, or provided with a continuous, entire or cleft dorsomarginal ridge or wing that is complete around the base or across the back of the nutlet, or both armed and ridged

            10 Nutlets more or less widely spreading in fruit, the rather large scar apical or apicolateral, not extending below the middle of the fruit, never elongate and narrow (Cynoglosseae)

                11 Nutlets armed with uncinate bristles; small, slender, white-flowered annuals     PECTOCARYA

                11 Nutlets armed with glochidiate prickles; coarse biennials or perennials; flowers mostly blue to red     CYNOGLOSSUM

            10 Nutlets erect or incurved to somewhat spreading; scar distinctly otherwise (Eritricheae in part)

                12 Nutlets essentially unarmed, the dorsomarginal ridge entire or merely toothed or lacerate

                    13 Scar medial; fornices conspicuous, exserted, outcurved; fruiting pedicels reflexed     DASYNOTUS

                    13 Scar basilateral; fornices not outcurved; fruiting pedicels erect to spreading

                        14 Dorsomarginal ridge complete around the base of the nutlet; plants sometimes caespitose, but not pulvinate     MYOSOTIS

                        14 Dorsomarginal ridge complete across the back of the nutlet well above the base, the part enclosed by the ridge (or wing) set at an angle to the basal part; plants pulvinate-caespitose     ERITRICHIUM

                12 Nutlets armed with glochidiate prickles at least along the dorsomarginal ridge

                  15 Nutlets narrowly attached to the elongate gynobase along the well-developed, median, ventral keel, free near the base; pedicels erect or ascending in fruit; "racemes" evidently bracteate; plants mostly annual     LAPPULA

                  15 Nutlets medially attached to the broad, low gynobase, the scar more or less

rounded, not elongate; pedicels recurved or deflexed in fruit; "racemes" naked or nearly so; plants mostly perennial (2 of our species annual or biennial)     HACKELIA
9 Nutlets smooth or variously roughened, but without hooked or glochidiate prickles or bristles (except for some very small dorsal ones in species of Plagiobothrys), sometimes
sharp-edged, especially distally, but not with the ridge complete across the back or around
the base of the nutlet (Eritricheae in part)
  16  Fruiting calyx greatly enlarged and prominently veiny; nutlets obliquely compressed,
      with a small, ventromarginal scar above the middle; blue-flowered, scrambling climb-
      ers with weak, retrorsely prickly-hispid stems     ASPERUGO
  16  Fruiting calyx not greatly enlarged, seldom evidently veiny; nutlets otherwise, though
      sometimes obliquely compressed; flowers various; habit otherwise
    17  Corolla blue or occasionally pink (in rare individuals white), tubular or tubular-fun-
        nelform to campanulate, never salverform; perennials, never pungently hairy
                                                                              MERTENSIA
    17  Corolla white to yellow or orange, often salverform; annual or perennial, often pun-
        gently hairy
      18  Corolla white, sometimes with a yellow eye (wholly yellow only in a few extra-
          limital species), the throat more or less closed by the fornices; cotyledons en-
          tire
        19  Nutlets with a closed or narrowly open ventral groove-scar running most of
            their length, this often expanded below into a depressed areola; plants an-
            nual to perennial, often pungently hairy     CRYPTANTHA
        19  Nutlets with a ventral keel extending to the middle or to near the base, the
            attachment elevated, carunclelike, placed at the base of the keel, or rarely
            extending along the keel; ours annuals, except for P. mollis, and generally
            not pungently hairy     PLAGIOBOTHRYS
      18  Corolla yellow or orange, the fornices obsolete and the throat open except for
          A. lycopsoides; cotyledons deeply 2-cleft, thus seemingly 4; annuals, usually
          pungently hairy, with the nutlets about as in Plagiobothrys     AMSINCKIA

Amsinckia Lehm. Nom. Conserv. Fiddle-neck; Tarweed; Fireweed

Flowers borne in sympodial, helicoid, mostly naked false racemes or spikes which tend to
elongate in age; calyx cleft essentially to the base, but some of the segments sometimes partly or wholly connate so that there may appear to be fewer than 5 members; corolla yellow or
orange, often marked with vermilion in the throat; corolla throat, except in A. lycopsoides,
open, the fornices obsolete; stamens included, the filaments short; nutlets with a well-developed ventral keel extending from the tip to near or below the middle, often somewhat keeled dorsally as well, the scar small, placed at the end of the ventral keel, often elevated and carunclelike; gynobase mostly short-pyramidal; cotyledons deeply cleft, thus apparently 4; taprooted, bristly-hairy annuals; leaves rather small and often narrow, entire or sometimes erose-
dentate.
  About a dozen species, native to w. N. Am. and w. and s. A. Am., centering in Calif.
(Named for William Amsinck, burgomaster of Hamburg and patron of its botanical garden
during the early part of the 19th century.)
  Of little, if any, horticultural value, but often included in wild flower seed mixtures.
  The species are all very much alike in general appearance, but numerous segregates have
been proposed on the basis of minor and inconstant differences in the nutlets and other features, and Suksdorf considered that more than 200 species were involved. Among our species,
A. lycopsoides, A. spectabilis, and A. tessellata are sharply limited and technically well marked,
at least within our area. Our other three species are less well defined, but still ordinarily
distinguishable without difficulty. Except for A. spectabilis, our species are all weedy, with little or no present geographic or ecologic differentiation, and two or more may occur together.
The synonymy as given here is subject to revision, since several of the segregates have been
placed from the descriptions alone, and even purported duplicates of the types are scarcely to
be trusted in a group of habitally similar, weedy species.

The name tarweed is commonly used for this genus in e. Wash. In Calif. the name fireweed has been reported, from the irritating effect of the stiff hairs on the skin of harvest workers.

It has been demonstrated that the nutlets of one or more of the species of e. Wash. are poisonous to livestock when eaten in quantity.

References:

Macbride, J. F. A revision of the North American species of Amsinckia. Contr. Gray Herb. n. s. 49:1-16. 1917. (23 species recognized).

Suksdorf, Wilhelm. Untersuchungen in der Gattung Amsinckia. Werdenda 1:47-113. 1931. (about 235 species recognized).

1 Plants maritime; leaves tending to be erose-denticulate; two of the sepals usually partly united; corolla tube about 10-nerved below the insertion of the stamens A. SPECTABILIS
1 Plants not maritime; leaves entire
 2 Sepals of many or all of the flowers unequal in width and reduced in number (commonly 4, sometimes 2 or 3) by fusion, the broader one(s) often apically bidentate; corolla tube mostly about 20-nerved below the insertion of the stamens   A. TESSELLATA
 2 Sepals 5, essentially distinct, not very unequal in width; corolla tube about 10-nerved below the insertion of the stamens
  3 Corolla throat obstructed by the well-developed, hairy fornices; stamens inserted below the middle of the corolla tube (unique among our species in both of these respects) A. LYCOPSOIDES
  3 Corolla throat open, glabrous, the fornices scarcely developed; stamens inserted above the middle of the corolla tube
   4 Stem spreading-hispid and also evidently puberulent or strigose with shorter and softer, more or less retrorse hairs; pubescence of the leaves tending to be ascending instead of widely spreading; corolla 5-8 mm. long   A. RETRORSA
   4 Stem spreading-hispid, nearly or quite without shorter and softer hairs below the inflorescence; hairs of the leaves often widely spreading
    5 Corolla 7-10 mm. long, the tube well exserted   A. INTERMEDIA
    5 Corolla 4-7 mm. long, the tube scarcely exserted   A. MENZIESII

Amsinckia intermedia Fisch & Mey. Ind. Sem. Hort. Petrop. 2:26. 1836.
 A. media Krause ex Sturm, Fl. Deutschl. 2nd ed. 11:42. 1903. Benthamia intermedia Druce, Rep. Bot. Exch. Cl. Brit. Isl. 3:25. 1912. (Bodega Bay, Calif.)
 A. howellii Brand, Fedde Rep. Sp. Nov. 25:213. 1928. (Howell, prairies of Oreg. in 1880)
 A. arvensis Suksd. Werdenda 1:32. 1927. (Suksdorf 2007, Bingen, Klickitat Co., Wash.)
 A. macounii Brand, Fedde Rep. Sp. Nov. 25:211. 1928. (First specimen cited is Macoun 78654, Victoria, B.C.)
 ?A. menziesii var. microcarya Brand, Pflanzenr. IV. 252 (Heft 97):211. 1931. (Howell, fields of Oreg. in 1881; no original material seen)

Plants 2-8 dm. tall, simple or branched; herbage spreading-hispid, or the hairs of the leaves often ascending, the upper part of the stem commonly with some soft, short hairs intermingled with the longer bristles; leaves linear to oblong or lanceolate, up to 15 cm. long and 2 cm. wide, seldom persistently clustered at the base; sepals free, often somewhat unequal, elongating to 6-12 mm. in fruit; corolla orange or orange-yellow, marked with vermilion in the throat, 7-10 mm. long, the tube well exserted, the limb commonly 4-6 mm. wide; nutlets 2.5-4 mm. long, ovoid, greenish or gray to brown or black, rugose and muricate or tuberculate. N=15, 17, 19.

A weed of roadsides, fields, and waste places; Calif. and Baja Calif. to N.M., n. to Klickitat Co., Wash., and reputedly to Ida.; seldom collected in our range. Apr.-May.

Amsinckia lycopsoides Lehm. ex Fisch. & Mey. Ind. Sem. Hort. Petrop. 2:2. 1836.
 Benthamia lycopsoides Lindl. ex Druce, List Brit. Pl. 103. 1908. (Cultivated plants, from seeds collected by Douglas, "above the rapids of the Columbia") Not Lithospermum lycopsoides Lehm. 1830. (See Johnston, Journ. Arn. Arb. 16:199-202. 1935)

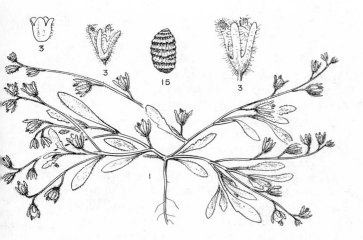

Phacelia tetramera

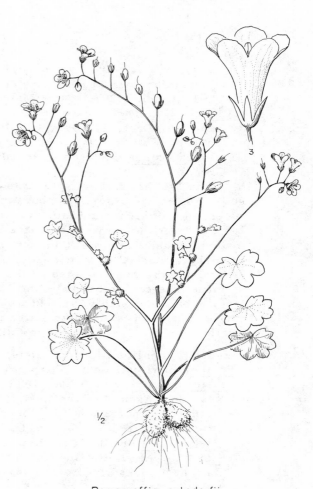

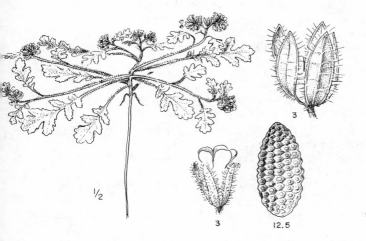

Phacelia thermalis

Romanzoffia suksdorfii

Romanzoffia sitchensis

Romanzoffia tracyi

JRJ

Amsinckia intermedia

Amsinckia barbata Greene, Erythea 2:192. 1894. Benthamia barbata Druce, Rep. Bot. Exch.
  Cl. Brit. Isl. 4:298. 1916. (Macoun 17093, Cameron Lake, Vancouver I., July 15, 1887)
Amsinckia arenaria Suksd. Deuts. Bot. Monats. 18:133. 1900. Benthamia arenaria Druce,
  Rep. Bot. Exch. Cl. Brit. Isl. 4:298. 1916. (Suksdorf 995, near Bingen, Klickitat Co.,
  Wash.)
Amsinckia simplex Suksd. Werdenda 1:33. 1927. (Suksdorf 3336, Lower Albina, Portland,
  Oreg.)

   Plants 1-6 dm. tall, simple or few-branched; stem spreading-hispid, also with shorter and
softer, often retrorse hairs above or throughout; leaves hispid-hirsute with mostly somewhat
ascending hairs, linear to linear-oblong or the upper more lanceolate, up to 10 cm. long and
1.5 cm. wide, often crowded at the base; sepals free, not very unequal, 6-10 mm. long at
maturity, their hairs white to somewhat fulvous; corolla yellow to more often orange-yellow
or orange, marked with vermilion in the throat, mostly 7-9 mm. long, the tube somewhat exserted, the throat obstructed by the well-developed, hairy fornices, the limb 3-6 mm. wide;
stamens inserted well down in the corolla tube, commonly below the middle; nutlets 2.5-3 mm.
long, ovoid, greenish to dark brown, tuberculate or muricate and often somewhat rugose. N=15.
   Dry, open slopes and flats, often in disturbed soil; s. B.C. to w. Mont., chiefly but not
wholly e. of the Cascades, s. to Calif. and reputedly Nev. Apr.-June (Aug.).

Amsinckia menziesii (Lehm.) Nels. & Macbr. Bot. Gaz. 61:36. 1916.
  Echium menziesii Lehm. Stirp. Pug. 2:29. 1830. (Menzies, N. W. coast of Am.)
Amsinckia micrantha Suksd. Deuts. Bot. Monats. 18:134. 1900. Benthamia micrantha Druce,
  Rep. Bot. Exch. Cl. Brit. Isl. 4:299. 1916. (Suksdorf 390, Klickitat Co., Wash.)
Amsinckia idahoensis M. E. Jones, Contr. West. Bot. 12:58. 1908. Benthamia idahoensis
  Druce, Rep. Bot. Exch. Cl. Brit. Isl. 4:298. 1916. (Jones, Weiser, Ida. in 1899)
A. microcalyx Brand, Fedde Rep. Sp. Nov. 25:211. 1928. (First collection cited is Macoun
66569, between Kettle and Columbia rivers, B.C.)
A. kennedyi Suksd. Werdenda 1:64. 1931. (Kennedy 1347, Truckee Pass, "Washoe Co.,
  Nev.")
A. borealis Suksd. Werdenda 1:71. 1931. (First collection cited is Eastwood 683, Lake Atlin,
  B.C.)
A. canadensis Suksd. Werdenda 1:77. 1931. (First collection cited is Macoun 78652, Esqui-
  mault, Vancouver I.,)
A. melanocarpa Suksd. Werdenda 1:89. 1931. (First collection cited is Otis 763, Embro,
  Wash.)
A. foliosa Suksd. Werdenda 1:90. 1931. (Macoun 78657, Nanaimo, Vancouver I.)
A. nephrocarpa Suksd. Werdenda 1:91. 1931. (First collection cited is Leiberg 809, near
  Prineville, Oreg.)

   Plants 1.5-7 dm. tall, simple to much branched, often softer and weaker, and with thinner
and broader leaves than in related species; herbage rather sparsely spreading-hispid, without
shorter and softer hairs except often in the inflorescence; leaves linear to more often oblong,
lance-elliptic, or even ovate, up to 12 cm. long and 2 cm. wide, not forming a persistent basal tuft; sepals free, not very unequal, 5-10 mm. long in fruit; corolla apparently rather light
yellow, 4-7 mm. long, the tube not much, if at all, exserted, the limb only 1-3 mm. wide;
nutlets ovoid, greenish or gray to brown or black, 2-3.5 mm. long, tuberculate or muricate
and often rugose.
   Fields and dry, open places, or not infrequently in moister soil with more luxuriant vegetation; throughout our range, on both sides of the Cascades, extending n. (perhaps only as an introduction) to s. Yukon and Alas., and rarely s. to Calif. and Nev. May-Sept.
   The use of the name Amsinckia menziesii for this species is based on examination of the
type specimen at the British Museum by Peter M. Ray. The specimen is not A. lycopsoides,
as might be inferred from the original description. An isotype at the British Museum is similar to the type.

Amsinckia retrorsa Suksd. Deuts. Bot. Monats. 18:134. 1900.
  Benthamia retrorsa Druce, Rep. Bot. Exch. Cl. Brit. Isl. 4:299. 1916. (Suksdorf 994, Bin-
    gen, Klickitat Co., Wash.)
Amsinckia hispidissima Suksd. Deuts. Bot. Monats. 18:133. 1900. Benthamia hispidissima
    Druce, Rep. Bot. Exch. Cl. Brit. Isl. 4:298. 1916. (Suksdorf 2316, Hood R., Oreg.)
  A. rugosa Rydb. Fl. Rocky Mts. 729. 1917. (Heller 2975, Lewiston, Ida.)
  A. helleri Brand, Fedde Rep. Sp. Nov. 25:212. 1928. (Heller & Kennedy 8850, Little Grizz-
    ly Creek, Plumas Co., Calif.)
  A. leibergii Brand, Fedde Rep. Sp. Nov. 25:214. 1928. (Leiberg 2119, Harper Ranch, Mal-
    heur Co., Oreg.)
  A. mollis Suksd. Werdenda 1:54. 1931. (Eastwood 13301, mouth of the Salmon R., Ida.)
    Plants simple or few-branched, 1-6 dm. tall; stem spreading-hispid and also evidently pu-
berulent or strigose throughout with shorter and softer, more or less retrorse hairs; leaves
hispid-hirsute with mostly ascending hairs, mostly linear or linear-oblong, sometimes wider,
up to 12 cm. long, seldom over 1 cm. wide, the basal ones often crowded and larger than the
others; sepals free, not very unequal, 5-12 mm. long at maturity; corolla orange or orange-
yellow, marked with vermilion in the throat, 5-8 mm. long, the tube not much, if at all, exserted, the limb mostly 1.5-3 mm. wide; nutlets ovoid, 2-3 mm. long, muricate-tuberculate
and often somewhat rugose. N=8, 13, 17.
    Roadsides, fields, and other dry, open places; s. B.C. and n. Ida. to Utah and s. Calif.,
chiefly but not wholly e. of the Cascades Apr.-July.

Amsinckia spectabilis Fisch. & Mey. Ind. Sem. Hort. Petrop. 2:26. 1836. (Bodega Bay,
  Calif.)
  Lithospermum lycopsoides Lehm. Stirp. Pug. 2:28. 1830. Amsinckia lycopsoides var. brac-
    teosa Gray, Syn. Fl. 2$^1$:198. 1878. Amsinckia scouleri Johnst. Journ. Arn. Arb. 16:202.
    1935. (Scouler, Strait of Juan de Fuca) Not Amsinckia lycopsoides Lehm., though in the
    past often confused with it. (See Johnston, Journ. Arn. Arb. 16:199-202. 1935)
    Plants up to 4 dm. tall, erect to often decumbent or prostrate, simple or branched; herbage
sparsely or moderately spreading-hispid; leaves slightly succulent, tending to be erose-denticulate, mostly lanceolate to lance-linear, lance-elliptic, or lance-oblong, or the lower more
oblanceolate, up to about 5 cm. long and 12 mm. wide, or perhaps larger; spikes often irregularly
leafy-bracteate below; sepals 4-8 mm. long at maturity, two of them generally connate below the
middle; corolla 4-8 mm. long, the limb 2.5-5 mm. wide; nutlets 2-2.5 mm. long, ovoid,
blackish, somewhat rugose or tuberculate, but much less strongly so than is usual for our
other species.
    Near or along the seashore, often on sandy beaches; Queen Charlotte Islands, B.C. (and the
Alaska panhandle?), to n. Baja Calif. June-July.
    Our plants, as here described, have smaller flowers than typical A. spectabilis, and may
prove to constitute a distinct variety. Small-flowered forms of A. spectabilis are also known
in California, however.

Amsinckia tessellata Gray, Proc. Am. Acad. 10:54. 1874.
  Benthamia tessellata Druce, Rep. Bot. Exch. Cl. Brit. Isl. 4:299. 1916. (The first of 5
    specimens cited is Brewer, near Mt. Diablo, Calif.)
Amsinckia hendersonii Suksd. Werdenda 1:107. 1931. (The first of 5 specimens cited is Hen-
    derson 8280, e. base of the Steens Mts., Oreg.)
  A. verucosissima Suksd. Werdenda 1:105. 1931. (Macbride 76, New Plymouth, Canyon Co.,
    Ida.)
  A. washingtonensis Suksd. Werdenda 1:110. 1931. (Suksdorf 9520 and 9518, both from Bingen,
    Klickitat Co., Wash., are cited in that order)
  A. densirugosa Suksd. Werdenda 1:111. 1931. (First of 7 collections cited is Suksdorf 8718,
    Spangle, Spokane Co., Wash.)
    Plants 1.5-6 dm. tall, simple or moderately branched; herbage spreading-hispid throughout,
or the hairs of the leaves merely ascending, the upper part of the stem also with shorter and
softer, somewhat retrorse hairs; leaves linear or generally broader, often lance-oblong or

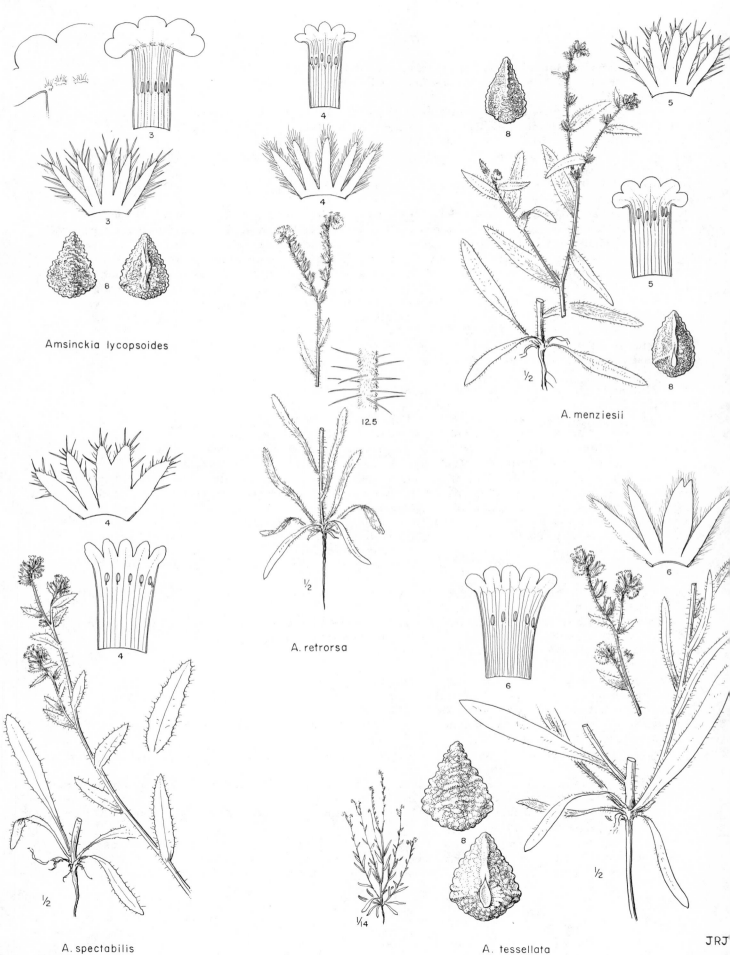

Amsinckia lycopsoides

12.5

A. menziesii

A. retrorsa

A. spectabilis

A. tessellata

JRJ

lance-ovate, up to 10 cm. long and 1.5 cm. wide; inflorescence often with some rufous or blackish bristles; sepals 7-14 mm. long at maturity, those of many or all of the flowers unequal in width and reduced in number (commonly 4, sometimes 2-3) by lateral fusion, the broader one(s) often apically bidentate; corolla yellow marked with orange, or more often golden or orange marked in the throat with vermilion, 7-12 mm. long, the limb mostly 3-5 mm. wide, the tube mostly about 20-nerved below the insertion of the stamens; nutlets ovoid, 2.5-3.5 mm. long, brownish or gray, tessellate-tuberculate and generally also somewhat rugose. N=12.

Roadsides and dry, open slopes and flats, often in disturbed soil; Wash. to n. Baja Calif., e. to Ida., Utah, and Ariz., chiefly but not wholly e. of the Cascades. Apr.-June.

Amsinckia vernicosa H. & A., a species chiefly of c. and s. Calif., with the calyx and corolla of A. tessellata, but with smooth and shining nutlets, is represented in Oreg. by a single collection (Leiberg 2234, Malheur Valley, near Harper Ranch) on which the name A. carinata Nels. & Macbr. Bot. Gaz. 62:145. 1916 was founded.

## Anchusa L. Alkanet

Flowers borne in terminal, sympodial, helicoid, bracteate, false racemes which tend to elongate and straighten in age; pedicels persistently erect or ascending; calyx shallowly to deeply cleft; corolla funnelform or salverform, the throat often poorly defined, not much, if at all, longer than the well-developed, more or less spreading, apically rounded lobes; filaments inserted shortly below the level of the fornices, the anthers extending into the corolla throat; nutlets with a stipelike basal attachment which fits into a pit in the otherwise flattish receptacle, the attachment surrounded by a prominent, thickened rim (the basal margin of the nutlet) which fits closely to the gynobase; annual or perennial leafy herbs, often pungently hairy.

About 40 species of the Old World, centering in the Mediterranean region. (Name from Greek anchousa, ancient name of a related plant, Alkanna tinctoria.)

1 Corolla limb 6-11 mm. wide; nutlets oblique, about 2 mm. high and 3-4 mm. long, the tip
     directed inward                                                              A. OFFICINALIS
1 Corolla limb 12-20 mm. wide; nutlets erect, 5-9 mm. high and 3-5 mm. thick
                                                                                 A. AZUREA

Anchusa azurea Mill. Gard. Dict. 8th ed. no. 9. 1768. (Crete)
  A. italica Retz. Obs. Bot. 1:12. 1779. (Italy)
  Taprooted perennial, 4-15 dm. tall, mostly single-stemmed, spreading-hispid throughout; basal leaves oblanceolate, petiolate, the others more lanceolate or ovate-lanceolate to oblong, sessile and often clasping, sometimes as much as 30 cm. long and 8 cm. wide; bracts narrow, resembling the calyx lobes; calyx 8-10 mm. long at anthesis, sometimes 15 mm. in fruit, cleft to well below the middle, the lobes slender; corolla showy, blue, the limb 12-20 mm. wide; nutlets erect, 5-9 mm. high and scarcely over half as thick, coarsely reticulate-ridged with mostly vertically elongate areolae. N=16.

Native of the Mediterranean region, cultivated and occasionally escaped; reported to be established at several localities in the Willamette Valley of Oreg. A good subject for the perennial garden.

Anchusa officinalis L. Sp. Pl. 133. 1753. (Europe)
  Taprooted perennial, 3-8 dm. tall, often several-stemmed, spreading-hirsute throughout; lower leaves oblanceolate, petiolate, mostly 6-20 cm. long (petiole included) and 1-2.5 cm. wide, the others gradually reduced, becoming sessile and more lanceolate; bracts lanceolate or lance-triangular; calyx 5-7 mm. long in flower, scarcely longer in fruit, the lanceolate or narrowly triangular lobes about equaling or a little longer than the tube; corolla blue, 6-11 mm. long, the limb 6-11 mm. wide; nutlets more or less rugose and tuberculate, oblique, about 2 mm. high and 3-4 mm. long, the tip directed inward. N=8.

Native of the Mediterranean region, now found as an occasional weed along roadsides and in other disturbed habitats here and there in the U.S.; e. of the Cascades in our range. May-July.

Two other species which might be confused with A. officinalis have been found in our range, but are probably not established. A. capensis Thunb. differs in its slightly smaller nutlets and particularly in its broad, blunt, somewhat fringed-ciliate calyx lobes that are distinctly short-er than the tube. A. arvensis (L.) Bieb. (=Lycopsis arvensis L.) is hispid rather than hirsute, and has the corolla tube bent near the middle and the limb oblique and slightly irregular.

## Asperugo L. Madwort

Flowers borne on short, stout, recurved pedicels in or near the axils of the leaves or bracts, and in the forks of the branches; calyx 5-lobed to about the middle, each lobe with a smaller tooth on each side near the base, the whole strongly accrescent, becoming compressed, firmly chartaceous, strongly reticulate-veiny, and shortly prickly-hispid with curved or hooked hairs; corolla small, blue, more or less campanulate, with well-developed fornices; anthers included; nutlets obliquely compressed, narrowly ovate, tessellate, attached to the elevated gynobase by a small scar just within the margin and above the middle; annual weeds with weak, climbing-scrambling, retrorsely prickly-hispid stems and opposite or subopposite to partly alternate or partly whorled leaves that are subentire and often remote.

A single species. (Name from the Latin asper, rough, referring to the pubescence.)

### Asperugo procumbens L. Sp. Pl. 138. 1753. (Europe)

Annual; stems 3-12 dm. long, glabrous except for the short, retrorse prickles which are borne chiefly along the angles; leaves thin, scabrous-hispid and irregularly hispid-ciliate, the lower oblanceolate, petiolate, rarely as much as 10 cm. long and 2.5 cm. wide, often de-ciduous, the others gradually reduced, often becoming more elliptic or subrhombic and sub-sessile; corolla 2-3 mm. long and wide; fruiting calyx 1-2 cm. wide; nutlets 2.5 mm. long, enveloped by the calyx. N=24.

A weed in fields, waste places, and disturbed sites, usually in fairly moist soil; native of Eurasia, now found here and there over much of n. U.S., and well established e. of the Cas-cades in our range. May-July.

## Borago L.

Flowers borne in loose, terminal, modified, sympodial cymes, these leafy-bracteate below, the elongate pedicels recurved in fruit; sepals narrow, elongate, distinct or nearly so; corolla rotate, blue, with 5 elongate acute lobes; fornices well developed; filaments inserted at the level of the fornices, each prolonged beyond the base of the anther into a prominent dorsal ap-pendage; anthers elongate, conspicuous, connivent around the style; nutlets with a well-devel-oped, swollen, stipelike basal attachment which fits into a pit in the otherwise flattish gyno-base, the attachment surrounded by a prominent, thickened rim (the basal margin of the nut-let) which fits closely to the gynobase; annual or perennial, broad-leaved, coarsely hairy an-nual or perennial herbs.

Three sharply distinct species of the Mediterranean region. (Medieval Latin name of un-known derivation.)

### Borago officinalis L. Sp. Pl. 137. 1753. (Europe)

Taprooted annual, 2-6 dm. tall, hispid-setose and hispidulous, especially the stem; lower leaves petiolate, with broadly elliptic or ovate to merely oblanceolate blade 3-11 cm. long and 2-6 cm. wide; cauline leaves progressively reduced and less petiolate, but still generally am-ple, the upper often sessile and clasping; pedicels 1-4 cm. long; sepals densely bristly, 1-1.5 cm. long in fruit; corolla 2 cm. wide; anthers dark, 5 mm. long, the linear appendages 3 mm. long; nutlets subcylindric, 4-5 mm. long, irregularly (mostly longitudinally) ridged and tuber-culate. N=8. Borage.

Native of Europe, sometimes cultivated, and established as a casual weed here and there in the U.S.; occasionally found w. of the Cascades in our range. July-Aug.

## Coldenia L.

Flowers small, essentially sessile in few-flowered, axillary clusters; calyx deeply (4)5-cleft; corolla white or pink, (4)5-lobed, with, or more often without, fornices; stamens included; ovary entire or more often 4-lobed, separating into 4 (or fewer by abortion) nutlets at maturity; style terminal, not gynobasic (though the ovary may be apically as well as laterally lobed), evidently 2-cleft, sometimes essentially to the base, each branch with a capitate stigma; annual or perennial herbs or subshrubs.

Perhaps 20 species, native mostly to the drier and warmer parts of the New World, one species in the Old World tropics. The genus consists of 3 or 4 sections that have sometimes been treated as distinct genera. (Named for Dr. Cadwallader Colden, colonial lieutenant-governor of New York and a correspondent of Linnaeus.)

Coldenia nuttallii Hook. Journ. Bot. & Kew Misc. 3:296. 1851.

Tiquiliopsis nuttallii Heller, Muhl. 2:239. 1906. (Nuttall, Rocky Mts.)

Tiquilia oregana Torr. Bot. Wilkes Exp. 2:411, pl. 12. 1874. (Wilkes Expedition 956, on the Walla Walla R.)

Prostrate, dichotomously branched, taprooted annual, forming rosettes up to 3 or 4 dm. across, short-hirsute throughout (or the stem more strigose) and commonly hispid-setose in the inflorescences, notably on the calyx lobes and on the petioles and margins of the subtending leaves; leaf blades rather broadly elliptic or ovate to subrotund, strongly few-veined, 3-8 mm. long, the petioles about as long; corolla minute, the limb only 1-2 mm. wide; fruiting calyx 3-4 mm. long; nutlets 1 mm. long, ovate, smooth, attached laterally near the base to the short, pyramidal gynobase, and connate along the ventral axis to about 3/4 of the way to the summit, the more or less deeply cleft style arising between the lobes; cotyledons so strongly sagittate at the base as to appear 2-cleft.

Dry, sandy places in desert regions; c. Wash. to Calif., e. to Wyo. and Ariz.; also in Argentina. May-Aug.

## Cryptantha Lehm.

Flowers ordinarily borne in a series of sympodial, helicoid, naked or bracteate, false spikes, these sometimes short and aggregated into a terminal thyrse, or rarely the flowers solitary in the axils; calyx cleft to the base or nearly so (except in C. circumscissa), more or less accrescent; corolla white, the well-developed fornices often yellow, the limb usually more or less rotately spreading, often very small; filaments short, attached below the middle of the corolla tube; nutlets 4, or 1-3 by abortion, affixed to the somewhat elongate gynobase for much of their length, the scar narrow and commonly appearing as an elongate, closed or narrowly open groove that is either forked at the base or opened into a basal areola; strigose to more often hirsute or partly hispid-setose herbs, annual or perennial, with mostly narrow leaves.

Perhaps 150 species, native to w. N. Am. and w. S. Am. (Name from the Greek kryptos, hidden, and anthos, flower, referring to the cleistogamous flowers of the original South American species.)

Species of Cryptantha are occasional components of wild flower seed mixtures, but ours are not particularly desirable.

The genus consists of several well-marked sections, some of which have frequently (and not without reason) been treated as distinct genera. The delimitation here adopted has been vigorously defended by Johnston and by Payson, the only students who have given the group serious monographic attention. The treatment of our annual species follows in the main the fundamentally sound revision by Johnston. With the exception of C. torreyana vs. C. ambigua, all of our annual species appear to be sharply defined in our region. The perennials are more difficult; the boundaries among the species even as here broadly defined are somewhat vague, while the more narrowly limited taxa of some previous students appear to me to be in several instances quite hopeless.

References:

Johnston, I. M. Studies in the Boraginaceae. -IV. The North American species of Cryptantha. Contr. Gray Herb. n. s. 74:1-114. 1923.

-------. Studies in the Boraginaceae. -VI. A revision of the South American Boraginoideae. Contr. Gray Herb. n. s. 78:1-118. 1927. (This reference for reduction of Oreocarya.)

Payson, E. B. A monograph of the section Oreocarya of Cryptantha. Ann. Mo. Bot. Gard. 14:211-358. 1927.

1 Plants perennial or sometimes merely biennial, relatively coarse, usually (C. salmonensis excepted) with a well-developed tuft of basal leaves; spikes aggregated into a dense, terminal, irregularly bracteate thyrse, often eventually elongating and distinct; corolla more or less conspicuous, the limb mostly 4-12 mm. wide; calyx persistent (Oreocarya)

  2 Nutlets smooth

    3 Corolla tube longer than the calyx at anthesis; leaves all relatively slender and tapering gradually to an acutish apex; plains of c. Wash. and adj. Oreg.

                                                  C. LEUCOPHAEA

    3 Corolla tube equaling or shorter than the calyx; lower leaves mostly more broadly rounded or obtuse at the tip; valley of the Salmon R. in c. Ida.        C. SALMONENSIS

  2 Nutlets more or less roughened, at least on the back

    4 Nutlets closely and unevenly rugose-reticulate on both sides; upper surfaces of the basal leaves uniformly short-hairy, or with some poorly developed and inconspicuous longer setae, in any case not pustulate; n. Malheur Co., Oreg., and adj. Ida.

                                                  C. PROPRIA

    4 Nutlets variously roughened, but scarcely as above; upper surfaces of the leaves usually provided with well-developed, mostly pustulate-based setae in addition to the fine, short hairs

      5 Corolla relatively large and showy, the limb mostly 8-12 mm. wide; basal leaves relatively broad, mostly broadly oblanceolate or spatulate, seldom conspicuously spreading-bristly; nutlets ovate or lance-ovate and more or less evidently roughened on both sides; biennial or perennial, occurring in the dry lands of e. Oreg. (northward from Grant Co.), Wash., and s. B. C., and in adj. Ida., thence to the foothills and intermontane valleys of Mont., and eastward onto the plains

                                         C. CELOSIOIDES

      5 Corolla smaller, the limb mostly 4-8 mm. wide (rarely to 10 mm. in C. interrupta); leaves various, often narrower and markedly spreading-bristly, especially in C. interrupta; nutlets lanceolate or lance-ovate; perennial

        6 Nutlets roughened ventrally as well as dorsally; plants of the plains, valleys, and foothills, in our range known from Mont., the more easterly intermontane valleys n. of the Snake R. plains in Ida., and rarely in e. Wash. and Oreg.

                                         C. INTERRUPTA

        6 Nutlets smooth or nearly so ventrally, slightly or moderately roughened dorsally; plants mostly of high altitudes in the mountains (C. thompsonii sometimes descending to the foothills)

          7 Plants dwarf, seldom as much as 15 cm. tall, absent from Wash., and in our range occurring wholly e. of the Cascades; leaves relatively short, the basal ones not over about 3.5 cm. long (petiole included); scar usually closed

                                         C. NUBIGENA

          7 Plants taller, 1-3 dm. tall, of the Wenatchee Mts. of Wash.; leaves more elongate, the larger basal ones commonly 4-7 cm. long; scar of the nutlets evidently open for most of its length        C. THOMPSONII

1 Plants annual, relatively slender, without any conspicuous tuft of basal leaves; corollas, except in C. intermedia, not over about 2.5 mm. wide

  8 Calyx circumscissile a little below the middle, the persistent cupulate basal portion scarious and obviously of different texture from the more herbaceous, deciduous portion; plants low, much branched when well developed, forming cushions 1-6 cm. high; flowers solitary in the closely crowded upper axils, not forming elongate spikes (Greeneocharis, Piptocalyx)        C. CIRCUMSCISSA

8 Calyx divided essentially to the base, not circumscissile, and without any abrupt change of
    texture, in most species eventually deciduous intact; plants not cushion-forming, most
    species erect and with a more or less evident central axis, or sometimes more bushy-
    branched; flowers borne in naked, eventually elongate, unilateral helicoid "spikes," these
    not closely aggregated (Krynitzkia)
  9 Corolla relatively large, the limb 4-8 mm. wide                    C. INTERMEDIA
  9 Corolla small, the limb 0.5-2.5 mm. wide
    10 Nutlets all smooth, or finely and inconspicuously granular, not at all tuberculate or
        spiculate-papillate
      11 Nutlets strictly solitary; hairs of the calyx tending to be uncinate or arcuate
        12 Nutlets broadly truncate at the base, the scar broadened below the middle into
            a definite open areola; style reaching to the middle of the nutlet or commonly
            beyond; rare                                               C. ROSTELLATA
        12 Nutlets more or less pointed or very narrowly truncate at the base, the scar
            closed or nearly so, not forming a definite areola; style scarcely reaching
            to the middle of the nutlet; common                       C. FLACCIDA
      11 Nutlets ordinarily 4, in any case more than 1; hairs on the calyx straight or near-
          ly so
        13 Nutlets symmetrical, the scar median on the ventral face
          14 Margins of the nutlets prominent, sharply angled, especially above
                                                                      C. WATSONII
          14 Margins of the nutlets rounded or obtuse, not prominent
            15 Nutlets lanceolate, 0.5-0.7 mm. wide; plants of sand dunes and very
                sandy soil, very local in our area                    C. FENDLERI
            15 Nutlets ovate, 0.8-1.2 mm. wide; plants common and widespread in
                our area                                               C. TORREYANA
        13 Nutlets obliquely compressed, with a distinctly excentric scar near one margin
                                                                      C. AFFINIS
    10 Nutlets, or some of them, rough, with evident tubercles or spiculate papillae on the
        dorsal surface
      16 Nutlets (except one) with conspicuously winged margins      C. PTEROCARYA
      16 Nutlets not at all wing-margined
        17 Nutlets distinctly heteromorphous, one nearly smooth and somewhat larger
            and more firmly attached than the other three, which are evidently tubercu-
            late                                                      C. KELSEYANA
        17 Nutlets all alike in size and texture when normally developed
          18 Nutlets lanceolate, 0.5-0.7 mm. wide                     C. SCOPARIA
          18 Nutlets ovate, (0.7)0.9-1.5 mm. wide
            19 Herbage closely strigose, essentially without spreading hairs
                                                                      C. SIMULANS
            19 Herbage spreading-hirsute, at least in part
              20 Nutlets with scattered, low tubercles, generally also granular;
                  throughout our range e. of the Cascades   C. AMBIGUA
              20 Nutlets densely spiculate-papillate; c. Ida.   C. ECHINELLA

Cryptantha affinis (Gray) Greene. Pitt. 1:119. 1887.
  Krynitzkia affinis Gray, Proc. Am. Acad. 20:270. 1885. (Lyall, e. side of the Cascades, in
    Wash.; lectotype by Johnston)
  Cryptantha confusa Rydb. Bull. Torrey Club 36:679. 1909. (Rydberg & Bessey 4884, Upper
    Madison Canyon, Yellowstone Nat. Pk., Wyo.)
  Slender annual 0.5-3 dm. tall, simple or branched; herbage strigose and somewhat hirsute;
leaves linear or narrowly oblong, scattered; spikes tending to be paired, naked except often for a
few leafy bracts at the base; calyx 2.5-4 mm. long at maturity, hispid and hirsute; corolla 1-
2 mm. wide; nutlets 4, hardly 2 mm. long, smooth or very finely granular, ovate, obliquely
compressed so that the scar is distinctly excentric, lying near one margin; style reaching to
well beyond the middle of the nutlets.

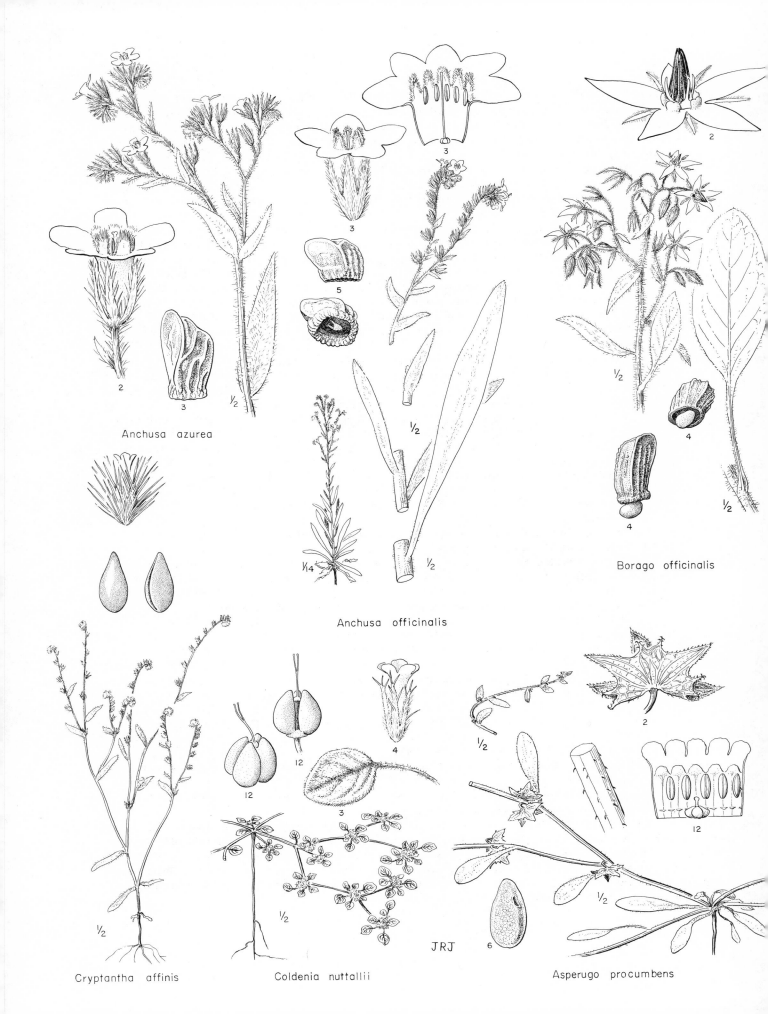

Anchusa azurea

Anchusa officinalis

Borago officinalis

Cryptantha affinis

Coldenia nuttallii

Asperugo procumbens

JRJ

Dry or sometimes moist, open or thinly wooded slopes from the foothills to moderate elevations in the mountains; Wash. and s. B. C., e. of the Cascade summits, to n. Mont., s. to s. Calif., n. Nev., and s. Wyo. June-July.

**Cryptantha ambigua** (Gray) Greene, Pitt. 1:113. 1887.

  Eritrichium muriculatum var. ambiguum Gray, Syn. Fl. $2^1$:194. 1878. Krynitzkia ambigua Gray, Proc. Am. Acad. 20:273. 1885. (Wilkes Expedition, "Nisqually," Wash., probably actually taken farther e. in the state)

  Cryptantha multicaulis A. Nels. Bot. Gaz. 30:194. 1900. (Nelson 6440, Snake R., Yellowstone Nat. Pk.)

  C. ambigua f. robustior Brand, Pflanzenr. IV. 252 (Heft 97) :69. 1931. (Specimens cited from several western states)

  Annual, 0.5-3 dm. tall, commonly much branched and often without a well-defined central axis; herbage evidently spreading-hirsute, and often strigose as well; leaves linear or linear-oblong; spikes seldom paired, and usually not markedly elevated above the main mass of the plant; fruiting calyx 4-6 mm. long, hirsute-strigose and conspicuously setose-hispid; corolla 1-2 mm. wide; nutlets 4, ovate, 1.2-2.0 mm. long, 0.7-1.1 mm. wide, somewhat granular, and with scattered (sometimes obscure) tubercles, the scar closed or nearly so; style nearly or quite equaling the nutlets.

  Dry, open places from the lowlands to moderate elevations in the mountains; e. Wash. and adj. s. B. C. to w. Mont., s. to Calif. and Colo. June-July.

  There appears to be some introgression between C. ambigua and C. torreyana, so that there is no sharp break in the gradient for either nutlets or habit, and some plants may have the nutlets of one and the habit of the other. Since the precise morphological position of the line of demarcation must in any case be arbitrary, and since determination of the texture of the nutlets is perhaps less subjective than that of habit, it seems convenient to stress the character of the nutlets, which are so commonly of value elsewhere in the genus and family. Plants with the nutlets essentially smooth are therefore referred to C. torreyana; those with fruit on which tubercles can be seen are referred to C. ambigua.

**Cryptantha celosioides** (Eastw.) Pays. Ann. Mo. Bot. Gard. 14:299. 1927.

  Oreocarya celosioides Eastw. Bull. Torrey Club 30:240. 1903. (Howell, banks of the Columbia R., e. Wash., July, 1881)

  Cynoglossum glomeratum Pursh, Fl. Am. Sept. 729. 1814. Myosotis glomerata Nutt. Gen. Pl. 1:112. 1818. Rochelia glomerata Torr. Ann. Lyc. N. Y. 2:226. 1827. Eritrichium glomeratum A. DC. Prodr. 10:131. 1846. Krynitzkia glomerata Gray, Proc. Am. Acad. 20: 279. 1885. Oreocarya glomerata Greene, Pitt. 1:58. 1887. Cryptantha bradburyana Pays. Ann. Mo. Bot. Gard. 14:307. 1927. (Bradbury, "Upper Louisiana," supposed to be about the Big Bend of the Missouri) Not Cryptantha glomerata Lehm.

  Oreocarya affinis Greene, Pitt. 3:110. 1896. Krynitzkia pustulata Blankinship, Mont. Agr. Coll. Stud. Bot. 1:96. 1905. (Greene, Red Buttes, Wyo., July 5, 1896) Not Cryptantha affinis (Gray) Greene.

  Oreocarya sericea sensu Piper, Contr. U. S. Nat. Herb. 11:482. 1906. Not Cryptantha sericea (Gray) Pays.

  Oreocarya sheldonii Brand, Fedde Rep. Sp. Nov. 19:73. 1925. Cryptantha sheldonii Pays. Ann. Mo. Bot. Gard. 14:301. 1927. (Sheldon 8315, Deep Creek, Wallowa Co., Oreg.)

  Biennial or short-lived perennial, with or without a branched caudex; stems 1-several, 1-5 dm. tall, often relatively robust, the central one frequently the largest; plants finely strigose or strigose-sericeous and pustulate-bristly, most of the bristles of the basal leaves generally appressed, those above more spreading; basal leaves tufted and conspicuous, spatulate or rather broadly oblanceolate, usually broadly rounded or even subretuse at the summit, rarely acutish, 2-8 cm. long and 4-15 mm. wide; cauline leaves more or less reduced; spikes aggregated into a terminal thyrse, often elongate and distinct at maturity; corolla relatively broad and showy, the limb mostly 8-12 mm. wide, the tube about equaling the calyx; nutlets ovate or lance-ovate, 3-5 mm. long, evidently roughened (generally rugose and tuberculate) on both sides, the scar essentially closed; style clearly surpassing the nutlets.

Dry, open places in the valleys, plains and foothills, occasionally ascending to moderate elevations in the mountains; e. Oreg. from Grant Co. to n. Wash. and s. B. C., e. through the lower parts of n. Ida. to Mont., and thence to s. Alta., N. D., and Neb. Apr. -July.

Plants from the Great Plains appear to be chiefly or wholly biennial, while those from our range are either biennial or more often short-lived perennial. In the absence of other apparent differences, no segregation seems tenable.

Cryptantha circumscissa (H. & A.) Johnst. Contr. Gray Herb. n. s. 68:55. 1923.
  Lithospermum circumscissum H. & A. Bot. Beechey Voy. 370. 1838. Piptocalyx circum-
    scissus Torr. in Wats. Bot. King Exp. 240. 1871. Eritrichium circumscissum Gray, Proc.
    Am. Acad. 10:58. 1874. Krynitzkia circumscissa Gray, Proc. Am. Acad. 20:275. 1885.
    Wheelerella circumscissa Grant, Bull. So. Calif. Acad. Sci. 5:28. 1906. Greeneocharis
    circumscissa Rydb. Bull. Torrey Club 36:677. 1909. (Tolmie, Snake Ft., Snake Country,
    Ida. )
  Cryptantha depressa A. Nels. Bot. Gaz. 34:29. 1902. (Merrill & Wilcox 873, w. of St. An-
    thony, Ida. )
Dwarf annual, much branched when well developed, forming cushions 1-6 cm. high; herbage coarsely strigose-hirsute with subappressed or partly ascending-spreading hairs; leaves narrowly linear, 5-13 mm. long, densely crowded toward the ends of the branches, the lower usually more remote; flowers solitary in the axils of the upper leaves; fruiting calyx 2-3 mm. long, circumscissile a little below the middle, the persistent, cupulate basal portion scarious, obviously of different texture from the more herbaceous deciduous portion, the sinuses between the lobes not reaching the line of dehiscence; corolla 0. 5-1. 5 mm. wide; nutlets ordinarily 4, nearly or quite alike, smooth or nearly so, rather narrowly triangular-ovate, 1.0-1. 3 mm. long, the scar closed; style nearly or quite equaling the nutlets.

Dry, open, usually sandy places in the lowlands; c. Wash. (and reputedly s. B. C.) to Baja Calif., largely e. of the Cascade-Sierra axis, extending e. throughout the Snake R. plains of Ida., and to Utah and Ariz.; also in Chile and Argentina. Apr. -July.

Our plants as here described are var. circumscissa. A more conspicuously spreading-hairy phase of the Sierra Nevada has been distinguished as var. hispida (Macbr.) Johnst.

Cryptantha echinella Greene Pitt. 1:115. 1887. (Sonne, Mt. Stanford, Calif., in 1886)
Annual, 0. 5-3 dm. tall, simple or branched; stem strigose and spreading-hirsute; leaves linear or linear-oblong, hirsute or strigose-hirsute with subappressed to somewhat spreading hairs; spikes naked, rarely paired; fruiting calyx 5-6 mm. long, strigose-hirsute and setose-hispid; corolla 1-2 mm. wide; nutlets ovate, 1. 8-2. 2 mm. long, 0. 9-1. 3 mm. wide, rather densely spiculate-papillate, the scar closed or narrowly open; style nearly equaling the nutlets.

Open places and dry woods at moderate to rather high elevations in the mountains; Sierra Nevada of Calif. to Nev. and s. w. Ida., and in c. Ida. July-Aug.

Cryptantha fendleri (Gray) Greene, Pitt. 1:120. 1887.
  Krynitzkia fendleri Gray, Proc. Am. Acad. 20:268. 1885. (Fendler, N. M. in 1847)
Annual, 1-4 dm. tall, subsimple or often much branched; stem strigose and spreading-hirsute; leaves linear or nearly so, acute, often rather numerous, hispid-hirsute with spreading or ascending, conspicuously pustulate-based hairs of varying size; spikes naked or nearly so; fruiting calyx 4-6 mm. long, subciliately strigose-hirsute and conspicuously pustulate-hispid; corolla inconspicuous, about 1 mm. wide; nutlets 4, lanceolate, 1. 5-2 mm. long, 0. 5-0. 7 mm. wide, nearly or fully 3 times as long as wide, smooth and shining, the margins rounded or broadly obtuse; scar opening at the base into an areola; style equaling or slightly surpassing the nutlets.

Sand dunes or very sandy soil; Franklin Co., Wash., and Umatilla Co., Oreg., irregularly to Sask., Neb., N. M., and Nev.; no records known to me from the parts of Ida. and Mont. lying within our range. May-July.

Cryptantha flaccida (Dougl.) Greene, Pitt. 1:115. 1887.
  Myosotis flaccida Dougl. in Lehm. Stirp. Pug. 2:22. 1830. (Douglas, "N. W. Coast in dry
    plains")

Eritrichium oxycaryum Gray, Proc. Am. Acad. 10:58. 1874. Krynitzkia oxycarya Gray,
    Proc. Am. Acad. 20:269. 1885. (Typification obscure)
Cryptantha multicaulis Howell, Fl. N.W. Am. 487. 1901. C. howellii A. Nels. Bot. Gaz.
    34:30. 1902. (Howell 502, Deschutes R. at Sherar's Bridge, Oreg.)
C. lyallii Brand, Fedde Rep. Sp. Nov. 24:57. 1927. (Lyall, Dalles of the Columbia)
C. flaccida var. minor Brand. Pflanzenr. IV. 252 (Heft 97):61. 1931. (Specimens cited
    from Wash. and Ida.)

Slender annual, 1-4 dm. tall, simple or branched; herbage strigose; leaves linear, scattered, only the lowermost opposite; spikes naked, tending to be geminate or ternate; calyx appressed to the rachis, closely enveloping the nutlet, swollen only near the base, 3-4 mm.
long at maturity, beset below with stout, spreading bristles that tend to be uncinate or arcuate;
corolla 1-2.5 mm. wide; nutlet solitary, 2 mm. long, smooth, lanceolate, acuminate, only
slightly, if at all, compressed, with rounded margins, somewhat pointed or very narrowly
truncate at the base, the scar closed or nearly so; style not reaching the middle of the nutlet.

Dry, open slopes and flats at lower elevations; c. and s. e. Wash. to s. Calif., and along
the w. fringe of Ida., as far e. as Elmore Co. May-June.

Cryptantha intermedia (Gray) Greene, Pitt. 1:114. 1887.
    Eritrichium intermedium Gray, Proc. Am. Acad. 17:225. 1882. Krynitzkia intermedia Gray,
        Proc. Am. Acad. 20:273. 1885. (Nevin, Los Angeles, Calif.)
    Allocarya hendersonii A. Nels. Erythea 7:69. 1899. Cryptantha hendersonii Piper ex J. C.
        Nels. Torreya 20:44. 1920. C. intermedia var. hendersonii Jeps. & Hoover in Jeps. Fl.
        Calif. 3:339. 1943. (Henderson, Potlatch R., Ida., May 31, 1895) Rough nutlet phase.
    Cryptantha monosperma Greene, Pitt. 5:53. 1902. C. hendersonii var. monosperma Brand,
        Pflanzenr. IV. 252. (Heft 97):71. 1931. (Suksdorf 180, Columbia R., w. Klickitat Co.,
        Wash.) Rough nutlet phase.
    C. muriculata sensu Piper, Contr. U.S. Nat. Herb. 11:484. 1906, but not (A. DC.) Greene.
    CRYPTANTHA INTERMEDIA var. GRANDIFLORA (Rydb.) Cronq. hoc loc. C. grandiflora Rydb.
        Bull. Torrey Club 36:679. 1909. C. torreyana var. grandiflora Nels. & Macbr. Bot. Gaz.
        61:43. 1916. (Sandberg et al. 10, valley of the Clearwater R., Ida.) Smooth nutlet phase.
    C. grandiflora var. annulata Brand, Pflanzenr. IV. 252 (Heft 97):59. 1931. (Elmer 156,
        Pullman, Wash.)
    C. hispidula var. elmeri Brand, Pflanzenr. IV. 252 (Heft 97):60. 1931. (Elmer 775, Wawa-
        wai, Whitman Co., Wash.)

Annual, 0.5-5 dm. tall, simple or branched, strigose and spreading-hirsute; leaves scattered, linear to narrowly oblong or linear-oblanceolate; spikes naked, tending to be geminate or
ternate; fruiting calyx 3-5(6) mm. long, evidently spreading-hispid at least below; corolla relatively large and showy, the limb 4-8 mm. wide, white with a yellow eye; nutlets 1-4, ovate
or lance-ovate, 2-3 mm. long, typically minutely granular and sparsely low-tuberculate, less
often smooth and shining; style slightly shorter than, to slightly surpassing, the nutlets.

Open, usually dry slopes at lower elevations on both sides of the Cascades from s. B.C. to
Calif., extending e. to the w. fringe of Ida. May-Sept.

Our plants, as described above, are var. grandiflora (Rydb.) Cronq. In n. Calif. they pass
into the more southern var. intermedia, with larger, sharper, and more numerous tubercles
on the nutlets, and with slightly more stiffly hairy calyx. The form of var. grandiflora with
smooth, shining, more often only 1 or 2 nutlets is neither geographically nor ecologically segregated from the form with rough and more often 3-4 nutlets, nor is the correlation between
texture and number of nutlets very strong. I am fully in agreement with Johnston and with Jepson that no segregation should be attempted here on these bases.

Cryptantha interrupta (Greene) Pays. Ann. Mo. Bot. Gard. 14:296. 1927.
    Oreocarya interrupta Greene, Pitt. 3:111. 1896. (Greene, e. of Wells, Nev., July 16, 1896)
    O. spiculifera Piper, Contr. U.S. Nat. Herb. 11:481. 1906. Cryptantha spiculifera Pays.
        Ann. Mo. Bot. Gard. 14:298. 1927. (Sandberg & Leiberg 164, Ritzville, Wash.)
    Oreocarya cilio-hirsuta Nels. & Macbr. Bot. Gaz. 55:378. 1913. (Nelson & Macbride 1799,
        Minidoka, Ida.)

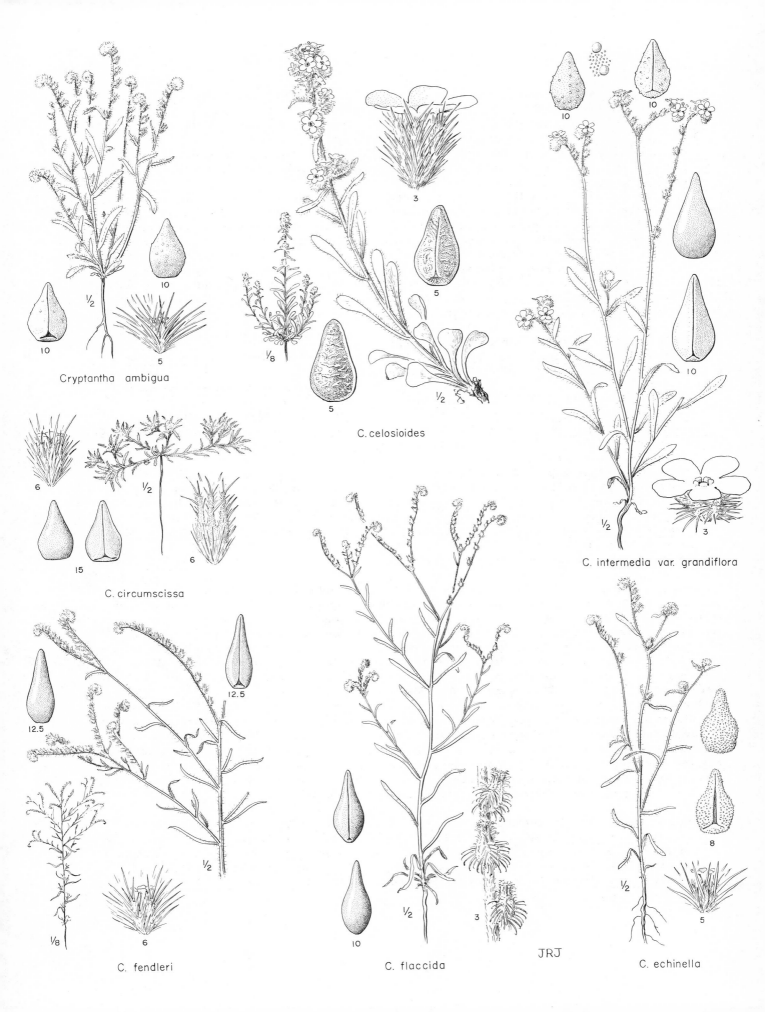

Cryptantha ambigua

C. celosioides

C. intermedia var. grandiflora

C. circumscissa

C. fendleri

C. flaccida

C. echinella

JRJ

O. macouni Eastw. Bull. Torrey Club 40:480. 1913. Cryptantha macouni Pays. Ann. Mo.
   Bot. Gard. 14:303. 1927. (Macoun, Moose mountain Creek, Sask., July 6, 1880)
   Perennial with solitary or more often several or numerous stems 0.5-4 dm. tall, the taproot surmounted by a branched caudex or simple crown; herbage more or less strigose or
strigose-sericeous and pustulate-setose, the setae spreading or those near the base often appressed; basal leaves densely tufted, oblanceolate, mostly rather narrowly so, acutish or obtuse to occasionally rounded at the apex, 1-6 cm. long, 1.5-8 mm. wide; cauline leaves more
or less reduced, not very numerous; spikes aggregated into a terminal thyrse, often more
elongate and distinct at maturity; corolla tube about equaling the calyx, the limb 5-8(10) mm.
wide; nutlets lanceolate or lance-ovate, 2-4 mm. long, evidently roughened on the back, commonly both rugose and tuberculate, sometimes merely one or the other, the ventral surface
similarly or less strongly roughened; scar closed or narrowly open; style distinctly surpassing
the nutlets.
   Dry, open slopes and flats in the plains, valleys, and foothills; common on the Snake R.
plains of Ida., extending s. to n. Nev., e. and n. to Baker Co., Oreg., and (rarely) Adams
Co., Wash.; passing through the valleys of the Lost R. and adjacent streams at the upper end
of the Snake R. plains to the valley of the Salmon R. (Stanley to Salmon), and across the continental divide into the intermontane valleys of Mont., thence n.e. to s. Sask. May-July.
   Plants from the type region of C. interrupta lie near the southern geographic limit of the
species as well as being morphologically peripheral. Both the appressed pubescence of the
lower parts of the plant and the sparsely tuberculate, scarcely rugose nutlets of C. interrupta
sens. strict. can be matched among plants from the nearby Snake R. plains, however, so that
there is evidently no good line to be drawn. Since C. interrupta sens. strict. is admittedly allied to the adjacent more northern population that has been treated as C. spiculifera, and since
each of the two features which supposedly distinguish C. interrupta from C. spiculifera can
readily be found in varying combination and degree among the population of C. spiculifera, it
does not seem necessary or useful to attempt to maintain the southern fragment as distinct.
The plants from Mont. and Sask. have customarily been treated as another separate species,
C. macounii, but I can see no consistent differences nor even an aspect on which to distinguish
them. C. interrupta as here defined appears to intergrade with C. celosioides in Mont. Elsewhere the two seldom occur close together, and characteristic specimens are obviously different.


Cryptantha kelseyana Greene, Pitt. 2:232. 1892. (Greene, Elliston, Mont., in 1889)
   Annual, 0.5-2.5 dm. tall, more or less bushy-branched and commonly without a strong
central axis; stem strigose and spreading-hirsute; leaves linear, hirsute with pustulate-based
hairs of varying size; spikes naked, seldom geminate; fruiting calyx 5-7 mm. long, hirsute-
strigose and coarsely spreading-hispid, the segments with somewhat thickened midrib; corolla
1-2 mm. wide; nutlets 4, lance-ovate, the axial one evidently though not greatly larger than
the others, and more firmly attached, very finely and obscurely granular, appearing dull and
very nearly smooth, 2.0-2.6 mm. long, the others evidently tuberculate; scar closed or nearly
so except for a small basal areola; style about equaling the 3 shorter nutlets.
   Dry, open plains, often in sandy soil; Sask. and Mont. (w. to Helena) to Colo. and Utah, extending into e. Ida. (Clark and Fremont cos.). June-July.
   The related C. minima Rydb., of the Great Plains, differing sharply in its prominently
bracteate spikes, has been taken as far w. as Great Falls, Mont., and may prove to occur
within our range.


Cryptantha leucophaea (Dougl.) Pays. Ann. Mo. Bot. Gard. 14:262. 1927.
   Myosotis leucophaea Dougl. in Lehm. Stirp. Pug. 2:22. 1830. Eritrichium leucophaeum A.
      DC. Prodr. 10:129. 1846. Krynitzkia leucophaea Gray, Proc. Am. Acad. 20:280. 1885.
      Oreocarya leucophaea Greene, Pitt. 1:58. 1887. (Douglas, "arid barrens of the Columbia,
      and of its northern and southern tributaries")
   Perennial with clustered stems 1.5-4 dm. tall; herbage mostly sericeous-strigose below,
becoming hispid-setose only upward; setae of the leaves mostly appressed and inconspicuous,
those of the upper surface of the lower leaves often poorly developed and scarcely pustulate,

or wanting; leaves relatively narrow and elongate, all tapering to an acute or acutish tip, the lower 5-10 cm. long (poorly defined petiole included) and 3-6 mm. wide, the cauline ones similar but becoming sessile and sometimes 1 cm. wide; corolla tube evidently surpassing the calyx at anthesis, the limb 5-9 mm. wide; nutlets ovate, smooth and shining, gray, 4 mm. long, sharp-edged, the scar closed; style moderately or strongly surpassing the nutlets.

Dry, often sandy places near the Columbia and lower Yakima rivers, from Wenatchee, Wash., to The Dalles, Oreg.; reputedly also in s. B. C. May-June.

The most sharply defined of our perennial species, not to be confused with anything else in our region.

Cryptantha nubigena (Greene) Pays. Ann. Mo. Bot. Gard. 14:265. 1927.
  Oreocarya nubigena Greene, Pitt. 3:112. 1896. (Chestnut & Drew, Cloud's Rest, Mariposa Co., Calif., July 10, 1889)
  Cryptantha sobolifera Pays. Ann. Mo. Bot. Gard. 14:305. 1927. (Jones, Upper Marias Pass, Mont., Sept. 10, 1909)
  C. subretusa Johnst. Journ. Arn. Arb. 20:393. 1939. Oreocarya subretusa Abrams, Ill. Fl. Pac. St. 3:599. 1951. (Thompson 12206, Crater Lake, Oreg.)
  Cryptantha hypsophila Johnst. Journ. Arn. Arb. 20:295. 1939. (Thompson 14129, head of Boulder Creek, Blaine Co., Ida.)
  C. andina attr. to Johnst. by Peck, Man. High. Pl. Oreg. 601. 1941, without Latin diagnosis or type.
Dwarf perennial with several or many stems 3-15 cm. tall from a taproot and branched caudex; herbage sparsely to densely sericeous-strigose and more or less pustulate-bristly, the bristles of the basal leaves sometimes appressed and poorly developed; basal leaves strongly tufted, oblanceolate or spatulate, with rounded to acutish tip, up to 3.5 cm. long (petiole included) and 8 mm. wide; cauline leaves, except the lower, more or less reduced, often few; spikes small, aggregated into a terminal thyrse, not becoming elongate; corolla tube about equaling the calyx, the limb 4-8 mm. wide; nutlets lanceolate or lance-ovate, 3-4 mm. long, slightly to moderately rugulose-roughened (or partly tuberculate) dorsally, smooth or nearly so ventrally, the scar generally closed or nearly so; the style equaling or somewhat surpassing the nutlets.

Dry, open, often rocky places at high elevations in the mountains, often above timber line; c. Ida. and w. Mont. to e. Oreg. (Wallowa, Steens, Warner, and s. Cascade mts.), s. to the Sierra Nevada of Calif. July-Aug.

The species shows some variation in pubescence, shape of the leaves, and details of sculpturing of the nutlets, but the variation is continuous, and the geographic correlations are so weak as to render any segregation difficult and arbitrary.

Cryptantha propria (Nels. & Macbr.) Pays. Ann. Mo. Bot. Gard. 14:317. 1927.
  Oreocarya propria Nels. & Macbr. Bot. Gaz. 62:145. 1916. (Leiberg 2049, Vale, Malheur Co., Oreg.)
  Krynitzkia fulvocanescens var. idahoensis M. E. Jones, Contr. West. Bot. 13:6. 1910. (Jones 6474, near Weiser, Ida.)
Perennial with several stems 1-3 dm. tall from a taproot and branched caudex; pubescence mostly appressed, except in the inflorescence; basal leaves conspicuously tufted, oblanceolate or spatulate, with rounded to acutish tip, mostly 3-10 cm. long (petiole included) and 5-10 mm. wide; cauline leaves few and reduced, becoming sessile; hairs of the upper side of the leaves all fine and about alike, or with some inconspicuous and poorly developed appressed setae, these not pustulate; setae of the lower surfaces of the leaves better developed and more evidently pustulate; spikes aggregated into a terminal, thyrsoid inflorescence; corolla tube about equaling the calyx, the limb 8 mm. wide; nutlets ovate, 3 mm. long, closely and irregularly rugose-reticulate on both sides, the scar rather narrowly open for most of its length.

Dry, open hillsides in n. Malheur Co., Oreg., and in adj. Ida. Apr.-June.

Cryptantha pterocarya (Torr.) Greene, Pitt. 1:120. 1887.
  Eritrichium pterocaryum Torr. Bot. Mex. Bound. 142. 1859. Krynitzkia pterocarya Gray,

Proc. Am. Acad. 20:276. 1885. (Pickering & Brackenridge 1047, Walla Walla, Wash.; lectotype by Johnston)

Annual, 1-4 dm. tall, simple or branched; herbage moderately hirsute or partly strigose, often pustulate; leaves linear; spikes tending to be paired, naked; fruiting calyx 4-5 mm. long and nearly as broad, in life angular, closed, and pyramidal, the segments lanceolate, hirsute-strigose and sometimes also hispid; corolla 0.5-2 mm. wide; nutlets 4, muricate or tuberculate on the back, about 3 mm. long, with lance-ovate body, the axial one wingless, the other three with broad (nearly 1 mm.), often toothed or lobed wing margins; scar dilated below into an open, excavated areola, often also more narrowly open upward; style about equaling the nutlets.

Dry, open slopes in the valleys and plains e. of the Cascades, from c. Wash. s. through e. Oreg. and s.w. Ida. to the Great Basin, thence to s. Calif., Tex., and Colo. Apr.-June.

Our plants, as here described, are var. pterocarya. The var. cycloptera (Greene) Macbr., with all four nutlets winged, occurs mostly to the s. of the Great Basin.

Cryptantha rostellata Greene, Pitt. 1:116. 1887.

Krynitzkia rostellata Greene, Bull. Calif. Acad. Sci. 1:203. 1886. (Curran, Leesburg, Calif., in 1884)

K. suksdorfii Greenm. Bot. Gaz. 40:146. 1905. Cryptantha suksdorfii Piper, Contr. U.S. Nat. Herb. 11:484. 1906. C. rostellata var. suksdorfii Brand, Pflanzenr. IV. 252 (Heft 97) 59. 1931. (Suksdorf 1495, near Rockland, Wash.)

Closely allied to C. flaccida, differing as follows: lower several pairs of leaves commonly opposite; calyx becoming somewhat spreading at maturity; nutlet averaging slightly larger (to 3 mm.) and proportionately a little broader, more distinctly compressed, often with distinctly angular margins, broadly truncate at the base, the scar evidently expanded below into an open areola; style reaching or commonly shortly surpassing the middle of the nutlet.

Scattered stations in dry, open places from Klickitat Co., Wash., through e. Oreg. to c. Cal. Apr.-June.

Cryptantha salmonensis (Nels. & Macbr.) Pays. Ann. Mo. Bot. Gard. 14:263. 1927.

Oreocarya salmonensis Nels. & Macbr. Bot. Gaz. 61:43. 1916. (Kirtley s.n., Salmon, Ida., June, 1896)

Perennial with clustered, simple or branched, often lax stems 1.5-4 dm. tall; herbage evidently but not densely spreading-bristly and pustulate, often with even the shorter fine hairs of the lower leaves spreading; lower leaves oblanceolate, generally broadly so, rounded or obtuse at the tip, commonly 4-10 cm. long (petiole included) and 6-12 mm. wide; usually no dense tuft of basal leaves present; middle and upper leaves scattered, broad, gradually reduced, becoming sessile; corolla tube about equaling the calyx at anthesis, the limb 8-11 mm. wide; nutlets lance-ovate, smooth and shining, 3-4 mm. long, sharp-edged, the scar closed; style evidently surpassing the nutlets.

Dry hillsides, washes, and shaly cliffs and slopes near the Salmon R., from shortly above Challis, Ida., to Salmon. June-Aug.

Cryptantha salmonensis in its typical form is sharply marked, but it evidently hybridizes with C. interrupta, so that otherwise typical specimens (or colonies?) may resemble C. interrupta in habit, or in flower size, or in sculpture of the nutlets.

Cryptantha scoparia A. Nels. Bot. Gaz. 54:144. 1912. (Nelson & Macbride 1311, Minidoka, Ida.)

Slender annual, 0.5-3 dm. tall, usually freely branched and often without a well-defined central axis; herbage strigose, or with a few more-spreading hairs, the hairs of the leaves pustulate-based; leaves linear; spikes naked, tending to be paired; fruiting calyx narrow, 4-6 mm. long, closely ascending, hirsute-strigose and evidently bristly-hispid; corolla about 1 mm. wide; nutlets ordinarily 4, lanceolate, 1.7-2.1 mm. long, 0.5-0.7 mm. wide, provided with numerous rather fine, forwardly directed, spiculate papillae, especially distally; scar closed or narrowly open; style about equaling the nutlets.

Dry, open slopes and flats, commonly among sagebrush; Snake R. plains of Ida., extending

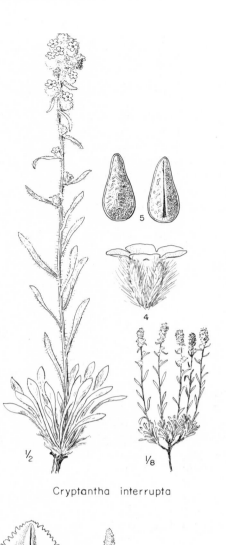

5

4

1/2

1/8

Cryptantha interrupta

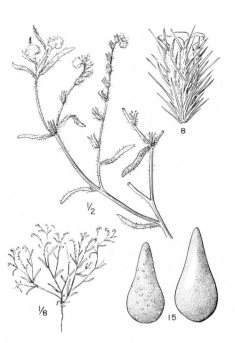

1/2

1/8

8

15

C. kelseyana

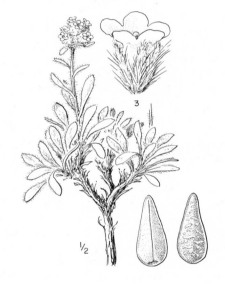

3

1/2

C. nubigena

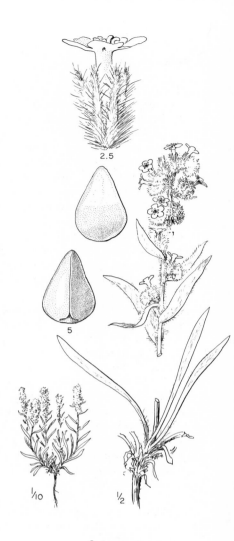

2.5

5

5

1/10

1/2

C. leucophaea

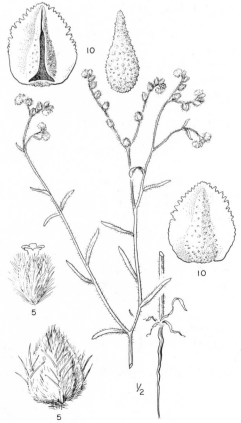

10

10

5

5

1/2

C. pterocarya

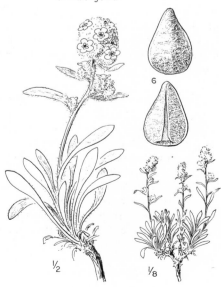

6

1/2

1/8

C. propria

JRJ

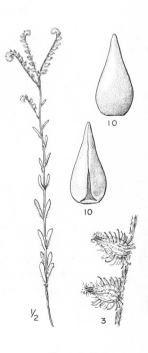

10

10

1/2

3

C. rostellata

into adj. Oreg., n.e. Nev., and s.w. Wyo.; also along the Salmon R. near and below Challis, Ida., and apparently in Yakima Co., Wash. May-July.

**Cryptantha simulans** Greene, Pitt. 5:54. 1902.

C. ambigua f. simulans Brand, Pflanzenr. IV. 252 (Heft 97): 68. 1931. (Hansen 1903, Amador Co., Calif.)

Openly branched or subsimple annual, 1.5-5 dm. tall; herbage strigose, often canescent, essentially without spreading hairs; leaves scattered, linear or linear-oblong; spikes relatively few, naked, tending to be paired; fruiting calyx 4-6 mm. long, strigose-hirsute and commonly with some stiffer but not very long bristles that may be somewhat curved; corolla 1-2 mm. wide; nutlets 4, ovate, 2-3 mm. long, 1.1-1.5 mm. wide, coarsely granular and with scattered, low tubercles, the scar closed; style a little surpassed by the nutlets.

In open ponderosa pine forests; s. Calif. to c. Wash. (Chelan Co.), wholly e. of the Cascade summits in our range; reported also by Johnston from Moscow Mt., Latah Co., Ida. June-July.

**Cryptanthe thompsonii** Johnst. Contr. Arn. Arb. 3:88. 1932.

Oreocarya thompsonii Abrams, Ill. Fl. Pac. St. 3:600. 1951. (Thompson 7663, crest of Iron Mt., Mt. Stuart region, Kittitas Co., Wash.)

Perennial with several or many stems 1-3 dm. tall from a stout taproot and branching caudex; herbage pustulate-bristly and more or less strigose-tomentulose, the bristles spreading above, often appressed below, those of the upper surface of the basal leaves sometimes scarcely pustulate; basal leaves tufted, oblanceolate, with relatively elongate blade and short petiole, the better developed ones mostly 4-7 cm. long (petiole included) and 5-10 mm. wide; cauline leaves becoming narrower and sessile, but often fairly well developed and elongate even in the inflorescence; spikes aggregated into a terminal, thyrsoid inflorescence; corolla tube equaling or shorter than the calyx, the limb 4-8 mm. wide; nutlets lance-ovate, 4 mm. long, somewhat tuberculate and rugulose-roughened dorsally, smooth ventrally, the scar evidently open for most of its length.

Open or sparsely wooded slopes and cliffs, often on talus, especially serpentine, from the ponderosa pine belt to rather high elevations in the Wenatchee Mts. of Wash. May-July.

**Cryptantha torreyana** (Gray) Greene, Pitt. 1:118. 1887.

Krynitzkia torreyana Gray, Proc. Am. Acad. 20:271. 1885. (Torrey 337, Yosemite Valley, Calif.; lectotype by Johnston)

K. torreyana var. calycosa Gray, Proc. Am. Acad. 20:271. 1885. Cryptantha torreyana var. calycosa Greene, Pitt. 1:119. 1887. C. calycosa Rydb. Mem. N.Y. Bot. Gard. 1:331. 1900. C. torreyana ssp. calycosa Piper, Contr. U.S. Nat. Herb. 11:484. 1906. C. torreyana var. genuina subvar. calycosa Brand, Pflanzenr. IV. 252 (Heft 97): 58. 1931. (Watson, East Humboldt Mts., Nev.)

Cryptantha affinis var. flexuosa A. Nels. Bot. Gaz. 30:195. 1900. C. flexuosa Nels. in Coult. & Nels. New Man. Bot. Rocky Mts. 416. 1909. (Nelson 6546, Jackson Lake, Wyo.)

C. torreyana var. genuina subvar. bracteata Brand, Pflanzenr. IV. 252. (Heft 97): 58. 1931. (Sheldon 8415, Crow Creek, Wallowa Co., Oregon.)

C. eastwoodae St. John, Fl. S.E. Wash. 342. 1937. (Eastwood and St. John 13219, Winona, Whitman Co., Wash.) A form resembling C. ambigua in habit.

Annual, 1-4 dm. tall, simple or branched; stem strigose and spreading-hirsute; leaves more uniformly hirsute with subappressed to somewhat spreading hairs, linear to narrowly oblong; spikes naked, typically somewhat elevated above the more leafy portion of the plant and tending to be paired, varying to occasionally as in characteristic C. ambigua; fruiting calyx 4-8 mm. long, the segments stiffly strigose-hirsute and spreading-hispid; corolla inconspicuous, about 1 mm. wide; nutlets 4, ovate, acute, with rounded or obtuse margins, mostly 1.5-2.3 mm. long and 0.8-1.2 mm. wide, smooth and shining, rarely very finely granular; scar closed; style somewhat surpassed by the nutlets.

Dry, open or lightly wooded slopes and flats, from the plains and foothills to moderate elevations in the mountains; Wash. (e. of the Cascade summits) and s. B.C. to w. Mont., s. to

Calif., n. Utah, and Wyo.; occasionally introduced elsewhere. May-July.

Our plants as described above are var. torreyana. Plants from Calif. with shorter calyx or with longer style have been recognized as distinct varieties. See comment under C. ambigua.

Cryptantha watsonii (Gray) Greene, Pitt. 1:120. 1887.

Krynitzkia watsoni Gray, Proc. Am. Acad. 20:271. 1885. (Watson 858, Wasatch Range, Utah)
Cryptantha vinctens Nels. & Macbr. Bot. Gaz. 62:143. 1916. (Leiberg 2235, Malheur ["Mathew"] Valley near Harper Ranch, Oreg.)

Annual, 1-3 dm. tall, subsimple or often much branched; herbage moderately spreading-hirsute, the stem sometimes also strigose; leaves linear or nearly so, mostly blunt-tipped; spikes naked, single or often paired; fruiting calyx 2-4.5 mm. long, hirsute-strigose and usually also spreading-hispid, the segments a little broader than in allied species; corolla inconspicuous, 1-2 mm. wide; nutlets 4, smooth, lanceolate, sometimes rather broadly so, 1.5-2.1 mm. long, 0.7-0.9 mm. wide, the margins prominent and sharply angled, especially toward the tip; scar opened at the base into a small areola; style nearly or fully equaling the nutlets.

Open slopes in the plains, valleys, and foothills; c. Wash. to w. Mont., s. to Nev. and Colo. May-Aug.

Seldom collected in our range.

## Cynoglossum L. Hound's tongue

Flowers borne in sympodial, naked or subnaked, false racemes, or in terminal, naked, mixed panicles, the pedicels ascending to more often spreading or even reflexed at maturity; calyx deeply cleft; corolla blue or violet to reddish, funnelform or salverform, with short broad tube and more or less spreading limb, the fornices well developed and often exserted; anthers included or borne at the corolla throat; nutlets more or less widely spreading at maturity, attached to the broad, low gynobase by the apical end, the broad scar not extending below the middle of the ventral surface, the whole surface covered with short, stout, glochidiate prickles, or these sometimes restricted to the distal region of the free portion; entire-leaved, taprooted annuals, biennials, or more often perennial herbs, moderately to very robust.

A cosmopolitan genus of perhaps 80 species. (Name from the Greek kyon, kynos, dog, and glossa, tongue, from the appearance of the leaves of C. officinale.)

Several species are choice garden perennials; our best is C. grande, which does well in both open and shady spots.

1 Inflorescence of numerous false racemes axillary to leaves or terminating short axillary branches; dorsal surface of the nutlets flattened, surrounded by a raised margin; introduced, weedy biennial                                                    C. OFFICINALE
1 Inflorescence terminal, with a naked common peduncle; dorsal surface of the nutlets broadly rounded, without a raised margin; native, not at all weedy perennials
  2 Plants with all the leaves (or all but the reduced uppermost ones) petiolate; stem glabrous or nearly so; corolla limb mostly 1-1.5 cm. wide                          C. GRANDE
  2 Plants with only the lower leaves petiolate, the others sessile and often clasping; stem evidently spreading-hairy; corolla limb less than 1 cm. wide
    3 Sepals 1-2.5 mm. long in flower and fruit; style 1-2 mm. long; nutlets 3.5-5 mm. long (unique among our species in all of these respects)                          C. BOREALE
    3 Sepals 4-5 mm. long in flower, sometimes 1 cm. in fruit; style 4-9 mm. long; nutlets 7-10 mm. long                                                                   C. OCCIDENTALE

Cynoglossum boreale Fern. Rhodora 7:250. 1905. (Williams, Collins, & Fernald, Little Cascapedia R., Que., July 17, 1905, is the first of 13 specimens cited)

Perennial, 4-8 dm. tall; stem spreading-hirsute, leaves more strigose; basal and lowermost cauline leaves long-petiolate, with elliptic or elliptic-ovate blade 7-20 cm. long and 2.5-7 cm. wide, the others sessile and clasping, becoming more oblong or oblong-ovate; inflorescence conspicuously naked-pedunculate, consisting of 3-several false racemes which greatly elongate in fruit; pedicels eventually arcuate-recurved; flowers small, the sepals only 1-2.5 mm. long in flower and fruit; corolla blue, funnelform, the short tube slightly or scarcely sur-

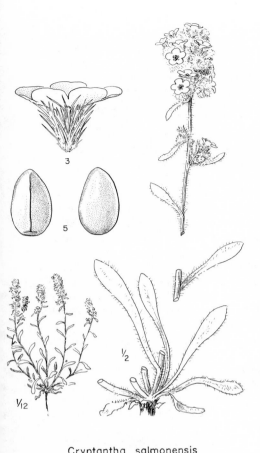

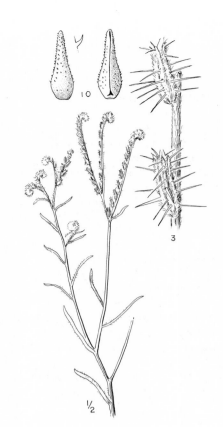

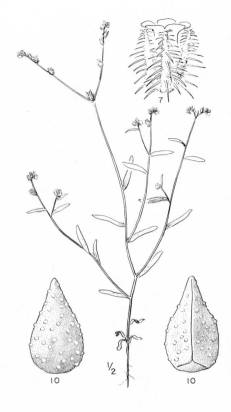

Cryptantha salmonensis

C. scoparia

C. simulans

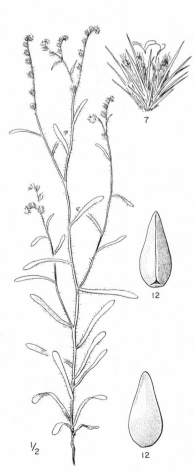

C. thompsonii

C. torreyana

JRJ

C. watsonii

passing the calyx, the limb only 5-8 mm. wide, the rounded-oblong fornices shortly exserted from the throat; anthers included; style inconspicuous, 1-2 mm. long; nutlets obovoid, 3.5-5 mm. long, prickly all over, free from the style, radially spreading so that the rather large ventro-apical scar might seem to be basal instead.

  Woodlands; Newf. to Conn., westward to Man. and n. Mich., and in s. B.C. June-July.

  Further study may necessitate the reduction of C. boreale to a variety of the more southern C. virginianum L., a more robust plant with larger flowers and fruits.

Cynoglossum grande Dougl. ex Lehm. Stirp. Pug. 2:25. 1830. (Douglas, N. Am.)

  Perennial, 2-8 dm. tall; stems glabrous or nearly so, often solitary; leaves confined to the lower half or third of the stem, all (except the scaly near-basal bracts) long-petiolate, with ovate to elliptic blade mostly 8-18 cm. long and 3-11 cm. wide, broadly rounded to truncate or shallowly cordate at the base, the petioles and lower surfaces more or less hirsute, the upper surfaces inconspicuously strigose or glabrous; inflorescence a mixed panicle, at first condensed, later open, the mature pedicels elongate and ascending; corolla blue or violet, almost salverform, the limb mostly 1-1.5 cm. wide, the fornices exserted, obcordately emarginate; anthers seated at the corolla throat; nutlets obovoid-globose, nearly 1 cm. long, radially spreading or somewhat ascending, so that the attachment appears basal rather than apical, free from the well-developed style, often some of them abortive; prickles often confined to the outer half of the nutlets. Bluebuttons.

  Woods at lower elevations w. of the Cascades (and extending through the Columbia gorge); extreme s. B.C. to San Luis Obispo and Tulare cos., Calif. Mar.-Apr.

Cynoglossum occidentale Gray, Proc. Am. Acad. 10:58. 1874. (Burgess, Sierra Nevada, and Lemmon, Sierra Co., Calif., are cited in that order)

  Perennial with clustered stems, 2-6 dm. tall, leafy to above the middle, coarsely hairy throughout; lower leaves oblanceolate, tapering very gradually to the petiole, 7-20 cm. long (petiole included) and 1-4 cm. wide, the others more oblong or lance-elliptic, becoming sessile and generally clasping; inflorescence a short-pedunculate, rather compact, mixed panicle, not much expanded in fruit, the pedicels eventually arcuate-recurved; calyx 4-5 mm. long in flower, up to 1 cm. in fruit; corolla dull red, funnelform, the limb not much expanded, 5-9 mm. wide, the fornices exserted and subentire; anthers seated at the corolla throat; style elongate, 4-9 mm. long; nutlets 1-4, obovoid-globose, 7-10 mm. long, radially spreading or somewhat ascending, so that the attachment appears basal rather than apical, free from the style, usually prickly all over.

  Open ponderosa pine woods, from Black Butte, Jefferson Co., Oreg., southward to the Sierra Nevada and n. Coast Ranges of Calif. May-June.

Cynoglossum officinale L. Sp. Pl. 134. 1753. (Europe)

  Coarse, single-stemmed biennial, 3-12 dm. tall, leafy to the top, villous or villous-hirsute throughout; lowermost leaves oblanceolate or narrowly elliptic, tapering to the petiole, 1-3 dm. long overall, 2-5 cm. wide, the others sessile and more oblong or lanceolate, numerous, only gradually reduced upward; inflorescence of numerous false racemes arising in the upper axils, or terminating short axillary branches, the mature pedicels curved-spreading; sepals broad, blunt, 5-8 mm. long in fruit; corolla dull reddish-purple, nearly salverform, the limb about 1 cm. wide or a little less, the fornices exserted, broadly rounded; anthers seated about at the corolla throat; nutlets 5-7 mm. long, ovate, descending-spreading, forming a broad low-pyramidal fruit, remaining attached to the style above even after splitting from the gynobase, the dorsal surface flattened, surrounded by a raised margin, the scar relatively long, extending about to the middle of the ventral surface. N=12. Common hound's tongue.

  A weed in disturbed sites, especially along roadsides; native of Europe, now well established in N. Am. and found more or less throughout our range. May-July.

Dasynotus Johnst.

Flowers borne in lax, few-flowered, false racemes that are basally leafy-bracteate but

Cynoglossum boreale

C. grande

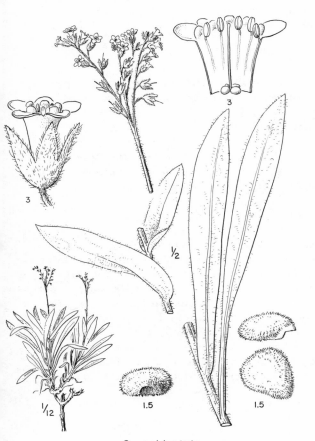

C. occidentale

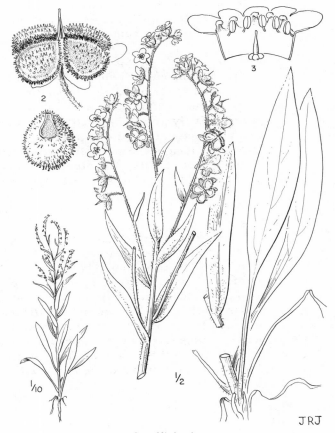

C. officinale

JRJ

otherwise naked; pedicels elongate, more or less deflexed in fruit; calyx deeply 5-cleft; corolla white, subsalverform or broadly funnelform, with short tube and relatively large limb; fornices conspicuous, exserted, recurved, apically retuse; anthers borne near the summit of the corolla tube; nutlets somewhat spreading below, compressed, with an evident, erect, continuous, unarmed dorsomarginal ridge, short-hairy on the back, ventrally keeled toward the tip, the rather large scar medial or slightly supra-medial; gynobase broadly pyramidal; erect perennial herbs with numerous stems and wholly cauline leaves.

A single species. (Name from the Greek dasys, hispid or woolly, and noton, back, referring to the dorsally hairy nutlets.)

Dasynotus daubenmirei Johnst. Journ. Arn. Arb. 29:234. 1948. (Daubenmire 46289, near Walde Mt. lookout, n. of Lowell, Idaho Co., Ida.)

Perennial from a stout but rather soft rhizome which may be shortened into a branching caudex, the rather sparsely spreading-hairy or subglabrous stems accordingly solitary or clustered, 3-6 dm. tall; lowermost leaves reduced, the others numerous, oblanceolate, sessile or nearly so, 9-17 cm. long, 1.5-3.5 cm. wide, coarsely strigose or hirsute-strigose on both sides; pedicels 1-3 cm. long, up to 7 cm. in fruit; calyx 6-8 mm. long in flower, twice as long in fruit; corolla tube 4-5 mm. long and wide, the limb 2-2.5 cm. wide; fornices 2-4 mm. long; nutlets 5-6 mm. long.

Openings in forests, at moderate to rather high elevations in the mountains; known only from the immediate vicinity of the type locality. June-July.

## Echium L.

Flowers borne in a series of sympodial, helicoid, bracteate cymes; calyx deeply cleft; corolla blue to purple or red (white), funnelform, irregular, the upper side evidently the longer, the lobes unequal; fornices wanting, the throat open; filaments slender, more or less unequal, some or all of them strongly exserted; gynobase flat or nearly so; nutlets more or less roughened, attached at the base, the large scar sometimes surrounded by a low rim; style exserted; annual, biennial, or perennial herbs or subshrubs.

Perhaps 50 species, native to Europe, the Mediterranean region, and Atlantic islands, and s. Africa. (From Greek echion, name for various members of the Boraginaceae. Echion, in turn, comes from echis, viper.)

Echium vulgare L. Sp. Pl. 139. 1753. (Europe)

Erect, taprooted biennial, 3-8 dm. tall, hirsute or rough-puberulent and (especially upward) more or less strongly spreading-bristly, the bristles often pustulate at the base; basal leaves more or less oblanceolate, 6-25 cm. long, (petiole included), 0.5-3 cm. wide; cauline leaves progressively smaller, becoming sessile; helicoid cymes numerous and often short, more or less aggregated into an elongate, often virgate inflorescence; corolla bright blue (pink or white), 12-20 mm. long; 4 filaments long-exserted, the fifth scarcely so; style hairy. N=8, 16. Blueweed.

Roadsides, fields, and other disturbed sites; native of s. Europe, now well established as a weed in e. U.S., and occasionally found in our range, notably about Spokane. June-Aug.

## Eritrichium Schrad.

Flowers borne in condensed, cymose clusters which may become elongate into sympodial, naked or somewhat leafy-bracteate, false racemes or spikes terminating the often very short stems, the fruiting pedicels more or less erect; calyx cleft essentially to the base; corolla blue, often with a yellow eye, or in occasional individuals white, salverform, with short, fairly narrow tube and abruptly spreading, evidently 5-lobed limb, the fornices well developed; filaments attached well down in the corolla tube, the anthers included; nutlets 1-4, basilaterally attached to the broad, low gynobase, provided with a prominent, ascending, entire to conspicuously toothed or lacerate dorsomarginal flange which is complete across the back of the nutlet well above the base, the flanged portion of the body tending to be set at an angle to the portion below the flange, more conspicuously so when the nutlets are only 1 or 2; dwarf, more or less strongly pulvinate-caespitose perennials, often acaulescent, with small, more or less hairy leaves densely crowded on the numerous short shoots or toward the base of the more elongate flowering stems.

About 4 closely allied species of Eurasia and w. N. Am., when strictly limited. By some authors more or less expanded, and then of uncertain limits and characters. (Name from the Greek erion, wool, and trichos, hair, referring to the woolly pubescence of E. nanum, the original species.)

Our species are attractive rock garden subjects, but are difficult to grow.

Reference:

Wight, William F. The genus Eritrichium in North America. Bull. Torrey Club 29:407-14. 1902.

1 Mature leaves densely and coarsely silvery-strigose, the surface ordinarily hidden by the coarse, mostly straight hairs, which do not form a pronounced apical tuft or fringe
E. HOWARDII

1 Mature leaves loosely long-hairy, the surface readily visible between the hairs, which are often more numerous distally and tend to form an apical tuft or fringe    E. NANUM

Eritrichium howardii (Gray) Rydb. Mem. N.Y. Bot. Gard. 1:327. 1900.
  Cynoglossum howardi Gray, Syn. Fl. 2$^1$:188. 1878. Omphalodes howardi Gray, Proc. Am. Acad. 20:263. 1885. (Howard, Rocky Mts. of Mont.)
  Densely pulvinate-caespitose, long-lived perennial, essentially acaulescent or with more or less erect, strigose, slender stems up to 1 dm. tall; leaves linear-oblanceolate to linear-elliptic or linear, tending to be acute, up to 2 cm. long and 2 mm. wide, densely and permanently silvery-strigose, the surface ordinarily hidden by the coarse, mostly straight hairs, which do not form a pronounced apical tuft or fringe, the hairs largely persistent even on dead leaves of past seasons; corolla limb 5-9 mm. wide; nutlets 1-4, often hispidulous on the back, the flange low and entire.
  Dry, open, often rocky places, often on limestone, from the foothills to high elevations in the mountains; w.c. and s.w. Mont. (e. of the continental divide) to the Big Horn Mts. of Wyo.; reported, doubtless through confusion in labeling, from the Washington Cascades. May-July.

Eritrichium nanum (Vill.) Schrad. Asperif. 16. 1820.
  Myosotis nana Vill. Prosp. 21. 1779. Omphalodes nana Gray, Proc. Am. Acad. 20:263. 1885. Lappula nana Car. in Parl. & Car. Fl. Ital. 6:861. 1886. (Presumably from the Alps)
  Myosotis villosa Ledeb. Mém. Acad. St. Pétersb. 5:516. 1812. Anchusa villosa Roem. & Schult. Syst. Veg. 4:775. 1819. Eritrichium villosum Bunge, Mém. Sav. Acad. St. Pétersb. 2:531. 1836. E. nanum var. villosum Kurtz, Engl. Bot. Jahrb. 19:471. 1894. E. nanum ssp. villosum Brand, Pflanzenr. IV. 252 (Heft 97):189. 1931. An Asiatic variety marked by its relatively large leaves and more robust habit; "villosum" may not be the oldest applicable epithet in varietal status.
  Myosotis aretioides Cham. Linnaea 4:443. 1829. Eritrichium aretioides A. DC. Prodr. 10:125. 1846. E. villosum var. aretioides Gray, Proc. Acad. Phila. 1863:73. 1864. E. nanum var. aretioides Herder, Acta. Hort. Petrop. 1:535. 1873. Omphalodes nana var. aretioides Gray, Proc. Am. Acad. 20:263. 1885. (Chamisso, St. Lawrence I. and St. Lawrence Bay)
  Eritrichium chamissonis A. DC. Prodr. 10:125. 1846. E. nanum var. chamissonis Herder, Acta, Hort. Petrop. 1:535. 1873. Omphalodes nana var. chamissonis Gray, Proc. Am. Acad. 20:263. 1885. (Chamisso, St. Lawrence Bay) = var. aretioides
  ERITRICHIUM NANUM var. ELONGATUM (Rydb.) Cronq. hoc loc. E. aretioides var. elongatum Rydb. Mem. N.Y. Bot. Gard. 1:327. 1900. E. elongatum Wight, Bull. Torrey Club 29:408. 1902. (Rydberg & Bessey 4891, Spanish Basin, Mont.)
  E. argenteum Wight, Bull. Torrey Club 29:411. 1902. E. elongatum var. argenteum Johnst. Contr. Gray Herb. n.s. 70:53. 1924. E. nanum ssp. villosum var. villosum f. argenteum Brand, Pflanzenr. IV. 252 (Heft 97): 191. 1931. (Crandall & Cowan 361, incorrectly cited by Wight as 36, near Como, Colo.)
  Pulvinate-caespitose, long-lived perennial, essentially acaulescent or with more or less erect, slender, villous to loosely strigose stems up to 1 dm. tall; leaves mostly oblong or nar-

rowly oblong to ovate, often obtuse, up to 1 cm. long and 2 mm. wide, greener and less firm than in E. howardii, loosely long-hairy, the surface readily visible at maturity between the hairs, which are often more numerous toward the tip, where they tend to form a fringe or tuft; corolla limb 4-8 mm. wide; nutlets 1-4, glabrous.

Open, rocky places at high altitudes in the mountains; irregularly from the Alps of Europe across Asia to Alas. and adj. Yukon, and in the Rocky Mts. of the U.S. from Mont. to n. N.M., extending w. to the Wallowa Mts. of Oreg. June-Aug.

Eritrichium nanum is here considered to be a widespread and variable but sharply limited species; a similar view was taken by Brand in Das Pflanzenreich. Plants of the U.S. constitute the rather weak var. elongatum (Rydb.) Cronq., apparently differing from typical European E. nanum only in being usually more densely pubescent; the less hairy specimens from our range, notably those from the Wallowa Mts., appear to be identical with the European plants in pubescence as well as in other characters. Plants of our range have the flange of the nutlets entire, while many of those from the s. Rocky Mts. have the flange conspicuously toothed and have therefore often been treated as a distinct species, variety, or form (argenteum); parallel variations occur in var. nanum.

Plants of Yukon, Alas., and e. Siberia, differing from the argenteum phase of var. elongatum in the much more elongate teeth of the nutlet, have often been treated as one or two distinct species, but are here regarded as E. nanum var. aretioides (Cham.) Herder. One or more additional varieties are to be recognized in Asia.

## Hackelia Opiz. Stickseed; Wild Forget-me-not

Flowers borne in sympodial, mostly naked or only basally leafy-bracteate false racemes which tend to elongate in age, the pedicels deflexed in fruit; calyx cleft nearly or quite to the base; corolla blue or white (sometimes ochroleucous or greenish-tinted), often with a yellow eye, salverform, with short tube and abruptly spreading, evidently 5-lobed limb, the fornices well developed; stamens included; nutlets surpassing the style, medially attached by a rather large scar to the broad, low gynobase, strongly glochidiate-prickly along the continuous dorso-marginal ridge and sometimes also across the back, the marginal prickles sometimes connate below into a cupulate border; taprooted perennials, or a few species biennial or even annual.

About 30 species, of wide geographic distribution, centering in w. N. Am. (Named for Joseph Hackel, 1783-1869, Czech botanist.)

Many of the species are attractive plants for arid regions, several of them being small enough for rock gardens. Our species are difficult or impossible w. of the Cascades.

The characters separating Hackelia from Lappula are scarcely stronger than those separating Allocarya from Plagiobothrys, but Johnston maintains that in the present instance the similarities are due to independent parallel evolution from cynoglossoid ancestors. The possibility of a closer relationship between Hackelia and Lappula is suggested by a specimen (Christ and Ward 8120, 18 miles e. of Riggins, Ida.) which may be a hybrid between H. cinerea and L. redowskii. The plant is chiefly sterile, the few nutlets being definitely like those of Hackelia. The mature pedicels are recurved, as in Hackelia, but the inflorescence is copiously and rather conspicuously bracteate and the flowers are rather crowded, as in Lappula. The corollas are intermediate in size between the suspected parents.

References:

Johnston, I. M. Studies in the Boraginaceae. 1. Restoration of the genus Hackelia. Contr. Gray Herb. n. s. 68:43-48. 1923.

Piper, C. V. Notes on the biennial and perennial West American species of Lappula. Bull. Torrey Club 29:535-49. 1902.

1 Corolla limb only 1.5-3 mm. wide; dorsal surface of the nutlets 2-3 mm. long; plants annual or biennial                                                                      H. DEFLEXA

1 Corolla limb mostly 4-20 mm. wide; dorsal surface of the nutlets mostly 3-5 mm. long; plants perennial, except for the mostly biennial H. floribunda

  2 Corolla limb blue, sometimes withering pink

    3 Marginal prickles of the nutlet united for at least a third of their length, forming a distinct cupulate border; basal leaves seldom and cauline leaves never over 1 cm. wide;

c. and n. Wash., e. of the Cascade summits                           H. CILIATA
3 Marginal prickles, or many of them, free nearly or quite to the base, not forming a distinct cupulate border; leaves often well over 1 cm. wide
    4 Plants robust, 3-10 dm. tall, the stem commonly (2)3-8 mm. thick toward the base,
      the larger leaves often over 1.5 cm. wide; fornices only very minutely papillate;
      widespread species
      5 Plants relatively short-lived, biennial or barely perennial, often single-stemmed
        from a taproot and simple crown; intramarginal prickles wanting, or occasionally
        1 or 2                                                        H. FLORIBUNDA
      5 Plants perennial, ordinarily with several or many stems from a taproot and branch-
        ing caudex; intramarginal prickles present, usually several. (Occasional plants of
        H. diffusa would be sought here, except for the evidently papillate-puberulent for-
        nices)                                                        H. JESSICAE
    4 Plants small and slender, 2-4 dm. tall, the stem mostly 1-2(3) mm. thick toward the
      base, the larger leaves not over about 1.5 cm. wide; fornices evidently papillate-
      puberulent
      6 Stem appressed-hairy; middle and upper cauline leaves narrow, seldom as much as
        1 cm. wide, scarcely clasping, mostly 4-10 times as long as wide; c. Oreg. to
        n. e. Calif.                                                  H. CUSICKII
      6 Stem spreading-hairy, at least below the middle; middle and upper cauline leaves
        markedly broad-based and clasping, 2-4 times as long as wide, many of them 1
        cm. wide or more; c. Ida.                                     H. DAVISII
2 Corolla limb white to ochroleucous or greenish-tinted, sometimes marked or very lightly
  washed with pale blue
  7 Corolla relatively very large and showy, the limb mostly 13-20 mm. wide; plants 2-4 dm.
    tall; Chelan Co., Wash.                                          H. VENUSTA
  7 Corolla smaller, the limb mostly 4-12 mm. wide; plants 2-10 dm. tall
    8 Intramarginal prickles well developed, some of them generally over 1 mm. long and
      more than half as long as the marginal ones; basal and lowermost cauline leaves re-
      duced or scarcely developed, the largest leaves borne shortly above the base of the
      stem; s. Jefferson Co., Oreg., and southward            H. CALIFORNICA
    8 Intramarginal prickles less well developed, not over about 1 mm. long, less than half
      as long as the marginal ones; basal leaves well developed and often persistent, com-
      monly larger than the cauline ones
      9 Corolla marked with light blue; pubescence of the stem largely or wholly appressed;
        s. w. Mont. and c. Ida. to n. Malheur Co., Oreg., c. Utah, and n. e. Nev.
                                                                      H. PATENS
      9 Corolla not marked with blue, though occasionally more evenly and very lightly
        washed with blue; pubescence of the stem appressed or often spreading; more
        northern or western plants, only H. cinerea overlapping the range of H. patens in
        c. Ida.
        10 Marginal prickles ordinarily united for at least 1/3 of their length, forming a
          distinct cupulate border to the nutlet
          11 Fornices essentially glabrous; corolla ochroleucous or greenish-tinted, the
            limb mostly 4-5 mm. wide; vicinity of the Snake River Canyon, and in the
            Grand Coulee                                          H. HISPIDA
          11 Fornices evidently papillate-puberulent; corolla white, the limb mostly (5)7-
            12 mm. wide; c. Wash. to n. w. Mont. (w. of the continental divide), but
            not in the Snake River Canyon or Grand Coulee region      H. CINEREA
        10 Marginal prickles or many of them free nearly or quite to the base, not forming
          a distinct, cupulate border (or tending to form such a border in forms of H.
          arida transitional to H. cinerea)
          12 Larger leaves rarely as much as 1 cm. wide; pubescence of the stem wholly
            appressed to partly spreading; c. Wash.           H. ARIDA
          12 Larger leaves mostly over 1 cm. wide; pubescence of the middle and lower
            part of the stem largely spreading; near the Columbia R. from near Blalock,

Oreg., to well into the gorge, and irregularly northward to s. B.C.    H. DIFFUSA

Hackelia arida (Piper) Johnst. Contr. Gray Herb. n. s. 68:48. 1923.
  Lappula arida Piper, Bull. Torrey Club 28:44. 1901. Echinospermum aridum K. Schum. Just
    Bot. Jahresb. 29[1]:564. 1903. (Piper 2676, Ellensburg, Wash.)
  Lappula hendersoni Piper, Bull. Torrey Club 29:539. 1902. Hackelia hendersonii Brand,
    Pflanzenr. IV. 252 (Heft 97): 132. 1931. (Henderson, Clemens Mt., Yakima Co., Wash.,
    June 14, 1892.)
  Lappula cottoni Piper, Bull. Torrey Club 29:549. 1902. Hackelia cottonii Brand, Pflanzenr.
    IV. 252. (Heft 97): 132. 1931. (Cotton 360, n. slope of Rattlesnake Hills, Yakima Co.,
    Wash.)
    Perennial, 2-8 dm. tall; stems several or sometimes solitary, antrorsely strigose above,
retrorsely strigose or more often softly spreading-hairy below; leaves strigose or hirsute-strigose to densely villous-puberulent, narrow, all linear or nearly so, the basal ones well developed
and often persistent, petiolate, 5-20 cm. long and 2-10(15) mm. wide, the cauline ones similar but mostly sessile and progressively smaller; corolla white with a yellow eye, the limb 6-12
(15) mm. wide, the fornices evidently papillate-puberulent; marginal prickles of the nutlets
generally free nearly or quite to the base, much larger than the intramarginal ones.
    Open or lightly wooded, dry slopes, often with sagebrush or ponderosa pine; Douglas,
Chelan, Grant, Kittitas, and Yakima cos., Wash. May-July.
    Hackelia arida and H. cinerea intergrade so fully in the Wenatchee region that they might
with good reason be considered phases of a single species. Such an expansion of specific limits
would however render the exclusion of H. diffusa difficult on morphological grounds, although
it is biologically apparently quite separate. It is therefore convenient to maintain the traditional treatment in which all three taxa are recognized as species.

Hackelia californica (Gray) Johnst. Contr. Gray Herb. n. s. 68:47. 1923.
  Echinospermum californicum Gray, Proc. Am. Acad. 17:225. 1882. Lappula californica
    Piper, Bull. Torrey Club 29:546. 1902. (Pringle, Mt. Shasta, Calif.; lectotype by Piper)
  Hackelia elegans Brand, Pflanzenr. IV. 252 (Heft 97): 128. 1931. (Cusick 2680, Black Butte,
    "Crook" Co., Oreg.)
    Perennial, 4-10 dm. tall; stems several or numerous, strigose or puberulent to shortly vil-
lous-hirsute, the hairs mostly antrorse above and retrorse or retrorsely spreading below;
leaves hirsute-puberulent to merely strigose, the largest ones oblanceolate and often obtuse,
petiolate, 5-15 cm. long, 1-3 cm. wide, borne above the base of the stem, the lowermost cauline leaves reduced and the truly basal ones apparently wanting; middle and upper cauline
leaves sessile and often clasping, becoming lanceolate or lance-oblong, gradually reduced upward; corolla pure white, including the fornices, which are shortly papillate-puberulent, the
limb 7-12 mm. wide; marginal prickles of the nutlets free to the base or nearly so, the intramarginal ones well developed, at least some of them generally over 1 mm. long and more than
half as long as the marginal ones.
    Forest openings in and near the Cascades, at elevations of 4000-7000 ft.; Black Butte, Jef-
ferson Co., Oreg., southward to Plumas Co., Calif. June-July.

Hackelia ciliata (Dougl.) Johnst. Contr. Gray Herb. n. s. 68:46. 1923.
  Cynoglossum ciliatum Dougl. ex Lehm. Stirp. Pug. 2:24. 1830. Echinospermum ciliatum
    Gray, Proc. Am. Acad. 17:225. 1882. Lappula ciliata Greene, Pitt. 2:182. 1891. (Douglas,
    "Kettle Falls and Spokan River," Wash.)
    Perennial, 3-9 dm. tall; stems generally several, retrosely strigose below, antrorsely so
above, and often coarsely spreading-hairy as well; leaves closely strigose to often densely villous-puberulent, and frequently with some coarser, loose or appressed bristles; basal leaves
well developed and often persistent, oblanceolate, petiolate, 5-15 cm. long, 3-12 mm. wide;
cauline leaves nearly linear, mostly sessile, up to 10 cm. long and 6 mm. wide; corolla light
blue, probably with a yellow eye, often withering pink, the limb (5)8-12 mm. wide, the fornices shortly papillate-hairy or merely papillate; marginal prickles of the nutlets united below,

1/2

1/14

1.5

Dasynotus daubenmirei

10    10

3

1/2

Eritrichium nanum

Hackelia arida

1/14    12    3    3

8    8

3

1/2

Eritrichium howardii

1.5

1/2

1/10    8    8    1/2

Echium vulgare

JRJ

usually for 1/3-1/2 their length, to form a shallow cup; intramarginal prickles several or rather many, evidently smaller than the marginal ones and sometimes poorly developed.

Dry, open slopes and flats in sagebrush and ponderosa pine lands; Spokane and Stevens cos. to Okanogan and Kittitas cos., Wash. May-June.

Hackelia cinerea (Piper) Johnst. Contr. Gray Herb. n.s. 68:46. 1923.
   Lappula cinerea Piper, Bull. Torrey Club 29:544. 1902. (Henderson 3006, Salmon R. bluffs, Ida.)
   Perennial, 2-8 dm. tall; stems several or sometimes solitary, strigose and generally also more or less spreading-bristly; leaves hirsute or hispid-hirsute with spreading or subappressed hairs, and often strigose or strigose-sericeous especially beneath, the basal ones well-developed and commonly persistent, petiolate, oblanceolate or narrowly elliptic, 5-20 cm. long and 4-15 mm. wide, the cauline ones mostly sessile and progressively smaller, lanceolate to linear or linear-oblong; corolla white with a yellow eye, the limb (5)7-12 mm. wide, the fornices evidently papillate-hairy; marginal prickles of the nutlets united for about 1/3-1/2 their length to form an evident cupulate border; intramarginal prickles small.
   Open or lightly forested places, especially on cliffs or talus, or loose stream banks, from the valleys and foothills to moderate elevations in the mountains; n.w. Mont. (w. of the continental divide) and n. and c. Ida. (s. to Boise) to Spokane Co., Wash.; also in Kittitas and Chelan cos., Wash. May-July.
   Hackelia cinerea overlaps the range of H. patens in Custer and Lemhi cos., Ida., but there is no evidence of intergradation. Where it overlaps the range of H. arida in c. Wash., H. cinerea tends to occur at the higher elevations, but no clear morphological or altitudinal line is evident between the two. See further comment under H. arida.

Hackelia cusickii (Piper) Brand, Pflanzenr. IV. 252 (Heft 97): 131. 1931.
   Lappula cusickii Piper, Bull. Torrey Club 29:542. 1902. L. arida var. cusickii Nels. & Macbr. Bot. Gaz. 61:41. 1916. Hackelia arida var. cusickii Johnst. Contr. Gray Herb. n.s. 68:48. 1923. (Cusick 2623, Logan Mts., e. Oreg.)
   Slender perennial, 2-4 dm. tall; stems several, antrorsely strigose throughout, or retrorsely so near the base, 1-2(3) mm. thick toward the base; leaves strigose, the basal ones oblanceolate or narrowly elliptic and petiolate, relatively well developed and persistent, mostly 4-14 cm. long and 4-12 mm. wide, acute; cauline leaves evidently smaller, all but the lowermost sessile and lanceolate or narrower, slightly or not at all clasping, seldom as much as 1 cm. wide, the middle and upper ones mostly 4-10 times as long as wide; corolla blue with a yellow or whitish eye, the limb 5-12 mm. wide, the fornices evidently papillate-puberulent; marginal prickles of the nutlets free nearly or quite to the base; intramarginal prickles several (about 10), much shorter than the marginal ones.
   In the shelter of junipers on the drier mountains from Crook and Harney cos., Oreg., to Modoc Co., Calif. May-July.
   A well-marked species of sharply limited habitat.

HACKELIA DAVISII Cronquist, hoc loc. (Type: Ray J. Davis 3046, Long Tom Camp, Range 16 East, Twp. 23 North, near the Salmon R. in Lemhi Co., Ida., May 15, 1941; N.Y. Bot. Gard.)
   Slender perennial with several lax, curved stems 2-3 dm. tall; stems 1-2 mm. thick toward the base, evidently spreading-hirsute, becoming more strigose above the middle or in the inflorescence; leaves hirsute, the basal ones oblanceolate, petiolate, up to about 10 cm. long and nearly 1.5 cm. wide; cauline leaves mostly sessile, the middle and upper ones broad-based and strongly clasping, 2-5 cm. long, 6-15 mm. wide, 2-4 times as long as wide; flowers relatively few and long-pedicellate, blue with a yellow eye, the limb 10-12 mm. wide, the fornices evidently papillate-puberulent; marginal prickles of the nutlets free to the base or nearly so; intramarginal prickles several, much shorter than the marginal ones.
   Herba perennis gracilis 2-3 dm. alta, caulibus pluribus patenti-hirsutis saltem sub medio, supra plerumque strigosis, foliis inferioribus oblanceolatis petiolatis, mediis superioribus-

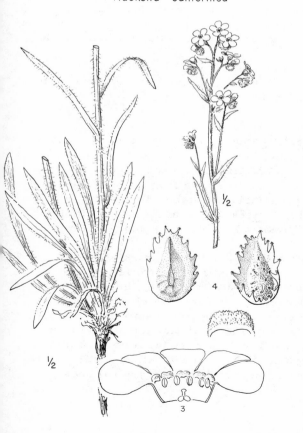

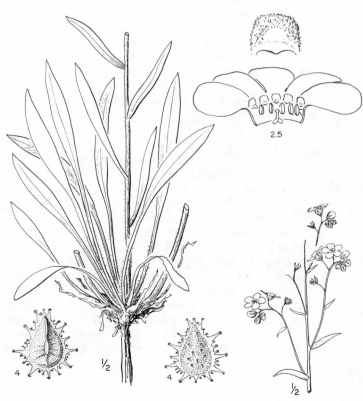

Hackelia californica

H. cinerea

H. ciliata

H. cusickii

JRJ

que sessilibus amplexicaulibus 2-5 cm. longis 6-15 mm. latis, corollis caeruleis 10-12 mm. latis ocula flavida, nuculis aculeis marginalibus plus minusve distinctis ornatis.

Moist rock crevices on the n. side of the Salmon R. shortly above the mouth of the Middle Fork; known only from the type locality (two collections, flower and fruit). May.

This species is probably allied to H. diffusa, from which it differs in its smaller size and more slender stems, more strongly clasping cauline leaves, blue flowers, and disjunct distribution. The name Hackelia davisii has already appeared in print, attributed to Ivan M. Johnston, but without Latin diagnosis. Dr. Johnston now tells me that he does not wish to assume formal responsibility for the publication of the name.

Hackelia deflexa (Wahlenb.) Opiz in Bercht. Fl. Böhm. 2²: 147. 1839.
  Myosotis deflexa Wahlenb. Svensk. Vet. Akad. Handl. 31:113. 1810. Echinospermum deflexum
    Lehm. Asperif. 120. 1818. Rochelia deflexa Roem. & Schult. Syst. Veg. 4:109. 1819.
    Cynoglossum deflexum Roth, Enum. 1:589. 1827. Lappula deflexa Garcke, Fl. Deutschl.
    6th ed. 275. 1863. (Europe)
  Echinospermum deflexum var. americanum Gray, Proc. Am. Acad. 17:224. 1882. Lappula
    deflexa var. americana Greene, Pitt. 2:183. 1891. L. americana Rydb. Bull. Torrey Club
    24:294. 1897. Hackelia deflexa var. americana Fern. & Johnst. Rhodora 26:124. 1924. H.
    americana Fern. Rhodora 40:341. 1938. (Bourgeau, Sask.; fide Fernald)
  Lappula leptophylla Rydb. Mem. N. Y. Bot. Gard. 1:329. 1900. Hackelia leptophylla Johnst.
    Contr. Gray Herb. n. s. 68:46. 1923. (Tweedy 223, Bozeman, Mont., is the first of four
    specimens cited)
  Lappula besseyi Rydb. Bull. Torrey Club 31:636. 1905. Hackelia leptophylla var. besseyi
    Brand, Pflanzenr. IV. 252 (Heft 97): 127. 1931. (Bessey s.n., Cheyenne Canyon, Colo.,
    July 25, 1895)
  Annual or biennial, 1.5-10 dm. tall; stems few or mostly solitary, strigose above, generally spreading-hairy below; leaves hirsute or strigose, largely or wholly cauline, often very thin, mostly 2-15 cm. long and 0.5-3 cm. wide, the lower oblanceolate and subpetiolate, the others generally more lanceolate or narrowly elliptic and sessile; corolla very small and inconspicuous, blue or sometimes white, the limb only 1.5-3 mm. wide; dorsal surface of the nutlets 2-3 mm. long; marginal prickles distinct nearly or quite to the base; intramarginal prickles few (2-3) and poorly developed, or more often none.

Thickets and open woods; circumboreal; in America from B. C. and n. Alta. to Que., southward to Wash., Ida., Colo., Iowa, and Vt. June-Aug.

The American plants as here described constitute the var. americana (Gray) Fern. & Johnst., from which the Eurasian var. deflexa differs in its mostly larger corollas (limb commonly 3-5 mm. wide) and in some minor statistical tendencies.

Hackelia diffusa (Lehm.) Johnst. Contr. Gray Herb. n. s. 68:48. 1923.
  Echinospermum diffusum Lehm. Stirp. Pug. 2:23. 1830. Lappula diffusa Greene, Pitt. 2:182.
    1891. (Douglas, N. W. America)
  Lappula trachyphylla Piper, Bull. Torrey Club 29:540. 1902. Hackelia hendersonii var. trachyphylla Brand, Pflanzenr. IV. 252 (Heft 97): 132. 1931. ("A single fragmentary specimen
    . . . collected by Winslow J. Howard in Montana"; presumably mislabeled)
  Lappula saxatilis Piper, Bull. Torrey Club 29:541. 1902. Hackelia saxatilis Brand, Pflanzenr. IV. 252 (Heft 97): 133. 1931. (Suksdorf 592, Klickitat R., Wash.) A blue-flowered
    form.
  Perennial, 2-7 dm. tall; stems several or sometimes solitary, 2-5 mm. thick toward the base, the middle and lower part evidently spreading-hirsute (sometimes retrorsely so below), the upper part more strigose (antrorsely); leaves hirsute or hirsute-strigose, the basal ones well developed and generally persistent, petiolate, oblanceolate or narrowly elliptic, mostly 6-18 cm. long and 8-25 mm. wide; cauline leaves well developed, mostly sessile and the upper sometimes somewhat clasping, lanceolate or lance-oblong to lance-elliptic, mostly 5-12 cm. long and 8-18 mm. wide; corolla white or rarely blue, with a yellow eye, the limb 7-12 mm. wide, the fornices evidently papillate-puberulent; marginal prickles of the nutlets distinct nearly or quite to the base; intramarginal prickles 8-30, much shorter than the marginal ones.

Cliffs and talus slopes along and near the Columbia R. in Oreg. and Wash. from Blalock to well into the gorge; irregularly northward to the Thompson and Fraser rivers in s. B.C. June-Aug.

The name H. diffusa is retained for the present plant in deference to Johnston (Contr. Gray Herb. n. s. 68:48. 1923), subject to correction when the type can be examined. Gray's earlier identification of H. diffusa with the plant now known as H. jessicae was based on study of original material, and has yet to be proven wrong. Lehmann's description applies less well to either H. jessica or H. hendersonii than to H. californica, a species which Douglas apparently had no opportunity to collect.

Hackelia floribunda (Lehm.) Johnst. Contr. Gray Herb. n. s. 68:46. 1923.
   Echinospermum floribundum Lehm. Stirp. Pug. 2:24. 1830. E. deflexum var. floribundum
      Wats. Bot. King Exp. 246. 1871. Lappula floribunda Greene, Pitt. 2:182. 1891. (Drum-
      mond, "Lake Pentanguishene to the Rocky Mts.")
   Robust biennial or short-lived perennial, 3-10 dm. tall; stems few or often solitary, an-trorsely strigose above, spreading-hirsute or occasionally retrorsely strigose below, mostly 2-8 mm. thick toward the base; leaves strigose or hirsute, the basal ones oblanceolate, petio-late, sometimes well developed and persistent, more often scarcely as large as the cauline ones and soon withering; cauline leaves well developed, more or less numerous, mostly 4-20 cm. long and 5-30 mm. wide, the lower oblanceolate and petiolate, the others more lanceolate or lance-elliptic and sessile, gradually reduced upward; flowers numerous in a mostly elon-gate and narrow inflorescence with ascending branches; corolla blue with a yellow eye, the limb 4-7 mm. wide, the fornices only minutely papillate; body of the nutlets 3-4 mm. long; marginal prickles free nearly or quite to the base; intramarginal prickles wanting, or rarely 1-2.

Thickets, meadows, stream banks, and other moist places from the foothills to moderate or rather high elevations in the mountains; s. B.C. to Sask. and reputedly w. Ont. and Minn., southward to Calif., s. Nev., and n. N.M., chiefly in the Rocky Mt. region, seldom collected in Wash., Oreg., and Calif. June-Aug.

Occasional plants of this species, particularly those from the s. Rocky Mts., show a tenden-cy to have the prickles fused toward the base to form a shallow cup.

Hackelia hispida (Gray) Johnst. Contr. Gray Herb. n. s. 68:46. 1923.
   Echinospermum diffusum var. hispidum Gray, Proc. Am. Acad. 17:225. 1882. E. hispidum
      Gray, Proc. Am. Acad. 20:259. 1885, nom. nud. Syn. Fl. 2nd ed. 2$^1$:422. 1886. Lappula
      hispida Greene, Pitt. 2:182. 1891 (Cusick, e. Oreg., in 1880 and 1881)
   Differing from the related H. cinerea as follows: plant greener, strigose chiefly in the in-florescence, the pubescence otherwise largely coarse and spreading, the stem sometimes nearly glabrous; corolla ochroleucous or greenish-tinged, the limb mostly 4-5 mm. wide, the fornices essentially glabrous.

Cliffs and talus slopes; Snake River Canyon and vicinity, in Oreg., Ida., and extreme s.e. Wash.; also in the Grand Coulee of Wash. May-July.

Plants from the Grand Coulee approach H. cinerea in having some shorter and finer, some-times subappressed hairs intermingled with the coarse, spreading hairs of the stem, but are otherwise apparently similar to typical H. hispida. Further collecting might warrant the es-tablishment of the Grand Coulee plants as a separate variety.

Hackelia jessicae (McGregor) Brand, Pflanzenr. IV. 252 (Heft 97): 132. 1931.
   Lappula jessicae McGregor, Bull. Torrey Club 37:262. 1910. L. floribunda var. jessicae
      Jeps. & Hoover in Jeps. Fl. Calif. 3:307. 1943. (McGregor 71, Half Moon Lake, near Ta-
      hoe, Calif.)
   Robust perennial from a taproot and branching caudex, 3-10 dm. tall; stems several or nu-merous, spreading-hirsute or occasionally retrorsely strigose below, antrorsely strigose or strigose-puberulent above, 3-8 mm. thick toward the base; leaves strigose or hirsute, the basal ones oblanceolate or narrowly elliptic, petiolate, generally well developed and persist-ent, up to 35 cm. long and 4 cm. wide, or sometimes smaller and soon deciduous; cauline

Hackelia davisii

H. deflexa var. americana

H. diffusa

H. hispida

H. floribunda

JRJ

leaves well developed, several or many, mostly 5-20 cm. long and 7-20 mm. wide, the lower oblanceolate and petiolate, the others more lance-elliptic to oblong and sessile, gradually reduced upward; flowers fairly numerous, the inflorescence tending to be shorter, broader, and looser than in H. floribunda; corolla blue with a yellow or whitish eye, the limb (5)7-11 mm. wide, the fornices only minutely papillate; marginal prickles of the nutlets free nearly or quite to the base; intramarginal prickles several, or sometimes only 1 or 2, much smaller than the marginal ones.

Forest openings and meadows, less often in thickets, along stream banks, or on drier, open slopes, from the foothills to moderate or fairly high elevations in the mountains; s. B.C., southward along the Cascades to the Sierra Nevada of Calif., eastward to Alta., Mont., w. Wyo., and Utah; much commoner in Oreg. and Wash. than H. floribunda. June-Aug.

Although all of the characters which distinguish H. jessicae from H. floribunda are subject to some failure or overlap, as will be noted by a comparison of the descriptions, the populations themselves appear to be quite distinct.

Hackelia patens (Nutt.) Johnst. Contr. Arn. Arb. 16:194. 1935.
  Rochelia patens Nutt. Journ. Acad. Phila. 7:44. 1834. (Wyeth, "near the Flat-Head River"; probably actually taken farther s. on Wyeth's route, in Ida., since the species is not otherwise known to occur in the vicinity of the Flathead R.) Identity fide Johnston.
  Echinospermum subdecumbens Parry, Proc. Davenport Acad. Sci. 1:148. 1876, nom. provis.
  Lappula subdecumbens Nels. in Coult. & Nels. New Man. Bot. Rocky Mts. 412. 1909.
  (Parry, Wasatch Mts. of Utah)
  Lappula coerulescens Rydb. Mem. N.Y. Bot. Gard. 1:328. 1900. L. subdecumbens (var.) coerulescens Garret, Spring Fl. Wasatch Reg. 78. 1911. Hackelia diffusa var. coerulescens Johnst. Contr. Gray Herb. n.s. 68:48. 1923. H. coerulescens Brand, Pflanzenr. IV. 252 (Heft 97): 130. 1931. (Rydberg & Bessey 4899, Bridger Mts., Mont.)
  Perennial, 2-8 dm. tall; stems generally several, 1.5-5 mm. thick toward the base, evidently pubescent, the hairs mostly or all appressed, retrorse below, antrorse above; leaves strigose or hirsute-strigose and sometimes bristly-ciliate, the basal ones well developed and generally persistent, petiolate, oblanceolate or narrowly elliptic, 5-30 cm. long and 7-30 mm. wide; cauline leaves evidently smaller, mostly sessile, more or less reduced upward; corolla limb mostly 6-11 mm. wide, white with a yellow eye, marked with pale blue (commonly with 10 bluish marks toward the center, and often lightly blue-veined in part, or even very lightly washed with blue), the fornices evidently villous-puberulent; marginal prickles of the nutlets free nearly or quite to the base, the intramarginal ones much smaller, 15 or fewer.

Dry, open places, often with sagebrush, from the valleys and foothills to moderate or sometimes high elevations in the mountains; s.w. Mont. (e. of the continental divide) to the mountains at the head of the e. fork of the Salmon R. in c. Ida., southward to w. Wyo., c. Utah, and n.e. Nev. June-Aug.

Hackelia venusta (Piper) St. John, Res. Stud. State Coll. Wash. 1:104. 1929.
  Lappula venusta Piper, Proc. Biol. Soc. Wash. 37:93. 1924. (Otis 895, between Tumwater and Drury, Chelan Co., Wash.)
  Stems numerous from a perennial taproot, 2-4 dm. tall; herbage bristly-hirsute, becoming more strigose above; leaves chiefly cauline, rather numerous, not much reduced upward, mostly 2-5 cm. long and 3-11 mm. wide, the lower oblanceolate and subpetiolate, the others more lanceolate or linear-oblong and sessile; flowers long-pedicellate, relatively large and showy, white, the corolla limb mostly 13-20 mm. wide, the fornices papillate; marginal prickles of the nutlets united for about 1/3-1/2 their length into an evident border; intramarginal prickles about 15, well developed but evidently smaller than the marginal ones.

Rocky slopes with ponderosa pine, at about 1000 ft.; so far known only from the immediate vicinity of the type locality in Chelan Co., Wash. May-June.

A well-marked species, apparently allied to H. cinerea.

## Heliotropium L.

Flowers mostly in terminal, naked or bracteate, helicoid, false spikes or racemes; calyx shallowly to deeply cleft; corolla mostly blue or white, salverform or funnelform, often with 5 small teeth alternating with the lobes; fornices wanting; anthers included, often connivent; ovary entire or merely shallowly lobed, the style terminal (or wanting and the stigma sessile); stigma with a broad, disklike base (often as broad as the ovary) that is commonly surmounted by a mostly short, entire or 2-cleft cone; fruit separating at maturity into 4 nutlets, or the nutlets cohering in pairs; herbs or shrubs.

More than 200 species, chiefly of tropical and warm-temperate regions. (Name slightly modified from the ancient Greek name of some of the species. The word is derived from helios, the sun, and tropos, turn.)
Reference:
Ewan, J. A review of the North American weedy heliotropes. Bull. So. Calif. Acad. Sci. 41:51-57. 1942.

Heliotropium curassavicum L. Sp. Pl. 130. 1753. (Tropical America)
H. curassavicum var. obovatum DC. Prodr. 9:538. 1845. (Douglas, Columbia R.)
H. spathulatum Rydb. Bull. Torrey Club 30:262. 1903. (Williams 542, Great Falls, Mont.)
Glabrous, succulent, taprooted annual or short-lived perennial, prostrate or ascending, the stems 1-6 dm. long; leaves wholly cauline, the lowermost reduced and scaly, the others oblanceolate to narrowly obovate, short-petiolate or subsessile, mostly 2-6 cm. long and 6-18 mm. wide; spikes 1-several at the end of a short, naked, common peduncle, up to 6(10) cm. long at maturity, naked; calyx 2-3 mm. long, the lobes persistently erect; corolla white or faintly tinged with blue, the limb mostly 5-9 mm. wide; stigma sessile, expanded and fully as broad as the ovary, the cone very short; nutlets 1.5-2 mm. long, tardily separating. N=13, 14.

Saline places at low elevations, often in the beds of dried ponds, or maritime; tropical and subtropical America, chiefly along the seacoast, and in suitable habitats over most of w. U.S., extending into s. Can.; in our range wholly e. of the Cascades. Introduced in the tropics of the Old World. June-Sept.

Only the var. obovatum DC. as described above, of interior w. U.S., occurs in our range.
Tournefourtia sibirica L. (=Messerschmidia sibirica), a low, rhizomatous perennial with white or bluish flowers in small, corymbiform cymes, and with subglobose, corky fruits about 7 mm. high that only tardily, if at all, separate into segments, is either repeatedly introduced or persistent on ballast at Portland, Oreg.

## Lappula Gilib. Stickseed

Flowers borne in sympodial, more or less conspicuously bracteate, terminal false racemes which may become elongate in age, the short pedicels erect or ascending in fruit; calyx cleft nearly or quite to the base; corolla blue or white, mostly small and relatively inconspicuous, more or less funnelform, with definite fornices; stamens included; nutlets usually surpassed by the style, narrowly attached to the elongate gynobase along the well-developed, median ventral keel (the lowermost part rounded and free), bearing one or more rows of glochidiate prickles along the continuous dorsomarginal ridge or cupulate border; entire-leaved, taprooted annuals or winter annuals or rarely biennials, reputedly short-lived perennial in some extralimital species.

About a dozen species, of wide distribution, chiefly in the Northern Hemisphere. Several species often included in Lappula are here referred to the segregate genus Hackelia. (Name a diminutive of the Latin lappa, a bur).
1 Marginal prickles of the nutlets in 2(3) rows, slender, not confluent at the base
                                                                          L. ECHINATA
1 Marginal prickles of the nutlets in a single row, often swollen and confluent toward the base
     so as to form a cupulate border to the nutlet                        L. REDOWSKII

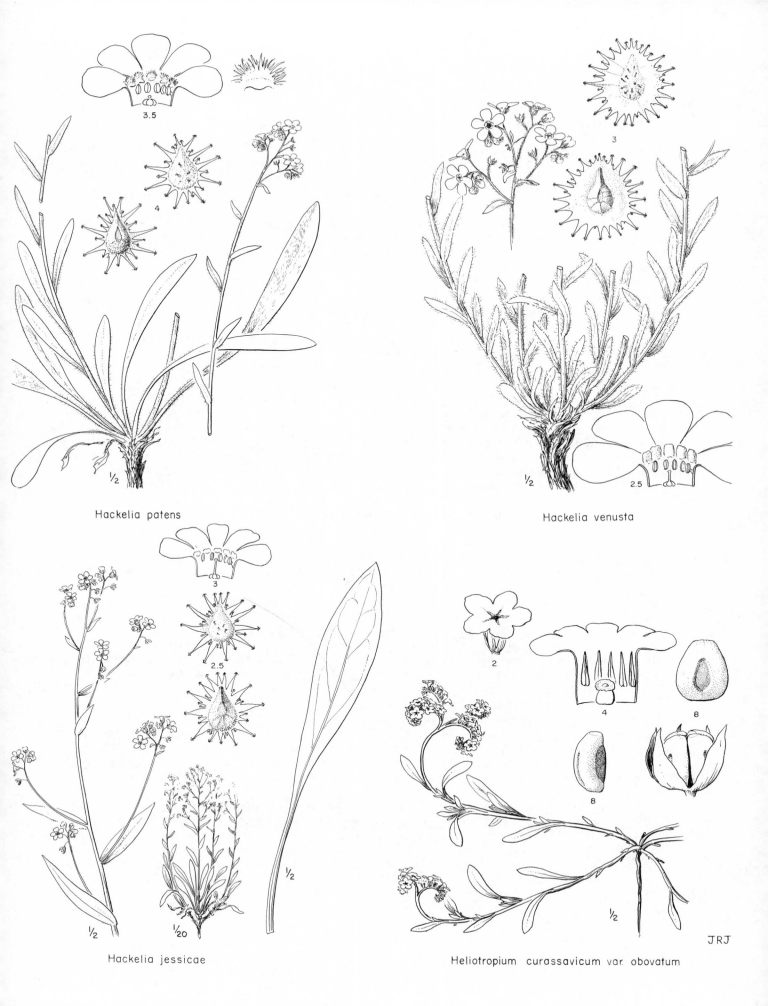

Hackelia patens

Hackelia venusta

Hackelia jessicae

Heliotropium curassavicum var. obovatum

JRJ

Lappula echinata Gilib. Fl. Lithuan. 1:25. 1781.

  Myosotis lappula L. Sp. Pl. 131. 1753. Lappula myosotis Moench. Meth. 417. 1794. Echi-
    nospermum lappula Lehm. Asperif. 121. 1818. Lappula lappula Karst. Deuts. Fl. 979.
    1882. (Europe)

Echinospermum fremontii Torr. Pac. R.R. Rep. 12:46. 1860. Lappula fremontii Greene,
    Pitt. 4:96. 1899. (Fremont 844, Pass Creek, near the s. end of the Sierra Nevada; thought
    by Johnston to be mislabeled and actually of n. Rocky Mt. origin)

Lappula erecta A. Nels. Bull. Torrey Club 27:268. 1900. (Nelson 424, Uva, Wyo., is first
    specimen cited)

    Similar to L. redowskii, sometimes a little more robust and larger-leaved; corolla more
consistently blue and averaging larger, the limb up to about 4 mm. wide; marginal prickles
of the nutlets in 2(3) rows, slender, not confluent at the base, those of the outer row commonly shorter than those of the inner. N=24.

    A weed in dry to moderately moist, usually more or less disturbed sites, as along roadsides
or on overgrazed ranges; Eurasia, where certainly native, and in much of the U.S. and s. Can.,
where perhaps also native in the northern Rocky Mt. region; wholly e. of the Cascades in our
region. June-Aug.

    Some botanists distinguish several varieties in the Old World on minor technical differences
which also occur among plants introduced in America.

    Plants with relatively large nutlets, which have some of the papillae along the median dorsal
line enlarged and somewhat glochidiate, appear to be native in the northern Rocky Mt. region,
and have been described as L. fremontii (Torr.) Greene, but are not sharply separable from
the introduced forms with more nearly uniform, nonglochidiate dorsal papillae. Monographic
study might support a varietal segregation, but no new combination is here proposed. The apparent disjunction in natural range between L. fremontii and L. echinata proper becomes less
significant in view of a similar gap in the range of L. redowskii, some of the native American
forms of which are considered by current students (including Johnston) to be taxonomically
identical with plants from Eurasia.

Lappula redowskii (Hornem.) Greene, Pitt. 4:97. 1899.

  Myosotis redowskii Hornem. Hort. Bot. Hafn. 1:174. 1813. Echinospermum redowskii Lehm.
    Asperif. 127. 1818. (Russia)

Echinospermum texanum Scheele, Linnaea 25:260. 1852. Lappula texana Britt. Mem. Torrey
    Club 5:273. 1894. L. redowskii var. texana Brand, Pflanzenr. IV. 252 (Heft 97): 150. 1931.
    (Roemer, San Antonio, Tex.) = var. cupulatum.

Echinospermum redowskii var. occidentale Wats. Bot. King Exp. 246. 1871. Lappula re-
    dowskii (var.) occidentalis Rydb. Contr. U.S. Nat. Herb. 3:170. 1895. L. occidentalis
    Greene, Pitt. 4:97. 1899. Echinospermum occidentale K. Schum. Just Bot. Jahresb. 27[1]:
    522. 1901. (Watson 861, "from the Sierras to the Wahsatch") = var. redowskii.

E. redowskii var. cupulatum Gray, Bot. Calif. 1:530. 1876. Lappula cupulata Rydb. Bull
    Torrey Club 28:31. 1901. Echinospermum cupulatum K. Schum. Just Bot. Jahresb. 29[1]:
    564. 1903. Lappula redowskii var. cupulata M. E. Jones, Bull. U. Mont. Biol. 15:44. 1910.
    (Watson 862, Trinity Mts., Nev.)

? Echinospermum brachycentrum var. brachystylum Gray, Proc. Am. Acad. 21:413. 1886.
    Lappula brachystyla Macbr. Contr. Gray Herb. n. s. 48:40. 1916. (Fletcher 1553, Spence's
    Bridge, Thompson R., B.C.) An unusual plant with the style surpassed by (or merely equal-
    ing) the nutlets, and the prickles nearly obsolete; twice re-collected, once (Thompson 199B)
    apparently in mixture with typical L. redowskii, which has been taken repeatedly at
    Spence's Bridge.

Lappula desertorum Greene, Pitt. 4:95. 1899. Echinospermum desertorum K. Schum. Just
    Bot. Jahresb. 27[1]:522. 1901. Lappula redowskii var. desertorum Johnst. Contr. Arn. Arb.
    3:93. 1932. (Greene, near Holborn, Nev., July 16, 1896) = var. cupulata.

L. heterosperma Greene, Pitt. 4:94. 1899. L. texana var. heterosperma Nels. & Macbr.
    Bot. Gaz. 61:41. 1916. (Baker, Earle, & Tracy 826, near Mancos, Colo.) = var. cupulata.

L. montana Greene, Pitt. 4:96. 1899. Echinospermum montanum K. Schum. Just Bot.
    Jahresb. 27[1]:522. 1901 (Kelsey, Helena, Mont., in 1887) = var. redowskii.

L. infelix Greene, Pitt. 4:235. 1901. (Cusick 1945 in part, Malheur R., Oreg.) = var. cupulata.

L. columbiana A. Nels. Bot. Gaz. 34:28. 1902. L. texana var. columbiana Johnst. Contr. Gray Herb. n. s. 70:50. 1924. (Piper 1703, Almota, Wash., is the first of the three specimens cited)

?L. anoplocarpa Greene, Ott. Nat. 16:39. 1902. (Macoun 17038, Spence's Bridge, B.C.)
The form previously described by Gray as Echinospermum brachycentrum var. brachystylum.

Annual or winter annual, probably occasionally biennial, simple or variously branched, 0.5-4(7) dm. tall, puberulent or short-hirsute (or partly strigose) throughout; leaves numerous, up to about 6 cm. long and 1 cm. wide, the basal ones oblanceolate and often deciduous, the cauline ones oblanceolate to more often linear or linear-oblong, gradually or abruptly reduced to the often more lanceolate bracts of the inflorescence, these seldom over 1(1.5) cm. long; corolla inconspicuous, blue or white, only slightly, if at all, surpassing the calyx, 2-4 mm. long, the limb mostly 1.5-2.5 mm. wide; marginal prickles of the nutlets in a single row, slender or swollen toward the base, distinct or united below to form a cupulate, often inflated border.

A weed in dry to moderately moist, usually more or less disturbed sites, as along roadsides or on overgrazed ranges; Eurasia and w. N. Am., wholly e. of the Cascades in our range. May-July.

The species exists in numerous named and nameless forms, both native and introduced. The var. redowskii, with the marginal prickles distinct, occurs essentially throughout the range of the species, and is the commoner phase in our range. The var. cupulata (Gray) M. E. Jones, with the prickles on (2)3 or all 4 of the nutlets of each flower fused below to form a prominent, cupulate, often swollen border, centers in s.w. U.S., and occasionally extends northward into our range in Mont., Ida., Oreg., and s.e. Wash. The extreme form of var. cupulata, including the type, with the prickles small and seeming to be merely seated on the strongly swollen border, is sometimes segregated as a distinct species, L. texana (Scheele) Britt.; the less extreme form, with the cupulate border obviously formed by the fused bases of the nutlets, is then treated as L. redowskii var. desertorum (Greene) Johnst. The difference is strictly technical, and the separation wholly arbitrary.

## Lithospermum L. Stoneseed

Flowers borne in modified, leafy-bracteate cymes, or solitary in or near the upper axils, often heterostylic; fruiting pedicels mostly erect or ascending; calyx deeply cleft; corolla yellow or yellowish, less often white (bluish-white), or in some extralimital species blue or purple, mostly funnelform or salverform, the fornices present or absent; anthers included or partly exserted; gynobase flat or depressed; nutlets smooth to pitted or wrinkled, basally attached, the large scar often surrounded by a sharp rim, sometimes only 1 nutlet maturing; annual to more often perennial herbs, seldom (ours never) pungently hairy.

About 75 species of wide distribution, mostly in temperate or mountainous regions. (Name from the Greek lithos, a stone, and sperma, seed, referring to the bony nutlets.)

The genus is composed of several more or less well-defined groups which have sometimes (notably by Johnston, Journ. Arn. Arb. 35:1-81; 158-66. 1954) been segregated as distinct genera, but Johnston's own treatments in closely successive papers present conflicting interpretations. Inasmuch as the only segregate genus (Buglossoides) occurring in our region is admittedly closely allied to Lithospermum, it is apparently not unnatural to maintain the genus in its customary sense as regards our species.

Reference:

Johnston, I. M. Studies in the Boraginaceae, XXIII. A survey of the genus Lithospermum. Journ. Arn. Arb. 33:299-366. 1952.

1 Plants annual; corolla white or bluish-white, 5-8 mm. long, the limb 2-4 mm. wide; introduced weedy species                                                              L. ARVENSE

1 Plants perennial, heavy-rooted; corolla yellowish or yellow, 8-40 mm. long, the limb 7-20 mm. wide; native species, not weedy

2  Corolla pale yellowish, often greenish-tinted, the tube 4-6 mm. long, the limb 7-13 mm.
      wide, the lobes entire or nearly so; common and widespread species  L. RUDERALE
2  Corolla bright yellow, the tube (12)15-30 mm. long, the limb 10-20 mm. wide, the lobes
      evidently erose; plains species, extending into the intermontane valleys of Mont., and in
      s. B. C.                                                                        L. INCISUM

Lithospermum arvense L. Sp. Pl. 132. 1753.
   Buglossoides arvense Johnst. Journ. Arn. Arb. 35:42. 1954. (Europe)
   Strigose annual, 1-7 dm. tall, simple or sparsely branched, with 1-several stems from the
base, the central one generally the largest; lowermost leaves oblanceolate and soon deciduous,
the others generally oblong, linear-oblong, or lanceolate, sessile or nearly so, mostly 1.5-6
cm. long and 2-15 mm. wide; flowers barely pedicellate, obliquely set in the axils of the more
or less reduced upper leaves, the inflorescence crowded at first, elongate and open in fruit;
corolla white or bluish-white, 5-8 mm. long, funnelform, the limb 2-4 mm. wide, the tube
bearing 5 hairy lines within; fornices wanting; nutlets gray-brown, 3 mm. long, wrinkled, pitted, and sometimes tuberculate, with a prominent ventral keel. N=14.
   Roadsides, fields, and other disturbed sites; native of Eurasia, now established as a weed
over most of the U.S.; wholly e. of the Cascades in our range. Apr.-June.

Lithospermum incisum Lehm. Asperif. 303. 1818.
   Batschia longiflora Pursh, Fl. Am. Sept. 132. 1814. Lithospermum longiflorum Spreng.
      Syst. 1:544. 1825, not of Salisb. in 1796. Pentalophus longiflorus A. DC. Prodr. 10:86.
      1846. (Nuttall, banks of the Missouri)
   Lithospermum angustifolium Michx. Fl. Bor. Am. 1:130. 1803, not of Forsk in 1775. Cy-
      phorina angustifolia Nieuwl. Am. Midl. Nat. 3:194. 1914. (Ohio R.)
   Batschia decumbens Nutt. Gen. Pl. 1:114. 1818. Lithospermum mandanense Spreng. Syst.
      1:544. 1825. Pentalophus mandanensis A. DC. Prodr. 10:87. 1846. Cyphorina mandanen-
      sis Nieuwl. Am. Midl. Nat. 4:515. 1916. (Nuttall, Ft. Mandan on the Missouri) Not Lithos-
      permum decumbens Vent. 1800.
   Lithospermum linearifolium Goldie, Edinb. New Phil. Journ. 6:322. 1822. (Head of Lake On-
      tario)
   Strigose perennial from a stout, woody taproot, 0.5-3 dm. tall; leaves wholly cauline, the
lowermost reduced and chaffy or rarely well developed and oblanceolate, the others linear-
oblong to narrowly lanceolate or linear, rather numerous, 2-6 cm. long and 2-6 mm. wide;
well-developed flowers crowded in the uppermost axils, short-pedicellate; corolla bright yellow, salverform, the tube (12)15-30 mm. long, with more or less evident fornices, the limb
1-1.5(2) cm. wide, with evidently erose, sometimes almost fimbriate lobes, these flowers
long-styled and seldom producing much fruit; cleistogamous, highly fertile, short-styled flowers commonly developed later in the season farther down on the stem, the plant becoming
slenderly much branched; nutlets gray, shining, sparsely pitted, 3-3.5 mm. long, with prominent ventral keel, the scar sunken and bearing a nearly central projection that is attached by
a ridge to the dorsal part of the prominent collar. N=12.
   Dry, open plains and foothills; plains of c. U.S. and adj. Can. and Mex., extending westward to Utah, the intermontane valleys of Mont., and s. B. C. May-July.
   Lithospermum caroliniense (Walt.) Macm. (L. gmelini (Michx.) Hitchc.), a species chiefly e.
and c. U.S., is known from Mont., and may eventually be found to occur at the e. edge of our
range. It differs from L. incisum in its more orange-yellow flowers with shorter (7-14 mm.)
tube and entire corolla lobes, in its smooth nutlets, and often also in being more robust and
broader-leaved. L. caroliniense is heterostylic, with some associated differences between
plants of different style length, but lacks cleistogamous flowers.

Lithospermum ruderale Dougl. ex Lehm. Stirp. Pug. 2:28. 1830. (Douglas, banks of the Co-
   lumbia and Multnomah rivers)
   L. pilosum Nutt. Journ. Acad. Phila. 7:43. 1834. Batschia pilosa G. Don, Gen. Syst. 4:372.
      1837. (Wyeth, Flathead R., Mont.)
   Lithospermum decumbens Torr. Ann. Lyc. N.Y. 2:225. 1827. L. torreyi Nutt. Journ. Acad.

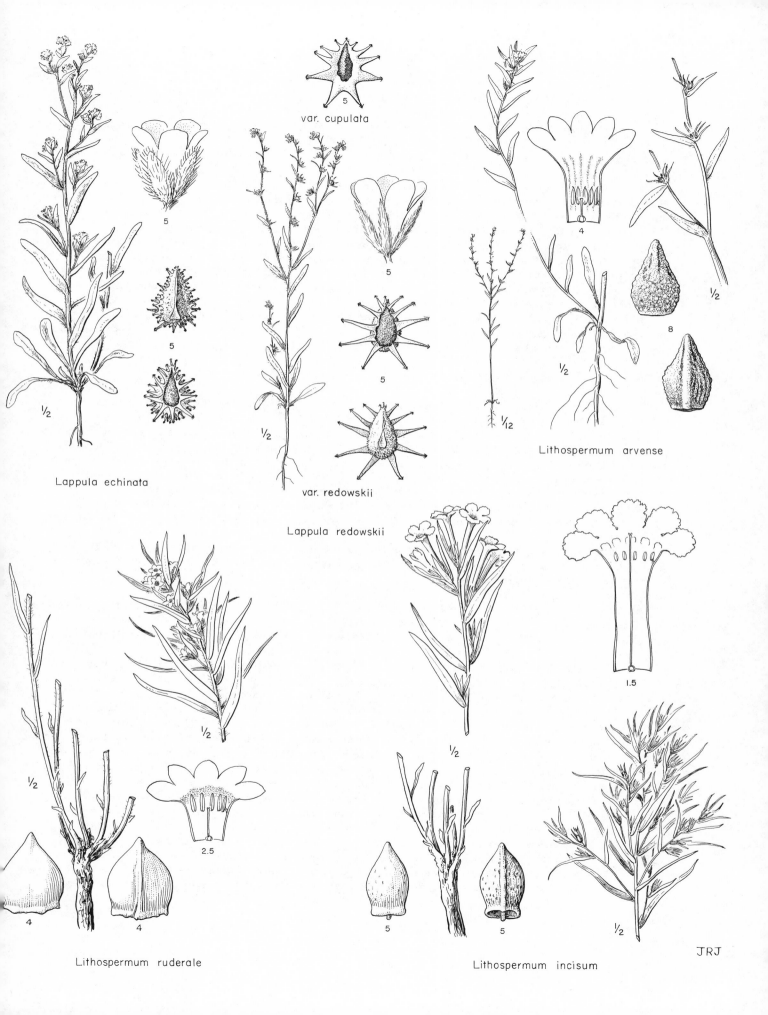

var. cupulata

Lithospermum arvense

Lappula echinata

var. redowskii

Lappula redowskii

Lithospermum ruderale

Lithospermum incisum

JRJ

Phila. 7:44. 1834. Batschia torreyi G. Don, Gen. Syst. 4:372. 1837. Lithospermum rude-
rale var. torreyi Macbr. Contr. Gray Herb. n. s. 48:55. 1916. (James 294, Rocky Mts.)
L. lanceolatum Rydb. Mem. N.Y. Bot. Gard. 1:333. 1900. L. ruderale var. lanceolatum A.
Nels. Bot. Gaz. 52:272. 1911. (Heller 3092, Lewiston, Ida.)
L. ruderale var. macrospermum Macbr. Contr. Gray Herb. n. s. 48:55. 1916. (Macbride
110, Big Willow, Canyon Co., Ida.)

Perennial, strigose to partly spreading-hirsute (especially on the stem), 2-6 dm. tall, the
stems clustered on the coarse, woody taproot; leaves wholly cauline, the lowermost ones reduced, the others numerous, sessile, lanceolate to linear, 3-10 cm. long, 2-10 mm. wide;
flowers in small, leafy-bracteate clusters in the upper axils, or more or less axillary to the
leafy bracts of the short, slender, upper branches; corolla light yellowish, often tinged with
green, 8-13 mm. long, the tube 4-6 mm. long, the limb 7-13 mm. wide, the lobes entire or
nearly so; corolla tube glandular within near the tip, but without definite fornices; style short;
nutlets smooth, shining, gray, 3.5-6(8) mm. long, with a more or less evident ventral keel,
often only 1 or 2 developed. Puccoon.

Open, fairly dry places from the foothills and adjacent lowlands to moderate elevations in
the mountains; s. B.C. and Alta. to n. Calif. and s. Colo.; in our range chiefly e. of the Cascades, but occasionally found in drier places in the Puget Sound region. Apr.-June.

## Mertensia Roth Nom. Conserv. Bluebells

Flowers in modified, bractless, mostly small cymes terminating the stem and branches;
calyx generally (except M. campanulata) cleft at least to the middle, often to the base; corolla
blue, or in occasional individuals white or pink, shallowly 5-lobed, generally abruptly expanded at the throat and thus evidently divided into a tube and limb (M. bella excepted), the fornices usually evident; filaments attached at or below the level of the fornices, often conspicuously expanded; nutlets attached laterally to the gynobase at or below the middle, generally
rugose; glabrous to strigose or hirsute perennial herbs, the hairs not pungent.

About 35 or 40 species, native to extratropical Eurasia and N. Am. Williams recognized 24
species in N. Am., most of them in w. U.S. (Named for F. C. Mertens, 1764-1831, German
botanist).

Nearly all of the species are useful for the native garden or perennial border, but the ones
found e. of the Cascades are mostly suitable only for that region. M. bella is a particularly
attractive species, as yet not cultivated.

References:
Johnston, I. M. Studies in the Boraginaceae, IX. 2. Notes on various borages of the western
United States. Contr. Arn. Arb. 3:83-98. 1932.
Macbride, J. F. The true Mertensias of western North America. Contr. Gray Herb. n. s.
48:1-20. 1916.
Williams, L. A monograph of the genus Mertensia in North America. Ann. Mo. Bot. Gard.
24:17-159. 1937.

Previous authors have differed in their evaluation of the taxonomic significance of the presence or absence of a ring of hairs in the corolla tube of the low species of Mertensia. Macbride considered it of subsectional importance, whereas Johnston and (later) Williams accorded it little significance. An examination of several hundred specimens of the group convinces
me that the character should not be disregarded. It is true that in w. Mont. and c. Ida. plants
which differ in this feature may be otherwise very much alike. It is also true that even in that
area the plants with the ring of hairs tend to occur at higher altitudes and to have the corolla
limb proportionately longer than do those without the ring of hairs, thus resembling many of
the plants of the s. Rocky Mts. which Williams has referred to M. viridis or M. bakeri. I
can, in fact, find no character to separate the n. Rocky Mt. plants with hairy tube that have
been referred by Williams to varieties of M. oblongifolia from others in the s. Rocky Mts.
that he refers to varieties of M. viridis and M. bakeri. If one refers these higher-altitude,
hairy-tubed plants of the n. Rockies to M. viridis (of which M. bakeri is here considered a
synonym), and grants that the cleavage between M. viridis and M. oblongifolia is blurred in

that area, some order emerges from the confusion, and the presence or absence of a ring of hairs in the corolla tube is revealed as a specific character in the low Mertensias.

1 Corolla campanulate, flaring from near the base, not sharply divided into a tube and limb, only 1 cm. long or less; leaves with evident lateral veins; filaments slender; plants of the Klamath region, extending n. to the Cascades of Lane Co., Oreg., and also in Idaho and Clearwater cos., Ida.                                        M. BELLA

1 Corolla either sharply divided into a tube and limb, or well over 1 cm. long, or often both; leaves and filaments diverse

  2 Plants relatively tall and robust (4-15 dm. tall when well developed), with evident lateral veins in the cauline leaves (except commonly in M. campanulata, marked by its short calyx lobes); plants blooming in late spring and summer, growing typically along streams and in wet meadows

    3 Calyx lobes distinctly shorter than the well-developed tube (unique among our species in this regard); c. Ida.                                        M. CAMPANULATA

    3 Calyx lobes distinctly longer than the mostly very short tube

      4 Anthers 1.3-2.2 mm. long (dry); corolla limb mostly 0.8-1.2(1.5) times as long as the tube; principal cauline leaves mostly sessile or subsessile (except the lower) and tending to taper to the base; Rocky Mt., Great Basin, and Sierra species, extending n. to Grant Co., Oreg., Boise and Lemhi cos., Ida., and Ravalli and Meagher cos., Mont.                                        M. CILIATA

      4 Anthers 2.2-5.0 mm. long; corolla limb mostly (1.0)1.2-1.6 times as long as the tube; principal cauline leaves mostly rounded toward the base and more or less petiolate

        5 Anthers 2.2-3.4 mm. long; boreal species, extending s. to Mineral Co., Mont., Idaho, Elmore, and Adams cos., Ida., and c. and s.w. Oreg., not extending w. of the Cascades in our region except in the Olympic Mts. of Wash.
                                                    M. PANICULATA

        5 Anthers 3.4-5.0 mm. long; species of the Willamette Valley and s. Puget Sound region, extending westward, at low altitudes, to the coast        M. PLATYPHYLLA

  2 Plants smaller, seldom as much as 4 dm. tall, usually without evident lateral veins in the cauline leaves (some forms here treated under M. oblongifolia are up to 6 dm. tall, with somewhat veiny leaves), blooming as soon as snow and temperature permit, growing typically on open or lightly shaded slopes or ridges, less often in meadows

    6 Filaments short, scarcely if at all over 1 mm. long, the base of the anthers not elevated beyond the fornices at the corolla throat; corolla 7-14 mm. long, the limb not much if at all shorter than the tube and tending to be widely flared; alpine plants of Mont., c. Ida., and southward

      7 Anthers included in the corolla tube, their tips barely reaching the fornices; corolla tube glabrous within                                        M. ALPINA

      7 Anthers exserted from the corolla tube, their bases about at the level of the fornices; corolla tube with a ring of hairs inside below the middle, or sometimes with the hairs scattered over much of the inner surface                M. PERPLEXA

    6 Filaments longer and more conspicuous, broad and flattened, 1.5-3 mm. long; base of the anthers elevated well above the fornices; corolla diverse; plants not alpine in our area, except sometimes M. viridis

      8 Corolla tube bearing a ring of hairs inside below the middle (or sometimes the hairs scattered over much of the inner surface), seldom much longer than the limb; plants mostly of moderate to high elevations in the mountains, from Mont., c. Ida., and c. Oreg. southward                                        M. VIRIDIS

      8 Corolla tube glabrous inside, obviously longer than the limb; plants of the valleys and foothills to moderate elevations in the mountains, occurring nearly throughout that portion of our area that lies e. of the Cascade summits

        9 Stems 1-2(5) from a short, tuberous-thickened, easily detached, shallow root; basal leaves rarely developed on flowering plants; cauline leaves mostly 1.5-4 times as long as wide; corolla tube commonly (1.7)2-3 times as long as the limb
                                       M. LONGIFLORA

9   Stems clustered on a stouter and more firmly attached, deeper-seated and scarcely tuber-
      ous root; basal leaves mostly well developed; cauline leaves mostly 2.5-7 times as long
      as wide; corolla tube mostly 1.3-2 times as long as the limb        M. OBLONGIFOLIA

Mertensia alpina (Torr.) G. Don, Gen. Syst. 4:372. 1837.
   Pulmonaria alpina Torr. Ann. Lyc. N.Y. 2:224. 1827. Cerinthodes alpinum Kuntze, Rev.
      Gen. 2:436. 1891. (James, Rocky Mts.)
   Mertensia tweedyi Rydb. Mem. N.Y. Bot. Gard. 1:336. 1900. (Rydberg & Bessey 4867, Old
      Hollowtop, Pony Mts., Mont.)
   Perennial from a branching caudex and one or several strong roots, the several or numerous, subprostrate to loosely erect, glabrous stems 0.5-2.5 dm. long; leaves strigose on the
upper surface, glabrous on the lower, the basal ones with elliptic or lanceolate blade mostly
1.5-5 cm. long and 7-15 mm. wide, on a well-developed petiole; cauline leaves well developed
but soon becoming sessile, often narrower than the basal ones, gradually reduced upward,
mostly longer than the internodes, the lateral veins obscure; calyx 2-3 mm. long, cleft essentially to the base; corolla 7-11 mm. long, the tube slender, about 2 mm. wide, the limb nearly or quite as long as the tube and abruptly expanded, 5-11 mm. wide; filaments inconspicuous,
shorter than the anthers, these 0.9-1.3 mm. long, with the tips barely or scarcely reaching
the level of the fornices; style surpassed by or barely surpassing the anthers.
   Open slopes and drier meadows at high altitudes in the mountains, often above timber line;
s.w. Mont. and adj. Ida. (Fremont Co.) to Colo. and n. N.M. July-Aug.
   A natural hybrid with M. ciliata has been collected in Carbon Co., Mont.

Mertensia bella Piper, Proc. Biol. Soc. Wash. 31:76. 1918. (Peck 5811, Horse Pasture Mt.,
   10 miles s.w. of McKenzie Bridge, Lane Co., Oreg.)
   Stems solitary from a shallow, ellipsoid-globose corm or cormose root 0.5-2 cm. thick,
slender and weak, 1-7 dm. tall, openly branched above in larger plants; leaves thin, evidently
veined, glabrous or strigulose, not very numerous, the lowermost one or more reduced to a
scarious sheath, the others with mostly ovate, acute, basally rounded blade 3-8 cm. long and
1-5 cm. wide, borne on progressively shorter petioles, or the upper sessile and often opposite; inflorescence mostly rather open; calyx 2-3 mm. long, strigulose, cleft to near the base;
corolla 6-10 mm. long, campanulate, flaring from near the base, not sharply divided into a
tube and limb, the fornices scarcely developed, the rounded lobes 2-3 mm. long; filaments
slender, attached about 1 mm. from the base of the corolla, a little longer than the 1 mm. anthers; style surpassing the anthers, but not exserted.
   Wet meadows, moist slopes, and springy banks at middle altitudes in the mountains; Josephine and Jackson cos., Oreg., to Lane Co. (2 stations), and in Clearwater and Idaho cos., Ida.
(1 collection each). May-July.
   A very distinctive species, apparently without close allies.

Mertensia campanulata A. Nels. Bot. Gaz. 54:150. 1912. (Woods 328, cited by Nelson as 325,
   Hailey, Blaine Co., Ida.)
   Perennial from a branching caudex, forming large clumps, glabrous throughout and somewhat glaucous; stems numerous, 3-10 dm. tall; basal leaves with the blade evidently veined,
lanceolate to narrowly or rather broadly elliptic or ovate-elliptic, 8-25 cm. long and 1.5-9
cm. wide, tapering to the well-developed petiole; cauline leaves rather numerous and narrow,
not over about 2.5 cm. wide, acute, thick and obscurely veined, more or less strongly reduced upward, only the lower petiolate; inflorescence tending to be elongate and often narrow; calyx tubular-campanulate, 5-7 mm. long, the rather blunt, deltoid lobes only 1-2.5 mm. long;
corolla 1.5-2 cm. long, glabrous within, the limb a little shorter than the tube and less sharply differentiated from it than in most species of the genus, the fornices poorly developed; anthers 2.9-3.4 mm. long, equaling or longer than the fairly broad filaments; style exserted
from the corolla.
   Meadows and open slopes in the valleys of the mountains of c. Ida. (Blaine, Camas, Custer,
and Elmore cos.); June-July.
   A sharply defined, local species, not to be confused with anything else.

Mertensia ciliata (Torr.) G. Don, Gen. Syst. 4:372. 1837.
  Pulmonaria ciliata Torr. Ann. Lyc. N.Y. 2:224. 1827. (James, Rocky Mts. of Colo.)
Mertensia subpubescens Rydb. Bull. Torrey Club 30:261. 1903. M. ciliata var. subpubescens
  Macbr. & Pays. Contr. Gray Herb. n. s. 49:67. 1917. (Rydberg & Bessey 4876, Spanish
  Basin, Gallatin Co., Mont.) The form with the leaves hairy beneath.
M. pallida Rydb. Bull. Torrey Club 36:680. 1909. (Rydberg 2777, Lima, Mont.) A white-
  flowered form.
M. incongruens Macbr. & Pays. Contr. Gray Herb. n. s. 49:66. 1917. (Macbride & Payson
  3759, Smoky Mts., Blaine Co., Ida.)
M. ciliata var. subpubescens f. candida Macbr. & Pays. Contr. Gray Herb. n. s. 49:67. 1917.
  (Macbride & Payson 3272, Bear Creek, below Parker Mt., Ida.) A white-flowered form.
  Stems numerous from a branched, woody caudex, 1.5-15 dm. tall; herbage glabrous, or the
leaves often strigose, especially beneath; leaves more or less evidently veined, the basal ones,
when present, with the blade elliptic to ovate or lance-ovate, generally not at all cordate, long-
petiolate; cauline leaves well developed and only gradually reduced upward, the blade narrowly
elliptic or lance-elliptic to rather narrowly ovate, 3-15 cm. long, 1-5 cm. wide, generally tapering to the base, or the lower sometimes more rounded, only the lower evidently petiolate;
inflorescence branched and open in well-developed plants; calyx 1-3 mm. long, cleft nearly or
quite to the base, the lobes with broadly rounded to acutish tip; corolla 10-17 mm. long, the limb
mostly 0.8-1.2(1.5) times as long as the tube, the tube with, or more often without, a ring of hairs
below the middle within; filaments attached at or shortly below the level of the fornices, broad
and conspicuous, 1.5-3 mm. long; anthers 1.2-2.2 mm. long, typically a little under 2 mm.;
styles elongate, often shortly exserted from the corolla. N=12, 24.
  Stream banks, wet meadows, damp thickets, and wet cliffs, from the foothills to high elevations in the mountains; Colo. and n. N.M. to Calif., n. to Meagher and Ravalli cos., Mont.,
Lemhi and Boise cos., Ida., and c. Oreg. (Grant Co.). June-Aug.
  Our plants as described above belong to the var. ciliata. In the Sierra Nevada the var. ciliata is replaced by var. stomatechoides (Kell.) Jeps., with longer calyx and more conspicuously
exserted style.

Mertensia longiflora Greene, Pitt. 3:261. 1898. (Sandberg & Leiberg, Medical Lake, Wash.,
  May, 1893)
M. pulchella Piper, Contr. U.S. Nat. Herb. 11:478. 1906. M. longiflora var. pulchella
  Macbr. Contr. Gray Herb. n. s. 48:17. 1916. (Sandberg et al. 75, lower Clearwater R., Ida.)
M. pulchella ssp. glauca Piper, Contr. U.S. Nat. Herb. 11:479. 1906. (Whited 1010, w. of
  Wenatchee, Wash.)
M. horneri Piper, Contr. U.S. Nat. Herb. 11:479. 1906. M. longiflora var. horneri Macbr.
  Contr. Gray Herb. n. s. 48:17. 1916. (Horner 366, Waitsburg, Wash.)
  Stems 1-2(5) from a shallow, tuberous-thickened, easily detached root, 0.5-2.5 dm. tall;
herbage glabrous or strigose; basal leaves rarely developed in flowering plants, petiolate and
with broadly elliptic, rounded blade when present; cauline leaves rather few, the lowermost 1
or several strongly reduced, sometimes to a mere petiolar sheath, the others well developed
and commonly surpassing the internodes, mostly sessile or nearly so, 2-6 cm. long and 0.5-3
cm. wide, mostly 1.5-4 times as long as wide, obtuse or rounded at the tip; inflorescence congested, often subcapitate; calyx 3-6 mm. long, cleft to the middle or beyond; corolla (1)1.5-2.5
cm. long, the tube glabrous within, (1.7)2-3 times as long as the not much expanded limb; filaments attached at or just below the fornices, broad and conspicuous, 1.5-3 mm. long, surpassing the 1.2-2.2 mm. anthers; style barely included or shortly exserted from the corolla.
  Open or lightly shaded places in the plains and foothills, often with sagebrush or with ponderosa pine, seldom ascending as high as 5000 ft.; s. B.C. to c. Oreg., e. of the Cascades,
and rarely to n. e. Calif., e. to n. w. Mont. (w. of the continental divide) and to Boise, Ida.
Apr.-June.
  Mertensia longiflora appears to be sharply distinct from its only near relative, M. oblongifolia.

Mertensia oblongifolia (Nutt.) G. Don, Gen. Syst. 4:372. 1837.
  Pulmonaria oblongifolia Nutt. Journ. Acad. Phila. 7:43. 1834. Cerinthodes oblongifolium

Kuntze, Rev. Gen. 2:436. 1891. (Wyeth, "towards the sources of the Columbia," perhaps along the Flathead R. in Mont.)

?Mertensia umbratilis Greenm. Erythea 7:118. 1899. M. arizonica var. umbratilis Macbr. Contr. Gray Herb. n. s. 48:9. 1916. (Cusick 1886, Sparta, near the Snake R. in Oreg.) See comment following description of M. oblongifolia.

M. tubiflora Rydb. Bull. Torrey Club 26:544. 1899. (Tweedy 119, headwaters of the Tongue R., Big Horn Mts., Wyo.)

?M. intermedia Rydb. Mem. N. Y. Bot. Gard. 1:335. 1900. (Rydberg & Bessey 4873, Bridger Range, Mont.) See comment under M. umbratilis following description of M. oblongifolia.

M. stenoloba Greene, Pl. Baker. 3:20. 1901. (Flodman 752, Bridger Range, Mont.)

M. nutans Howell, Fl. N. W. Am. 491. 1901. (Howell, near Goldendale, Wash.)

M. nevadensis A. Nels. Proc. Biol. Soc. Wash. 17:96. 1904. M. foliosa var. nevadensis Macbr. Contr. Gray Herb. n. s. 48:19. 1916. M. oblongifolia var. nevadensis Williams, Ann. Mo. Bot. Gard. 24:125. 1937. (Kennedy & True 711, near Reno, Nev.)

?M. ambigua Piper, Contr. U.S. Nat. Herb. 11:477. 1906. (Vasey, Cascade Range of c. Wash. in 1879) See comment under M. umbratilis following description of M. oblongifolia.

?M. infirma Piper, Contr. U.S. Nat. Herb. 11:476. 1906. (Whited 307, Ellensburg, Wash.) See comment under M. umbratilis following description of M. oblongifolia.

M. pubescens Piper, Contr. U.S. Nat. Herb. 11:479. 1906, not of Willd. M. foliosa var. pubescens Macbr. Contr. Gray Herb. n. s. 48:19. 1916. (Whited 1214, near Waterville, Douglas Co., Wash.)

M. nutans ssp. subcalva Piper, Contr. U.S. Nat. Herb. 11:479. 1906. M. foliosa var. subcalva Macbr. Contr. Gray Herb. n. s. 48:18. 1916. (Cotton 328, Rattlesnake Mts., Wash.)

M. eplicata Macbr. Contr. Gray Herb. n. s. 48:16. 1916. (Macbride 856, Dry Buck, Boise Co., Ida.)

M. foliosa var. nimbata Macbr. Contr. Gray Herb. n. s. 53:18. 1918. (Gottschalck, Bozeman, Mont., May 18, 1893)

M. oblongifolia var. amoena sensu Williams, Ann. Mo. Bot. Gard. 24:130. 1937, in large part, but not as to type, which is M. viridis.

Perennial with several or numerous more or less erect stems 1-4 dm. tall arising from a fairly stout root which is often surmounted by a branching caudex; herbage glabrous to strigose; basal leaves with elliptic to oblanceolate or lance-ovate blade 2-15 cm. long and 0.7-6 cm. wide borne on a well-developed petiole; cauline leaves well developed, commonly longer than the internodes and often rather numerous, acute or obtuse, mostly 2.5-7 times as long as wide, only the lower petiolate; calyx 2.5-6 mm. long, cleft to near the base; corolla 1-2 cm. long, the tube glabrous within, mostly 1.3-2 times as long as the only slightly expanded limb; filaments broad and conspicuous, 1.5-4 mm. long, attached at or just below the level of the fornices, commonly longer than the 1.2-2.2 mm. anthers; style barely included or shortly exserted from the corolla.

Open slopes and drier meadows, often among sagebrush, from the plains and foothills to moderate elevations in the mountains, ascending to alpine stations in Nev.; e. side of the Cascades in Wash. to the edge of the plains e. of the mountains in Mont., s. to c. Nev., n. Utah (Wasatch and Bear River ranges), and n. Wyo. Apr.-May, and to July at higher elevations.

Status of M. umbratilis Greenm.: There appears to have been some hybridization and subsequent introgression between M. paniculata and M. oblongifolia in the area from c. Wash. to n. e. Oreg., and a considerable number of specimens from that region present one or another combination of the characters of these very different species. The most common form among these intermediates occurs in the contact zone of the habitats of the presumed parents or ancestors, and is almost exactly intermediate in habit, being 3-6 dm. tall, with somewhat veiny leaves; the flowers, however, are almost exactly those of M. oblongifolia. The name M. umbratilis Greenm. applies to such plants, as do also M. ambigua Piper and M. infirma Piper. "M. umbratilis" intergrades more freely with M. oblongifolia than with M. paniculata, suggesting that it might perhaps properly be attached to the former as a variety; however, since it remains to be demonstrated that M. umbratilis has a real existence as a self-perpetuating natural population (as contrasted to a series of continually re-created hybrids and hybrid progeny), it seems unwise to disturb the nomenclature at this time. Presumed hybrids between M. oblongfolia and one or another of the tall species are certainly not restricted to the M. umbratilis area, and the type of M. intermedia, from Mont. , has the appearance of a cross between M. oblongifolia and M. ciliata. The combination of characters presented by M. umbratilis is such that the plants might well key to the chiefly more southern M. ciliata, except for their relatively longer corolla tube, but M. ciliata does not seem to be involved in their origin.

Mertensia paniculata (Ait. ) G. Don, Gen. Syst. 4:318. 1837.
   Pulmonaria paniculata Ait. Hort. Kew. 1:181. 1789. Lithospermum paniculatum Lehm. Asperif. 2:289. 1818. Casselia paniculata Dum. Com. Bot. 22. 1822. Platynema paniculata Schrad. Ind. Sem. Hort. Goett. 1835. Cerinthodes paniculatum Kuntze, Rev. Gen. 2:436. 1891. (Garden specimens, coming eventually from Hudson Bay, Can. )
   Mertensia membranacea Rydb. Bull. Torrey Club 28:33. 1901. (MacDougal 3, Priest River Valley, Ida. )
   M. laevigata Piper, Contr. U.S. Nat. Herb. 11:477. 1906. M. paniculata var. laevigata G. N. Jones, U. Wash. Pub. Biol. 5:219. 1936, (Piper 2116, Mt. Rainier, Wash. )
   M. brachycalyx Piper, Contr. U.S. Nat. Herb. 11:477. 1906. M. laevigata var. brachycalyx Macbr. Contr. Gray Herb. n. s. 48:10. 1916. (Sandberg & Leiberg 678, Nason Creek, Chelan Co. , Wash. )
   M. leptophylla Piper, Contr. U.S. Nat. Herb. 11:478. 1906. M. paniculata var. subcordata f. leptophylla Macbr. Contr. Gray Herb. n. s. 48:7. 1916. (Elmer 2826, Olympic Mts. , Clallam Co. , Wash. )
   M. pratensis var. borealis Macbr. Contr. Gray Herb. n. s. 48:8. 1916. M. paniculata var. borealis Williams, Ann. Mo. Bot. Gard. 24:49. 1937. (Leiberg 1217, Clearwater-St. Joe divide, Ida. )
Stems numerous (or sometimes apparently solitary) from a mostly multicipital caudex or stout rhizome, 2-15 dm. tall, glabrous or hairy; leaves evidently veined, often very thin, the basal ones, when present, long-petiolate and more or less cordate; cauline leaves numerous, well developed, mostly short-petiolate, with lanceolate to broadly ovate, apically acute or acuminate blade 3-14 cm. long and 1.5-7 cm. wide, mostly rounded at the base, the upper only gradually reduced and eventually subsessile; inflorescence generally branched and open; calyx 2-6 mm. long, cleft to below the middle or to the base; corolla 9-16 mm. long, the limb (1.0)1.2-1.6 times as long as the tube, the tube with, or more often without, a ring of hairs below the middle within; filaments broad and conspicuous, attached at the level of the fornices, 1.5-3.5 mm. long; anthers 2.2-3.4 mm. long; style elongate, often shortly exserted.

   Stream banks, wet meadows, damp thickets, and wet cliffs, from the foothills to high elevations in the mountains; Alas. to Que. , s. to Ont. , Iowa, Mont. (Mineral Co. ), Ida. (Elmore and Adams cos. ), and c. and s. w. Oreg. ; not occurring w. of the Cascades in our region except in the Olympic Mts. May-Aug.

   Over most of its range, M. paniculata has the leaves strigose above and hirsute or coarsely strigose beneath. A less hairy phase, with the leaves glabrous at least above, and sometimes even glaucous, occurs from n. w. Mont. and adj. B. C. to Oreg. and Wash. , and wholly replaces the typical phase in Oreg. and all but the n. e. corner of Wash. These less hairy plants may be distinguished as var. borealis (Macbr. ) Williams. A third variety, with narrow leaves, occurs to the n. of our range.

Mertensia perplexa Rydb. Bull. Torrey Club 31:639. 1904.
   M. alpina var. perplexa Macbr. Contr. Gray Herb. n. s. 48:20. 1916. (Osterhout 2439, mountains s. of Ward, Boulder Co. , Colo. )
   M. viridis sensu Williams, Ann. Mo. Bot. Gard. 24:110. 1937, in part; sensu Weber, Handb. Colo. Fr. Range, 156. 1953, not of A. Nels. Bull. Torrey Club 26:244. 1899.
   Resembling M. alpina (compare illustrations), averaging a little more robust and broader-leaved; corolla 9-14 mm. long, the tube bearing a ring of hairs near the base within, the limb often less flared and sometimes a little shorter than in M. alpina; differing sharply from M. alpina in the position of the anthers, these 1-2 mm. long and so placed that the base is at the level of the fornices.

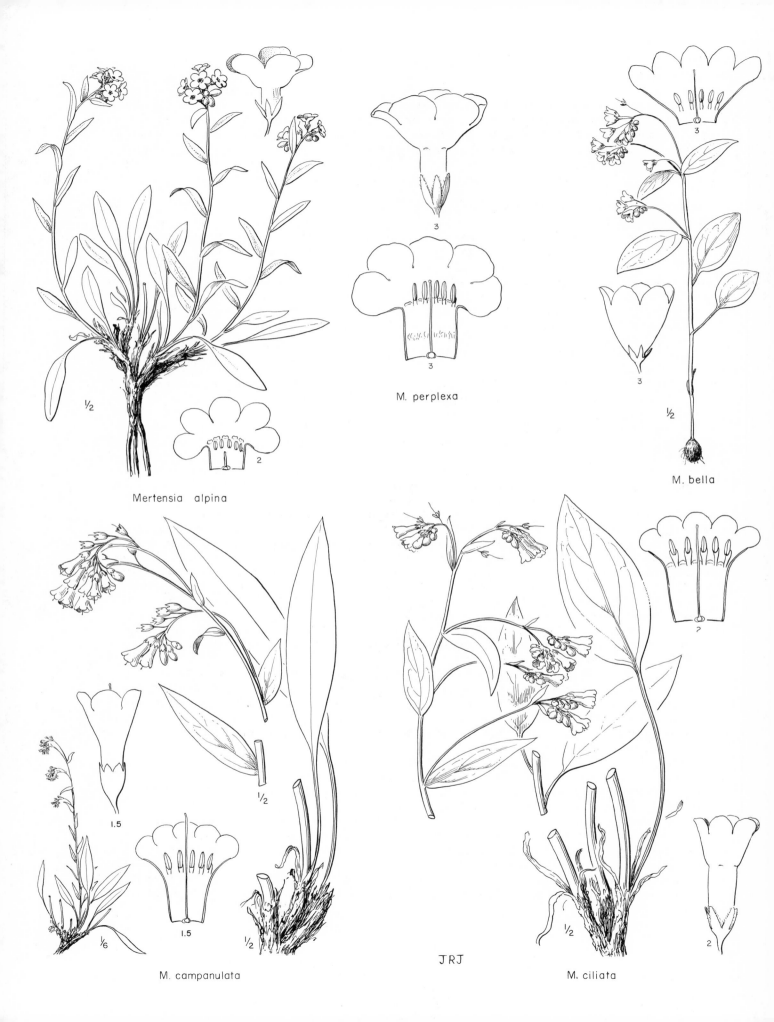

M. perplexa

Mertensia alpina

M. bella

M. campanulata

JRJ

M. ciliata

Open, often rocky slopes and summits at high altitudes in the mountains; s. w. Mont. (Park, Deerlodge, Beaverhead cos.) to Colo.; seldom collected in our range. July-Aug.

Mertensia perplexa, as here defined, is a technically fairly well-marked species which seems to diverge from the obviously related M. alpina in the direction of M. lanceolata Pursh. See further discussion under M. viridis.

Mertensia platyphylla Heller, Bull. Torrey Club 26:548. 1899.

M. paniculata var. platyphylla G. N. Jones, U. Wash. Pub. Biol. 5:220. 1936. (Heller 3872, near Montesano, Wash.)

M. subcordata Greene, Pitt. 4:89. 1899. M. paniculata var. subcordata Macbr. Contr. Gray Herb. n. s. 48:7. 1916. M. platyphylla var. subcordata Williams, Ann. Mo. Bot. Gard. 24:60. 1937. (Howell s. n., Roseburg, Oreg., May 3, 1897)

Perennial with the stems mostly arising singly from a stout, branching rhizome; otherwise similar to M. paniculata (though more consistently broad-leaved) except for the longer (3.4-5.0 mm.) anthers; leaves scabrous above, coarsely and rather sparsely strigose or hirsute-strigose to glabrous beneath.

Stream banks and moist, low woods at lower elevations from the w. base of the Cascades to the coast, from the s. Puget Sound region to Douglas and Curry cos., Oreg. May-July.

Perhaps M. platyphylla is better considered a variety of M. paniculata. Plants of M. platyphylla from Oreg. tend to have the calyx shorter (2.5-4 mm.) than those from Wash. (4.5-7 mm.), and Williams has therefore distinguished the Oregon plants as a separate variety (subcordata). Unfortunately, the type of var. subcordata, from Roseburg, Oreg., has the calyx elongate as in the Washington specimens. Until further evidence is forthcoming in support of the segregation, it seems better to leave the species undivided.

Mertensia viridis A. Nels. Bull. Torrey Club. 26:244. May, 1899.

M. lanceolata var. viridis A. Nels. Bull. Wyo. Exp. Sta. 28:158. 1896. (Nelson 1608, Laramie Peak, Wyo.)

M. paniculata var. nivalis Wats. Bot. King Exp. 239. 1871. M. nivalis Rydb. Mem. N. Y. Bot. Gard. 1:336. 1900. (Watson 844, Bear River Canyon, Uintah Mts., Utah)

M. foliosa A. Nels. Bull. Torrey Club 26:243. May, 1899. (Nelson 2951, Evanston, Wyo.)

M. bakeri Greene, Pitt. 4:90. Dec. 1899. (Baker, Earle, & Tracy 576, Hayden Peak, Colo.)

M. amoena A. Nels. Bot. Gaz. 30:195. 1900. M. bakeri var. amoena Nels. in Coult. & Nels. New Man. Bot. Rocky Mts. 422. 1909. M. foliosa var. amoena Johnst. Contr. Arn. Arb. 3:85. 1932. M. oblongifolia var. amoena Williams, Ann. Mo. Bot. Gard. 24:130. 1937. (Nelson 5413, Monida, Mont.)

M. cusickii Piper, Bull. Torrey Club 29:643. 1902. M. foliosa var. amoena f. cusickii Johnst. Contr. Arn. Arb. 3:85. 1932. (Cusick 2582, cited by Piper as 2532, Steens Mts., Oreg.)

M. bakeri var. subglabra Macbr. Contr. Gray Herb. n. s. 49:66. 1917. M. foliosa var. subcalva f. macbridei Johnst. Contr. Arn. Arb. 3:84. 1932. (Macbride & Payson 3544, Josephus Lakes, Custer Co., Ida.)

Perennial with few to many lax stems 0.5-4 dm. tall arising from a fairly stout root which may be surmounted by a branching caudex; herbage glabrous to strigose; basal leaves with oblanceolate to more often elliptic or lance-ovate blade 2-10 cm. long and 0.7-3 cm. wide borne on a well-developed petiole; cauline leaves well developed but mostly sesssile, few or rather numerous, mostly longer than the internodes, narrow or broad, gradually or scarcely reduced upward, the lateral veins obscure; calyx 2.5-5 mm. long, cleft to well below the middle or commonly to the base; corolla 9-17 mm. long, the tube 2-3 mm. wide, about equaling or usually a little longer than the somewhat flared limb, and bearing a ring of hairs internally below the middle (or the hairs sometimes scattered over much of the inner surface of the tube); filaments attached at or just below the level of the fornices, conspicuous, as broad as, and often longer than, the anthers, mostly 1.5-3 mm. long; anthers 1.3-3 mm. long, the base elevated well above the level of the fornices; style surpassed by or surpassing the anthers.

Open, often rocky places at moderate to high elevations in the mountains, sometimes above

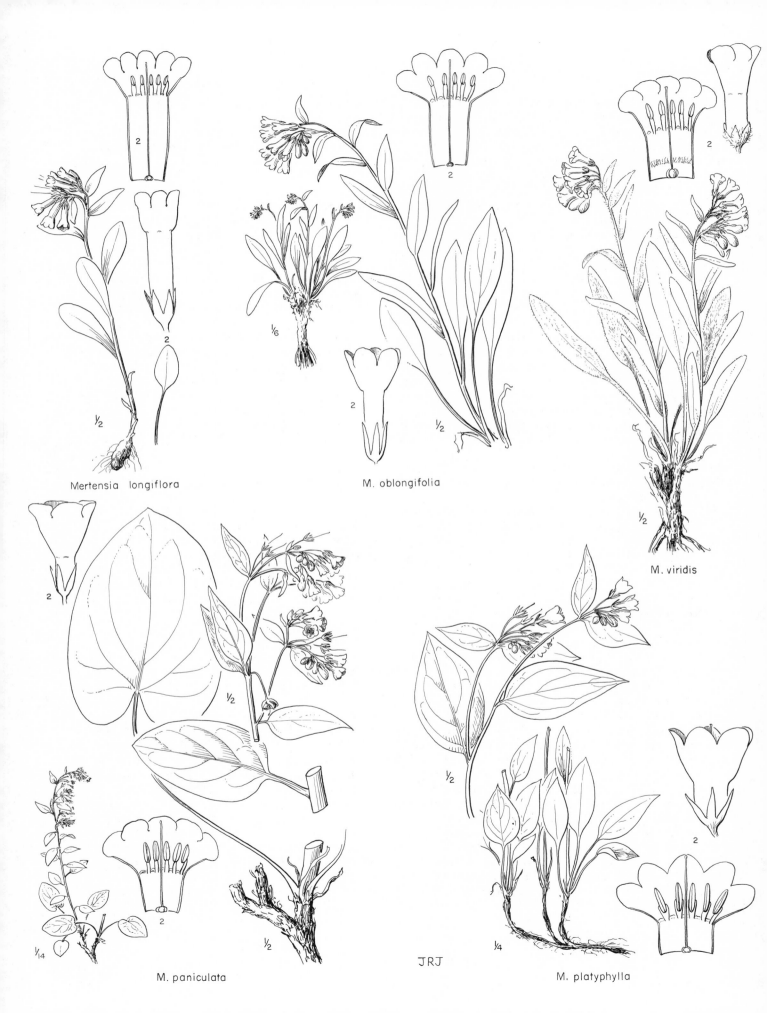

Mertensia longiflora

M. oblongifolia

M. viridis

M. paniculata

JRJ

M. platyphylla

timber line; Mont. and c. Ida. s. to Colo., Utah, and n. N. M.; also in c. and s. e. Oreg. and adj. Ida., Nev., and Calif. June-Aug.

Mertensia viridis is an ill-defined species. In Colo. it seems to be scarcely more than a subalpine ecotype of M. lanceolata Pursh, but in Mont. and c. Ida. it passes equally freely into M. oblongifolia. At higher altitudes it often closely resembles M. perplexa, but a fairly sharp distinction is furnished by the nature of the filaments and resultant position of the anthers. An isolated fragment of the population of M. viridis, occurring at relatively low elevations in c. and s. e. Oreg. and adj. Ida., Nev., and Calif., has been described as M. cusickii Piper. The specimens are rather uniform among themselves, but are scarcely to be distinguished morphologically from the larger and more hairy specimens of the principal population of the species, which itself may extend well down into the foothills as in s. w. Wyo.

## Myosotis L.

Flowers borne in terminal, naked, helicoid, eventually more or less elongate, sympodial false racemes, or the lower ones scattered among the leaves, the fruiting pedicels erect or spreading; calyx 5-lobed, with a distinct tube; corolla mostly blue, less often white, rarely yellow, salverform or broadly funnelform, with abruptly spreading, evidently 5-lobed limb, the fornices usually well developed; anthers mostly included; nutlets 4, attached by a small, basilateral scar to the broad, low gynobase, smooth and shining, with an evident raised margin all the way around; annual to perennial herbs, glabrous to hirsute, but not hispid or setose.

Perhaps 30 or 40 species, of wide distribution in temperate and boreal regions. (Name from the Greek mus, mouse, and ous, ear, from the appearance of the leaves of some species.)

Several of the species are attractive plants for moist areas, especially along streams, although they have seldom been cultivated.

1 Calyx closely strigose, the hairs neither spreading nor uncinate
  2 Corolla limb 2-5 mm. wide; style distinctly surpassed by the nutlets; stem often lax and
      decumbent at the base, but scarcely creeping, and not at all stoloniferous    M. LAXA
  2 Corolla limb 5-10 mm. wide; style equaling or usually surpassing the nutlets; stem often
      with creeping or stoloniferous base                  M. SCORPIOIDES
1 Calyx tube with some loose or spreading, uncinate hairs
  3 Corolla rather showy, the limb 4-8 mm. wide; plants perennial, montane or alpine, not
      at all weedy                               M. SYLVATICA
  3 Corolla not showy, the limb 1-4 mm. wide; plants annual or biennial, occurring from the
      lowlands to moderate elevations in the mountains, often weedy
    4 Calyx asymmetrical, subbilabiate, 3 of the lobes somewhat shorter than the other 2;
        corolla white; fruiting pedicels evidently to barely shorter than the calyx
                                        M. VERNA
    4 Calyx symmetrical, with essentially similar lobes; corolla commonly blue, sometimes
      white in M. arvensis, or (in M. discolor) yellow when young
      5 Fruiting pedicels equaling or generally surpassing the calyx    M. ARVENSIS
      5 Fruiting pedicels distinctly shorter than the calyx
        6 Plants floriferous nearly to the base, the lower flowers scattered among the
          leaves; style distinctly surpassed by the nutlets    M. MICRANTHA
        6 Plants floriferous to not much if at all below the middle, the inflorescence ordi-
          narily essentially naked; style somewhat surpassed by to more often equaling or
          distinctly surpassing the nutlets         M. DISCOLOR

Myosotis arvensis (L.) Hill, Veg. Syst. 7:55. 1764.
  M. scorpioides var. arvensis L. Sp. Pl. 131. 1753. (Europe)
Biennial or less commonly annual, tending to be fibrous-rooted, 1-4 dm. tall, simple or branched especially above, strigose to evidently hirsute-puberulent throughout; leaves mostly 1-6 cm. long and 3-16 mm. wide, the lower mostly oblanceolate, the upper often more oblong or lanceolate; racemes slightly, if at all, longer than the leafy portion of the plant; fruiting pedicels ascending or spreading, equaling or generally surpassing the 3-5 mm. fruiting calyx; calyx hirsute-puberulent and shortly uncinate-hispid; corolla blue or occasionally white, the

limb 2-4 mm. wide; nutlets brown or nearly black, distinctly surpassing the style. 2N=about 48, about 54.

Roadsides and other disturbed sites, in wet or dry soil, less often in woodlands; native of Europe, now established as a weed in n. e. U. S. and adj. Can., and occasionally found in our range. June-Aug.

See nomenclatural discussion under M. scorpioides.

Myosotis discolor Pers. in Murr. Syst. Veg. 15th ed. 190. 1797. (Europe)
    M. arvensis var. versicolor Pers. Syn. 1:156. 1805. M. lutea var. versicolor Macbr. Contr. Gray Herb. n. s. 49:19. 1917. (Europe)
    M. versicolor J. E. Smith, Engl. Bot. sub pl. 2558. 1814. (England) Published without reference to M. arvensis var. versicolor Pers.
    Slender, sparsely leafy annual or perhaps sometimes biennial, tending to be fibrous-rooted, 1-5 dm. tall, simple or moderately branched, evidently hirsute-puberulent to obscurely strigose; leaves mostly 1-4 cm. long and 2-8 mm. wide, the lowermost commonly oblanceolate, the others scattered, generally oblong or narrowly elliptic to nearly linear; racemes naked or with one or two leafy bracts near the base, not much, if at all, longer than the leafy portion of the plant; fruiting pedicels 1-3 mm. long, evidently shorter than the 3-5 mm. calyx, ascending or somewhat spreading; calyx-tube shortly uncinate-hispid, the lobes merely strigose or strigose-puberulent; corolla at first yellow or yellowish, ordinarily changing to blue, the limb mostly 1-2 mm. wide, not completely flat; nutlets generally dark brown or blackish, distinctly shorter than or merely equaling the style, or occasionally surpassing it.

Roadsides and moist ground; native of Europe, introduced in n. e. U. S. and adj. Can., and occasionally found w. of the Cascades in our range. May-Aug.

Myosotis laxa Lehm. Asperif. 1:83. 1818.
    M. caespitosa var. laxa DC. Prodr. 10:105. 1846. M. palustris var. laxa Gray, Man. 5th ed. 365. 1867. M. scorpioides ssp. laxa Hegi, Fl. Mittel-Eur. 5³:2165. 1927. (N. Am.)
    Plants fibrous-rooted, short-lived, sometimes annual, slender and weak, often curved and decumbent at the base, but scarcely creeping, and not at all stoloniferous, 1-4 dm. tall, inconspicuously strigose; leaves mostly 1.5-8 cm. long and 3-15 mm. wide, the lower mostly oblanceolate, the middle and upper more oblong or narrowly elliptic to lanceolate; inflorescence terminal, essentially naked, becoming loose and open, seldom appreciably longer than the leafy part of the stem; fruiting pedicels spreading, equaling or mostly longer than the 3-5 mm. calyx; calyx closely but not densely strigose, the lobes equal or unequal, shorter or a little longer than the tube; corolla blue, the limb 2-5 mm. wide; style distinctly surpassed by the brown to more often blackish nutlets. 2N=about 80.

In moist soil and shallow water; interruptedly circumboreal, but not at particularly high altitudes or latitudes; n. e. U. S. and adj. Can.; s. B. C. to n. w. Calif., from the e. base of the Cascades to the coast; Chile. June-Sept.

Myosotis laxa is in the main well set off from M. scorpioides, but there are a few doubtful specimens which may perhaps represent hybrids.

Myosotis micrantha Pall. ex Lehm. Neue Schrift. Ges. Nat. Halle 3(2):24. 1817. (Pallas, Wolgam, in s. Russia)
    M. stricta Link ex Roem. & Schult. Syst. Veg. 4:104. 1819. (Fields, presumably in Europe)
    Annual or winter annual up to 2 dm. tall, simple or more often branched from near the base, more or less hirsute-puberulent throughout; leaves seldom over 2 cm. long and 7 mm. wide, the lowermost commonly oblanceolate or broader, the others more oblong or elliptic; plants floriferous nearly to the base, the lower flowers scattered among the leaves (i. e., the racemes irregularly leafy-bracteate below); fruiting pedicels ascending or somewhat spreading, less than 2 mm. long, evidently shorter than the 3-5 mm. calyx; calyx strigose at least above, the tube also shortly uncinate-hispid; corolla blue, the limb 1-2 mm. wide, not completely flat; mature nutlets mostly brown, sometimes paler, distinctly surpassing the style.

Roadsides, stream banks, and moist or rather dry, open places, somewhat weedy; native of Eurasia, now more or less established over much of the n. U. S. and s. Can., and not uncommon in our range, chiefly or wholly e. of the Cascades. Apr.-June.

For the proper application of this name see: G. Stroh. Myosotis micrantha Pallas. Ein Beitrag zur Nomenklaturfrage. Notizb. Bot. Gart. Berl. 12:471-73. 1935.

Myosotis scorpioides L. Sp. Pl. 131. 1753.
 M. scorpioides var. palustris L. Sp. Pl. 131. 1753. M. palustris Lam. Fl. Fr. 2:283. 1778. (Europe) See comment following specific description.

Plants fibrous-rooted, apparently perennial, 2-6 dm. tall, often creeping at the base, and commonly stoloniferous as well, inconspicuously strigose; leaves mostly 2.5-8 cm. long and 7-20 mm. wide, the lower commonly oblanceolate, the others more oblong or elliptic to lance-elliptic; inflorescence terminal, essentially naked, becoming loose and open, seldom much, if at all, longer than the leafy part of the stem; fruiting pedicels spreading, about equaling or somewhat surpassing the 3-5 mm. calyx; calyx closely but not densely strigose, the lobes equaling or shorter than the tube, sometimes unequal; corolla blue, the limb 5-10 mm. wide, flat; style equaling or more often distinctly surpassing the mostly blackish nutlets, or rarely surpassed by them. N=32.

In shallow water and wet soil; native of Europe, now widespread in N. Am., and occasionally found in our range, from the lowlands to moderate elevations in the mountains. June-Aug.

Myosotis scorpioides L. was composed of two co-equal elements, the var. α arvensis and the var. β palustris. The first author to restrict the name M. scorpioides to one of these two elements was Hill, who used the name M. arvensis for M. scorpioides var. arvensis L., and M. scorpioides for M. scorpioides var. palustris L. Authors and synonyms were consistently omitted from this work of Hill's, so that technically his M. arvensis might be considered a new species rather than the elevation of Linnaeus' variety, but by long established custom this and many other names used by Hill are considered to be founded on the Linnaean names of the Species Plantarum. Recent European usage is divided between that here presented, and the complete suppression of M. scorpioides in favor of the two original varieties at specific status. The latter practice is contrary to Article 53 of the Rules (1956 edition).

Myosotis sylvatica Hoffm. Deutschl. Fl. 61. 1791.
 M. arvensis var. sylvatica Pers. Syn. 1:156. 1805. M. perennis var. sylvatica DC. in Lam. & DC. Fl. Fr. 3rd ed. 3:629. 1805. (Ehrhart Herb. 31, Europe)
 M. alpestris F. W. Schmidt, Fl. Boëm. 3:26. 1794. M. perennis var. alpestris DC. in Lam. & DC. Fl. Fr. 3rd ed. 3:629. 1805. M. sylvatica var. alpestris Koch, Syn. Deuts. Fl. 504. 1838. M. pyrenaica var. alpestris Schinz & Thell. Viert. Nat. Ges. Zürich 53:558. 1908. M. sylvatica ssp. alpestris Hegi, Fl. Mittel-Eur. 5$^3$:2168. 1927. (Bohemia)
 M. alpestris ssp. asiatica Vestergr. in Hultén, Fl. Kamtch. 4:80. 1930. M. alpestris var. asiatica attr. to Vestergr. by auct. (Kamchatka; several specimens cited)

Fibrous-rooted perennial with several or many stems 0.5-4 dm. tall from a short, branched caudex, hirsute or hirsute-puberulent to hirsute-strigose throughout; basal leaves oblanceolate or elliptic, up to 13 cm. long (petiole included) and 13 mm. wide; cauline leaves several, smaller, seldom over 6 cm. long, mostly oblong to lance-elliptic and sessile; inflorescence at first compact and rather showy, eventually elongate and open, naked; fruiting pedicels ascending-spreading, about equaling or somewhat surpassing the 3-5 mm. calyx; calyx hirsute or hirsute-strigose, and bearing some spreading, uncinate hairs on the tube; calyx lobes distinctly longer than the tube; corolla blue, rarely white, the flat limb mostly 4-8 mm. wide; mature nutlets surpassing the style, black or blackish. N=12, 24, about 36.

Meadows and moist, open slopes at moderate to high elevations in the mountains; Eurasia; Alas. to s. B.C., c. Ida., n. Wyo., and the Black Hills of S.D. June-Aug.

Our plants, as decribed above, belong to the var. alpestris (F. W. Schmidt) Koch, with essentially the range of the species. The var. sylvatica, a European plant of lower elevations, is distinguished chiefly by its less caespitose and less strongly perennial habit. The cultivated forget-me-not belongs to this species, being derived chiefly or wholly from var. sylvatica.

The name M. sylvatica may have to give way to M. pyrenaica Pourret (1788) when the taxonomy and typification of the taxa involved are given monographic study.

The related species M. azorica H. C. Wats. has been reported as a casual weed w. of the

Myosotis arvensis

M. discolor

M. laxa

M. scorpioides

M. micrantha

M. sylvatica var. alpestris

Cascades. It is decumbent at the base, without a prominent cluster of basal leaves, and the stem is conspicuously retrorse-hairy below.

Myosotis verna Nutt. Gen. pl. 2: Add. 1818. (N.J.)

?Lycopsis virginica L. Sp. Pl. 139. 1753. Myosotis virginica B.S.P. Prelim. Cat. N.Y. Pl. 37. 1888. (Clayton, Va.) See comment following specific description.

Plants annual or winter annual, tending to be fibrous-rooted, 0.5-4 dm. tall, simple or moderately branched, hirsute-puberulent or hirsute-strigose throughout; leaves mostly 1-5 cm. long and 2-10 mm. wide, the lowermost commonly oblanceolate and somewhat petiolate, the others more oblong or narrowly elliptic and sessile; inflorescence seldom comprising much more than the upper half of the plant, often irregularly leafy-bracteate below; fruiting pedicels ascending or suberect, nearly equaling or more often distinctly shorter than the 4-7 mm. calyx; calyx rather densely and shortly uncinate-hairy below, stiffly ascending-hirsute above, asymmetrical, subbilabiate, three of the lobes more or less evidently shorter than the other two; corolla inconspicuous, white, the limb 1-2 mm. wide, not wholly flat; mature nutlets distinctly surpassing the short style, tending to be closely pale-stippled on a darker (commonly brown) background.

In open, wet or sometimes rather dry places in the foothills and lowlands; e. U.S. and adj. Can.; wanting from most of the plains and Rocky Mt. states; reappearing in s. B.C., Wash., Oreg., and Ida. May-July.

This is the species that has been generally known for half a century as M. virginica (L.) B.S.P., based on Lycopsis virginica L. Fernald pointed out in 1941 (Rhodora 43:636-37) that Lycopsis virginica L. was based solely on a polynomial of Gronovius (Fl. Virg.) which indicated that the plant had blue flowers. Since the present species has white flowers rather than blue, Fernald rejected the name M. virginica (L.) B.S.P. as being of doubtful application. Myosotis virginica (L.) B.S.P. can also be rejected on other grounds as illegitimate. The original edition of the Species Plantarum included a Myosotis virginiana (now considered to be a species of Hackelia). In both the second (1762) and the third (1764) editions of the Species Plantarum, this same species was called Myosotis virginica. The fact that the same form was used in two consecutive editions indicates that the change was deliberate. If it be maintained that Myosotis virginica L. of the second and third editions of the Species Plantarum is merely an orthographic variant of M. virginiana L., then obviously M. virginica (L.) B.S.P. is also an orthographic variant of M. virginiana L., and the B.S.P. name is therefore illegitimate. On the other hand, if it be maintained that M. virginica of the second and third editions of the Species Plantarum is not an orthographic variant of M. virginiana of the first edition, then M. virginica L. 1762 is a new name, with the epithet "virginica" as an illegitimate but validly published substitute for "virginiana." M. virginica L. 1762, though itself illegitimate, would therefore be an earlier homonym of M. virginica (L.) B.S.P. 1888, which latter would thus be rendered illegitimate under Article 64 of the Rules (1956 edition). The name M. virginica (L.) B.S.P. is therefore in any case illegitimate and cannot be used under the Rules.

Onosmodium Michx.

Flowers borne in leafy-bracteate, helicoid terminal cymes which tend to elongate and straighten in age, precociously sexual, the style exserted and the anthers dehiscent well before the corolla is mature, the style remaining exserted; calyx deeply 5-cleft, the narrow, often unequal segments sometimes eventually disarticulate at the base; corolla externally hairy, white to yellow, commonly with somewhat greenish lobes, nearly tubular, with rather narrow, erect, pointed lobes and thickened, basally inflexed sinuses; fornices wanting; anthers barely or only partly included; nutlets turgid, without a prominent ventral keel, smooth or merely pitted, broadly attached at the base to the flattish or depressed gynobase, commonly only 1 or 2 maturing; rather coarse perennial herbs, often rough-hairy, with largely or wholly cauline leaves that have several prominent veins.

Five sharply defined species occurring e. of the continental divide in the U.S. and adjacent parts of Can. and Mex. (Named for its resemblance to Onosma, an Old World genus of the same family.)

References:

Johnston, I. M. Studies in the Boraginaceae, XXIV. Further revaluations of the genera of the Lithospermeae. Journ. Arn. Arb. 35:1-81. 1954.

Mackenzie, K. K. Onosmodium. Bull. Torrey Club 32:495-506. 1906.

Onosmodium molle Michx. Fl. Bor. Am. 1:133. 1803.

Lithospermum molle Muhl. Cat. Pl. 19. 1813. Purshia mollis Lehm. Asperif. 383. 1818.
Onosmodium carolinianum var. molle Gray, Syn. Fl. 2$^1$:206. 1878. (Nashville, Tenn.)
ONOSMODIUM MOLLE var. BEJARIENSE (DC.) Cronq. hoc loc. O. bejariense DC. Prodr.
10:70. 1846. (Berlandier 1681, near Bejar, Rio de la Trinidad, "Mexico boreali")
ONOSMODIUM MOLLE var. SUBSETOSUM (Mack. & Bush) Cronq. hoc loc. O. subsetosum
Mack. & Bush ex Small, Fl. S.E.U.S. 1001. 1903. (Bush 135, Eagle Rock, Mo.)
ONOSMODIUM MOLLE var. HISPIDISSIMUM (Mack.) Cronq. hoc loc. O. hispidissimum
Mack. Bull. Torrey Club 32:500. 1905. (Numerous specimens are cited, from N.Y. to Neb.,
La., and Tex.)
O. occidentale Mack. Bull. Torrey Club 32:502. 1905. O. molle var. occidentale Johnst.
Contr. Gray Herb. n.s. 70:18. 1924. (Numerous specimens are cited, from Man. and Alta.
to N.M., Tex. and Ill.)
O. hispidissimum var. macrospermum Mack. & Bush, Bull. Torrey Club 32:502. 1905.
(Mackenzie, Atherton, Jackson Co., Mo., Sept. 7, 1895) A form nearest var. hispidis-
simum, but intermediate toward var. occidentale in size and texture of the nutlets, and the
epithet therefore abandoned.
O. occidentale var. sylvestre Mack. Bull. Torrey Club 32:504. 1905. (Engelmann, opposite
St. Louis, in Ill., July 21, 1861) A form perhaps nearest var. occidentale, but intermediate
toward var. hispidissimum in habit, pubescence, and nutlet size, and the epithet, "sylvestre,"
therefore abandoned.
Several-stemmed perennial from a woody root, mostly 3-7 dm. tall, coarsely and loosely
hairy throughout; lower leaves reduced and deciduous, the others rather numerous and uniform,
sessile, lanceolate or rather narrowly ovate, mostly 3-8 cm. long and 1-2 cm. wide; corolla
12-16 mm. long, the sharply acute lobes 2-4 mm. long; nutlets smooth and shining, mostly
3.5-5 mm. long, broadly ovoid, without a collar at the base.

Open, moderately dry places; U.S. and adj. Can. from the Appalachian region to the e. base
of the Rocky Mts., and extending into the more easterly intermontane valleys of Mont. in our
range. June-July.

The species consists of five intergradient geographic varieties which it has been customary
for the past half-century to dignify with specific rank. Only the var. occidentale (Mack.)
Johnst., as described above, occurs in our range. The several varieties may be distinguished,
more or less, by the following key:

1 Stem evidently hairy to the base, some or all of the hairs (except sometimes in var. molle)
   spreading
   2 Nutlets constricted just above the base to form an evident collar, mostly dull and smooth
     or nearly so, 2.5-3.5 mm. long; plants robust, often 1 m. tall or more, more coarse-
     ly and conspicuously hairy than the other varieties, the leaves sometimes as much as
     15 cm. long and 4 cm. wide; Appalachian region to the w. edge of the deciduous forest
     region                                                            var. hispidissimum (Mack.) Cronq.
   2 Nutlets without a collar; plants smaller, commonly 3-7 dm. tall, the leaves up to about
     8 cm. long and 2 cm. wide, or sometimes larger in var. bejariense
     3 Corolla lobes acuminate; nutlets 3-4 mm. long, pitted to sometimes smooth; s. Tex.
                                                            var. bejariense (DC.) Cronq.
     3 Corolla lobes merely acute
       4 Nutlets smooth and shining, 3.5-5 mm. long; pubescence relatively coarse and loose;
         plains and prairie region of c. U.S. and adj. Can.
                                                            var. occidentale (Mack.) Johnst.
       4 Nutlets dull and pitted, 2.5-3 mm. long; barrens of Ky. and Tenn., extending to s.e.
         Mo.                                                var. molle
1 Stem strigose above, glabrous or nearly so below; pubescence wholly appressed; nutlets ap-

proximately of var. molle, habit more nearly of var. hispidissimum; Ozark region of Mo. and Ark.                                        var. subsetosum (Mack. & Bush) Cronq.

Pectocarya DC.

Flowers borne in terminal, naked or irregularly bracteate, sympodial false racemes, or the lower scattered among the leaves, the short pedicels ascending to recurved; calyx deeply cleft, the sepals distinct or nearly so; corolla minute, white; anthers included; nutlets radially spreading from the broad, low gynobase at maturity, with relatively broad, apicolateral scar, uncinate-bristly at least in part, and with raised or winged margins; low, slender, taprooted strigose annuals with small, narrow leaves.

Perhaps 10 species, of w. N. Am. and S. Am. (Name from the Greek pektos, combed, and karyon, nut, from the row of bristles on the nutlet.)

1 Nutlets isometrically spreading; margins lineate or scarcely developed; sepals uncinate-bristly toward the tip                                              P. PUSILLA
1 Nutlets spreading in pairs, at least one of each pair evidently wing-margined; hairs of the calyx not uncinate
   2 Nutlets equaling or more often surpassing the sepals, oblong, with upcurved to inflexed margins which are cleft (at least distally) into uncinate bristles; plants tending to be prostrate or weakly ascending                                              P. LINEARIS
   2 Nutlets surpassed by the sepals, orbicular-obovate, with more or less scattered uncinate bristles, one nutlet of each pair with spreading wing margins, the other nearly margin-less and partly hidden by the margined one; plants tending to be ascending or erect                                                                                         P. SETOSA

Pectocarya linearis (R. & P.) DC. Prodr. 10:120. 1846.
   Cynoglossum lineare R. & P. Fl. Peruv. 2:6. 1799. (Santiago, Chile)
   C. penicillatum H. & A. Bot. Beechey Voy. 371. 1838. Pectocarya penicillata A. DC. Prodr. 10:120. 1846. P. linearis var. penicillata M. E. Jones, Proc. Calif. Acad. Sci. II. 5:709. 1895. (Douglas, Calif.)
   Pectocarya lateriflora var. nuttallii (Spreng.) Brand sensu Brand, Pflanzenr. IV. 252 (Heft 78): 95. 1921; but not as to the type, fide Johnst. Contr. Gray Herb. n. s. 70:37. 1924.
   PECTOCARYA LINEARIS var. PLATYCARPA (Munz & Johnst.) Cronq. hoc loc. P. gracilis var. platycarpa Munz & Johnst. Contr. Gray Herb. n.s. 70:36. 1924. P. platycarpa Munz & Johnst. Contr. Gray Herb. n. s. 81:81. 1928. (Pringle s. n., in part, near Camp Lowell, Ariz., April 16, 1881) The new combination here proposed represents a return of this taxon to the position originally intended by Munz & Johnston, since the type of P. gracilis proves, according to Johnst. (Contr. Gray Herb. n. s. 81:81. 1928), to be a Plagiobothrys.
   P. linearis var. ferocula Johnst. Contr. Arn. Arb. 3:95. 1932. (Munz & Crow 11846, Lady Harbor, Santa Cruz I., Calif.) = var. linearis.

Prostrate or weakly ascending annual, often freely branched at the base and forming a rosette; leaves linear, up to 2 cm. long and 1 cm. wide; corolla limb about 1 mm. wide or less; nutlets spreading in pairs, oblong, 1.5-2 mm. long, about equaling or more often surpassing the sepals, the prominent margins upturned, often inflexed near the middle, lacerate (at least distally) into uncinate bristles; smaller uncinate bristles often present near the tip of the ventral surface of the nutlet, and the remainder of the body often very finely uncinate-puberulent.

In dry, often sandy soil at lower elevations, commonly among sagebrush; s. B. C. to Baja Calif., e. to Salmon, Ida., s. w. Wyo., n. e. Nev., s. e. Utah, w. Ariz., and Sonora, Mex.; also in Chile and Argentina. Wholly e. of the Cascades in our range. Apr. -June.

Three geographic varieties may be distinguished on technical characters of the nutlets, as follows:

1 Margins of the nutlets cleft into uncinate bristles chiefly near the tip, the central region generally smooth, the proximal portion (nearest the gynobase) sometimes with a few bristles; s. B. C. to n. Baja Calif., e. to Salmon, Ida., s. w. Wyo., and n. e. Nev.
                                                    var. penicillata (H. & A.) M. E. Jones

1 Margins of the nutlets cleft into uncinate bristles along the sides as well as near the tip,
  the ones along the sides commonly stouter than those near the tip
  2 Margins of the nutlets very narrow, the bristles nearly distinct; cismontane and insular
    s. w. Calif., and in w. S. Am.                          var. linearis
  2 Margins of some or all of the nutlets relatively broad and conspicuous, the bristles evi-
    dently confluent below; desert regions of s. w. Utah, s. Nev., w. Ariz., s. Calif., and
    n. w. Mex.                                       var. platycarpa (Munz & Johnst.) Cronq.

Pectocarya pusilla (A. DC.) Gray, Proc. Am. Acad. 12:81. 1876.
  Gruvelia pusilla A. DC. Prodr. 10:119. 1846. (Poeppig 276, Chile)
  Pectocarya pusilla var. flagellaris Brand, Pflanzenr. IV. 252 (Heft 78): 96. 1921. (Speci-
    mens cited from Wash., Oreg., and Calif.) Name intended to apply to the N. Am. phase
    of the species, here considered (as by Johnston) taxonomically identical with the S. Am.
    phase.
  Erect or ascending, simple or loosely branched, slender annual 0.5-2 dm. tall; leaves linear, not very numerous, up to 1.5 cm. long and 1.5 mm. wide, the lower opposite; sepals
with forward-pointing, uncinate bristles near the tip; corolla limb about 1 mm. wide or less;
nutlets isometrically spreading, obovate, 2.5 mm. long, surpassed by the sepals, beset with
uncinate bristles all along the slightly (if at all) raised margin, and often very finely uncinate-
puberulent over the surface.
  Dry, open places at lower elevations; Calif., northward occasionally to Klickitat Co., Wash.,
wholly e. of the Cascades in our range; also in Chile. May.

Pectocarya setosa Gray, Proc. Am. Acad. 12:81. 1876. (Palmer 379, upper Mojave R., Calif.)
  More or less branched, mostly erect or ascending annual up to 1 dm. tall, stouter and firmer than our other species, the herbage often more bristly; leaves linear or nearly so, up to 2 cm.
long and 2 mm. wide, the lower opposite; sepals conspicuously and pungently spreading-setose,
as well as strigulose; corolla limb about 1 mm. wide or less; nutlets obovate-orbicular, 2-3
mm. long, surpassed by the sepals, spreading in pairs, one of each pair with well-developed,
spreading wing margins, the other nearly marginless and partly concealed by the margined one;
uncinate bristles more or less scattered over the dorsal surface of the nutlet, tending to be
concentrated distally and near the margins, which may be finely uncinulate-ciliate, but are not
at all cleft.
  Dry, open places in the lowlands, commonly with sagebrush; Yakima Co., Wash., southward to Baja Calif., eastward to Boise, Ida., Utah, and Ariz. Rarely collected in our range.
May-June.

## Plagiobothrys F. & M.

  Flowers borne in a series of sympodial, helicoid, naked or irregularly bracteate, false racemes or spikes, these sometimes condensed into glomerules, but elongating with age in most
species; calyx cleft to below the middle or near the base, sometimes moderately accrescent;
corolla white, the well-developed fornices sometimes yellow, the limb usually more or less
rotately spreading, not large; stamens included, the filaments short; nutlets 4, or 1-3 by
abortion, tending to be keeled on the back, and with a well-developed ventral keel extending
from the tip to near the middle or near the base; scar generally elevated and carunclelike,
mostly small, lateral to basal, placed at the end of the ventral keel (rarely extending along it);
gynobase mostly short and broad; strigose or subglabrous to coarsely spreading-hairy herbs,
annual or perennial, mostly rather small; leaves mostly narrow, the lower often opposite.
  Perhaps 50 species, native to w. N. Am. and S. Am., with one outlying species in e. Asia
and another in Australia. (Name from the Greek, plagios, placed sideways, and bothros, pit
or excavation, referring to the position of the nutlet scar.)
  Plagiobothrys is composed of 2 well-marked sections or subgenera that often (and not
without reason) have been treated as distinct genera. The circumscription here adopted has
been vigorously defended by Johnston, who further points out that the Allocarya group, if recognized as a genus, must take the older name Maccoya. The treatment here presented con-

forms more nearly to Johnston's first (1923) revision than to his second (1932), in which he
recedes toward the stand earlier taken by Piper that minor technical variations in the nutlets,
unsupported by other characters or by geography, furnish sufficient basis for the establish-
ment of species. The species as recognized in the present treatment are sharply defined, at
least as to our region, and are readily recognizable in flower as well as in fruit.

References:

Johnston, I. M. Studies in the Boraginaceae. 4. A synopsis and redefinition of Plagiobothrys. Contr. Gray. Herb. n. s. 68:57-80. 1923.

-------. Studies in the Boraginaceae. IX. 1. The Allocarya section of Plagiobothrys in the
western United States. Contr. Arn. Arb. 3:5-82. 1932.

Piper, C. V. A Study of Allocarya Contr. U.S. Nat. Herb. 22:79-113. 1920.

1 Lower cauline leaves, like the others, ordinarily alternate; scar lateral, near the middle
of the nutlet; plants occurring mostly in well-drained soils, definitely taprooted; herbage
more or less markedly spreading-hairy  (Plagiobothrys)
  2 Nutlets 4, thick-cruciform (unique among our species in this regard); corolla limb mostly 2-4 mm. wide; habit of P. nothofulvus; widespread           P. TENELLUS
  2 Nutlets 1-4, not at all cruciform
    3 Corolla minute, the limb not over about 1.5 mm. wide; nutlets mostly 1-2, horizontal;
plants bushy-branched when well developed, the stem evidently leafy and the basal
leaves mostly deciduous at maturity, the inflorescences short and glomerate; Basin
species, barely reaching our range in c. Oreg.           P. HISPIDUS
    3 Corolla larger, the limb mostly 4-9 mm. wide; plants not at all bushy-branched, or
somewhat so in P. harknessii; basal leaves, except commonly in P. harknessii, tufted and persistent
      4 Scar short, scarcely longer than wide, placed at the end of the ventral keel; calyx
eventually circumscissile; inflorescences becoming elongate; nutlets 1-4; e. end
of the Columbia gorge, and irregularly southward w. of the Cascades
P. NOTHOFULVUS
      4 Scar narrow and elongate, extending along the crest of the ventral keel; calyx not
circumscissile; inflorescences more or less glomerate, not elongating; nutlets
3-4; Basin species, perhaps reaching our range in c. Oreg.
P. HARKNESSII
1 Lower cauline leaves, or some of them, opposite; scar basal, basilateral, or lateral near
the base; plants occuring mostly in poorly drained soils, tending to be fibrous-rooted; species, except for the perennial P. mollis, with the herbage mostly appressed-hairy or subglabrous  (Allocarya)
  5 Plants perennial, evidently spreading-hairy at least in part; Basin species, perhaps
reaching our range in c. Oreg.           P. MOLLIS
  5 Plants annual, mostly appressed-hairy or subglabrous
    6 Corolla relatively large, the limb mostly 5-10 mm. wide; plants erect; spikes naked,
tending to be paired; scar lateral, near but often not reaching the base; w. of the
Cascades           P. FIGURATUS
    6 Corolla smaller, the limb 1-4 mm. wide; plants prostrate to erect; spikes seldom
paired, often irregularly bracteate below
      7 Scar basilateral, commonly more lateral than basal; calyx lobes neither elongate
nor much thickened, more or less symmetrically disposed; plants prostrate to ascending or erect; widespread           P. SCOULERI
      7 Scar very nearly basal; calyx lobes becoming elongate and slightly thickened, tending
all to be directed towards the same side of the fruit; plants mostly prostrate; e.
of the Cascades in Oreg., extending eastward and southward
P. LEPTOCLADUS

Plagiobothrys figuratus (Piper) Johnst. ex Peck, Man. High. Pl. Oreg. 609. 1941.
Allocarya figurata Piper, Contr. U.S. Nat. Herb. 22:101. 1920. Plagiobothrys hirtus var.
figuratus Johnst. Contr. Arn. Arb. 16:193. 1935. (J. C. Nelson 1509, Illahe, Curry Co.,
Oreg.)

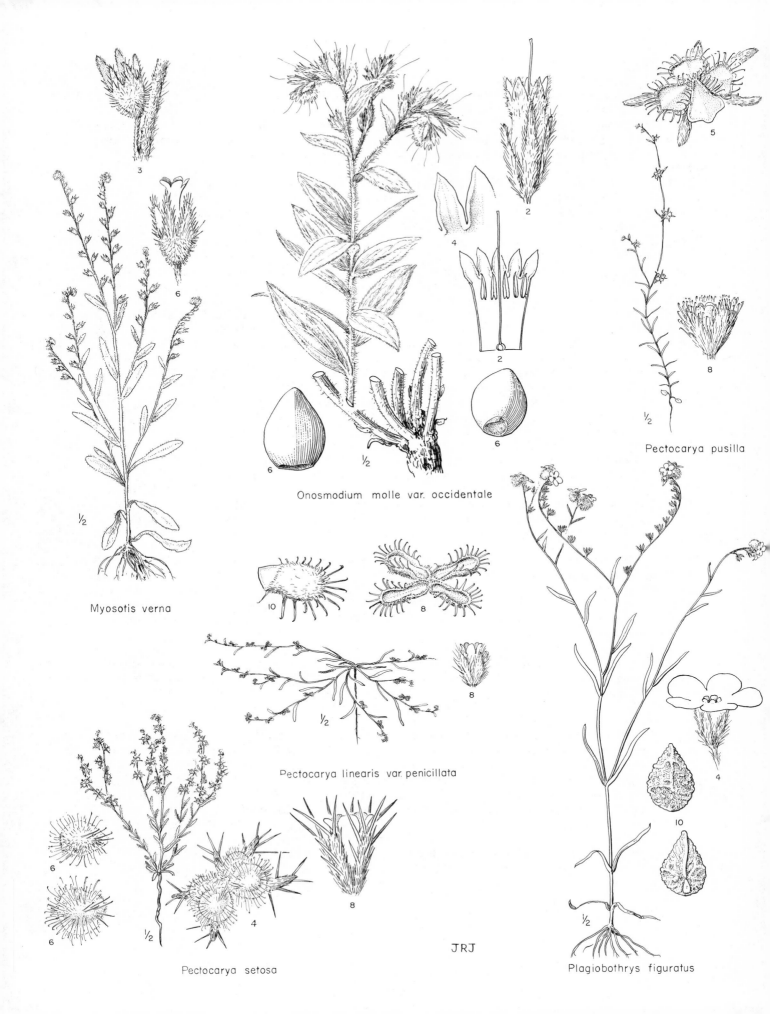

Onosmodium molle var. occidentale

Myosotis verna

Pectocarya pusilla

Pectocarya linearis var. penicillata

Pectocarya setosa

Plagiobothrys figuratus

JRJ

Allocarya dichotoma Brand, Fedde Rep. Sp. Nov. 18:313. 1922. (Howell 46, Oreg., June,
   1881)

Erect, fibrous-rooted annual 1-4 dm. tall, simple or moderately branched; herbage sparsely to moderately strigose; leaves essentially all cauline, linear, sometimes elongate, the lower 2-4 pairs opposite, the others alternate; spikes naked, tending to be paired, eventually
elongate; calyx 3-4 mm. long at maturity, more densely and loosely hairy than the herbage,
the hairs often fulvous; flowers fragrant, white with a yellow eye, relatively large and showy,
the corolla limb 5-10 mm. wide; nutlets 4, ovate, 1.2-1.7 mm. long, rugose or rugose-tuberculate on both sides, often more finely tuberculate in the areolae, the scar small, distinctly
lateral, near but often not quite reaching the base.

Meadows, low ground, and moist fields; s. Vancouver I. to s.w. Oreg., w. of the Cascades,
chiefly in the Puget trough, especially the Willamette Valley, and extending through the Columbia gorge to Klickitat Co., Wash. May-July.

Further collecting may necessitate the subordination or complete reduction of P. figuratus to
P. hirtus (Greene) Johnst., which differs in its coarse, spreading pubescence and slightly
larger nutlets with more subdued sculpture, and which is so far known only from the type locality, near Drain, Douglas Co., Oreg.

The name Allocarya scouleri (H. & A.) Greene, or Plagiobothrys scouleri (H. & A.) Johnst.,
has often been misapplied to the present species. An apparent isotype at N.Y. of P. scouleri
bears out Johnston's comments (Journ. Arn. Arb. 16:192-93. 1935) concerning the application
of the name.

Plagiobothrys figuratus has often been confused with the superficially similar but technically
very distinct Cryptantha intermedia var. grandiflora (=C. hendersonii), which however is definitely taprooted rather than fibrous-rooted, is more spreading-hairy, shows slight, if any,
tendency toward the development of opposite lower leaves, and has merely tuberculate or
smooth, not at all rugose-reticulate nutlets with the characteristic Cryptantha-type scar.

Plagiobothrys harknessii (Greene) Nels. & Macbr. Bot. Gaz. 62:143. 1916.
   Sonnea harknessii Greene, Pitt. 1:23. 1887. Plagiobothrys kingii var. harknessii Jeps. Man.
   Fl. Pl. Calif. 856. 1925. (Harkness, near Mono Lake, Calif., June, 1886)

Taprooted annual (or biennial?) up to 25 cm. tall, coarsely spreading-hirsute throughout, well-
developed plants with 2-several stems from the base; leaves alternate, the lower well developed, narrowly oblanceolate, up to 6 cm. long and nearly 1 cm. wide, tending to be crowded toward the base, but scarcely forming a basal tuft; remaining leaves linear or nearly so and
often evidently smaller; inflorescences compact and rather few-flowered, somewhat glomerate,
not elongating, often surpassed by their subtending leaves; fruiting calyx 3-4 mm. long; corolla limb 4-6 mm. wide; nutlets 3-4, ovate, incurved, 2-3 mm. long, coarsely rugose-reticulate,
the scar elongate, narrow, extending along the ventral keel.

Dry, open slopes in sagebrush and juniper regions; Nev. and adj. Calif. to s.e. Oreg., and
perhaps entering our range. May-July.

Plagiobothrys kingii (Wats.) Gray, a related but more robust and broader-leaved species,
with the inflorescences tending to elongate in age, appears to be wholly more southern.

Plagiobothrys hispidus Gray, Proc. Am. Acad. 20:286. 1885.
   Sonnea hispida Greene, Pitt. 1:22. 1887. (Curran, between Carson City and Virginia City,
   Nev.)

Taprooted annual, 5-20 cm. tall, bushy-branched, at least when well developed, evidently
spreading-hirsute throughout; leaves alternate, well scattered along the stems, not clustered
at the base, mostly 1-3.5 cm. long, the middle and lower mostly linear-oblong, the upper often more lanceolate or lance-ovate; inflorescences numerous, small and few-flowered, compact and somewhat glomerate, often surpassed by their subtending leaves; calyx 2-3 mm. long
at maturity; corolla minute, the limb not over about 1.5 mm. wide; nutlets mostly 1-2(4) and
horizontal, broadly ovoid, 1-2 mm. long, tuberculate or rugose-tuberculate and finely papillate, the scar relatively large and situated at or a little above the middle, the ventral keel
short.

Dry, open places, often in sandy soil, from Deschutes and Crook cos., Oreg., s. to Nev. and Calif. July-Sept.

Plagiobothrys leptocladus (Greene) Johnst. Contr. Arn. Arb. 3:38. 1932.
  Allocarya leptoclada Greene, Pitt. 3:109. 1896. (Greene, Pine Creek, Eureka Co., Nev., July 20, 1896)
  Eritrichium californicum var. subglochidiatum Gray, Bot. Calif. 1:526. 1876. Krynitzkia californica var. subglochidiata Gray, Proc. Am. Acad. 20:266. 1885. Allocarya californica var. subglochidiata Jeps. Fl. W. Middle Calif. 443. 1901. A. subglochidiata Piper, Contr. U.S. Nat. Herb. 11:485. 1906. (Watson 851, Clover Mts., Elko Co., Nev.; lectotype by Johnston)
  Allocarya orthocarpa Greene, Pitt. 4:235. 1901. Plagiobothrys orthocarpus Johnst. Contr. Gray Herb. n. s. 68:78. 1923. (Mulford 147 in part, Cache Valley, Utah)
  A. oricola Piper, Contr. U.S. Nat. Herb. 22:92. 1920. (Nelson & Macbride 1170, Shoshone, Lincoln Co., Ida.)
  A. wilcoxii Piper, Contr. U.S. Nat. Herb. 22:93. 1920. (Wilcox, near Boise, Ida., in 1883)
  A. tuberculata Piper, Contr. U.S. Nat. Herb. 22:95. 1920. (Leiberg 166, Pine Creek, Gilliam Co., Oreg.)
  Taprooted or fibrous-rooted annual, rather sparsely strigose or subglabrous, with several or many prostrate stems 5-25 cm. long from the base; leaves linear or narrowly linear-oblanceolate, the lower up to 6 cm. long and 5 mm. wide, 1 or more pairs near the base opposite; stems commonly floriferous to near the base, each forming an elongate, loosely flowered false raceme or spike that may be irregularly leafy-bracteate below; calyx somewhat accrescent, becoming 4-7 mm. long, the linear lobes slightly thickened and tending all to be directed toward the same side of the fruit; corolla minute, the limb only 1-2 mm. wide; nutlets lanceolate, 1.5-2.5 mm. long, rugose-tuberculate, generally also granulate, and often with minute, distally branched bristles; ventral keel extending almost to the base, the scar very nearly basal, visible even from the rear.
  Moist clay flats and beds of drying pools, tolerant of alkali; e. of the Cascades in Oreg., s. to Baja Calif., e. to n. Utah, the Snake R. plains of Ida., and a single specimen (Coues in 1874) reputedly from Mont. Apr.-July.

Plagiobothrys mollis (Gray) Johnst. Contr. Gray Herb. n. s. 68:74. 1923.
  Eritrichium molle Gray, Proc. Am. Acad. 19:89. 1883. Krynitzkia mollis Gray, Proc. Am. Acad. 20:267. 1885. Allocarya mollis Greene, Pitt. 1:20. 1887. (Lemmon, Sierra Valley, Calif., in 1883)
  Fibrous-rooted perennial; stems lax, 5-25 cm. long, or sometimes arching-trailing, rooting at the nodes, and more elongate; herbage, especially the stems, more or less densely and softly pubescent with long spreading hairs, the stoloniform portions more glabrate; leaves all, or nearly all, cauline, broadly linear or narrowly oblong, mostly 2-7 cm. long and 3-8 mm. wide, the lower 2-several pairs opposite; racemes terminating the stems and branches, naked above the base, becoming somewhat elongate; calyx 4-6 mm. long at maturity; corolla limb mostly 5-10 mm. wide; nutlets ovate, 1.5-2 mm. long, more or less rugose-reticulate; ventral keel extending to the middle or commonly a little below, the scar scarcely elevated, lateral, rather narrowly triangular, not quite reaching the base.
  Moist fields and pastures, meadows, etc., tolerant of alkali; n. Calif. and w. Nev. to n. Harney Co., Oreg., and perhaps barely entering our range. June-Aug.

Plagiobothrys nothofulvus Gray, Proc. Am. Acad. 20:285. 1885.
  Eritrichium nothofulvum Gray, Proc. Am. Acad. 17:227. 1882. (Douglas, Calif.)
  Taprooted annual 1.5-5 dm. tall, with one or several simple or few-branched stems from the base; herbage spreading-hairy, the hairs of the stem obscurely viscid, shorter, softer and more curled than those of the leaves; leaves apparently all alternate, the basal ones tufted and persistent, 2-10 cm. long, 5-20 mm. wide, carrying their width for much of their length, and scarcely petiolate; cauline leaves few and progressively reduced; racemes becoming elongate, often paired; calyx more or less fulvous-hirsute, especially distally, eventually circumscissile,

leaving an expanded, scarious, saucer-shaped base; corolla relatively large and showy, the limb mostly 5-9 mm. wide; nutlets 1-4, 2-3 mm. long, loosely rugose-reticulate and somewhat granular-tuberculate, broad-based, more or less horizontally bent over the broad, low gynobase, the abruptly narrowed tips intercrossed; scar somewhat enlarged, medial. Popcorn flower.

Open slopes, fields, and roadsides; common at the e. end of the Columbia R. gorge in Oreg. and Wash.; rarely and irregularly w. of the Cascades in Oreg.; throughout the length of Calif. to n. Baja Calif. March-May.

Plagiobothrys scouleri (H. & A.) Johnst. Contr. Gray Herb. n. s. 68:75. 1923.
  Myosotis scouleri H. & A. Bot. Beechey Voy. 370. 1838. Erithrichium scouleri A. DC.
    Prodr. 10:130. 1846. Krynitzkia scouleri Gray, Proc. Am. Acad. 20:267. 1885. Allocarya
    scouleri Greene, Pitt. 1:18. 1887. (Scouler, n. w. coast)
  ?Myosotis californica Fisch. & Mey. Ind. Sem. Hort. Petrop. 2:42. 1835. Erithrichium
    californicum A. DC. Prodr. 10:130. 1846. Krynitzkia californica Gray, Proc. Am. Acad.
    20:266. 1885. Allocarya californica Greene, Pitt. 1:20. 1887. Plagiobothrys reticulatus var.
    rossianorum Johnst. Contr. Arn. Arb. 3:79. 1932. (Ft. Ross, Sonoma Co., Calif.) Not P.
    californicus Greene, 1887. Probably represents a distinct, chiefly Californian, subcoastal
    variety of P. scouleri, but no new combination is here proposed.
  Allocarya scopulorum Greene, Pitt. 1:16. 1887. Plagiobothrys scopulorum Johnst. Contr.
    Gray Herb. n. s. 68:79. 1923. (Greene s. n., Denver, Colo., June 15, 1870) = var. penicil-
    latus.
  Allocarya cusickii Greene, Pitt. 1:17. 1887. Plagiobothrys cusickii Johnst. Contr. Arn. Arb.
    3:63. 1932. (Cusick, Union Co., Oreg., in 1883) = var. penicillatus.
  Allocarya hispidula Greene, Pitt. 1:17. 1887. Plagiobothrys hispidulus Johnst. Contr. Arn.
    Arb. 3:71. 1932. (Parish 1470, Bear Lake, San Bernardino Mts., Calif.) = var. penicil-
    latus.
  PLAGIOBOTHRYS SCOULERI var. PENICILLATUS (Greene) Cronq. hoc loc. Allocarya
    penicillata Greene, Pitt. 1:18. 1887. A. hispidula var. penicillata Jeps. Man. Fl. Pl.
    Calif. 853. 1925. (Greene, Donner Lake, Calif., in 1883)
  Allocarya nelsonii Greene, Erythea 3:48. 1895. Plagiobothrys nelsonii Johnst. Contr. Gray
    Herb. n. s. 68:77. 1923. (Nelson 1198, Silver Creek, Wyo.) = var. penicillatus.
  Allocarya cognata Greene, Pitt. 4:235. 1901. Plagiobothrys cognatus Johnst. Contr. Arn.
    Arb. 3:59. 1932. (Mulford 147 in part, Cache Valley, Utah) = var. penicillatus.
  Allocarya ramosa Piper, Contr. U. S. Nat. Herb. 22:100. 1920. (Leiberg 318, Crooked R.,
    near Prineville, Oreg.) = var. penicillatus.
  A. media Piper, Contr. U. S. Nat. Herb. 22:107. 1920. Plagiobothrys medius Johnst. Contr.
    Arn. Arb. 3:58. 1932. (Flett 3378, Pt. Angeles, Wash.) = var. scouleri.
  Allocarya divaricata Piper, Contr. U. S. Nat. Herb. 22:107. 1920. (Wm. Palmer, Victoria,
    B. C., June 6, 1905) = var. penicillatus.
  A. conjuncta Piper, Contr. U. S. Nat. Herb. 22:109. 1920. A. charaxata var. debilis Brand,
    Pflanzenr. IV. 252 (Heft 97): 165. 1931. (Copeland 3046, Chico, Butte Co., Calif.) = var.
    penicillatus.
  A. granulata Piper, Contr. U. S. Nat. Herb. 22:109. 1920. Plagiobothrys granulatus Johnst.
    Contr. Arn. Arb. 3:57. 1932. (Nelson 1338, Salem. Oreg.) = var. scouleri.
  Allocarya insculpta Piper, Contr. U. S. Nat. Herb. 22:109. 1920. (Piper 3869, Coulee City,
    Wash.) = var. penicillatus.
  A. fragilis Brand, Fedde Rep. Sp. Nov. 18:312. 1922. (Suksdorf 2207, Bingen, Klickitat
    Co., Wash.) = var. scouleri.
    Taprooted to fibrous-rooted annuals with several or numerous prostrate to ascending or sub-
erect stems up to 20 cm. long, or single-stemmed and erect, especially when crowded among
other low vegetation; herbage more or less evidently strigose; leaves essentially all cauline,
linear, up to 6. 5 cm. long and 5 mm. wide, the lower 1-4 pairs opposite, the others alternate;
stems terminating in an elongate, loosely flowered false raceme or spike that is often irregularly leafy-bracteate below, often floriferous to near the base; fruiting calyx 2-4 mm. long,
with lanceolate, symmetrically disposed segments; corolla small, the limb only 1-4 mm. wide;

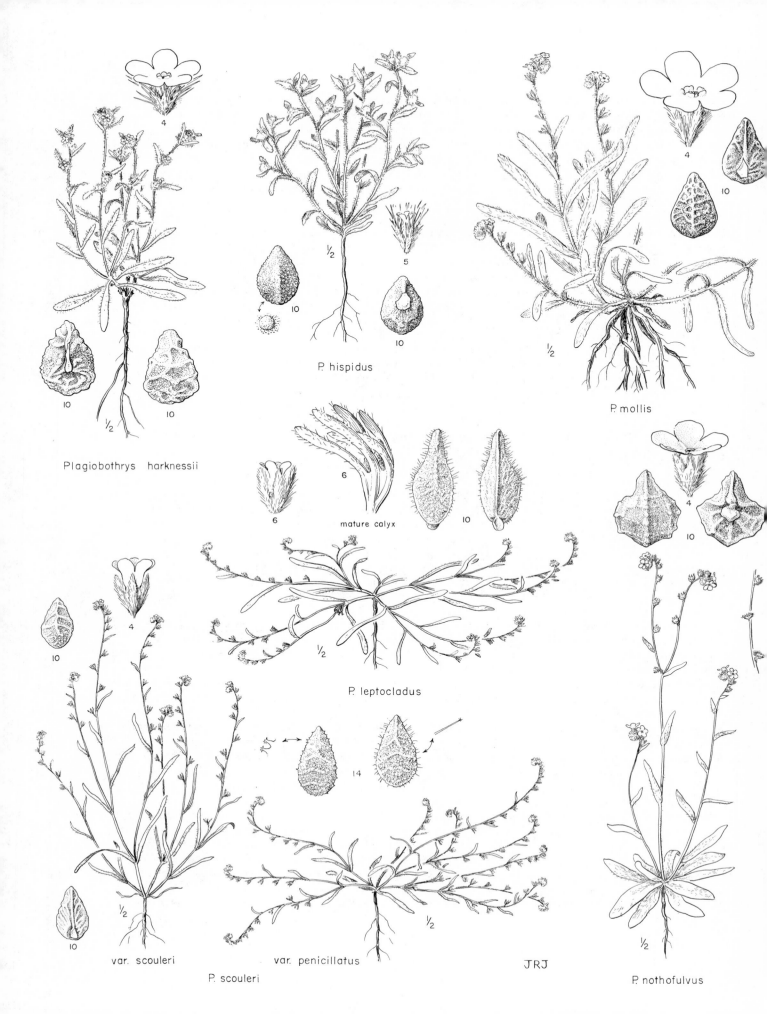

4

P. hispidus

5

10

10

10

Plagiobothrys harknessii

10

10

½

10

10

P. mollis

6

6

mature calyx

10

4

10

P. leptocladus

½

14

10

4

½

var. scouleri

var. penicillatus

½

JRJ

½

P. scouleri

P. nothofulvus

nutlets ordinarily 4, 1 sometimes more persistent than the others, ovate or lance-ovate, 1.5-2 mm. long, more or less rugose and commonly also tuberculate, with or without minute, distally branched bristles; scar small, basilateral, more lateral than basal.

Moist, often poorly drained soil, from the lowlands to moderate elevations in the mountains, seldom in alkaline places; B.C. to Sask. and Man., s. to Calif. and n. N.M.; introduced in Alas. May-Aug.

Our plants may be divided, with some difficulty, into var. scouleri, a more often ascending or erect plant, occurring w. of the Cascade summits, with the corolla limb mostly 2-4 mm. wide, and var. penicillatus (Greene) Cronq., a more often prostrate to ascending plant, occurring chiefly e. of the Cascade summits, with the corolla limb mostly 1-2 mm. wide. The species shows a good deal of variation in the ornamentation of the nutlets, but at least in our range these variants are neither clearly definable nor correlated with other characters, and plants belonging to different "species" as defined by Piper or in 1932 by Johnston frequently occur in the same colony without other obvious differences. The relationship of P. scouleri to some very similar plants in S. Am. remains to be elucidated.

Plagiobothrys tenellus (Nutt.) Gray, Proc. Am. Acad. 20:283. 1885.
    Myosotis tenella Nutt. ex Hook. Kew Journ. Bot. 3:295. 1851. Eritrichium tenellum Gray,
        Proc. Am. Acad. 10:57. 1874. (Geyer, "mountains along the Coeur d'Aleine River," Ida.)
    Plagiobothrys echinatus Greene, Pitt. 3:262. 1898. P. tenellus var. echinatus Brand, Pflan-
        zenr. IV. 252 (Heft 97): 108. 1921. (Macoun, Cedar Hill, Vancouver I., May 16, 1887)
    P. asper Greene, Pitt. 3:262. 1898. ("northern California to Washington"; no type designated)
    Taprooted annual, 0.5-2.5 dm. tall, with one or (more often) several erect or ascending, simple or few-branched stems arising from the base; herbage spreading-hirsute, the stem often also retrorsely strigose; leaves apparently all alternate, the basal ones tufted and persistent, 0.7-3 cm. long, 2-8 mm. wide, sessile or nearly so; cauline leaves few and scattered, mostly lanceolate, seldom as large as the basal ones; racemes tending to become elongate, sometimes paired; calyx 3-5 mm. long at maturity; corolla limb mostly 2-4 mm. wide; nutlets ordinarily 4, 1.5-2.5 mm. long, thick-cruciform, tuberculate in rows, especially along the dorsal keel and on the transverse ridges near the ends of the lateral arms; scar set just beneath the middle of the nutlet.

Dry, open places at lower elevations; Wash. and adj. s. B.C. and w. Ida., s. to Baja Calif., Nev., and reputedly Utah, chiefly e. of the Cascades. Apr.-June.

### Symphytum L. Comfrey

Flowers borne in several or many small, sympodial, somewhat helicoid, naked, modified cymes; calyx shallowly to deeply cleft; corolla white or ochroleucous to pink or blue, tubular-campanulate, the well-defined throat much longer than the short, erect or apically spreading lobes; fornices narrow and elongate, erect; filaments inserted at the level of the fornices; anthers included; style elongate, shortly exserted; nutlets incurved, ventrally keeled, smooth or finely wrinkled or tuberculate, with a well-developed, stipelike, basal attachment which fits into a pit in the otherwise flattish receptacle, the attachment surrounded by a prominent, toothed rim (the basal margin of the nutlet) which fits closely to the gynobase; broad-leaved perennial herbs.

About 25 species, native to Europe and adj. Asia. (Name from the Greek symphyton, growing together, ancient name of S. officinale.)
    Reference:
    Bucknall, C. A revision of the genus Symphytum. Journ. Linn. Soc. 41:491-556. 1913.
1 Stem evidently winged by the conspicuously decurrent leaf bases        S. OFFICINALE
1 Stem not winged, the leaf bases not decurrent, or only shortly and inconspicuously so
                                                                        S. ASPERUM

Symphytum asperum Lepech. Nov. Act. Acad. Sci. Petrop. 14:442. 1805. (Mountains of the Caucasus)
    S. asperrimum Donn, Curtis' Bot. Mag. 24: pl. 929. 1806. (Caucasus)

Similar to S. officinale; hairs of the stem and inflorescence stouter, flattened, often recurved; leaves not decurrent, or only shortly (up to 1 cm.) and narrowly so; corolla averaging a little larger, blue (pink before anthesis). N=20.

Native of the Caucasus region; escaped from cultivation and more or less established along roadsides and in other disturbed habitats here and there in the U.S.; w. of the Cascades in our range. May-July.

Symphytum officinale L. Sp. Pl. 136. 1753. (Europe)

Taprooted perennial 3-12 dm. tall; stem and inflorescence somewhat hispid-hirsute with spreading or recurved, subterete hairs; leaves large, the basal ones petiolate, with ovate or lance-ovate blade mostly 15-30 cm. long and 7-12 cm. wide; cauline leaves gradually reduced and less petiolate but still ample, the upper commonly sessile; stem evidently winged by the conspicuously decurrent bases of the leaves; calyx 5-7 mm. long, cleft to below the middle; corolla ochroleucous or dull blue, about 1.5 cm. long; nutlets brownish-black, slightly wrinkled, 4 mm. long. N=18.

Native of Europe; escaped from cultivation and more or less established along roadsides and in other disturbed habitats in much of e. U.S., and recorded from Wash. and Mont. in our range. May-Aug.

## VERBENACEAE. Verbena or Vervain Family

Flowers gamopetalous, usually perfect; calyx gamosepalous, persistent, typically 4- to 5-lobed; corolla regular or more often irregular, the limb mostly 4- to 5-lobed and often bilabiate; stamens (2)4(5), epipetalous, mostly didynamous; ovary superior, scarcely or not at all lobed apically, 2(4-5)-carpellary, or 1-carpellary by abortion, each carpel ordinarily partitioned into 2 uniovulate segments; placentation axile; style slender, mostly 2-cleft at the tip, the short branches often unequal; endosperm usually scanty or wanting; fruit indehiscent and often drupaceous, or more commonly a dry schizocarp separating into 4 nutlets; herbs, shrubs, trees, or woody vines, with opposite (rarely alternate or whorled), exstipulate, simple or compound leaves, usually square stems, and diverse sorts of determinate or indeterminate inflorescences.

A large, cosmopolitan family, best developed in the tropics, composed of about 90 genera and nearly 3000 species. Our species are without horticultural value.

### Verbena L. Verbena; Vervain

Flowers borne in terminal, often densely many-flowered spikes, each flower solitary in the axil of a narrow bract; calyx 5-angled and unequally 5-toothed, only slightly enlarged in fruit; corolla salverform or funnelform, with flat, unequally 5-lobed, weakly bilabiate limb; only one lobe of the style stigmatic; ovary entire or very shallowly 4-lobed; fruit mostly enclosed in the calyx, dry, readily separating into 4 linear-oblong nutlets; endosperm wanting; herbs (ours) or shrubs with mostly toothed to dissected leaves.

More than 200 species, native chiefly to the warmer parts of the New World. Perry recognized 51 species in N. Am. in 1933; Moldenke (in Gleason's Illustrated Flora) apparently considered in 1952 that 69 species occurred in the U.S. (The classical Latin name for any of certain sacred boughs; derivation obscure.)

Reference:

Perry, Lily M. A revision of the North American species of Verbena. Ann. Mo. Bot. Gard. 20:239-362. 1933.

Each of our three species will hybridize with either of the other two. The hybrid of V. hastata and V. stricta has been named V. rydbergii Moldenke, Rev. Sudam. Bot. 4:19. 1937 (a strictly synonymous nomenclatural substitute for the apparently quite legitimate name V. paniculato-stricta Engelm. Am. Journ. Sci. 46:100. 1844, which in turn was based on Geyer, banks of the Mississippi opposite St. Louis); the hybrid of V. stricta and V. bracteata has been named V. deamii Moldenke, Rev. Sudam. Bot. 4:18. 1937, without description or type; and the hybrid of V. hastata and V. bracteata was described as V. bingenensis Moldenke, Phytologia

2:145. 1946 (Suksdorf, Bingen, Wash., July 9, 1898). V. bingenensis was mistakenly considered by Moldenke to be a hybrid of V. bracteata and V. lasiostachys Link.; the latter species is not known to occur n. of approximately the Willamette-Umpqua divide, 150 miles s.w. of Bingen, Wash.

Verbena officinalis L., a more or less erect, European annual with pinnatifid leaves and short, inconspicuous bracts, is introduced in e. U.S. and has been collected on ballast at Portland, Oreg.; it gives no evidence of becoming established as a permanent member of our flora.

1 Bracts of the inflorescence inconspicuous, barely as long as the calyx, or shorter; plants erect; leaves mostly merely toothed or very shallowly lobed
  2 Leaves sessile or subsessile, broadly elliptic or ovate; calyx 4-5 mm. long; corolla limb 7-11 mm. wide; spikes typically few or solitary, 6-30 cm. long     V. STRICTA
  2 Leaves evidently petiolate, mostly lanceolate or lance-ovate; calyx 2-3 mm. long; corolla limb 2.5-5 mm. wide; spikes panicled at the top of the stem and generally more or less numerous, 3-10(15) cm. long     V. HASTATA
1 Bracts of the inflorescence elongate, conspicuously surpassing the calyx; plants generally prostrate or decumbent; leaves, or many of them, more or less deeply cleft
                                 V. BRACTEATA

Verbena bracteata Lag. & Rodr. Anal. Cienc. Nat. 4:260. 1801. (A specimen cultivated at the Bot. Gard. of Madrid)
  V. bracteosa Michx. Fl. Bor. Am. 2:13. 1803. ("In the region of Illinois and in the town of Nashville," Tenn.)

Taprooted annual or more often perennial, usually with numerous prostrate or decumbent, sparsely spreading-hirsute stems 1-6 dm. long, small plants rarely single-stemmed and erect; leaves coarsely strigose, mostly wing-petiolate, the blade commonly 2-5 cm. long and 1-2.5 cm. wide, irregularly toothed and cleft, often with one or two pairs of pinnately disposed lower segments well differentiated from the larger, terminal segment; racemes terminating the main stem and branches, up to 1.5 dm. long; leaves passing abruptly into the entire, lance-linear bracts of the raceme, these spreading, 5-15 mm. long, conspicuously surpassing the 2.5-4 mm. calyx; corolla inconspicuous, almost hidden by the bracts, bluish or pinkish to rarely white, the tube 4 mm. long, the limb 2-3 mm. wide.

Roadsides and other disturbed habitats; B.C. to Me., s. to Calif., Mex., and Fla., probably not native in the n.e. part of its range. May-Sept.

Verbena hastata L. Sp. Pl. 20. 1753. (Can.)

Fibrous-rooted perennial apparently from a short caudex, erect, usually single-stemmed, (2)5-15 dm. tall; stem strigose or spreading-hirsute; leaves with short, distinct petioles mostly (0.5)1-2 cm. long; blades lanceolate or lance-ovate, acute, scabrous-hirsute, 4-15 cm. long, 1-4 cm. wide, sharply serrate, occasionally shallowly lobed, rarely with a pair of deeper, spreading lobes at the base; spikes 3-10(15) cm. long, panicled at the summit of the stem, often forming a flat-topped cluster, the stem simple below; bracts inconspicuous, subulate, 2-2.5 mm. long, surpassed by the (2)2.5-3 mm. calyx; corolla blue or violet, the tube 3-4 mm. long, the limb 2.5-5 mm. wide.

Ditch banks and moist low ground; B.C. to N.S., s. to Calif., Ariz., and Fla. July-Sept.

Verbena stricta Vent. Descr. Pl. Jard. Cels. 53. 1800. (Cultivated plants from seeds collected by Michaux in Ill.)

Short-lived, perhaps sometimes annual, plants with one or several erect stems 3-12 dm. tall arising from a taproot; herbage densely and conspicuously spreading-hairy, or the hairs of the upper surfaces of the leaves often appressed; leaves wholly cauline, narrowed to a sessile or subpetiolar base rarely as much as 1 cm. long, the blade broadly elliptic or ovate, 4-11 cm. long, 2-5 cm. wide, coarsely (often doubly) serrate, firm and evidently rugose-veiny; spikes elongate, 6-30 cm. long, typically few or solitary, terminating the stem and branches, more numerous when the stem is more branched; bracts lance-subulate, 4 mm. long, equaling or slightly shorter than the 4-5 mm. calyx; corolla deep blue or purple, the tube 6-7 mm. long, the limb 7-11 mm. wide.

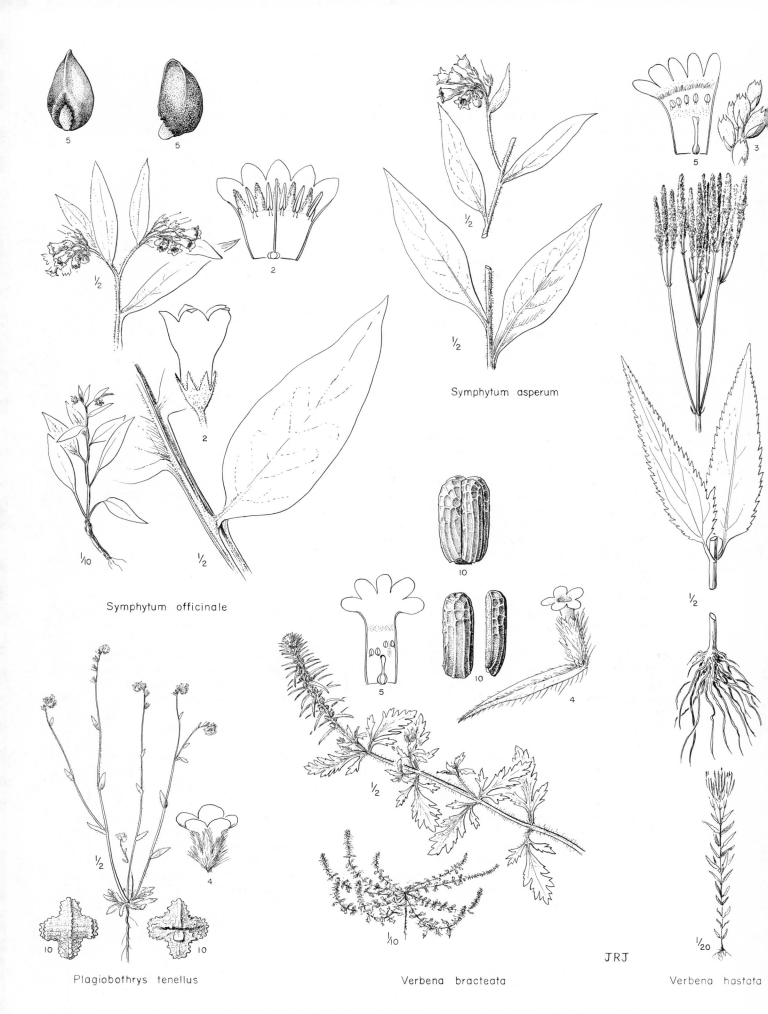

5      5

2

Symphytum asperum

Symphytum officinale

3

5

½

½

½

½

10

10     10

5

4

2

1/10      ½

½

½

½

1/10      4

10     10

½

1/10

JRJ

1/20

Plagiobothrys tenellus

Verbena bracteata

Verbena hastata

Roadsides and other dry open places; a characteristic species of the prairies and plains of c. U.S., extending w. through n. Ida. to n.e. Wash., and introduced eastward to the Atlantic. June-Sept.

## LABIATAE. Mint Family

Flowers gamopetalous, perfect (some plants sometimes pistillate); calyx gamosepalous, regular or irregular, often bilabiate, 5(10)-lobed, or the lobes rarely suppressed; corolla more or less irregular, usually bilabiate, 5-lobed, or 4-lobed by fusion of the two lobes of the upper lip; stamens epipetalous, 4, didynamous, or 2 by abortion of either the upper or the lower pair; ovary superior, basically 2-carpellary, but more or less deeply cleft into 4 uniovulate segments which ripen into hard nutlets, the style typically basal and surrounded by the basally attached, otherwise essentially distinct nutlets; style slender, mostly 2(4)-cleft at the tip, the short branches often unequal; ovules erect; seeds nearly or quite without endosperm; aromatic or sometimes inodorous herbs or shrubs, rarely trees or vines, with opposite (rarely whorled), exstipulate, entire or toothed to dissected leaves, ordinarily square stems, and diverse sorts of basically cymose (determinate) inflorescences, the flowers very often verticillate, or in small cymes in the axils of leaves or bracts.

A large family, of more than 3000 species and at least 150 genera, these often poorly defined; nearly cosmopolitan in distribution, but seldom in arctic or alpine regions. Monardella and Scutellaria are our only genera with native species of much horticultural interest.

1 Corolla seemingly 1-lipped, or 2-lipped with the upper lip cleft to the base; ovary merely 4-lobed, not cleft to the base, the nulets laterally attached
  2 Lobes of the upper lip of the corolla adjacent, often slightly displaced toward the lateral lobes of the lower lip, but the sinus between them scarcely more prominent than the lateral sinuses, and the central lobe of the lower lip less than twice as long as any of the other 4 corolla lobes; ours taprooted annuals     TRICHOSTEMA
  2 Lobes of the upper lip of the corolla well separated, displaced onto the lateral margins of the lower lip; central lobe of the lower lip much larger than any of the other 4 corolla lobes; ours a rhizomatous perennial     TEUCRIUM
1 Corolla either evidently 2-lipped or nearly regular, not at all 1-lipped nor with the upper lip cleft to the base; ovary cleft to the base, the nutlets basally attached
  3 Corolla only obscurely or scarcely bilabiate, either subequally 5-lobed with the lobes all about alike, or subequally 4-lobed with one of the lobes usually broader than the others and sometimes emarginate
    4 Stamens 4; plants aromatic; inflorescences axillary or terminal
      5 Inflorescence a dense terminal head subtended by conspicuous leafy or (in our species) dry bracts; corolla 5-lobed     MONARDELLA
      5 Inflorescence of axillary verticels, or terminal, spikelike, and inconspicuously bracteate; corolla usually 4-lobed, 5-lobed only in occasional individuals     MENTHA
    4 Stamens 2; plants odorless; inflorescences wholly axillary; corolla mostly 4-lobed     LYCOPUS
  3 Corolla more or less strongly bilabiate
    6 Stamens 2
      7 Connective much elongate, jointed to and somewhat resembling the relatively short filament, the upper arm bearing a single pollen sac at the end, the shorter lower arm often sterile and sometimes wholly suppressed; herbs or shrubs (elongate connective unique among our genera)     SALVIA
      7 Connective merely expanded at the end of the filament, bearing the two pollen sacs adjacent to each other, end to end; herbs
        8 Upper lip of the corolla narrow, galeate, in our species elongate and arcuate; inflorescence (ours) a single, terminal, basally leafy-bracteate head; ours with leaves mostly over 1 cm. wide     MONARDA
        8 Upper lip of the corolla short, straight, flat; inflorescence a series of axillary verticels; ours with the leaves well under 1 cm. wide     HEDEOMA

6  Stamens 4 (or obsolete in pistillate plants)
  9  Calyx bilabiate, with entire lips, bearing a prominent transverse external appendage on the upper side; lower stamens each with only one functional pollen sac
                                                   SCUTELLARIA
  9  Calyx regular or irregular, sometimes bilabiate, but in any case evidently toothed, and lacking any transverse appendage; each stamen with 2 pollen sacs, these sometimes confluent in dehiscence
    10  Inflorescence appearing terminal, the verticels tending to be crowded and mostly subtended by mere bracts that are evidently differentiated from the foliage leaves
      11  Stamens evidently exserted, readily visible without dissection of the flower; upper lip of the corolla not at all galeate
        12  Calyx evidently bilabiate, the upper lip with 3 short teeth, the lower cleft to the base into two longer lobes; lower stamens equaling or longer than the upper; introduced plants                                              THYMUS
        12  Calyx (at least in our species) regular or nearly so, with equal or subequal teeth
          13  Flowers solitary (or paired) in the axils of small but evident bracts, crowded into dense, trichotomously branched, headlike clusters at the ends of the stem and branches; lower stamens longer than the upper; introduced plants                                                ORIGANUM
          13  Flowers verticillate, the verticels crowded into a dense, spikelike inflorescence (in our species); upper stamens longer than the lower, and in our species declined between them; native plants          AGASTACHE
      11  Stamens ascending under the more or less galeately rounded upper lip of the corolla, scarcely or not at all exserted
        14  Flowers borne singly in the axils of small bracts, forming elongate terminal racemes                                                DRACOCEPHALUM
        14  Flowers verticillate in dense, sometimes interrupted inflorescences
          15  Calyx 5- to 10-nerved; lower stamens longer than the upper
            16  Calyx bilabiate, the upper lip broad and shallowly 3-toothed, the lower with two longer, narrow segments; filaments apically bidentate, the anther borne on the lower tooth                  PRUNELLA
            16  Calyx subequally 5-toothed; filaments normal          STACHYS
          15  Calyx 15-nerved; upper stamens longer than the lower
            17  Calyx regular or nearly so, with subequal teeth          NEPETA
            17  Calyx strongly irregular, ours with the upper tooth much broader than the other 4                                        MOLDAVICA
    10  Inflorescences appearing axillary, the flowers or verticels of flowers mostly subtended by more or less well developed leaves that are separated (except often in Lamium) by internodes of approximately normal length
      18  Calyx lobes recurved-hooked at the tip, 10 in our species; stamens included in the corolla tube                                          MARRUBIUM
      18  Calyx lobes not at all hooked, 5; stamens mostly ascending under the upper lip of the corolla
        19  Lower lip of the corolla constricted at the base of the enlarged, cleft or emarginate central lobe, the lateral lobes broad and low, seeming to arise from the corolla throat, each terminating in a short, divergent tooth          LAMIUM
        19  Lower lip of the corolla not constricted at the base of the central lobe, the lateral lobes directed more or less forward.
          20  Lower lip of the corolla with a pair of nipples projecting internally near the base; annual                                        GALEOPSIS
          20  Lower lip of the corolla without nipples; our species perennial
            21  Calyx evidently bilabiate, the upper lip nearly flat, shortly 3-toothed, the lower more deeply cleft into two segments; our species erect, 3-10 dm. tall                                          MELISSA
            21  Calyx in our species scarcely bilabiate, though the teeth may be unequal

22 Calyx lobes firm, spinulose; calyx 5-nerved; principal leaves palmately lobed; erect plants,
    up to 15 dm. tall                                                                 LEONURUS
22 Calyx lobes merely acute, scarcely spinulose; calyx in our species 12- to 15-nerved;
    leaves merely toothed; our species lax or trailing, not over about 4 dm. tall
    23 Upper pair of stamens longer than the lower; flowers verticillate, ours short-pedicel-
        late; introduced plants                                             GLECOMA
    23 Lower pair of stamens equaling or longer than the upper; ours mostly native plants with
        the flowers solitary in the axils and borne on pedicels 5-15 mm. long   SATUREJA

## Agastache Clayton

Flowers verticillate, the verticels crowded into a dense, spikelike inflorescence (in all our
species), or the inflorescence sometimes more open; calyx with 15 or more prominent veins,
equally 5-toothed, or the 3 upper teeth more or less connate, often whitish or tinged with pink
or blue; corolla pink to violet or white, bilabiate, with short lobes; stamens 4, usually exserted, the upper pair longer than the lower, and in our species declined between them; pollen sacs
parallel or nearly so; perennial herbs with petiolate or subsessile, entire or more often toothed, mostly ovate to deltoid or subcordate leaves.

Twenty species, native to N. Am. and s. e. Asia. (Name from the Greek agan, much, and
stachys, ear of grain, referring to the inflorescence. )

Reference:

Lint, Harold, and Carl Epling. A revision of the genus Agastache. Am. Midl. Nat. 33:207-
   30. 1945.

1 Plants dwarf, only 1-2 dm. tall, the leaf blades mostly 1-2 cm. long          A. CUSICKII
1 Plants robust, 4-15 dm. tall, many of the leaf blades over 3 cm. long
   2 Leaves very finely and closely tomentose-puberulent beneath; Cascades of Wash.
                             A. OCCIDENTALIS
   2 Leaves glabrous beneath, or often very minutely and obscurely hirtellous-scabrous,
     varying to occasionally loosely villous-puberulent; widespread, but absent from the
     Washington Cascades                                             A. URTICIFOLIA

Agastache cusickii (Greenm.) Heller, Muhl. 1:59. 1904.
  Lophanthus cusickii Greenm. Erythea 7:119. 1899. (Cusick 2001, Steens Mts., Oreg.)
  Agastache cusickii var. parva Cronq. Madrono 7:81. 1943. (Cronquist 3200, n. e. of Dickey,
    Custer Co., Ida.)
Stems numerous from a woody taproot and branching caudex, 1-1.5(2) dm. tall, simple or
branched, tending to be somewhat woody at the base; stem, leaves, bracts, and calyces finely hirtellous-puberulent; leaf blades ovate, deltoid-ovate, or a little narrower, crenate, mostly 1-2 cm. long and 6-15 mm. wide, borne on petioles up to 1 cm. long; inflorescence mostly
1.5-4 cm. long (exclusive of any remote verticels), the bracts and calyces tinged with lavender-purple; calyx teeth 2-5 mm. long, lance-subulate, obscurely veined, or only the midrib
evident; corolla 8-12 mm. long, whitish; stamens conspicuously exserted.

In dry, rocky places at moderate elevations in the mountains; c. Ida. to n. Nev. and s. e.
Oreg. (Steens Mts.). July-Aug.

The plants of Ida. and Nev., as described above, belong to the var. parva Cronq. The var.
cusickii, known only from the Steens Mts., is slightly taller (to 4 dm. or perhaps more) and
tends to have longer petioles.

Agastache occidentalis (Piper) Heller, Muhl. 1:4. 1900.
  Vleckia occidentalis Piper, Erythea 6:31. 1898. (Elmer 396, s. w. of Ellensburg, Wash.)
Very similar to A. urticifolia, averaging smaller (4-8 dm.) and less branched; leaf blades
up to 6 cm. long and 4(5) cm. wide, on petioles up to 5 cm. long; differing constantly in the
very short, fine, and close puberulent tomentum of the lower surfaces of the leaves.

Open, often rocky slopes and ledges, from the e. base to near the summit of the Cascades;
Chelan Co. to Yakima Co., Wash. June-Aug.

Agastache occidentalis is evidently allied to A. urticifolia, and perhaps better treated as a geographic variety.

Agastache urticifolia (Benth.) Kuntze, Rev. Gen. 2:511. 1891.
  Lophanthus urticifolius Benth. Bot. Reg. 15: sub pl. 1282. 1829. Vleckia urticifolia Raf. Fl.
    Tellur. 3:89. 1837. (Douglas, n. w. coast of America)
  AGASTACHE URTICIFOLIA var. GLAUCIFOLIA (Heller) Cronq. hoc loc. A. glaucifolia
    Heller, Muhl. 1:32. 1900. (Heller 5792, Knight's Valley, Sonoma Co., Calif.)
  Fibrous-rooted perennial from a branching, woody caudex; stems numerous, 4-15 dm. tall, simple or often branched, very finely retrorse-puberulent or retrorse-strigillose to glabrous; leaf blades rather coarsely crenate or crenate-serrate, ovate to more often narrowly to broadly deltoid or subcordate, the better developed ones mostly 4-10 cm. long and 2-8 cm. wide, on petioles 1-5 cm. long, glabrous or scaberulous above, glabrous or often very shortly and finely hirtellous-scabrous beneath; inflorescence 3-15 cm. long; calyx teeth 3-5 mm. long, narrowly triangular-acuminate, evidently 3-veined, usually tinged with lavender-purple; corolla whitish, 10-14 mm. long; stamens conspicuously exserted.
  Open slopes and draws, from the foothills to rather high elevations in the mountains; s. e. B. C. and w. Mont. to Colo. and Calif.; in Wash. not known to extend w. of Grand Coulee Dam, but in Oreg. sometimes reaching the e. foot of the Cascades, and in Calif. extending to the coast. June-Aug.
  In the Sierra Nevada and Coast Ranges of Calif., the otherwise widespread var. urticifolia (as described above) is largely replaced by the var. glaucifolia (Heller) Cronq., with the leaves evidently and loosely puberulent, and transitional specimens have been reported to extend into our range in Oreg.


Dracocephalum L.


  Flowers borne singly in the axils of small bracts, forming elongate, terminal racemes; calyx 10-nerved, equally or unequally 5-toothed; corolla pink or purple to white, elongate, bilabiate, the upper lip subgaleate and scarcely or barely notched, the lower lip about as long as the upper, spreading, 3-lobed; stamens 4, ascending under the upper lip of the corolla, the lower pair the longer; pollen sacs nearly parallel; glabrous or finely puberulent, fibrous-rooted, perennial herbs with toothed or sometimes entire leaves, recalling some of the larger species of Penstemon in aspect.
  About 7 species, native to N. Am. (Name from the Greek drakon, a dragon, and kephale, head, from a fancied resemblance in the shape of the flower.)
  The name Dracocephalum was used by Linnaeus in the Species Plantarum and in the fifth (1754) edition of the Genera Plantarum to cover species which are now considered to belong to two distinct genera. Each of these two elements has been provided with a generic name, and the name Dracocephalum has by different authors been restricted to each. Linnaeus' description and comments in the Genera Plantarum clearly indicate that D. virginianum must be considered the type species. When the genus in the Linnaean sense is subdivided, the name Moldavica [Tourn.] Adans. (1763) must be maintained, and the name Physostegia Benth. (1829, based on D. virginianum) must be reduced to Dracocephalum. McClintock's review of the nomenclatural history (Leafl. West. Bot. 5:171-72. 1949) is clear and pertinent, but her proposal to retypify Dracocephalum on D. ruyschianum cannot stand in the absence of action by an international botanical congress to conserve Physostegia Benth., or of some other unanticipated alteration of the Rules of Nomenclature.

Dracocephalum nuttallii Britt. in Britt. & Brown Ill. Fl. 2nd ed. 3:117. 1913.
  Physostegia parviflora Nutt. ex Gray, Proc. Am. Acad. 8:371. 1872. Physostegia nuttallii
    Fassett, Rhodora 41:525. 1939. (Nuttall, Oreg.)
  Stems arising singly from a rhizome or rhizomelike vertical base, 2-10 dm. tall, simple, or branched at the base of the inflorescence; lower leaves reduced and deciduous, the others 3-10 cm. long, 5-20 mm. wide, serrate or subentire, sessile, linear-oblong to elliptic-oblong, those directly beneath the inflorescence becoming lanceolate or lance-ovate, broadest

near the base; racemes 2-8 cm. long, closely flowered; flowers subsessile, the finely glandu-lar-puberulent, shortly and unevenly toothed calyx 4-6 mm. long, much surpassing the mostly entire, more or less ovate bracts; corolla lavender-purple, 12-16 mm. long, the lips only 3-5 mm.

Shores of streams and lakes, marshes, and other moist, low places in the valleys and foot-hills; s. B.C. and Wash. (e. of the Cascade summits), and through the Columbia gorge to the vicinity of Portland, Oreg.; e. through most of Ida. and thence to s. Sask., N.D., and ex-treme w. Minn. and w. Neb. July-Sept.

The related, more eastern species, D. virginianum L., with larger flowers (1.5-3 cm.) and with the uppermost leaves more slender, narrow-based, and often relatively small, has been collected as far west as Great Falls, Mont. [D. formosius (Lunell) Rydb.]

### Galeopsis L. Hemp Nettle

Flowers sessile, numerous in remote or crowded axillary verticels, the leaves subtending the upper verticels more or less reduced; calyx 5-15-nerved, with 5 equal or unequal, firm and apically subspinescent teeth; corolla bilabiate, the upper lip galeately rounded and entire; lower lip spreading, 3-lobed, with the central lobe sometimes cleft, bearing a pair of prom-inent projections or "nipples" on the upper side near the base; stamens 4, ascending under the upper lip, the lower pair the longer; pollen sacs contiguous in a nearly straight, transverse line at the summit of the apically expanded filament, the expanded portion representing the connective; each pollen sac transversely 2-valved, the upper valve ciliate, the lower larger and glabrous; annuals with mostly toothed leaves.

About half a dozen species, native to Europe and temperate Asia. (The classical Latin name for some plants with bilabiate corollas.)

Galeopsis tetrahit L. Sp. Pl. 579. 1753. (Europe)
 G. bifida Boenn. Prodr. Fl. Monast. 178. 1824. G. tetrahit var. bifida LeJeune & Court.
  Comp. Fl. Belg. 2:241. 1831. G. tetrahit ssp. bifida Fries, fide Hegi, Fl. Mittel-Eur.
  5$^4$:2467. 1927. (Westphalia)
Annual, 1.5-7 dm. tall, branched when well developed, the stem more or less hispid with retrorsely spreading hairs, and often with shorter but still prominent, gland-tipped hairs, especially just beneath the upper nodes; leaves petiolate, with rhombic-elliptic to more or less ovate, coarsely blunt-serrate, basally rounded or cuneate, more or less strigose or hirsute blade 3-10 cm. long and 1-5 cm. wide; verticels dense, generally several, the lower often re-mote; calyx about 1 cm. long at anthesis, somewhat larger in fruit, the erect spinescent lobes about equaling the tube; corolla mostly 15-23 mm. long, purple or pink (white) with darker markings, the middle lobe of the lower lip nearly square, not emarginate. N=16.

Meadows and other moist places; native of Eurasia, apparently only introduced in America, although often occurring in undisturbed habitats and appearing to be native. Known in our range from n. Ida., w. Wash., and s. B.C. July-Aug.

Galeopsis tetrahit is known to have originated from G. pubescens Bess. and G. speciosa Mill. by hybridization and polyploidy, and has been experimentally synthesized in complex fashion, but the F$_1$ between the two parent species does not closely resemble the eventual tetra-ploid. The species may well have arisen more than once, and may owe some of its variability in technical characters to the multiplicity of origin. There are two principal intergradient phases of more or less concurrent distribution that have sometimes been dignified with separate binomials. Our plant, as here described, is var. tetrahit. The var. bifida (Boenn.) LeJeune & Court. differs in its somewhat smaller corollas (seldom much over 1.5 cm. long) with the central lobe of the lower lip emarginate or cleft; other differences have been alleged, but au-thors are not in full agreement about them.

### Glecoma L. Ground Ivy

Flowers in few-flowered verticels axillary to normal leaves; calyx 15-nerved, oblique at the mouth, unequally 5-toothed; corolla mostly blue or purple, bilabiate, the upper lip 2-lobed,

2

2

½

½

½

Verbena stricta

Galeopsis tetrahit

½

2

3

½

½

Agastache cusickii

½

½

3

½

1/20

Agastache occidentalis

JRJ

3

2

2

3

½

1/12

½

Dracocephalum nuttallii

3

3

1/20

½

Agastache urticifolia

somewhat concave, the lower lip spreading, with large central and smaller lateral lobes; stamens 4, the upper pair the longer, ascending under the upper lip of the corolla and scarcely exserted; pollen sacs divergent at about right angles, opening by separate slits; diffuse or creeping, perennial herbs with toothed leaves.

Half a dozen species, native to Eurasia, related to Nepeta and sometimes included in that genus, but sharply set off by the inflorescence and habit. (Name from the Greek glechon, the classical name for a species of Mentha.) The name has also been spelled Glechoma, but Glecoma is the correct form under the rules of nomenclature.

Glecoma hederacea L. Sp. Pl. 578. 1753.
  Nepeta glechoma Benth. Lab. Gen. & Sp. 485. 1834. Nepeta hederacea Trevis. Prosp. Fl.
    Eug. 26. 1842. (n. Europe)
Fibrous-rooted perennial from slender stolons or superficial rhizomes, the stems lax, 1-4 dm. tall, retrorsely scabrous (or hirsute) to subglabrous, pilose at the nodes; leaves wholly cauline, all about alike, petiolate, the blades glabrous or hirsute, rotund-cordate to cordate-reniform, strongly crenate, 1-3 cm. long; flowers short-pedicellate; calyx narrow, 5-6 mm. long, hirtellous-scabrous, the upper teeth the longer; corolla blue-violet, purple-maculate, 13-23 mm. long, or in forms with reduced anthers only about 1 cm. long or less.

Moist woods and thickets, often in disturbed habitats; native of Eurasia, now well established across the U.S., and occasionally found in our range. Apr.-June.

A good ground cover, but too aggressive for most areas. The small-flowered, pistillate plants, which are commoner in America (and in our range) than the normal ones, have sometimes been segregated as var. micrantha Trevis. or var. parviflora (Benth.) Druce, but they are considered by modern European students to be without taxonomic significance.

### Hedeoma Pers.

Flowers verticillate in the axils of ordinary leaves; calyx about 13-nerved, mostly distended near the base and usually more or less bilabiate, the three upper teeth more or less connate, the two lower ones longer, narrower, and distinct; corolla bilabiate, the upper lip short, straight, flat, emarginate or 2-lobed, the lower spreading and 3-lobed; stamens 2, ascending under the upper lip of the corolla and often slightly exserted beyond it; pollen sacs separately dehiscent, contiguous in a nearly straight transverse line (or at a very wide angle) at the summit of the apically expanded filament, the expanded portion representing the connective; small, wiry-stemmed annual or perennial herbs with entire or obscurely toothed, mostly subsessile leaves.

About 28 species, native to the New World, chiefly N. Am. (Name from the Greek hedyosmon [hedys, sweet, and osme, odor], used for some other species of the family.)
  Reference:
    Epling, Carl, and W. S. Stewart. A revision of Hedeoma, with a review of allied genera.
      Fedde Rep. Sp. Nov. Beih. 115: 1-49. 1939.
1 Plants perennial; calyx lobes subconnivent in fruit, the whole calyx seeming to taper gradually from near the base to (near) the tips of the lobes          H. DRUMMONDII
1 Plants annual; upper calyx lobes more or less strongly upcurved and divergent from the lower ones          H. HISPIDUM

Hedeoma drummondii Benth. Lab. Gen. & Sp. 368. 1834. (Berlandier, near Monterrey, Nuevo Leon, Mex.)
Taprooted perennial, somewhat woody at the base, 1-2.5 dm. tall, evidently puberulent throughout, the several, mostly branching stems retrorsely so; leaves linear to rather narrowly elliptic, entire, 5-15 mm. long, the lower mostly deciduous, the middle and upper mostly subtending few-flowered verticels; calyx 6 mm. long, strongly ribbed-sulcate, slender, tapering gradually from near the base, the teeth subconnivent in fruit, the upper ones slightly or scarcely upcurved and evidently surpassed by the slender, upwardly arcuate, hispid-ciliate lower ones; corolla nearly or fully 1 cm. long, the stamens slightly surpassing the short lips.

Dry, open places at lower elevations; w. Minn. to the more easterly intermontane valleys of Mont., s. to Okla., Tex., Ariz., and n. Mex. June-July.

Hedeoma hispidum Pursh, Fl. Am. Sept. 414. 1814.
   Cunila hispida Spreng. Syst. Veg. 1:54. 1825. Ziziphora hispida Roem. & Schult. Syst. Mant. 1:179. 1822. (Banks of the Missouri)
   Similar in general appearance to H. drummondii, but annual, and with consistently narrow, often less hairy leaves up to 2 cm. long, and with the hairs of the stem longer and more conspicuous; calyx 5 mm. long, stouter, with broader internerves; upper calyx teeth about equaling the lower, more or less strongly upcurved and divergent from them.
   Dry, open places at lower elevations; s. Alta. and the more easterly intermontane valleys of Mont., s. to Colo. and e. Tex., and e. nearly to the Atlantic. June-Aug.

<h3 style="text-align:center">Lamium L. Dead Nettle</h3>

Flowers verticillate in the axils of ordinary leaves or modified but well-developed and leafy bracts; calyx inconspicuously nerved, the 5 pointed teeth equal or the upper one the largest; corolla purple to white or rarely yellow, bilabiate, the upper lip entire or 2-lobed, galeately rounded, the lower spreading, constricted at the base of the enlarged, cleft or emarginate central lobe, the lateral lobes broad and low, terminating in a short tooth, seemingly borne on the corolla throat; stamens 4, the lower pair the longer, ascending under the upper lip; pollen sacs more or less divergent or divaricate, conspicuously hairy; nutlets angularly 3-sided, truncate at the summit; annual or perennial herbs with toothed or pinnatifid, mostly cordate leaves.
   About 40 species, native to temperate Eurasia and n. Africa. (The ancient Latin name.)
1 Upper lip of the corolla 7-12 mm. long; plants perennial; leaves all petioled
                                                                        L. MACULATUM
1 Upper lip of the corolla 2-5 mm. long; plants annual
   2 Leaves all petiolate                                                L. PURPUREUM
   2 Leaves subtending the flower clusters mostly or all sessile and clasping
                                                                        L. AMPLEXICAULE

Lamium amplexicaule L. Sp. Pl. 579. 1753. (Europe)
   Annual from a short taproot, generally branched at the base, the several weak stems decumbent below; herbage inconspicuously hirsute or strigose to subglabrous; proper leaves restricted to the lower part of the stem, petiolate, with broad, rounded, more or less cordate, coarsely crenate or lobulate blade seldom as much as 1.5(2) cm. long; leaves subtending the flower clusters (bracts) sessile, broad-based, clasping, often 1.5(2.5) cm. long, surpassing the calyces but usually surpassed by the corollas; verticels few and (except sometimes the upper) mostly well spaced, the lowest fully developed one often borne at or below the middle of the stem; lowest verticel sometimes few-flowered and subtended by petiolate leaves; calyx hirsute, 5-8 mm. long, the narrow, erect lobes about equaling the tube; corolla purplish, 12-18 mm. long, glabrous inside, hairy outside, the hairs of the galea purple; upper lip 3-5 mm. long; occasional plants produce small, cleistogamous flowers. N=9.
   A weed in fields and waste places; native to Eurasia and n. Africa, now well established in N. Am., and occasionally found throughout our range. Apr.-July.

Lamium maculatum L. Sp. Pl. 2nd ed. 809. 1763. (Italy)
   Fibrous-rooted perennial from a creeping base, decumbent below, 1.5-7 dm. tall, somewhat hirsute; leaves all petiolate, with cordate-ovate or triangular-ovate, crenate-serrate blade 1.5-6 cm. long and 1-5 cm. wide, generally with an irregular broad white stripe along the midvein; verticels several, less crowded than in L. purpureum, but often more so than in L. amplexicaule, the lower subtended by full-sized leaves, the leaves subtending the upper verticels becoming somewhat reduced and short-petiolate; calyx mostly 5-10 mm. long, with firm, unequal, somewhat divaricate slender lobes equaling or shorter than the tube; corolla pink-purple, rarely white, mostly 2-2.5 cm. long, curved, hairy outside and with a ring of hairs inside; galea 7-12 mm. long. N=9.

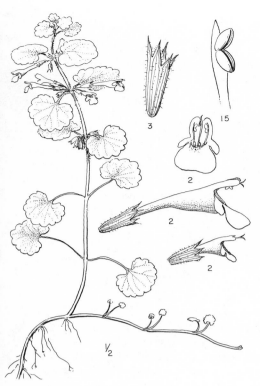

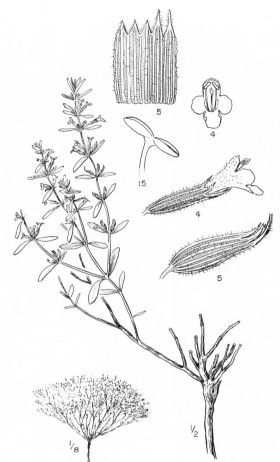

Glecoma hederacea

Hedeoma drummondii

Hedeoma hispidum

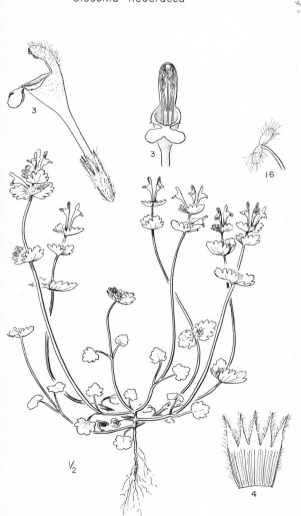

Lamium amplexicaule

JRJ

Lamium maculatum

Native of Europe and adj. Asia, now established in e. U.S., and reported to be escaped from cultivation and established in the Willamette Valley of Oreg. Apr. -Aug.

**Lamium purpureum L. Sp. Pl. 579. 1753. (Europe)**

Annual from a short taproot, simple or more often branched at the base, decumbent or sub-prostrate, inconspicuously hairy, the stems 1-4 dm. long; leaves all petiolate, crenate or crenate-serrate, the lower with rotund-cordate or broader blade 0.7-3 cm. long and separat-ed by 1-3 very elongate internodes from the several more ovate-cordate, closely crowded, shorter-petiolate, and eventually smaller upper pairs which subtend the verticels of flowers; calyx 5-6 mm. long, with firm, unequal, somewhat divaricate, slender lobes about as long as the tube; corolla pink-purple, mostly 1-1.5 cm. long, hairy outside and with a ring of hairs inside near the base; galea 2-4 mm. long. N=9.

Native of Eurasia; now irregularly established in America and known from several stations in our range as a weed. Apr. -July.

## Leonurus L. Motherwort

Flowers verticillate in the axils of somewhat reduced leaves; calyx 5(10)-nerved, with 5 narrow, firm, spinulose, subequal lobes; corolla pink or white, bilabiate, the tube scarcely exserted from the calyx, the upper lip entire and galeately rounded, the lower spreading and 3-lobed, with obcordate central lobe and oblong lateral lobes; stamens 4, ascending under the upper lip and scarcely exserted, the lower pair equaling or longer than the upper; pollen sacs mostly parallel; nutlets triquetrous, truncate above; erect perennial herbs with the lower leaves palmately cleft, the upper gradually smaller, narrower, and less cleft, but still well surpassing their axillary verticels of flowers.

About 10 species, native mostly to temperate Eurasia. (Name from the Greek leon, lion, and oura, tail.)

**Leonurus cardiaca L. Sp. Pl. 584. 1753. (Europe)**

Fibrous-rooted perennial with the stems more or less clustered on a stout, branching rhi-zome or caudex, 4-15 dm. tall, the stem retrorsely strigose-puberulent on the angles; leaves wholly cauline, the lower soon deciduous, the principal ones borne somewhat above the base, with palmately veined, palmately cleft and again coarsely toothed blade 5-10 cm. long and wide, loosely short-hairy beneath and inconspicuously strigose above, the petiole about equal-ing the blade; middle and upper leaves gradually smaller, narrower, less cleft, and shorter-petiolate, the uppermost often entire; flowers subsessile; calyx tube firm, 5-angled, 3-4 mm. long; calyx lobes about equaling the tube, the 2 lower ones deflexed; corolla 1 cm. long, pale pink, the upper lip conspicuously white-villous. N=9.

Native of Asia, formerly cultivated as a home remedy, and now established as a casual weed over much of the U.S.; in our range known only from the more easterly intermontane valleys of Mont. and about Bellingham, Wash. June-Sept.

## Lycopus L. Bugleweed; Water Horehound

Flowers sessile, verticillate in the axils of scarcely reduced leaves; calyx evidently to ob-scurely 5-nerved, sometimes with additional lesser nerves, the 5 teeth about equal; corolla small, the tube short, internally hairy at the throat, the limb nearly regularly 4-lobed; upper corolla lobe formed by the fusion of the two lobes of the upper lip, tending to be broader than the other lobes, and often apically emarginate; fertile stamens 2, slightly exserted, the upper pair represented by small staminodes, or obsolete; pollen sacs parallel; nutlets widened up-ward, bearing a corky ridge along the lateral angles and often across the top, the outer surface smooth and nearly plane, the inner convex and commonly glandular; rhizomatous, perennial, scarcely aromatic herbs with wholly cauline leaves.

About a dozen species, native to the N. Temp. regions, one in Australia. (Name from the Greek lykos, wolf, and pous, foot, translating the French vernacular name, patte du loup.)

Our species usually can be recognized on vegetative characters, but study of the calyx and/or the nutlets may be necessary to confirm the identification.

1 Calyx lobes narrow, firm, slenderly subulate-pointed, distinctly surpassing the mature nutlets

  2 Nutlets mostly 1.6-2.1 mm. long by 1.4-1.8 mm. wide, the outer apical margin truncate and often irregularly toothed; leaves sessile or nearly so, with abruptly contracted to occasionally more tapering base and rather coarsely but fairly evenly serrate margins
                                                              L. ASPER

  2 Nutlets mostly 1.0-1.4 mm. long by 0.8-1.2 mm. wide, the outer apical margin smooth and broadly rounded; leaves tapering to a short petiole or petiolar base, irregularly incised-toothed or subpinnatifid, rarely merely serrate        L. AMERICANUS

1 Calyx lobes broader, nearly ovate, soft, merely acutish, scarcely surpassing the mature nutlets; leaves mostly tapering to a short petiole, sometimes more abruptly contracted and subsessile, the margins merely toothed        L. UNIFLORUS

Lycopus americanus Muhl. Cat. Pl. Sept. 3. 1813, nom. subnud.; ex Barton, Fl. Phila. Prodr. 15. 1815. Phytosalpinx americanus Lunell, Am. Midl. Nat. 5:2. 1917. (Pa.)
    Rhizomes elongate, not tuberiferous; plants 2-8 dm. tall, simple or branched, hairy at the nodes, especially upward, otherwise generally glabrous or nearly so; leaves numerous, gradually or scarcely reduced upward, mostly 3-8 cm. long and 1-3.5 cm. wide, tapering to a short petiole or narrow, subpetiolar base, the margins coarsely and irregularly incised-toothed or subpinnatisect, the lower teeth the larger; calyx lobes narrow, firm, slenderly subulate-pointed, with evident midnerve, distinctly surpassing the mature nutlets; corolla white, 2-3 mm. long, barely if at all surpassing the calyx; staminodes small, clavellate; nutlets mostly 1.0-1.4 mm. long by 0.8-1.2 mm. wide, the outer apical margin smooth and broadly rounded, the corky lateral ridges confluent around the tip. N=11.
    Marshes and moist low ground along streams, in the foothills and lowlands; B. C. to Newf., s. to Calif. and Fla. June-Aug.

Lycopus asper Greene, Pitt. 3:339. 1898.
  Phytosalpinx asper Lunell, Am. Midl. Nat. 5:2. 1917. (Sheldon, Battle Lake, Minn., in 1892)
  Lycopus lucidus sensu American authors. Not Turcz. ex Benth.
  L. lucidus var. americanus Gray, Proc. Am. Acad. 8:286. 1870. (Specimens cited from Sask., Neb., and Kans.)
  L. maritimus Greene, Pitt. 3:340. 1898. (Greene, Suisun marshes, Calif., in 1889)
    Rhizomes often (regularly?) tuberiferous; plants 2-8 dm. tall, simple or branched, the stem spreading-hairy on the angles, at least above, or more uniformly so throughout; leaves glabrous or frequently scabrous to hirsute-puberulent especially on the upper surface, numerous, gradually or scarcely reduced upward, mostly 3.5-10 cm. long and 0.6-3.5 cm. wide, broadbased and sessile or subsessile, more rarely tapering gradually to a narrow base, the margins rather coarsely but fairly evenly serrate; calyx lobes narrow, firm, slenderly subulate-pointed, with evident midnerve, distinctly surpassing the mature nutlets; corolla white, 3-5 mm. long, only slightly (if at all) surpassing the calyx; staminodes small, clavate; nutlets mostly 1.6-2.1 mm. long by 1.4-1.8 mm. wide, the outer apical margin truncate and often irregularly toothed.
    Marshes and shores of streams and lakes, at lower elevations, tolerant of alkali; s. B. C. to Calif., e. to Colo., Iowa, Minn., Sask., and more rarely Mich. and Ont. In our range found wholly e. of the Cascades. June-Aug.
    Lycopus asper has frequently been confused with the related, chiefly e. Asian species, L. lucidus, which however has consistently narrower nutlets that tend to be broadly rounded above.

Lycopus uniflorus Michx. Fl. Bor. Am. 1:14. 1803. (Lake St. John and Lake Mistassini, in e. Can.)
  L. communis Bicknell ex Britt. Man. 803. 1901. (Van Cortlandt, N. Y.; no collector cited)
  Stems of the season 1-4(7) dm. tall, simple or occasionally branched, puberulent or stri-

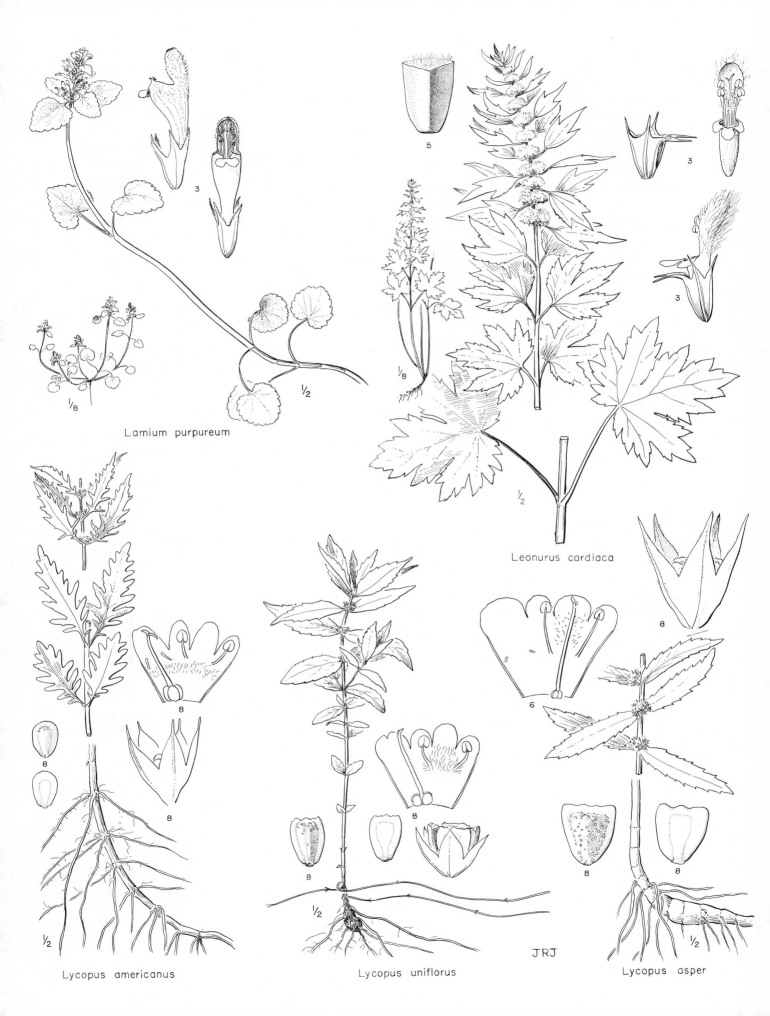

Lamium purpureum

Leonurus cardiaca

Lycopus americanus

Lycopus uniflorus

Lycopus asper

JRJ

gose-puberulent to finely scaberulous, arising singly from distinct tubers and producing very slender, elongate rhizomes and/or stolons near the base; leaves obscurely scabrous-puberulent or glabrous, not much reduced upward, mostly 2-8 cm. long and 0.6-3 cm. wide, gradually or occasionally more abruptly narrowed to the short petiole or subpetiolar base, the margins rather coarsely and often somewhat irregularly serrate-dentate; calyx small, soft, obscurely or scarcely nerved, the relatively broad, nearly ovate teeth merely acutish, about equaling the mature nutlets; corolla white or pinkish, 2.5-3.5 mm. long, distinctly surpassing the calyx; staminodes minute or obsolete; nutlets 1.1-1.8 mm. long, 0.8-1.2 mm. wide, the outer apical margin subtruncate and toothed.

Stream banks, marshes, and peat bogs, chiefly in the mountains, but descending to near sea level w. of the Cascades; Alas. to Newf., s. to n.w. Calif., n. Ida., Mont., Neb., Ark., and N.C. July-Sept.

## Marrubium L. Horehound

Flowers in dense whorls axillary to normal leaves; calyx 5- to 10-nerved, the 5- to 10 mostly pungent teeth commonly recurved-hooked at the tip, at least in age; corolla white or purplish, or rarely yellowish, bilabiate, the short tube included or barely exserted from the calyx, the upper lip flat or subgaleately rounded, entire or shortly bifid, the lower spreading, 3-lobed, the larger central lobe often emarginate; stamens 4, the lower pair the longer, included in the corolla tube; pollen sacs divaricate, soon confluent; perennial herbs, generally woolly, with toothed or incised, rugose leaves.

About three dozen species, native to the Old World. (The classical Latin name for this and probably some other species of Labiatae, perhaps eventually derived from the Hebrew marrob, a bitter juice.)

Marrubium vulgare L. Sp. Pl. 583. 1753. (N. Europe)
Perennial from a stout taproot; stems generally several, nearly prostrate to suberect, 3-10 dm. long, conspicuously white-woolly; leaves wholly cauline, canescent-woolly or partly subglabrate, not much reduced upward, petiolate, with broadly elliptic to rotund-ovate or nearly flabellate, evidently crenate blade 2-5.5 cm. long and often nearly or quite as wide; verticels compact; calyx stellate and often more or less long-hairy, bearing a ring of exserted long hairs within the throat; calyx tube 4-5 mm. long, the 10 narrow, firm teeth somewhat shorter, eventually widely spreading, sometimes unequal, their spinulose tips recurved from the first; corolla whitish, slightly exserted, the lips about equal, the upper erect and narrowly 2-lobed, the lower spreading and with broadly rounded central lobe. N=17, 18.

A casual weed along roadsides and in other disturbed habitats; native of Europe, now widespread elsewhere in the world and found throughout our range. June-Oct.

## Melissa L. Balm

Flowers borne in small verticels axillary to well-developed leaves; calyx bilabiate, strongly 13-nerved, the upper lip nearly flat, shortly 3-toothed, the lower more deeply 2-cleft; corolla white to yellowish or light pink or blue, bilabiate, the upper lip nearly flat, emarginate, the lower spreading and 3-lobed; stamens 4, ascending under the upper lip, the lower pair the longer; pollen sacs more or less strongly divaricate; herbs with toothed leaves.

Four species, native to Eurasia. (Name from the Greek melissa, a honey bee, the plant being used as a source of honey.)

Melissa officinalis L. Sp. Pl. 592. 1753. ("in montibus Genevensibus, Allobrogocis, Italicis")
Lemon-scented, fibrous-rooted perennial from a woody rhizome or elongate caudex, 3-10 dm. tall, commonly branched, finely canescent (at least on the upper part of the stem) and generally with some long spreading hairs and long-stipitate glands upward; leaves petiolate, the lower soon deciduous, the middle cauline ones ovate to nearly deltoid, coarsely blunt-serrate, 4-9 cm. long, 2.5-5 cm. wide, the upper and rameal ones gradually reduced, and more tapering to the shorter petiole; flowers short-pedicellate; calyx 7-9 mm. long, the teeth firm,

with subspinulose tip; corolla white to pinkish or light blue, 10-15 mm. long, with short lips. N=16.

Roadsides and other disturbed habitats; native of Eurasia, long in cultivation in Europe, and sometimes grown also in America; now found here and there over much of the U.S. and adj. Can. June-Aug.

### Mentha L. Mint

Flowers in verticels subtended by leaves or more or less reduced bracts, the verticels crowded or remote, often forming spikelike inflorescences; calyx 10-nerved, regular or subbilabiate, the 5 teeth equal or unequal; corolla with short tube and nearly regularly 4-lobed limb, the upper lobe formed by the fusion of the two lobes of the upper lip, tending to be broader than the other lobes, and often apically emarginate; stamens 4, equal, usually exserted; pollen sacs parallel; aromatic, rhizomatous perennial herbs with wholly cauline leaves.

More than 200 binomials exist; of these, perhaps as few as 15 represent original species, one circumboreal, the rest equally divided between Eurasia and Australia. The European species, long in cultivation, have given rise through hybridization, polyploidy, and other kinds of chromosomal aberration to numerous, more or less stabilized, additional populations which have become established in the wild as well as being retained in cultivation, thus weakening or even obliterating the specific distinctions. Many species produce small-flowered pistillate plants as well as normal perfect-flowered ones. (Name from the Latin menta, Greek minthe, mint, perhaps eventually from the Greek nymph Minthe.)

1 Verticels of flowers axillary, subtended by ordinary leaves and separated by internodes of
    normal length                                      M. ARVENSIS
1 Verticels of flowers crowded into terminal, inconspicuously bracteate, spikelike inflorescences
  2 Leaves sessile or subsessile, the petioles, if any, not over about 3 mm. long; calyx 1-2 mm. long; spikes relatively slender, mostly 0.5-1 cm. thick at anthesis
    3 Leaves mostly 1-2 times as long as wide, acutish to more often obtuse or rounded, evidently woolly-villous on both sides, especially beneath    M. ROTUNDIFOLIA
    3 Leaves mostly 2-3.5 times as long as wide, more or less acute, glabrous, or commonly hirsute along the main veins beneath    M. SPICATA
  2 Leaves evidently petiolate, many of the petioles 4 mm. or more (to 15 mm.) long; calyx 2.5-3 mm. long; spikes relatively stout, mostly 1-1.5 cm. thick at anthesis
                                  M. PIPERITA

Mentha arvensis L. Sp. Pl. 577. 1753. (Europe)
M. canadensis L. Sp. Pl. 577. 1753. M. canadensis var. villosa Benth. Lab. Gen. & Sp. 181. 1833. M. arvensis var. canadensis Kuntze, Rev. Gen. 2:524. 1891. M. arvensis var. villosa S. R. Stewart, Rhodora 46: 333. 1944. (Kalm, Can.)
M. borealis Michx. Fl. Bor. Am. 2:2. 1803. M. canadensis var. glabrata Benth. Lab. Gen. & Sp. 181. 1833. M. canadensis var. glabrior Hook. Fl. Bor. Am. 2:111. 1838. M. arvensis var. borealis Kuntze, Rev. Gen. 2:524. 1891. M. canadensis ssp. borealis Piper, Contr. U.S. Nat. Herb. 11:492. 1906. M. arvensis var. glabrata Fern. Rhodora 10:86. 1908. M. glabrior Rydb. Bull. Torrey Club 36:686. 1909. M. arvensis var. villosa f. glabrata S. R. Stewart, Rhodora 46:333. 1944. (Michaux, Hudson Bay)
M. arvensis var. glabra Benth. Lab. Gen. & Sp. 179. 1833. M. arvensis var. typica f. glabra S. R. Stewart, Rhodora 46:331. 1944. (Sole, presumably from England, surely not from the New World)
M. arvensis var. penardii Briq. Bull. Herb. Boiss. 3:215. 1895. M. penardii Rydb. Bull. Torrey Club 33:150. 1906. (Penard 297, Boulder, Colo.)
M. rubella Rydb. Mem. N.Y. Bot. Gard. 1:337. 1900. (Rydberg & Bessey 4900, Lower Geyser Basin, Yellowstone Nat. Pk.)
M. arvensis var. lanata Piper, Bull. Torrey Club 29:223. 1902. M. canadensis ssp. lanata Piper, Contr. U.S. Nat. Herb. 11:492. 1906. M. lanata Rydb. Bull. Torrey Club 36:687.

1909. M. arvensis var. typica f. lanata S. R. Stewart, Rhodora 46:332. 1944. (Lake & Hull
603, Parrotts P. O., Wash.)

M. occidentalis Rydb. Bull. Torrey Club 36:687. 1909. M. arvensis var. occidentalis G. N.
Jones, U. Wash. Pub. Biol. 5:222. 1936. (Heller 3486, Forest, Nez Perce Co., Ida.)

M. arvensis var. typica f. puberula S. R. Stewart, Rhodora 46:332. 1944. (Heller & Halbach
641, McCall's Ferry, York Co., Pa.)

M. arvensis var. villosa f. lanigera S. R. Stewart, Rhodora 46:333. 1944. (Eastwood 1666,
Ft. Bragg to Glen Blair, Mendocino Co., Calif.)

M. arvensis var. villosa f. brevipilosa S. R. Stewart, Rhodora 46:334. 1944. (Heller 11561,
near Chico, Butte Co., Calif.)

Perennial from creeping rhizomes; stems ascending or erect, 2-8 dm. tall, pubescent with
few to numerous, short and retrorse to longer and more spreading hairs, often glabrous between the angles; leaves short-petiolate, the blade 2-8 cm. long and 6-40 mm. wide, glabrous
or hairy, serrate, acuminate, with several pairs of lateral veins, rather narrowly ovate or
elliptic-ovate to more often somewhat rhombic-elliptic, those of the inflorescence, at least,
tending to be cuneately tapered to the petiole; verticels of flowers compact, borne in the axils
of the scarcely reduced upper (or middle and upper) leaves, and separated by internodes of ordinary length; calyx pubescent, 2.5-3 mm. long; corolla white to light purple or pink, 4-7 mm.
long. N=6, 27, 30-31, 32, 36, about 45, 46. (European forms.)

Moist places, especially along streams and shores, from the lowlands to moderate elevations in the mountains; circumboreal, in America extending s. to Calif., N.M., Mo., and Va.
July-Sept.

The native American and e. Asian plants, as described above, constitute the var. glabrata
(Benth.) Fern. The European var. arvensis, with the leaves of the inflorescence relatively
broader, more ovate, and somewhat broadly rounded to the petiole, is introduced in e. U.S.
and Can., but is not yet known in our range, although extreme specimens of var. glabrata have
been confused with it. Both var. glabrata and var. arvensis are variable in the quantity,
quality, and distribution of the pubescence, and numerous specific and infraspecific segregates
have been founded on these characters. In var. glabrata, at least, these pubescence types are
not only wholly confluent, but show little if any geographic or ecologic coherence, so that no
taxonomic segregation is desirable.

The name M. arvensis var. villosa (Benth.) S. R. Stewart has been used to include var. glabrata, but M. canadensis var. villosa Benth., on which M. arvensis var. villosa was based,
is illegitimate under articles 26 and 64 of the International Rules of Botanical Nomenclature
(1956 edition), since it was evidently intended to apply to the nomenclaturally typical phase of
M. canadensis.

Mentha pulegium L., pennyroyal, a densely velvety native of Europe, is occasionally found
in the Willamette Valley and southward. It is distinguished from M. arvensis by its relatively
small leaves that have only 2 or 3 pairs of lateral veins, the leaves of the inflorescence tending
to be deflexed and often scarcely surpassing the flower clusters.

Mentha piperita L. Sp. Pl. 576. 1753. (England)

Perennial from creeping rhizomes, 3-10 dm. tall, glabrous or glandular, or the leaves often
hirsute along the main veins beneath; leaves petiolate, the petioles of the principal leaves
mostly 4-15 mm. long, the blade ovate or lance-ovate to somewhat elliptic, serrate, acute,
3-6 cm. long, 1.5-3 cm. wide; verticels of flowers crowded into dense terminal spikes (sometimes interrupted below) 2-7 cm. long and 1-1.5 cm. wide at anthesis; calyx 2.5-3 mm. long,
the lobes hispid-ciliate, the tube without hairs; corolla 3.5-5 mm. long, pink-lavender to
white. 2N=36, 64, 66, 68, 70, 72, 84. Peppermint.

Banks of streams and ditches, bottom lands, and moist roadsides; native of Europe, apparently originating through hybridization between M. spicata and M. aquatica L., and more nearly resembling the former than the latter. Now widely established in America and occasionally
found in our range. Aug.-Sept.

Mentha citrata Ehrh., a similar form thought to have been derived also from hybridization
between M. spicata and M. aquatica, but more nearly resembling M. aquatica, sometimes is
found as an introduction w. of the Cascades in our range and southward. It differs from M.

piperita in its somewhat broader leaves, slightly longer calyx with scarcely ciliate teeth, stouter, shorter spikes that often have the lower verticel remote and axillary, and in its characteristic lemon odor.

Mentha rotundifolia (L.) Huds. Fl. Angl. 221. 1762.
 M. spicata var. rotundifolia L. Sp. Pl. 576. 1753. (N. Europe)
 Perennial from creeping rhizomes, the aerial stems erect or in part creeping, 2-10 dm. tall, loosely viscid-villous; leaves sessile, broad-based and often clasping, broadly elliptic-oblong to broadly ovate or subrotund, acutish to more often obtuse or rounded, 2-5 cm. long, 1-3.5 cm. wide, 1-2 times as long as wide, crenate-serrate, strongly rugose-reticulate, strongly villous-tomentose beneath, less so above; verticels of flowers crowded into elongate terminal spikes (sometimes interrupted below) 3-15 cm. long and 0.5-1 cm. wide at anthesis; calyx 1-2 mm. long, the tube and teeth hispidulous; corolla 2-3 mm. long, white or nearly so. 2N=18, 24, 54.
 Roadsides and waste places; native of Europe, now sparingly established across s. U.S., and reported to extend into our range in w. Oreg. Aug.-Sept.
 Mentha alopecuroides Muhl., a very similar plant thought to have been derived by hybridization between M. rotundifolia and M. longifolia L. or perhaps M. spicata, has been reported as an escape from cultivation in w. Oreg. and Wash. It differs from M. rotundifolia in its more sharply toothed, less reticulate and only slightly or scarcely rugose, often less densely pubescent and commonly somewhat larger leaves, and in its more canescent and slightly stouter spikes.

Mentha spicata L. Sp. Pl. 576. 1753. (N. Europe)
 Perennial from creeping rhizomes, 3-10 dm. tall, glabrous or subglabrous (often glandular), the leaves often hirsute along the main veins beneath; leaves sessile or subsessile, the petioles not over about 3 mm. long, the blade lance-ovate or elliptic, 2-7 cm. long, 0.8-2.5 cm. wide, 2-3.5 times as long as wide, serrate, more or less acute; verticels of flowers crowded into slender, terminal spikes (sometimes interrupted below) 3-12 cm. long and 0.5-1 cm. wide at anthesis; calyx 1.5-2 mm. long, the lobes generally hispid-ciliate, the tube without hairs; corolla 2-4 mm. long, pale lavender to sometimes white. 2N=36, 48, 84. Spearmint.
 Banks of streams and ditches, and other moist places; native of Europe, now widely introduced in America and occasionally found throughout our range. June-Aug.

## Moldavica Adans.

 Flowers verticillate in the axils of entire or more often spinose-dentate, small or leafy bracts; calyx 15-nerved, with oblique orifice, 5-toothed, the upper tooth much broader than the others, or the 3 upper teeth connate into a lip; corolla blue or purple to sometimes white, bilabiate, the upper lip emarginate and subgaleately rounded, the lower 3-lobed, with the large central lobe sometimes 2-cleft; stamens 4, ascending under the upper lip of the corolla, the upper pair the longer; pollen sacs divaricate at a wide angle, confluent in dehiscence; perennial or biennial (annual) herbs with toothed or cleft leaves.
 Three or four dozen species, native to the Northern Hemisphere, only one native to the New World. (Named for Moldavia.) See comment under Dracocephalum.

Moldavica parviflora (Nutt.) Britt. in Britt. & Brown, Ill. Fl. 2nd ed. 3:114. 1913.
 Dracocephalum parviflorum Nutt. Gen. Pl. 2:35. 1818. (Nuttall, Ft. Mandan, on the Missouri)
 Biennial or short-lived perennial (sometimes annual?) from a taproot, the solitary or more often clustered stems 1.5-8 dm. tall, simple or branched, inconspicuously strigose or hirtellous with retrorse hairs; leaves obscurely short-hairy, petiolate, the lower small, relatively broad, and often soon deciduous, the others with lance-elliptic or lance-oblong to broadly lance-triangular blade 2.5-8 cm. long and 1-2.5 cm. wide, coarsely serrate, the teeth often spine-tipped; inflorescence dense and spikelike, 1.5-3.5 cm. thick, often interrupted below, the terminal segment 2-10 cm. long; bracts subfoliaceous, mostly 1-3 cm. long, nearly or quite sessile, aristately few-toothed; flowers short-pedicellate, rather numerous in each ver-

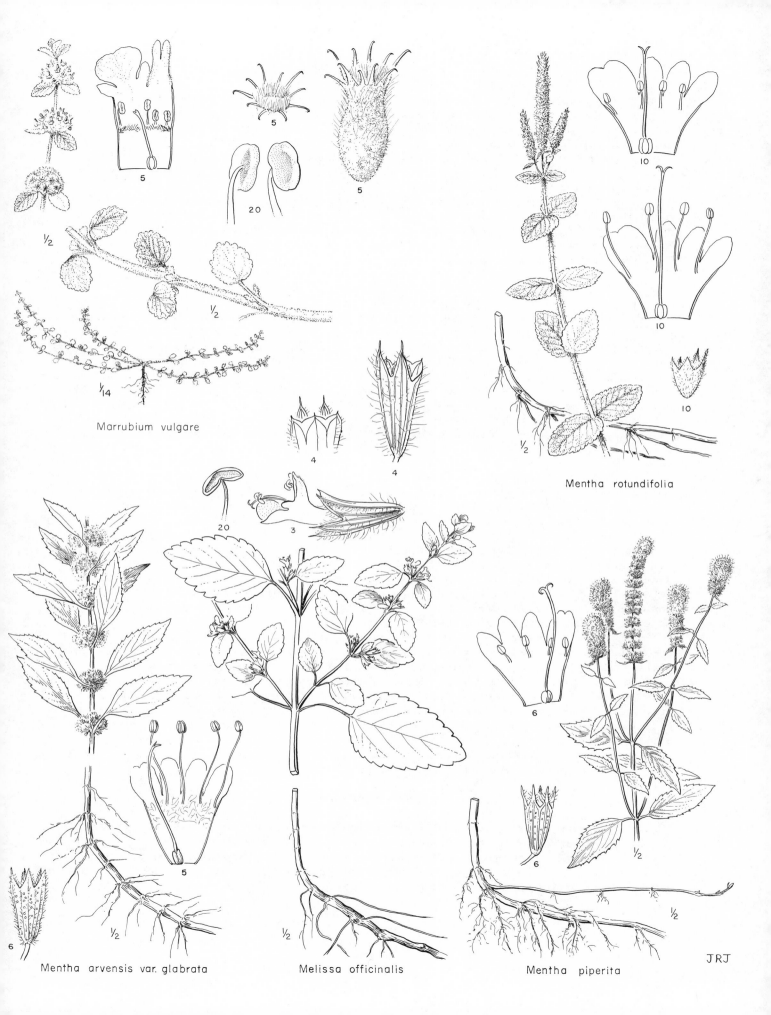

5

5

5

20

5

20

½

½

½

¼₄

Marrubium vulgare

4

4

10

10

10

10

½

Mentha rotundifolia

20

3

6

6

½

6

6

5

6

½

½

½

Mentha arvensis var. glabrata    Melissa officinalis    Mentha piperita

JRJ

ticel; calyx loosely hirsute or subglabrous, about 1 cm. long, the tube about equaling the aris-
tate-tipped lobes; upper lobe broadly ovate-oblong, conspicuously wider but not much longer
than the narrowly lance-triangular lateral and lower ones; corolla purplish, barely surpassing
the calyx, with short lips; anthers nearly equaling the upper lip. N=7.

Open, often rather moist places, from the foothills to moderate elevations in the mountains;
Alas. to Que., s. to n.e. Oreg., Ariz., N.M., S.D., Mo., and N.Y.; not known w. of the
Cascade summits in our range. June-Aug.

Moldavica thymiflora (L.) Rydb. (Dracocephalum thymiflorum L.), a smaller-leaved and
more slender plant with the bracts of the narrower and less compact inflorescence entire or
nearly so, and with the calyx mostly 6-8 mm. long, the tube obviously longer than the teeth,
has been collected here and there in the U.S., including one station at Henry's Lake, Fremont
Co., Ida. It is a native of Eurasia.

## Monarda L.

Flowers in one or several dense, leafy-involucrate, terminal heads; calyx 13-15-nerved, the
5 teeth about equal; corolla strongly bilabiate, the upper lip narrow, entire, galeate, the lower
spreading and 3-lobed; stamens 2, ascending under the upper lip of the corolla and usually exserted; pollen sacs end to end on the expanded summit of the filament (the connective), confluent in dehiscence; annual or more often perennial herbs with mostly toothed leaves.

About 15 species, native to N. Am. (Commemorating Nicolas Monardes, 1493-1588, Spanish
physician and botanist.)
Reference:
McClintock, Elizabeth, and Carl Epling. A review of the genus Monarda. U. Calif. Pub. Bot.
20:147-94. 1942.

Monarda fistulosa L. Sp. Pl. 22. 1753. (Can.)
M. menthaefolia Grah. Edinb. New Phil. Journ. 347. 1829. M. mollis var. menthaefolia
Fern. Rhodora 3:15. 1901. M. fistulosa var. menthaefolia Fern. Rhodora 46:495. 1944.
(Garden plants, from seeds collected by Drummond at "Norway House on the Saskatchewan")
Perennial from creeping rhizomes, 3-7 dm. tall, seldom branched, finely puberulent
throughout, rarely with longer spreading hairs; leaves wholly cauline, the lance-triangular to
rather narrowly ovate, more or less serrate blade 2.5-8 cm. long, 1-3 cm. wide; petioles
short, seldom 1 cm. long; flowers in a single, large, terminal, basally leafy-bracteate head;
calyx 7-11 mm. long, the teeth only 1 mm. long, the orifice markedly white-hairy within and
sometimes also without; corolla 2.5-3.5 cm. long, purple, puberulent, the long, arcuate,
upper lip surpassed by the stamens. N=16, 18.

Moist, open places (more rarely on open slopes) in the valleys and at moderate elevations in
the mountains; Que. to Vancouver I., s. to Ga., Ariz., and probably n. Mex., but seldom occurring w. of the Rocky Mt. region, and not known from Wash. or Oreg. June-Aug.

Worthy of a place in the native garden e. of the Cascades, but a poor prospect in more humid
regions. Our plants as here described belong to the var. menthaefolia (Grah.) Fern., the cordilleran and northern plains phase of the species. The typical variety of the species (including
M. mollis L.), with longer-petiolate leaves and often more branched stems, occurs in the deciduous forest region of e. U.S.

Monarda didyma L., a more robust species with bright red, mostly larger flowers, has
been collected as an escape from cultivation in w. Wash.

## Monardella Benth.

Flowers borne in dense, terminal heads subtended by conspicuous, leafy or dry bracts; calyx 10- to 15-nerved, with short, subequal teeth; corolla mostly pink-purple, obscurely bilabiate, with 5 well-developed, slender lobes; stamens 4, rather shortly exserted, subequal or
the lower pair the longer; pollen sacs divergent or subparallel; annual or perennial herbs with
small, entire or serrate, sessile or petiolate leaves.

About 20 species, native to w. U.S. and n. Mex., most numerous in Calif. (Name a diminutive of <u>Monarda.</u>)

Good rock garden subjects in arid regions, but not persistent w. of the Cascades.

Reference:

Epling, Carl. Monograph of the genus Monardella. Ann. Mo. Bot. Gard. 12: 1-106. 1925.

Monardella odoratissima Benth. Lab. Gen. & Sp. 332. 1834.

<u>Madronella</u> <u>odoratissima</u> Greene, Leafl. 1:168. 1906. (<u>Douglas</u>, above Kettle Falls on the Columbia)

Monardella discolor Greene, Pitt. 2:24. 1889. <u>Madronella</u> <u>discolor</u> Greene, Leafl. 1:169. 1906. <u>Monardella</u> <u>odoratissima</u> ssp. <u>discolor</u> Epling, Ann. Mo. Bot. Gard. 12:60. 1925. M. odoratissima var. <u>discolor</u> St. John, Res. Stud. State Coll. Wash. 1:64. 1929. (<u>Greene</u>, Yakima R. near "Clealum," Wash., Aug. 13, 1889)

Monardella <u>glauca</u> Greene, Pitt. 4:321. 1901. <u>Madronella</u> <u>glauca</u> Greene, Leafl. 1:169. 1906. <u>Monardella</u> <u>odoratissima</u> ssp. <u>glauca</u> Epling, Ann. Mo. Bot. Gard. 12:62. 1925. <u>Monardella</u> <u>odoratissima</u> var. <u>glauca</u> St. John, Res. Stud. State Coll. Wash. 1:64. 1929. (<u>Cusick</u> <u>1956</u>, deserts of e. Oreg.)

Monardella <u>nervosa</u> Greene, Pitt. 4:322. 1901. <u>Madronella</u> <u>nervosa</u> Greene, Leafl. 1:169. 1906. (<u>Sandberg</u> & <u>Leiberg</u>, arid region of e. Wash. in 1893, probably <u>413</u> from Alkali Lake, Douglas Co.) = var. discolor.

Monardella <u>elegantula</u> Gand. Bull. Soc. Bot. France 65:67. 1918. (<u>Cusick</u> <u>1956</u>; presumably an isotype of <u>M.</u> <u>glauca</u>)

Monardella <u>odoratissima</u> [var. <u>odoratissima</u>] f. <u>alba</u> St. John, Res. Stud. State Coll. Wash. 1:64. 1929. (<u>St.</u> <u>John</u> et al. <u>9658</u>, Clayton Spring, Columbia Co., Wash.)

Stems numerous from a stout taproot and branching caudex, 1-5 dm. tall, simple or nearly so, loosely ascending or erect, rather slender, becoming somewhat woody below; leaves shortly or scarcely petiolate, with lanceolate to lance-ovate or narrowly elliptic, entire, firm blade 1-3.5 cm. long and 3-12 mm. wide; bracts conspicuous, 7-15 mm. long, forming a distinct involucre beneath the head, rather dry, veiny, more or less purplish-tinged, long-ciliate on the margins; heads 1-4 cm. wide; corolla pink-purple to whitish, 1-2 cm. long, the lips subequal, about half as long as the tube, or a little longer; corolla lobes slender, apically narrowly rounded, the three lower distinct to the base of the lip, the 2 upper connate for about the lower half; filaments short-hairy below.

Open, wet or dry, often rocky places, from the plains to moderate elevations in the mountains; e. of the Cascade summits in Wash. and adj. n. Ida., s. to s. Calif. and e. to Colo. and N.M. May-Aug.

The species is divisible into about half a dozen poorly defined but geographically significant varieties. Most of our plants fall into two of these: var. <u>discolor</u> (Greene) St. John, occurring in the dry lands of c. Wash. and c. Oreg., westward onto the e. slopes of the Cascades, has the leaves more or less densely short-hairy beneath. Var. <u>odoratissima</u>, with the leaves glabrous or nearly so, occurs in the mountains of c. and n.e. Oreg. and adj. w. Ida., n. to n.e. Wash. and n. Ida., and perhaps rarely also in the Cascades. The floral bracts of var. <u>odoratissima</u> are typically ovate to subrotund, and evidently hirsute-puberulent on the back, but extreme specimens approach or match the var. <u>glauca</u> (Greene) St. John, the characteristic Basin and Klamath form of the species, which typically has the more elliptic or oblong bracts only very finely puberulent on the back. Epling's report of var. <u>glauca</u> from Mont. was based on a collection (<u>Shear</u> <u>3164</u>) actually taken near Logan, Utah. The name var. <u>glauca</u> may eventually have to be replaced by var. <u>ovata</u> (Greene) Jeps., depending on what disposition is made of Greene's type of <u>M.</u> <u>ovata</u> (<u>Brown</u> <u>381</u>, near Sisson, Siskiyou Co., Calif.), which appears to be morphologically and geographically transitional between var. <u>glauca</u> and the otherwise mostly southern variety which has been treated by Epling as ssp. <u>pallida</u> (Heller) Epling.

## Nepeta L.

Flowers in dense verticels crowded to form a terminal, inconspicuously bracteate inflorescence, or sometimes borne in open, branched, paniculiform inflorescences; calyx 15-nerved, commonly curved, oblique at the mouth, 5-toothed, the teeth often unequal and sometimes forming two lips; corolla blue, white, or rarely yellow, bilabiate, the upper lip somewhat concave and often subgaleate, entire or bifid, the lower lip spreading, 3-lobed, the large central lobe sometimes notched; stamens 4, the upper pair the longer, ascending under the upper lip of the corolla and scarcely exserted; pollen sacs widely divaricate, confluent in dehiscence; annual or more often perennial herbs, mostly more or less erect, with toothed or cleft leaves.

About 150 species, native to Eurasia (The classical Latin name of Nepeta cataria.)

Nepeta cataria L. Sp. Pl. 570. 1753. (Europe)

Taprooted perennial, commonly several-stemmed, branched upward, more or less erect, 3-10 dm. tall, canescent throughout, especially on the lower surfaces of the leaves, with soft, loosely spreading, not very long hairs; leaves wholly cauline, scarcely reduced upward, petiolate, with coarsely crenate-serrate, triangular-ovate, basally cordate blade mostly 2.5-7 cm. long and 1.5-5 cm. wide; flowers in short, dense, spikelike clusters mostly 2-8 cm. long and 1.5-2.5 cm. wide at the ends of the branches; calyx 5-6 mm. long, the slender, acuminate teeth shorter than the tube, not very unequal; corolla whitish, commonly dotted with purple, shortly exserted from the calyx, short-hairy outside like the calyx, the upper lip 2-lobed, the broad central lobe of the lower lip crenulate. N=16, 18. Catnip.

Roadsides, stream banks, and waste places, less often in relatively undisturbed habitats; native of Eurasia, now widely naturalized in America and found throughout our range. June-Sept.

### Origanum L. Wild Marjoram

Flowers solitary (or paired) in the axils of small but evident bracts, crowded into dense, trichotomously branched, headlike clusters which terminate the stem and its upper branches, forming a compact, corymbose-paniculiform inflorescence; calyx about 13-nerved, subequally 5-toothed, bearded internally at the throat; corolla bilabiate, the upper lip flat or nearly so, emarginate or cleft, the lower lip spreading and 3-lobed; stamens 4, the upper pair about equaling the corolla, the lower pair exserted and divergent; pollen sacs more or less strongly divergent, separately dehiscent; perennial herbs with rather small, toothed or entire leaves.

Half a dozen species, native to Eurasia and n. Africa; sometimes expanded to include two related groups, Marjorana and Amaracus. (Name from the ancient Greek name, origanon, for several species of Labiatae.)

Origanum vulgare L. Sp. Pl. 590. 1753. (Europe)

Perennial from creeping rhizomes, 3-7 dm. tall, the stem generally simple below the inflorescence except for the axillary fascicles of small leaves; stem and the margins and lower surfaces of the leaves loosely hairy; leaves wholly cauline, well distributed along the stem and only gradually reduced toward the inflorescence, mostly ovate or deltoid-ovate and 1.5-3 cm. long, entire or nearly so, rather short-petiolate; bracts mostly elliptic or rhombic to obovate, 3-5 mm. long, evidently purple-tipped; calyx 2-2.5 mm. long, the lance-triangular or nearly ovate teeth shorter than the tube and generally purplish; corolla 5 mm. long; small-flowered, pistillate plants occur. N=15, 16.

Native of Eurasia; now found as a roadside weed over much of the n.e. U.S., and reported to be well established near Estacada, Clackamas Co., Oreg. July-Sept.

### Prunella L. Self-heal; All-heal

Flowers in verticels, crowded into a dense, evidently bracteate, terminal spike, the bracts sharply differentiated from the leaves; calyx irregularly 10-nerved, bilabiate, the upper lip broad, shallowly 3-toothed, the lower deeply cleft, with two narrow segments; corolla blue or purple to white, bilabiate, the upper lip galeate and entire or nearly so, the lower shorter and 3-lobed; stamens 4, ascending under the galea, the lower pair the longer, scarcely exserted; filaments more or less bidentate at the tip, the anther borne on the lower tooth; pollen sacs

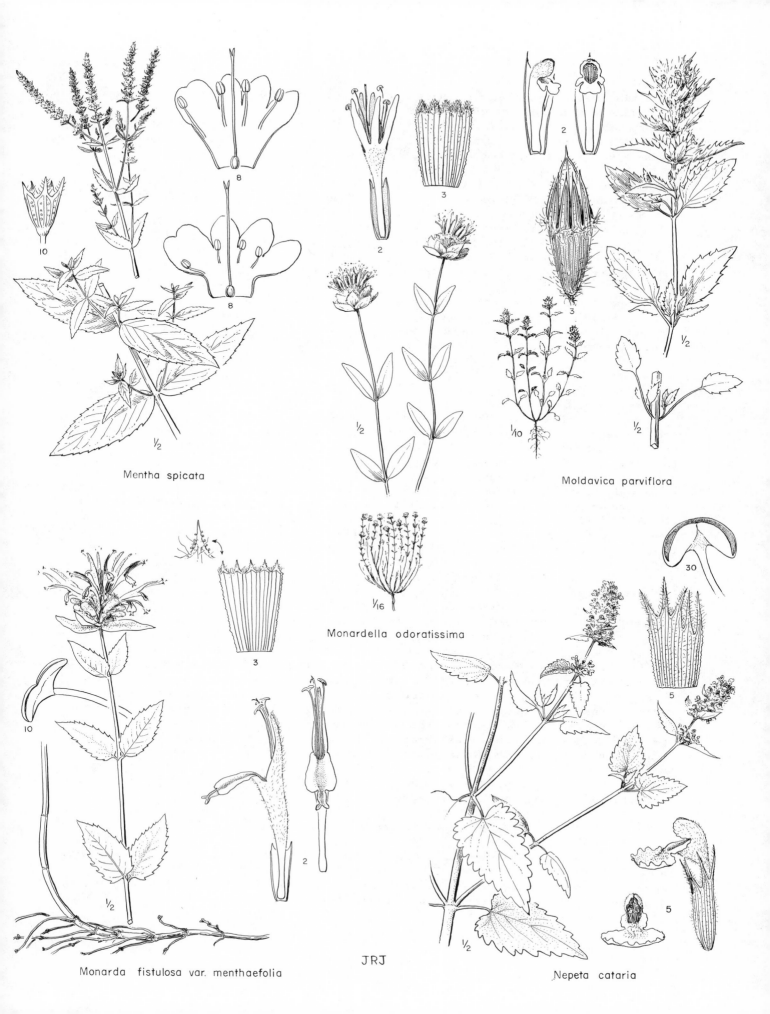

Mentha spicata

Moldavica parviflora

Monardella odoratissima

JRJ

Monarda fistulosa var. menthaefolia

Nepeta cataria

subparallel on the somewhat expanded connective, separately dehiscent; perennial herbs, erect to spreading, with entire to pinnatifid leaves.

About half a dozen species, one nearly cosmopolitan, the others native to Europe and adj. parts of Asia and Africa. (Name of doubtful origin.) The name has also been spelled <u>Brunella</u>, but the spelling here used is correct under the rules.

Prunella vulgaris L. Sp. Pl. 600. 1753. (Europe)

  P. parviflora Poir. Voy. Barb. 2:188. 1789. <u>P. vulgaris</u> var. <u>parviflora</u> Benth. Lab. Gen. & Sp. 417. 1834. (<u>Poiret</u>, Barbary Coast)

  P. pennsylvanica var. lanceolata Barton, Comp. Fl. Phila. 2:37. 1818. <u>P. vulgaris</u> var. lanceolata Fern. Rhodora 15:183. 1913. <u>P. vulgaris</u> ssp. lanceolata Hultén, Fl. Alas. 9: 1364. 1949. (Phila.)

  P. vulgaris var. elongata Benth. Lab. Gen. & Sp. 417. 1834. (<u>Douglas</u>, presumably in n. w. U.S.)

  P. vulgaris var. major Hook. Fl. Bor. Am. 2:114. 1838. (Specimens from w. U.S. and Can. by <u>Drummond</u>, <u>Douglas</u>, <u>Scouler</u>, and <u>Tolmie</u> are cited in that order)

  P. vulgaris var. calvescens Fern. Rhodora 15:185. 1913. (<u>Rosendahl</u> & <u>Brand</u> 1, Renfrew Dist., Vancouver I.) A form of var. lanceolata with the bracts only obscurely ciliate.

  P. vulgaris var. calvescens f. erubescens Henry, Fl. So. B.C. 258. 1915. (Cameron Lake, Vancouver I.)

  P. vulgaris var. calvescens f. alba J. C. Nels. Am. Bot. 24:84. 1918. (<u>J. C. Nelson</u> 1619, near Salem, Oreg.)

Fibrous-rooted perennial from a caudex or rather short rhizome, the solitary or clustered, erect to decumbent or even prostrate stems 1-5 dm. long, loosely villous-puberulent in lines or nearly throughout, or practically glabrous; leaves few, petiolate, basal and cauline or wholly cauline, entire or obscurely toothed, with lanceolate or elliptic to broadly ovate blade, 2-9 cm. long and 0.7-4 cm. wide, the lower ones tending to be broader and to have a more rounded base than the upper; spikes short and dense, 2-5 cm. long, commonly 1.5-2 cm. thick, the depressed-ovate, abruptly short-acuminate bracts about 1 cm. long, more or less strongly ciliate; calyx 7-10 mm. long, green or purple, the lips longer than the tube, the teeth spinulose-tipped; corolla blue-violet or occasionally pink or white, mostly 1-2 cm. long (or smaller in plants with reduced anthers), the tube equaling or surpassing the calyx, the lips short. N=14, 16.

In moist places, from sea level to moderate elevations in the mountains. May-Sept.

A nearly cosmopolitan species, with several regionally more or less differentiated phases that can with some difficulty be recognized as varieties. Typical European <u>P. vulgaris</u> (var. <u>vulgaris</u>) tends to have relatively broad leaves, the middle cauline ones about half as wide as long, with broadly rounded base. In America the var. <u>vulgaris</u> is known only as an occasional introduced weed in lawns, along roadsides, and in other disturbed sites. In such situations it may become dwarfed or prostrate, but under proper conditions it assumes the more vigorous, suberect habit of the native plants. The native American and e. Asian plants, var. <u>lanceolata</u> (Barton) Fern., typically have the middle cauline leaves about 1/3 as wide as long, with tapering base. They occur in moist, open or partly shaded places at various elevations, in both disturbed and undisturbed habitats. The broader-leaved specimens, when growing in disturbed sites, would doubtless pass as the introduced var. <u>vulgaris</u>, just as the narrower-leaved specimens of var. <u>vulgaris</u> from Europe appear to be fully characteristic of the American var. <u>lanceolata</u>. The small-flowered, pistillate plants of var. <u>vulgaris</u> have sometimes been segregated as var. <u>parviflora</u> (Poir.) Benth., but they are considered by modern European students to be without taxonomic significance.

## Salvia L. Sage

Flowers in more or less interrupted spikes (series of verticels) or in panicles or racemes; calyx 10- to 15-nerved, bilabiate, the lips toothed or cleft to entire; corolla bilabiate, the lower lip 3-lobed, the upper lip tending to be subgaleate and with the two lobes sometimes suppressed; stamens 2, ascending under the upper lip, sometimes exserted; filaments short;

connective elongate, articulate near the middle to the filament and bearing a pollen sac at each end, or the lower pollen sac or the whole lower arm of the connective more or less suppressed; herbs or shrubs of diverse habit.

The largest genus of the family, cosmopolitan in distribution, best developed in tropical and subtropical America. Epling recognized 468 species of a single subgenus (the largest) in 1939. (The classical Latin name of S. officinalis L., the cultivated sage.)

Salvia officinalis L., the garden sage, with elliptic or lanceolate, rarely few-lobed, strongly canescent leaves, is rarely found in our range as an escape from cultivation.

1 Shrubs; leaf blades 1-3.5 cm. long                                S. DORRII
1 Coarse herbs; blades of the lower leaves 6-25 cm. long
  2 Corolla pale yellow; herbage floccose-woolly, eventually partly glabrate   S. AETHIOPIS
  2 Corolla blue to sometimes white, or marked with yellow; herbage more coarsely hairy
    to subglabrate, not at all floccose-woolly
    3 Upper lip of the calyx with short, approximate, inconspicuous teeth less than 1 mm.
      long; bracts inconspicuous, mostly less than 1 cm. long    S. PRATENSIS
    3 Upper lip of the calyx with the two lateral teeth prominent, aristate, 1.5-3 mm. long,
      and well separated from the smaller central tooth; bracts conspicuous, 1-3 cm. long
                                        S. SCLAREA

Salvia aethiopis L. Sp. Pl. 27. 1753. (Illyria, Greece, and Africa)

Coarse biennial, 2-7 dm. tall from a stout taproot, more or less floccose-woolly throughout when young, eventually partly glabrate (especially the upper surfaces of the leaves); basal and lowermost cauline leaves distinctly petiolate, with large, rather broadly ovate-oblong or subcordate to sometimes merely elliptic blade 6-25 cm. long, pinnately rather shallowly lobed, or in part merely coarsely toothed, the lobes or large teeth finely dentate or erose; stem freely branched to the middle or below, the verticels of flowers scattered along the branches in an open, paniculiform inflorescence; cauline leaves abruptly reduced to mere sessile foliaceous bracts, those subtending the verticels subrotund or depressed-ovate, abruptly aristate, 1-2 cm. long; calyx 1 cm. long, floccose-woolly; corolla pale yellow, 1.5-2.5 cm. long, the upper lip arched and galeate, about equaling the tube; stamens shortly or scarcely exserted; upper arm of the filamentlike connective much longer than the proper filament; lower arm of the connective shorter than the upper and without a well-developed pollen sac. N=12.

A weed of roadsides and waste places, native to the Mediterranean region, now found over much of s.w. U.S., and extending into our range in e. Oreg. June-July.

Salvia dorrii (Kell.) Abrams, Ill. Fl. Pac. St. 3:639. 1951.
  Audibertia dorrii Kell. Proc. Calif. Acad. Sci. 2:190. 1863. (Dorr, presumably near Virginia City, Nev.)
  SALVIA DORRII var. CARNOSA (Dougl.) Cronq. hoc loc. Audibertia incana Benth. Bot. Reg. 17: pl. 1469. 1831. Salvia carnosa Dougl. ex Benth. loc. cit. as a synonym; ex Greene, Pitt. 2:235. 1892. Audibertiella incana Briq. Bull. Herb. Boiss. 2:73. 1894. Ramona incana Briq. Bull. Herb. Boiss. 2:440. 1894. Salvia dorrii ssp. carnosa Abrams, Ill. Fl. Pac. St. 3:639. 1951. (Cultivated plants, from seeds collected by Douglas "on plains of the Columbia, near the Priest's Rapid, and . . .near the Big Birch") Not Salvia incana Mart. & Gal. 1844.

Much-branched shrubs, 2-5 dm. tall, often broader than high, the branches rigid and often spinescent; leaves numerous, often fascicled, silvery with a close, mealy pubescence (or partly glabrate in age) and often coarsely atomiferous-glandular, the blade oblanceolate or elliptic to spatulate or obovate, entire and apically acutish to more often rounded, (1)1.5-3(3.5) cm. long and 4-15 mm. wide, gradually or abruptly narrowed to the rather short, petiolar base; flowers in a series of dense, approximate or somewhat remote, conspicuously bracteate verticels at the ends of many of the branches; bracts broadly elliptic to obovate or subrotund, 7-12 mm. long, rather dry and often suffused with anthocyanin, ciliate-margined, granular on the back; corolla bright blue-violet to rarely white, about 1 cm. long, the spreading lower lip evidently longer than the short, 2-lobed, not at all galeate upper one; stamens long-exserted; connective obliquely jointed to and about equaling the filament, the lower arm suppressed.

Dry, open places in the plains and foothills, often associated with sagebrush; Wash. and Oreg., e. of the Cascades, to Calif., s. w. Ida., Utah, and Ariz. May-June.

Our plants, as described above, belong to the var. carnosa (Dougl.) Cronq. Several other varieties, all with somewhat smaller leaves, and sometimes with more hairy bracts, occur well to the s. of our range.

Salvia pratensis L. Sp. Pl. 25. 1753. (Europe)

Similar to S. sclarea, but perennial and averaging somewhat smaller; differing sharply in the much smaller bracts and minute, crowded teeth of the upper lip of the calyx, as indicated in the key. N=9.

Native of Europe, occasionally found in the U.S. (including e. Wash., n. Ida.) along roadsides and in other disturbed habitats, often with S. sclarea. June-Aug.

Salvia sclarea L. Sp. Pl. 27. 1753. ("Syria, Italia")

Coarse, spreading-hairy biennial, 5-15 dm. tall, many of the hairs (at least in the inflorescence) gland-tipped; lowermost leaves long-petiolate, with rugose, ovate to ovate-oblong, basally subcordate, toothed or doubly toothed blade 7-20 cm. long, sometimes deciduous in larger plants; cauline leaves progressively smaller and less petiolate, often few, especially in smaller plants; stem freely branched above, the branches with scattered verticels of flowers subtended by conspicuous, caudate-acuminate, often dry and anthocyanic bracts 1-3 cm. long; calyx glandular and coarsely hairy, the upper lip with well-developed, aristate, lateral teeth 1.5-3 mm. long and well separated from the shorter, central tooth; corolla blue to sometimes white, or marked with yellow, 1.5-3 cm. long, the upper lip strongly arched, more or less galeate, longer than the tube and evidently surpassing the lower lip; stamens exserted; upper arm of the filamentlike connective much longer than the proper filament; lower arm of the connective shorter than the upper, without a well-developed pollen sac. N=11. Clary.

Native of the Mediterranean region, occasionally found in the U.S. (incl. e. Wash., n. and w. Ida.) along roadsides and in other disturbed habitats. June-Aug.

The European S. glutinosa L., differing in its larger flowers (3-4.5 cm. long) with the upper calyx lip entire, and in its subhastately cordate leaves and less conspicuous bracts, has been collected near Montesano, Wash.

## Satureja L.

Flowers verticillate (or solitary) in the axils of ordinary leaves, or the leaves subtending the upper (rarely all) whorls reduced; calyx 5- to 15-nerved, the 5 teeth subequal or arranged in two lips; corolla purple or yellow to white, bilabiate, the upper lip entire or emarginate, flat or nearly so; lower lip spreading, 3-lobed, the central lobe often larger than the others and sometimes emarginate; stamens 4, ascending under the upper lip, the lower pair often longer than the upper; pollen sacs parallel or divergent, opening by separate slits; herbs or semishrubs with small, entire or toothed leaves.

A loosely knit genus of perhaps 150 species, of wide geographic distribution. (The classical Latin name of one of the species, S. hortensis L.)

Satureja douglasii (Benth.) Briq. in Engl. & Prantl. Nat. Pflanzenf. IV. 3a:300. 1897.

Thymus douglasii Benth. Linnaea 6:80. 1831. Micromeria douglasii Benth. Lab. Gen. & Sp. 372. 1834. (Douglas [?], Calif.)

Thymus chamissonis Benth. Linnaea 6:80. 1831. Micromeria chamissonis Greene, Man. Bay Reg. 289. 1894. (Chamisso & Schlechtendahl, Calif.)

Inconspicuously puberulent perennial from a woody rhizome, the stems prostrate and frequently rooting, sometimes 1 m. long, often with short ascending branches; leaves shortpetiolate or subsessile, ovate to subrotund, 1-3.5 cm. long, generally with a few blunt teeth; flowers solitary in the axils on slender pedicels 5-15 mm. long; calyx (4)5 mm. long, 12- to 15-ribbed-costate, with short, acute, subequal teeth; corolla white or purple-tinged, 7-10 mm. long, externally hirtellous-puberulent, with short lips; pollen sacs nearly parallel.

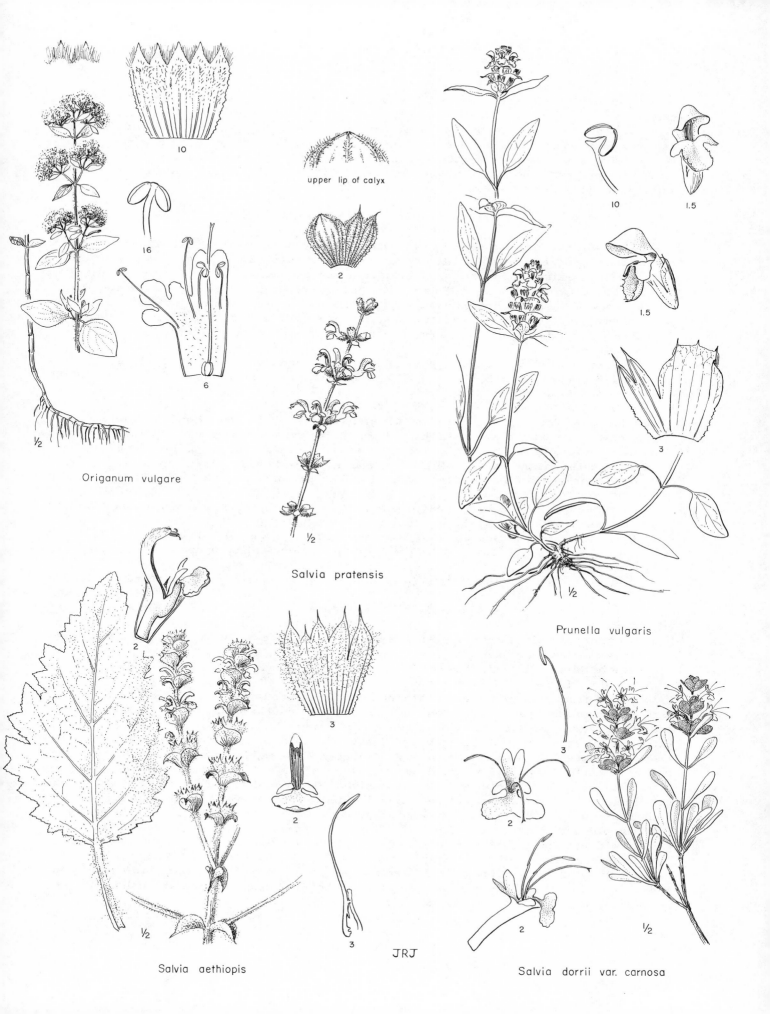

10

16

6

Origanum vulgare

upper lip of calyx

2

Salvia pratensis

10          1.5

1.5

3

Prunella vulgaris

2

Salvia aethiopis

3

2

3

JRJ

3

2

2

Salvia dorrii var. carnosa

In coniferous woods from s.w. B.C. to s. Calif., chiefly w. of the Cascade summits, but extending e. to n. Ida. June-July.

Satureja acinos (L.) Scheele [Acinos arvensis (Lam.) Dandy], a native of Europe, has been reported to be established in grassy openings in Whatcom Co., Wash. It differs from S. douglasii, among other respects, in being annual or nearly so, branched from the base, 1-2 dm. tall, with smaller, relatively narrower leaves, and in having the bluish flowers borne in small axillary clusters on pedicels less than 5 mm. long.

## Scutellaria L. Skullcap

Flowers solitary (or less often verticillate) in the axils of normal or reduced leaves, sometimes forming more or less definite racemes; calyx obscurely nerved, bilabiate, with entire lips, bearing a raised and mostly prominent transverse appendage on or proximal to the upper lip; corolla variously colored, bilabiate, the upper lip generally more or less galeate, the lateral lobes of the lower lip somewhat separated from the broad central lobe and marginally partly connate with the upper lip; stamens 4, the upper pair normal, the lower pair longer than the upper, each with a single functional pollen sac, the other abortive; perennial (rarely annual) herbs or low shrubs.

A large, cosmopolitan, sharply defined genus, the species often closely allied. Epling recognized 113 species in the New World. (Name from the Latin scutella, tray, referring to the appendage of the calyx.)

Reference:

Epling, Carl. The American species of Scutellaria. U. Calif. Pub. Bot. 20: 1-146. 1942.

1 Flowers in elongate racemes which are borne in the axils or at the ends of axillary branches (the stem also terminating in a raceme); corolla 6-8 mm. long     S. LATERIFLORA
1 Flowers paired at the nodes (solitary in the axils); corolla well over 1 cm. long
   2 Palate merely papillate; leaves mostly truncate-subcordate at the base, the larger ones mostly (2)2.5-5 cm. long     S. GALERICULATA
   2 Palate evidently long-hairy; leaves with more or less tapering or rounded base, the larger ones mostly 1-2.5(3.5) cm. long
     3 Corolla pale; pubescence of the stem down-curved; plants 0.5-1 dm. tall    S. NANA
     3 Corolla deep blue-violet; pubescence of the stem spreading or upcurved; plants (0.5)1-3 dm. tall
       4 Corollas as pressed (13)15-18(22) mm. long, measured to the tip of the galea
                                       S. ANTIRRHINOIDES
       4 Corollas as pressed (22)24-28(32) mm. long     S. ANGUSTIFOLIA

Scutellaria angustifolia Pursh, Fl. Am. Sept. 412. 1814. (Lewis, along the Clearwater near the present site of Kamiah, Ida.)

S. veronicifolia Rydb. Bull. Torrey Club 36:681.1909. (Sandberg et al. 115, Peter Creek, Nez Perce Co., Ida.)

Perennial from slender or somewhat moniliform-thickened rhizomes, 1-3 dm. tall, the slender, erect stems often clustered, or branched near the base, strigulose or strigulose-puberulent with upward-pointed hairs, sometimes nearly glabrous, or occasionally the hairs a little longer, spreading, and gland-tipped; leaves short-petiolate or the upper subsessile, the blade usually entire, often 3- to 5-nerved from near the base, lance-elliptic to oblong or nearly ovate, mostly 1.5-2.5 (3.5) cm. long and 3-10(16) mm. wide, obtuse or rounded at the apex, the lower often a little broader and longer-petiolate than the others and often somewhat toothed, but commonly deciduous; hairs of the leaves like those of the stem, or a little looser; flowers solitary in the axils of slightly reduced leaves; calyx 3.5-5.5 mm. long, slightly larger in fruit; corolla deep blue-violet, (22)24-28(32) mm. long as pressed (measured to the tip of the galea), with open throat and somewhat spreading lower lip, the palate provided with more or less numerous, long, flattened, white hairs.

In a variety of open, moist or dry, often rocky habitats in the foothills and lowlands; Wash. and s. B.C. to Calif., chiefly e. of the Cascades (but occurring rarely in the Willamette Valley of Oreg.), e. to w. and n. Ida. May-June.

Well worth a place in the drier rock garden. Our plants as described above belong to the nomenclaturally typical variety. Some other rather poorly marked varieties, which have sometimes been considered as species, occur in Calif. In our range glandular plants are known only in the Snake R. drainage near the Oregon-Washington boundary, where they sometimes occur with the nomenclaturally and biologically typical, eglandular plants.

**Scutellaria antirrhinoides** Benth. Bot. Reg. 18: sub pl. 1493. 1832. (<u>Scouler,</u> Columbia R. near Ft. Vancouver)

Very similar to <u>S.</u> angustifolia, sometimes slightly smaller, differing principally in the smaller flowers (see key) with the throat reputedly nearly closed by the palate.

In a variety of open, moist or dry, often rocky habitats from the lowlands to moderate elevations in the mountains; c. Ida. to Utah, Calif., and s. Oreg., seldom extending into n. Oreg. (Wasco. Clackamas, and Washington cos.), the type locality (where not since re-collected) representing the northern extremity of the range. June-July.

The bulk of the population of <u>S. antirrhinoides</u> is clearly set off as a distinct taxon from the bulk of the population of <u>S. angustifolia,</u> but there are a few specimens which, at least in the herbarium, appear to be intermediate.

**Scutellaria galericulata** L. Sp. Pl. 599. 1753. (Europe)

<u>S. epilobiifolia</u> Hamilt. Ann. Soc. Linn. Lyon. 1:32. 1832. <u>S. galericulata</u> var. <u>epilobiifolia</u> Jordal, Rhodora 53:158. 1951. (<u>Bigelow,</u> without locality)

<u>S. galericulata</u> var. <u>pubescens</u> Benth. Lab. Gen. & Sp. 437. 1834. (No type designated, but evidently intended to apply to American plants)

<u>S. galericulata</u> var. <u>glaberrima</u> Benth. Lab. Gen. & Sp. 437. 1834. (<u>Douglas,</u> n.w. Am.)

Perennial from slender rhizomes, the stems simple or branched, weak but mostly erect, 2-8 dm tall, strigose-puberulent especially along the angles with descending hairs, to less often glabrous or glandular; leaves scarcely petiolate, the blade lanceolate to narrowly ovate-oblong, pinnately veined, mostly blunt-toothed, more or less acute, 2-5 cm. long and 6-20 mm. wide, more or less truncate-subcordate at the base, glabrous above, puberulent or rarely sub-glabrous beneath; flowers solitary in the axils of slightly reduced leaves; calyx 3.5-4.5 mm. long, slightly larger in fruit; corolla blue marked with white, 1.5-2 cm. long, the palate papillate, not hairy. N=about 16.

Wet meadows and similar habitats; circumboreal, extending s., not at particularly high elevations, to Del., Mo., Ariz., and Calif. June-Aug.

Although the American plants of <u>S. galericulata</u> differ somewhat from those of the Old World in the average expression of certain characteristics such as pubescence and flower size, the overlap is so great that any infraspecific segregation seems unprofitable unless further differences, or correlations of known differences, are discovered.

**Scutellaria lateriflora** L. Sp. Pl. 598. 1753. (<u>Clayton,</u> Va.)

Perennial from slender creeping rhizomes, the stems arising singly, 2-8 dm. tall, commonly puberulent in lines, or glabrous; petioles 0.5-2.5 cm. long; leaf blades thin, pinnately veined, with a few hairs along the main veins beneath, otherwise mostly glabrous, ovate or lance-ovate, with broadly rounded or subcordate base, crenate-dentate, 3-8 cm. long, 1.5-5 cm. wide; racemes mostly axillary or at the ends of axillary branches, 2-12 cm. long, fairly closely flowered, bracteate, but scarcely leafy except sometimes near the base; calyx 1.5-2.5 mm. long, slightly larger at maturity; corolla blue, 6-8 mm. long, the short galea evidently surpassed by the lower lip.

Moist bottom lands; Newf. to s. B.C., s. to Ga. and Calif. July-Sept.

**Scutellaria nana** Gray, Proc. Am. Acad. 11:100. 1876. (<u>Lemmon,</u> Winnemucca Valley near Pyramid Lake, Nev.)

<u>S. footeana</u> Mulford, Bot. Gaz. 19:118. 1894. (<u>Mulford</u> s.n., Black Canyon, Gem Co., Ida.)

Described as having the corolla yellow, becoming orange on the lobes; otherwise apparently not differing significantly from the usual form except for the slightly larger leaves, which are up to 3 cm. long.

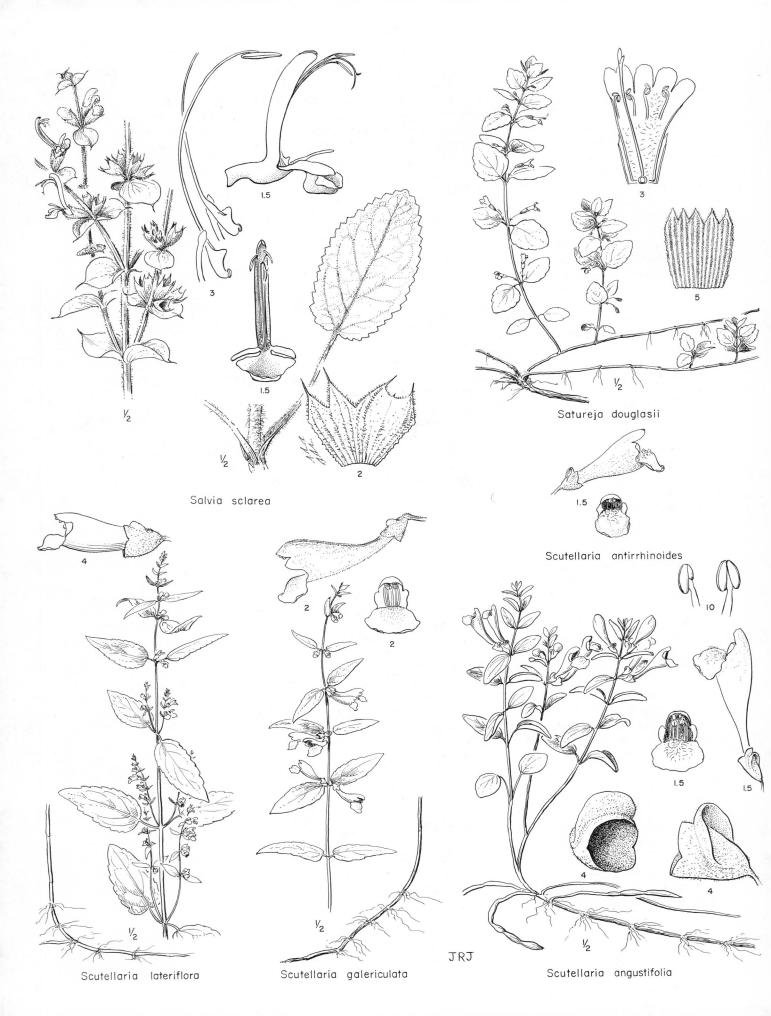

Salvia sclarea

Satureja douglasii

Scutellaria antirrhinoides

Scutellaria lateriflora

Scutellaria galericulata

Scutellaria angustifolia

JRJ

Perennial from irregularly thickened or somewhat tuberiferous rhizomes, branched at or near the ground level, 5-10 cm. tall, densely cinereous-puberulent, the hairs of the stem, pedicels, and calyx downcurved, those of the leaves loosely spreading; leaves crowded, entire, mostly oblanceolate or narrowly elliptic and subpetiolate, 1-2 cm. long and 3-7 mm. wide, or the lower varying to rhombic-ovate and more petiolate, but no larger; flowers solitary in the axils; calyx 3.5-5 mm. long, slightly larger in fruit; corolla 14-21 mm. long, very pale in effect, the lower lip creamy, the upper generally dull pale-purplish; palate white-hairy as in allied species.

Open, dry, often rather barren soil, frequently associated with basalt or andesite, in the foothills and plains; Jefferson Co., Oreg., e. to Gem Co., Ida., s. to Calif. and Nev. May-June.

Our plants belong to the var. nana. Plants of e. Nev. have blue flowers and have been described as var. sappharina Barneby.

## Stachys L. Hedge Nettle

Flowers verticillate in the axils of the reduced upper leaves or bracts, commonly forming terminal, often interrupted spikes; calyx 5- to 10-nerved, the 5 subequal teeth commonly spinulose-tipped; corolla bilabiate, with entire or emarginate, subgaleate upper lip and spreading 3-lobed lower lip, the central lobe much the largest; stamens 4, ascending under the upper lip, not much, if at all, exserted; pollen sacs strongly divergent, often confluent in dehiscence; annual or perennial herbs with toothed or entire leaves.

Perhaps as many as 200 species, widespread in the N. Temperate Zone, and extending also to S. Am. and s. Africa. Epling recognized 77 species in the New World. (Name a transliteration of the Greek word for ear of grain, referring to the inflorescence.)

Reference:

Epling, Carl. Preliminary revision of American Stachys. Fedde Rep. Sp. Nov. Beih. 80: 1-75. 1934.

1 Leaves all evidently petiolate, the middle cauline ones with petioles mostly 1.5-4.5 cm. long
  2 Flowers relatively large, the corolla tube 15-25 mm. long; inland as well as coastal plants          S. COOLEYAE
  2 Flowers smaller, the corolla tube (dry) 8-13 mm. long; mostly coastal and subcoastal plants, occasionally extending as far inland as Portland, Oreg. S. MEXICANA
1 Middle and upper cauline leaves sessile, or some of them on short petioles not over 1 cm. long
  3 Leaves becoming evidently petiolate toward the base of the stem, some of the lower ones with petioles at least 1 cm. (to 4 cm.) long; Vancouver, Wash., s. on both sides of the Cascades to Calif.          S. RIGIDA
  3 Leaves all sessile or nearly so, the petioles, if any, mostly well under 1 cm. long; widespread northern species, occurring nearly throughout our range      S. PALUSTRIS

Stachys cooleyae Heller, Bull. Torrey Club 26:590. 1899. (Cooley, "Nanimo, Vancouver Island, B.C.," July 18, 1891)

S. caurina Piper, Proc. Biol. Soc. Wash. 32:42. 1919. (Elmer 2543, Olympic Mts., Clallam Co., Wash.)

S. confertiflora Piper, Proc. Biol. Soc. Wash. 32:42. 1919. (Piper, near Corvallis, Oreg., Aug. 7, 1918)

S. ciliata ssp. macrantha Piper, Proc. Biol. Soc. Wash. 32:43. 1919. (Macoun 54685, Chilliwack Valley, B.C.)

Rhizomatous perennial, 7-15 dm. tall, the mostly simple stem shortly retrorse-bristly on the angles; leaves pubescent on both sides with long, straight, often appressed hairs, evidently petiolate, equally distributed and only gradually reduced and shorter-petiolate upward, the middle ones commonly with the narrowly deltoid-ovate to cordate-ovate, rather coarsely crenate blades 6-15 cm. long and 2.5-8 cm. wide, and borne on petioles 1.5-4.5 cm. long; inflorescence a series of verticels, the lower often axillary to slightly reduced leaves, the others to progressively reduced bracts; calyx 8-11(13) mm. long, glandular-villous, with broadly lance-triangular, spinulose-tipped teeth much shorter than the tube; corolla deep red-purple, relatively large, the tube 15-25 mm. long, the lower lip 8-14 mm. long; internal ring of hairs near the base of the corolla tube nearly transverse, scarcely marked externally; anthers explanate.

Swamps and moist low ground, from the e. slope of the Cascades to the coast, ascending to at least 4000 ft., from s. B.C. to s. Oreg. June-Aug.

This species has commonly been called S. ciliata, but the name does not properly apply. See the discussion under S. mexicana.

Stachys chamissonis Benth., of coastal Calif., differs from S. cooleyae chiefly in its more densely long-hairy leaves and calyces, and in its slightly larger calyx (up to 15 mm. long); further study may well necessitate the reduction of S. cooleyae to a variety of S. chamissonis, in spite of an apparent gap between the ranges of the two.

Stachys mexicana Benth. Lab. Gen. & Sp. 541. 1834. (Mocino, "Nova Hispania," doubtless collected during the expedition to Nootka)
S. ciliata Dougl. ex Benth. Lab. Gen. & Sp. 539. 1834. (Douglas, Columbia R.)
?S. ciliata var. pubens Gray, Syn. Fl. 2$^1$:388. 1878. S. pubens Heller, Bull. Torrey Club 25:581. 1898. (Holmes, Fraser R., B.C.; lectotype by Heller) An isotype at New York has short-petiolate leaves and conspicuously long-hairy calyx, suggesting that it may be a hybrid between S. palustris var. pilosa and either S. mexicana or S. cooleyae.
S. emersonii Piper, Erythea 6:31. 1898. (Lamb 1138, near Hoquiam, Wash.)
Habitally similar to S. cooleyae, but smaller, only 3-8 dm. tall; hairs of the stem often longer and sometimes more spreading; calyx 5-9 mm. long, pubescent with long, spreading hairs, and sometimes also glandular; corolla pink or pink-purple, distinctly paler than in S. cooleyae, and smaller, the tube 8-13 mm. long.

Swamps and moist woodlands, chiefly near the coast, extending inland as far as Portland, Oreg.; s. B.C. to n. Calif. June-Aug.

The application of the name S. mexicana to the present species is based on Epling's examination of an isotype, as reported by him in Madroño 4:272. 1938. This somewhat misleading name is here chosen instead of the more familiar S. ciliata Benth. because the latter name has customarily been misapplied to the related, larger-flowered species here treated as S. cooleyae. A specimen at New York of the paratype collection (Scouler) of S. ciliata belongs to the population here treated as S. mexicana; a possible isotype, received by Torrey from Hooker and bearing the data, "Stachys ciliata Flor. Bor. Am.," is also the present species, and was annotated as S. emersonii by Epling in 1931. Furthermore, the statement in the original description of S. ciliata that the calyx is "vix 2 lin. longus," while true of smaller-flowered specimens of the present species, would be grossly inaccurate for S. cooleyae, which has the calyx mostly 4-6 lines long.

Stachys palustris L. Sp. Pl. 580. 1753. (Europe)
S. pilosa Nutt. Journ. Acad. Phila. 7:48. 1834. S. palustris ssp. pilosa Epling, Fedde Rep. Sp. Nov. Beih. 80:63. 1934. S. palustris var. pilosa Fern. Rhodora 45:474. 1943. (Wyeth, valleys of the Rocky Mts., presumably in Mont. or Ida.)
S. scopulorum Greene, Pitt. 3:342. 1898. (Osterhout, New Windsor, Colo.; lectotype by Epling)
S. leibergii Rydb. Bull. Torrey Club 36:682. 1909. (Leiberg 1328, Blue Creek, Coeur d'Alene Mts., Ida.)
Rhizomatous perennial, simple or branched, 2-7 dm. tall, evidently hairy throughout, often glandular as well, the stem typically with long, coarse, spreading or somewhat retrorse hairs along the angles, and shorter, more slender, frequently viscid or gland-tipped hairs along the sides or all around; leaves sessile or some of the ones near or shortly below the middle of the stem borne on short petioles mostly well under 1 cm. long; lowermost leaves small and deciduous, the others lance-triangular or lance-ovate to elliptic, broadly rounded to truncate-subcordate at the base, 3.5-9 cm. long, 1-4 cm. wide, crenate; inflorescence a series of verticels, the lower often axillary to foliage leaves, the others to progressively reduced bracts; ca-

lyx (6)7-9 mm. long, pubescent with slender, gland-tipped hairs and long, stout, glandless ones, the narrow lobes a little shorter than the tube and tapering to a slender, firm point; corolla purplish, white-maculate, 11-16 mm. long, the tube only slightly (if at all) surpassing the calyx, abruptly expanded on the lower side at the level of the base of the oblique internal ring of hairs, and often with a small saccate gibbosity; anthers explanate. 2N=about 64, 102.

Along the shores of streams and lakes, in meadows, and in other moist places; circumboreal, but rarely or scarcely montane, in America occurring from Alas. to Oreg., Nev., Ariz., and N.M., and e. to New England. June-Aug.

Our plants as here described belong to the var. pilosa (Nutt.) Fern., which with the poorly defined, more eastern, more uniformly hairy var. homotricha Fern. might be considered to constitute an American subspecies (pilosa) of S. palustris. Typical var. palustris, with shortly glandular-hairy calyx, also occurs in e. U.S., apparently as an introduction from Europe.

Stachys rigida Nutt. ex Benth. in DC. Prodr. 12:472. 1848.
  S. ajugoides var. rigida Jeps. & Hoover in Jeps. Fl. Calif. 3:426. 1943. (Nuttall, Columbia R.)
  S. rivularis Heller, Muhl. 1:33. 1904. S. rigida ssp. rivularis Epling, Fedde Rep. Sp. Nov. Beih. 80:60. 1934. (Heller 7114, near Truckee, Calif.) The shorter-petiolate form of S. rigida var. rigida, possibly representing an infusion of genes from S. palustris; not here given taxonomic status.

Very similar to S. palustris var. pilosa, but with the middle and lower leaves progressively more petiolate, some of the lower ones ordinarily with petioles 1-4 cm. long; calyx 5-6 mm. long, less conspicuously long-hairy, the teeth relatively short and broad, approaching a deltoid shape; corolla 9-14 mm. long, the tube often evidently surpassing the calyx.

Stream banks and moist bottom lands; Calif. and adj. Nev., n. to c. Oreg. and rarely to Vancouver, Wash. July-Aug.

Our plants as here described belong to the var. rigida. One or two other varieties occur to the s. of our range. S. rigida is not sharply separated from the Californian S. ajugoides Benth., and it is possible that the two should be combined under the latter name, as was done by Jepson & Hoover. Plants treated as S. bullata in Piper's Flora of Washington probably represent S. rigida.

## Teucrium L. Germander; Wood Sage

Flowers in terminal, bracteate spikes or racemes, or solitary in the axils of modified upper leaves; calyx 10-nerved, with 5 scarcely to evidently unequal teeth; corolla seemingly one-lipped, the upper lip represented only by its two lobes, which are separated and displaced so as to arise from the lateral margins of the well-developed, declined, otherwise 3-lobed lower lip, of which the central lobe is much the largest; stamens 4, exserted, the lower pair the longer; pollen sacs strongly divergent, often confluent in dehiscence; ovary merely 4-lobed, the nutlets laterally attached and almost completely united; herbs or rarely shrubs with entire to deeply lobed leaves.

About 100 species, cosmopolitan in distribution, centering in the Mediterranean region. Only 8 species are known in the New World. (Name from Teukrion, the ancient Greek name for species of this or a similar genus.)
  Reference:
  McClintock, Elizabeth, and Carl Epling. A revision of Teucrium in the New World . . .. Britt. 5:491-510. 1946.

Teucrium canadense L. Sp. Pl. 564. 1753. ("Canada")
  T. occidentale Gray, Syn. Fl. 2$^1$:349. 1878. T. canadense var. occidentale McClintock & Epling, Britt. 5:499. 1946. (Fendler 617, "west of Vegas," N.M.; lectotype by McClintock & Epling)
  T. occidentale ssp. viscidum Piper, Contr. U.S. Nat. Herb. 11:487. 1906. (Kreager 482, Mission, Stevens Co., Wash.)
  Rhizomatous perennial with solitary, erect stems 2-10 dm. tall, spreading-hairy throughout,

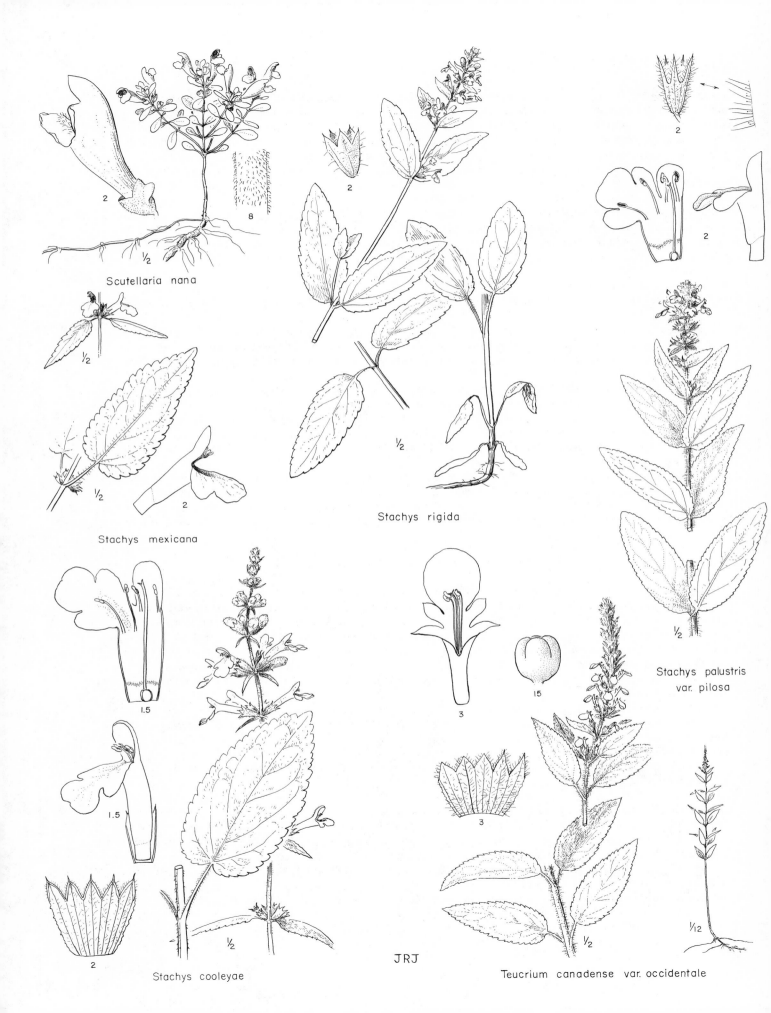

Scutellaria nana

Stachys mexicana

Stachys rigida

Stachys palustris
var. pilosa

Stachys cooleyae

JRJ

Teucrium canadense var. occidentale

or the hairs of the upper leaf surfaces appressed, many of the hairs, especially in the inflorescence, gland-tipped; lower leaves deciduous, the others short-petiolate, with lance-ovate to narrowly elliptic or lance-oblong, serrate blade 3-10 cm. long and 1-4 cm. wide; inflorescence a crowded, spiciform raceme 5-20 cm. long, with slender bracts about 1 cm. long or less; calyx 5-7 mm. long, the three upper teeth deltoid or broader and acutish to obtuse, the two lower longer and lance-subulate; corolla purplish, 11-18 mm. long, the lower lip longer than the tube and with broad, declined central lobe, the corolla cleft on the upper side to the mouth of the campanulate calyx.

Stream banks and moist bottom lands; throughout most of the U.S. and adj. parts of Can. and Mex., apparently wanting from Ida. and Mont., but occasionally collected in Oreg., Wash., and s. B.C. June-Aug.

Our plants, as described above, belong to the var. occidentale (Gray) McClintock & Epling. Two other varieties, both eglandular, occur in e. and s. U.S.

## Thymus L. Thyme

Flowers verticillate in the axils of the upper leaves, these fully developed or often reduced to bracts; calyx 10- to 13-nerved, villous at the throat within, bilabiate, the broad upper lip 3-toothed, the lower lip more deeply cleft into two narrow lobes; corolla bilabiate, the upper lip nearly flat, the lower spreading and 3-lobed; stamens 4, mostly exserted, subequal or the lower pair the longer; pollen sacs parallel or divergent, the connective expanded; small shrubs or subshrubs with small, entire leaves.

Perhaps three dozen species of Eurasia and n. Africa; many more have been described. (Name from thymos, the classical Greek name of T. vulgaris L.)

Thymus serpyllum L. Sp. Pl. 590. 1753. (Europe)

Low perennial with branching, widely divaricate or partly creeping, slender, elongate stems, puberulent in lines or all around, woody especially below; leaves small, entire, narrowly or broadly elliptic or nearly ovate, seldom over 1 cm. long, short-petiolate or subsessile; verticels mostly crowded to form a terminal spicate-capitate or spiciform inflorescence, sometimes more remote, the subtending leaves of the upper ones commonly reduced; calyx 3 mm. long, the rigidly ciliate lower lobes about equaling the tube; corolla purple, 4-5 mm. long.

Native of Eurasia; now well established as a casual weed of roadsides and waste places in n.e. U.S., and occasionally found especially w. of the Cascades in our range. June-Sept.

A very good and widely used ground cover for exposed sites. A complex species, consisting of numerous races (variously treated as species, subspecies, varieties, forms, or wholly ignored) that differ in chromosome number and minor morphological details. When the whole population is considered, these races are morphologically confluent, and no attempt is here made to recognize any of them.

## Trichostema L.

Flowers borne in cymose axillary clusters; calyx 10-nerved, regular (in our species) or the 3 upper teeth partly connate and the calyx then inverted; corolla irregular, the lowest lobe the largest, declined, the other 4 about equal and more than half as long as the lowermost one, the sinus between the two upper lobes scarcely deeper and not much, if at all, more pronounced than the lateral sinuses; stamens 4, exserted, the filaments often arcuate; pollen sacs divaricate, often confluent in dehiscence; ovary rather deeply 4-lobed, the sculptured nutlets laterally attached, united about 1/3 their length; annual or perennial herbs or near-shrubs, with mostly entire leaves.

Sixteen species, distributed in 5 well-marked sections, native to N. Am., best developed in s.w. U.S. (Name from the Greek trichos, hair, and stema, stamen, from the elongate, exserted stamens.)

Reference:

Lewis, Harlan. A revision of the genus Trichostema. Britt. 5:276-303. 1945.

Ajuga reptans L., a low, stoloniferous perennial somewhat resembling Prunella vulgaris in appearance, but with the technical characters nearly of Trichostema (differing in that the upper lip of the corolla is not cleft to the base), has been collected in Wash. A native of Eurasia, the species is known in gardens as bugle.

1 Filaments 2.5-6 mm. long; corolla tube slightly arcuate, scarcely or not at all exserted
    from the calyx; leaves relatively broad (see text), scarcely crowded, pinnately veined
                                                                        T. OBLONGUM
1 Filaments 10-20 mm. long; corolla tube evidently exserted from the calyx, abruptly bent
    upward at a right angle near the tip; leaves relatively narrow (see text) and crowded,
    strongly 3- to 5-ribbed from below the middle                    T. LANCEOLATUM

Trichostema lanceolatum Benth. Lab. Gen. & Sp. 659. 1835. (Douglas, "prope ... Vancouver
  ... et in Nova California")
    Taprooted annual, 1-5(10) dm. tall, often branched near the base, evidently pubescent throughout, usually with intermingled glands and long and short hairs; leaves numerous, crowded, mostly well surpassing the internodes, sessile or nearly so, lanceolate or rather narrowly lance-elliptic to lance-ovate, distinctly acute and often cuspidate, 2-7 cm. long and 4-20 mm. wide, mostly 3.5-7 times as long as wide, prominently (3)5-nerved, one pair of lateral veins arising near the base, the other below the middle; flowers borne in numerous small, axillary, racemelike cymes; calyx regular, 2.5-4 mm. long at anthesis; corolla tube slender, 5-10 mm. long, strongly exserted from the calyx, abruptly bent upward at right angles near the tip, the unexpanded limb forming a conspicuous knob before anthesis; lowest lobe of the corolla (central lobe of the lower lip) 4-8 mm. long; filaments exserted between the lobes of the "upper lip," conspicuously arcuate, 1-2 cm. long; nutlets 1.5-3 mm. long, reticulate-ridged, sparsely tuberculate, and finely hirtellous. N=7.
    Fields, waste places, and moist, open, low ground; Willamette Valley, s. to Baja Calif. July-Aug.

Trichostema oblongum Benth. Lab. Gen. & Sp. 659. 1835. (Douglas, near Ft. Vancouver,
  Wash.)
    Taprooted annual 1-5 dm. tall, simple or branched especially below, evidently pubescent throughout with intermingled short and notably long spreading hairs, sometimes glandular as well; leaves not much crowded, the principal ones slightly, if at all, surpassing the internodes, short-petiolate, the mostly rather broadly elliptic blade (narrower in depauperate plants) rounded to acutish at the tip, commonly 1.5-5 cm. long and 5-20 mm. wide, mostly 1.7-3.5 times as long as wide, with normal pinnate venation; flowers in short, compact, secund, racemelike axillary cymes; calyx regular, 1.5-3 mm. long at anthesis; corolla tube 2-3.5 mm. long, slightly upcurved, scarcely or not at all exserted from the calyx; lowest lobe of the corolla 2-3.5 mm. long; filaments shortly exserted between the lobes of the "upper lip," somewhat arcuate, 2.5-6 mm. long; nutlets averaging slightly smaller than in T. lanceolatum. N=7.
    Moist, open places, often in disturbed soil; e. Wash. and adj. Ida., through the Columbia gorge to the Willamette Valley of Oreg., southward to Calif. and adj. Nev.; not known in w. Wash., except the type collection, nor in e. Oreg. except near the Cascades. July-Aug.

## SOLANACEAE. Potato or Nightshade Family

Flowers gamopetalous, perfect, mostly 5-merous as to the calyx, corolla and androecium; corolla usually regular, elongate to rotate; stamens epipetalous, alternate with the corolla lobes; ovary superior, ordinarily 2-carpellary and 2-locular, with axile placentation, occasionally falsely 4-locular, rarely with several carpels and locules; style solitary, with capitate or slightly bilobed stigma; fruit a capsule or berry; seeds with well-developed endosperm and subperipheral, often curved embryo; herbs, less often shrubs, vines, or even trees, with mostly alternate, extipulate, simple to pinnately dissected leaves and diverse types of inflorescences that appear to be mostly or wholly of eventually determinate origin.

Nearly 100 genera and perhaps 3000 species, cosmopolitan, best developed in tropical America. Our native species are of little or no horticultural value.

1 Fruit capsular, dehiscent; flowers relatively large, the corolla funnelform to salverform,
   in ours 2-10 cm. long; heavy-scented, narcotic herbs, ours annual or biennial
   2 Flowers essentially terminal and solitary in origin, soon appearing to be in the forks of
      the branches; calyx circumscissile near the base; corolla (ours) 6-10 cm. long; fruit
      usually spiny                                                            DATURA
   2 Flowers borne in terminal, sometimes leafy, inflorescences (panicles, mixed panicles,
      racemes, or spikes); calyx not circumscissile; corolla (ours) 2-6 cm. long; fruit not
      spiny
      3 Fruiting calyx conspicuously accrescent, concealing the fruit; capsule operculate;
         leaves mostly sessile; flowers in a terminal, secund raceme or spike that is leafy-
         bracteate at least below; corolla funnelform                    HYOSCYAMUS
      3 Fruiting calyx scarcely accrescent, not concealing the fruit; capsule dehiscent by api-
         cal valves; ours with mostly petiolate leaves and salverform flowers borne in a ter-
         minal, sparsely or scarcely bracteate, mixed panicle            NICOTIANA
1 Fruit a fleshy or dry berry, indehiscent; flowers smaller, ours (except in Atropa) less than
   2 cm. long, funnelform or campanulate to rotate, or with reflexed lobes; plants of diverse
   habit, ours not markedly heavy-scented
   4 Corolla lobes much longer than the short tube, sooner or later reflexed; anthers connivent,
      longer than the filaments, opening by terminal pores                 SOLANUM
   4 Corolla lobes not much, if at all, longer than the tube, not reflexed except casually in
      Lycium, or the corolla scarcely lobed; anthers free, about equaling or shorter than the
      filaments, dehiscent by longitudinal slits
      5 Shrubs, often thorny; corolla with flaring, basally slender tube and well-developed
         lobes more than half as long as the tube                          LYCIUM
      5 Herbs, never thorny; corolla more or less campanulate to subrotate, the lobes short
         or obscure, much less than half as long as the tube
         6 Corolla mostly 2.5-3 cm. long, distinctly longer than wide; robust, entire-leaved,
            perennial herbs up to 15 dm. tall, with pedicellate, mostly solitary flowers at the
            upper axils                                                     ATROPA
         6 Corolla less than 2 cm. long, scarcely, if at all, longer than wide; annual or peren-
            nial herbs, ours up to about 6 dm. tall
            7 Fruiting calyx bladdery-inflated, nearly closed, loosely investing and completely
               concealing the berry; corolla in ours light yellow or greenish-yellow, marked
               with purple or brown in the throat                          PHYSALIS
            7 Fruiting calyx rather closely investing the berry, wide-mouthed, so that the top
               of the berry is exposed; corolla white or pale creamy, sometimes marked with
               purple or greenish                                          CHAMAESARACHA

Atropa L.

Flowers pedicellate and mostly solitary at the upper axils; calyx herbaceous, deeply 5-cleft,
somewhat accrescent and spreading in fruit; corolla broadly tubular-campanulate, distinctly
longer than wide, dirty purplish to greenish or yellowish, shallowly 5-lobed; filaments
elongate, attached to the base of the corolla; anthers short, longitudinally dehiscent; fruit a
many-seeded black berry; seeds flattened, with strongly curved embryo; robust perennial herbs
with entire, alternate leaves, or the upper leaves in alternate pairs, the members of each pair
unequal and both on the same side of the stem.

Two species, native to Europe, n. Africa, and w. and c. Asia. (Named for Atropos, the
Fate who severs the thread of life.)

Atropa belladonna L. Sp. Pl. 181. 1753. (Europe)
Perennial, mostly 6-15 dm. tall, generally with several erect, branching stems from a
short, stout, quickly deliquescent root, usually somewhat glandular-hairy above, otherwise
glabrous or nearly so; leaves short-petiolate, ovate to elliptic, acute or acuminate, mostly
8-16 cm. long and 4-8 cm. wide, or a smaller leaf often also present alongside each normal
upper leaf; pedicels 1-2.5 cm. long; calyx segments ovate or broadly lanceolate, acuminate,

1-1.5 cm. long in flower, slightly larger in fruit; corolla 2.5-3 cm. long, dirty purplish to somewhat greenish or yellowish, the broad lobes 5-8 mm. long; berry subglobose, 1-2 cm. thick. 2N=50(?), 72. Belladonna; Deadly nightshade.

Native of Europe, widely cultivated especially in the Old World as a source of the narcotic drug atropine, and occasionally found as a weed in our range w. of the Cascades. June-Aug.

The berries are highly poisonous.

## Chamaesaracha Gray

Flowers actually terminal and solitary (or in small clusters), but appearing axillary or eventually in the forks of the branches, because of the adjacent shoots developed from one or a pair of somewhat displaced axillary buds; calyx campanulate, 5-toothed or -lobed, somewhat accrescent, but wide-mouthed, so that the rather closely invested berry is exposed above; corolla subrotate, white or pale-creamy, sometimes marked with purple, often greenish in the throat, woolly at the base within, slightly (if at all) 5-lobed; anthers equaling or shorter than the filaments, longitudinally dehiscent; fruit a many-seeded berry; seeds flattened, with strongly curved embryo; low perennials with entire to pinnatifid leaves.

About half a dozen species, native from w. U.S. to n. S. Am. (Name originally used as a section of the tropical American genus Saracha, the latter genus named for Isadore Saracha, a Spanish monk, with the addition of a Greek prefix meaning on the ground.)

Chamaesaracha nana Gray, Bot. Calif. 1:540. 1876.
  Saracha nana Gray, Proc. Am. Acad. 10:62. 1874. (Kellogg & Harford 719, and Lemmon s.n., both from California; syntypes)
  Perennial from a branching system of creeping rhizomes, seldom as much as 1.5 dm. tall, the herbage and calyx strigose-puberulent; leaves petiolate, with narrowly to broadly ovate or rhombic, entire or subentire blade 1.5-5 cm. long; corolla 1.5-2 cm. wide; fruiting pedicels arching-recurved; berry about 1 cm. thick.

Dry, open places and open woods, especially in sandy soil; Deschutes and Crook cos., Oreg., s. to Calif. and s. Nev. June-July.

## Datura L. Jimson Weed

Flowers terminal and solitary in origin, but appearing (at least at maturity) to be borne in the forks of the branches, because of the shoots produced from a pair of somewhat displaced axillary buds; calyx cylindric or prismatic, generally circumscissile near the base, leaving a persistent flaring collar under the fruit; corolla elongate-funnelform, very large, the lobes well developed or represented by slender projections; stamens more or less included, the anthers longitudinally dehiscent, much shorter than the filaments; fruit a 2-carpellary, 4-celled, mostly spiny capsule, generally dehiscing by 4 apical valves; seeds numerous, flattened, with curved embryo; narcotic, poisonous herbs, shrubs, or trees with large, entire to toothed or lobed leaves.

About two dozen species, now more or less cosmopolitan, probably originally all from tropical or warm-temperate regions. (Name somewhat altered from the Hindustani and Arabic cognate names for one or more of the species.)

Datura stramonium L. Sp. Pl. 179. 1753. ("America, nunc vulgaris per Europam")
  D. tatula L. Sp. Pl. 2nd ed. 256. 1762. D. stramonium var. tatula Torr. Fl. N. Mid. U.S. 232. 1824. (Origin unknown to Linnaeus)
  Coarse, heavy-smelling, inconspicuously puberulent annual, often divaricately branched, up to 1.5 m. tall; leaves petiolate, with large, coarsely few-toothed or sublobate blade up to 2 dm. long and 1.5 dm. wide; calyx 3.5-5 cm. long; corolla 6-10 cm. long, the limb 3-5 cm. wide, shallowly lobed and with slender projecting teeth up to 1 cm. long; capsule ovoid, 3-5 cm. long. N=12.

In dry soil and waste places; widely distributed in temperate and warm-temperate regions of

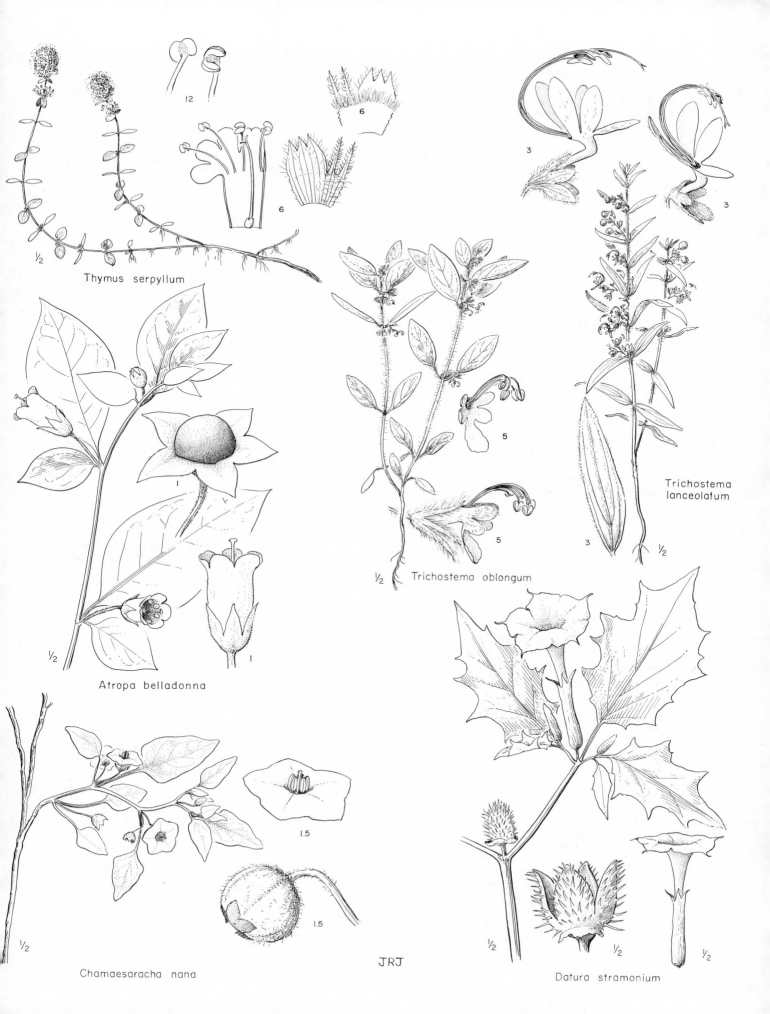

½  Thymus serpyllum

12

6  6

½  Atropa belladonna

1  1

5  5  ½  Trichostema oblongum

3  3  3  ½  Trichostema lanceolatum

½  1.5  1.5  Chamaesaracha nana

JRJ

½  ½  ½  Datura stramonium

the world, perhaps of American origin; occasionally found in our range, but more common southward. June-Aug.

This species is one of the sources of the hypnotic and sedative drug hyoscyamine. The seeds are poisonous.

Two varieties occur. Var. stramonium has green stems and white flowers, and usually has the lower spines of the fruit shorter than the upper. Var. tatula (L.) Torr. (apparently the more common phase in our area) has purplish stems and purplish or partly purplish flowers, and has the lower spines of the fruit generally similar to the upper. Rare individuals of either variety may have smooth fruits.

### Hyoscyamus L. Henbane

Flowers showy, borne in terminal, mostly secund and more or less leafy-bracteate racemes or spikes, the lower merely axillary; calyx campanulate or urceolate, 5-toothed, accrescent and enclosing the fruit; corolla funnelform, with an oblique, 5-lobed, slightly irregular limb; stamens mostly exserted, the anthers longitudinally dehiscent, much shorter than the filaments; capsule more or less 2-locular, circumscissile well above the middle; seeds numerous, flattened, tuberculate or roughened; embryo strongly curved; narcotic, poisonous annual or perennial herbs with ample, mostly toothed to incised-pinnatifid leaves.

About 15 species, occurring chiefly in the Mediterranean region. (Name from the ancient Greek hyoskyamos, from hys, sow, and kyamos, bean.)

Hyoscyamus niger L. Sp. Pl. 179. 1753. (Europe)

Coarse, strong-scented biennial or annual up to 1 m. tall, conspicuously viscid-villous, especially the stem; leaves large, sessile, 5-20 cm. long, 2-14 cm. wide, rather shallowly pinnatilobate, with unequal, triangular segments; mature calyx 2.5 cm. long, urceolate, dry, reticulate-veiny; corolla 2.5-4.5 cm. long and nearly or quite as wide at the top, prominently purple-reticulate on a pale, often greenish-yellow background, more distinctly purple in the throat; fruit 1-1.5 cm. long, with strongly thickened lid. N=17.

A weed along roadsides and in waste places; native of Europe, now casually established over much of the U.S. May-Aug.

This species is one of the sources of the hypnotic and sedative drugs hyoscyamine and scopolamine. Both the herbage and the seeds are poisonous, but so unpalatable as to be seldom eaten by stock or children.

### Lycium L. Matrimony Vine

Flowers axillary, 1-4 (many) in an axil; calyx 3- to 6-lobed, campanulate to tubular, ruptured by the growing fruit; corolla tubular to funnelform, 4- to 7-lobed; anthers longitudinally dehiscent, much shorter than the slender filaments; fruit a fleshy or dry berry; seeds 2-many, somewhat compressed, with strongly curved embryo; shrubs or small trees, usually thorny, with entire or minutely toothed, often fascicled leaves.

Perhaps 100 species, occurring in drier areas throughout the world; 14 species in N. Am. (Name from the ancient Greek lykion, the name of some thorny plant.)
Reference:
Hitchcock, C. L. A monographic study of the genus Lycium of the western hemisphere. Ann. Mo. Bot. Gard. 19:179-274. 1932.

Lycium halimifolium Mill. Gard. Dict. 8th ed. 1768. (Cultivated plants, from seeds of Chinese origin)
L. chinense Mill. Gard. Dict. 8th ed. 1768. (Cultivated plants, from seeds of Chinese origin)

Glabrous shrub with long, weak, generally sparsely thorny, arched or climbing branches, 1-6 m. tall; leaves short-petiolate, entire, dull, elliptic to lanceolate, ovate, or oblanceolate, up to 7 cm. long and 3.5 cm. wide on vigorous young shoots, or only 1.5 cm. long and 3 mm. wide on older ones; pedicels 1-3 per axil, 0.7-2 cm. long; corolla lavender or purplish, 9-

14 mm. long, with (4)5 broad, spreading lobes shorter than, or about equaling, the tube; berry ellipsoid or ovoid, 1-2 cm. long, red, 10- to 20-seeded. N=12

Native of Asia and s. e. Europe, commonly cultivated in the U.S., and occasionally escaping. June-Sept.

Fairly attractive in fruit, but not one of our choicer species. Plants cultivated as Chinese matrimony vine (L. chinense Mill.), supposedly differing in their less spiny or unarmed habit, broader, greener leaves, more acute calyx teeth, and slightly more deeply lobed corollas, appear to fall within the range of variability of L. halimifolium, and any correlations among the supposedly diagnostic characters are obscure.

## Nicotiana L. Tobacco

Flowers borne in terminal panicles, mixed panicles, or racemes; calyx toothed or cleft; corolla funnelform or salverform, with mostly spreading limb; stamens included or exserted, with long filaments and short, longitudinally dehiscent anthers; fruit capsular, ordinarily 2-locular and dehiscent by 2-4 apical valves; seeds numerous, scarcely flattened, with straight or curved embryo; narcotic herbs or shrubs with entire or merely toothed leaves.

More than 50 species, native to N. Am. and S. Am., Australia, and some Pacific islands. The cultivated tobacco is N. tabacum L. (Named for Jean Nicot de Villemain, who was concerned with the introduction of tobacco into Europe in the sixteenth century.)

Nicotiana multivalvis Lindl. Bot. Reg. 13: pl. 1057. 1827. (N. quadrivalvis var. multivalvis Gray, Bot. Calif. 1:546. 1876), described from plants grown from seeds taken by Douglas from plants cultivated by Indians along the Columbia R., apparently does not occur in the wild state. It is more robust than N. attenuata, with larger flowers, and differs especially in its large, multilocular fruits (2.5-3 cm. high, 6- to 8-carpellary) and in having usually more than 5 lobes to the calyx and corolla. Like N. quadrivalvis Pursh, which was based on plants cultivated by the Indians of the Great Plains, it has been thought to be eventually derived from the otherwise more southern native species, N. bigelovii (Torr.) Wats.

1 Longer calyx lobes 2-4 mm. long, distinctly shorter than the tube; corolla limb 8-14 mm.
    wide; widespread native plant                           N. ATTENUATA
1 Longer calyx lobes 5-7 mm. long, equaling, or longer than, the tube; corolla limb 12-22 mm.
    wide; rare introduced plant                           N. ACUMINATA

Nicotiana acuminata (Grah.) Hook. Curtis' Bot. Mag. 56: pl. 2919. 1829.
    Petunia acuminata Grah. Edinb. New Phil. Journ. 1828:378. 1828. (Garden plants from seeds.
      sent by Gillies from Mendoza, Argentina)
    Nicotiana multiflora Phil. Linnaea 33:197. 1864. N. acuminata var. multiflora Reiche, Anal.
      Univ. Chil. 125:460. 1910. (Landbeck, Choapa, Coquimbo, Chile)
    N. caesia Suksd. Werdenda 1:37. 1927. (Several Suksdorf specimens from Bingen, Wash.,
      are cited)

Similar to N. attenuata, but more densely and persistently glandular-villous; leaves sometimes broader; calyx lobes much narrower and longer, nearly linear, the longer ones 5-7 mm. long and equaling or exceeding the tube; corolla often longer (up to 6 cm.), and with broader (12-22 mm.) limb. N=12.

Sandy stream banks; native to S. Am., now established in Calif., and known in our range from Suksdorf's collections on which the name N. caesia was founded. Aug.-Oct.

Our plants, as described above, have been referred to the var. multiflora (Phil.) Reiche.

Nicotiana attenuata Torr. ex Wats. Bot. King Exp. 276. 1871.
    N. torreyana Nels. & Macbr. Bot. Gaz. 61:43. 1916. (Torrey 354, Lake Washoe, Nev.; lec-
      totype)

Glandular-pubescent, heavily odorous annual 3-10 dm. tall; leaves entire, the lowermost petiolate and with lance-ovate to elliptic blade 2.5-12 cm. long and 1-5 cm. wide, the others progressively reduced and relatively narrower upward; flowers more or less numerous in an elongate, subnaked, mixed panicle; calyx 5-toothed, the teeth triangular, unequal, the longer ones 2-4 mm. long, much shorter than the tube; corolla vespertinal, dirty white, salverform,

(2)2. 5-3. 5 cm. long, the limb 8-14 mm. wide when expanded; capsule ovoid, 2-locular, 4-valved, about 1 cm. long, partly enclosed by the slightly accrescent calyx tube. N=12. Wild tobacco, coyote tobacco.

Dry, sandy bottom lands, and in other dry, open places; s. B. C. and n. Ida. to Baja Calif., Sonora and Tex., wholly e. of the Cascades. June-Sept.

### Physalis L. Ground Cherry

Flowers actually terminal and mostly solitary, but appearing axillary (or eventually in the forks of the branches) because of the adjacent shoots developed from one or a pair of somewhat displaced axillary buds; calyx fairly small and 5-toothed or -lobed at anthesis, becoming bladdery-inflated, 5- to 10-angled, and dry at maturity, loosely enclosing the fruit, closed or barely open at the summit, the teeth remaining relatively small; corolla broadly campanulate to subrotate, shallowly or scarcely 5-lobed, typically (including our species) pale yellow or greenish yellow, with purplish or brownish throat; anthers longitudinally dehiscent, about equaling, or shorter than, the filaments; fruit a many-seeded berry; seeds flattened, with curved embryo; annual or perennial herbs with entire or toothed to rarely pinnatifid leaves. X=12.

Perhaps as many as a hundred species, more or less cosmopolitan, well developed in tropical America. (Greek word for bladder, referring to the inflated fruiting calyx.)

1 Perennial from a running rhizome; stem glabrous or inconspicuously strigose
                                                                 P. LONGIFOLIA
1 Annual from a taproot; stem conspicuously spreading-villous     P. PRUINOSA

Physalis longifolia Nutt. Trans. Acad. Phila. II. 5:193. 1834.
    P. lanceolata var. longifolia Trelease, Rep. Ark. Geol. Surv. 1888. 4:207. 1891. (Nuttall, sandy banks of the Arkansas, near Belle Point)
    P. lanceolata var. laevigata Gray, Proc. Am. Acad. 10:68. 1874. (Based partly on P. longifolia Nutt.)
    PHYSALIS LONGIFOLIA var. SUBGLABRATA (Mack. & Bush) Cronq. hoc loc. P. subglabrata Mack. & Bush, Trans. Acad. Sci. St. Louis 12:86. 1902. (Mackenzie, Sheffield, Mo.)
Perennial from a running rhizome, branched, 2-6 dm. tall, commonly somewhat pubescent with short, opaque, mostly appressed, only obscurely septate hairs on the upper part of the stem and on the pedicels, on the veins of the lower side of the leaves, and especially along the 10 principal nerves of the young calyx, often glabrate at maturity; leaves petiolate, lanceolate, elliptic-lanceolate, or narrowly rhombic and generally tapering to the petiole (in var. longifolia), or thinner, broader, more ovate, and more abruptly narrowed to the somewhat longer petiole (in var. subglabrata), entire or coarsely and irregularly blunt-toothed, the blade 4-10 cm. long; corolla 11-17 mm. long and wide, shortly 5-toothed; filaments flattened and somewhat dilated, nearly or quite 1 mm. wide; fruiting calyx broadly ovoid, 2. 5-4 cm. long, often somewhat sunken at the base; berry 1 cm. thick. N=12.

A weed in cultivated fields and waste places; native from the Great Plains (and the Great Basin?) to the Atlantic, the natural range overlapping our area in Mont.; introduced elsewhere in our region, chiefly or wholly e. of the Cascades. June-Sept.

Most of our plants belong to the plains phase of the species, the var. longifolia; a few represent the var. subglabrata (Mack. & Bush) Cronq., which has a natural range in the deciduous forest region of e. U.S.

The closely related plains species, P. lanceolata Michx., has also been reported as an introduction in our range. It has the hairs of the calyx longer, looser, more generally distributed, and conspicuously septate, and often has more evidently rough-pubescent herbage.

Physalis pruinosa L. Sp. Pl. 184. 1753. (America)
Taprooted, generally branching annual, 1-6 dm. tall, the stem gray with soft, loosely spreading, septate, viscid hairs; leaves similarly or often more shortly pubescent, or later glabrate, petiolate, the blade relatively broad, coarsely and unevenly toothed or occasionally subentire, obliquely cordate or subcordate at the base, 3-10 cm. long and often nearly as wide;

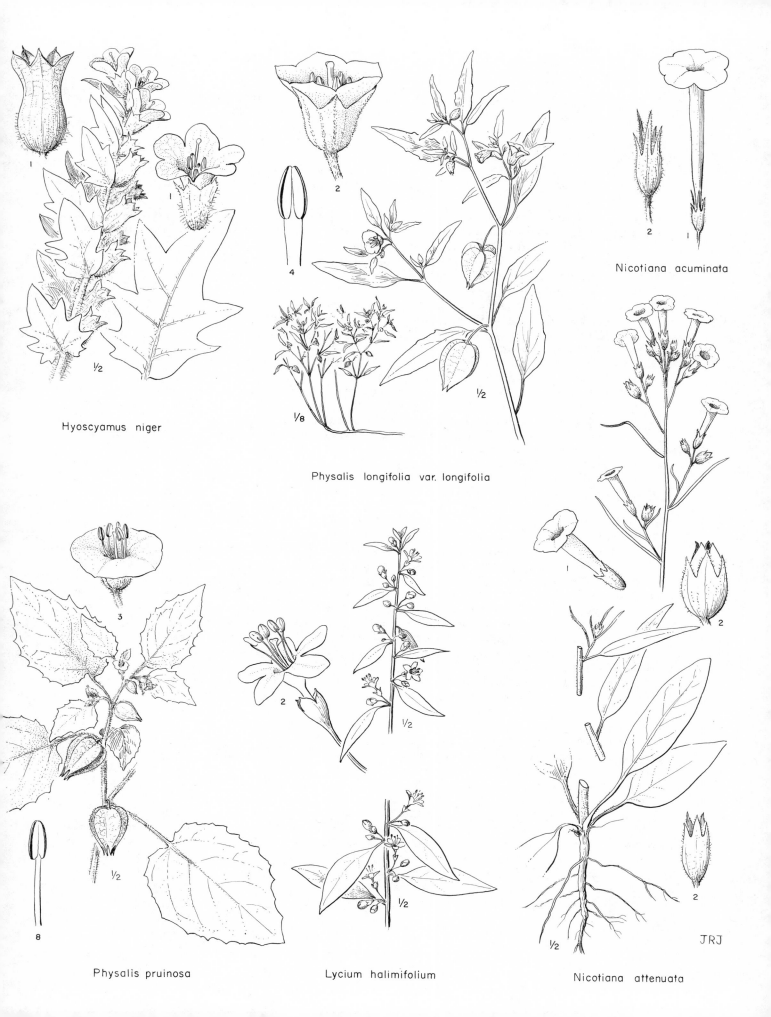

Hyoscyamus niger

Physalis longifolia var. longifolia

Nicotiana acuminata

Physalis pruinosa

Lycium halimifolium

Nicotiana attenuata

JRJ

corolla scarcely 1 cm. long and wide; filaments slender, 0. 2-0. 3 mm. wide; fruiting calyx villous-puberulent, broadly ovoid, 2-3 cm. long; berry about 13 mm. thick. N=12.

An occasional weed in cultivated ground and waste places, often in moist, sandy soil; Me. to Fla., w. to Wis., Kan., Mo., and Ala., and introduced w. of the Cascades in Oreg. and Wash. July-Sept.

## Solanum L. Nightshade

Flowers in diverse sorts of inflorescences, which in all our species are extra-axillary, arising opposite or between the leaves; calyx campanulate or rotate, toothed or cleft, sometimes accrescent; corolla with short tube and more or less spreading, eventually reflexed, mostly pointed lobes; anthers connivent, usually longer than the filaments, opening by terminal pores; fruit a berry; seeds numerous, flattened, with annular embryo; herbs, shrubs, or vines with simple to bipinnatifid leaves.

One of the larger genera of angiosperms, containing perhaps 1000 species, cosmopolitan in distribution, best developed in tropical and subtropical America. (The classical Latin name for some plant.)

1 Plants unarmed; pubescence never stellate
  2 Rhizomatous perennial, woody below, tending to climb or scramble; inflorescences
      branched, 10 to 25-flowered; fruits bright red       S. DULCAMARA
  2 Taprooted annuals, not climbing; inflorescences simple, 2- to 8-flowered; fruits not red
    3 Leaves evidently pinnatilobate       S. TRIFLORUM
    3 Leaves entire to merely toothed or wavy-margined
      4 Calyx accrescent, cupping the lower half of the greenish or yellowish fruit; stems
        spreading-hairy       S. SARRACHOIDES
      4 Calyx scarcely accrescent, not cupping the fruit; ours with black fruits and glabrous
        or somewhat appressed-hairy stems       S. NIGRUM
1 Plants spiny; pubescence of the leaves stellate       S. ROSTRATUM

Solanum dulcamara L. Sp. Pl. 185. 1753. (Europe)

Rhizomatous perennial, becoming shrubby below, tending to climb or scramble on other vegetation to a height of 1-3 m.; herbage moderately short-hairy to glabrous; leaves petiolate, some (rarely all?) simple and with rather broadly ovate-subcordate blade 2.5-8 cm. long by 1.5-5 cm. wide, others with a pair of smaller basal lobes or leaflets, the terminal one then frequently not at all cordate; peduncles 1.5-4 cm. long, 10- to 25-flowered, the inflorescence 3-8 cm. wide, jointed, bractless, often subdichotomously branched, centripetally flowering; corollas blue or light violet, the lobes 5-9 mm. long, soon reflexed; anthers conspicuous, yellow; fruit bright red, ellipsoid-globose, 8-11 mm. long. N=12. Bittersweet.

Thickets, clearings, and open woods, often in moist soil; native of Eurasia, now widely introduced in the U.S. and s. Can., and occasionally found throughout our range. May-Sept.

The herbage and fruits are mildly poisonous to livestock and to humans, but are seldom eaten in sufficient quantity to be fatal.

Solanum nigrum L. Sp. Pl. 186. 1753. (Europe)
  S. nigrum var. virginicum L. Sp. Pl. 186. 1753. (Va.)
  S. americanum Mill. Gard. Dict. 8th ed. no. 5. 1768. S. nigrum var. americanum O. E.
   Schulz, Symb. Ant. 6:160. 1909. (Va.)
  S. nodiflorum Jacq. Coll. Bot. 2:288. 1786. S. nigrum var. nodiflorum Gray, Syn. Fl. 2$^1$:
   228. 1878. (Mauritius I.)

Branching annual 1.5-6 dm. tall, glabrous, or somewhat strigose especially above; leaves petiolate, the blade ovate to deltoid, irregularly blunt-toothed or subentire, mostly 2-8 cm. long and 1-5.5 cm. wide; peduncles numerous, ascending, 0.5-2.5 cm. long, few-flowered, the pedicels subumbellately clustered, mostly deflexed, at least in fruit; mature calyx 2-3 mm. long, the lobes often unequal, sometimes reflexed; corolla white or faintly bluish, 5-10 mm. wide when expanded; fruits globose, black, about 8 mm. thick, poisonous when young, becoming edible at maturity, at least in some races. Black nightshade, wonderberry (a cultivated strain).

A weed in fields, along roadsides, and in other disturbed sites, mostly in rather moist soil; cosmopolitan. July-Oct.

The herbage and immature fruits are poisonous, but are so unpalatable that they are seldom consumed in dangerous quantities. The quantity of the toxic principle, solanine, in the fruits gradually decreases to insignificant amounts as they ripen, and mature fruits, at least of some cultivated strains, are palatable and harmless.

A highly variable species, with numerous more or less definable races, forming a polyploid series on a base number of 12. Typical S. nigrum, a European hexaploid, is apparently only casual in N. Am.; one collection has been made, among other places, at Portland, Oreg. The morphologically very similar, apparently native American plants, occurring in slightly vary-ing forms over most of the U.S. (including our range) and in the tropics of both hemispheres, are diploid; the name var. virginicum L. is available for these, but it is doubtful that any taxonomic segregation is necessary or desirable. Our plants are referred by Stebbins & Pad-dock (The Solanum nigrum complex in Pacific North America. Madroño 10:70-81. 1949) to the pantropical segregate S. nodiflorum Jacq., which they consider to be probably native only in the New World, and to be only introduced in our area. The var. douglasii (Dunal) Gray, a native w. American diploid differing from our plants (as described above) in its more robust, mostly perennial habit and somewhat larger flowers, occurs wholly to the south of our area. S. furcatum Dunal, a South American hexaploid morphologically much like var. douglasii, was reported by Stebbins & Paddock from ballast near Portland, Oreg. It does not appear to be more than varietally distinct from S. nigrum, but no new combination is here proposed. Records from our area of S. nigrum var. villosum L., a European tetraploid differing from var. nigrum in its more pubescent herbage and yellow to red fruits, are based largely or whol-ly on the related but apparently quite distinct S. sarrachoides (q.v.)

Solanum rostratum Dunal, Hist. Sol. 234. 1813.
   Androcera rostrata Rydb. Bull. Torrey Club 33:150. 1906. (Garden specimens, presumably from seeds collected by Nuttall on the banks of the Missouri)
   Solanum heterandrum Pursh, Fl. Am. Sept. 156. 1814. Androcera lobata Nutt. Gen. Pl. 1: 129. 1818. (Nuttall, banks of the Missouri)
   Coarse, erect, branching annual 3-10 dm. tall, the herbage and flowers stellate-pubescent, the stems, calyces, and to a lesser extent the leaves provided with yellow spines 3-12 mm. long; leaves 6-20 cm. long (petiole included), 2-7 cm. wide, deeply pinnatilobate or in part pinnatifid, with broad, round-tipped lobes that may be again more shallowly lobed; racemes short-pedunculate, 3- to 15-flowered, condensed at first, elongating to as much as 1.5 dm. at maturity; corolla light yellow, 2-3 cm. wide when expanded; one anther much longer than the others. Buffalo bur.

A weed along roadsides and in other disturbed sites; native of the Great Plains, now widely introduced throughout most of the U.S., and occasionally found in our range. June-Sept.

The somewhat similar, mostly more southern, likewise annual weed S. sisymbriifolium Lam. has been collected near Portland, Oreg. It has more pointed leaf lobes, coarser, more flattened spines, bluish flowers, and equal anthers, and has the stellate hairs largely con-fined to the leaves, the stem being viscid-villous. S. carolinense L., another chiefly more southern weed which may eventually be found to occur in our range, is a rhizomatous peren-nial, often less spiny, with the leaves more shallowly or scarcely lobed, the flowers white or light bluish, and the calyx nonaccrescent and usually unarmed.

Solanum sarrachoides Sendt. in Mart. Fl. Bras. 10:16. 1846. (Sellow 281, Villa das Minos, Uruguay)
   S. nigrum var. villosum sensu American authors. Not L.
   Similar to S. nigrum; stem softly spreading-hairy, the hairs flattened, viscid, often gland-tipped; leaves evidently hairy along the main veins beneath, sometimes over one or both sur-faces as well; calyx viscid-pubescent, accrescent, at maturity 4-6(9) mm. long and cupping the lower half of the greenish or yellowish fruit, the lobes evidently connate below. N=12.

A weed in fields, along roadsides, and in other disturbed sites, mostly in rather moist soil; native to S. Am., now widely introduced elsewhere in the world, and found throughout our range. May-Oct.

It is possible that monographic study will disclose an earlier name for this species.

Solanum triflorum Nutt. Gen. Pl. 1:128. 1818. (Nuttall, near Ft. Mandan, N. D.)

Fetid annual, branched from the base, the principal branches 1-6 dm. long, decumbent at least below; herbage sparsely or moderately short-hairy, or eventually glabrate; leaves rather short-petiolate, evidently pinnatilobate, the blade mostly 2-5 cm. long and 1-3 cm. wide, the rachis seldom wider than the length of the lobes; peduncles stout, ascending, few-flowered, the pedicels subumbellately clustered and becoming deflexed at least in fruit; calyx accrescent, the tube short, the rather narrow lobes often unequal, up to about 6 mm. long at maturity, frequently reflexed at least at the tip; corolla white, 5-9 mm. wide when expanded; fruit globose, greenish, 9-14 mm. thick.

A weed in fields, along roadsides, and in other disturbed sites, sometimes also occurring in relatively undisturbed places at moderate and low elevations; apparently native in the drier areas from the Great Plains to the Cascades, and introduced farther eastward. July-Aug.

The fruits are reported to be mildly poisonous, but are seldom eaten.

## SCROPHULARIACEAE. Figwort Family

Flowers gamopetalous, perfect; calyx of 4 or 5 distinct or united, similar or dissimilar sepals, or fewer by reduction or fusion; corolla mostly 4- to 5-lobed, tubular and bilabiate, less often rotate or regular or both, rarely wanting; stamens generally epipetalous, typically 4 and didynamous, sometimes only 2, or 5, with the fifth one sterile or rarely (Verbascum) fertile, basically alternate with the corolla lobes; ovary superior, 2-carpellary, 2-celled; placentation axile; style solitary, with distinct or united stigmas; fruit capsular; seeds with well-developed endosperm, numerous or less often few; herbs, or occasionally shrubs or vines, with opposite or alternate, exstipulate, simple to sometimes dissected leaves and diverse types of mostly indeterminate inflorescences.

A large family, of wide geographic distribution, comprising nearly 200 genera and perhaps 3000 species. One of the most important sources of horticultural herbs and low shrubs, with promising native and introduced species in the genera Antirrhinum, Collinsia, Digitalis, Linaria, Mimulus, Penstemon, Synthyris, and Veronica.

1 Corolla galeate, i. e. the upper lip forming a hood or beak (galea) that tends to enclose the anthers, the teeth of the upper lip short or obsolete (galeate condition least marked in Euphrasia); inflorescence of leafy-bracted spikes, spikelike racemes, or heads; stigmas wholly united; sepals more or less connate at least below (sometimes in groups)
  2 Leaves opposite (leafy bracts subtending the flowers sometimes alternate); ours annual
    3 Calyx somewhat inflated at anthesis, accrescent and very conspicuously inflated in fruit; seeds flattened, winged        RHINANTHUS
    3 Calyx not much, if at all, inflated; seeds turgid, wingless
      4 Leaves entire (though the leafy bracts may have a few teeth); capsule asymmetrical, curved, dehiscent only along the convex margin; seeds few, commonly 1 or 2 per locule        MELAMPYRUM
      4 Leaves (in our species) evidently toothed; capsule symmetrical or nearly so in form and dehiscence; seeds numerous
        5 Corolla in ours yellow and 1.5-2 cm. long; introduced, weedy plants        PARENTUCELLIA
        5 Corolla white or pinkish, in ours well under 1 cm. long; native, boreal plants        EUPHRASIA
  2 Leaves alternate, sometimes nearly all basal; plants annual or perennial
    6 Pollen sacs similar in size and position; calyx lobes typically 5, sometimes 4 or only 2; leaves toothed to dissected, never entire, often basal as well as cauline        PEDICULARIS
    6 Pollen sacs unequally set, one medifixed and appearing terminal on the filament, the

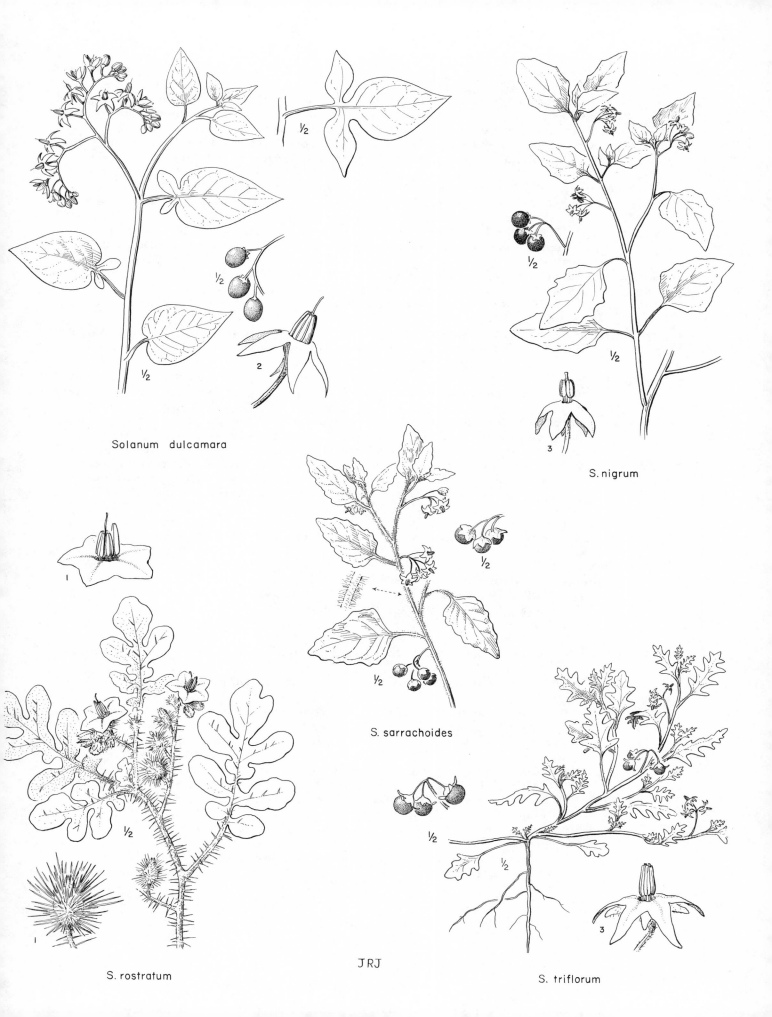

Solanum dulcamara

S. nigrum

S. sarrachoides

S. rostratum

S. triflorum

JRJ

other (sometimes reduced or obsolete) attached by its apex and pendulous or lying alongside the upper part of the filament; calyx lobes 2-4; leaves wholly cauline, entire or cleft, seldom toothed or definitely pinnatifid with broad segments

  7 Galea more or less strongly surpassing the lower lip; plants perennial, except for C. exilis, in which the galea is very conspicuously longer than the lower lip
                                                                            CASTILLEJA

  7 Galea only slightly, or not at all, surpassing the lower lip; plants annual
    8 Calyx not cleft to the base, the 4 lobes subequal or often partly connate in pairs to form two lateral segments; inflorescences many-flowered    ORTHOCARPUS
    8 Calyx cleft to the base into a dorsal and a ventral segment, or the lower segment (in extralimital species) obsolete; inflorescences in our species with about 5 or fewer flowers                                        CORDYLANTHUS

1 Corolla not galeate, though often bilabiate, the upper lip, if differentiated, not forming a hood or beak; inflorescences, sepals, and stigmas diverse
  9 Filaments 5 (the fifth a mere projecting knob or scale on the upper lip in Scrophularia); stigmas wholly united
    10 Anthers 5; leaves alternate; corolla rotate, only slightly irregular    VERBASCUM
    10 Anthers 4; leaves opposite (the cauline ones sometimes much reduced); corolla tubular, more or less strongly bilabiate
      11 Sterile filament reduced to a mere projecting knob or scale on the upper lip; leave all cauline, petiolate; corolla yellowish green or greenish purple to partly maroon, firm and not very showy                          SCROPHULARIA
      11 Sterile filament elongate, attached well down in the corolla tube; leaves often basal as well as cauline, the cauline ones petiolate or more often sessile; corolla blue or purple to less commonly red, white, or yellow, more or less showy
        12 Inflorescence thyrsoid to open-paniculate, or often of condensed verticillasters, in a few species racemose with 2 flowers per node; cauline leaves fairly well developed in most species; calyx cleft essentially to the base
                                              PENSTEMON
        12 Inflorescence racemose, with 1 flower per node; cauline leaves much reduced and inconspicuous; calyx with a distinct tube, not cleft to the base                CHIONOPHILA

  9 Filaments 2-4
    13 Anthers 2; sepals distinct, except in Besseya
      14 Leaves wholly cauline, opposite (the bracts sometimes leafy and alternate)
        15 Corolla tubular, bilabiate; stigmas distinct
          16 Lower filaments wanting, or present as mere linear vestiges near the base of the corolla                            GRATIOLA
          16 Lower filaments developed as projecting knobs on the lower lip below the sinuses                                LINDERNIA
        15 Corolla subrotate, slightly irregular, 4-lobed; stigmas wholly united
                                              VERONICA
      14 Leaves chiefly basal, the cauline ones, if present, reduced and usually alternate; stigmas wholly united
        17 Corolla present, unequally 4-lobed, not bilabiate; sepals distinct
                                            SYNTHYRIS
        17 Corolla wanting or vestigial in our species, bilabiate in others; sepals more or less connate below                                BESSEYA
    13 Anthers 4 (one pair somewhat reduced in Mimetanthe); sepals diverse
      18 Sepals distinct
        19 Corolla evidently bilabiate and either spurred, pouched-saccate, or strongly gibbous at the base; stigmas wholly united
          20 Corolla with a slender, more or less elongate spur at the base; capsule symmetrical                              LINARIA

20 Corolla shortly pouched-saccate or strongly gibbous at the base; capsule asymmetrical, one cell larger and wholly in front of the pedicel      ANTIRRHINUM

19 Corolla barely, if at all, bilabiate, sometimes ventricose beneath, but not at all spurred or saccate; stigmas distinct

   21 Leaves alternate; flowers in racemes; terrestrial plants    DIGITALIS
   21 Leaves opposite; flowers axillary; ours aquatic, with the tips of the stems floating      BACOPA

18 Sepals connate below

   22 Leaves all basal; corolla small and inconspicuous, essentially regular    LIMOSELLA
   22 Leaves well distributed along the stem (chiefly or entirely opposite), as well as sometimes basal; corolla diverse, often showy and irregular

      23 Flowers with the central lobe of the lower lip keeled-saccate, forming a pouch which encloses the stamens; annuals      COLLINSIA
      23 Flowers with the central lobe of the lower lip flat or nearly so, not forming a pouch; annuals or perennials

         24 Corolla subrotate, slightly bilabiate, with short tube; stigmas wholly united; seeds mostly 1-2 in each cell       TONELLA
         24 Corolla tubular, more or less evidently bilabiate; stigmas mostly distinct; seeds more or less numerous

            25 Flowers borne in terminal, minutely bracteate racemes; corolla blue-violet; calyx regular, the lobes equaling or surpassing the tube and with more or less evident midrib        MAZUS
            25 Flowers axillary to the upper leaves, these sometimes reduced, but the inflorescence still generally more or less leafy (flowers falsely terminal-solitary in one species of Mimulus); corolla yellow to red, purple, or pink, but not blue-violet; calyx otherwise

               26 Calyx strongly 5-angled, the midribs to the equal or unequal segments raised and prominent; deepest sinus of the calyx extending much less than halfway to the base       MIMULUS
               26 Calyx not angled, the midribs to the unequal segments obscure; ours with the deepest sinus of the calyx extending more than halfway to the base      MIMETANTHE

### Antirrhinum L. Snapdragon

Flowers borne in terminal racemes, or merely axillary; calyx of 5 essentially distinct sepals; corolla blue or purple to white or partly yellow, bilabiate, the lower lip with a prominent palate, the tube shortly pouched-saccate or strongly gibbous ventrally near the base; stamens 4, didynamous; stigma punctiform; capsule asymmetrical, one cell larger and wholly in front of the pedicel, the smaller cell opening by a single terminal pore, the larger one by 2 terminal pores, or the 2 pores confluent; style persistent and deflexed; seeds numerous; annual or perennial herbs, often glandular, with shortly or scarcely petiolate leaves, the lower opposite, the upper generally alternate.

About 30 species of the N. Hemisphere, most numerous in the Mediterranean region and California. (From the Greek anti, like, and rhinos, snout, in reference to the corolla.)

Two species of Antirrhinum have been found in our range, but neither can as yet be considered an established member of our flora. A. majus L., the garden snapdragon, blooms the first year, but is potentially perennial, and bears large, typically purple and yellow corollas about 3.5 cm. long that are several times as long as the more or less ovate sepals. A. orontium L., a native of Europe, is annual and has smaller, mostly pink-purple corollas about 1.5 cm. long that are often equaled or exceeded by the elongate, linear sepals; it has been collected along the Willamette R. in Marion Co., Oreg., and also at Victoria on Vancouver I.

### Bacopa Aubl.

Flowers axillary; sepals 5, distinct, dissimilar; corolla tubular-campanulate, 5-lobed, regular or nearly so, the upper lobes external in bud; stamens 4; style shortly cleft at the summit, with distinct, terminal stigmas, rarely nearly entire; capsules septicidal and loculicidal; seeds numerous; aquatic or subaquatic herbs with opposite, mostly sessile leaves.

Perhaps 60 species, mostly of tropical and warm-temperate regions, especially in the New World. (Latinized form of an aboriginal name among the Indians of French Guiana. )

Bacopa rotundifolia (Michx. ) Wettst. in Engl. & Prantl, Nat. Pflanzenf. IV. 3b:76. 1891.
  Monniera rotundifolia Michx. Fl. Bor. Am. 2:22. 1803. Herpestis rotundifolia Pursh, Fl.
    Am. Sept. 418. 1814. Macuillamia rotundifolia Raf. Autik. Bot. 44. 1840. Ranapalus ro-
    tundifolius Pennell, Proc. Acad. Phila. 71:242. 1920. (Ill. )
  Fibrous-rooted perennial; stems succulent, branched, 1-4 dm. long, distally floating and spreading-hairy; leaves opposite, sessile, entire, palmately several-nerved, broadly rotund-obovate to suborbicular, 1-3 cm. long, glabrous; pedicels stout, 0.5-2 cm. long; sepals 3-5 mm. long, the outer one broad, rotund-elliptic, the others much narrower; corolla white with yellow throat, 5-10 mm. long, the lobes equaling, or a little shorter than, the tube.

In mud-bottomed shallow pools; chiefly in the Mississippi Valley and Great Plains, but extending at least to Great Falls, Mont. , and reported from Ida. ; closely related, perhaps identical plants in Nev. and Calif. June-Sept.

Besseya Rydb.

Flowers borne in terminal spikes or racemes; calyx of 2-4 segments, connate below; corolla wanting or vestigial in our species, in others purple, yellow, or white, evidently bilabiate, the upper lip much larger than the lower; stamens 2; stigma capitate; capsule somewhat compressed, loculicidal; seeds numerous; fibrous-rooted perennial herbs with petiolate basal leaves and erect, leafy-bracteate stems, the bracts alternate.

About 7 species, native to the U.S. and adj. Can. , chiefly in the Rocky Mt. region. Closely allied to Synthyris, and perhaps better restored to that genus. (Named for Charles E. Bessey, 1845-1915, American botanist. )
1 Calyx (3)4-lobed, the base cupulate and surrounding the ovary and stamens          B. RUBRA
1 Calyx 2(3)-lobed, abaxial (external) to the ovary and stamens, and not surrounding them at
    the base                                                                                             B. WYOMINGENSIS

Besseya rubra (Dougl. ) Rydb. Bull. Torrey Club 30:280. 1903.
  Gymnandra rubra Dougl. ex Hook. Fl. Bor. Am. 2:103. 1838. Synthyris rubra Benth. in DC.
    Prodr. 10:455. 1846. Wulfenia rubra Greene, Erythea 2:83. 1894. Lunellia rubra Nieuwl.
    Am. Midl. Nat. 3:189. 1914. (Douglas, "banks of M'Gillivary's and Flathead rivers, near
    the Kettle Falls of the Columbia, and in the valleys of the Rocky Mts.")
  Perennial from a short caudex or crown, villous-puberulent throughout when young, later partly or wholly glabrate, 2-6 dm. tall, the inflorescence at first rather compact, later much elongate, the lower fruits becoming remote; basal leaves with elliptic-ovate to subrotund, coarsely and often doubly crenate-dentate, basally deltoid to subcordate blade 4-12 cm. long, the petiole a little shorter or a little longer; bracts of the inflorescence distinctly narrowed toward the base; calyx (3)4-lobed, often unequally so, cupulate at the base and surrounding the ovary and stamens; corolla wanting or vestigial; filaments 4-6 mm. long, dark red; capsules 5-6 mm. high and about as wide, slightly notched, glabrous or hairy.

Open slopes and dry meadows in the foothills and lowlands; Wash. (e. of the Cascades) to c. Oreg. , e. to w. Mont. Apr. -May.

A hybrid with Synthyris missurica has been reported.

Besseya wyomingensis (A. Nels. ) Rydb. Bull. Torrey Club 30:280. 1903.
  Wulfenia wyomingensis A. Nels. Bull. Torrey Club 25:281. 1898. Synthyris wyomingensis
    Heller, Muhl. 1:5. 1900. (Nelson 2142, Laramie Hills, Wyo. )
  Wulfenia gymnocarpa A. Nels. Bull. Torrey Club 25:282. 1898. Synthyris gymnocarpa Heller,
    Muhl. 1:5. 1900. Besseya gymnocarpa Rydb. Bull. Torrey Club 30:280. 1903. Synthyris

wyomingensis var. gymnocarpa Nels. in Coult. & Nels. New Man. Bot. Rocky Mts. 450.
1909. (Nelson 2959, Evanston, Wyo.)

Besseya cinerea (Raf.) Pennell, sensu Pennell, Proc. Acad. Phila. 85:104. 1933; the name
based on Veronica cinerea Raf. New Fl. N. Am. 4:39. 1838, which was stated to have the
corolla "dark blue or purple (almost brown in my specimen)"; Rafinesque's name therefore
cannot apply to the present species, in which the corolla is lacking.

Similar to B. rubra, but smaller, only 1-4 dm. tall, often more evidently white-hairy, and
with more compact inflorescence, this rarely as much as 1.5 dm. long, and with the fruits
generally all contiguous; basal leaf blades 1.5-7 cm. long, less coarsely toothed, seldom at all
cordate; bracts of the inflorescence seldom much narrowed below; filaments averaging longer,
up to 12 mm.; differing sharply from B. rubra in the calyx, as indicated in the key.

Open slopes, from the foothills and high plains to rather high elevations in the mountains;
s. Alta., s. through w. Mont. and the e. half of Ida. to n. Utah., e. to S.D., Neb., and n.
Colo. Apr.-July.

### Castilleja Mutis ex L. f. Indian Paintbrush

Flowers borne in short or elongate, prominently bracteate terminal spikes; calyx 4-cleft,
somewhat accrescent in fruit, the lobes subequal and in most of our species partly to completely connate in lateral pairs; corolla elongate and narrow, bilabiate, the upper lip (galea) beaklike, its lobes united to the tip and enclosing the anthers, the lower lip somewhat saccate-inflated or reduced and more or less rudimentary, sometimes as long, or nearly as long, as the
galea, usually much shorter, and in some species less than 1/5 the galea length, 3-toothed,
external in bud; stamens 4, attached near or above the middle of the corolla tube; pollen sacs
unequally placed; stigma capitate, entire or 2-lobed, penicillate; capsule loculicidal; seeds numerous; perennial or rarely annual (these only distantly related to Orthocarpus) herbs, with
alternate, entire to dissected, wholly (in our species) cauline leaves. N=12, 24.

A genus of possibly some 150-200 species, occurring chiefly in w. N. Am., but represented
in e. N. Am. (2 species), n. Asia (1 species), and Andean S. Am. (several species). (Named
for Domingo Castillejo, Spanish botanist.)

Castilleja is one of our most difficult genera. Its limits, however, are easily defined, and
with the exception of a few species which some authors with good reason have placed with the
related Orthocarpus, there has been unanimity in this regard. These few species are perennials with short Orthocarpus-type flowers, and undoubtedly stand near the common ancestor of
both genera. They are here included in Castilleja because the only satisfactory line which can
be drawn between the two genera is this one. These perennial Orthocarpus-like species cannot
be separated generically from Castilleja, whereas Orthocarpus, as here defined, is a homogeneous natural taxon without the related perennial species.

The delimitation of Castilleja species is quite another matter. Although there are some well-
defined species, the majority of them do not have clearly defined limits. This variability is of
a peculiar sort. Ordinarily, it is in the direction of another well-differentiated species with
which hybridization conceivably is possible. Some very probable hybrids between morphologically distant species are known, and it seems likely that even more extensive hybridization
occurs between less distant species. This would explain much of the apparent intergradation
which makes species delimitation so difficult. The percentage of intermediate specimens,
however, is low, and unless one insists that every morphologically distinguishable specimen
be given a name, the number of species of Castilleja in our area, although large, is limited.
Inasmuch as it attempts to recognize as species most variants of more than local-populational
significance, the following treatment is by no means a conservative one. A more satisfactory
account, however, does not seem possible on the basis of present information. Castilleja badly
needs the attention of a competent monographer.

1 Root annual; stems usually solitary; leaves and bracts entire, linear-lanceolate, the latter
    much longer than the flowers, with only the distal 1/3 of the uppermost colored red
    (Stenanthae)                                                         C. EXILIS
1 Root perennial, usually woody; stems clustered, often decumbent and sometimes rooting at
    the base; bracts 3- to 9-lobed or -divided or, if entire, much broader

2 Galea short, rarely over 1/2 the length of the corolla tube; lower lip prominent, usually
    more than 1/3 the length of the galea (somewhat reduced and as little as 1/5 galea length in
    C. cervina, C. flava, C. levisecta, C. lutescens, and C. rustica); bracts usually yellow
    or yellowish, rarely purplish or reddish[*]
  3 Lower lip prominent, usually 1/2 or more the length of the galea (as little as 1/3 galea
     length in C. cusickii)
    4 Corolla tube greatly elongated, 3-4 cm. in length, arcuate (Sessiliflorae)
                                                  C. SESSILIFLORA
    4 Corolla tube shorter, mostly less than 2 cm., not strongly arcuate
      5 Calyx subequally cleft into 4 linear or triangular lobes
        6 Stems and leaves puberulent to villous-hirsute (Pilosae)
          7 Inflorescence puberulent and villous, not glandular    C. LONGISPICA
          7 Inflorescence glandular as well as villous          C. RUBIDA
        6 Stems and leaves tomentulose or arachnoid-lanate (Arachnoideae)
          8 Galea puberulent or only moderately pubescent with often crisped hairs; bracts
            arachnoid-lanate                                  C. ARACHNOIDEA
          8 Galea conspicuously shaggy-villous; bracts tomentose  C. NIVEA
      5 Calyx divided usually less than half as deeply laterally as sagittally
        9 Calyx lobes acute or, if obtuse or obtusish, the plants rather harshly puberulent
          or pubescent or growing in dry habitats (Pallescentes)
          10 Lower lip more or less strongly pouched, densely puberulent-pubescent; hairs
            on stems and leaves minute
            11 Plants densely and uniformly short-puberulent; inflorescence compact;
              cauline leaves usually crowded; c. and s. Ida. and n. Nev.
                                                C. INVERTA
            11 Plants puberulent, but also with some longer hairs, particularly in the in-
              florescence and on the lower stems; inflorescence ultimately elongate;
              leaves more remote
              12 Calyx lobes triangular, usually not more than 3 times as long as broad;
                Mont., Wyo., and adj. Ida.         C. PALLESCENS
              12 Calyx lobes linear; e. Oreg. and adj. Ida.   C. ORESBIA
          10 Lower lip little pouched, not or obscurely puberulent; stems and leaves vil-
            lous or hispid; Wash. and s. B. C.        C. THOMPSONII
        9 Calyx lobes obtuse or, if acute, the plants rather softly pubescent and growing in
          meadows or other moist situations (Chrysanthae)
          13 Bracts mostly as long as, or longer than, the flowers, obtuse or rounded;
            plants mostly of medium and lower elevations
            14 Bracts deeply divided, with broad lateral lobes; John Day drainage near
              Clarno, Oreg.                           C. XANTHOTRICHA
            14 Bracts entire or little divided, the lateral lobes short and much narrower
              than the mid-blade; widespread         C. CUSICKII
          13 Bracts mostly shorter than the flowers, at least the lower ones acute or acut-
          ish; plants of alpine and subalpine habitats
           15 Calyx lobes acute; stigma not or tardily exserted; galea densely puberulent;
            Mt. Rainier                              C. CRYPTANTHA
           15 Calyx lobes broadly obtuse or rounded; stigma exserted
            16 Lobes of lower lip and margins of galea colored red; galea glandular
              puberulent; Wallowa Mts.          C. OWNBEYANA
            16 Lobes of lower lip yellow (or rarely purple or purplish in the Rocky
             Mt. C. pulchella); galea margins yellowish or purplish, never red
             17 Galea margins, calyces, or bracts usually purple or purplish,

---

[*]The natural separation of the species at this point is very difficult because of the variability
of these features even within a single taxon. If the lower lip is between 1/5 and 1/3 the galea
length, it is suggested that both branches of the key be explored.

rarely yellow or yellowish; galea usually strongly puberulent, rarely obscurely so; pouch of lower lip green and thickened; stems mostly less than 1 dm. tall, the inflorescence remaining rather compact; Rocky Mts.          C. PULCHELLA

17 Galea margins, calyces, and bracts yellow or yellowish, rarely faintly purplish; galea obscurely puberulent; pouch of lower lip not notably green and thickened; stems mostly 1-2(5) dm. tall, the inflorescence at length rather elongate; Wallowa and Blue Mts.          C. CHRYSANTHA

3 Lower lip somewhat reduced, less than 1/2, and sometimes as little as 1/5, the galea length, proximally more or less thickened and green

18 Bracts of inflorescence yellow or, if red or reddish, the color not extending to galea margins and lobes of lower lip, and the herbage more or less densely cinereous-puberulent or -pubescent

19 Calyx lobes acute; interior (Flavae)

20 Calyx less deeply cleft above than below

21 Bracts and herbage greenish, glabrous or with only a minute crisped puberulence; northern          C. CERVINA

21 Bracts and herbage grayish; pubescence usually longer, often of retrorse hairs; southern          C. FLAVA

20 Calyx subequally cleft above and below

22 Pubescence soft; leaves smooth to the touch          C. RUSTICA

22 Pubescence harsh; leaves rough to the touch          C. LUTESCENS

19 Calyx lobes obtuse; flowers mostly hidden by the bracts; w. of Cascade Range (Chrysanthae)          C. LEVISECTA

18 Bracts, calyces, galea margins, and lobes of lower lip red; Wallowa Mts. (Fraternae)          C. FRATERNA

2 Galea usually longer and more slender, at times equaling or exceeding corolla tube length; lower lip usually reduced, dark green and thickened, in most species 1/5 galea length or less (more prominent and as much as 1/3 galea length in C. occidentalis, C. parviflora, C. rhexifolia, and C. sulphurea); bracts showy, red or purple, rarely yellowish or whitish except in C. glandulifera, C. gracillima, C. occidentalis, C. sulphurea, and a variety of C. parviflora

23 Calyx much more deeply cleft below than above, usually more showy than the bracts (Linariaefoliae)          C. LINARIAEFOLIA

23 Calyx subequally cleft above and below or more deeply so above; bracts usually more showy than the calyx

24 Stems and leaves pubescent with branched, dendritic hairs (Pruinosae)          C. PRUINOSA

24 Stems and leaves glabrate to pubescent with simple hairs

25 Leaves ordinarily all entire (rarely the uppermost somewhat lobed); bracts entire or lobed, but rarely deeply divided; calyx subequally cleft above and below, or sometimes a little more deeply so below (Septentrionales)

26 Bracts predominantly yellow or yellowish (but varying to red or purple in 2 of the 3 species)

27 Stems clumped, rather stout, erect or ascending from a woody base

28 Plants 1-2 dm. tall, unbranched, strongly viscid-villous, at least in the inflorescence; alpine situations          C. OCCIDENTALIS

28 Plants mostly 2-5 dm. tall, often branched, glabrous to viscid-villous, but not strongly so; moist slopes and meadows, mostly below alpine zone          C. SULPHUREA

27 Stems mostly solitary, although several may be attached to a remote caudex, slender, decumbent or creeping and often rooting at the base; wet meadows, middle altitudes          C. GRACILLIMA

26 Bracts predominantly red or purple, yellow only in rare individuals

29 Stems mostly 1-3 dm. tall, usually unbranched; bracts crimson (drying purple), rarely scarlet; flowers mostly 20-30 mm. long; alpine and subalpine zones

30    Plants strongly viscid-villous, at least in the inflorescence; bracts mostly entire;
  Wenatchee and n. Cascade mts., Wash.    C. ELMERI
30    Plants glabrate, or only obscurely viscid-villous in the inflorescence; bracts commonly lobed; widespread       C. RHEXIFOLIA
 29 Stems mostly more than 3 dm. tall, often branched; bracts scarlet, rarely crimson;
  flowers mostly 30-40 mm. long; plants of lower elevations
   31  Bracts rather broad and rounded; stems more or less decumbent or ascending;
    leaves often ovate, pubescent; Oregon coast  C. LITORALIS
   31  Bracts mostly narrower and more strongly lobed; stems mostly erect; leaves usual-
    ly lanceolate or narrower, glabrous to pubescent; widely distributed, but not
    coastal in Oregon       C. MINIATA
25 Upper leaves and bracts ordinarily deeply cleft into 3-7 linear, spreading lobes (rarely all
 entire); calyx cleft subequally above and below, or more deeply so above (Parviflorae)
 32 Plants hispid with long multicellular hairs
   33  Mid-blade of leaves lanceolate to ovate-lanceolate, the lateral lobes much narrower
                C. HISPIDA
   33  Mid-blade of leaves linear to linear-lanceolate, the lateral lobes often nearly as
    broad
     34  Inflorescence narrow, becoming greatly elongate; bracts purplish (rarely yellow)
              C. ANGUSTIFOLIA
     34  Inflorescence broader and more compact; bracts bright red, scarlet, or yellow
              C. CHROMOSA
 32 Plants more villous than hispid, the hairs slender and entangled, sparse, sometimes
  glabrate (see also C. hispida var. hispida)
   35  Stems and leaves not or scarcely glandular-viscid
    36  Leaves and bracts mostly 5(7)-parted; mid-blade narrow, usually not much
     wider than the lateral lobes
      37  Galea 1/2-2/3 the length of the corolla tube, not or inconspicuously exceed-
       ing the bracts; Rocky Mts.   C. COVILLEANA
      37  Galea about equaling or longer than the corolla tube, conspicuously exserted;
       Cascade Range     C. RUPICOLA
    36  Leaves and bracts mostly 3(5)-parted; mid-blade broad, usually much wider
     than the lateral lobes
      38  Stems clustered on a woody caudex
       39  Stems numerous, slender, 1-3 dm. tall; leaves close-set, mostly 3(5)-
        lobed; corolla 15-25 mm. long; Cascade and Coast ranges
             C. PARVIFLORA
       39  Stems few, stout, 2-5 dm. tall; leaves remote, mostly entire; corolla
        25-40 mm. long; Rocky Mts.  C. CRISTA-GALLI
      38  Stems usually solitary; base slender, rhizomatous C. SUKSDORFII
   35  Stems and leaves more or less strongly glandular-viscid
    40  Bracts scarlet (rarely yellow), deeply cleft into long, rather blunt lobes; corol-
     la 25-35 mm. long, with the galea nearly as long as the tube C. APPLEGATEI
    40  Bracts yellow or reddish, cleft into shorter, more acute lobes; corolla 15-
     25 mm. long, with the galea shorter than the tube C. GLANDULIFERA

Castilleja angustifolia (Nutt.) G. Don, Gen. Syst. 4:616. 1838.
 Euchroma angustifolia Nutt. Journ. Acad. Phila. 7:46. 1834. (Wyeth, Little Goddin R., now
  Little Lost R., Butte Co., Ida.)
 E. bradburyi Nutt. Journ. Acad. Phila. 7:47. 1834. Castilleja bradburyi Don, Gen. Syst. 4:
  616. 1838. C. angustifolia var. bradburyi Fern. Erythea 6:48. 1898. (Wyeth, Little Goddin
  R., now Little Lost R., Butte Co., Ida.)
 C. buffumii Nels. in Coult. & Nels. New Man. Bot. Rocky Mts. 459. 1909. (Buffum [A. Nel-
  son 9307], Slick Creek Badlands, Worland, Washakie Co., Wyo.)
 C. bennittii Nels. & Macbr. Bot. Gaz. 55:380. 1913. (Nelson & Macbride 1714, Shoshone
  Falls, Twin Falls Co., Ida.)

Perennial; stems clustered, erect or ascending from a woody base, 1-4 dm. tall, often branched, finely puberulent with interspersed long hispid hairs, usually purplish; leaves puberulent and hispid , lower ones linear, entire, upper ones usually divided into 3-5 divergent lobes, the lateral lobes not much narrower than the mid-blade; inflorescence conspicuous, crimson, pale to deep rose, or occasionally yellow, rather narrow and becoming greatly elongate; bracts puberulent and villous, 3- to 5-parted, the lobes blunt; calyx 15-25 mm. long, subequally cleft above and below or less deeply so below, its primary lobes again divided 1-4 mm. into 2 usually rounded segments, corolla 20-25 mm. long, little longer than the calyx, its puberulent to pubescent galea usually a little shorter than the tube and 5 or more times the length of the dark green, thickened, lower lip, the last generally hidden in the calyx tube.

Dry hills and plains, usually associated with sagebrush, n. w. Wyo. , c. and s. w. Ida. , and extreme e. Oreg. May-July.

Castilleja angustifolia has been a little known and completely misunderstood species, almost all of the material so named in herbaria belonging to either C. chromosa or C. hispida. Although it is related to these species, it is a well-defined natural unit of co-ordinate value. There is evidence of hybridization with C. rustica.

Castilleja applegatei Fern. Erythea 6:49. 1898. (Applegate 87, summit of Mt. Scott, Klamath Co. , Oreg. )

C. angustifolia var. adenophora Fern. Erythea 6:48. 1898. (Applegate 413, "bald hills," near Ashland, Jackson Co. , Oreg. ) = var. applegatei.

C. pinetorum Fern. Erythea 6:50. 1898. (Applegate 415, Swan Lake Valley, Klamath Co. , Oreg. ) = var. applegatei.

CASTILLEJA APPLEGATEI var. VISCIDA (Rydb. ) Ownbey, hoc loc. C. viscida Rydb. Bull. Torrey Club 34:38. 1907. (Rydberg & Carlton 6593, near headwaters of Big Cottonwood Creek, Wasatch Mts. , Salt Lake Co. , Utah)

C. rhexifolia var. pubens Nels. & Macbr. Bot. Gaz. 55:380. 1913. C. confusa var. pubens Nels. & Macbr. Bot. Gaz. 61:45. 1916. (Nelson & Macbride 2023, Jack Creek Canyon, Jarbidge, Elko Co. , Nev. ) = var. applegatei.

C. wherryana Pennell, Proc. Acad. Phila. 99:180. 1947. (Pennell 15454, Dooley Mt. , s. of Salisbury, Baker Co. , Oreg. ) = var. applegatei.

Perennial; stems clustered, erect or ascending from a woody base, 1-5 dm. tall, branched or unbranched, glandular-villous to -hirsute; leaves glandular-puberulent or -pubescent, the lower ones usually entire and linear-lanceolate, the upper ones broader and usually with a pair of lateral lobes, these narrower than the mid-blade, rarely leaves all divided or not uncommonly all entire; inflorescence conspicuous, bright red, scarlet, or occasionally yellow; bracts glandular-villous, deeply 3- to 5-parted, mostly as long as, or longer than, the flowers; calyx 15-23 mm. long, deeply and subequally cleft above and below, its primary lobes again divided 3-8 mm. into 2 narrow, acute segments; corolla 20-35 mm. long, its puberulent or pubescent galea not or little shorter than the tube and 5 or more times the length of the dark green, thickened, lower lip.

Dry mountain slopes and summits at medium and high elevations, n. Utah and w. Wyo. , across s. and c. Ida. to e. Oreg. , thence s. to Nev. and c. Calif. May-Aug.

This is a complex and variable species, but its limits are well defined. Its only near relative in our area is C. glandulifera, which seems satisfactorily distinct. These two species clearly are members of the Parviflorae, being unique in this alliance only in their well-developed glandular pubescence. The plants of the mountains of Utah, Ida., and Wyo. seem sufficiently different from those of Oreg. and southward to deserve varietal status. They may be distinguished by the following key:

1 Stems 1-2(3) dm. tall, mostly unbranched; leaves mostly divided

var. viscida (Rydb. ) Ownbey

1 Stems (2)3-5 dm. tall, mostly branched; leaves mostly entire      var. applegatei

Castilleja arachnoidea Greenm. Bot. Gaz. 53:510. 1912.

Orthocarpus pilosus var. arachnoideus Jeps. Man. Fl. Pl. Calif. 940. 1925. (Butler 422, Marble Mt. , Siskiyou Co. , Calif. )

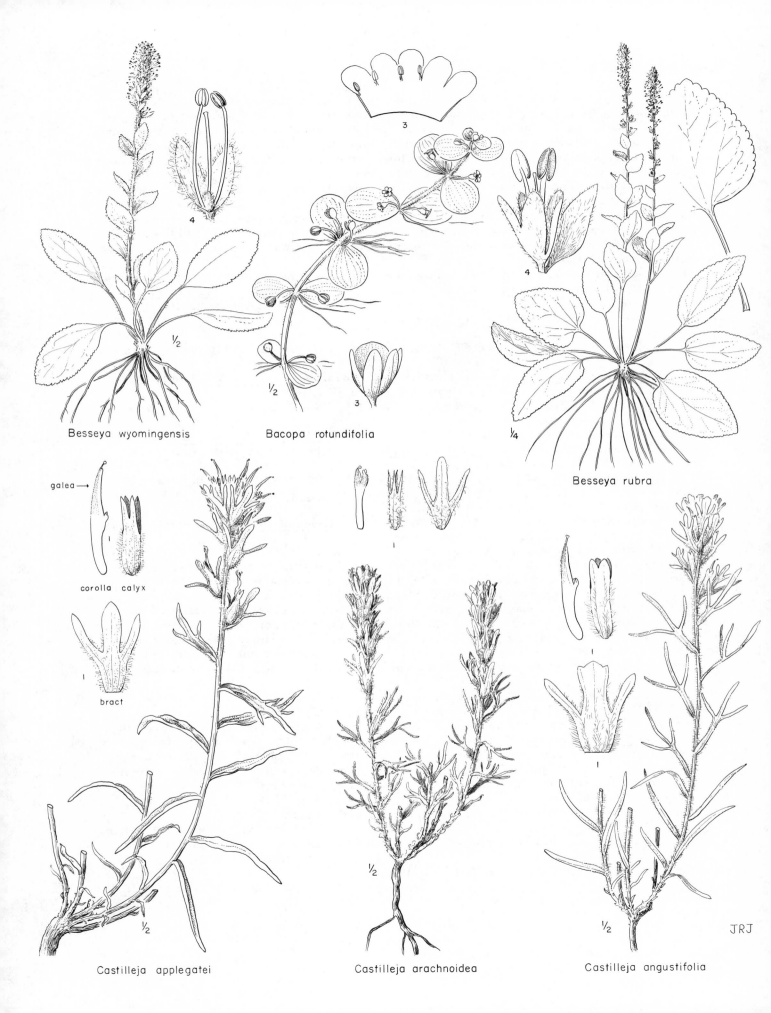

Besseya wyomingensis

Bacopa rotundifolia

3

½

3

Besseya rubra

½

galea →

corolla  calyx

bract

½

Castilleja applegatei

½

Castilleja arachnoidea

Castilleja angustifolia

JRJ

?Castilleja filifolia Eastw. Leafl. West. Bot. 2:243. 1940. (J. T. Howell 6900, 8 mi. n. of Diamond Lake, Douglas Co., Oreg.)

C. payneae Eastw. Leafl. West. Bot. 2:245. 1940. (Payne 126, Mt. Warren, Warner Mts., Modoc Co., Calif.)

C. eastwoodiana Pennell, Not. Nat. Acad. Phila. 74:1. 1941. (Pennell 26140, Elijah Peak, s. of Oregon Caves, Josephine Co., Oreg.)

C. floccosa Pennell ex Peck, Man. High. Pl. Oreg. 664. 1941. (Siskiyou Mts., Oreg.)
An unintentional renaming of C. eastwoodiana; without Latin diagnosis

C. pumicicola Pennell, Not. Nat. Acad. Phila. 74:3. 1941. (Pennell 15577, lower slopes of Mt. Scott, Crater Lake Nat. Pk., Oreg.)

C. arachnoidea ssp. shastensis Pennell, Proc. Acad. Phila. 99:177. 1947. (Cooke 16188, Mt. Shasta at 7000 ft., Calif.) The nom. nud. C. shastensis Eastw. ex Baker, W. Am. Pl. 3:4. 1904, cited as a synonym by Pennell, is without nomenclatural significance.

Perennial; stems clustered, erect or ascending, sometimes branched, 1-2.5 dm. tall, arachnoid-lanate throughout; leaves subequally 3-parted to about the middle, the lobes linear, divergent; bracts dull red or yellowish, mostly 3-parted with short lateral lobes, the middle lobe usually much broader and rounded at the apex; flowers mostly hidden by the bracts, about 12-15 mm. long; calyx and corolla subequal in length, the former divided somewhat less than halfway into 4 equal segments; lower lip prominent, pouched, about as long as the galea and like it more or less strongly puberulent or pubescent with often crisped hairs. N=12 (C. payneae).

Open summits and slopes of the s. Cascade and Klamath mts., to Mt. Lassen, and the Warner Mts.; s.w. Oreg. and n. Calif.; reaching our area in the vicinity of Three Sisters Peaks. June-Aug.

This is a rather constant and easily recognized species, and none of the above segregates seems to have morphological or geographical significance. However, Pennell has confused it with the wholly different C. schizotricha Greenm., which is sympatric with it in the Klamath area. C. schizotricha is distinguished by its lower stature, often entire leaves and bracts, little pouched lower lip, and abundant and conspicuous indumentum of branched tangled hairs, these extending even to the galea and lobes of the lower lip.

Castilleja cervina Greenm. Bot. Gaz. 25:269. 1898. (Dawson, Lower Arrow Lake, B.C.; lectotype)

Perennial; stems clustered, erect, usually branched above, 3-6 dm. tall, crisp-puberulent or glabrous, sometimes purplish; leaves crisp-puberulent (rarely glabrous), lower ones linear, entire, upper ones usually with a pair of linear, divaricate lobes; bracts broader than the leaves, 3- to 5-parted, crisp-puberulent, yellowish; flowers rather remote and not hidden by the bracts; calyx 15-20 mm. long, more deeply cleft below than above, its primary lobes again divided 2-3 mm. into 2 slender, acute segments; corolla 18-25 mm. long, usually exserted from the calyx, not strongly curved, its lower lip reduced, 1/2-1/4 the length of the galea, with puberulent lobes.

Grasslands and open coniferous woods, s. B.C., adj. n. Ida., and n. Wash. To be expected in n.w. Mont. June-July.

This species is closely related only to C. flava, from which it appears to be satisfactorily distinct.

Castilleja chromosa A. Nels. Bull. Torrey Club 26:245. 1899.

C. miniata chromosa Garrett, Spring Fl. Wasatch Reg. 87. 1911. (A. Nelson 4577, Leroy, Uinta Co., Wyo.; lectotype)

C. collina A. Nels. Bull. Torrey Club 28:231. 1901. C. angustifolia collina Garrett, Spring Fl. Wasatch Reg. 87. 1911. (A. Nelson 6995, Sand Creek, Albany Co., Wyo.)

C. angustifolia var. dubia A. Nels. Bull. Torrey Club 29:404. 1902. C. dubia Nels. in Coult. & Nels. New Man. Bot. Rocky Mts. 460. 1909. (E. Nelson 4898a, Indian Grove Mts., Carbon Co., Wyo.; lectotype) A yellow-bracted form.

Perennial; stems clustered, erect or ascending from a woody base, 1-4 dm. tall, mostly unbranched, finely puberulent with interspersed long hispid hairs, occasionally somewhat viscid; leaves puberulent and hispid, lower ones linear-lanceolate, entire, upper ones deeply divided into 3-5 divergent lobes, the lateral lobes not much narrower than the mid-blade; inflorescence conspicuous, bright red, scarlet, or not infrequently yellow, at first short and broad, but later elongating; bracts puberulent and villous, 3- to 5-parted, the lobes blunt; calyx 15-25 mm. long, subequally cleft above and below or more deeply so above, its primary lobes again divided 1-4 mm. into 2 acute or rounded segments; corolla 20-30(40) mm. long, its puberulent to pubescent galea about equaling the tube in length and more than 5 times as long as the dark green, thickened, lower lip, the last generally hidden in the calyx tube.

Dry soils, usually associated with sagebrush, c. and s. Wyo., w. Colo., n.w. N.M., across s. Ida., Utah, n. Ariz., and Nev. to e. Oreg., trans-Sierran and s. Calif. Definitely known in our area only from Jefferson and Crook cos., Oreg., and apparently from the Salmon River Canyon, n. of Riggins, Ida. Apr.-Aug.

This is closely related to and perhaps only varietally distinct from C. angustifolia, in which position the varietal name dubia is available. The differences between the two, however, are conspicuous and relatively clear-cut, and there is little, if any, overlap in their respective geographical ranges. There is evidence of hybridization with C. flava, C. linariaefolia, and perhaps C. miniata.

Castilleja chrysantha Greenm. Bot. Gaz. 48:146. 1909. (Cusick 3200b, head of West Eagle Creek, Wallowa Mts., Union Co., Oreg.)
    C. indecora Piper, Proc. Biol. Soc. Wash. 31:76. 1918. (Peck 4282, 10 mi. n. of Cornucopia, Wallowa Mts., Wallowa Co., Oreg.)
    Perennial; stems clustered, erect or from a decumbent base, usually unbranched, 0.5-5 (mostly 1-2) dm. tall, softly viscid-villous; leaves viscid-villous or somewhat hispidulous, lower ones linear-lanceolate, entire, upper ones broader and commonly with a pair of short lateral lobes; bracts broader than the leaves, ovate and acute, sometimes entire, but more commonly with a pair of short lateral lobes, viscid-villous, pale yellow or rarely purplish; inflorescence strict and ultimately rather elongate, the flowers mostly not hidden by the bracts; calyx 12-20 mm. long, deeply and subequally cleft above and below, its primary lobes broad and shallowly notched at the apex into two rounded segments; corolla about equaling or a little longer than the calyx, its lower lip prominent, not notably greenish and thickened, 1/2-2/3 the length of the obscurely puberulent galea; stigma exserted.

Alpine and subalpine meadows, Wallowa and Blue mts., n.e. Oreg. Late June-Sept.

Castilleja chrysantha seems most closely related to C. pulchella, and the numerous small differences indicated in the key are not without exceptions. However, there is no difficulty in separating the two taxa on a geographic basis, and when so separated the two species are entirely comparable with the other species of the Chrysanthae. The plants from the Blue Mts. ordinarily are taller and have longer spikes than those from the Wallowa Mts. Ultimately it may be desirable to distinguish two varieties of C. chrysantha on this basis.

Castilleja wallowensis Pennell (Not. Nat. Acad. Phila. 74:7. 1941.) almost certainly is a hybrid between C. chrysantha and C. rhexifolia, with which two species the type (Pennell 21109, mountain s. of Ice Lake, Wallowa Mts., Wallowa Co., Oreg.) was found growing. In general habit, pubescence, prominent lower lip, and short galea, C. wallowensis is like C. chrysantha. The bracts, however, and margins of the galea are colored as in C. rhexifolia. It is rare.

Castilleja covilleana Henderson, Bull. Torrey Club 27:353. 1900. (Henderson 3388, Soldier Mt., Blaine Co., Ida.)
    C. multisecta A. Nels. Bot. Gaz. 54:148. 1912. (Nelson & Macbride 1261, Ketchum, Blaine Co., Ida.)
    Perennial; stems clustered, erect or ascending from a woody base, 1-3(4) dm. tall, mostly unbranched, finely villous with crisped hairs, not at all hispid; leaves finely villous, all except sometimes the lowermost deeply divided into 3-7 divergent lobes, the lateral segments not much narrower than the mid-blade; inflorescence conspicuous, bright red, scarlet, or occasionally pale yellow to orange, at first short and compact, but later elongating greatly, the fruiting calyces becoming remote; bracts puberulent and villous, deeply 5- to 7-parted,

mostly as long or longer than the flowers; calyx 15-25 mm. long, deeply cleft above, less deeply so below, its primary lobes again divided 1-5 mm. into 2 acute segments; corolla 20-35 mm. long, its minutely puberulent galea 1/2-2/3 the length of the tube and usually 5 or more times as long as the dark green, thickened, lower lip.

Mountain slopes and summits, c. Ida. and adj. Mont. June-Aug.

This is a well-defined species with clear affinities in the Parviflorae, especially with C. rupicola.

Castilleja crista-galli Rydb. Mem. N.Y. Bot. Gard. 1:355. 1900. (Rydberg & Bessey 4950, Bridger Mts., Gallatin Co., Mont.)

Perennial; stems clustered, erect or ascending from a woody base, 2-5 dm. tall, usually unbranched, puberulent to obscurely villous, rarely glabrous; leaves puberulent or sometimes glabrous, the lower ones linear-lanceolate, entire, the upper ones commonly with a pair of lateral lobes, these narrower than the mid-blade; inflorescence conspicuous, bright red or scarlet, rather lax; bracts puberulent and villous, deeply 3- to 5-parted, about as long as the flowers; calyx (20)25-35 mm. long, deeply cleft below, usually less deeply so above, its primary lobes again divided into 2 slender, acute segments 1-10 mm. long; corolla (25)30-40 mm. long, its puberulent to pubescent galea usually about 3/5 the length of the tube and about 5 times as long as the dark green, thickened, lower lip.

Dry mountainsides, extreme n.w. Wyo. and adj. Mont., westward (but specimens atypical) to Lemhi and Custer cos., Ida. Late June-Aug.

The specific status of C. crista-galli is open to question. It is here maintained only to dispose of a number of collections which cannot be placed elsewhere with any degree of confidence. Apparently, the characteristics of these collections are of populational significance, and they happen to include those features which ordinarily are reliable for separating the Parviflorae from the Septentrionales. It is possible that C. crista-galli is only a variant of C. miniata, but it departs from that species in the direction of C. linariaefolia, and it could with equal reason be considered a variant of the latter. Perhaps, it is ultimately of hybrid origin involving these two species, which are not, however, known to be sympatric within the range of C. crista-galli at the present time. Certainly, its relationships, as a whole, are not with any other member of the Parviflorae, but with C. miniata (Septentrionales), C. linariaefolia (Linariaefoliae), or both.

Castilleja cryptantha Pennell & Jones, Proc. Biol. Soc. Wash. 50:208. 1937. (Pennell & Danner 21173, Yakima Park, Mt. Rainier, Wash.)

Perennial; stems clustered, erect or ascending, unbranched, 1.0-1.5 dm. tall, softly viscid-villous; leaves viscid-villous, lower ones linear, upper ones broader, with or without a pair of short lateral lobes; bracts broader than the leaves, lanceolate and acute, with a pair of rather broad lateral lobes, viscid-villous, purplish; inflorescence strict, not greatly elongate, the flowers not hidden by the bracts; calyx yellow, 12-15 mm. long, deeply and subequally cleft above and below; its primary lobes broad and again divided at the apex into 2 triangular, acute segments 1 mm. long; corolla about equaling or a little longer than the calyx, its lower lip prominent, about 2/3 the length of the densely puberulent galea; stigma not or tardily exserted.

Subalpine meadows, Mt. Rainier Nat. Park, Wash. July-Aug.

This is a quite constant highly localized species with clear affinities with C. chrysantha and C. pulchella from which it differs in the technical characters of the key.

Castilleja cusickii Greenm. Bot. Gaz. 25:267. Apr. 1898 (Cusick 1700, Sumpter Valley, Blue Mts., Baker Co., Oreg.)

C. pallida var. camporum Greenm. Bot. Gaz. 25:266. 1898. C. camporum Howell, Fl. N.W. Am. 532. 1901. (Suksdorf 423 [=691], low prairies, Spokane Co., Wash., probably near Spangle; lectotype)

C. lutea Heller, Bull. Torrey Club 25:268. May, 1898. (Heller 3267 [=3090], near the mouth of Potlatch R., Nez Perce Co., Ida.)

C. villosa Rydb. Mem. N.Y. Bot. Gard. 1:361. 1900. (Canby 261, Blackfoot R., Mont.)

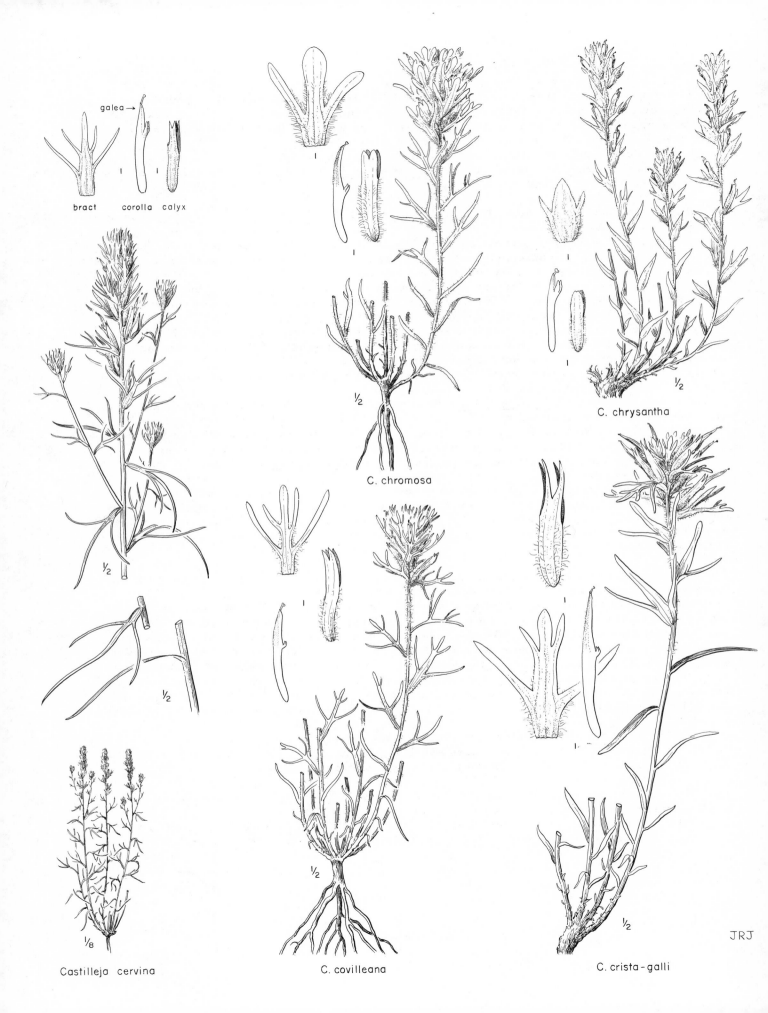

galea→

bract    corolla  calyx

C. chromosa

C. chrysantha

Castilleja cervina

C. covilleana

C. crista-galli

JRJ

C. pilifera A. Nels. Bull. Torrey Club 31:246. 1904. (A. & E. Nelson 5878, Soda Butte,
 Yellowstone Park, Wyo.)

C. pannosa Eastw. Leafl. West. Bot. 3:116. 1942. (Eastwood & Howell 3241, near Anatone,
 Asotin Co., Wash.)

 Perennial; stems clustered, erect or from a decumbent base, usually unbranched, 1-6 dm.
tall, softly viscid-villous or rarely hispidulous; leaves viscid-villous or sometimes hispidulous,
lower ones linear-lanceolate, entire, upper ones broader and usually with 1-3 pairs of linear
divergent lobes, mostly from above the middle; bracts broader than the leaves, mostly oblong
and obtuse, entire or with 1-2 pairs of short lateral lobes, viscid-villous or puberulent, pale
yellow or occasionally purplish; inflorescence strict and ultimately elongate, the flowers remote and mostly hidden by the overlapping bracts; calyx 20-30 mm. long, deeply and subequally cleft above and below, its primary lobes broad and usually rounded at the apex, entire or
shallowly cleft; corolla sometimes shorter, sometimes longer, than the calyx, its lower lip
prominent, 1/3 to nearly as long as the densely glandular-puberulent galea.

 Meadows, mostly at medium elevations (but to 10,000 ft. in the Sawtooth Mts.), n. and w.
Wyo., w. Mont., s.w. Alta., Ida., e. Wash., and n.e. Oreg. Apr.-Aug.

 This is a common and variable species. The type collection in particular differs from most
of the other specimens referred here in its almost complete lack of long hairs on the leaves,
which accordingly are cinereous-puberulent rather than villous. It is otherwise well within the
limits of the taxon as here delimited and is thought to represent no more than a local race, influenced perhaps through hybridization with one of the cinereous-puberulent-leaved species of
the Pallescentes. Clear-cut instances of natural hybridization are known between C. cusickii
and C. miniata, and between C. cusickii and C. rhexifolia, intermediates having been found
with specimens of both parental species at a number of stations. C. miniata and C. rhexifolia
are not otherwise to be considered closely related to C. cusickii. In the Beartooth Range and
perhaps elsewhere, there is complete intergradation between what looks like a subalpine ecotype of C. cusickii and C. pulchella. Because of this situation, these plants, which certainly
are not typical C. cusickii, are included in C. pulchella. Similar plants from the high mountains of Idaho, outside the range of typical C. pulchella, are included in C. cusickii with little
hesitation.

Castilleja elmeri Fern. Erythea 6:51. 1898. (Elmer 457, Wenatchee Mts., 19.3 km. s. [=n.]
 of Ellensburg, Wash.)

C. angustifolia whitedii Piper, Bull. Torrey Club 27:399. 1900. (Whited 1141, Wenatchee
 [Mts. ?], Wash.)

 Perennial; stems clustered, erect or ascending from a woody base, 1.5-3 dm. tall, rarely
branched, viscid-villous; leaves linear-lanceolate, ordinarily entire, but sometimes with a
pair of lateral lobes, viscid-puberulent or -villous; inflorescence conspicuous, crimson,
scarlet, or sometimes yellowish; bracts oblong-ovate, entire and rounded or rarely somewhat
lobed, viscid-villous; calyx 15-25 mm. long, subequally cleft above and below, its primary lobes
again divided into 2 blunt segments 1-4 mm. long; corolla 20-30 mm. long, its puberulent galea
nearly as long as the tube and mostly 4-5 times as long as the dark green, thickened, lower lip.

 Mountain slopes and meadows, mostly above 5000 ft., Wenatchee Mts. and northward, Cascade Range, Wash. June-Aug.

 Castilleja elmeri includes a rather distinct series of specimens with clear affinities with C.
miniata, but differing in their low stature, mostly entire bracts, and glandular pubescence.
Intermediates are rare. It is more difficult to distinguish from the similar high altitude predominantly Rocky Mt. C. rhexifolia, which may be no more than varietally distinct. A few collections of apparently typical C. rhexifolia from the n. Cascade Range in Wash. and B.C.,
however, show no intermediacy toward C. elmeri.

Castilleja exilis A. Nels. Proc. Biol. Soc. Wash. 17:100. 1904.

C. stricta Rydb. Mem. N.Y. Bot. Gard. 1:354. 1900, but not of Benth. in DC. in 1846.
 (Watson 809, Ruby Valley, Nev.)

 Annual; stem erect, usually unbranched, glandular-villous, 3-8 dm. tall; leaves linear-lanceolate, entire, glandular-puberulent or -pubescent; inflorescence becoming greatly elongate, the fruiting calyces remote; bracts lanceolate, entire, glandular, much longer than the flowers, the distal third of the upper ones scarlet; calyx 15-20 mm. long, deeply and subequally cleft above and below, or sometimes more deeply above, its primary lobes again divided 2-3 mm. into linear, acute segments; corolla yellowish, 15-25 mm. long, usually not much longer than the calyx, its galea short and blunt, puberulent, about 4-5 times as long as the inconspicuous lower lip.

Alkaline marshes and meadows, mostly at lower elevations, throughout the intermountain region from c. Wash. and e. Oreg. to s. Mont. and w. Wyo., south to n. w. N. M., n. Ariz., Nev., and adj. Calif. Late June-Sept.

This, our only annual species of Castilleja, is without close relatives in the Pacific Northwest. Three closely allied annual species, however, occur in Calif., Ariz., and N. M.

Castilleja flava Wats. Bot. King Exp. 230. 1871. (Watson 813, Bear River Valley, near the Uintas [s. w. Wyo. or adj. Utah])
  C. breviflora Gray, Am. Journ. Sci. 34:338. 1862, but not of Benth. in DC. in 1846.
    Euchroma breviflora Nutt. ex Gray, loc. cit. as a synonym. Castilleja brachyantha Rydb. Mem N. Y. Bot. Gard. 1:360. 1900. (Rocky Mts., Nuttall, Parry 243) According to Pennell (unpublished), Nuttall's plant is the present species, and Parry's is the one later described by Rydberg in 1904 as C. puberula. Gray's description evidently was drawn from both specimens. In reducing C. brachyantha to synonymy under C. flava, Harrington (Man. Pl. Colo. 506. 1954) has in effect typified it with the Nuttall specimen and established precedent which may well be followed.
  ?C. linoides Gray, Syn. Fl. 2$^1$:299. 1878. (Watson, Clover [present East Humboldt] Mts., Elko Co., Nev.)
  C. curticalix Nels. & Macbr. Bot. Gaz. 55:380. 1913. (Nelson & Macbride 2099, Gold Creek, Elko Co., Nev.)
  Perennial; stems clustered, erect or ascending, generally branched above, 2-5 dm. tall, finely pubescent with soft crisped hairs, often purplish; leaves densely crisp-puberulent, lower ones linear, entire, upper ones commonly with a pair of linear, divaricate lobes; bracts broader than the leaves, usually 3(5)-parted, villous or hispid, yellowish or occasionally reddish; flowers rather remote and not hidden by the bracts; calyx 12-18 mm. long, much more deeply cleft below than above, its primary lobes again divided into 2 acute segments 1-3 mm. long; corolla 15-25 mm. long, often curved and exserted through the anterior cleft in the calyx; galea 6-10 mm. long; lower lip reduced, 1/3-1/5 the length of the galea, its lobes, but usually not its pouch, puberulent.

Dry soils, frequently associated with sagebrush, Mont. (Park and Meagher cos.), Wyo. (in the Big Horn and Laramie mts., and westward), n. Colo., n. Utah, e. and s. Ida. (mostly s. of the Snake River Plain, but n. of the Snake R. in Fremont and Teton cos.), and n. e. Nev. June-July.

This species is habitally similar to C. rustica which replaces it in the mountains of s. w. Mont., c. Ida., and n. e. Oreg. The two taxa are technically and geographically distinct, but seem to be closely related. Rare intermediates suggest that C. flava can hybridize with C. chromosa and an extensive hybrid swarm connecting it with C. linariaefolia is described under that species. A specimen from Park Co., Mont., has the pouch of the lower lip pubescent as in C. pallescens which occurs there. Essentially glabrous specimens from n. e. Nev. seem to be referable to C. linoides, but both C. linariaefolia and C. flava occur in this area, and since they hybridize, the difference is of doubtful significance.

Castilleja fraterna Greenm. Bot. Gaz. 48:147. 1909. (Cusick 3125, near extreme source of Imnaha R., Wallowa Mts., Oreg.; specific locality from the collector's label)
  Perennial; stems clustered, more or less flexuous and ascending, unbranched, 1-1.5 dm. tall, softly viscid-villous; leaves viscid-villous or -puberulent, lower ones linear-lanceolate, entire, upper ones broader and sometimes with a pair of short lateral lobes; bracts relatively inconspicuous, shorter and broader than the leaves, ovate, acute, with a pair of short lateral lobes, viscid-villous, greenish or the upper ones red-tipped; calyx conspicuous, bright red, 15-20 mm. long, deeply and subequally cleft above and below, the primary lobes broad and again divided into 2 rounded or acute segments 1-3 mm. long; corolla 20-30 mm. long, the

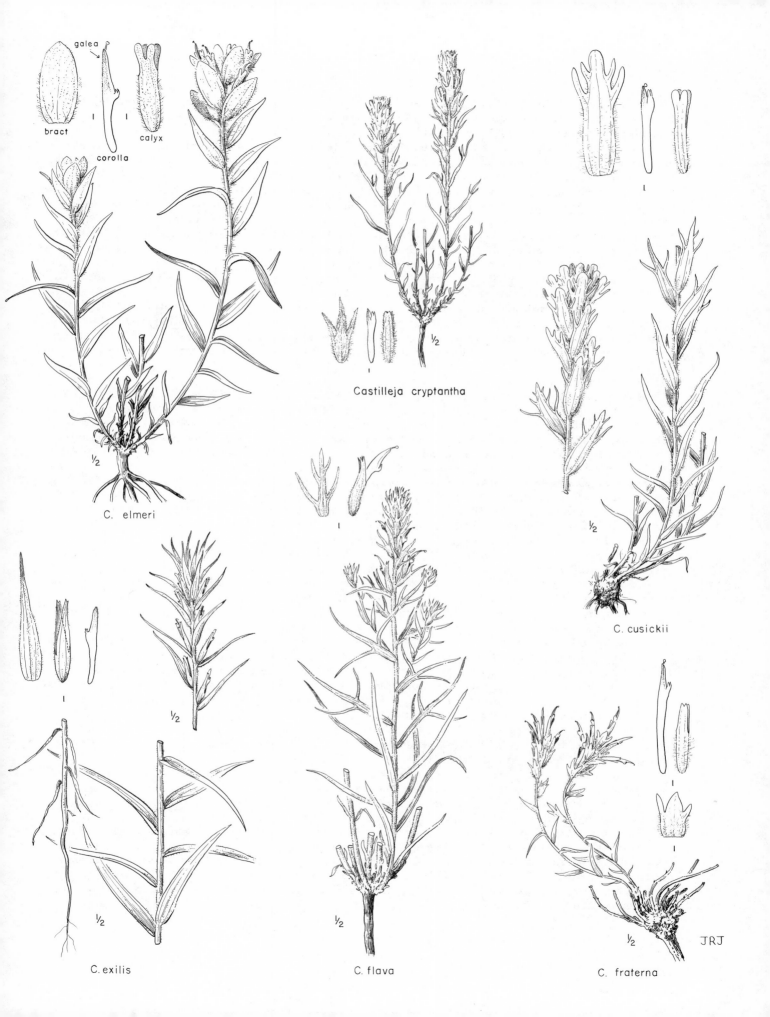

bract

galea

corolla

calyx

C. elmeri

Castilleja cryptantha

C. cusickii

C. exilis

C. flava

C. fraterna

JRJ

tube more than twice the length of the short galea, lower lip prominent though reduced, about
1/3 the length of the galea, its lobes and the margins of the galea bright red.

Alpine meadows, in "thickets" (mats) of dwarf willow, Wallowa Mts., Oreg. Aug.

This species is known only from the original collection which was ample and widely distributed. In an unpublished manuscript entitled: "A list, from memory, of the plants to be seen near
at hand in crossing the canyon of the Middle Fork of the Imnaha near its source," Cusick de-
scribes its type locality in greater detail. "Near the base of the cliff, a fine colony of the alpine
willow, Salix petrophylla [sic!], a few inches in height, finds a congenial home in the moist
sand. With it, and doubtless parasitic on it, is Castilleja fraterna, its crimson spikes quite
over-topping the willow." Apparently, the species is common in this area, for he lists it as
occurring at the starting point, going into the canyon from the west, on the upper slopes on this
side, and on the lower slopes of the east wall, the last evidently the spot described above. The
only other species of Castilleja mentioned are C. rustica on the cliffs of the west wall and C.
chrysantha on the valley floor.

Inasmuch as it combines characteristics of the Chrysanthae and of the Septentrionales, C.
fraterna is a well-marked species. Its nearest relative doubtless is C. ownbeyana, also of the
high Wallowas, but it differs from that species in its much longer corollas, brighter red bracts
and calyces, and more obscurely pubescent stems and leaves. See also C. wallowensis under
C. chrysantha.

Castilleja glandulifera Pennell Not. Nat. Acad. Phila. 74:8. 1941. (Pennell 21138, Anthony
  Peak [=Lake, according to an isotype], Elkhorn [Blue] Mts., Baker Co., Oreg.)

Perennial; stems clustered, erect or ascending from a woody base, 1-3 dm. tall, not uncommonly branched, glandular-hirsute; leaves glandular-puberulent, the lower ones usually
entire and linear-lanceolate, the upper ones broader and usually with a pair of short lateral
lobes, these narrower than the mid-blade, rarely leaves all entire or all divided; inflorescence
inconspicuous, greenish yellow or the bracts sometimes tipped with dull red; bracts glandular-
villous, 3- to 5-parted, mostly shorter than the flowers; calyx 15-20 mm. long, subequally
cleft above and below, its primary lobes again divided 2-6 mm. into 2 acute segments; corolla
20-25 mm. long, its puberulent to pubescent galea about 2/3 as long as the tube and 5 or more
times the length of the dark green, thickened, lower lip.

Dry mountainsides in gravelly soil, e. Oreg. (in the Blue, Wallowa, and Steens mts., and
in the Cascade Range near Bend) and s. w. Ida. Late June-Aug.

This species is clearly a near relative of C. applegatei through which it is connected to the
other members of the Parviflorae. Its relationship to the earlier and little-known C. viscidula
Gray is an unsolved problem. The two species are remarkably similar in their glandular pubescence, divided leaves, and yellowish inflorescences. They look alike and frequently have not
been distinguished. However, the specimens of C. viscidula available are much smaller than
those of C. glandulifera. The flowers are only 14-17 mm. long, and the shorter galea is only
3-4 times the length of the more prominent lower lip. If these characteristics are constant in
the Nevada species, it is obvious that our plants as described above are distinct. It is possible,
however, that the known collections of C. viscidula are not representative, and that C. glandulifera will prove to be synonymous with it.

Castilleja gracillima Rydb. Bull. Torrey Club 34:39. 1907. (Rydberg & Bessey 4964, Lower
  Geyser Basin, Yellowstone Park, Wyo.)
  ?C. ampliflora Rydb. Bull. Torrey Club 34:39. 1907. (Vreeland 995, divide between
    McDonald and Camas lakes, Flathead Co., Mont.)
  C. ardifera Macbr. & Pays. Contr. Gray Herb. n. s. 49:69. 1917. (Macbride & Payson 3524,
    Cape Horn, Custer Co., Ida.)

Perennial; stems slender, erect or ascending, 2-5 dm. tall, from a remote woody caudex,
more or less creeping and often rooting at the base, usually unbranched, villous to glabrate;
leaves linear-lanceolate, ordinarily all entire, villous-puberulent to glabrate; inflorescence
conspicuous, rather narrow and elongating in fruit, predominantly yellow, but varying in col-
or to orange or even red; bracts oblong, entire and rounded or sometimes with a pair of lateral lobes, villous-puberulent; calyx 15-22 mm. long, deeply and subequally cleft above and be-

low, its primary lobes again divided into 2 acute segments 2-4 mm. long; corolla 20-30 mm.
long, its densely puberulent galea usually shorter than the tube and about 5 times the length of
the dark green, thickened lower lip.

Wet meadows, n.w. Wyo. and adj. Mont. to c. Ida. and possibly e. B.C. June-Aug.

In its typical form, C. gracillima seems to be a well-marked and easily distinguishable taxon with affinities with both C. sulphurea and C. miniata. Where the three grow in the same
general region, they form rather uniform natural populations ecologically and morphologically,
with no clear evidence of hybridization. Outside of this area, however, the characters of C.
gracillima occur sporadically and independently in both related species, and occasional specimens are difficult to place. Deserving special comment in this connection are plants from the
Rocky Mountain trench near the head of the Columbia R. in e. B.C. These are certainly not
C. miniata. They agree with C. gracillima in the characters of the key, and adequate reasons
scarcely exist for considering them other than conspecific with it. However, the lower lip in
some but not in all of them is more prominent than that described above.

Castilleja hispida Benth. in Hook. Fl. Bor. Am. 2:105. 1838.
    C. angustifolia var. hispida Fern. Erythea 6:47. 1898. (Douglas, Ft. Vancouver, Clark Co.,
      Wash.; lectotype)
    C. angustifolia var. abbreviata Fern. Erythea 6:49. 1898. C. hispida ssp. abbreviata Pennell
      in Abrams, Ill. Fl. Pac. St. 3:841. 1951. (Piper 2175, Olympic Mts., Wash.) = var. his-
      pida.
    C. remota Greene, Pitt. 4:2. 1899. (Macoun, Goldstream, Vancouver I., B.C.) = var. his-
      pida.
    C. desertorum Geyer ex Rydb. Fl. Rocky Mts. 790. 1917. (Mont.-Ida.-Wash. No type des-
      ignated) = var. acuta. This is nomenclaturally independent of the nom. subnud. C. deserto-
      rum Geyer mentioned by Hooker (Journ. Bot. and Kew Misc. 5:258. 1853) as a probable
      synonym of C. hispida, the type of which (Geyer 511, "desert of Upper Colorado" [Wyo.])
      surely is C. chromosa rather than C. hispida. It must, accordingly, be typified by a speci-
      men of the species which Rydberg described rather than by Geyer 511.
    CASTILLEJA HISPIDA var. ACUTA (Pennell) Ownbey, hoc loc. C. hispida ssp. acuta Pennell,
      Not. Nat. Acad. Phila. 74:11. 1941. C. hispida var. acuta Pennell, ex Peck, Man. High.
      Pl. Oreg. 669. 1941, without a Latin diagnosis. (Pennell 21074, Adams Creek, Wallowa
      Mts., Wallowa Co., Oreg.)
    C. taedifera Pennell, Not. Nat. Acad. Phila. 74:6. 1941. (Peck 18338, Buckhorn Springs,
      Wallowa Co., Oreg.) = var. acuta.
Perennial; stems clustered, erect or ascending from a woody base, 2-6 dm. tall, mostly
unbranched, finely villous to more or less hispid; leaves lanceolate or broader, finely villous
or occasionally somewhat viscid- to hispid-villous, lower ones entire and reduced, upper ones
with usually 1-2(3) pairs of lateral lobes, these much narrower than the mid-blade, occasionally all entire; inflorescence conspicuous, bright red or scarlet, sometimes yellow, at first
short and broad, but later elongating; bracts broad and rather deeply 3- to 5-lobed, more or
less puberulent and villous; calyx 15-30 mm. long, deeply and subequally cleft above and below,
its primary lobes again divided 1-7 mm. into 2 rounded or acute segments; corolla 20-40 mm.
long, its puberulent to pubescent galea about equaling the tube in length and 5 or more times
the length of the dark green, thickened, lower lip.

Grassy slopes and forest openings at medium and low elevations, s.w. Alta. and n.w. Mont.
to Vancouver I., s. along the coast to Benton Co., Oreg., and in the interior to Grant Co.,
Oreg., and Payette Co., Ida. Late Apr.-Aug.

This is a common, complex and variable species. It is most closely related to C. chromosa
which replaces it along the southern border of our area, but there is little difficulty in separating the two on the characters of the key. In e. Oreg., in particular, it intergrades with C. miniata in a manner suggesting natural hybridization. C. peckiana was set up to include these
intermediates, but this passes imperceptibly into C. miniata on one hand and into C. hispida
(and perhaps C. chromosa) on the other. There is a definite break, however, in the middle of
the series of intermediates, and it is here that the line between the species must be drawn.

Within C. hispida, two fairly well-defined geographical races may be separated by the following key:

1  Pubescence, on the whole, more villous than hispid; calyx lobes obtuse and rounded; Cascade Range and westward, extending inland on both sides of the international boundary to n. w. Mont. and s. w. Alta.                                  var. hispida

1  Pubescence usually strongly hispid-villous; calyx lobes generally acute; more robust plants with thicker, more often entire leaves; mountains of e. Oreg. to s. e. Wash., across Ida. to Missoula Co., Mont.                        var. acuta (Pennell) Ownbey

These varieties are not sharply delimited, but intermediates are much less common than the extremes. In general, the correlation of morphology and geography is good, but individuals and probably populations closely resembling var. hispida are found at higher altitudes within the range of var. acuta as far south as the Blue Mts. in Wash. and the Salmon-Clearwater divide in Ida. Conversely, hispid individuals and presumably populations strongly suggestive of var. acuta occur within the range of var. hispida, particularly along the lower Columbia R. in Wash. and Oreg. and in the vicinity of Wenatchee, Wash.

Castilleja inverta (Nels. & Macbr.) Pennell & Ownbey in Davis, Fl. Ida. 614. 1952.
    C. fasciculata var. inverta Nels. & Macbr. Bot. Gaz. 55:381. 1913. C. pilosa var. inverta Nels. & Macbr. Bot. Gaz. 61:44. 1916. (Nelson & Macbride 1915, Rattlesnake Springs, Owyhee Co., Ida.)
    Perennial; stems clustered, erect or ascending, unbranched, mostly less than 1 dm. tall, densely and uniformly puberulent, without long hairs, usually purplish; leaves densely short-puberulent, rather crowded, lower ones linear, entire, upper ones with a pair of linear, divaricate lobes; bracts broader than the leaves, 3- to 5-parted, densely puberulent, not ciliate, yellowish, prominently veined; calyx 15-30 mm. long, purplish, strongly nerved, subequally cleft above and below or more deeply so above, its primary lobes again divided 1-5 mm. into 2, triangular or linear-triangular, acute segments; corolla sometimes shorter, sometimes longer, than the calyx, its lower lip prominent, pouched, pubescent, nearly as long as the galea or a little longer.
    Dry plains and hills, often associated with sagebrush, c. and s. w. Ida. to n. Nev., and probably s. e. Oreg. June-Aug.
    This is a rather constant taxon closely related only to C. pallescens and C. oresbia, from both of which it may be distinguished by its shorter, more uniform puberulence, lower stature, more strongly nerved bracts, and different distribution. The morphological intermediates which exist are excluded from C. inverta as here defined.

Castilleja levisecta Greenm. Bot. Gaz. 25:268. 1898. (Howell 279, Mill Plain [Ft. Vancouver], Wash.; lectotype)
    Perennial; stems several, more or less decumbent or creeping at the base, usually unbranched, 1-5 dm. tall, softly viscid-villous; leaves viscid-villous to hispidulous, lower ones linear-lanceolate, entire, upper ones oblong-ovate or -obovate, with mostly 1-3 pairs of short lateral lobes from the distal 1/3 of the blade; bracts about the width of the upper leaves, oblong, obtuse, entire or more usually with 1-3 pairs of short lateral lobes from near the apex, puberulent and more or less viscid-villous, golden yellow; inflorescence strict and ultimately elongate, the flowers remote and mostly hidden by the overlapping bracts; calyx 15-18 mm. long, deeply and subequally cleft above and below, its primary lobes again rather deeply divided into 2 linear obtuse segments; corolla 20-23 mm. long, its galea rather slender, puberulent, about 3-4 times the length of the unpouched lower lip.
    Meadows and prairies at low elevations west of the Cascade Range, Vancouver I. to Linn Co., Oreg. Apr.-Sept.
    Castilleja levisecta is a well-marked species with clear affinities in the Chrysanthae, especially with C. cusickii.

Castilleja linariaefolia Benth. in DC. Prodr. 10:532. 1846.
    C. affinis var. linariaefolia Zeile in Jeps. Man. Fl. Pl. Calif. 938. 1925. (Fremont, Rocky

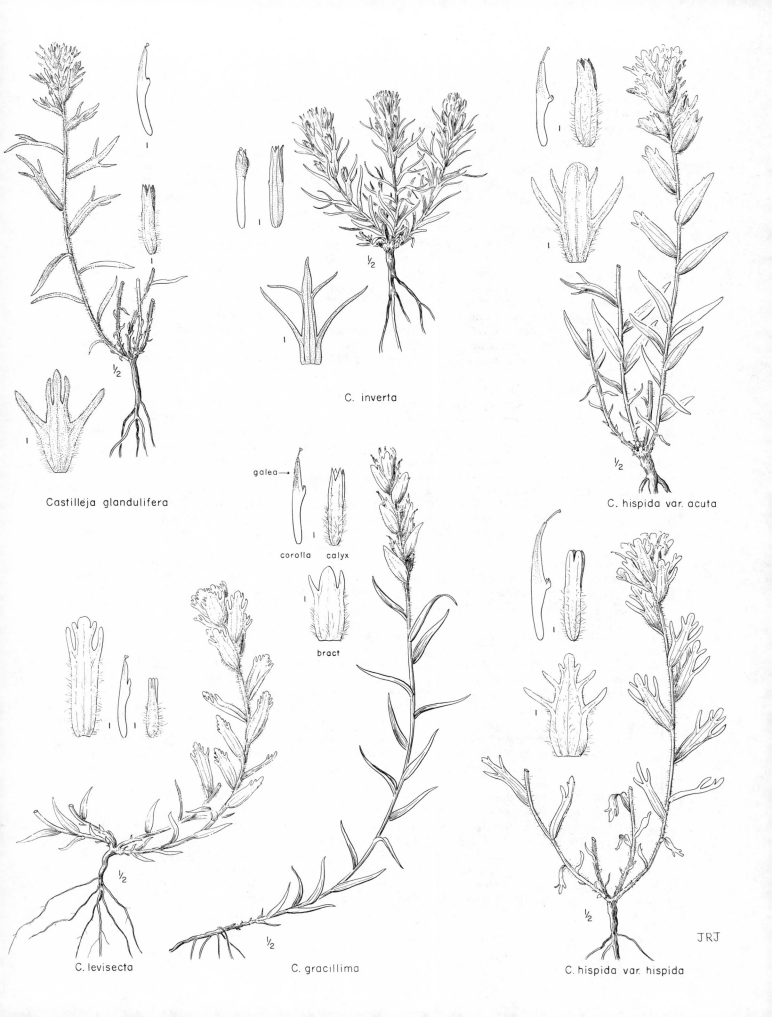

Castilleja glandulifera

C. inverta

galea→

corolla    calyx

bract

C. hispida var. acuta

C. levisecta

C. gracillima

C. hispida var. hispida

JRJ

Mts., actually at n. end of Laramie Mts. in Converse or Platte Co., Wyo., according to
Pennell [unpublished])

C. arcuata Rydb. Bull. Torrey Club 34:35. 1907. (Rydberg & Carlton 7508, s. end of Fish
Lake, Sevier Co., Utah)

Perennial; stems erect or ascending from a woody base, 3-7(10) dm. tall, commonly
branched, usually more or less hispid below, otherwise glabrous, occasionally glabrous or
pubescent throughout; leaves mostly entire and linear, but the upper ones commonly with a
pair of lateral lobes, and sometimes most of them divided, usually glabrous, rarely puberulent or pubescent; inflorescence conspicuous, bright red, scarlet, or occasionally yellow,
the color mostly confined to the calyces and upper bracts, the lower flowers mostly well separated; bracts mostly 3-lobed, commonly shorter than the calyx and relatively inconspicuous,
finely villous-tomentulose; calyx conspicuously colored, puberulent to tomentulose, 20-35 mm.
long, deeply cleft below, much less deeply so above, its primary lobes again divided into 2
slender, acute segments 2-4 mm. long; corolla (25)30-45 mm. long, usually curved and exserted through the anterior cleft of the calyx, its puberulent galea sometimes shorter, sometimes longer, than the tube and many times the length of the dark green, thickened, lower lip.

Dry plains and hills, usually associated with sagebrush, and in the mountains in suitable
habitats to 10,000 ft. elevation or more, s. Mont., Wyo., w. Colo., and n.w. N.M., across
s. Ida., Utah, Nev., and Ariz. to c. Oreg. and Calif., e. of the Sierra Nevada. Late May-
Aug.

This is our most spectacular species of Castilleja and the state flower of Wyo. In its usual
and typical form it is very easily recognized, but like other widespread species of the genus,
it tends to intergrade with equally distinct species wherever there is opportunity for hybridization. On the west face of the Big Horn Mts., e. of Kane, Wyoming, there is an extensive
hybrid swarm involving C. linariaefolia and C. flava. This population consists of thousands of
individuals grading imperceptibly from the former species at its lower limits to the latter at the
top of the mountain. Similar intermediates from elsewhere, particularly from n.e. Nev. and s.w.
Oreg., indicate that this situation is by no means unique. Ordinarily, however, these two species
do not grow together, and the effect of hybridization is limited. Similarly, there is some evidence
of occasional hybridization between C. linariaefolia and C. chromosa as well as with C. miniata. For a puzzling series of intermediates between the first and the last, which replaces
C. linariaefolia from s.w. Mont. to c. Ida., see C. crista-galli.

Among our species, C. linariaefolia is most closely related to C. flava. These are, however, very dissimilar species, and this relationship is perhaps not so close as that between
C. linariaefolia and a large series of similar species extending to the Andes of S. Am. and including C. fissifolia, the type species of the genus.

Castilleja litoralis Pennell, Proc. Acad. Phila. 99:183. 1947. (Pennell 15651, Bandon, Coos
Co., Oreg.)

Perennial; stems few, ascending or more or less decumbent, 1-9 dm. long, commonly
branched, finely hispid-villous with spreading hairs; leaves lanceolate to ovate, the broader
ones commonly rounded at the apex, ordinarily entire, but sometimes with a pair of lateral lobes,
more or less hispidulous or villous; inflorescence conspicuous, scarlet to crimson, elongating in
fruit; bracts broad and rounded, entire or with 1-2 pairs of short lateral lobes, villous; calyx
17-25 mm. long, deeply and subequally cleft above and below, its primary lobes again divided
1-5 mm. into 2 usually blunt segments; corolla 25-40 mm. long, its villous galea about 3/4
as long as the tube and usually 5 or more times the length of the dark green, thickened, lower
lip.

Seaside bluffs, along the coast of Oreg. and n.w. Calif. Reported from Grays Harbor Co.,
Wash., but the similar seashore plants there (C. dixoni) pertain to C. miniata. Apr.-Sept.

Castilleja litoralis includes a diverse series of specimens intermediate between the Californian C. latifolia (sens. lat.) and C. miniata. At the present time, C. latifolia and C. miniata
are nowhere in contact, but the existence of the variably intermediate C. litoralis suggests that
at one time they may have been. The possibility that C. litoralis is ultimately of hybrid origin
is strengthened by the existence within C. miniata, as here delimited, of a poorly marked seaside ecotype (C. dixoni), some individuals of which could almost as well be C. litoralis. No

satisfactory criteria have been found for distinguishing C. dixoni from C. miniata, and the line between them and C. litoralis is an arbitrary geographic one drawn at the mouth of the Columbia R.

Castilleja longispica A. Nels. Bull. Torrey Club 26:480. 1899. (Nelson 900, in part, Gros Ventre R., Wyo.)

   Orthocarpus psittacinus Eastw. Bull. Torrey Club 29:78. 1902. Castilleja psittacina Pennell in Abrams, Ill. Fl. Pac. St. 3:825. 1951. (Bruce 2240, Warner Mts., Oreg.)

   Castilleja pratensis Heller, Muhl. 2:139. 1906. (Heller 8079, Gazelle, Siskiyou Co., Calif.)

   C. robiginosa Macbr. & Pays. Contr. Gray Herb. n. s. 49:68. 1917. (Macbride & Payson 3387, Robinson Bar, Custer Co., Ida.)

   C. steenensis Pennell, Not. Nat. Acad. Phila. 74:4. Feb. 1941. (Peck 14076, Steens Mts., 5 mi. s. of Wild Horse Creek, Harney Co., Oreg.)

   C. ochracea Eastw. Leafl. West. Bot. 3:91. Nov. 1941. (Eastwood & Howell 8016, 7 mi. s. w. of Canby, Modoc Co., Calif.)

   Perennial; stems clustered, often decumbent at the base and sometimes branched above, 1-3 dm. tall, purplish, puberulent to villous-hirsute; lower leaves linear, entire, upper ones with 1 or 2 pairs of slender, divergent lobes, densely short-pubescent; bracts yellowish (rarely purplish), 3- to 9-parted, the lobes shorter and narrower than the mid-blade; flowers rather remote and usually not hidden by the bracts; calyx 10-20 mm. long, subequally cleft into 4 linear, linear-triangular, or sometimes deltoid segments; corolla 15-20 mm. long, generally exserted well beyond the calyx, its lower lip prominent, pouched, sometimes purplish, often nearly equaling or even exceeding the galea in length and like it more or less strongly puberulent.

   Generally associated with sagebrush, often in meadowy areas, mostly at lower elevations, n. e. Calif., across e. Oreg. and c. Ida., to w. Wyo. and s. w. Mont.; also in the Clearwater River drainage in n. Ida. June-July.

   This relatively constant and easily recognized species on one hand has been confused with the related C. pilosa (Wats.) Rydb. of Nev. and Calif., and on the other has been segregated into the numerous proposals listed above, none of which seems to have any great morphological or geographical significance. The calyx lobes are subject to considerable variation in shape, but plants in which they are triangular or deltoid are sporadic in occurrence. The most extreme forms in this respect occur in the Clearwater River drainage, an area in which C. longispica would not ordinarily be expected.

Castilleja lutescens (Greenm.) Rydb. Mem. N. Y. Bot. Gard. 1:359. 1900.

   C. pallida var. lutescens Greenm. Bot. Gaz. 25:265. 1898. (Suksdorf 424 [=690?], probably near Spangle, Spokane Co., Wash.; lectotype)

   Perennial; stems clustered, erect (sometimes decumbent at the base), stout, often branched above, 3-6 dm. tall, puberulent to obscurely villous or hispid, sometimes purplish; leaves linear or linear-lanceolate, scabrous-puberulent, mostly entire, but the upper ones sometimes with 1 or rarely 2 pairs of short lateral lobes; bracts yellowish, densely puberulent, broader than the leaves, obtuse or obtusish, entire or more commonly 3- to 7-lobed, the lobes usually short; calyx 15-25 mm. long, deeply and subequally cleft above and below, its primary lobes again divided into 2, triangular or linear, acute segments 1-7 mm. long; galea usually exserted 1-10 mm. beyond the calyx; lower lip reduced, 1/2-1/5 the length of the galea, its lobes commonly puberulent and not incurved-thickened.

   Grasslands and open coniferous woods, s. w. Alta., n. w. Mont., Ida. n. of Salmon R., n. e. Oreg., e. Wash., and s. B. C. May-Aug.

   This is an easily recognized species without close relatives. Apparently it is related to C. rustica, and through that species with C. pallescens and its allies.

Castilleja miniata Dougl. ex Hook. Fl. Bor. Am. 2:106. 1838.

   C. pallida var. miniata Gray, Am. Journ. Sci. 34:337. 1862. (Douglas, Blue Mts.; lectotype)

   C. confusa Greene, Pitt. 4:1. 1899. (Type not designated, but based perhaps wholly on Baker, Earle, & Tracy 369, 578, and 755 from the mountains of s. w. Colo.)

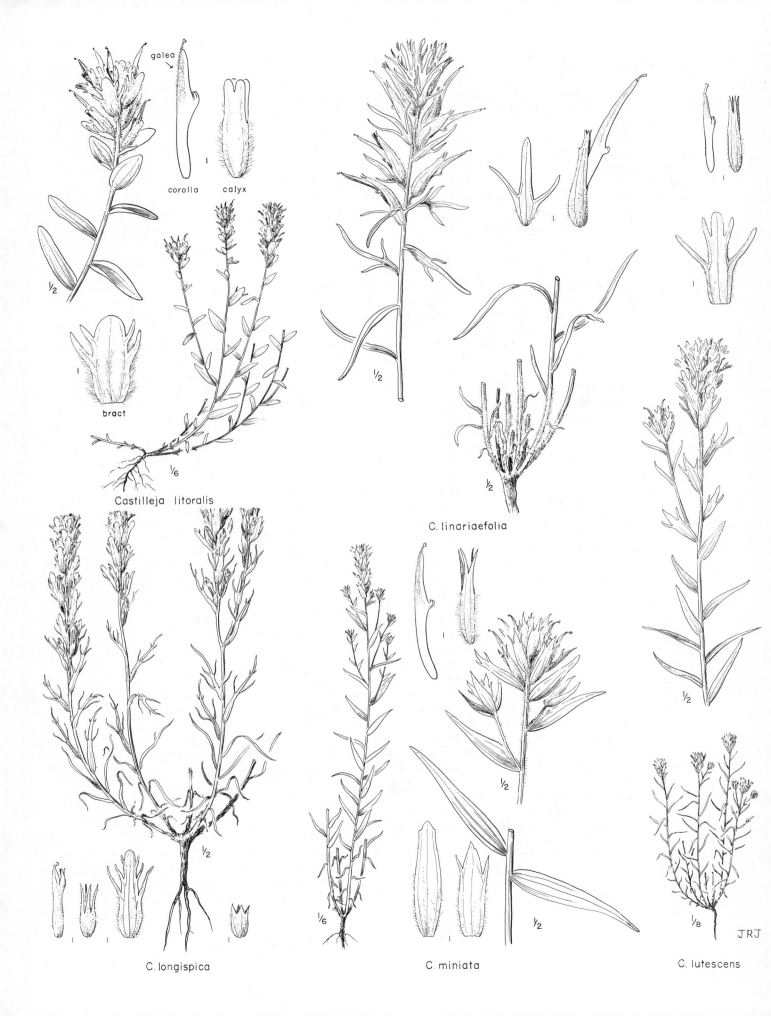

galea

corolla    calyx

½

bract

Castilleja litoralis

⅙

C. linariaefolia

½

½

½

C. longispica

½

⅙

C. miniata

½

½

C. lutescens

⅛

JRJ

C. dixoni Fern. Erythea 7:122. 1899. C. miniata var. dixoni Nels. & Macbr. Bot. Gaz. 65:70. 1918. (Dixon, Quinaitt [=Quinault] Indian agency, Wash.). A poorly defined seashore ecotype.

C. lanceifolia Rydb. Mem. N.Y. Bot. Gard. 1:357. 1900. (Rydberg & Bessey 4954, Spanish Basin, Madison Co., Mont.) A glabrate form.

C. tweedyi Rydb. Mem. N.Y. Bot. Gard. 1:358. 1900. (Rydberg & Bessey 4962, Jack Creek, Madison Co., Mont.). A pubescent form.

C. trinervis Rydb. Bull. Torrey Club 28:26. 1901. (Rydberg & Vreeland 5620, headwaters of Sangre de Cristo Creek, Costillo Co., Colo.) A pubescent form.

C. crispula Piper, Contr. U.S. Nat. Herb. 11:516. 1906. C. miniata var. crispula Nels. & Macbr. Bot. Gaz. 61:45. 1916. (Coville 768, Mt. St. Helens, Skamania Co., Wash.) A pubescent form.

C. magna Rydb. Bull. Torrey Club 34:36. 1907. (Shaw 205, above Carbonate, Kootenay Dist., B.C.) A glabrous form.

C. variabilis Rydb. Bull. Torrey Club 34:37. 1907. (Rydberg 6773, Big Cottonwood Canyon, below Silver Lake, Salt Lake Co., Utah) A pubescent form.

C. vreelandii Rydb. Bull. Torrey Club 34:38. 1907. (Vreeland 1000, divide between McDonald and Camas lakes, Flathead Co., Mont.) A glabrous form.

C. peckiana Pennell, Not. Nat. Acad. Phila. 74:9. 1941. (Pennell 15503, 3 mi. n. of Sisters, Deschutes Co., Oreg.). A pubescent form.

Perennial; stems few, erect or ascending from a woody base (rarely creeping and rooting at the base or decumbent), 2-8 dm. tall, often branched, glabrous to short-pubescent or somewhat viscid-villous, particularly above; leaves linear or lanceolate, sometimes broader, ordinarily all entire, but sometimes a few lobed, glabrous to puberulent or finely villous with simple hairs; inflorescence conspicuous, bright red or scarlet, occasionally crimson or rarely yellow, at first short and broad, but often elongating in fruit; bracts oblong-ovate, more or less toothed or cleft, with acute segments, rarely entire, puberulent and villous, often viscid; calyx 15-30 mm. long, deeply and subequally cleft above and below, its primary lobes again divided into 2 usually linear segments 3-9 mm. long; corolla 20-40 mm. long, its puberulent to pubescent galea 3/4 to about as long as the tube and 5 or more times the length of the dark green, thickened, lower lip. N=24.

Meadows and slopes, mostly at medium and lower elevations in the mountains of w. N. Am., occurring in every state and province from Alas. to Calif., Ariz. and N.M. May-Sept.

This is our most common species of Castilleja. It is complex and variable and overlaps in distribution most of our other species. Morphological intermediates between many of these and C. miniata are not uncommon, but first generation hybrids and hybrid swarms appear to be rare. In spite of its variability, therefore, C. miniata, as a whole, is a readily recognizable taxon. See notes under C. cusickii, C. gracillima, C. hispida, C. litoralis, and C. rhexifolia.

Castilleja nivea Pennell & Ownbey, Not. Nat. Acad. Phila. 227:2. 1950. (Pennell, Cotner, & Schaeffer 23920, s. side of Rock Creek Valley, Beartooth Mts., Carbon Co., Mont.)

Perennial; stems clustered, erect or ascending, mostly unbranched, 0.5-1.5 dm. tall, tomentulose; lower leaves linear, entire, upper ones broader, 3-parted to about the middle, with linear lobes, tomentulose; bracts lanceolate, acute, 3-parted, snowy-tomentose; inflorescence congested, but the flowers not hidden by the bracts, yellow or yellowish; calyx 15-20 mm. long, subequally cleft into 4 broadly linear segments; corolla 18-25 mm. long, generally exserted 3-5 mm. beyond the calyx; lower lip prominent, strongly pouched, villous-lanate, about half the length of the shaggy-villous galea.

Arctic-alpine in the mountains of c. and s. Mont. and adj. Wyo. Now known from the Big Snowy, Tobacco Root, Crazy and Beartooth ranges. July-Aug.

This remarkable species is without near relatives.

Castilleja occidentalis Torr. Ann. Lyc. N.Y. 2:230. 1828.

C. pallida var. occidentalis Gray, Bot. Calif. 1:575. 1876. (Collected by Dr. Edwin James on the Long Expedition, probably on James' [now Pikes] Peak, Colo., which he climbed on July 14, 1820)

C. pallida var. alpina Porter, Syn. Fl. Colo. 96. 1874. (Porter, summit of Pikes Peak, Colo.)

Perennial; stems clustered, erect or ascending from a woody base, 1-2 dm. tall, unbranched, mostly purplish, more or less viscid-villous; leaves linear-lanceolate (rarely broader), usually all entire, but the upper ones not uncommonly with a pair of lateral lobes, viscid-villous or -puberulent; inflorescence short and compact, predominantly yellow, but varying to red and purple with all conceivable intermediates; bracts ovate-oblong, entire and rounded or with 1-2 pairs of lateral lobes, strongly viscid-villous; calyx 15-20 mm. long, deeply and subequally cleft above and below, its primary lobes again divided into 2 mostly blunt segments 1-4 mm. long; corolla 18-25 mm. long, its minutely puberulent galea much shorter than the tube and only 3-4 times the length of the rather prominent, but scarcely saccate, lower lip.

Alpine slopes and meadows, higher mountains of Colo. and Utah, appearing again in the n. Rocky Mts. from Glacier Nat. Pk., Mont., northward. Not known elsewhere in Mont., Wyo., and Ida. July-Aug.

Castilleja occidentalis forms a connecting link between the Septentrionales and the boreal C. pallida complex. The latter is not known to occur in our area. Among our species, C. occidentalis shows a particularly close relationship only with C. rhexifolia, from which it is ordinarily sharply distinct.

Castilleja oresbia Greenm. Bot. Gaz. 48:147. 1909. (Cusick 3201a, Kettle Creek, Wallowa Mts., Baker Co., Oreg.)

Perennial; stems clustered, erect or ascending, usually unbranched, 1-2 dm. tall, densely puberulent to finely villous or hispidulous, purplish; leaves densely puberulent, remote, lower ones linear, entire, upper ones with 1-2 pairs of linear, divergent lobes; bracts broader than the leaves, 3- to 5-parted, puberulent and more or less ciliate, yellowish, mostly not strongly veined; calyx 10-25 mm. long, yellowish or sometimes purplish, deeply and subequally cleft above and below, its primary lobes again divided usually into 2, linear, usually acute segments 5-10 mm. long; corolla sometimes shorter, sometimes longer, than the calyx, its lower lip prominent, pouched, pubescent, 2/3 to fully as long as the galea.

Dry hills and plains, often associated with sagebrush, n. e. Oreg. and adj. Ida. May-Aug.

This somewhat variable taxon connects C. pallescens with C. inverta. From the first, it is usually separable by its long calyx lobes and softer pubescence, but occasional specimens of either may approach the other in these respects. From C. inverta, it may usually be distinguished by its longer calyx lobes, longer pubescence, and obscurely nerved bracts, but none of these features provides an absolute criterion for distinction. There is apparently no difficulty, however, in drawing the geographical limits here indicated, and most of the specimens so separated agree in the characteristics enumerated.

Castilleja ownbeyana Pennell, Proc. Acad. Phila. 99:179. 1947. (Cusick 2354, "moist subalpine Wallowa Mts., near the lake," n. e Oreg.) Cusick's unpublished notes indicate that he knew this plant only from Keystone Basin.

Perennial; stems clustered, mostly decumbent or flexuous, unbranched, 0.5-1.5 dm. tall, softly viscid-villous; leaves viscid-villous, linear-lanceolate, entire, or the upper ones with a pair of short lateral lobes; bracts broader than the leaves, broadly ovate, acute or acutish, with a pair of short lateral lobes, viscid-villous, dull red or purplish, especially above; inflorescence strict and at length rather elongate, the flowers mostly not hidden by the bracts; calyx 12-18 mm. long, deeply and subequally cleft above and below, its primary lobes broad and shallowly notched at the apex into two rounded segments; corolla about equaling or a little longer than the calyx, lower lip prominent, about 2/3 the length of the galea, its lobes red as are also the broad margins of the minutely puberulent galea; stigma exserted.

Subalpine meadows, Wallowa Mts., n. e. Oreg. July-Aug.

Castilleja ownbeyana is a little-known taxon, distinguished from C. chrysantha, the common species of the high Wallowas, by its flexuous stems, red or reddish bracts, calyces, galea margins, and lobes of the lower lip. In these respects, it resembles C. fraterna, but it is well differentiated in other ways. The bracts and calyces in C. ownbeyana are not nearly so red as in C. fraterna, the flowers are much shorter, and the pubescence is much more conspicuous.

Castilleja pallescens (Gray) Greenm. Bot. Gaz. 25:266. 1898.
   Orthocarpus pallescens Gray, Am. Journ. Sci. 34:339. 1862. Euchroma pallescens Nutt. ex
      Gray loc. cit. as a synonym. (Nuttall, Rocky Mts.)
   Orthocarpus parryi Gray, Am. Nat. 8:214. 1874. (Parry 218, Pacific Springs, n.w. Wyo.
      Pennell [unpublished] cites W. A. Jones, Reconn. N.W. Wyoming 312. 1875, as indicating
      that this specimen was collected within the limits of the present Yellowstone Nat. Pk.)
   C. fasciculata A. Nels. Bull. Torrey Club 26:133. 1899. (E. Nelson 4998 [=4898], Indian
      Grove Mts., Carbon Co., Wyo.)
   Perennial; stems clustered, erect or ascending, usually unbranched, 1-2(3) dm. tall, densely puberulent with retrorse hairs or somewhat hispidulous, usually purplish; leaves densely puberulent, remote, lower ones linear, entire, upper ones usually with 1-2 pairs of linear, divergent lobes; bracts broader than the leaves, 3- to 7-parted, puberulent and ciliate, yellowish or purplish, not strongly veined; calyx 12-25 mm. long, deeply and subequally cleft above and below, its primary lobes again divided into 2, triangular or linear-triangular, usually acute segments 1-5 mm. long; corolla shorter than the calyx or but little longer, its lower lip prominent, pouched, pubescent, 2/3 to fully as long as the galea.
   Dry plains and hills, commonly associated with sagebrush, c. and w. Wyo., w. and s. Mont., and n.e. Ida. May-Aug.
   This species is closely related only to C. inverta and C. oresbia, from which it is distinguished on trivial morphological and geographical criteria. Perhaps the three should be considered to be geographical varieties of C. pallescens, but this disposition is not clearly required at the present time.

Castilleja parviflora Bong. Mém. Acad. St. Pétersb., Ser. VI, 2:158. 1833. (Mertens, Island
   of Sitka, Alaska)
   CASTILLEJA PARVIFLORA var. OREOPOLA (Greenm.) Ownbey, hoc loc. C. oreopola
      Greenm. Bot. Gaz. 25:264. 1898. C. miniata var. alpina Suksd. Deuts. Bot. Monats. 18:
      155. 1900, not accepted by its author, hence invalid. (Suksdorf 2046, Mt. Adams [s.e.
      side], Yakima Co., Wash.; lectotype)
   ?C. henryae Pennell, Proc. Acad. Phila. 86:534. 1934. (Henry 303, head of Besa R., B.C.)
   CASTILLEJA PARVIFLORA var. OLYMPICA (G. N. Jones) Ownbey, hoc loc. C. olympica
      Jones, U. Wash. Pub. Biol. 5:231. 1936. (Jones 3808, Mt. Angeles, Clallam Co., Wash.)
   C. oreopola ssp. olympica Pennell, Proc. Acad. Phila. 99:188. 1947. (Pennell & Meyer
      21237, Bogachiel Ridge, above Sol Duc Hot Springs, Clallam Co., Wash.)
   CASTILLEJA PARVIFLORA var. ALBIDA (Pennell) Ownbey, hoc loc. C. oreopola ssp.
      albida Pennell, Proc. Acad. Phila. 99:188. 1947. (Thompson 14353, Mt. Pugh, Snohomish
      Co., Wash.)
   Perennial, strongly blackening in drying; stems clustered, erect or ascending from a woody base, 1.5-3(5) dm. tall, unbranched, glabrate to obscurely villous, not at all hispid; leaves glabrate to somewhat villous, all except sometimes the lowermost divided into 3-5 divergent lobes, the lateral segments usually much narrower than the mid-blade; inflorescence conspicuous, rose-colored to crimson or, in var. albida, whitish, at first short and compact, but later elongating; bracts villous, 3- to 5-parted, about as long as the flowers; calyx 12-20 mm. long, deeply and subequally cleft above and below, its primary lobes again divided into 2 obtuse or acute segments up to 6 mm. long, rarely undivided; corolla 12-25 mm. long, its minutely puberulent galea about as long as the tube and 3 or more times the length of the dark green, thickened, lower lip.
   Subalpine meadows at 4000-7500 ft. elevation in the Cascade and Olympic mts., n. to s. Alas. (at 2000-3000 ft.) and the Rocky Mts. of B.C. and Alta. Late June-Sept.
   Three well-marked geographical varieties of C. parviflora may be distinguished in our area, as follows:

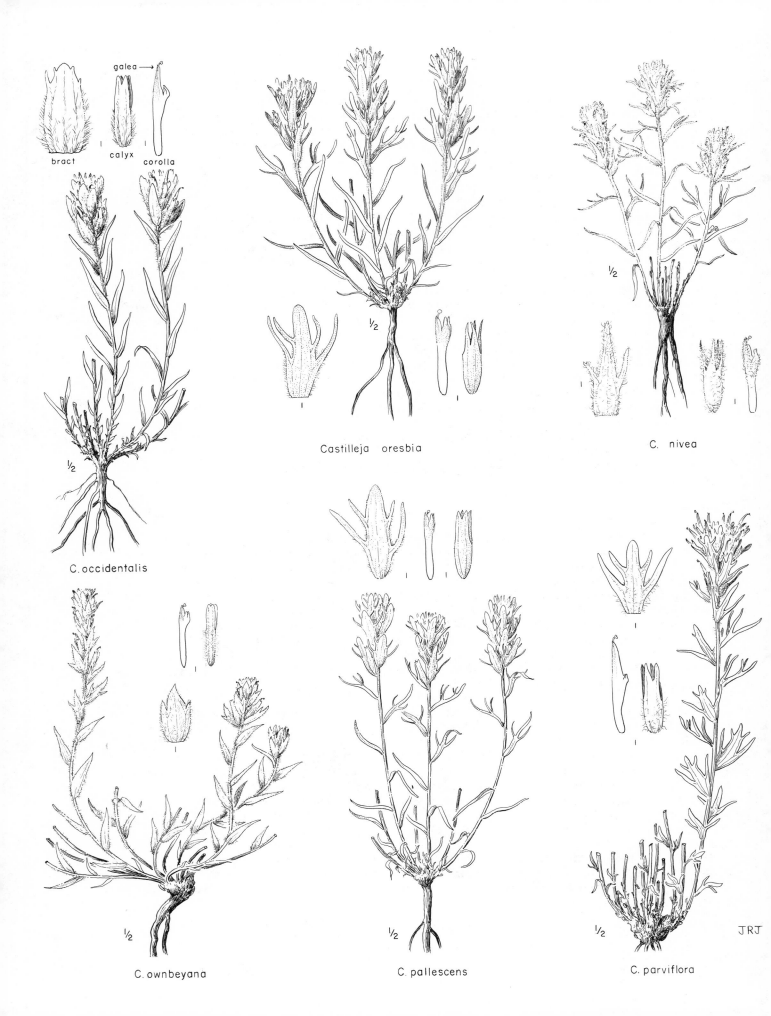

bract    galea →

calyx    corolla

Castilleja oresbia

C. nivea

C. occidentalis

C. ownbeyana

C. pallescens

C. parviflora

JRJ

1 Corolla 20-25 mm. long; bracts deep rose to crimson, rarely whitish; calyx lobes mostly
  obtuse or obtusish (but sometimes acute); Cascade Range from Mt. Rainier, Wash., to the
  vicinity of Three Sisters Peaks, Oreg.      var. oreopola (Greenm.) Ownbey
1 Corolla 12-20 mm. long
  2 Bracts deep rose to crimson; calyx lobes obtuse or obtusish (but sometimes acute); Olympic Mts., Wash.        var. olympica (G. N. Jones) Ownbey
  2 Bracts usually white or pinkish (but rose-colored in the vicinity of Harts Pass); calyx
   lobes often acute or acutish (but sometimes obtuse); Cascade Range from the vicinity of
   Cascade Pass, Wash. (reported from Mt. Stuart) to s. B. C.
            var. albida (Penn.) Ownbey

  These varieties of C. parviflora appear about as distinct as some of the taxa accorded specific rank elsewhere in this account. For instance, C. hispida vars. hispida and acuta, C.
chromosa and C. angustifolia could be considered as varieties of an expanded C. angustifolia.
C. inverta, C. oresbia, and C. pallescens could be included in a more comprehensive C. pallescens. C. miniata, C. rhexifolia, C. elmeri, C. gracillima, and C. sulphurea could be in-
corporated into an enlarged C. septentrionalis. There are two reasons for maintaining this
inconsistency for the moment: (1) There is considerable precedent for essentially the present
treatment. (2) It is difficult to determine the practical limits to which the reduction of species
to varieties could or should be carried. Even more inclusive species would in many instances
be as difficult to delimit. Specifically, with regard to C. parviflora, there has been near unanimity in that our plants represent a single species which has been called C. oreopola. This
species is easily recognized and is without near relatives. That it must also include the earlier C. parviflora from Alaska and probably C. henryae of the Canadian Rockies has been overlooked, but this conclusion seems to be inescapable.

**Castilleja pruinosa** Fern. Erythea 6:50. 1898. (Applegate 416, Swan Lake Valley, Klamath
  Co., Oreg.)
  Perennial; stems clustered, erect or ascending from a woody base, (2)3-6 dm. tall, commonly branched, more or less villous with a variable mixture of simple and branched hairs,
the latter always present, but sometimes sparse; leaves mostly entire and linear-lanceolate,
but the upper ones not infrequently with 1-2 pairs of lateral lobes, and sometimes most of
them divided, variably pubescent with branched and simple hairs, commonly pruinose with
most of the hairs branched, not or obscurely glandular; inflorescence conspicuous, bright red,
scarlet, or occasionally yellow; bracts 3- to 5-lobed, mostly shorter than the flowers, villous
and puberulent, usually somewhat viscid; calyx 15-25 mm. long, deeply and subequally cleft
above and below or more deeply so above, its primary lobes again divided into 2 slender, acute
segments 3-7 mm. long; corolla 20-35 mm. long, its finely pubescent galea equaling or usually longer than the tube and many times the length of the dark green, thickened, lower lip.
  Forested slopes at medium elevations, mountains of s.w. Oreg. and n. Calif., s. to Tuo-
lumne Co. Reported, perhaps erroneously, to extend northward in the Cascade Range to Mt.
Jefferson. See below. Late Apr.-Aug.
  This is a common and characteristic species of s.w. Oreg., from Klamath to Douglas cos.
southward, as well as of n. Calif. It is recorded from our area only from a single collection
(Leach 4601, Hunts Cove, Mt. Jefferson, Linn Co., Oreg., in Herb. Acad. Phila.), the identity, but not the source, of which has been verified. That this unexpected record may not be
valid is supported by the fact that the collection does not bear the collector's original label.
  The relationship of C. pruinosa is obscure, although it is probably a derivative of the Parviflorae. It is unique among our species in the possession of branched hairs, and its galea is relatively longer than that of any other species of our area.

**Castilleja pulchella** Rydb. Bull. Torrey Club 34:40. 1907. (Rydberg & Bessey 4967, mountains
  near Indian Creek, Madison Co., Mont.)
  C. pulchella var. acutina Kelso, Rhodora 37:227. 1935. (Kelso 3506, Electric Peak, Yellowstone Park, Mont.)
  Perennial; stems clustered, erect or from a decumbent base, simple, 0.5-2.0 (mostly less
than 1.0) dm. tall, softly viscid-villous; leaves viscid-villous or somewhat hispidulous, lower

ones linear, entire, upper ones broader and usually with a pair of short lateral lobes; bracts broader than the leaves, ovate and at least the lower ones acute, sometimes entire, but usually with a pair of divergent lateral lobes, viscid-villous and puberulent, varying in color from yellow to deep purple; inflorescence relatively short and broad, the flowers and bracts close together; calyx usually purple or purplish, rarely entirely yellow, 12-25 mm. long, deeply and subequally cleft above and below, its primary lobes broad and either shallowly notched at the apex into 2 rounded (rarely acute) segments or sometimes entire and rounded; corolla at anthesis a little longer than the calyx, its lower lip prominent, 1/2-2/3 the length of the usually strongly puberulent galea, the pouch green and thickened, and the lobes, like the galea margins, yellow to deep purple; stigma exserted.

Alpine and subalpine meadows, high mountains of s. w. Mont., w. Wyo., and n. e. Utah. Late June-Aug.

Castilleja pulchella is a variable species. In its typical and usual form, it seems most closely related to C. chrysantha of the alpine and subalpine regions of the Wallowa and Blue mts. However, it is more difficult to separate from reduced high-altitude forms of C. cusickii, and no entirely satisfactory means has been found for doing this. Hybridization may be involved, but it is not certain that present ranges are in contact, although the general range of C. cusickii includes much of that of C. pulchella. It is apparently absent from the high mountains of Ida., being replaced by alpine ecotypes of C. cusickii, but intermediate specimens from the Sawtooth Mts., now referred to C. cusickii, may actually belong with C. pulchella.

Castilleja rhexifolia Rydb. Mem. N. Y. Bot Gard. 1:356. Feb. 1900. (Rydberg & Bessey 4951, Cedar Mountain, Madison Co., Mont.)

   C. oreopola var. subintegra Fern. Erythea 6:45. 1898. (Piper 2322, Powder River [Wallowa] Mts., n. e Oreg.; lectotype)

   C. lauta A. Nels. Bull. Torrey Club 27:269. May, 1900. (A. & E. Nelson 6708, Dunraven Peak, Yellowstone Park, Wyo.)

   C. obtusiloba Rydb. Bull. Torrey Club 31:644. 1904. (Cowen, Leroux Park, Delta Co., Colo.)

   C. purpurascens Greenm. Bot. Gaz. 42:146. 1906. (Farr 568, near Field [Kicking Horse R.], Kootenay Dist., B.C.)

   C. humilis Rydb. Bull. Torrey Club 34:37. 1907. (A. Nelson 7919, Medicine Bow Mts., Albany Co., Wyo.)

   C purpurascens Rydb. Bull. Torrey Club 34:38. 1907. C. subpurpurascens Rydb. Fl. Rocky Mts. 791, 1066. 1917. (Peterson 11, flood plains of Kicking Horse [R.], Kootenay Dist., B.C.)

   C. oregonensis Gand. Bull. Soc. Bot. France 66:217. 1919. (Cusick 2443, East Eagle-Cornucopia trail, Wallowa Mts., Baker Co., Oreg.)

Perennial; stems clustered, erect or ascending from a woody base, 1-3(4) dm. tall, usually unbranched, glabrate, villous, or sometimes viscid-villous; leaves linear-lanceolate or sometimes broader, ordinarily all entire, but the upper ones sometimes lobed, glabrous to puberulent, finely villous, or viscid-villous; inflorescence conspicuous, generally crimson (drying purple), but varying to scarlet and occasionally yellow, at first short and broad, but often elongating in fruit; bracts oblong-ovate, entire and rounded or sometimes acute, commonly with 1-2 pairs of short lateral lobes, villous-puberulent; calyx 15-25 mm. long, subequally cleft above and below, its primary lobes again divided into 2 usually blunt segments 3-6 mm. long; corolla 20-35 mm. long, its puberulent galea usually much shorter than the tube and 4-5 times as long as the dark green, thickened, lower lip.

Alpine and subalpine meadows and slopes, Rocky Mts. from Alta. and B. C. to Colo. and n. Utah, westward through Ida. to the Wallowa and Blue mts. of n. e. Oreg., and apparently in the n. Cascade Range, from Lake Chelan, Wash., northward to B. C. June-Aug.

Castilleja rhexifolia is the common high-altitude species throughout the eastern half of our area. It forms extensive relatively uniform populations which are clearly distinct from those of the related C. miniata occurring ordinarily toward the lower edge of the forest belt on the mountains. But where the forest is not continuous, it is apparently possible for the two species to meet and completely obliterate the distinctions between them through hybridization. On the whole, however, indeterminably intermediate specimens comprise only a small percentage of the total. In addition, some very probable hybrids are known between C. rhexifolia and C. cusickii, and between it and C. chrysantha, the latter having been described as C. wallowensis. The relationship of C. rhexifolia to C. elmeri is noted elsewhere.

Castilleja rubida Piper, Bull. Torrey Club 27:398. 1900. (Cusick 2094, Wallowa Mts., Keystone Basin according to Cusick's unpublished notes, 9000 ft., n. e. Oreg.)
 ?C. lapidicola Heller, Muhl. 8:49. 1912. (Heller 9390, head of middle canyon of the South Fork of the Humboldt R., above Lee post office, Ruby Mts., Elko Co., Nev.)
 Perennial; stems clustered, often decumbent, unbranched, 0.5-1 dm. tall, purplish, villous-hirsute and somewhat glutinous; lower leaves linear, entire, upper ones with 1 or 2 pairs of short lateral lobes, villous-puberulent and usually glutinous; bracts broader than the leaves, red-purple, with 1 or 2 pairs of linear lobes; inflorescence congested, but the flowers usually not hidden by the bracts; calyx 10-12 mm. long, subequally cleft into 4 broadly linear or linear-triangular segments; corolla 12-15 mm. long, glandular-puberulent, generally exserted about 3 mm. beyond the calyx, its lower lip prominent but not strongly pouched, a little shorter than the galea, red-purple with red-purple lobes.
 Arctic-alpine in the Wallowa Mts., n. e. Oreg., and possibly in the high mountains of n. e. Nev. and adj. Utah. July-Aug.
 In our area, this is a well-marked and highly restricted species. The plants from Utah and Nev. are practically identical with ours and may be conspecific. However, they differ slightly in the shape of the bracts and in that the lobes of the lower lip are yellow instead of red-purple. Until more adequate collections are available, it cannot be determined if these differences are constant and significant.

Castilleja rupicola Piper, Erythea 6:45. 1898, (Piper, Paradise Valley, Mt. Rainier, Lewis Co., Wash.)
 Castilleja andrewsii Henderson, Madroño 3:31. 1935. (Andrews 233, Horsepasture Mt., e. Lane Co., Oreg.)
 Perennial; stems clustered, erect or ascending from a woody base, 1-2(3) dm. tall, unbranched, obscurely villous with crisped hairs, not at all hispid; leaves finely villous, all divided into 3-5(7) divergent lobes, or the lowermost rarely entire, the lateral segments not much narrower than the mid-blade; inflorescence conspicuous, bright scarlet or crimson, rather short and few-flowered; bracts puberulent and villous, mostly 5-parted, much shorter than the flowers at anthesis; calyx 15-25 mm. long, deeply and subequally cleft above and below, its primary lobes again divided into 2 obtuse or acute segments 1-5 mm. long; corolla 25-35(45) mm. long, its minutely puberulent galea about equaling or longer than the tube and many times as long as the dark green, thickened, lower lip.
 Perpendicular cliffs and rocky slopes at 4000-7000 ft., Cascade Range, from c. Oreg. (e. Lane Co.) to s. B. C. June-Aug.
 This is a well-defined species with clear affinities in the Parviflorae, especially with C. covilleana.

Castilleja rustica Piper, Bull. Torrey Club 27:398. 1900. (Cusick, Wallowa R., Wallowa Mts., Oreg.)
 C. pecten Rydb. Bull. Torrey Club 34:41. 1907. (Shear 3041, Beaver Canyon, Clark Co., Ida.)
 C. subcinerea Rydb. Bull. Torrey Club 40:484. 1913. C. angustifolia var. subcinerea Nels. & Macbr. Bot. Gaz. 61:45. 1916. (A superfluous name for C. pecten independently based on the same type)
 Perennial; stems clustered, erect or ascending, generally branched above, 2-4 dm. tall, appressed puberulent to finely villous or hispidulous, often purplish; leaves densely puberulent, lower ones linear, entire, upper ones usually with 1-2 pairs of linear, divergent lobes; bracts broader than the leaves, 3- to 5-parted, puberulent or villous, yellowish or occasionally reddish; flowers rather remote and not hidden by the bracts; calyx 12-23 mm. long, deeply and subequally cleft above and below, its primary lobes again divided into 2 triangular, acute seg-

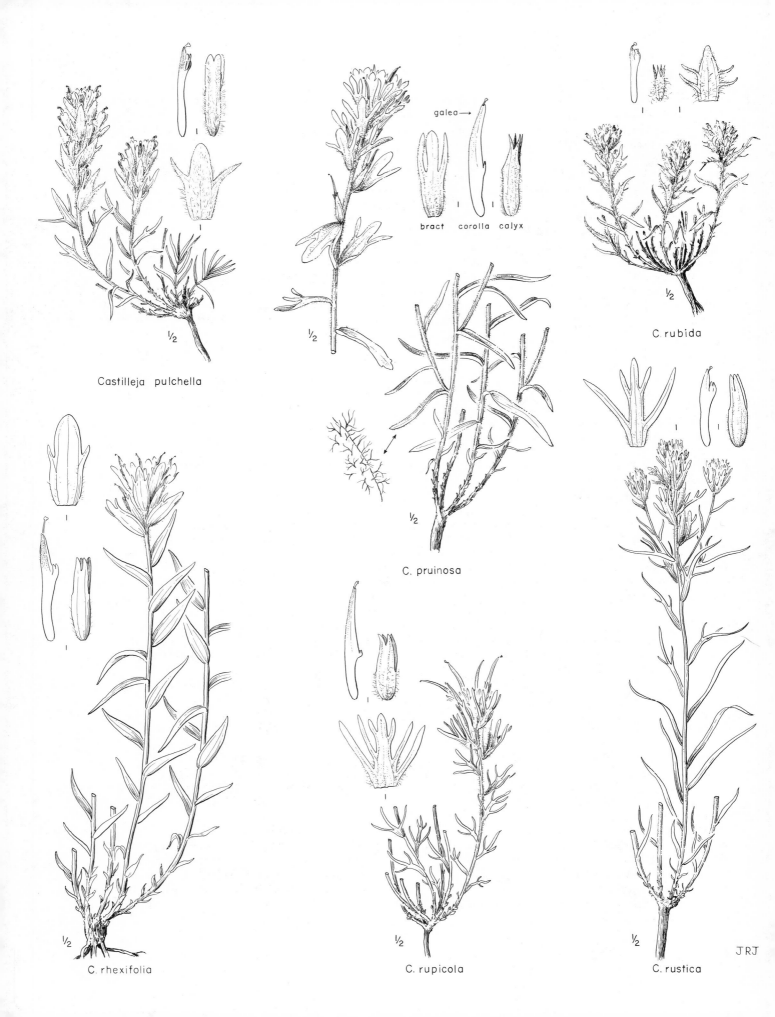

galea →

bract    corolla    calyx

Castilleja pulchella

C. rubida

C. pruinosa

C. rhexifolia

C. rupicola

C. rustica

JRJ

ments 2-6 mm. long; corolla 15-25 mm. long, usually longer than the calyx; lower lip reduced, 1/2-1/5 the length of the galea, its lobes and sometimes its pouch puberulent.

Dry soils, frequently associated with sagebrush, s. w. Mont. (Madison and Beaverhead cos.), c. Ida., and n. e. Oreg. (Wallowa Mts.) June-Aug.

This species is related to C. pallescens and C. flava. From the first, it differs in its more robust habit and shorter lower lip; from the second, in its subequally cleft calyx. Individuals with reddish bracts or calyces suggest frequent hybridization with C. angustifolia, and pubescence on the pouch of the lower lip of some specimens may indicate introgression from C. pallescens.


Castilleja sessiliflora Pursh, Fl. Am. Sept. 738. 1814. (Bradbury, "Upper Louisiana," i. e., along the Missouri R., probably in N. D. or S. D.)

Euchroma grandiflora Nutt. Gen. 2:55. 1818. Bartsia grandiflora Spreng. in Ersch. & Grub. Allg. Encyc. 7:461. 1821. Castilleja grandiflora Spreng. Syst. Veg. 2:775. 1825. ("On the plains of the Missouri from the confluence of the river Platte to the [Rocky] Mountains; also near the Prairie du Chien, [on the] Mississippi")

C. sessiliflora var. or ssp. betheli Cockerell, Torreya 18:181. 1918. (Bethel, Boulder, Colo.)

Perennial; stems clustered, erect or ascending, sometimes branched below, 1-4 dm. tall, villous-tomentose; lower leaves linear, entire, upper ones mostly broader and 3- to 5-parted, with linear, divergent lobes, densely puberulent; bracts leaflike, green, puberulent or lanulose, mostly 3-parted to below the middle; flowers 4-5 cm. long, exserted far beyond the bracts, the inflorescence becoming elongate in fruit; calyx 25-40 mm. long, yellowish, glandular-puberulent, subequally cleft above and below or more deeply so above, its primary lobes again divided 5-15 mm. into linear segments; corolla strongly curved, greenish yellow, pinkish, or purplish, minutely villous to glandular puberulent, its galea 10-12 mm. long, and 2-3 times the length of the prominent lower lip with its 3 spreading lobes.

Dry soils; throughout the Great Plains from n. w. Mont., s. Man., and s. Sask. to w. Tex. and s. Ariz., thence e. on the central prairies, through s. Minn. and Iowa to s. Wisc. and n. Ill. In our area, occurring only along the eastern front of the Rocky Mts. in Mont. May-July.

This is a very distinct species without near relatives in our area. Some related taxa, however, occur in w. Tex. and n. Mex.


Castilleja suksdorfii Gray, Proc. Am. Acad. 22:311. 1887. (Suksdorf, Mt. Adams, Wash., at 6000-7000 ft. elevation, probably s. e. side, in Yakima Co.)

Perennial; stems usually solitary, erect, 3-5(8) dm. tall, from a slender creeping base, unbranched, glabrate to obscurely villous or sometimes almost hispid; leaves glabrate to somewhat villous or hispidulous, sometimes all entire and linear lanceolate, but usually at least the upper ones with 1-2 pairs of lateral lobes, these much narrower than the mid-blade; inflorescence conspicuous, at first short and thick, but later elongating, the bracts and calyces with a yellow band below the red tips; bracts villous and puberulent, rather broad, mostly 5-parted, shorter than the flowers; calyx 20-30 mm. long, deeply and subequally cleft above and below, its primary lobes again divided 8-12 mm. into 2 linear, acute segments; corolla 30-50 mm. long, its puberulent to pubescent galea about the length of the tube and many times as long as the dark green, thickened, lower lip.

Wet subalpine meadows, particularly about springs, Cascade Range from Mt. Adams, Wash., to Crater Lake, Oreg. Late June-Sept.

This is a very distinct species with affinities, however, with the Parviflorae, particularly with C. rupicola.


Castilleja sulphurea Rydb. Mem. N. Y. Bot. Gard. 1:359. 1900. (Rydberg & Bessey 4966, Electric Peak, Yellowstone Park, Mont.)

C. luteovirens Rydb. Bull. Torrey Club 28:26. Jan. 1901. (Rydberg & Vreeland 5616, Sangre de Cristo Creek, Costilla Co., Colo.)

C. wyomingensis Rydb. Bull. Torrey Club 28:502. Sept. 1901. (Tweedy 2341, Big Horn Mts., Sheridan Co., Wyo.) A narrow-leaved, puberulent form.

Perennial; stems clustered, erect or ascending from a woody base, 1.5-5 dm. tall, often branched above, usually glabrous or glabrate below, commonly viscid-villous above, occasionally puberulent throughout; leaves linear to lance-ovate, ordinarily all entire, but the upper ones sometimes lobed, glabrous to puberulent or finely villous; inflorescence conspicuous, pale yellow, at first short and broad, but often elongating in fruit; bracts oblong-ovate, generally entire and rounded, but sometimes acute or with 1-2 pairs of short lateral lobes, puberulent and villous; calyx 15-25 mm. long, deeply and subequally cleft above and below, its primary lobes again notched or cleft into 2 blunt or acute segments 1-3 mm. long; corolla 18-30 mm. long, its minutely puberulent galea shorter than the tube and 3-4 times the length of the dark green, thickened, lower lip.

Moist meadows and slopes at medium and high elevations, mountains of w. Mont. and s. Alta., through Wyo. and adj. Ida. to Colo., Utah, and n. N.M.; also in the Black Hills of S. D. Reported (as C. septentrionalis) from Mt. Stuart, Wash., but the record perhaps based on C. elmeri. June-Sept.

Castilleja sulphurea commonly has been referred to the eastern C. septentrionalis Lindley, and certainly the two are closely related. The eastern and western plants, however, are not identical, ours on the whole being taller, with stouter stems, broader leaves, and larger bracts and flowers. Rarely, if ever, are the bracts of ours other than yellow, whereas reddish or purplish bracts are common in the eastern ones. C. septentrionalis was described originally from Labrador, and its range extends from the mountains of n. New England, through e. Can., to Newf. and Baffin I. Westward, it occurs on the Keeweenaw Peninsula, Isle Royale, and the n.w. shore of Lake Superior. The Rocky Mt. distribution of C. sulphurea is indicated above. This is not the only near relative of C. septentrionalis in our area. C. gracillima and C. occidentalis are almost equally close. If they deserve specific rank, so does C. sulphurea.

Castilleja thompsonii Pennell, Proc. Acad. Phila. 99:178. 1947. (Thompson 11512, Coulee City, Grant Co., Wash.)

C. villicaulis Pennell & Ownbey, Proc. Acad. Phila. 99:177. 1947. (Suksdorf 5767, Mt. Adams [s.e. side], Yakima Co., Wash.)

Perennial; stems clustered, erect or ascending, often branched above, 1-4 dm. tall, hispid or villous, sometimes purplish; lower leaves linear, entire, upper ones with 1-2 pairs of linear, divergent lobes, hispid or villous and sometimes glandular; bracts broader than the leaves, 3- to 5-parted, puberulent and ciliate, yellowish; calyx often purplish, 12-25 mm. long, deeply and subequally cleft above and below, its primary divisions usually again divided into 2, triangular or linear, acute segments 1-3 mm. long, occasionally merely notched or the segments obtuse; galea included in the calyx or little exserted, its lower lip prominent, scarcely pouched, usually not or obscurely puberulent, 2/3-4/5 the length of the galea.

Dry soils, frequently associated with sagebrush, widespread in the arid interior of Wash. and s. B.C.; local on open slopes and bald summits of the surrounding mountains to about 7000 ft. May-July (Sept.)

This is a complex species which may be resolved ultimately into several varieties. In general, the mountain plants, besides being smaller and flowering later, tend to be less hispid and more villous and glandular. However, there seems to be a different race on almost every mountain where it occurs, each appearing to have been derived independently from the main body of the species at lower elevations. The race on Mt. Adams (C. villicaulis) is perhaps distinct enough to deserve varietal status, but until the other mountain races are as well known, there is little to be gained by recognizing this one. The limits of the inclusive species are geographically and morphologically well marked.

Castilleja xanthotricha Pennell, Not. Nat. Acad. Phila. 74:5. 1941. (Peck 10017, s.e. Wasco Co., near Clarno, Oreg.)

Perennial; stems clustered, erect or ascending, usually unbranched, 1-2 dm. tall, softly viscid-villous; leaves viscid-villous, with some shorter gland-tipped hairs, lower ones linear, entire, upper ones broader, with mostly 1 pair of linear, divergent lobes from below the middle, these nearly as broad as the mid-blade; bracts broader than the leaves, deeply divided, the mid-blade and 1-2 pairs of broad lateral lobes enlarged distally and blunt or rounded at

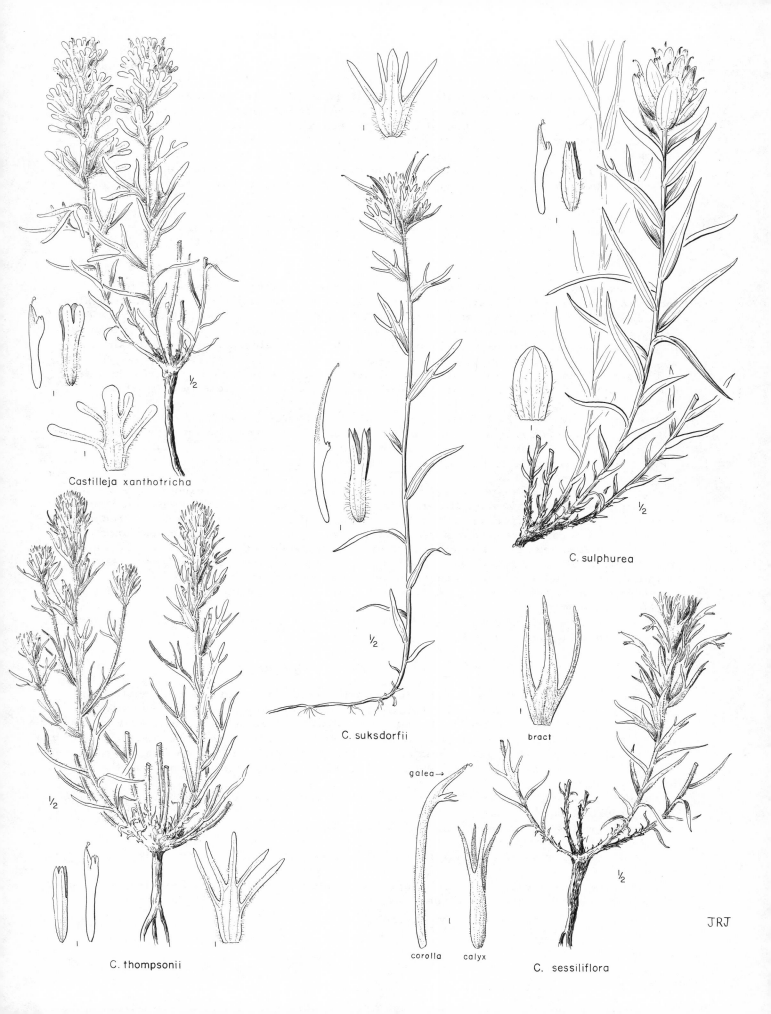

Castilleja xanthotricha

C. suksdorfii

C. sulphurea

C. thompsonii

bract

galea→

corolla    calyx

C. sessiliflora

JRJ

the apex, viscid-villous and glandular-puberulent, pale yellow or sometimes pinkish distally; inflorescence rather broad and loose, the flowers mostly not hidden by the bracts; calyx yellowish or pinkish, 15-22 mm. long, deeply and subequally cleft above and below, its primary lobes broad and again divided at the apex into 2 rounded segments 2-5 mm. long; corolla at anthesis longer than the calyx, in fruit, shorter, its lower lip prominent, about 1/2 as long as the minutely puberulent galea, lip and margins of the galea pinkish or yellowish.

Rocky slopes and sagebrush flats, John Day drainage in the vicinity of Clarno, Oreg. Apr. - May (July).

This is a remarkably distinct and localized species. It is clearly a member of the Chrysanthae, but it is well differentiated from the other members of this section.

## Chionophila Benth.

Flowers borne in a terminal, bracteate, secund raceme or spike; calyx with 5 teeth or lobes; corolla tubular, bilabiate, the upper lip external in bud; fertile stamens 4, the pollen sacs distally confluent; sterile filament slender, attached near the base of the corolla, evidently shorter than the fertile ones; stigma capitate; seeds numerous, with an ariliform, reticulate coat; small, fibrous-rooted perennials with well-developed, petiolate, basal leaves and more or less reduced opposite cauline leaves.

The genus consists of the following species and C. jamesii Benth., of the s. Rocky Mts. (Name from the Greek chion, snow, and philos, beloved, referring to the alpine habitat.)

Chionophila tweedyi (Canby & Rose) Henderson, Bull. Torrey Club 27:352. 1900.
 Penstemon tweedyi Canby & Rose, Bot. Gaz. 15:66. 1890. Penstemonopsis tweedyi Rydb. Fl.
   Rocky Mts. 717. 1917. (Tweedy 35, Beaverhead Co., Mont.)
 Penstemon albrightii A. Nels. Bot. Gaz. 65:69. 1918. (Macbride & Payson 3570, Josephus
   Lakes, Custer Co., Ida.)
Slender, mostly single-stemmed perennial from a short rhizome or caudex, 0.5-2.5 dm. tall, stipitate-glandular in the inflorescence, otherwise glabrous; basal leaves oblanceolate, mostly 2.5-9 cm. long (petiole included) and 3-13 mm. wide; cauline leaves much reduced and distant, 1-3 pairs, seldom 2 cm. long, the stem appearing nearly naked; racemes mostly 4- to 10-flowered, the bracts small, opposite, the short-pedicellate flowers mostly solitary at each node; calyx 3.5-5 mm. long, the lobes about equaling the tube; corolla pale lavender, 9-14 mm. long; sterile filament about half as long as the others.

Open slopes, meadows, and talus, near timber line in the mountains of c. Ida. and adj. s. w. Mont. June-Aug.

This would be considered a rock garden triumph, but it has not yet been grown successfully, and may prove impossible in cultivation.

## Collinsia Nutt. Blue-eyed Mary

Flowers 1-several in the axils of normal or more or less reduced upper leaves (bracts); calyx with 5 subequal lobes; corolla strongly bilabiate, the upper lip 2-lobed, distally recurved, external in bud; lower lip 3-lobed, the central lobe shorter than the lateral ones, keeled-saccate, forming a pouch that encloses the style and stamens; stamens 4, didynamous; pollen sacs confluent at the tip; stigma capitate or slightly 2-lobed; capsule dehiscing along 4 sutures; seeds 1-many in each cell; annuals with opposite (or partly whorled), mostly entire or toothed leaves.

About 17 species of temperate N. Am., especially Calif. (Named for Zacheus Collins, 1764-1831, American botanist.)

Several species are fine annuals in wild flower seed mixtures.

Reference:

Newsom, V. M. A revision of the genus Collinsia (Scrophulariaceae). Bot. Gaz. 87:260-301.
   1929.

1 Upper filaments pubescent; capsule subglobose, 4-6 mm. wide; seeds flattened, wing-margined, evidently cellular-reticulate                                    C. SPARSIFLORA

1 Upper filaments, like the lower, glabrous except sometimes at the point of insertion; capsule ellipsoid, 2-4 mm. wide; seeds, except commonly in C. rattanii, turgid, with thickened, inrolled margins, not evidently cellular-reticulate

  2 Corolla tube only slightly, if at all, bent; many of the hairs of the inflorescence with expanded, glandular tips                                                    C. RATTANII

  2 Corolla tube evidently bent near the base so as to stand at an angle to the calyx and pedicel; hairs of the inflorescence eglandular or somewhat glandular, but seldom with expanded gland tips

    3 Corolla 4-7(10) mm. long, the tube bent at an oblique angle to the calyx; widespread
                                                                     C. PARVIFLORA

    3 Corolla 9-17 mm. long, the tube bent at about a right angle to the calyx; w. of the Cascade summits (and to the e. end of the Columbia gorge)                C. GRANDIFLORA

Collinsia grandiflora Lindl. Bot. Reg. 13: pl. 1107. 1827. (Garden specimens, from seeds collected by Douglas on the banks of the Columbia, a hundred miles or more from the ocean) ?Antirrhinum tenellum Pursh, Fl. Am. Sept. 421. 1814. Linaria tenella F. G. Dietr. Vollst. Lexik. Gaertn. Nachtr. 4:408. 1818. Collinsia tenella Piper, Contr. U.S. Nat. Herb. 11: 496. 1906. (Lewis, "on the banks of the Missouri"; no certain specimen known to be extant) Not C. tenella Benth. 1846.

Collinsia grandiflora var. nana Gray, Proc. Am. Acad. 8:394. 1872. (Hall 366, Oreg.)

Collinsia multiflora Howell, Fl. N. W. Am. 506. 1901. (Howell, Willamette Valley, Oreg.)

C. diehlii M. E. Jones, Contr. West. Bot. 12:68. 1908. C. parviflora var. diehlii Pennell in Abrams, Ill. Fl. Pac. St. 3:778. 1951. (Diehl, Oregon City, Oreg.)

Much like C. parviflora, tending to be stouter and more erect, often with somewhat better developed and more toothed lower leaves (these with the blade up to 1.5 cm. long); flowers averaging shorter-pedicellate, the inflorescence often with an interrupted-thyrsoid aspect; calyx 5-8 mm. long; corolla 9-17 mm. long, the tube abruptly bent near the base at about a right angle to the calyx, shortly spur-pouched at the bend.

Open, moist or rather dry slopes and flats, from the valleys to moderate elevations w. of the Cascade summits, extending up the Columbia to Hood R.; s. B.C. to Calif. Apr-June.

A very attractive annual with considerable promise for the native garden. Probably hybridizes with C. parviflora.

Collinsia parviflora Lindl. Bot. Reg. 13: pl. 1082. 1827. (Garden specimens, from seeds collected by Douglas on the Columbia R.)

C. minima Nutt. Journ. Acad. Phila. 7:47. 1834. C. parviflora var. minima M. E. Jones, Contr. West. Bot. 12:69. 1908. (Wyeth, Flathead R., Mont.)

C. grandiflora var. pusilla Gray, Syn. Fl. 2$^1$:256. 1878. C. pusilla Howell, Fl. N. W. Am. 506. 1901. C. grandiflora ssp. pusilla Piper, Contr. U.S. Nat. Herb. 11:496. 1906. (Austin, Plumas Co., Calif.; lectotype by Newsom)

C. parviflora f. rosea Warren, Proc. Biol. Soc. Wash. 41:197. 1928. (Warren 745, near Pullman, Wash.) A form with the flowers pink and white.

Plants 0.5-4 dm. tall, simple or branched, lax and weak when elongate; stem and pedicels minutely pubescent with short, thick, often spreading or retrorse hairs (those of the inflorescence often glandular), or rarely glabrous; leaves glabrous, or short-hairy like the stem, entire or nearly so, the lower small, petiolate, spatulate to rotund, commonly deciduous, the others narrow and becoming sessile, narrowly elliptic or oblong to nearly linear, seldom as much as 5 cm. long and 12 mm. wide; upper leaves often whorled; flowers long-pedicellate, the lower mostly solitary in the axils, the upper more often clustered; calyx 3-6 mm. long, the narrow, slenderly pointed lobes longer than the tube; corolla blue, with white or whitish upper lip, 4-7(10) mm. long, the tube abruptly bent near the base, forming an oblique angle with the calyx and strongly gibbous on the upper side at the bend; capsule ellipsoid, shorter than the calyx, 3-5 mm. long, 2-3.5 mm. wide; seeds ellipsoid, about 2 mm. long, turgid, with thickened, inrolled margins.

In a variety of vernally mesic habitats, from the valleys to rather high elevations in the mountains; Alaska panhandle to s. Calif., e. to Ont., Mich., and Colo. March-July, depending partly on the altitude.

Occasional specimens from well beyond the range of C. grandiflora (such as in n. Ida.) have relatively large corollas up to 1 cm. long, suggesting the possibility that Pleistocene remnants of C. grandiflora have been absorbed into the population of C. parviflora.

Collinsia rattanii Gray, Proc. Am. Acad. 15:50. 1879.

C. torreyi var. rattani Jeps. Man. Fl. Pl. Calif. 905. 1925. (Rattan, Trinity R. in n. w. Calif.)

C. glandulosa Howell, Fl. N. W. Am. 506. 1901. C. rattani ssp. glandulosa Pennell in Abrams, Ill. Fl. Pac. St. 3:780. 1951. (Howell, Cold Camp, John Day Valley, Oreg.)

Resembling C. parviflora, but more erect and less frequently branched; leaves more often linear; spreading, evidently gland-tipped hairs intermingled with the shorter coarse hairs of the pedicels and at least the upper part of the stem; calyx lobes shorter and less acute, often merely equaling the tube; corolla tube only slightly (if at all) bent, but gibbous on the upper side near the base.

Mostly in open woods; s. Wash. (w. of the Cascades) to n. Calif. passing through the Columbia gorge and extending (rarely) to the John Day Valley in Oreg. Apr.-June.

Plants from our range have the seeds flattened, thin-margined or wing-margined, and evidently cellular-reticulate; to the southward, the seeds are more variable, passing to the turgid type with thickened, inrolled margins as in C. parviflora. Further investigation may support the presently dubious segregation of our plants as ssp. glandulosa (Howell) Pennell.

Collinsia sparsiflora Fisch. & Mey. Ind. Sem. Hort. Petrop. 2:33. 1836.

C. parviflora var. sparsiflora Benth. in DC. Prodr. 10:319. 1846. (Near the present Ft. Ross, Calif.)

C. brucae M. E. Jones, Contr. West. Bot. 12:69. 1908. C. sparsiflora var. brucae Newsom, Bot. Gaz. 87:285. 1929. (Bruce 2063, Little Chico, Butte Co., Calif.)

Plants 0.5-2 dm. tall, simple or often branched, erect, glabrous or minutely spreading-hirtellous; leaves opposite throughout, the lower petiolate, with broadly elliptic or ovate to subrotund, often few-toothed blade about 1 cm. long or less, often deciduous, the others narrow and becoming sessile, commonly linear to linear-oblong or linear-lanceolate, mostly entire, up to about 3 cm. long and 5 mm. wide; flowers long-pedicellate, 1-3 at each of the upper nodes, their subtending leaves more or less reduced; calyx 5-11 mm. long, the lanceolate to narrowly lance-triangular, acute or acutish lobes prominent, firm-foliaceous, much longer than the tube, commonly concealing much of the corolla tube; corolla blue-lavender or often white, 8-11 mm. long, the tube abruptly bent near the base, forming an oblique angle with the calyx and strongly gibbous on the upper side at the bend; keel generally somewhat hairy externally near the tip; upper pair of filaments shortly spreading-hairy over most of their length; capsule subglobose, 4-6 mm. wide; seeds flattened, irregularly wing-margined, evidently cellular-reticulate, 3-4 mm. long.

Open slopes and swales; Klickitat Co., Wash., to the Snake River Canyon in Wallowa Co., Oreg., s. to Calif. Apr.

Our plants as described above belong to the var. bruciae (Jones) Newsom. Two additional varieties, differing chiefly in the size and proportions of the calyx, corolla, and fruits, occur in Calif.

Cordylanthus Nutt. ex Benth. in DC. Nom. Conserv.

Flowers borne in small heads or compact spikes, or some of them singly; calyx cleft to the base into dorsal and ventral segments, or the lower segment obsolete; corolla tubular, bilabiate, the upper lip (galea) hooded and enclosing the anthers, with the lobes suppressed; lower lip about equaling the galea, a little inflated, external in bud, the 3 lobes short or obsolete; stamens 4 or 2, attached above the middle of the corolla tube; pollen sacs unequally set, one medifixed and appearing terminal on the filament, the other (sometimes reduced or obsolete) attached by its apex and pendulous or lying alongside the upper part of the filament; stigmas

Scrophulariaceae

329
wholly united; capsule loculicidal; seeds numerous to rather few; annual with alternate, entire to more often cleft or dissected, wholly cauline leaves. (Adenostegia.)

About 25 species, native to w. N. Am. (Name Latin from the Greek kordyle, club, and anthos, flower.)

Reference:

Ferris, R. S. Taxonomy and distribution of Adenostegia. Bull. Torrey Club 45:399-423. 1918.

1 Stamens 4, each with 2 pollen sacs; upper calyx segment 5- to 7-nerved like the lower, and only minutely bidentate (to less than 1 mm.) at the apex; seeds numerous (about 20)
C. RAMOSUS

1 Stamens 2, each with a single pollen sac; upper calyx segment (the one next the galea) obviously only 2-nerved and obviously bifid (to 2-5 mm.) at the apex; seeds about 8
C. CAPITATUS

Cordylanthus capitatus Nutt. ex Benth. in DC. Prodr. 10:597. 1846.

Adenostegia capitata Greene, Pitt. 2:180. 1891. (Nuttall, "Nova California")

Cordylanthus bicolor A. Nels. Bot. Gaz. 54:416. 1912. Adenostegia bicolor Rydb. Fl. Rocky Mts. 797, 1066. 1917. (Nelson & Macbride 1239, Ketchum, Ida.)

Plants 1-6 dm. tall, ordinarily more or less branched; herbage spreading-hairy, many of the hairs gland-tipped; leaves 1-5 cm. long, linear, those of the branches mostly entire; inflorescences capitate, terminating the branches, mostly 2- to 5-flowered, subtended by trifid bracts which resemble the foliage leaves; calyx tending to be purplish, scarious between the nerves, the lower segment several-nerved, nearly or quite as long as the 1-1.5 cm. purplish corolla; upper calyx segment somewhat shorter, 2-nerved, bifid, the teeth 2-5 mm. long; stamens 2, each with only a single pollen sac; seeds relatively few, about 8.

Open slopes and dry woods from the lowlands to moderate elevations in the mountains, often among sagebrush; Kittitas Co., Wash., to n.e. Calif., e. to s.w. Mont. and n.e. Nev. June-Sept.

Cordylanthus ramosus Nutt. ex Benth. in DC. Prodr. 10:597. 1846.

Adenostegia ramosa Greene, Pitt. 2:180. 1891. (Nuttall, Rocky Mts.)

Adenostegia ciliosa Rydb. Bull. Torrey Club 34:35. 1907. (Tweedy 545, Spread Creek, Wyo.)

Plants 1-3 dm. tall, much branched, cinereous-puberulent throughout, and sometimes slightly glandular as well; leaves 1-4 cm. long, narrowly linear and entire or more often 3- to 5-cleft, with linear-filiform segments; inflorescences capitate, terminating the branches, (1) 3- to 5-flowered, subtended by 5- to 7-cleft bracts which resemble the foliage leaves, the bracts often with some longer, flattened spreading hairs; calyx segments equal, somewhat scarious between the 5-7 nerves, often somewhat purple-tinged, the upper segment inconspicuously bidentate, the apical teeth about 0.5 mm. long; corolla dull yellowish, 12-17 mm. long, only slightly, if at all, surpassing the calyx; stamens 4, each with 2 pollen sacs; seeds numerous, about 20.

Dry, open places, especially among sagebrush; Wasco Co., Oreg., to n.e. Calif., e. to Wyo. and Colo. June-Aug.

Digitalis L. Foxglove

Flowers borne in terminal, bracteate racemes; calyx of 5 essentially distinct sepals; corolla tubular-campanulate, open, somewhat ventricose beneath, the lobes much reduced with the lowest one the largest, the upper external in bud; stamens 4, didynamous; stigmas 2, flattened; capsule septicidal and loculicidal; seeds numerous; perennial or biennial herbs with alternate leaves.

About 30 species, native to Eurasia. (Name Latin, pertaining to a finger, or fingerlike, referring to the flowers of D. purpurea.)

Digitalis purpurea L. Sp. Pl. 621. 1753. (S. Europe)

Robust biennial, 5-18 dm. tall, puberulent throughout (or in part glabrate), becoming viscid or glandular upward; leaves somewhat basally disposed, the lower ones 1.5-5 dm. long (in-

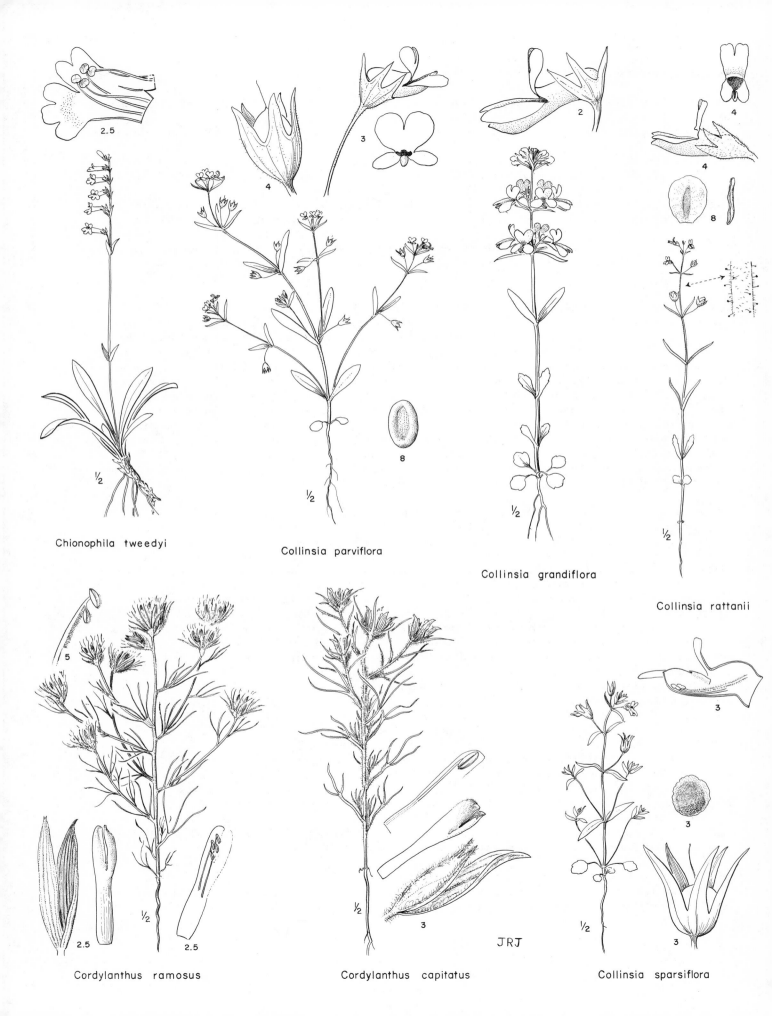

Chionophila tweedyi

Collinsia parviflora

Collinsia grandiflora

Collinsia rattanii

Cordylanthus ramosus

Cordylanthus capitatus

JRJ

Collinsia sparsiflora

cluding the petiole) and 3-12 cm. wide, crenate or crenate-serrate, the cauline ones well developed but more or less reduced upward; raceme elongate, secund, leafy-bracteate, the
bracts entire and sharply differentiated from the foliage leaves; pedicels 6-20 mm. long; calyx
foliaceous, 1-1.8 cm. long; corolla 4-6 cm. long, pink-purple, the lower side paler and mottled or spotted; seeds minute. N=28.

Native of Europe, now well established along roadsides and in other mostly disturbed sites,
from s. B.C. to Calif., especially w. of the Cascades, and occasionally found farther east.
Mostly June-July.

The drug digitalis is derived from this species.

## Euphrasia L.

Flowers borne in condensed or elongate, prominently leafy-bracteate terminal spikes; calyx
4-cleft, the lobes sometimes partly connate in lateral pairs; corolla bilabiate, weakly galeate,
the upper lip only slightly hooded and often shortly 2-lobed; lower lip about equaling or even
surpassing the upper, evidently 3-lobed, external in bud; stamens 4, didynamous, converging
under the galea; pollen sacs 2, equal, one more strongly acuminate-spurred at the base than
the other; stigma entire; capsule loculicidal, symmetrical in form and dehiscence; seeds mostly numerous; slender annual or perennial herbs, reputedly root parasites; leaves opposite,
wholly cauline, entire to toothed or cleft, passing into the well-developed, often alternate bracts.

A fairly large, widespread genus, best developed in Eurasia. Wettstein recognized 96 species in his monograph in 1896, and Pennell (Scrophulariaceae of eastern temperate North
America) estimated 200 species in 1935. These figures, however, reflect an inordinate sub-
division of the section Euphrasia into scarcely definable microspecies which still do not nearly
provide for all the combinations of characters which exist; this section, at least, is better reduced to a few more broadly defined, readily recognizable, if still not always sharply limited,
species. (Name Greek, meaning delight, referring probably to reputed medicinal qualities.)

Euphrasia arctica Lange ex Rostrup, Bot. Tidskr. 4:47. 1870. (Colmaster, Labrador)
   EUPHRASIA ARCTICA var. DISJUNCTA (Fern. & Wieg.) Cronq. hoc loc. E. disjuncta Fern.
      & Wieg. Rhodora 17:190. 1915. (Fernald et al. 6169, Grand Falls, Newf.)
   E. subarctica Raup, Rhodora 36:87. 1934. (Raup 4633, Lake Athabaska, Alta.)
   Annual, 3-20 cm. tall, simple or few-branched, the stem retrorsely puberulent; leaves few,
1 cm. long or less, prominently few-toothed, the teeth rounded to acute; bracts similar to, or
larger than, the leaves, sometimes more sharply toothed, sparsely hispidulous-puberulent
and finely (sometimes obscurely) glandular; flowers small and inconspicuous, the whitish corolla 3-5 mm. long and only slightly, if at all, surpassing the calyx; capsule more or less hairy.

Stream banks, bogs, and other wet places; Scandinavia, Greenl., Can., and Alas., barely
extending into the U.S. in Me., Mich., Minn., and Mont. July-Aug.

Our plants belong to the transcontinental American var. disjuncta (Fern. & Wieg.) Cronq.
The var. arctica, with larger flowers (corolla mostly 5-8 mm. long), is the sole representative of the species in Scandinavia, Greenl., and probably Labrador; in Newf. and Que. the two
varieties mingle and merge.

The closely related E. mollis (Ledeb.) Wettst., of Kamchatka and the Aleutian Islands, may
eventually have to be included as a third variety of E. arctica.

## Gratiola L.

Flowers pedicellate and solitary in the axils of the more or less reduced upper leaves; calyx
of 5 distinct sepals, often closely subtended by a pair of sepal-like bracteoles; corolla tubular,
bilabiate, the upper lip entire or shallowly lobed, external in bud, the lower 3-lobed; antheriferous stamens 2, the others wanting or represented by a pair of vestigial filaments near the
base of the corolla tube on the lower side; connective much enlarged and flattened, bearing the
parallel pollen sacs on one face; stigmas 2, flattened; capsule loculicidal and generally also
septicidal (thus dehiscing by four valves); seeds numerous; annual or perennial herbs with opposite, sessile, entire or toothed leaves.

About 20 species of wide distribution, mostly of subaquatic habitats. (Name diminutive of the Latin gratia, favor, in allusion to its reputed healing properties.)

1 Pedicels bearing at the summit a pair of sepal-like bracteoles, the sepals thus apparently 7
                                                    G. NEGLECTA

1 Pedicels without bracteoles, the sepals evidently 5         G. EBRACTEATA


Gratiola ebracteata Benth. in DC. Prodr. 10:595. 1846. (Nuttall, Oreg.)
   Similar to G. neglecta, averaging a little smaller, glabrous or only obscurely glandular upward; leaves more pointed, often acuminate; bracteoles wanting; sepals more elongate and more pointed, commonly somewhat acuminate, often well over 1 cm. long; corolla 5-7 mm. long; capsule more globose, not pointed, 4-5 mm. long.
   In shallow water and on muddy shores, and in other wet places in the valleys and plains; s. B.C. to Calif., e. to w. Mont. May-Aug.


Gratiola neglecta Torr. Cat. Pl. N.Y. 89. 1819. (Torrey, vicinity of New York City)
   Fibrous-rooted annual, simple or branched, erect or often diffuse, 1-3 dm. tall, generally more or less stipitate-glandular upward; leaves entire or often with a few, small, sharp teeth, 1-4 cm. long, 2-15 mm. wide, obtuse or moderately acute, pedicels 1-3 cm. long; calyx 3-7 mm. long, closely subtended by a pair of sepal-like bracteoles, these and the sepals narrow but seldom very strongly acute; corolla 7-10 mm. long, with white or yellowish tube and small purplish limb; capsule broadly ovoid, pointed, 5-7 mm. long.
   In shallow water and on muddy shores, and in other wet places in the valleys and plains; throughout most of the U.S. and s. Can. June-Aug.
   This species has sometimes been confused with the more strictly eastern, short-pedicellate G. virginiana L.


## Limosella L.

   Flowers borne on slender scapes which resemble mere elongate pedicels; calyx 5-toothed; corolla white or pinkish, inconspicuous, essentially regular, the 5 spreading lobes shorter than the tube, the upper lobes external in bud; stamens 4, nearly equal, the cells of the anther confluent; stigma capitate; capsule septicidal, distally 1-celled, the septum not extending to the tip; seeds numerous; glabrous, perennial, scapose herbs, the leaves entire and mostly long-petiolate.
   About a dozen species, widely distributed. (Name from the Latin limus, mire, and sella, seat, referring to the habitat of the plants.)


Limosella aquatica L. Sp. Pl. 631. 1753. (N. Europe)
   L. aquatica var. americana (with f. terrestris and f. natans) Glueck, Notizb. Bot. Gart. Berl. 12:75. 1934. (C. and w. N. Am.; no type given)
   Diminutive, fibrous-rooted perennial from a small crown; leaf blades rather narrowly elliptic, 5-18 mm. long, (1)2-7 mm. wide, much shorter than the slender petioles, these mostly 1-8 cm. long; scapes very slender, lax, 8-30 mm. long, bearing a solitary flower, or sometimes bearing at the summit a whorl of leaves and a cluster of short-pedicellate flowers; calyx 2-3 mm. high; corolla tube nearly equaling the calyx, the lobes short and spreading; capsule 2-3.5 mm. high; seeds transversely finely rugulose. N=20.
   In shallow water or on wet mud in the valleys and plains; widespread in the Northern Hemisphere, and occurring throughout our range. June-Oct.
   What seems to be L. subulata Ives, a related, maritime, chiefly N. Atlantic American species with the very narrow leaves not differentiated into blade and petiole, was collected many years ago at Alberni, Vancouver I.


## Linaria Mill. Toadflax

   Flowers borne in terminal racemes; calyx of 5 essentially distinct sepals; corolla yellow to blue or white, ventrally spurred at the base, bilabiate, the upper lip external in bud, the lower lip often raised into a palate; stamens 4, didynamous; stigmas punctiform or capitate; capsule cylindric or subglobose, rupturing irregularly across the distal width of each cell; seeds mostly numerous; annual to perennial herbs; leaves sessile, alternate, or the lowermost opposite.

More than 100 species, 2 native to N. Am., the others from Eurasia. (Name from the Latin linum, flax, because of the vegetative resemblance of L. vulgaris to flax.)

Cymbalaria muralis Gaertn. (Linaria cymbalaria), the Kenilworth ivy, a twining, often climbing annual or biennial with petiolate, palmately lobed leaves, and long-pedicellate, axillary, violet flowers about 1 cm. long, is occasionally found in our range as an escape from cultivation.

Kickxia elatine (L.) Dumort, a native of Europe that has become established in parts of California, has recently been collected near Hillsboro, Washington Co., Oreg. It is a viscid-villous annual with slender, prostrate stems, small hastate leaves, and pedicellate axillary purple and yellow flowers about 1 cm. long or less and resembling Linaria in structure. The capsule opens by large pores with deciduous lids.

1 Flowers yellow; plants perennial, more or less robust
  2 Leaves linear, not clasping                       L. VULGARIS
  2 Leaves ovate or lance-ovate, clasping        L. DALMATICA
1 Flowers blue, or blue and white; plants annual or winter annual, slender
                                           L. CANADENSIS

Linaria canadensis (L.) Dumont, Bot. Cult. 2:96. 1802.
  Antirrhinum canadense L. Sp. Pl. 618. 1753. (Va., Can.; Kalm, s. N.J., fide Pennell)
  Linaria texana Scheele, Linnaea 21:761. 1848. L. canadensis (var.) texana Pennell, Proc.
  Acad. Phila. 73:502. 1921. (Between Houston and Austin, Tex.)

Slender annual or winter annual from a short taproot, 1-5 dm. tall, essentially glabrous throughout, somewhat glaucous in life, producing a rosette of short, prostrate stems with assurgent tips, the leaves of which are mostly opposite or ternate and shorter and relatively broader than those of the erect stems; main stems 1-several from the base, simple, erect, sparsely leafy, the leaves linear, 1-3.5 cm. long and 1-2.5 mm. wide, the lowermost often opposite or ternate, the others alternate; racemes nearly naked; corolla rather light blue, with paler, scarcely raised palate, 8-12 mm. long exclusive of the slender, downwardly and forwardly arcuate spur; capsules ellipsoid-globose, 2.5-4 mm. high; seeds prismatic-angled.

Moist, sandy places; widespread in the less arid parts of the U.S. and adj. Can. and Mex. Apr.-June.

Two fairly well-marked varieties may be recognized, as follows:
1 Corolla mostly 8-10 mm. long, exclusive of the 2-6 mm. spur; seeds smooth or nearly so; chiefly northern and eastern, and occasionally found w. of the Cascades in our area as an apparent introduction                              var. canadensis
1 Corolla mostly 10-12 mm. long, exclusive of the 5-9 mm. spur (the final flowers of an inflorescence, or those on depauperate plants, often smaller); seeds densely tuberculate; chiefly southern and western, and apparently native w. of the Cascades in our area
                                     var. texana (Scheele) Pennell

Linaria dalmatica (L.) Mill. Gard. Dict. 8th ed. Linaria no. 13. 1768.
  Antirrhinum dalmaticum L. Sp. Pl. 616. 1753. (Crete)

Stout, erect perennial, branched above, 4-12 dm. tall, spreading by creeping roots; herbage glaucous; leaves numerous, ovate or lance-ovate, sessile and clasping, commonly 2-5 cm. long and 1-2 cm. wide (probably sometimes larger), entire; flowers in elongate racemes, short-pedicellate or subsessile, bright yellow, with well-developed, orange-bearded palate, 2.5-4 cm. long, including the well-developed spur, which about equals the rest of the corolla; capsules broadly ovoid-cylindric, 7-8 mm. high; seeds irregularly wing-angled. N=6.

Roadsides and other disturbed sites; native of the e. Mediterranean region, now established at several widely scattered localities on both sides of the Cascades in our range. Summer.

Linaria genistifolia (L.) Mill., a related species differing in its smaller flowers (1.5-2 cm. long) and mostly somewhat narrower leaves, has been collected in w. Oreg.

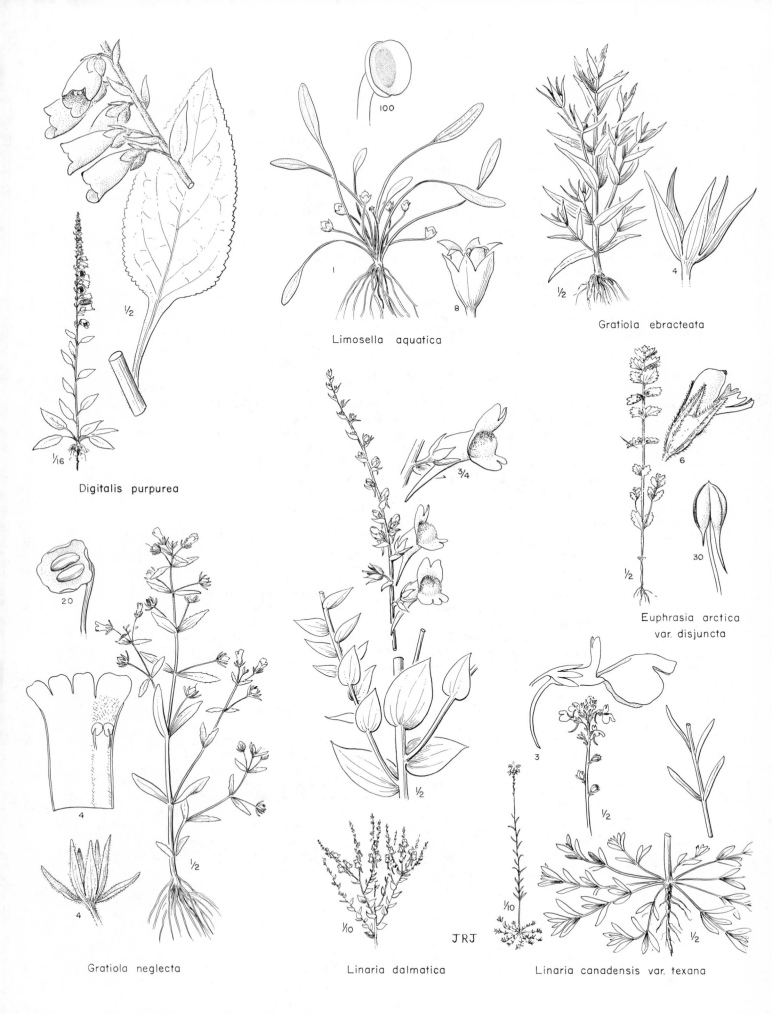

Digitalis purpurea

Limosella aquatica

Gratiola ebracteata

Euphrasia arctica
var. disjuncta

Gratiola neglecta

Linaria dalmatica

Linaria canadensis var. texana

JRJ

Linaria vulgaris Hill, Brit. Herb. 109. 1756.

   Antirrhinum linaria L. Sp. Pl. 616. 1753. Linaria linaria Karst. Deuts. Fl. 947. 1880-83. (Eu-
      rope)

   Perennial from creeping roots, 2-8 dm. tall, ill-smelling, glabrous throughout and somewhat
glaucous in life; leaves numerous, linear, 2-10 cm. long, 1-5 mm. wide, often narrowed toward
the base; racemes at first dense, in age more elongate; flowers bright light yellow, with well-developed, bearded, orange palate, 2-3.5 cm. long including the nearly straight spur, which is about
as long as the rest of the corolla; capsules broadly cylindric, 1 cm. high; seeds flattened, winged.
N=6. Butter-and-eggs.

   Roadsides, pastures, and waste places; native of Eurasia, now widely naturalized in temperate N. Am. June-Sept.


                                        Lindernia All.


   Flowers pedicellate, solitary in the axils of the leaves; sepals 5, distinct; corolla blue-violet, bilabiate, the upper lip with short, acutish lobes, external in bud, the lower much longer
and 3-lobed; filaments 4, didynamous, the upper short and antheriferous, the lower adnate to
the corolla tube for most of their length, but visible as hairy ridges, the tips free and projecting upward, lacking anthers in our species; stigmas distinct, flattened; capsule septicidal,
the septum persistent; seeds numerous; branching annuals with opposite, entire or toothed,
wholly cauline, often sessile leaves, most or all of which subtend flowers.

   Perhaps 50 species, mostly in the warmer parts of the Old World. (Named for Franz
Balthazar von Lindern, 1682-1755, German physician and botanist.)

1 Pedicels (except sometimes the lowermost ones) conspicuously surpassing their subtending
      leaves; leaves all, or all except the lowermost, broadly rounded at the base
                                        L. ANAGALLIDEA
1 Pedicels surpassed by, or only slightly surpassing, their subtending leaves; at least the
      lower leaves narrowed to the base          L. DUBIA


Lindernia anagallidea (Michx.) Pennell, Monog. Acad. Phila. 1:152. 1935.

   Gratiola anagallidea Michx. Fl. Bor. Am. 1:6. 1803. Ilysanthes anagallidea Raf. Autik. Bot.
      46. 1840. (Carolina)

   ?Gratiola inaequalis Walt. Fl. Carol. 61. 1788. Ilysanthes inaequalis Pennell, Torreya 19:
      149. 1919. (?South Carolina) Nom. dub.

   Ilysanthes gratioloides var. depressa Suksd. W. Am. Sci. 15:61. 1906. (Suksdorf 2192,
      Bingen, Wash.)

   Low, branching annuals, 0.5-2 dm. tall, glabrous or glandular; leaves entire or obscurely
few-toothed, 6-15 mm. long, 3-10 mm. wide, all broadly rounded at the base, or the lowermost a little narrowed; pedicels slender, mostly 1-2.5 cm. long, all except sometimes the
lowermost ones conspicuously surpassing their subtending leaves, often more than twice as
long; sepals narrow, 3-nerved, 3-4 mm. long; corolla 6-9 mm. long; seeds mostly brownish
yellow, 1.5-2 times as long as wide.

   Moist banks, especially along the larger rivers; widespread in N. Am. and S. Am. but infrequently collected in our range. July-Oct.

   The European species L. pyxidaria L., which has been collected as a waif in Wash., differs
in having the leaves more narrowed to the base, and in having all 4 stamens fertile.


Lindernia dubia (L.) Pennell, Monog. Acad. Phila. 1:141. 1935.

   Gratiola dubia L. Sp. Pl. 17. 1753. Capraria gratioloides L. Syst. Nat. 10th ed. 1117. 1759.
      Ilysanthes gratioloides Benth. in DC. Prodr. 10:419. 1846. I. dubia Barnh. Bull. Torrey
      Club 26:376. 1899. (Clayton 164, Va.)

   Ilysanthes pyxidaria var. major Pursh, Fl. Am. Sept. 419. 1814. Lindernia dubia ssp. ma-
      jor Pennell, Monog. Acad. Phila. 1:146. 1935. L. dubia var. major Deam, Fl. Ind. 845.
      1940. (Probably Pursh s.n., from Greenbriar, W. Va.)

   Very similar to L. anagallidea; leaves averaging larger, up to 3 cm. long, at least the lower ones narrowed to the base; pedicels shorter, mostly 5-15 mm. long, surpassed by, or only

slightly surpassing, their subtending leaves; corollas 7-10 mm. long, those of the later flowers mostly falling unopened (cleistogamous); seeds pale yellow, mostly 2-3 times as long as wide.

Moist banks, especially along the larger rivers; widespread in e. and c. U.S., and extending to S. Am.; seldom collected in our range and perhaps not native.

## Mazus Lour.

Flowers in terminal, nearly naked racemes; calyx 5-cleft to about the middle or a little beyond, the narrow segments equal, with more or less evident midrib; corolla blue-violet, bilabiate, the short upper lip subacutely 2-toothed, external in bud, the lower lip much larger and 3-lobed; filaments 4, didynamous, all antheriferous; stigmas 2, flattened; fruit loculicidal, dehiscing across the septum; seeds numerous; diffuse annuals or biennials with basal and opposite leaves, or the uppermost leaves alternate, as are the minute or obsolescent bracts of the raceme.

Perhaps 30 species, mostly of s.e. Asia. (Name from the Greek mazos, papilla, referring to the tubercles in the mouth of the corolla.)

Mazus japonicus (Thunb.) Kuntze, Rev. Gen. 2:462. 1891.
   Lindernia japonica Thunb. Fl. Jap. 253. 1784. (Japan)
   Mazus rugosus Lour. Fl. Cochinch. 385. 1790. (Indochina)
   Apparently annual, commonly branched from the base, up to 15 cm. tall, the stem and sometimes the leaves with short, retrorse-spreading hairs; basal leaves spatulate or broader, up to 4 cm. long and 1.5 cm. wide, irregularly toothed; cauline leaves often smaller; raceme longer at maturity than the proper stem, openly 4- to 10-flowered; calyx 4-5 mm. long at anthesis, up to 9 mm. in fruit; corolla 7-10 mm. long, blue-violet, marked with yellow and white.

Lawns and wet bottom lands; native of e. Asia, introduced in e. U.S. and about Portland, Oreg. May-Nov.

## Melampyrum L.

Flowers in terminal, leafy-bracteate spikes or racemes (or solitary in the upper axils); calyx deeply 4-cleft; corolla bilabiate, galeate, the upper lip hooded and nearly lobeless, lower lip about equaling the upper, 3-lobed and with a well-developed palate at the base, external in bud; stamens 4, didynamous; anthers enclosed in the galea; pollen sacs equal, awned at the base; stigma entire; capsule loculicidal, curved, asymmetrical, dehiscent only on the convex margin; seeds few, commonly only 1 or 2 per locule, hard; annuals with opposite, wholly cauline, entire leaves, the bracts often with divergent slender teeth near the base.

About 15 species, one North American, the others from Eurasia. (Name Latin from the Greek melas, black, and pyros, wheat, referring to the color of the seeds.)

Melampyrum lineare Desr. in Lam. Encyc. Meth. 4:22. 1796. (Fraser, "Carolina")
   Slender annual, 1-3 dm. tall; stem simple or few-branched, glandular, especially upward; leaves (and bracts) opposite throughout, short-petiolate, linear or lanceolate, 2-5 cm. long, 1-8 cm. wide, glandular or glabrous; bracts scarcely differentiated from the proper leaves, but sometimes with a few, divergent, slender teeth near the base; flowers short-pedicellate; corolla 5-10 mm. long, white or pinkish with yellow palate; seeds 3 mm. long.

Rich woods and wet meadows; Newf. and Que. to n. Ga., w. to Alta., c. B.C. and Vancouver I., and n.e. Wash. July-Aug.

Our plant is the widespread, northern var. lineare. Two or three other varieties, with broader leaves or more prominently toothed bracts, occur in e. U.S.

## Mimetanthe Greene

Flowers axillary; calyx somewhat 5-sulcate, but not angled, the midribs to the well-developed, unequal segments obscure; corolla small, yellow or purplish, slightly bilabiate, the upper lobes external in bud; stamens 4, didynamous, the anthers of the lower pair tending to be reduced, or wanting; stigmas distinct; capsule loculicidal, and splitting across the septum (dividing the placenta) in the upper half or third; seeds numerous; annual with opposite, narrow, sessile or subsessile leaves.

The genus, consisting of the following and one other California species, has sometimes been included in Mimulus, where it is a discordant element. (Name from the Greek mimetes, an imitator, and anthos, flower, from its resemblance to Mimulus.)

Mimetanthe pilosa (Benth.) Greene, Bull. Calif. Acad. Sci. 1:181. 1886.
  Herpestis pilosa Benth. Comp. Bot. Mag. 2:57. 1836. Mimulus pilosus Wats. Bot. King
    Exp. 225. 1871. (Douglas, North California)
  Mimulus exilis Dur. & Hilg. Pac. R.R. Rep. 5:12. 1855. (Heerman, Posé Creek, Calif.)
    Glandular and spreading-hairy, unpleasantly odorous, simple or much-branched annual, 1-
4 dm. tall; leaves entire or nearly so, 1-5 cm. long, up to 1.5 cm. wide, the lower mostly oblanceolate and often subpetiolate, the others mostly oblong to broadly linear or narrowly elliptic and sessile, or the upper sometimes reduced and more lanceolate; pedicels mostly longer than the calyx; calyx glandular-pubescent like the herbage, 5-6 mm. long at anthesis, the upper lobe the longest, the deepest sinus extending more than halfway to the base; corolla yellow, often with some maroon dots, 5-9 mm. long; ovary, style, and fruit rather finely and shortly stipitate-glandular.

Stream banks, dry stream beds, and other low, moist places; dry regions from e. Wash. to Baja Calif., e. to Ida. (King Hill), Utah, and Ariz. June-Aug.

### Mimulus L. Monkey Flower

Flowers axillary; calyx strongly 5-angled, the midveins to the mostly rather short lobes prominent and often raised; corolla slightly to strongly bilabiate, yellow to purple or red, the upper lobes external in bud; stamens 4, didynamous, all with well-developed anthers; stigmas mostly distinct, or sometimes marginally connate into a funnelform structure; capsule loculicidal, sometimes splitting across the septum, the halves of the placenta then adherent to the respective valves; seeds numerous; annual or perennial herbs, or some extralimital species shrubs, with opposite, entire or toothed leaves.

A rather large genus, best developed in w. N. Am., especially California, but native to parts of Asia, Africa, and Australia as well. Grant recognized 114 species in 1924. Pennell (in Abrams' Flora) recognized 107 species for the three Pacific states alone in 1952, but this latter figure, at least, is doubtless too high. (Name a diminutive of the Latin mimus, a mimic actor.)

Several species are desirable garden subjects for moist or wet soil, especially M. guttatus. M. lewisii is perhaps our most showy species, but does not do well in cultivation. M. moschatus is readily established, but apt to be too aggressive. M. nanus is worth trying in well-drained soil e. of the Cascades.

Reference:

Grant, Adele. A monograph of the genus Mimulus. Ann. Mo. Bot. Gard. 11:99-388. 1924.

1 Plants perennial from rhizomes or stolons; flowers long-pedicellate, the pedicels longer
    than the calyx; corollas dropping before withering; septum of the capsule remaining intact
    at maturity, or splitting only above the middle
  2 Corollas pink-purple, marked with yellow, 3-5.5 cm. long; calyx 1.5-2.5(3) cm. long;
      erect montane plants with the leaves sessile and several-nerved from the base
                                                                M. LEWISII
  2 Corollas yellow, sometimes marked with red or maroon, or in part washed with pale
      purplish; calyx, habit, and habitat diverse
    3 Upper calyx tooth conspicuously larger than the others; leaves palmately or subpalmately veined, the 3-7 main veins arising at or very near the base; corolla strongly
        bilabiate, and with broad, strongly flaring throat
      4 Corolla throat open; lateral and lower calyx teeth blunt and mostly very short, the
          lower not folded upward; corolla mostly 1-2 cm. long; stems weak, mostly decumbent to creeping or floating; plains and southern species, extending westward into
          our range in Mont.                                      M. GLABRATUS

4 Corolla throat nearly closed by the well-developed palate; lateral and lower calyx teeth
   more or less acute, the lower tending to fold upward in fruit and partially close the
   orifice; corolla (1)2-4 cm. long; stems erect, or if weak and more or less creeping,
   then the flowers mostly over 2 cm. long; cordilleran species, occurring throughout
   our range
     5 Plants with definite, creeping, often sod-forming rhizomes, often stoloniferous as
       well; flowers few (mostly 1-5), large, the corolla mostly 2-4 cm. long; low plants,
       commonly 2 dm. tall or less, of high altitudes in the mountains     M. TILINGII
     5 Plants with stolons, but only rarely with definite creeping rhizomes; flowers often
       more than 5, and usually less than 2 cm. long when few; plants usually over 2 dm.
       tall; sea level to moderate elevations in the mountains     M. GUTTATUS
3 Upper calyx tooth not evidently larger than the others, or if so, then the leaves pinnately
   veined; corolla various
     6 Plants leafy-stemmed; flowers axillary; calyx tube pubescent, at least on the nerves
       7 Calyx lobes 2-5 mm. long, acute or acuminate; rhizomes not producing cormlike
         resting buds, though sometimes moniliform in M. moschatus; leaves pinnately
         veined
           8 Herbage somewhat hirsute, but scarcely viscid; calyx tube hirsute along the 5
             ribs only; corolla 2.5-4 cm. long, strongly bilabiate, with broad, evidently ex-
             panded throat; w. of the Cascades     M. DENTATUS
           8 Herbage viscid-villous and tending to be slimy; calyx tube viscid-villous over the
             surface as well as along the ribs; corolla 1-3 cm. long, only slightly irregular,
             and with relatively narrow, not much expanded, scarcely differentiated throat;
             widespread     M. MOSCHATUS
       7 Calyx lobes 1-2 mm. long, abruptly apiculate to rounded-mucronulate; rhizomes
         very slender, producing cormlike resting buds from which the stems arise; leaves
         palmately or subpalmately veined; herbage and calyx viscid-villous as in M.
         moschatus; local at and near the e. end of the Columbia R. gorge
                                    M. JUNGERMANNIOIDES
     6 Plants mat-forming; leaves crowded at or near the ground; flowers borne mostly sin-
       gly on axillary pedicels which are generally longer than the stem, and which appear
       to be nearly terminal; calyx tube glabrous; leaves 3(5)-nerved from the base; wide-
       spread, but apparently absent w. of the Cascades     M. PRIMULOIDES
1 Plants annual, without stolons or rhizomes; pedicels, corollas, and capsules diverse
   9 Corollas yellow, often marked with red or maroon; pedicels elongate, mostly longer than
     the calyx; corollas (except sometimes in M. suksdorfii) generally deciduous before with-
     ering; septum of the capsule remaining intact at maturity, or splitting only above the
     middle
      10 Corolla strongly bilabiate, the lower lip evidently longer than the upper and strongly
        deflexed from it; corolla not less than 8 mm. long; principal leaves of fairly broad
        form and usually petiolate
         11 Upper calyx tooth obviously larger than the others; widespread     M. GUTTATUS
         11 Upper calyx tooth about the same size as the others, or smaller
           12 Corolla pubescent on the palate, sometimes finely red-dotted, but without a
             prominent blotch; calyx teeth all about alike, mostly acute; e. of the Cas-
             cades     M. WASHINGTONENSIS
           12 Corolla glabrous, bearing a prominent maroon blotch at the base of the lower
             lip; calyx teeth unlike, the three upper acute, the 2 lower longer and round-
             ed; in and w. of the Cascades     M. ALSINOIDES
      10 Corolla only slightly bilabiate, the lower lip only slightly longer than the upper and
        not much deflexed from it; corolla often less than 8 mm. long; leaves diverse
         13 Leaves abruptly contracted to the petiole; herbage generally evidently viscid-
          pubescent, or sometimes only shortly and inconspicuously so
                             M. FLORIBUNDUS
         13 Leaves tapered to the petiolar or sessile base; herbage finely glandular-puberu-
          lent

14   Corolla 8-16 mm. long, mostly 2-3 times as long as the calyx; pedicels commonly 1-
      1.5 cm. long at anthesis, tending to become arcuate-spreading or strongly divergent
      in fruit                                                          M. PULSIFERAE
14   Corolla 4-8 mm. long, slightly surpassing the calyx; pedicels shorter, less than 1 cm.
      long at anthesis
   15   Leaves mostly rather narrowly elliptic or rhombic-elliptic, commonly short-petio-
           late; calyx teeth more or less acute; fruiting pedicels tending to be loosely ascend-
           ing                                                          M. BREVIFLORUS
   15   Leaves narrower, linear to narrowly oblong or oblanceolate, sessile or the lower
           short-petiolate; calyx teeth tending to be rounded-mucronulate; fruiting pedicels
           tending to be widely spreading, with suberect tips   M. SUKSDORFII
9   Corollas purple or red, commonly marked in the tube or throat with yellow or white; pedi-
     cels, except often in M. breweri, mostly shorter than the calyx; corollas tending to per-
     sist for some time after withering (the tendency least marked in M. breweri)
  16   Flowers small and slender, 5-10 mm. long; septum splitting only above the middle; ped-
           icels from a little shorter to evidently longer than the calyx         M. BREWERI
  16   Flowers larger and more showy, 1-5 cm. long; septum splitting to the base at maturity,
           the halves of the placenta adherent to the respective valves; pedicels shorter than the
           calyx (sometimes not much shorter in M. clivicola)
   17   Capsule symmetrical, dehiscent; corolla 1-3.5 cm. long; e. of the Cascades
    18   Leaves notably broad, evidently 3- to 5-nerved, the better developed ones com-
                monly 1-2.5(3) cm. wide, at least the upper ones broadly ovate or broader,
                with strongly acute to acuminate tip; corolla 2-3.5 cm. long   M. CUSICKII
    18   Leaves narrower and more obscurely nerved, rarely any of them as much as 1.2
                cm. wide, seldom at all ovate, all obtuse or merely acutish; corolla 1-2.5 cm.
                long
     19   Flowers subsessile, the pedicels only 1-3 mm. long; capsule ovate, obtuse;
                     plants of very dry places, becoming much branched when well developed;
                     widespread, but absent from all except perhaps the s. end of the range of
                     M. clivicola                                        M. NANUS
     19   Flowers evidently pedicellate, the pedicels becoming 3-7 mm. long in fruit;
                     capsule lance-linear; plants more mesophytic and mostly simple or nearly
                     so, occurring from n. Ida. and adj. Wash. s. to the s. end of the Snake
                     River Canyon                                       M. CLIVICOLA
   17   Capsule strongly oblique at the base, somewhat woody, only tardily or not at all de-
           hiscent; corolla 3-5 cm. long; s. Willamette Valley, and southward
                                                                        M. TRICOLOR

Mimulus alsinoides Dougl. ex Benth. Scroph. Ind. 29. 1835.
  M. alsinoides var. paniculatus Benth. Scroph. Ind. 29. 1835. Intended as the typical form of
  the species. (Douglas, Columbia R.)
  M. alsinoides var. minimus Benth. Scroph. Ind. 29. 1835. (Columbia R., collected by
  Douglas and by Scouler)
  Slender, glandular-puberulent to partly glabrous annual, 0.4-3 dm. tall, simple or freely
branched; leaves thin and rather small, the blade elliptic to deltoid or subrhombic, obscurely
to evidently toothed, obtuse or rounded to acutish, 3- to 5-nerved from the base, 0.5-2 cm.
long, longer or shorter than the petiole; flowers long-pedicellate; calyx glabrous or obscurely
glandular, 4-7 mm. long, the 3 upper teeth acute and very short, the 2 lower ones somewhat
longer and rounded; corolla yellow, with a conspicuous reddish-brown blotch at the base of the
lower lip, 8-14 mm. long, glabrous, evidently bilabiate, the central lobe of the lower lip longer than the others and deflexed.
  Moist, shady places, particularly in moss mats on cliffs; s. B.C. to n. Calif., in and w. of
the Cascades, at low elevations. Apr.-June.

Mimulus breviflorus Piper, Bull. Torrey Club 28:45. 1901. (Piper 1858, Pullman, Wash.)
 M. inflatulus Suksd. Werdenda 1:38. 1927. (Suksdorf cites several of his own collections,
 chiefly from Klickitat Co., Wash.)
 Slender, glandular-puberulent annual, 0.5-2 dm. tall, often much branched; leaves small
and often rather numerous, the blade mostly 5-20 mm. long, mostly rather narrowly elliptic
or rhombic-elliptic, entire or toothed, 3(5)-nerved from the base, tapering to the short petiole,
or some of them sessile; pedicels less than 1 cm. long at anthesis, often longer later, tending
to be loosely ascending in fruit; calyx glandular-puberulent, 3.5-5 mm. long at anthesis, the
short teeth more or less acute; corolla yellow, faintly spotted, narrow, 4-7 mm. long, up to
about half again as long as the calyx, slightly 2-lipped, the subequal lobes truncate or rounded,
the throat ventrally pubescent within, scarcely 2 mm. wide; style scarcely surpassing the calyx.
 Moist, open places in the valleys and plains; Wash. to c. Oreg., e. of the Cascades, e. to
n. Ida. and on at least the n. margin of the Snake R. plains (Martin, Butte Co., Ida.), and reputedly s. to n. Calif. May-July.

Mimulus breweri (Greene) Rydb. Mem. N.Y. Bot. Gard. 1:351. 1900.
 Eunanus breweri Greene, Bull. Calif. Acad. Sci. 1:101. 1885. (Brewer 2114, Wood's Peak,
 near Amador Pass, Sierra Nevada, Calif.)
 Slender, simple or sparingly branched annual, up to 1.5 dm. tall, the herbage and calyx copiously stipitate-glandular (or with short, spreading, gland-tipped hairs) throughout, and often anthocyanic; leaves linear to linear-oblanceolate or linear-elliptic, 1-2 cm. long, 1-4 mm. wide,
obscurely 3-nerved; pedicels from slightly shorter to evidently longer than the calyx, ascending
in fruit; calyx 3-6 mm. long, the short teeth about equal; corolla light purple to nearly red,
sometimes slightly marked with yellow, small and slender, 5-10 mm. long, scarcely 2 mm. wide
at the throat, somewhat bilabiate, obscurely pubescent ventrally within, tending to persist for
some time after withering; style exserted or included; septum usually splitting only near the tip.
 Moist to moderately dry slopes and meadows at moderate elevations in the mountains, in and
e. of the Cascades; s. B.C. to s. Calif., e. through Ida. to Yellowstone Nat. Pk. June-Aug.
 The species is allied to M. suksdorfii.

Mimulus clivicola Greenm. Erythea 7:119. 1899.
 Eunanus clivicola Heller, Muhl. 1:60. 1904. (Sandberg et al. 586, near the foot of Wiess-
 ner's Peak, Kootenai Co., Ida.)
 Glandular-pubescent, slender, mostly simple annuals up to 1.5 dm. tall; leaves inconspicuously or scarcely 3-nerved, obtuse or merely acutish, entire or with scattered small teeth,
oblanceolate or elliptic, mostly 0.5-3 cm. long, 2-12 mm. wide; flowers short-pedicellate at
first, the pedicel later elongating to 3-7 mm.; calyx 5.5-8 mm. long at anthesis, glandular-
pubescent, the sharp teeth 1-2 mm. long; corolla purple, conspicuously marked with yellow in
the throat and tube, 1.2-2 cm. long, evidently bilabiate, persistent for some time after withering; capsule lance-linear, curved, exserted, dehiscent; placenta splitting to the base at maturity, the halves adherent to their respective valves.
 Moist to moderately dry slopes in the foothills and valleys; n. Ida. and adj. Wash., s. to
the s. end of the Snake River Canyon (Pine Creek, Union Co., Oreg.) May-July.

Mimulus cusickii (Greene) Piper, Contr. U.S. Nat. Herb. 11:508. 1906.
 Eunanus cusickii Greene, Pitt. 1:36. 1887. (Cusick 1262, Malheur R., Oreg.)
 Mimulus bigelovii var. ovatus Gray, Syn. Fl. 2nd ed. 2$^1$:445. 1886. (Torrey 372, Lake
 Washoe, Nev.; lectotype by Grant)
 Low but often rather stout, simple or much-branched annual up to 2.5 (reputedly 4) dm. tall,
glandular-pubescent and somewhat mephitically aromatic; leaves broad, evidently 3- to 5-
nerved, entire, the lower tending to be somewhat obovate or oblanceolate and often obtuse, the
others generally broadly ovate or broader, at least the upper ones strongly acute or acuminate,
mostly (1)1.5-7 cm. long and 1-3 cm. wide, or the lower smaller; flowers subsessile; calyx
7-13 mm. long, strongly glandular-pubescent, the sharp teeth 2-4 mm. long; corolla rich magenta, marked in the throat with yellow and deeper red, 2-3.5 cm. long, evidently bilabiate
(but the lips about equal), persistent for some time after withering; capsule lanceolate, obtuse,

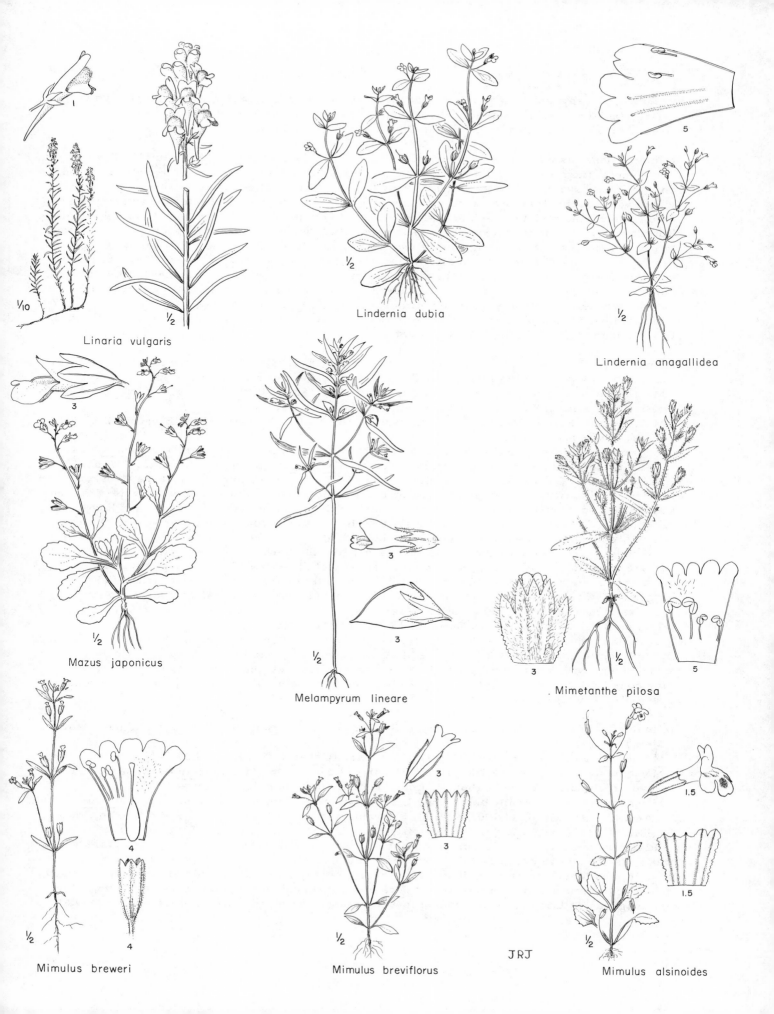

1/10

1/2

Linaria vulgaris

1/2

Lindernia dubia

5

1/2

Lindernia anagallidea

3

1/2

Mazus japonicus

3

3

1/2

Melampyrum lineare

3

1/2

5

. Mimetanthe pilosa

4

4

1/2

Mimulus breweri

3

3

1/2

Mimulus breviflorus

1.5

1.5

1/2

Mimulus alsinoides

JRJ

exserted; placenta splitting to the base at maturity, the halves adhering to the respective valves.

Dry, mostly loose and often sandy slopes in desert and semidesert regions; s. Wasco Co., Oreg., to Boise Co., Ida., s. to n.e. Calif. and n. Nev. May-Sept.

Mimulus dentatus Nutt. ex Benth. in DC. Prodr. 10:372. 1846. (Nuttall, Columbia R.)

Perennial from well-developed, shallow rhizomes; stems ascending or loosely erect, 1-4 dm. tall; herbage sparsely to evidently rough-hirsute with flattened, often spreading, scarcely viscid white hairs; leaves serrate, pinnately veined, but the principal lateral veins arising chiefly below the middle, the blade lance-elliptic to ovate, or the lower sometimes broadly elliptic to obovate, more or less acute, short-petiolate or, especially the upper, sessile, mostly 2-7 cm. long and 1-3.5 cm. wide; flowers few, long-pedicellate; calyx 8-16 mm. long, evidently spreading-hirsute along the 5 rib angles, otherwise essentially glabrous except for the ciliolate margins of the acute teeth, these 2-5 mm. long, the upper one somewhat the largest; corolla 2.5-4 cm. long, evidently bilabiate, yellow, the long, strongly flaring throat often red-dotted, the well-developed lobes sometimes washed with reddish purple.

Stream banks and other moist places; Wash. to n. Calif., w. of the Cascades, chiefly in the Coast Ranges and w. Olympics. May-Sept.

Mimulus floribundus Lindl. Bot. Reg. 13: pl. 1125. 1828. (Garden specimens, from seeds collected by Douglas "in the interior of the districts of the river Columbia")

M. peduncularis Dougl. ex Benth. Scroph. Ind. 29. 1835. M. floribundus var. minor Hook. Fl. Bor. Am. 2:99. 1838. (Douglas "North-West America")

M. serotinus Suksd. Deuts. Bot. Monats. 18:154. 1900. (Suksdorf 2185, Bingen, Klickitat Co., Wash.)

M. membranaceus A. Nels. Bot. Gaz. 34:30. 1902. M. floribundus var. membranaceus Grant, Ann. Mo. Bot. Gard. 11:221. 1924. (Nelson 1683, Medicine Bow Mts., Wyo.)

M. deltoideus Gand. Bull. Soc. Bot. France 66:218. 1919. (Cusick 2237, Pine Creek, Oreg.)

Erect to subprostrate annual, simple or more often branched, the stem 0.5-2.5(4) dm. long; herbage conspicuously to sometimes obscurely viscid-pubescent and glandular, tending to be slimy as well; leaves mostly deltoid-ovate to subcordate, callous-dentate or -denticulate, subpalmately or nearly pinnately veined, the blade (0.5)1-3 cm. long, the petiole usually shorter; flowers long-pedicellate; calyx viscid-pubescent like the herbage, or more finely and sparsely so, 4-8 mm. long, the acute, more or less equal teeth less than 2 mm. long; corolla yellow, often with some red dots, 6-14 mm. long, only slightly bilabiate, the lower lip only slightly longer than the upper and not much deflexed from it, the throat ventrally pubescent within.

Moist open places, chiefly e. of the Cascades, seldom ascending very high into the mountains; s. B.C. to Mont., s. to Calif. and n. Mex. May-Oct.

A variable species, from which several segregates have been proposed. Many of the plants in the Rocky Mts. area (Mont. and Ida. to Colo. and Utah) tend to be small, few-flowered, and thin-leaved, with relatively long petioles and inconspicuous pubescence. These plants, which may constitute a shade ecotype, may be distinguished, with some difficulty, as var. membranaceus (A. Nels.) Grant.

Mimulus glabratus H.B.K. Nov. Gen. & Sp. 2:370. 1817. (Humboldt & Bonpland, near Moran, Mex.)

M. geyeri Torr. Nicoll. Rep. Miss. 157. 1843. (Geyer 119, Devil's Lake, N.D.)

M. jamesii T. & G. ex Benth. in DC. Prodr. 10:371. 1846. M. glabratus var. jamesii Gray, Syn. Fl. 2nd ed. 2¹:447. 1886. (James, near the Missouri R., in Iowa)

M. jamesii var. fremontii Benth. in DC. Prodr. 10:371. 1846. M. glabratus var. fremontii Grant, Ann. Mo. Bot. Gard. 11:190. 1924. M. glabratus ssp. fremontii Pennell, Monog. Acad. Phila. 1:120. 1935. (Fremont, Wind River Mts., Wyo.) Not M. fremontii (Benth.) Gray, 1876.

Plants perennial, rhizomatous, glabrous or inconspicuously hairy; stems 0.5-6 dm. long, weak, decumbent to creeping (often rooting at the nodes) or floating; leaves short-petiolate or sessile, the blade irregularly denticulate (rarely dentate) or entire, palmately 3- to 7-nerved, subrotund to more or less reniform or occasionally rotund-ovate, mostly 1-3 cm. long; flowers long-pedicellate, their subtending leaves not much reduced; calyx accrescent, irregular, the upper tooth much the largest, the lateral and lower teeth blunt and mostly short, the lower not folded upward; corolla 1-2 cm. long, yellow, sparingly if at all red-dotted, evidently bilabiate, the throat open, the lower lip strongly bearded.

In very wet places or in shallow water, especially in calcareous situations; Man. to Mich. and Mont. (Missoula), s. through the plains region of c. N. Am. (and westward occasionally to Nev. and Ariz.) to Mex. and thence to S. Am. May-Aug.

Our plants, as described above, representing the common plains phase of the species, belong to the var. fremontii (Benth.) Grant. Other varieties exhibit minor differences in size of the flowers and in shape and margin of the leaves.

Mimulus guttatus DC. Cat. Hort. Monspel. 127. 1813.

M. langsdorfii var. guttatus Jeps. Fl. W. Middle Calif. 406. 1901. (Garden specimens, from seeds sent by Langsdorff from Unalaska)

M. luteus sensu auct., not L., this properly being a Chilean species.

M. langsdorfii Donn in Sims, Curtis' Bot. Mag. 36: pl. 1501. 1812, as a synonym of M. luteus; Greene, Journ. Bot. 33:6. 1895. (Garden specimens from seed sent by Langsdorff from Unalaska)

M. rivularis Nutt. Journ. Acad. Phila. 7:47. 1834. (Wyeth, valleys of the Rocky Mts.)

M. scouleri Hook. Fl. Bor. Am. 2:100. 1838. M. guttatus ssp. scouleri Pennell, Proc. Acad. Phila. 99:166. 1947. (Scouler, Columbia R.)

M. microphyllus Benth. in DC. Prodr. 10:371. 1846. M. luteus var. depauperatus Gray, Bot. Calif. 1:567. 1876. M. langsdorfii var. depauperatus Henry, Fl. So. B.C. 268. 1915. M. langsdorfii var. microphyllus Nels. & Macbr. Bot. Gaz. 61:44. 1916. M. guttatus var. depauperatus Grant, Ann. Mo. Bot. Gard. 11:170. 1924. (Douglas, banks of the Columbia)

M. nasutus Greene, Bull. Calif. Acad. Sci. 1:112. 1885. M. langsdorfii var. nasutus Jeps. Fl. W. Middle Calif. 407. 1901. M. guttatus var. nasutus Jeps. Man. Fl. Pl. Calif. 928. 1925. (Edwards s.n., Knight's Valley, Sonoma Co., Calif., in 1877) = var. depauperatus, at least as to our plants.

M. hallii Greene, Bull. Calif. Acad. Sci. 1:113. 1885. M. guttatus var. hallii Grant, Ann. Mo. Bot. Gard. 11:172. 1924. (Hall & Harbour 358, mountains of Colo.; lectotype by Grant) = var. depauperatus.

M. guttatus var. grandis Greene, Man. Bay Reg. 277. 1894. M. langsdorfii var. grandis Greene, Journ. Bot. 33:7. 1895. M. grandis Heller, Muhl. 1:110. 1904. (Greene s.n., Berkeley, Calif., in 1883)

M. thermalis A. Nels. Bull. Torrey Club 27:269. 1900. (Nelson 6285, Upper Geyser Basin, Yellowstone Nat. Pk.) = var. depauperatus.

M. hirsutus Howell, Fl. N. W. Am. 520. 1901. (Tualatin and Willamette rivers, Oreg.)

M. grandiflorus Howell, Fl. N. W. Am. 520. 1901. (W. Oreg. and Wash.)

M. longulus Greene, Leafl. 2:4. 1909. (Greene, Humboldt R. at Deeth, Nev. in 1896) = var. depauperatus.

M. langsdorfii var. minimus Henry, Fl. So. B.C. 268. 1915. (Paisley I., B.C.) = var. depauperatus.

M. puncticalyx Gand. Bull. Soc. Bot. France 66:219. 1919. (Suksdorf 2775, Bingen, Klickitat Co., Wash.) = var. depauperatus.

M. puberulus Gand. Bull. Soc. Bot. France 66:219. 1919. (Suksdorf 5016, Bingen, Klickitat Co., Wash.) Not M. puberulus Greene, 1906.

M. bakeri Gand. Bull. Soc. Bot. France 66:219. 1919. (Baker 2618, near Napa, Calif.)

M. guttatus var. decorus Grant, Ann. Mo. Bot. Gard. 11:173. 1924. M. decorus Suksd., Werdenda 1:37. 1927. (Lyon 59, near Oregon City)

M. laxus Pennell ex Peck, Man. High. Pl. Oreg. 655. 1941, without Latin diagnosis.

M. guttatus ssp. litoralis Pennell, Proc. Acad. Phila. 99:165. 1947. (Maguire 17304, near Otter Rock, Lincoln Co., Oreg.) = var. grandis.

M. guttatus var. gracilis G. Campbell, El Aliso 2:328. 1950. (Thurber 498, Napa Co., Calif.)

Plants annual and fibrous-rooted, or perennial by stout stolons, only rarely perennial from well-developed rhizomes, very variable in stature and vigor, sometimes dwarf and small-

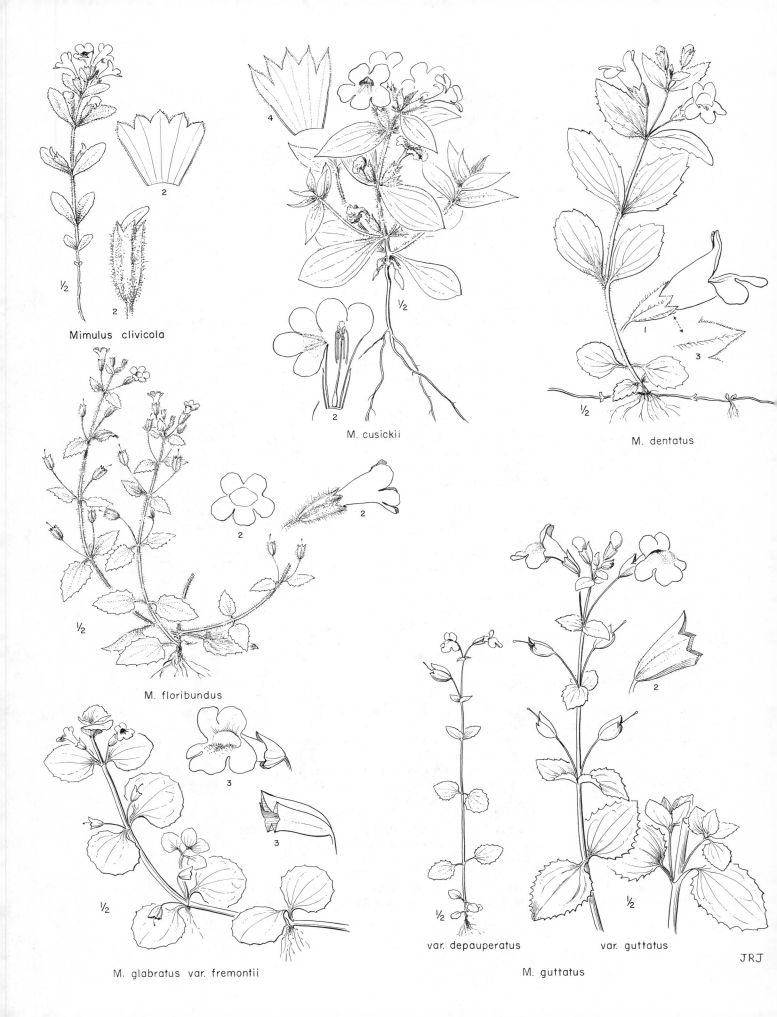

Mimulus clivicola

M. cusickii

M. dentatus

M. floribundus

M. glabratus var. fremontii

var. depauperatus

var. guttatus

M. guttatus

JRJ

leaved, sometimes robust and nearly a meter high, and with the leaf blades up to nearly 1 dm. long, glabrous or pubescent, soft and often somewhat succulent; leaves irregularly dentate, ovate to rotund or reniform-cordate, palmately or subpalmately veined, the 3-7 main veins all arising at or near the base; lower leaves petiolate, upper becoming sessile, those of the inflorescence reduced, relatively broad, tending to clasp the stem, and sometimes connate; flowers mostly several or many (up to about 2 dozen, or more when the inflorescence is branched) in terminal racemes, long-pedicellate, or sometimes solitary in small forms; calyx accrescent, irregular, the upper tooth much the largest, the two lower ones tending to fold upward in fruit and partly close the orifice; corolla 1-4 cm. long, strongly bilabiate, and with flaring throat, yellow, dotted or more heavily marked with maroon on or about the well-developed, pubescent palate. N=14.

In wet places, from sea level to moderate elevations in the mountains; cordilleran region of w. N. Am., from Alas. and Yukon to n. Mex.; introduced in Europe and in other temperate regions. Mar.-Sept.

A highly variable species, either very plastic, or consisting of numerous biotypes, or both, but not readily divisible into well-defined infraspecific taxa. The commoner forms are more or less robust, mostly 2-8 dm. tall, and often stoloniferous, with the leaves well developed (sometimes as much as 8 cm. long) and with the mostly several or many flowers rather large, the corolla 2-4 cm. long. These represent the var. guttatus. A very robust, stout-stemmed, evidently pubescent form of ocean bluffs, occurring from s.w. Wash. to Calif., has been distinguished as var. grandis Greene. More or less depauperate forms, seldom over 3 dm. tall, generally without stolons, and with the leaves 0.5-2 cm. long and the often fewer flowers only 1-2 cm. long, tend to occur in less distinctly hydric habitats than var. guttatus, and sometimes bloom very early in the spring. These have been segregated as var. depauperatus (Gray) Grant. Other more or less recognizable phases of the species occur beyond our borders, and one careful and competent student of the family (Pennell) believed that several additional forms in our range should be dignified with binomials.

The name M. guttatus var. gracilis Campbell, stated by Campbell to be based on M. luteus var. gracilis Gray in Torr. Bot. Mex. Bound. 115. 1859, has recently been used in part for some of the plants here treated as M. guttatus var. depauperatus. As clearly indicated by the typography (compare a more fully described but likewise unnamed variety of Gerardia heterophylla Nutt. on p. 119 of the same work), and by the fact that in subsequent writings Gray did not refer to a variety gracilis, the word "gracilis" was intended by Gray to be merely part of a three-word description of an unnamed variety of M. luteus.

Mimulus jungermannioides Suksd. Deuts. Bot. Monats. 18:154. 1900. (Suksdorf 1470, near Bingen, Wash.)

Perennial from slender rhizomes which produce cormlike resting buds, these giving rise to the weak, often prostrate or drooping stems 0.5-3 dm. long; herbage viscid-villous, especially the stems; leaves broadly ovate to reniform-cordate, irregularly dentate or denticulate, subpalmately veined, the blade up to 2.5 cm. long, the petiole mostly somewhat shorter; flowers long-pedicellate, the pedicels divaricate at least in fruit; calyx glandular-villous, 6-7 mm. long in fruit, the short (1-2 mm.), broad lobes abruptly apiculate or rounded-mucronulate; corolla yellow with some red dots, mostly 13-18 mm. long, 2-lipped, the throat ventrally pubescent within.

Moss mats on cliffs at the e. end of the Columbia R. gorge in Wash. and Oreg., and extending s. along the Deschutes R. to Maupin. May-June.

Seldom collected and insufficiently known.

Mimulus lewisii Pursh, Fl. Am. Sept. 427. 1814. (Lewis, head springs of the Missouri; stated by Pennell to be near Lolo Pass in Ida.)

M. lewisii var. alba Henry, Fl. So. B.C. 268. 1915. (Vancouver I.) A white-flowered form.
M. lewisii f. leuceruthrus Hardin, Mazama 11:88. 1929. (Hardin & English 1392, Bagley Creek, vicinity of Mt. Baker, Wash.) A pink-flowered form.

Perennial with stout stems 3-10 dm. tall clustered on stout, branching rhizomes; herbage viscid-villous; leaves sessile, strongly several-nerved from the base, irregularly callous-dentate to entire, the lowermost ones reduced, the others 3-7(10) cm. long, 1-3.5 cm. wide, lanceolate to ovate or elliptic, acute; pedicels 3-6(10) cm. long; calyx 1.5-2.5(3) cm. long, the sharp teeth about equal; corolla very showy, pink-purple, marked with yellow, 3-5.5 cm. long, evidently bilabiate. N=8

In and along streams, and in other wet places, at moderate to high elevations in the mountains; B.C. to Calif., e. to Alta., Mont., Wyo., and Utah; in all the higher mountains of our range. June-Aug.

Mimulus moschatus Dougl. in Lindl. Bot. Reg. 13: pl. 1118. 1828. (Garden specimens, from seeds collected by Douglas in the "margins of springs in the country about the River Columbia")
   M. moschatus var. longiflorus Gray, Syn. Fl. 2$^1$:278. 1876. (Calif.; no type specified)
   M. inodorus Greene, Bull. Calif. Acad. Sci. 1:119. 1885. M. moschatus var. sessilifolius Gray, Syn. Fl. 2nd ed. 2$^1$:447. 1886. (Calif. and Oreg.; no type specified)
   M. moschatus var. pallidiflorus Suksd. Deuts. Bot. Monats. 18:154. 1900. (Suksdorf 2320, near Chenowith, Skamania Co., Wash.)

Perennial from well-developed, sometimes moniliform rhizomes; stems lax, often prostrate, 0.5-7 dm. long, often freely branched; herbage viscid-villous with slender, flattened, shining white hairs, and tending to be slimy, strongly to scarcely musk-scented; leaves remotely and sometimes obscurely callous-dentate, pinnately veined, short-petiolate or sessile, the blade ovate to elliptic-ovate or lance-ovate, barely to strongly acute, 1-8 cm. long, 7-35 mm. wide; flowers long-pedicellate; calyx 7-13 mm. long, viscid-villous like the herbage, often more strongly so on the 5 rib angles than between them; calyx teeth acute or acuminate, mostly 2-4 mm. long, the upper tooth often a little larger than the others; corolla 1.5-3 cm. long, obscurely bilabiate, yellow, often with some dark lines or dots, the tube only slightly widened upward, without a well-defined throat, ventrally pubescent within.

Moist places, from near sea level to moderate elevations in the mountains; s. B.C. to Mont., s. to Calif., Utah, and Colo.; introduced in parts of Europe and e. U.S. May-Aug.

A variable species, from which several segregates have been proposed. Most of the plants from w. of the Cascades tend to be relatively robust, with the stems up to 7 dm. long, and with the relatively large, often sessile leaves not more than about half as wide as long. These may be distinguished, often with some difficulty, as var. sessilifolius Gray. The more widespread var. moschatus, occurring sparingly w. of the Cascades as well as throughout the rest of our range and elsewhere is smaller, with the stems seldom 3 dm. long, and has mostly petiolate, shorter and relatively broader leaves, the blade seldom over 4 cm. long and often more than half as wide as long. Plants of var. moschatus with relatively large flowers have been called var. longiflorus Gray, but do not form a clearly differentiated taxon.

Mimulus nanus H. & A. Bot. Beechey Voy. 378. 1838.
   Eunanus tolmiei Benth. in DC. Prodr. 10:374. 1846. Eunanus nanus Holz. Contr. U.S. Nat. Herb. 3:244. 1895. Mimulus tolmiei Rydb. Mem. N.Y. Bot. Gard. 1:351. 1900. (Tolmie, country of the Snake Indians)
   M. stamineus Grant, Ann. Mo. Bot. Gard. 11:302. 1924, as to the Washington specimens cited.
   M. microphyton Pennell, Proc. Acad. Phila. 99:169. 1947. (Thompson 8280, Tumwater Canyon, near Leavenworth, Chelan Co., Wash.)

Dwarf annual up to 1 dm. high, becoming much branched when well developed, rather finely glandular-puberulent; leaves rather inconspicuously (or scarcely) 3- to 5-nerved, entire, obtuse or merely acutish, the lower commonly oblanceolate, the upper elliptic or elliptic-oblong to rather narrowly ovate, up to about 3.5 cm. long, and rarely as much as 1 cm. wide; flowers subsessile, commonly crowded; calyx 5-8 mm. long, glandular-puberulent, the sharp teeth 1-3 mm. long; corolla rich magenta, marked in the throat with yellow and deeper red, 1-2.5 cm. long, evidently bilabiate (but the lips about equal), persisting for some time after withering; placenta splitting to the base at maturity, the halves adhering to the separate valves.

Dry, open, often sandy or gravelly places, in the plains, valleys, and foothills; Chelan Co., Wash., s. to n. Calif., e. to s.w. Mont., Yellowstone Nat. Pk., and n.e. Nev.; rarely collected in Wash. May-Aug.

Mimulus pulsiferae Gray, Proc. Am. Acad. 11:98. 1876. (Mrs. Pulsifer-Ames, Indian Valley, Calif.)

Slender, glandular-puberulent annual, 0.5-1.5 dm. tall, sometimes branched; leaves small, the blade mostly 5-12 mm. long, rhombic to elliptic or lance-ovate, toothed or entire, 3(5)-nerved from the base, tapering to the short petiole, or some of them sessile; pedicels commonly 1-1.5 cm. long at anthesis, tending to become arcuate-spreading or strongly divergent in fruit; calyx glandular-puberulent, 3.5-5 mm. long at anthesis, the short lobes acute; corolla yellow, commonly with some maroon dots, 8-16 mm. long, mostly 2-3 times as long as the calyx, slightly bilabiate, the lower lip not much deflexed from the upper, the throat ventrally pubescent within; style conspicuously surpassing the calyx.

Moist, open places in the foothills and valleys; Klickitat Co., Wash., s. along the e. base of the Cascades in Oreg., and thence to Mariposa Co., Calif. June-July.

Rarely collected in our range.


Mimulus primuloides Benth. Scroph. Ind. 29. 1835. (Douglas, Western North America)
M. pilosellus Greene, Erythea 4:22. 1896. M. primuloides var. pilosellus Smiley, U. Calif. Pub. Bot. 9:332. 1921. (Calif.; no type specified)
M. primuloides var. minimus Peck, Proc. Biol. Soc. Wash. 47:187. 1934. (Leach 4361, Roz Lake, Wallowa Mts., Oreg.)

Perennial with flagelliform rhizomes, forming dense mats, the leaves all crowded at or near the ground, occasionally more lax and the stem up to 6 cm. long; leaves villous-hirsute on one or both sides and often viscid, or glabrous, oblanceolate or nearly so, essentially sessile, entire or obscurely toothed, 3(5)-nerved from the base, 7-25 mm. long, 3-11 mm. wide; pedicels slender, usually only 1(3) from a stem, 2-10 cm. long, rarely shorter in alpine depauperates; calyx narrow, 4-8 mm. long, glabrous except for the often ciliolate margins of the short, acute or mucronulate, about equal teeth; corolla yellow, often dotted with maroon, 1-2 cm. long, obscurely bilabiate, with spreading, mostly notched lobes, the throat somewhat flaring from a slender tube.

Wet meadows and other moist, mostly open places at moderate to rather high elevations in the mountains; Wash. to Calif., e. to c. Ida. and Ariz., in our range not extending w. of the Cascades. June-Aug.


Mimulus suksdorfii Gray, Syn. Fl. 2nd ed. $2^1$:450. 1886. (Suksdorf 487, Mt. Paddo, Wash.)

Slender, often much-branched, finely glandular-puberulent annual 3-10 cm. tall; leaves small and often rather numerous, linear to narrowly oblong or the lower oblanceolate, sessile or the lower short-petiolate, entire or nearly so, 1- to 3-nerved from the base, up to about 2 cm. long; pedicels well under 1 cm. long at anthesis, seldom much longer later, tending to be widely spreading in fruit, with suberect tip; calyx glandular-puberulent, 3-5 mm. long at anthesis, the short teeth mostly rounded-mucronulate; corolla yellow, faintly spotted, narrow, 4-8 mm. long, slightly 2-lipped, the subequal lobes notched at the tip, the throat scarcely 2 mm. wide, ventrally pubescent within; style scarcely to evidently surpassing the calyx.

Open, moist or rather dry places, from the valleys and foothills to moderate or occasionally rather high elevations in the mountains; Mt. Adams, Wash. (not extending w. of the Cascades) to s. Calif., e. to Wyo. and Colo. May-July.


Mimulus tilingii Regel, Gartenfl. 18:321. 1869.
M. langsdorfii var. tilingii Greene, Journ. Bot. 33:8. 1895. (Garden specimens, from seeds collected by Tiling near Nevada City, Calif.)
M. luteus var. alpinus Gray, Proc. Acad. Phila. 1863:71. 1864. M. langsdorfii alpinus Piper, Mazama 2:99. 1901. M. alpinus Piper, Contr. U.S. Nat. Herb. 11:510. 1906. (Parry 135a, alpine region of Colo.)
M. langsdorfii var. argutus Greene, Journ. Bot. 33:7. 1895. (Presumably a Nuttall or a Tolmie collection from Oreg.)
M. scouleri var. caespitosus Greene, Pitt. 2:22. 1889. M. caespitosus Greene, Journ. Bot. 33:8. 1895. M. tilingii var. caespitosus Grant, Ann. Mo. Bot. Gard. 11:154. 1924. (Greene s.n. in 1889, Mt. Rainier, Wash.)

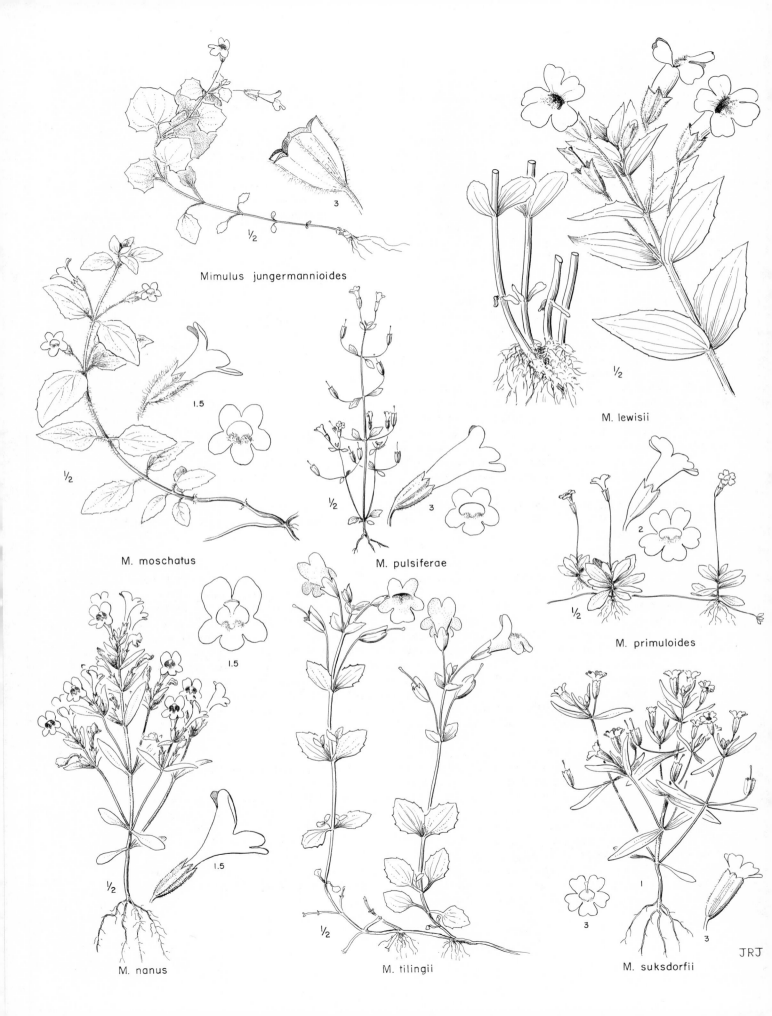

Mimulus jungermannioides

M. lewisii

M. moschatus

M. pulsiferae

M. primuloides

M. nanus

M. tilingii

M. suksdorfii

JRJ

M. minor A. Nels. Proc. Biol. Soc. Wash. 17:178. 1904. (Andrews 8, Arapahoe Pass, Boulder Co., Colo.)
M. veronicifolius Greene, Leafl. 2:7. 1909. (Piper 2177, Olympic Mts., Wash.)
M. lucens Greene, Leafl. 2:7. 1909. (Piper 2518 and 2519, Powder River Mts., Oreg.)
   Perennial from well-developed, often sod-forming, creeping rhizomes, sometimes stoloniferous as well, 0.5-2 dm. tall, glabrous or inconspicuously pubescent; leaves sessile or short-
petiolate, the blade seldom as much as 2.5 cm. long, rhombic to elliptic or ovate, or less often subrotund, not much reduced upward, commonly with a few irregular small teeth on the
margins, subpalmately veined as in M. guttatus; flowers few or solitary, mostly 1-5, long-
pedicellate, similar to those of M. guttatus, the corolla large for the size of the plant, commonly 2-4 cm. long. N=14.
   Wet places, particularly along or in small cold streams, at high altitudes in the mountains;
B.C. and Alta. to Calif., n. Baja Calif., and n. N.M.; also, according to Campbell, northward to Alas. July-Sept.
   Mimulus tilingii is closely allied to M. guttatus, and the two are sometimes difficult to distinguish, at least in the herbarium. Grant, Campbell, and Pennell in successive studies have
maintained M. tilingii as distinct, and strong support for their view is provided by the recent
extensive experimental work of R. K. Vickery, who finds (personal communication) an almost
complete genetic barrier between the two species. Two fairly definite varieties of M. tilingii
may be recognized, as follows:
1 Leaves small, seldom over 1 cm. long; plants freely branched, the branches more or less
      stoloniform; s. B.C., from Vancouver I. to the Selkirk Range, s. to the Olympic Mts. and
      Cascade Range of Wash., the Wallowa Mts. of Oreg., and the mountains of c. Ida.
                                                          var. caespitosus (Greene) Grant
1 Leaves larger, up to about 2.5 cm. long, the better developed ones mostly 1 cm. long or
      more; plants less branched, or the branches not stoloniform; nearly the range of the spe-
      cies, but apparently absent (or very rare) from the Washington Cascades     var. tilingii

Mimulus tricolor Hartw. ex Lindl. Journ. Hort. Soc. Lond. 4:222. June, 1849.
   Eunanus tricolor Greene, Bull. Calif. Acad. Sci. 1:99. 1885. (Garden specimens from seeds
      collected by Hartweg in the Sacramento Valley, Calif.)
   Eunanus coulteri "Harv. et Gr. ex A. Gray in litt." acc. to Benth. Pl. Hartw. 329. Aug.
      1849. (Hartweg, Sacramento Valley, Calif.) The Coulter plant, which would have been the
      type of the species had Gray himself published it, was later said by Gray to belong to an-
      other species.
   Simple or branching annual up to 1.5 dm. tall, glandular-puberulent; leaves entire or irregularly toothed, pinnately and longitudinally veined, up to 4 cm. long and 1 cm. wide, the lower tending to be oblanceolate and often subpetiolate, the others rather narrowly elliptic or elliptic-oblong and sessile; flowers very shortly pedicellate; calyx glandular-puberulent, 1-2
cm. long, the short lobes unequal, the upper the largest and most foliaceous; corolla principally purple, with a darker spot on each lobe, conspicuously marked in the throat with white
and (distally) yellow, 3-5 cm. long, the tube slender and elongate, the throat shorter and flaring, both lips well developed and with broad, rounded lobes; capsule 5-10 mm. long, firm,
tardily, or not at all, dehiscent, strongly oblique at the base, with one valve better developed
than the other; placenta splitting to the base at maturity, the halves adherent to the respective
valves.
   Wet clay soil, especially on sites of vernal pools; s. Willamette Valley (Corvallis) to c.
Calif. May.
   Mimulus douglasii Gray, a chiefly more southern species, known from Douglas Co., Oreg., to
c. Calif., has been reported from Corvallis. It is similar in most respects to M. tricolor,
though often smaller, but differs strikingly in the very much reduced, nearly obsolete lower
lip of the corolla.

Mimulus washingtonensis Gand. Bull. Soc. Bot. France 66:218. 1919. (Suksdorf 560, Bingen,
   Klickitat Co., Wash.)

M. ampliatus Grant, Ann. Mo. Bot. Gard. 11:214. 1924. (Heller 3330, Lake Waha, Nez Perce
Co., Ida.)

M. patulus Pennell, Proc. Acad. Phila. 99:162. 1947. (Elmer 752, Wawawai, Whitman Co.,
Wash.) A small-flowered extreme; an isotype at New York has the corollas 11-12 mm.
long, instead of 7-9 mm. as described by Pennell.

Slender, glandular-puberulent annual, 0.4-2 dm. tall, simple or freely branched; leaves
small, the blade ovate to subcordate, slightly toothed, acute or acutish, palmately or subpalmately veined, 6-14 mm. long, the petiole from a little shorter to a little longer; flowers
long-pedicellate; calyx glandular-puberulent, 4.5-8 mm. long, the short, mostly acute teeth
about equal; corolla yellow, often with some reddish-brown dots, (0.8)1-2 cm. long, strongly
bilabiate, the lower lip longer than the upper and evidently deflexed from it, the throat ventrally pubescent within.

Wet, open places, at low elevations; Klickitat Co., Wash., to Wheeler, Grant, and Wallowa
cos., Oreg., and Nez Perce Co., Ida. May-Sept.

The name M. peduncularis Dougl., has sometimes been applied to this species, but is properly a synonym of M. floribundus according to Grant.

## Orthocarpus Nutt.

Flowers borne in short or elongate, prominently bracteate, terminal spikes or spiciform
racemes; calyx 4-cleft, the lobes often partly connate in lateral pairs; corolla elongate and
narrow, bilabiate, the upper lip (galea) beaklike, its lobes united to the tip and enclosing the
anthers, the lower lip more or less saccate-inflated, nearly or quite as long as the galea,
usually 3-toothed at the tip, external in bud; stamens 4, didynamous, attached near the summit of the corolla tube; pollen sacs unequally placed, one apically attached and lying alongside
the upper part of the filament, the other medifixed and versatile, or the lower pollen sac obsolete; stigma entire, penicillate; capsule loculicidal; seeds numerous; slender annual with alternate, entire to dissected, wholly cauline leaves.

Twenty-five species, occurring chiefly in w. U.S., especially in Calif., one species in The
Andes of S. Am. (Name from the Greek orthos, straight, and karpos, fruit, referring to the
symmetrical capsule.)

Reference:

Keck, David D. A revision of the genus Orthocarpus. Proc. Calif. Acad. Sci. IV. 16:517-
71. 1927.

1 Corolla mostly 9-25 mm. long; anthers 2-celled
  2 Lower lip of the corolla simply saccate, or nearly so
    3 Bracts and calyces glandular-pubescent; upper leaves tending to pass gradually into
      the bracts; galea about equaling the lower lip, or surpassing it by less than 1 mm.
      4 Corolla yellow, mostly 9-14 mm. long, the lower lip with definite short teeth; e. of
        the Cascades                                                      O. LUTEUS
      4 Corolla pink-purple or occasionally white, mostly 12-20 mm. long, the lower lip
        without teeth; from the e. base of the Cascades, westward      O. BRACTEOSUS
    3 Bracts and calyces scarcely or not at all glandular; upper leaves passing rather ab-
      ruptly into the strongly differentiated bracts; galea commonly surpassing the lower
      lip by 1-1.5 mm.
      5 Leaves entire; upper bracts with pink-purple, somewhat petaloid tips; Olympic Mts.
        of Wash., Cascade Range of Oreg., and southward      O. IMBRICATUS
      5 Leaves (at least the upper) cleft, with elongate, narrow segments; e. of the Cas-
        cades
        6 Upper bracts with soft, pink-purple, somewhat petaloid, rounded or broadly ob-
          tuse to obscurely mucronate tips; widespread e. of the Cascades
                                    O. TENUIFOLIUS
        6 Bracts all green or yellowish-green throughout, with firm, mostly cuspidate, not
          at all petaloid tips; c. Wash.                    O. BARBATUS
  2 Lower lip of the corolla more or less trisaccate

7 Bracts green throughout (rarely somewhat anthocyanic as well as chlorophyllous); teeth of the lower lip poorly developed and inconspicuous, 0.2-0.8 mm. long; widespread species                                    O. HISPIDUS

7 Bracts (or some of them) tipped with white, yellow, or purple; teeth of the lower lip small but evident, slender, 1-3 mm. long; w. of the Cascades

    8 Corolla linear, the lower lip scarcely expanded; bracts only minutely tipped with white or yellowish (rarely purple), the slender spike scarcely showy; rachis and segments of the leaves, and segments of the bracts, very slender; plants not maritime

                                                  O. ATTENUATUS

    8 Corolla clavate, the lower lip somewhat expanded; bracts more evidently tipped with white, yellow, or purple, the stout spike rather showy; rachis and segments of the leaves and bracts mostly broader (or the segments of the leaves often narrow); plants maritime or submaritime                        O. CASTILLEJOIDES

1 Corolla mostly 4-6 mm. long; anthers 1-celled; w. of the Cascades          O. PUSILLUS

**Orthocarpus attenuatus** Gray, Pac. R.R. Rep. 4:121. 1856. (Bigelow, Corte Madera, Calif.)

Plants 1-3.5 dm. tall, seldom branched, the herbage spreading-hirtellous throughout; leaves very narrowly lance-linear, long-attenuate, 2-6 cm. long, entire or the upper often 3-cleft; inflorescence elongate and rather narrow, not showy, the bracts gradually differentiated from the leaves, becoming shorter and more cleft, with broader rachis, the upper ones with very short white, yellowish, or occasionally purplish tips; calyx subequally lobed; corolla 1-2.5 cm. long, linear, whitish or slightly pinkish, the narrowly trisaccate, scarcely expanded lower lip more yellowish and with some evident purple spots; teeth of the lower lip well developed, erect, slender, 1-1.5 mm. long, nearly equaling the galea.

Open, often grassy slopes, meadows, and pastures; s. Vancouver I. to Calif., wholly w. of the Cascades in our range, except for a station near Mackenzie Pass in Oreg. Apr.-June.

**Orthocarpus barbatus** Cotton, Bull. Torrey Club 29:574. 1902. (Sandberg & Leiberg 234, junction of Crab and Wilson creeks, Douglas, now Grant, Co., Wash.)

Plants 0.8-2.5 dm. tall, simple or with several erect branches, the herbage pubescent with mixed long and short, loosely spreading hairs, sometimes obscurely glandular upward; leaves 2-4 cm. long, narrow and entire or deeply 3- to 5-cleft with elongate, narrow lobes; inflorescence dense and rather short; bracts abruptly differentiated from the leaves, broader and more membranous, reticulate-veiny, mostly cuspidate, the upper progressively less cleft, green or yellowish-green throughout, not at all petaloid; calyx 2-cleft, with bifid segments; corolla yellow, 1-1.2 cm. long, the lower lip simply saccate, its teeth well under 1 mm. long; galea surpassing the lower lip by 1-1.5 mm.

Sagebrush slopes and flats; Grant Co. to s. Okanogan Co., Wash. June.

**Orthocarpus bracteosus** Benth. Scroph. Ind. 13. 1835. (Douglas, Columbia R.)

O. bracteosus var. albus Keck, Proc. Calif. Acad. Sci. IV. 16:554. 1927. (Leiberg 521, Big Meadows, Deschutes R., Oreg.)

Plants 1-4 dm. tall, simple or branched above; herbage pubescent throughout, the hairs of the stem mostly short and retrorse, of the leaves short and spreading or appressed; hairs of the inflorescence mostly longer, and some of them gland-tipped; leaves 1.5-3.5 cm. long, narrow, the lower entire, the upper trifid and passing into the broader, divergently trilobed bracts of the dense, often finally elongate inflorescence; calyx 2-cleft, with bifid segments; corolla 12-20 mm. long, pink-purple or occasionally white, gradually expanded to the swollen, simply saccate, toothless lower lip; galea short and broad, scarcely surpassing the lower lip.

Meadows at low elevations; s. B.C. to n. Calif., from the e. base of the Cascades, westward. June-Aug.

**Orthocarpus castillejoides** Benth. Scroph. Ind. 13. 1835. (Douglas, "California")

Plants 1-3.5 dm. tall, simple or often branched, the stem shortly spreading-hairy and obscurely viscid; leaves scabrous-puberulent and obscurely glandular, 1-5 cm. long, lanceolate to oblong or ovate, the lower mostly entire, the upper with 1 or 2 pairs of mostly short,

sometimes slender segments; bracts gradually differentiated from the leaves, becoming apically cleft into several short, more or less round-tipped, subequal segments, these white,
yellow, or purple, the spike fairly dense and rather showy; calyx deeply 2-cleft, the segments
with bifid tips that are colored like the tips of the bracts; corolla 14-25 mm. long, yellow with
some purple markings, the tube gradually expanded to the somewhat inflated, trisaccate lower
lip, which has well-developed, slender, erect, mostly purple apical teeth 1-3 mm. long; galea
straight or nearly so, surpassing the teeth of the lower lip by 1-2 mm.

Salt marshes and other saline soils along the coast (Puget Sound included), from s. Vancouver I. to Monterey Co., Calif. June-Sept.

Orthocarpus purpurascens Benth., a related California species with prominently purple-
tipped bracts, hooked galea, and very narrow leaf segments, occurring in nonmaritime habitats, has been collected as an introduction at Seattle and Tacoma.

Orthocarpus hispidus Benth. Scroph. Ind. 13. 1835.
  Triphysaria hispida Rydb. Bull. Torrey Club 40:484. 1913. (Douglas, Columbia R.)
  Orthocarpus rarior Suksd. Allg. Bot. Zeit. 12:27. 1906. (Suksdorf 2779, Falcon Valley,
    Klickitat Co., Wash.)
  Plants 1-4 dm. tall, simple or with a few erect branches; herbage spreading-hairy throughout, many of the hairs of at least the inflorescence and upper part of the stem attaining 1-2
mm. in length, some of the finer and shorter ones (at least in the inflorescence) commonly
gland-tipped; leaves 1-4 cm. long, the lower linear and entire, the upper a little wider and 3-
to 5-cleft, with slender segments; bracts gradually differentiated from the leaves, becoming
shorter, more cleft, and with broader rachis, green throughout, or occasionally slightly over-
laid with purple, but not at all showy; calyx 2-cleft, with bifid segments; corolla white or light
yellow, 12-20 mm. long, exserted, the lower lip inflated and trisaccate, with inconspicuous,
poorly developed teeth 0.2-0.8 mm. long; galea straight or nearly so, 1-2 mm. longer than
the lower lip.

In moist meadows, beds of vernal pools, and other moist sites, from the lowlands to moderate elevations in the mountains; s. B.C. (and introduced at Skagway, Alas.) to s. Calif., e. to
Ida. and Nev., and to be expected in n.w. Mont.; both sides of the Cascades, but more common eastward. May-Aug.

Orthocarpus imbricatus Torr. ex Wats. Bot. King Exp. 458. 1871. (Newberry, Cascade Mts.,
  Oreg.)
  O. olympicus Elmer, Bot. Gaz. 36:60. 1903. (Elmer 2574, Olympic Mts., Clallam Co.,
    Wash.)
  Plants 1-3.5 dm. tall, simple or corymbosely branched above; herbage sparsely puberulent
or subglabrous, sometimes with a few longer hairs in the inflorescence; leaves 1-4 cm. long,
linear to narrowly lanceolate, entire; bracts abruptly differentiated from the leaves, shorter,
and much broader and blunter, sometimes some of them with a pair of short, lateral lobes below the middle, at least the upper ones with pink-purple, somewhat petaloid tips, the spike
dense and showy; calyx short, 2-cleft with bifid segments; corolla 10-13 mm. long, purplish,
in large part covered by the bracts, the lower lip simply saccate or nearly so, with inconspicuous, short teeth; galea about 1 mm. longer than the lower lip, minutely hooked at the tip.

Meadows and open slopes in the mountains; Olympic Mts. of Wash.; Cascade Range from c.
Oreg. to n. Calif., and in the Klamath region of s.w. Oreg. and adj. Calif. July-Sept.

Orthocarpus luteus Nutt. Gen. Pl. 2:57. 1818. (Nuttall, on the plains of the Missouri)
  O. strictus Benth. Scroph. Ind. 13. 1835. (Douglas, Columbia R. and Red R.)
  Plants 1-4 dm. tall, simple or sometimes branched above; herbage spreading-hairy throughout (or finally glabrate below), the hairs of the leaves mostly shorter than those of the stem,
many of the hairs, at least in the inflorescence, gland-tipped; leaves 1.5-4 cm. long, linear
or lance-linear, entire or occasionally some of them trifid; bracts gradually differentiated
from the leaves, becoming shorter, broader, and more cleft; inflorescence finally elongate;
calyx short, subequally 4-lobed; corolla 9-14 mm. long, golden yellow, gradually expanded to

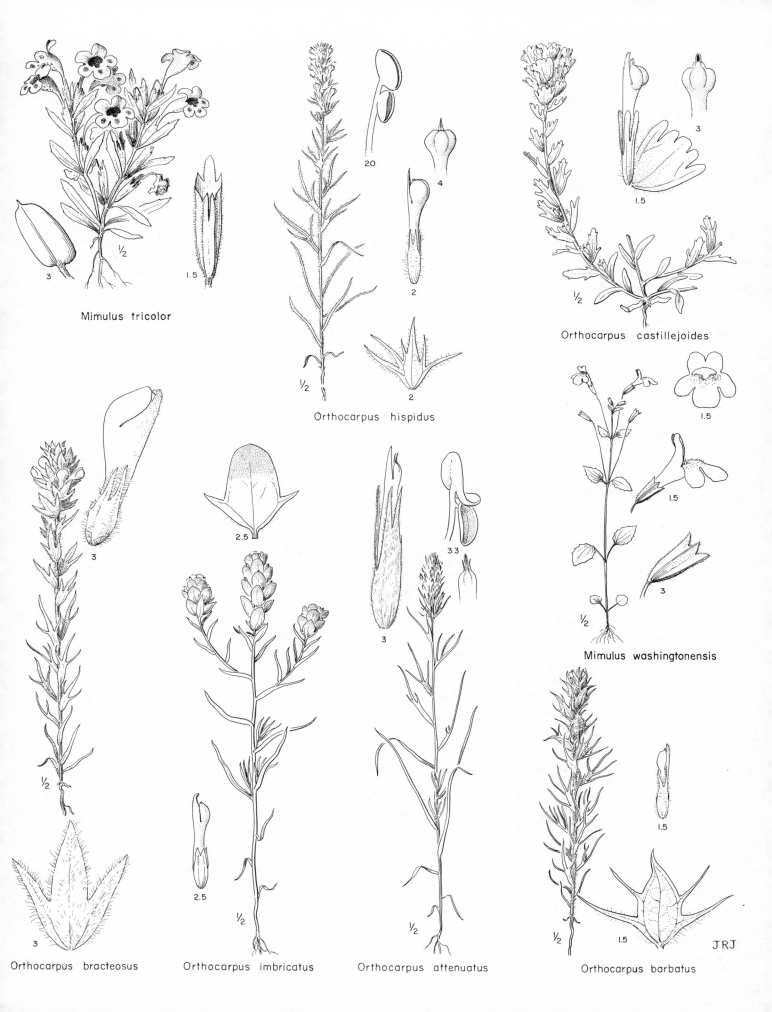

Mimulus tricolor

Orthocarpus hispidus

Orthocarpus castillejoides

Mimulus washingtonensis

Orthocarpus bracteosus

Orthocarpus imbricatus

Orthocarpus attenuatus

Orthocarpus barbatus

JRJ

the simply saccate, minutely 3-toothed, lower lip; galea short and broad, about equaling the lower lip.

Low ground, from the plains to moderate elevations in the mountains; B.C. to Calif., e. of the Cascades, e. to Man., Minn., Neb., and N.M. July-Aug.

Orthocarpus pusillus Benth. Scroph. Ind. 12. 1835. (Douglas, Calif.)
   O. densiusculus Gand. Bull. Soc. Bot. France 66:218. 1919. O. pusillus var. densiusculus Keck, Proc. Calif. Acad. Sci. IV. 17:569. 1927. (Suksdorf 5015, Bingen, Klickitat Co., Wash.)
   Plants 4-20 cm. tall, simple or more often branched from the base, very slender; herbage spreading-hispidulous; leaves 0.5-3 cm. long, with slender rachis and filiform lateral segments; spike elongate, often extending to near the base of the stem, the lower flowers becoming remote; bracts resembling the leaves, or more dissected; calyx subequally 4-lobed; corolla 4-6 mm. long, red-purple or sometimes yellow, the lower lip trisaccate, with inconspicuous teeth; galea slightly surpassing the lower lip, somewhat hooked at the tip; anthers 1-celled.
   Moist places in the lowlands; s. B.C. to Calif., w. of the Cascades. Apr.-June.
   Orthocarpus erianthus Benth., a related California species with similarly 1-celled anthers, but differing prominently in its larger corollas 1-2.5 cm. long, has been collected as an introduction at Seattle.

Orthocarpus tenuifolius (Pursh) Benth. Scroph. Ind. 12. 1835.
   Bartsia tenuifolia Pursh, Fl. Am. Sept. 429. 1814. (Lewis, "banks of Clark's River")
   Plants 1-3 dm. tall, simple or branched above; herbage puberulent; leaves 1-5 cm. long, the lower mostly narrowly linear and entire, the others with 1 or 2 pairs of slender lobes; bracts abruptly differentiated from the leaves, much broader and blunter, entire or the lower with a pair of slender lateral lobes, the lower commonly hispid-ciliate toward the base, the upper with conspicuous, pink-purple, somewhat petaloid tips; spike dense and showy; calyx 2-cleft, with bifid segments; corolla yellow, or purplish at the tip, 14-20 mm. long, the lower lip simply saccate, with short teeth; galea about 1 mm. longer than the lower lip, hooked at the tip.
   Open, moist or dry places from the valleys and plains to moderate elevations in the mountains; Wash. and adj. s. B.C. to c. Oreg., e. of the Cascades, e. to w. Mont. May-Aug.

## Parentucellia Viviani

Flowers borne in a terminal, leafy-bracteate, spiciform raceme; calyx 4-lobed; corolla galeate, the upper lip saccate and lobeless, enclosing the anthers, the lower lip spreading, 3-lobed, about as long as the galea, its lobes external to the galea in bud; stamens 4, didynamous, the pollen sacs equal, mucronate, woolly; stigma capitate, entire or nearly so; capsule loculicidal, the cells slightly unequal; seeds numerous, minute, turgid, smooth or nearly so; annual or biennial, glandular-hairy herbs with sessile, toothed, wholly cauline leaves, these opposite, or the upper alternate or offset.

Two species, native to the Mediterranean region. (Named for Tomaso Parentucelli, founder of the botanic garden at Rome)

Parentucellia viscosa (L.) Car. in Parl. & Car. Fl. Ital. 6:482. 1885.
   Bartsia viscosa L. Sp. Pl. 602. 1753. (Europe)
   Erect, fibrous-rooted annual 1-7 dm. tall, usually unbranched, rather coarsely spreading-hairy, becoming stipitate-glandular upward; leaves ovate or lanceolate, crenate-dentate, 1-4 cm. long, up to 2 cm. wide, the upper often alternate or offset; flowers subsessile, mostly alternate or offset; corolla yellow, 1.5-2 cm. long, well surpassing the eventually accrescent (but not inflated) calyx.
   A weed in moist low ground; native of the Mediterranean region, now occasionally found w. of the Cascades from Wash. to n. Calif., and also in w. S. Am. June-Aug.

## Pedicularis L. Lousewort

Flowers borne in a lax and elongate to capitate, usually spiciform raceme; calyx regularly, or often irregularly, cleft, with 5, or less commonly 4 or only 2, lobes; corolla yellow or white to purple or red, bilabiate, the upper lip or galea hooded, enclosing the anthers, often extended into a beak, the lower lip usually shorter, generally 3-lobed, external to the upper in bud; stamens 4, didynamous; stigma capitate; capsule glabrous, flattened, asymmetrical, loculicidal, mostly arcuate and opening chiefly or wholly on the upper side; seeds several, often slightly winged; perennial (rarely apparently annual) herbs with alternate, toothed to more often pinnatifid or pinnately dissected leaves.

A large genus of several hundred species, occurring chiefly in the n. temp. and boreal regions, extending also to the Andes of S. Am.; best developed in the Old World. (Name Latin, pertaining to lice, because of a superstition that the ingestion of these plants by stock promoted infestation with lice.)

Nearly all of the species would be valued members of the wild garden, especially in moist soil. However, they are difficult to grow, and they present a real challenge to the garden enthusiast.

1 Leaves merely toothed; calyx lobes 2                   P. RACEMOSA
1 Leaves, or many of them, pinnatilobate to bipinnatifid; calyx lobes 5 (except in P. parviflora)
  2 Stem branched; plants short-lived, perhaps annual; galea essentially beakless
                               P. PARVIFLORA
  2 Stem simple; plants evidently perennial; galea beaked or beakless
    3 Plants leafy-stemmed, the basal leaves, if present, not markedly larger than the cauline (though sometimes with longer petioles); bracts sharply differentiated from the leaves; galea beakless or with a very short beak     P. BRACTEOSA
    3 Plants with the leaves basally disposed, the cauline leaves mostly few and more or less reduced (this habit least marked in P. langsdorfii, which has the bracts, especially the lower ones, scarcely differentiated from the leaves)
      4 Galea essentially beakless, occasionally with an inconspicuous apiculation not over 1 mm. long
        5 Corolla yellow or ochroleucous, sometimes faintly tinged with pink or purple
          6 Pinnae of the leaves, or many of them, strongly incised
            7 Flowers large, the corolla (2)2.5-3.5 cm. long; galea much longer than the corolla tube; boreal species, extending s. to s.e. B.C.   P. CAPITATA
            7 Flowers smaller, the corolla about 1.5 cm. long; galea about equaling the corolla tube; Mt. Rainier, Wash.     P. RAINIERENSIS
          6 Pinnae mostly merely toothed; corolla 1.5-2.5 cm. long, the galea a little shorter than the tube; boreal species, extending s. to n.w. Wyo.
                              P. OEDERI
        5 Corolla purple
          8 Pinnae of the leaves, or many of them, deeply cleft; Mont. and Wyo.
           9 Bracts not strongly differentiated from the leaves, evidently pinnatifid or bipinnatifid at least distally; dwarf plants, less than 1 dm. tall
                           P. PULCHELLA
           9 Bracts sharply differentiated from the leaves, the distal portion elongate, narrow, and entire or merely toothed; taller plants, 1-4.5 dm. tall
                        P. CYSTOPTERIDIFOLIA
        8 Pinnae or lobes of the leaves mostly merely toothed
         10 Cauline leaves few and small, or none, the stem appearing naked or nearly so; rachis narrow, the leaves distinctly pinnatifid; boreal species, not yet known in our range (See comment under P. cystopteridifolia)
                         P. sudetica Willd.
         10 Cauline leaves rather numerous and not much, if at all, reduced, the stem appearing strongly leafy; rachis relatively broad, the leaves often merely

pinnatilobate; boreal species, extending s. to the Marble Mts. of B.C.

P. LANGSDORFII

4 Galea evidently beaked, the beak 2 mm. long or more (or sometimes only about 1 mm. long
in P. parryi)
  11 Beak straight, spreading, about 1-4 mm. long
    12 Leaves subbipinnatifid, the pinnae deeply cleft and again toothed; inflorescence cap-
       itate, sometimes with 1 or 2 smaller, separated, lower clusters; corolla mostly
       1-1.5 cm. long; beak of the galea mostly 2-4 mm. long; Cascade Range of Wash.,
       and northward                                           P. ORNITHORHYNCHA
    12 Leaves once pinnatifid, with the pinnae mostly merely toothed; inflorescence spi-
       cate-racemose, more evidently so in age; corolla mostly 1.5-2 cm. long; beak of
       the galea about (1)2 mm. long; Rocky Mts.             P. PARRYI
  11 Beak strongly curved, often more than 4 mm. long
    13 Beak lunately downcurved, not much, if at all, exserted from the well-developed
       lower lip; widespread species of wooded or open slopes and drier meadows in the
       mountains                                                P. CONTORTA
    13 Beak strongly upcurved, the living flowers (especially in P. groenlandica) reminis-
       cent of an elephant's head with trunk upraised; plants of wet meadows and other wet
       places in the mountains
      14 Plants essentially glabrous throughout; beak conspicuously exserted beyond the
         relatively small lower lip; widespread species   P. GROENLANDICA
      14 Plants evidently villous in the inflorescence; beak not extending much farther
         outward than the rather well-developed lower lip; Cascades and Sierra Nevada,
         from Mackenzie Pass, Oreg., southward              P. ATTOLLENS

Pedicularis attollens Gray, Proc. Am. Acad. 7:384. 1868.
  Elephantella attollens Heller, Muhl. 1:4. 1900. (Swamps in the Sierra Nevada; several col-
    lections cited)
  Similar to P. groenlandica, differing chiefly in the villous inflorescence, and slightly
smaller flowers with better developed lower lip and relatively shorter and less conspicuous
beak; upper calyx tooth evidently the shortest; corolla purple, or purple-spotted on a pale or
white background.
  Wet meadows and other wet places at moderate to high elevations in the mountains; Sierra
Nevada and s. Cascades, n. nearly to Mt. Hood, Oreg. July-Aug.

Pedicularis bracteosa Benth. in Hook. Fl. Bor. Am. 2:110. 1838. (Drummond, Rocky Mts.)
  PEDICULARIS BRACTEOSA var. CANBYI (Gray) Cronq. hoc loc. P. canbyi Gray, Syn. Fl.
    2nd ed. 2¹:454. 1886. (Canby 266, McDonald's Peak, Mission Range, Mont.)
  P. montanensis Rydb. Bull. Torrey Club 24:292. 1897. P. bracteosa var. montanensis M.
    E. Jones, Bull. U. Mont. Biol. 15:45. 1910. (Flodman 796, Little Belt Mts., Mont., 9
    miles from Barker) = var. bracteosa.
  PEDICULARIS BRACTEOSA var. SIIFOLIA (Rydb.) Cronq. hoc loc. P. siifolia Rydb. Bull.
    Torrey Club 34:35. 1907. (Elrod et al. 97, Grant Creek, Mont.)
  PEDICULARIS BRACTEOSA var. ATROSANGUINEA (Pennell & Thompson) Cronq. hoc loc.
    P. atrosanguinea Pennell & Thomps. Bull. Torrey Club 61:443. 1934. (Pennell & Thompson
    15822, Mt. Angeles, Clallam Co., Wash.)
  P. paddoensis Pennell, Bull. Torrey Club 61:444. 1934. (Pennell 15738, Mt. Adams, Wash.)
    A form, several times collected on Mt. Adams, most nearly resembling var. bracteosa,
    but apparently diverging in the direction of var. flavida or var. latifolia.
  PEDICULARIS BRACTEOSA var. FLAVIDA (Pennell) Cronq. hoc loc. P. flavida Pennell,
    Bull. Torrey Club 61:445. 1934. (Pennell 15545, Elk Lake, Deschutes Co., Oreg.)
  PEDICULARIS BRACTEOSA var. PACHYRHIZA (Pennell) Cronq. hoc loc. P. pachyrhiza
    Pennell, Bull. Torrey Club 61:445. 1934. (Pennell 15414, Blue Mts. n.w. of Elgin, Union
    Co., Oreg.)

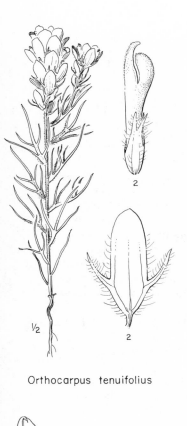

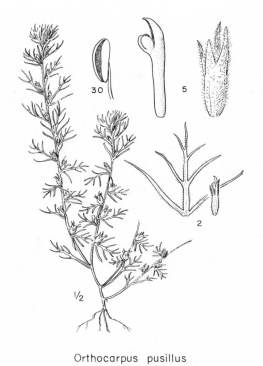

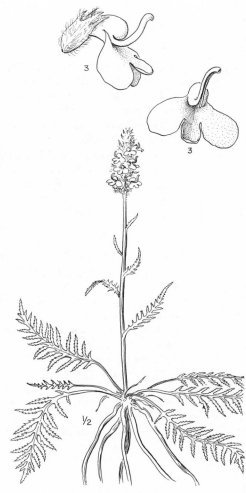

Orthocarpus tenuifolius

Orthocarpus pusillus

var. siifolia

Pedicularis attollens

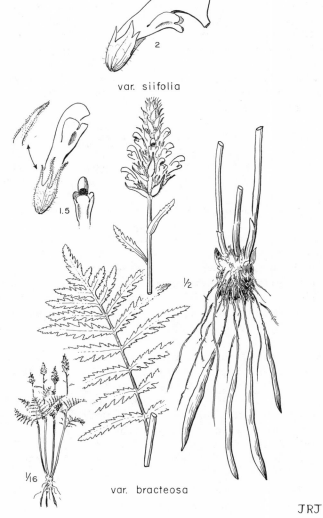

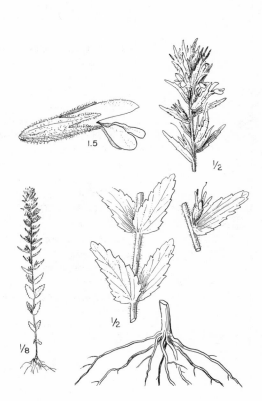

Orthocarpus luteus

Pedicularis bracteosa

var. bracteosa

Parentucellia viscosa

JRJ

PEDICULARIS BRACTEOSA var. PAYSONIANA (Pennell) Cronq. hoc loc. P. paysoniana
   Pennell, Bull. Torrey Club 61:446. 1934. (Payson & Armstrong 3724, near Cottonwood
   Lake, e. of Smoot, Lincoln Co., Wyo.)
P. thompsonii Pennell, Bull. Torrey Club 61:447. 1934. (Thompson 7142, near Wauconda,
   Okanogan Co., Wash.) A form resembling var. paysoniana except for its partly purple co-
   rolla; perhaps actually an intermediate between var. bracteosa and var. latifolia.
PEDICULARIS BRACTEOSA var. LATIFOLIA (Pennell) Cronq. hoc loc. P. latifolia Pennell,
   Bull. Torrey Club 61:448. 1934. (Pennell 15786, Mt. Rainier, Wash.)
   Erect perennials up to about 1 m. tall, coarsely fibrous-rooted, or often some of the roots
tuberous-thickened; herbage glabrous below the inflorescence; stems leafy, the leaves short-
petiolate or sessile, pinnatifid, the principal segments distinct or very nearly so, linear-
oblong to lanceolate, 1-7 cm. long, somewhat incised and again finely serrate, or sometimes
merely doubly serrate; uppermost leaf segments smaller and confluent; basal leaves similar
to the cauline, but more petiolate, or reduced or wanting; calyx lobes 5, the upper one much
the shortest, the others further partly connate into 2 lateral segments; corolla purple or red
to yellow, 13-21 mm. long, the galea beakless or nearly so, about as long as the tube.
   Woods, meadows, and moist open slopes in the mountains; B.C. and Alta. to n. Calif. and
Colo. June-Aug.
   The species as here defined is variable but sharply limited. It may be further broken down,
with some difficulty, into 8 varieties as follows:
1 Free tips of the lateral sepals very slender and elongate, almost filiform, evidently glandular; corolla red or purple to partly or occasionally wholly yellow
   2 Corolla dark blood-red to sometimes yellow; sepal tips tending to be evidently hairy as
      well as glandular, the glands dark, purple to blackish; Olympic Mts.
                                       var. atrosanguinea (Pennell & Thomps.) Cronq.
   2 Corolla purple to partly or sometimes wholly yellow; sepal tips mostly without evident
      glandless hairs, the glands mostly pale; B.C. and Alta. to n.w. Mont., n. Ida., and
      n.e. Wash., and apparently isolated, in a less pronounced form, on Mt. Adams, Wash.
                                       var. bracteosa
1 Free tips of the sepals somewhat wider (linear to lanceolate or triangular) and often shorter,
   only very finely, or not at all, glandular; corolla yellow or yellowish, or in var. latifolia
   sometimes partly, or even wholly, purple
   3 Galea slightly beaked
      4 Inflorescence evidently arachnoid-villous; w.c. and n.w. Mont., and n. Ida.
                                       var. canbyi (Gray) Cronq.
      4 Inflorescence glabrous or nearly so; s.w. Mont., n.c. Ida., and extreme s.e. Wash.
                                       var. siifolia (Rydb.) Cronq.
   3 Galea beakless
      5 Free tips of the lateral sepals evidently longer than the portion which is connate above
         the dorsal sinus
         6 Inflorescence evidently hairy; free tips of the lateral sepals commonly at least twice
            as long as the connate portion
            7 Galea not much raised above the lower lip, sometimes even partly enfolded by it;
               roots tending to be tuberous-thickened; Wallowa and Blue Mt. area of n.e. Oreg.
               and adj. s.e. Wash.              var. pachyrhiza (Pennell) Cronq.
            7 Galea strongly raised above the lower lip; roots generally not tuberous-thickened;
               Colo., Utah, Wyo., and e. Ida., extending occasionally into s.w. Mont.
                                       var. paysoniana (Pennell) Cronq.
         6 Inflorescence glabrous or nearly so; free tips of the lateral sepals less than twice
            as long as the connate portion; Klamath region, and the Cascades of Oreg., ap-
            parently passing into var. latifolia on Mt. Rainier
                                       var. flavida (Pennell) Cronq.
      5 Free tips of the lateral sepals mostly shorter than the portion which is connate above
         the dorsal sinus; inflorescence evidently hairy to sometimes nearly glabrous; Cas-
         cades of Wash., from Mt. Rainier northward, extending to the Marble Mts. of B.C.,
         and occasionally across n. Wash. to n. Ida.      var. latifolia (Pennell) Cronq.

Pedicularis capitata Adams, Nouv. Mém. Soc. Nat. Mosc. 5:100. 1817. (Islands at the mouth of the Lena R., Siberia)

Perennial from a creeping rhizome, 0.5-1.5 dm. tall, more or less puberulent, or the inflorescence villous; basal leaves 3-7 cm. long, the blade equaling or shorter than the petiole, 0.5-1.5 cm. wide, the pinnae, or many of them, sharply incised, the segments often again finely few-toothed; stem nearly naked, bearing 0-2 reduced leaves below the capitate inflorescence; bracts more or less foliose-serrate or -dissected, at least distally, the lowest ones often resembling the leaves, but sessile; calyx lobes 5, not very unequal, rather broad and foliaceous, sometimes distally expanded and toothed; corolla ochroleucous, perhaps sometimes tinged with pink or purple, (2)2.5-3.5 cm. long, the galea much longer than the tube, beakless and without evident subapical teeth; lower lip well developed, up to 1.5 cm. long.

Open slopes above timber line, and in boreal or subalpine forests; n. Asia and n. N. Am., extending s. at high altitudes to the mountains of s.e. B.C. July-Aug.

Pedicularis contorta Benth. in Hook. Fl. Bor. Am. 2:108. 1838. (Tolmie, Mt. Rainier)
P. ctenophora Rydb. Bull. Torrey Club 24:293. 1897. P. contorta var. ctenophora Nels. & Macbr. Bot. Gaz. 61:44. 1916. (Rydberg 2789, near Lima, Mont.)
P. lunata Rydb. Bull. Torrey Club 28:27. 1901. (Tweedy 2317, Big Horn Mts., Wyo.) = var. ctenophora.

Perennial from a stout caudex, glabrous throughout (except var. ctenophora), 1.5-6 dm. tall, the stems clustered; basal leaves 5-18 cm. long, the blade about equaling or longer than the petiole, 1-3.5 cm. wide, the pinnae narrow, not crowded, serrate; inflorescence elongate, seldom very dense, the bracts narrow, deeply trifid or pinnipalmately cleft; calyx tube pale and somewhat scarious between the 5 darker veins; calyx lobes 5, narrow, entire, the upper one the shortest; corolla ochroleucous or white, often finely marked with purple (wholly pink or purple in var. ctenophora), about 1 cm. long as pressed, the galea short, strongly arched, attenuate into an elongate, lunately downcurved beak which is only slightly, if at all, exserted from the well-developed lower lip.

Wooded or open slopes and drier meadows at moderate to high elevations in the mountains; B.C. and Alta. to n. Calif., c. Ida., and n. and w. Wyo. June-Aug.

In the Big Horn Mts. of Wyo. the typical variety, as chiefly described above, is replaced by the technically well-marked but essentially similar var. ctenophora (Rydb.) Nels. & Macbr., which differs in its pink or purplish corolla, and in having the calyx (often also the bracts) slightly villous at the base. The var. ctenophora is occasionally found as far w. as Lima, Mont., where the type was collected. Many of the plants from Idaho Co., Ida., have pink flowers, as in var. ctenophora, but are glabrous, as in var. contorta. The proper taxonomic disposition of these is as yet uncertain.

Pedicularis cystopteridifolia Rydb. Mem. N.Y. Bot. Gard. 1:365. 1900. (Rydberg & Bessey 4983, Cedar Mt., Mont.)

Fibrous-rooted perennial from a short, erect caudex, 1-4.5 dm. tall, glabrous below the villous inflorescence, or slightly puberulent; basal leaves 5-15 cm. long, the blade equaling or longer than the petiole, 0.5-2 cm. wide, the pinnae, or many of them, sharply incised, the segments often again toothed and tending to be somewhat callous-margined; cauline leaves several, progressively reduced and becoming sessile; inflorescence tending to elongate in age; bracts sharply differentiated from the leaves, occasionally bearing a pair of slender lateral lobes at the junction of the lance-elliptic to ovate base with the elongate, narrow, entire or merely dentate lip; calyx lobes 5, the upper the smallest, the others more or less connate below into 2 lateral pairs; corolla purple, 2-2.5 cm. long, the galea equaling, or shorter than, the tube, essentially beakless, and bearing a pair of small, divergent teeth near the tip.

Open slopes at moderate to high elevations in the mountains; n. and w. Wyo. and s.w. Mont. June-Aug.

The related, circumboreal species, P. sudetica Willd., has been collected as far s. as Barkerville, B.C., and may be found to occur in our range. It differs from P. cystopteridifolia in its more nearly naked stem, mostly less dissected leaves (the pinnae usually merely toothed), and in the usually toothed or lobulate distal portions of the bracts. An isolated southern fragment of the population of P. sudetica, occurring in the high mountains of Colo., tends to approach P. cystopteridifolia in all of these features except the dissection of the leaves, without being sharply separable from typical P. sudetica. These Colorado plants, which further differ from typical P. sudetica in the usual reduction or obsolescence of the subapical teeth of the galea, have commonly been known as P. scopulorum (Gray) Gray; they might perhaps better be treated as P. sudetica var. scopulorum Gray.

Pedicularis groenlandica Retz. Prodr. Fl. Scand. 2nd ed. 145. 1795.
  Elephantella groenlandica Rydb. Mem. N.Y. Bot. Gard. 1:362. 1900. (By implication, Greenl., but unknown there, and presumably actually from Labr.)
  Pedicularis surrecta Benth. in Hook. Fl. Bor. Am. 2:107. 1838. P. groenlandica var. surrecta Gray, Proc. Am. Acad. 8:396. 1872. P. groenlandica ssp. surrecta Piper, Contr. U.S. Nat. Herb. 11:512. 1906. (Douglas, N. West Interior)
  Coarsely fibrous-rooted perennial, sometimes with an evident caudex, 1.5-7 dm. tall, glabrous throughout (except for the often minutely ciliolate calyx lobes), the stems often clustered; basal leaves 5-25 cm. long, the blade equaling or exceeding the petiole, 0.5-4 cm. wide, the pinnae narrow, sharply serrate or subincised, often somewhat cartilaginous-margined; cauline leaves several, progressively reduced; inflorescence elongate, fairly dense; bracts mostly much shorter than the flowers, at least the lower more or less cleft into narrow segments; calyx lobes 5, short, entire, subequal; corolla pink-purple or almost red, 1-1.5 cm. long as pressed, the galea short and strongly hooded, tipped with a slender, elongate, conspicuously upturned beak; lower lip rather small. Elephant's head, little red elephants.
  Wet meadows, and in small, cold streams, at moderate to high elevations in the mountains; in the cordilleran region, from B.C. and Alta. to Calif. and N.M., and across c. Can. to Labrador. June-Aug.
  Plants from the U.S. tend to have the beak somewhat longer than do most of the Canadian specimens, and have been distinguished on this basis as a separate species, subspecies, or variety (surrecta), but no clear-cut taxonomic segregation seems possible.

Pedicularis langsdorfii Fisch. ex Steven, Nouv. Mém. Soc. Nat. Mosc. 6:49. 1822. (Unalaska and St. Lawrence I., fide Hultén Fl. Alas.)
  Perennial from a stout caudex, often with a stout taproot as well, 1-3 dm. tall, more or less woolly-villous, at least upward; stems solitary or several; leaves 4-7 cm. long, 6-12 mm. wide, the blade longer than the broad petiole, pinnatilobate or subpinnatifid, the rachis relatively broad, the lobes more or less toothed; cauline leaves relatively numerous and not much, if at all, smaller than the basal ones; inflorescence elongate, often longer than the vegetative part of the stem, the lower bracts greatly surpassing the flowers and scarcely differentiated from the foliage leaves, the upper bracts gradually reduced; calyx 5-lobed, the lateral lobes partly connate into 2 pair, with the upper of the pair the larger; corolla purple, 2-2.5 cm. long, the galea about equaling the tube, beakless, and bearing a pair of slender, divergent teeth near the tip; lower lip well developed.
  Alpine slopes in the mountains; Siberia and arctic N. Am., extending s. in the cordillera to the Marble Mts. of B.C. June-Aug.
  The habitally rather similar P. lanata Cham. & Schlecht., a closely related, also boreal species, may later be found to occur within our northern limits. It differs from P. langsdorfii in its more deeply cleft leaves, smaller bracts, and more pointed, equal calyx lobes, and in the absence of the subapical teeth of the galea.

Pedicularis oederi Vahl in Hornem. Dansk. Oekonom. Plantel. 2nd ed. 580. 1806. (Norway)
  Perennial from fibrous, or somewhat thickened, roots, 0.5-2 dm. tall, villous-puberulent at least in the inflorescence; basal leaves 3-10 cm. long, the blade from a little shorter to evidently longer than the petiole, 5-15 mm. wide; pinnae rather broad and crowded, spreading or

2

2

1/8

1/2

Pedicularis contorta

1.5

1.5

1/2

1/6

2.5

P. cystopteridifolia

2.5

1/2

P. capitata

1/2

1.5

P. langsdorfii

1.5

1/2

P. oederi

2

2

1/8

1/2

P. groenlandica

JRJ

retrorse, evidently, sometimes deeply, toothed; cauline leaves few and reduced; inflorescence spicate-racemose, the bracts, especially the lower ones, more or less foliose and toothed (or even pinnatifid) distally; calyx lobes 5, tending to be distally enlarged and coarsely few-toothed; corolla light yellow, 1.5-2.5 cm. long, the galea a little shorter than the tube, beakless and without subapical teeth.

Open slopes above timber line; circumboreal, extending s. in the Rocky Mt. region to the Beartooth Mts. of Wyo. July-Aug.

Pedicularis ornithorhyncha Benth. in Hook. Fl. Bor. Am. 2:108. 1838. (Tolmie, Mt. Rainier, Wash.)

More or less fibrous-rooted perennial with an evident caudex, 0.5-3 dm. tall, glabrous below the more or less villous inflorescence; basal leaves 3-12 cm. long, the blade exceeding or about equaling the petiole, 0.7-2 cm. wide, subbipinnatifid, the narrow pinnae deeply cleft and often again toothed; stem nearly naked, the cauline leaves few and poorly developed, or wanting; inflorescence capitate, sometimes with one or two smaller, separated, lower clusters; bracts more or less cleft or dissected; calyx lobes 5, rather short, subequal; corolla purple, mostly 1-1.5 cm. long, the galea produced into a distinct, straight, spreading beak mostly 2-4 mm. long.

Meadows and open slopes at alpine and subalpine stations in the mountains; Cascades of Wash., from Mt. Rainier (where abundant) northward, and extending n. to the Alaska panhandle. July-Sept.

Pedicularis parryi Gray, Am. Journ. Sci. II. 34:250. 1862. (Parry 251, alpine ridges in the Rocky Mts. of Colo.)

P. parryi var. purpurea Parry, Am. Nat. 8:214. 1874. (Parry 215, Yellowstone Lake, Wyo.)
P. hallii Rydb. Mem. N.Y. Bot. Gard. 1:364. 1900. (Hall, Yellowstone Nat. Pk., Wyo.) = var. purpurea.
P. anaticeps Pennell, Not. Nat. Acad. Phila. 95:10. 1942. (Pennell et al. 24186, Cascade Creek, Grand Teton Nat. Pk., Wyo.) = var. purpurea.

Perennial from somewhat thickened fibrous roots, 0.5-3 dm. tall, glabrous below the inflorescence; basal leaves 4-10 cm. long, the blade equaling or longer than the petiole, 0.5-1 cm. wide; pinnae strongly toothed and tending to be cartilaginous-margined; cauline leaves few and reduced; inflorescence spicate-racemose, sometimes short at first; bracts villous-ciliate on the margins, sometimes similarly hairy over the surface, commonly trifid with the central lobe the largest, or subpinnately cleft; calyx lobes 5, rather short, not very unequal, the ventral sinus the deepest; corolla purple, mostly 1.5-2 cm. long, the galea tapering to a short, straight, spreading, sometimes poorly defined beak about (1)2 mm. long.

Open slopes and meadows, from moderate to high elevations in the mountains; s.w. Mont. and c. Ida. to n. N.M. and Ariz. June-Aug.

Our plants as described above represent the var. purpurea Parry. Typical P. parryi, with yellow flowers and glabrous or subglabrous bracts, occurs chiefly in the s. Rocky Mts. (extending n. to the Big Horn Mts.) and probably does not enter our range.

Pedicularis parviflora Smith in Rees, Cycl. 26: Pedicularis no. 4. 1813. (Menzies, coast of w. N. Am.)

Short-lived, perhaps annual plants, 1-4 dm. tall, glabrous or nearly so below the inflorescence, with a single, branching stem arising from the slender taproot; leaves chiefly cauline, 1.5-5 cm. long, seldom 1 cm. wide, pinnatilobate or pinnatifid, with entire or toothed narrow segments; flowers capitately clustered at the ends of the branches, or the inflorescence often more elongate, with the lower flowers remote and axillary; bracts foliose-dissected; calyx 2-cleft, with irregularly lacerate or lobed segments; flowers slender; 11-14 mm. long, purple at least as to the galea, which is shorter than the tube, beakless, and usually without subapical teeth.

Wet places, commonly in muskeg; Alas. to B.C. and Sask., reputedly e. to Hudson Bay and Lake Mistassini, and reputedly also s. along the coast to Oreg., but I have seen no specimens from s. of about lat. 53 degrees (Barkerville, B.C.); a similar, perhaps separable plant in Siberia. July-Aug.

The habitally similar, likewise boreal species P. labradorica Wirsing (P. euphrasioides Steph.) extends s. at least as far as 53 degrees lat., and may be found to occur in our range. It differs from P. parviflora in its less dissected upper leaves and bracts, the latter being mostly merely toothed, in its less cleft calyx segments, and in its often slightly larger corolla, the galea of which bears a pair of prominent, slender teeth near the apex.

Pedicularis pulchella Pennell, Not. Nat. Acad. Phila. 95:7. 1942. (Pennell et al. 23713, high mountains s.w. of Anaconda, Mont.)

Dwarf perennial from an apparent taproot, less than 1 dm. tall, more or less villous; basal leaves 2-5 cm. long, less than 1 cm. wide, the blade mostly longer than the petiole, more or less bipinnatifid, the narrow pinnae deeply lobed and often again toothed; cauline leaves few and reduced; inflorescence compactly spicate-racemose; bracts pinnatifid or subbipinnatifid, not sharply differentiated from the leaves; calyx lobes 5, about equal, short, entire or toothed; corolla purple, about 2 cm. long, the galea about equaling the tube, obscurely or scarcely short-beaked, and bearing a pair of divergent, small, subapical teeth.

Open alpine slopes; mountains of s.w. Mont. and n.w. Wyo. July-Aug.

Pedicularis racemosa Dougl. ex Hook. Fl. Bor. Am. 2:108. 1838. (Douglas, "summit of the high mountains of the Grand Rapids of the Columbia")

PEDICULARIS RACEMOSA var. ALBA (Pennell) Cronq. hoc loc. P. racemosa ssp. alba Pennell, Proc. Acad. Phila. 99:176. 1947. (Pennell 20922, near Lolo Creek, Clearwater Co., Ida.)

Coarsely fibrous-rooted perennial from a woody caudex, 1.5-5 dm. tall, glabrous throughout, or obscurely hairy above; stems clustered; lowermost leaves much reduced, the others well distributed along the stems, short-petiolate or subsessile, lanceolate or lance-elliptic to lance-linear or linear-oblong, mostly 4-10 cm. long and 0.5-1.5 cm. wide, doubly serrate, the secondary teeth often inconspicuous; inflorescence mostly lax and elongate, the lower flowers (or pedunculate flower clusters) often axillary to scarcely altered leaves, the middle and upper bracts progressively smaller and less leafy; calyx deeply cleft below, and more shallowly above, into 2 oblique, broad-based segments, each with a dorsally oriented, acuminate tip; corolla 1-1.5 cm. long as pressed, the galea strongly arched and tapering into a slender, downcurved beak which approaches or touches the prominent lower lip.

Coniferous woods in the mountains, less commonly in dry meadows or on open slopes; B.C. and Alta. to c. Calif. and N.M. June-Sept.

The var. racemosa, occurring in the Cascade and n. Sierra Nevada region and westward, has pink to purplish flowers. The var. alba (Pennell) Cronq., occurring mostly e. of the Cascades and Sierra Nevada, has white or ochroleucous flowers, and slightly narrower (on the average) leaves.

Pedicularis rainierensis Pennell & Warren, Bull. Torrey Club 55:317. 1928. (Warren 537, Mt. Rainier, Wash.)

Coarsely fibrous-rooted perennial, 1.5-4 dm. tall, glabrous below the villous inflorescence, the stems mostly clustered; basal leaves 5-15 cm. long, the blade longer than the petiole, up to 3 cm. wide, the pinnae, or many of them, sharply incised and again toothed; cauline leaves several, but progressively reduced and becoming sessile; inflorescence shortly spicate-racemose; bracts distally somewhat foliose-serrate; calyx lobes 5, lance-attenuate, the upper the shortest, the others further partly connate into 2 lateral pairs; corolla ochroleucous, about 1.5 cm. long; galea about equaling the tube, beakless or with an inconspicuous apiculation, without subapical teeth.

Moist alpine meadows and open coniferous forest on Mt. Rainier, Wash. July-Aug.

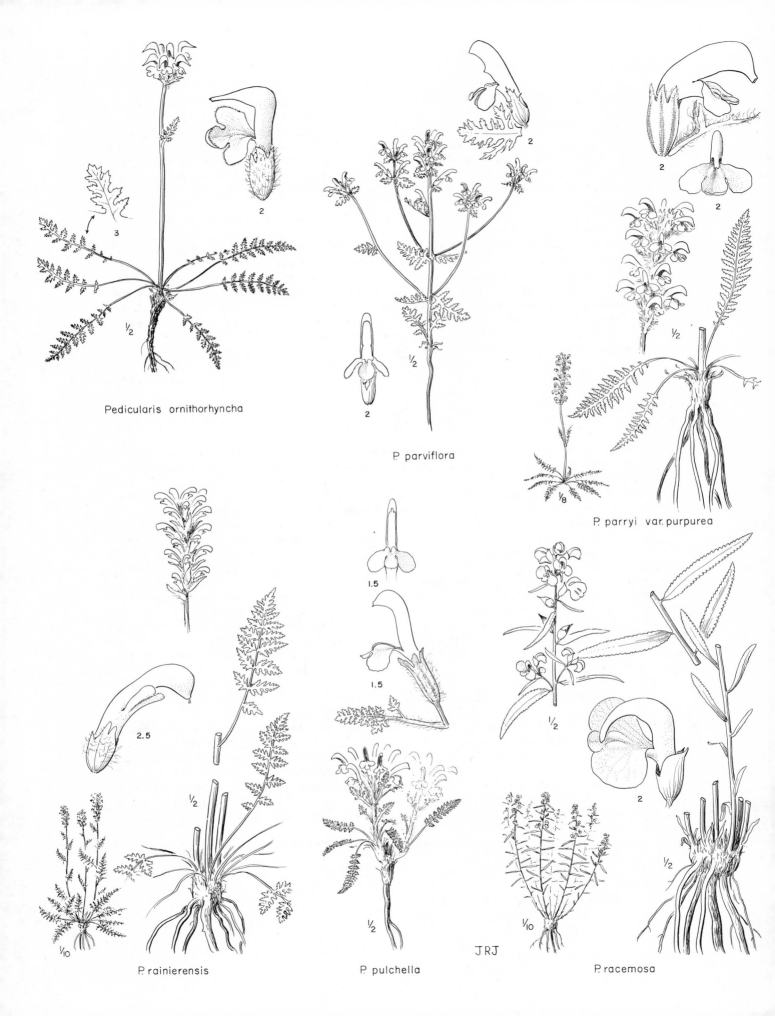

Pedicularis ornithorhyncha

P. parviflora

P. parryi var. purpurea

P. rainierensis

P. pulchella

JRJ

P. racemosa

## Penstemon Mitch. Beardtongue

Flowers typically borne in verticillasters, these sometimes expanded to form an open-paniculate inflorescence, or in a few species reduced to a pair of opposite flowers at each node of a then essentially racemose inflorescence; calyx 5-cleft essentially to the base; corolla tubular, slightly to strongly bilabiate, typically blue to purple or lavender, varying to pink, red, yellow, or white, the upper lobes external in bud; fertile stamens 4, paired; anthers deeply sagittate, the pollen sacs joined only near the tip; sterile filament well developed, longer or shorter than the fertile ones, often bearded, inserted well down in the corolla tube; stigma small, capitate; capsule septicidal; seeds numerous, with a reticulate, sometimes ariliform or winglike coat; perennial herbs or shrubs with opposite (rarely alternate, or partly ternate or quaternate), entire or toothed to rarely laciniate-pinnatifid leaves.

About 225 species, native to N. Am., especially w. U.S., one species in Kamchatka and Japan. (Name from the Greek pente, five, and stemon, thread.)

The genus includes some of our most beautiful native plants, and most of them take readily to cultivation. Among the most desirable shrubby species are P. rupicola, P. barrettiae, P. cardwellii, P. fruticosus, and P. davidsonii, all of which are easily propagated by cuttings. The more herbaceous species are all easily grown from seed.

The name has sometimes been spelled Pentstemon, but the original spelling, as here used, is philologically acceptable and must be maintained under the Rules.

Many of our species of Penstemon are poorly defined, and taxa which superficially look very different sometimes hybridize freely in nature. The species of the group represented by P. albertinus, P. attenuatus, P. elegantulus, P. humilis, P. ovatus, P. pruinosus, P. subserratus, and P. wilcoxii, in particular, are so different in their typical forms and yet so difficult of definition that a wholly satisfactory treatment does not seem possible on the basis of our present information.

The section Habroanthus (represented in our range by P. cyaneus, P. lemhiensis, P. payettensis, P. pennellianus, and P. speciosus) consists of about two dozen superficially similar but technically well-marked taxa which occupy distinctive and hardly overlapping geographic areas. Most of these have at one time or another been submerged in P. glaber Pursh (the oldest name in the group), but there are so few intermediates that it seems better to retain them in specific status.

The floral measurements in the following treatment are taken from dried herbarium specimens. Measurements of the calyx are taken at anthesis; the calyx tends to be more or less accrescent in fruit. Measurements of the pollen sacs are taken at full maturity, after dehiscence. When the pollen sacs open so widely after dehiscence as to form essentially a plane, they are said to be explanate. The pollen sacs are joined (by the connective) only at the apical end, and except in the subgenus Saccanthera they become more or less divergent at maturity. When the divergence is so strong that the two pollen sacs assume an end-to-end position in essentially a straight line, they are said to be opposite. The ends of the pollen sacs which remain in contact are here called the proximal ends, and the ends which become progressively more divergent are called the distal ends.

References:

Keck, David D. Studies in Penstemon. I. A systematic treatment of the section Saccanthera. U. Cal. Pub. Bot. 16:367-426. 1932.

-------. Studies in Penstemon. VI. The section Aurator. Bull. Torrey Club 65:233-55. 1938.

-------. Studies in Penstemon. VII. The subsections Gairdneriani, Deusti, and Arenarii of the Graciles, and miscellaneous new species. Am. Midl. Nat. 23:594-616. 1940.

-------. Studies in Penstemon. VIII. A cyto-taxonomic account of the section Spermunculus. Am. Midl. Nat. 33:128-206. 1945.

-------, and A. Cronquist. Studies in Penstemon. IX. Notes on northwestern American species. Britt. 8:247-50. 1957.

1 Anthers more or less densely long-woolly with tangled hairs (conspicuously so to the naked eye) (Dasanthera)

  2 Plants 3-8 dm. tall, herbaceous nearly or quite to the base; some or all of the axillary

peduncles generally branched and 2- to-several-flowered, the inflorescence thus more
or less paniculate; leaves all cauline, the lower ones reduced, the better developed ones
often more than 6 cm. long, or more than 2.5 cm. wide, or both

   3  Leaves distinctly short-petiolate, rather thin and sharply serrate, mostly (1)1.5-4 cm.
wide; corolla glandular-hairy externally, glabrous internally; staminode bearded;
Cascade region and westward                                                P. NEMOROSUS

   3  Leaves sessile or merely narrowed to an obscurely subpetiolar base, thicker and less
toothed, commonly remotely serrulate to subentire, mostly 3-10(15) mm. wide; corolla glabrous externally, densely woolly-villous along the two prominent ventral
ridges internally; staminode glabrous; e. of the Cascades                     P. LYALLII

2  Plants 0.5-4 dm. tall, usually more or less distinctly woody toward the base (especially
in the larger forms); axillary peduncles generally all simple and uniflorous, the inflorescence thus essentially racemose (often some of the peduncles branched and 2-flow-
ered in P. barrettiae); leaves not over about 6 cm. long and 2.5 cm. wide, often basal or
near-basal as well as cauline

   4  Leaves all cauline, the lower reduced, the sterile shoots few and similar to the fertile
ones, or more often wanting, the plants therefore without any tendency for the leaves
to be clustered toward the base; Valley Co., Ida., to Park Co., Mont., and southward                                                                          P. MONTANUS

   4  Larger leaves tending to be clustered near the base of the plant or near the base of the
season's growth, usually on short, sterile shoots as well as on the flowering stems

      5  Herbage and inflorescence glabrous and glaucous throughout, or the more or less
ovate calyx segments finely and inconspicuously glandular; leaves relatively large,
up to 8 cm. long and 2.5 cm. wide, the larger ones generally more than 1.5 cm.
wide; e. end of the Columbia gorge                                       P. BARRETTIAE

      5  Herbage generally more or less hairy or glandular at least on the stems or in the
inflorescence, seldom (except in P. rupicola) at all glaucous; leaves rarely over
1.5 cm. wide (to 18 mm. in P. ellipticus); calyx segments often narrower

         6  Plants forming dense mats on the surface of the substrate, with scattered, sim-
ple, and mostly erect flowering stems not over about 1.5 dm. tall; leaf blades
1-2.5 times as long as wide

            7  Leaves of the erect flowering stems relatively well developed, rarely less than
1 cm. long; calyx 8-15 mm. long, the conspicuously glandular-hairy seg-
ments lance-linear to linear-oblong; leaves glabrous, not, or scarcely, glaucous, finely and often irregularly serrulate, rarely quite entire; n. Rocky
Mt. region                                                              P. ELLIPTICUS

            7  Leaves of the erect flowering stems small and often bractlike, rarely as much
as 1 cm. long; other characters diverse, but not combined as in P. ellipticus;
Cascade region and westward

               8  Flowers pink or nearly red to pink-lavender or rose-purple; leaves evident-
ly glaucous and often shortly spreading-hairy                          P. RUPICOLA

               8  Flowers blue-lavender to purple-violet; leaves glabrous but not glaucous
P. DAVIDSONII

        6  Plants more or less ascending or erect, branched above the ground surface, the
clusters of leaves at the base of the season's growth more or less elevated
above the ground, not forming mats; plants often well over 1.5 dm. tall; leaf
blades mostly 2-10 times as long as wide

            9  Flowers bright purple to deep blue-violet; leaves mostly obtuse or rounded at
the tip and with serrulate margins; w. of the Cascade summits
P. CARDWELLII

            9  Flowers mostly blue-lavender to light purplish; leaves mostly acute or acutish,
entire or toothed; e. of the Cascade summits  P. FRUTICOSUS

1  Anthers glabrous, or inconspicuously hispidulous with short, mostly straight hairs

  10  Pollen sacs opening across their confluent apices, the free tips remaining saccate and
indehiscent, not becoming divaricate, the anther permanently more or less horseshoe-
shaped (Saccanthera)

11   Plants not at all glandular
 12   Staminode glabrous; corolla 15-24 mm. long; leaves entire, less than 10(14)mm.
  wide; Wheeler Co., Oreg. to Blaine Co., Ida. P. CUSICKII
 12   Staminode bearded; leaves evidently toothed or rarely subentire, often well over
  10 mm. wide
   13   Corolla lobes and fertile filaments glabrous; corolla 17-25(28)mm. long; Cas-
    cade region and westward P. SERRULATUS
   13   Corolla lobes ciliate; fertile filaments hairy above; corolla 25-38 mm. long; n.
    and w. Ida. to s.e. Wash. and n.e. Oreg. P. VENUSTUS
11   Plants evidently glandular in the inflorescence
 14   Plants evidently glandular nearly or quite throughout; leaves basal and cauline,
  usually toothed; corolla 28-40 mm. long; staminode glabrous; w. Ida. to the foot
  of the Cascades in c. Wash. and n. Oreg. P. GLANDULOSUS
 14   Plants glandular only in the inflorescence; leaves all cauline
   15   Leaves entire; staminode glabrous; corolla 15-25 mm. long; Wheeler Co.,
    Oreg., and southward P. LAETUS
   15   Leaves more or less toothed to laciniate-pinnatifid; staminode usually bearded,
    except in one var. of P. richardsonii
     16   Corolla 13-19 mm. long; upper lip cleft more than half its length
      17   Leaves irregularly arranged, many of them ternate, or quaternate, and
       often some of them scattered or alternate; low elevations along the
       Snake R. and its larger tributaries in w. Ida., n.e. Oreg., and s.e.
       Wash. P. TRIPHYLLUS
      17   Leaves all, or nearly all, opposite or subopposite; Missoula, Granite,
       and Beaverhead cos., Mont., to Idaho and Valley cos., Ida., and
       perhaps rarely in Whitman Co., Wash. P. DIPHYLLUS
     16   Corolla mostly 22-32 mm. long (15-20 mm. in the local var. curtiflorus);
      upper lip cleft less than half its length; B.C. to c. Oreg., e. of the Cas-
      cade summits (and in the Columbia gorge), not closely approaching the
      Ida. boundary P. RICHARDSONII
10   Pollen sacs opening throughout their length, or remaining indehiscent at the apex, almost
 always becoming divaricate (or even opposite) after dehiscence (Eupenstemon)
 18   Flowers white or whitish; plants more or less distinctly shrubby at the base
                 P. DEUSTUS
 18   Flowers either not whitish, or the plants scarcely shrubby at the base
  19   Plants not glandular (though sometimes glutinous in P. acuminatus); leaves entire
   or nearly so
   20   Seeds 1.8-4 mm. and capsules 7-13 mm. long; flowers (except in P. nitidus
    and P. acuminatus) large, the corolla 18-38 mm. long, the pollen sacs 1.1-
    3.0 mm. long; palate glabrous; caudex compactly branched, not creeping, of-
    ten surmounting a short taproot
     21   Pollen sacs essentially glabrous, 0.8-1.2 mm. long; corolla 13-21 mm.
      long; plants more or less strongly glaucous (Anularius)
      22   Bearded portion of the staminode mostly 1-1.5 mm. long, the hairs
       about 0.5 mm. long or less, or the staminode rarely glabrous; Wash.,
       Oreg., and w. Ida. (to Gooding Co.) P. ACUMINATUS
      22   Bearded portion of the staminode mostly 2-4 mm. long, the hairs often
       well over 0.5 mm. long, sometimes more than 1 mm. long; Mont. and
       adj. Ida. (Lemhi Co.), and eastward P. NITIDUS
     21   Pollen sacs evidently setose-dentate or dentate-ciliolate along the sutures,
      and sometimes hispidulous over the surface as well, 1.1-3.0 mm. long;
      corolla often well over 21 mm. long; plants often not glaucous (Habroan-
      thus)
      23   Pollen sacs 1.1-1.9 mm. long, straight or arcuate, becoming opposite
       or upwardly divaricate after dehiscence, glabrous except along the
       sutures, or in P. payettensis often obscurely short-hairy toward the

connective, often wholly dehiscent, especially in P. payettensis; corolla 18-28
mm. long; staminode bearded, or occasionally glabrous in P. payettensis
24 Middle and upper leaves relatively broad, especially toward the base, mostly
lance-ovate or broader and more or less clasping, (1)1.5-4 cm. wide; calyx
5-8 mm. long, the segments with narrow to fairly broad, scarcely to evidently scarious-margined base and more or less elongate, acuminate to subcaudate tip; c. Ida., from the Salmon R. axis to the n. margin of the Snake R.
plains, chiefly toward the w. part of the state, extending also to n. Owyhee
Co. and to the Wallowa Mts. of Oreg.            P. PAYETTENSIS
24 Middle and upper leaves narrow, mostly lance-linear to linear-oblong, about
1(1.5) cm. wide or less; calyx 3.5-5.5 mm. long, with broad, more or less
strongly scarious-margined, acutish to mucronate or abruptly short-acuminate segments; Snake R. plains of Ida., from Bingham Co. to Payette Co.,
scarcely entering our range               P. perpulcher A. Nels.
23 Pollen sacs 1.8-3.0 mm. long, tending to be sigmoidally twisted, downwardly divaricate (rarely becoming opposite in P. cyaneus), their proximal (apical) portions remaining indehiscent; corolla 25-38 mm. long
25 Leaves relatively broad, the middle and upper cauline ones commonly 2.5-4
cm. wide and 2-3 times as long; pollen sacs somewhat hispidulous toward the
connective; calyx 6-9 mm. long, the segments fairly narrow and evidently
acuminate, the scarious margins not very broad; staminode bearded; Blue
Mts., from Wallowa Co., Oreg., to Asotin, Garfield, and Columbia cos.,
Wash.                               P. PENNELLIANUS
25 Leaves (except for the wider-leaved forms of P. cyaneus) relatively narrow,
seldom any of them as much as 2.5 cm. wide, the middle and upper cauline
ones mostly 3.5-10 times as long as wide
26 Pollen sacs evidently hispidulous
27 Calyx 4-7 mm. long, the segments very broad and with prominently
erose-scarious margins, inconspicuously or scarcely pointed; staminode bearded; upper Snake R. plains, from Bingham and Butte cos.,
Ida. (also along the n. margin of the plains to Elmore Co.) e. to Madison Co., Mont., and Park Co., Wyo., and n. to Custer and s. Lemhi
cos., Ida.                        P. CYANEUS
27 Calyx 7-11 mm. long, the segments long-acuminate or subcaudate, less
prominently scarious; staminode glabrous; n. Lemhi Co., Ida., to s.
Ravalli Co. and n. Beaverhead Co., Mont.          P. LEMHIENSIS
26 Pollen sacs glabrous, except along the sutures; staminode glabrous or occasionally with a sparse, short beard; calyx 4-8 mm. long, the segments
mostly ovate and acutish to acuminate, with more or less prominently
erose-scarious margins; c. Wash., s. through e. Oreg. to Calif., and e.
to Nev. and n. e. Utah, entering Ida. only near the Oreg. and Nev. borders
                                    P. SPECIOSUS
20 Seeds 0.5-1.5 mm. and capsules 3-7 mm. long; flowers (except in P. globosus) relatively small, the corolla 6-16(18) mm. long, the pollen sacs 0.3-1.0 mm. long; palate bearded; caudex somewhat rhizomelike and tending to creep along the surface of
the ground (Spermunculus in part)
28 Flowers blue or purple, rarely ochroleucous in forms of P. procerus; bracts of
the inflorescence with entire, not evidently scarious margins
29 Flowers relatively very small, the corolla 6-11 mm. long, with narrow, more
or less declined tube mostly 2-3 mm. wide at the mouth; pollen sacs 0.3-0.7
mm. long, wholly dehiscent and eventually explanate
30 Leaves narrow, mostly linear or nearly so, rarely as much as 7 mm. wide,
often arcuate-recurved; calyx 1.5-2.5 mm. long, the segments mucronate
to subtruncate; dry foothills and lowlands from Deschutes Co., Oreg., to
n. Calif.                           P. CINICOLA

        30  Leaves mostly wider; calyx various; plants widespread, mostly occurring in
             moister habitats, or at higher altitudes, or both        P. PROCERUS
    29  Flowers larger, the corolla (10)11-20(22) mm. long, scarcely declined, commonly
        (except in P. laxus) with more expanded, ventricose tube mostly 3-7 mm. wide at
        the mouth; pollen sacs mostly more ovate or elliptic, 0.6-1.2 mm. long
        31  Herbage glaucous throughout; Cascade Range from Mt. Adams to the Three
             Sisters               P. EUGLAUCUS
        31  Herbage not glaucous; mostly not of the Cascades
           32  Pollen sacs opening fully, confluent after dehiscence, the glabrous sutures
               becoming widely separated; corolla mostly 11-15 (18) mm. long
               33  Leaves all cauline, the lower ones strongly reduced, the stem appear-
                    ing relatively densely leafy; calyx 2-3.5 mm. long; corollas tending to
                    be ascending, the tube relatively very narrow, commonly 2-3 mm.
                    wide at the mouth; n. of the Snake R. plains in Blaine, Elmore, and
                    Boise cos., Ida.         P. LAXUS
               33  Leaves both basal and cauline, the basal ones generally well developed
                    and often forming distinct rosettes, the stem appearing less leafy than
                    in P. laxus; calyx 3-7(9) mm. long; corollas generally more spread-
                    ing, the tube more expanded, commonly 3-5 mm. wide at the mouth;
                    widespread, but absent from the range of P. laxus     P. RYDBERGII
           32  Pollen sacs incompletely dehiscent, their distal (free) ends shortly pouched,
               the partition between their proximal (apical) ends remaining intact, the
               mostly setulose-dentate sutures not becoming very widely separated; co-
               rollas mostly 15-20(22) mm. long; c. and n. Ida., extending into Wallowa
               and Baker cos., Oreg.         P. GLOBOSUS
  28  Flowers ochroleucous or light yellow to slightly brownish; bracts of the inflorescence,
      or many of them, evidently scarious-margined and erose
    34  Flowers relatively small, the calyx 3-5 mm. long, the corolla 8-12 mm. long, its
        tube narrow, 2-3.5 mm. wide at the mouth, the pollen sacs mostly 0.4-0.7 mm.
        long; leaves relatively thin and light green, not blackening in drying; widespread
        e. of the Cascade summits, but apparently absent from the range of P. flavescens
                             P. CONFERTUS
    34  Flowers larger, the calyx 5-9 mm. long, the corolla 12-16 mm. long, the tube
        ventricosely expanded, 3.5-6 mm. wide at the mouth, the pollen sacs 0.7-0.9
        mm. long; leaves relatively thick and firm, tending to blacken in drying; Idaho
        Co., Ida., and adj. Ravalli Co., Mont.        P. FLAVESCENS
19  Plants more or less glandular in the inflorescence (or on the corollas), or occasionally
    essentially glabrous in forms of some species that generally have more or less toothed
    leaves
    35  Corolla glandular-puberulent near the mouth within, white, or with pale pink or bluish
        tube and white lobes; plains species, extending into the more easterly intermontane
        valleys of Mont.           P. ALBIDUS
    35  Corolla generally either glabrous within or bearded on the palate, glandular within only
        in the more western, darker-flowered P. gairdneri; corolla sometimes pale, but only
        rarely white (the following species, except P. eriantherus and P. pumilus, belong to
        section Spermunculus)
        36  Ovary and capsule ordinarily glandular-puberulent near the summit; calyx elongate,
            7-13 mm. long, with herbaceous and entire segments; corolla 18-40 mm. long,
            more or less inflated, 6-14 mm. wide at the mouth; staminode more or less ex-
            serted from the mouth of the corolla
           37  Plants essentially glabrous below the inflorescence; staminode bearded only
               near the tip, or glabrous; montane species of the s. Rocky Mt. region, bare-
               ly entering our range in Madison Co., Mont.   P. WHIPPLEANUS
           37  Plants more or less hairy below the inflorescence, at least in part; staminode
               usually bearded for much of its length; foothill and lowland species, wide-
               spread e. of the Cascade Range in our range   P. ERIANTHERUS

36 Ovary and capsule ordinarily glabrous; calyx seldom over 7(8) mm. long; corolla often
    narrower, or shorter, or both; staminode included in most species

    38 Stems arising singly or few together from a short, simple, subterranean caudex; co-
        rolla pale lilac or pale violet, whitish within; pollen sacs 1.0-1.5 mm. long, the su-
        tures not becoming widely separated after dehiscence; plains species, extending into
        the intermontane valleys of Mont.                                   P. GRACILIS

    38 Stems mostly arising from a more or less surficial, branched, woody caudex; corolla
        and anthers diverse, but not combined as in P. gracilis; cordilleran species

        39 Leaves all entire or nearly so

            40 Leaves (except for often being glandular-hairy in the inflorescence) all essen-
                tially glabrous, not at all cinereous (rarely a few of them subcinereous in P.
                aridus, marked by its narrow, basally disposed leaves)

                41 Corolla relatively very small, 8-12 mm. long, the narrow tube mostly 2-
                    3.5 mm. wide at the mouth; pollen sacs subrotund, mostly 0.4-0.6 mm.
                    long

                    42 Leaves narrow, mostly linear or nearly so, up to about 5 mm. wide, the
                        basal ones not much, if at all, developed, not forming rosettes; corol-
                        la pale purplish-blue to white; plants 2.5-7 dm. tall; Cascade Range
                        of Oreg.                                   P. PECKII

                    42 Leaves wider, many or all of them over 5 mm. (to 18 mm.) wide, the
                        basal ones commonly well developed and forming obvious rosettes; co-
                        rolla blue-purple, rarely ochroleucous; plants 1-2.5 dm. tall; Chelan
                        and Okanogan cos., Wash.              P. WASHINGTONENSIS

                41 Corollas (except in forms of P. attenuatus var. palustris from c. Oreg.)
                  either longer, or broader and with evidently ventricose tube, or commonly
                  both; pollen sacs more or less ovate, 0.6-1.2 mm. long

                    43 Leaves all narrow, linear or lance-linear to narrowly oblanceolate, sel-
                      dom any of them as much as 5 mm. wide, the basal ones numerous and
                      well developed, forming prominent rosettes; verticillasters relatively
                      loose and few-flowered; s.w. Mont. (barely extending into adj. Ida.)
                      and n. Wyo.                              P. ARIDUS

                  43 Leaves, or many of them, wider both in shape and measurement; verti-
                    cillasters dense and many-flowered (the entire-leaved extremes of P.
                    albertinus and P. subserratus would be sought here, except for their
                    looser and fewer-flowered verticillasters)

                    44 Plants 1-2.5 dm. tall, with well-developed basal rosettes; basal
                      leaves (petiole included) up to 6 cm. long, with more or less ellip-
                      tic blade; cauline leaves up to 3.5 cm. long; corolla 10-13 mm.
                      long, with evident guidelines in the throat; Wallowa Mts. of Oreg.
                                        P. SPATULATUS

                  44 Plants otherwise, commonly taller, or with larger leaves, or with
                    longer corollas, or often differing in all of these respects; guide-
                    lines obscure or wanting; widespread, but apparently absent from
                    the Wallowa Mts.                      P. ATTENUATUS

        40 Leaves (all, or at least some of the cauline ones) cinereous with numerous very
            short hairs; verticillasters relatively loose and few-flowered

            45 Corolla glandular-hairy near the mouth within; leaves alternate, irregularly
                scattered, or opposite, their bases generally remaining distinctly sepa-
                rated even when opposite; e. of the Cascades in Wash. and Oreg., extend-
                ing e. to w. Valley Co., Ida.              P. GAIRDNERI

            45 Corolla not glandular within, either glabrous or merely bearded on the pal-
                ate; leaves all or mostly opposite, their bases tending to meet around the
                stem, or to be joined by a raised line

                46 Corolla glabrous within; leaves narrow, not over 6 mm. wide

                    47 Cauline leaves not very numerous, mostly 1-5 pairs below the inflo-

rescence; plants less than 2 dm. tall, from a compactly branched woody caudex which sometimes surmounts a short taproot; c. Ida.     P. PUMILUS

  47 Cauline leaves numerous, mostly 6-12 pairs below the inflorescence; plants often taller, the stems arising from a more loosely branched caudex; Owyhee Co., Ida., to Harney and Jefferson cos., Oreg.     P. SEORSUS

46 Corolla bearded on the palate; leaves often more than 6 mm. wide

  48 Leaves all cauline, the lower ones strongly reduced; pollen sacs 0.8-1.1 mm. long; s.w. Mont. and adj. Ida. to n. Nev., n. Utah, and n. Colo.     P. RADICOSUS

  48 Leaves basal and cauline, the basal ones (or those on short, sterile shoots) more or less well developed and persistent, tending to form distinct rosettes; pollen sacs mostly 0.4-0.8 mm. long; widespread between the Cascade-Sierra crest and the continental divide     P. HUMILIS

39 Leaves, or many of them, more or less evidently toothed; verticillasters relatively loose and often few-flowered

  49 Corollas relatively small, mostly 11-16 mm. long; c. Wash. extending barely into adj. Oreg. (Hood River Co.), and n. along the Okanogan R. into s. B.C.

    50 Herbage essentially glabrous below the inflorescence, or the stem inconspicuously hirtellous-puberulent; leaves shallowly, irregularly, and rather inconspicuously toothed, or occasionally all entire; plants mostly 3-8 dm. tall; Yakima, Klickitat, and e. Skamania cos., Wash., and Hood River Co., Oreg.     P. SUBSERRATUS

    50 Herbage below the inflorescence glandular-hirsute to merely hirtellous-puberulent, less often essentially glabrous; leaves mostly sharply and conspicuously toothed; plants mostly 1-4 dm. tall; along the e. side of the Cascade Range (Kittitas Co., northward) in Wash. and adj. B.C., extending e. to Adams and Franklin cos., Wash.     P. PRUINOSUS

  49 Corollas larger, mostly (13)15-23 mm. long; plants of either more eastern or more western distribution

    51 Plants relatively robust, (3)4-10 dm. tall, with sharply toothed, mostly relatively broad leaves, the larger cauline ones mostly (0.8)1.5-4 cm. wide

      52 Plants occurring w. of the Cascade summits; stem spreading-hirsute, often shortly so; leaves hairy like the stem, at least along the midrib beneath     P. OVATUS

      52 Plants occurring from n.w. Mont. (wholly w. of the continental divide) to n. and c. Ida., barely extending into e. Wash. and n.e. Oreg., usually glabrous below the inflorescence, the stem occasionally and the leaves rarely spreading-hirsute     P. WILCOXII

    51 Plants smaller, commonly 1-4 dm. tall, with narrow, mostly rather irregularly and obscurely few-toothed leaves, the cauline ones up to about 1.5 cm. wide

      53 Cauline leaves, or some of them, finely hirtellous-puberulent, all under 1 cm. wide; Wallowa Co., Oreg., and adj. Nez Perce and Idaho cos., Ida.     P. ELEGANTULUS

      53 Cauline leaves, like the basal ones, glabrous, not at all hirtellous-puberulent, often over 1 cm. wide; w. Mont. (both sides of the continental divide) to s.w. Alta. and s.e. B.C., s. to c. Ida.; also in Owyhee Co., Ida.     P. ALBERTINUS

Penstemon acuminatus Dougl. ex Lindl. Bot. Reg. 15:pl. 1285. 1829. (Douglas, barren sandy plains of the Columbia)

Perennial, commonly with several rather stout stems arising from a short, branched caudex which may surmount a quickly deliquescent taproot, 1.5-6 dm. tall, wholly glabrous (except the staminode), the herbage glaucous and sometimes glutinous; leaves thick and firm, entire, the basal ones tufted, oblanceolate or a little broader, up to about 15 cm. long and 2 cm. wide; cauline leaves (except the lower) mostly sessile and clasping, the upper progressively shorter and broader, the bracts of the inflorescence often broader than long; inflorescence of rather numerous, slightly separated, rather dense verticillasters, the bracts subtending the lower verticillasters often prominent and leafy, the upper ones more reduced;

calyx 5-9 mm. long, the segments lance-attenuate to lance-ovate, scarcely scarious-margined, entire; corolla bright blue, mostly 14-21 mm. long, the tube expanded distally; pollen sacs essentially glabrous, dehiscent throughout and becoming opposite, but not at all explanate, 0.8-1.2 mm. long; staminode scarcely exserted, not much expanded distally, yellow-bearded for about 1-1.5 mm. toward the tip (or even glabrous), the hairs mostly about 0.5 mm. long or less; capsule 7-12 mm. long, excluding the conspicuous, slender beak which is up to 5 mm. long; seeds 1.8-3.5 mm. long. N=8.

Dry, open, commonly sandy places at low elevations, often on dunes; Grant Co., Wash., to Klickitat and Walla Walla cos., and across the Columbia R. in adj. n. Oreg.; Snake R. plains from Gooding and Owyhee cos., Ida., to Washington Co., extending w. to Harney Co., Oreg.; apparently absent from the Blue Mt. region. Apr.-June.

Penstemon albertinus Greene, Leafl. 1:167. 1906. (Macoun 11865, Sheep Mt., Alta.)
P. caelestinus Pennell, Not. Nat. Acad. Phila. 95:1. 1942. (Pennell 20536, Little Blackfoot R., 9 miles n.e. of Garrison, Powell Co., Mont.)

Stems clustered from a surficial branched, woody caudex, 1-4 dm. tall; plants evidently to obscurely glandular-hirsute (rarely quite glabrous) in the inflorescence, otherwise essentially glabrous, or the stem finely hirtellous-puberulent; leaves rather thick and firm, generally some of them, especially the cauline ones, with a few small, callous teeth; basal leaves well developed, tufted, with oblanceolate to more often elliptic or ovate blade mostly 1.5-5 cm. long and 0.4-2 cm. wide, longer or shorter than the petiole; cauline leaves well developed, mostly sessile by a broad, or less often narrow, base, mostly 2-5 cm. long and 3-15 mm. wide; inflorescence of several often confluent verticillasters, these less loose and open than in P. wilcoxii; calyx 3-5 mm. long, the segments broadly lanceolate to ovate, obscurely to evidently (but narrowly) scarious-margined and sometimes erose; corolla blue (pink), with paler throat and evident guidelines, (13)15-20 mm. long, the tube 4-7 mm. wide at the mouth; palate bearded; pollen sacs 0.6-0.9 mm. long, dehiscent essentially throughout and becoming opposite, glabrous except for the often minutely setulose-dentate sutures; staminode scarcely or not at all expanded toward the bearded, recurved tip; capsule about 5 mm. and seeds 0.8-1.5 mm. long. N=8.

Dry, open, commonly rocky places, from the foothills to fairly high elevations in the mountains; s.e. B.C. and extreme s.w. Alta. to w. Mont. (both sides of the continental divide) and c. Ida.; also in the mountains of Owyhee Co., Ida.; apparently not reaching Oreg. or Wash. May-July.

Replaced in the s. Rocky Mts. e. of the continental divide by the morphologically very similar, but geographically well removed, P. virens Pennell. P. albertinus shares much of its range with P. wilcoxii, and the two often hybridize. Where I have seen them hybridizing in Mont., P. albertinus was distinguished by its smaller size, thicker and less evidently toothed leaves, slightly smaller corollas, and less open inflorescence.

Penstemon albidus Nutt. Gen. Pl. 2:53. 1818.
Chelone albida Spreng. Syst. Veg. 2:813. 1825. (Nuttall, "plains of the Missouri, common, from the confluence of the river Platte to the Mountains")

Stems few or solitary from a rather short, simple or few-branched, largely subterranean caudex; plants 1-4 dm. tall, glandular-hirsute in the inflorescence, otherwise very finely cinereous-puberulent, or the lower leaves rarely subglabrate; leaves entire or with a few scattered, small teeth, the basal ones well developed, up to 10 cm. long and 1.5 cm. wide, or sometimes larger, petiolate, with oblanceolate or elliptic blade; cauline leaves mostly sessile and more lanceolate, but often not much reduced; inflorescence of several verticillasters; calyx 5-8 mm. long, the segments lanceolate or ovate, obscurely, if at all, scarious-margined, entire; corolla white, or with pale pinkish or bluish tube and white lobes, glandular-hirsute outside, densely glandular-puberulent at the throat and base of the lobes within, 1.5-2.5 cm. long, the tube flaring distally, the limb conspicuously expanded and spreading; pollen sacs dark, 0.9-1.3 mm. long, dehiscent throughout, becoming opposite and eventually explanate; sterile filament bearded for much of its length, scarcely expanded upward; capsule 7-12 mm. long; seeds 2-3 mm. long.

Open plains and hillsides, often in grasslands; Great Plains, s. Man. to Alta., s. to Okla., Tex., and n. e. N. M., extending into the more easterly intermontane valleys of Mont. June-July.

**Penstemon aridus** Rydb. Mem. N. Y. Bot. Gard. 1:348. 1900. (<u>Rydberg</u> & <u>Bessey</u> <u>4920</u>, Spanish Basin, Mont.)

Plants tufted from a compact, branching caudex, 0.5-2 dm. tall, glandular-hairy in the inflorescence; stems slender, finely scaberulous-puberulent or occasionally glabrous; leaves essentially glabrous, or a few of them obscurely scaberulous-puberulent (the margins and petioles sometimes more evidently scabrous-ciliolate), all entire or nearly so, narrow, mostly 1-5 cm. long and 1-5 mm. wide, the numerous, tufted, basal ones linear to linear-oblanceolate, the cauline ones linear to narrowly linear-oblong or lance-linear; inflorescence of several rather loose, few-flowered verticillasters; calyx 2.5-4 mm. long, the segments broadly ovate, scarious-margined and often erose, abruptly pointed, often anthocyanic; corolla blue-purple, glandular-hairy, 12-18 mm. long, the tube 3-6 mm. wide at the mouth; palate bearded, as also the slightly expanded tip of the staminode; pollen sacs glabrous, dehiscent throughout, becoming opposite and often explanate, 0.6-0.9 mm. long; capsule 5-6 mm. and seeds 1-1.5 mm. long.

Dry, open, often rocky places, from the intermontane valleys and plains to moderate elevations in the mountains; s.w. Mont., barely extending into adj. Ida. (Clark and Lemhi cos.); Big Horn Mts. region of Wyo. June-July.

<u>Penstemon</u> aridus is a well-marked species which rarely gives any trouble in identification, although it may perhaps rarely hybridize with <u>P</u>. <u>humilis</u> and <u>P</u>. <u>albertinus</u>.

**Penstemon attenuatus** Dougl. ex Lindl. Bot. Reg. 15: pl. 1295. 1830.

<u>P</u>. <u>digitalis</u> var. <u>attenuatus</u> Trautv. Bull. Acad. St. Pétersb. 5:345. 1839. <u>P</u>. <u>confertus</u> var. <u>attenuatus</u> M. E. Jones, Contr. West. Bot. 12:62. 1908. (<u>Douglas</u>, "mountains of Lewis and Clark's River")

PENSTEMON ATTENUATUS var. PSEUDOPROCERUS (Rydb.) Cronq. hoc loc. <u>P</u>. <u>pseudoprocerus</u> Rydb. Mem. N. Y. Bot. Gard. 1:346. 1900. <u>P</u>. <u>procerus</u> var. <u>pseudoprocerus</u> Nels. in Coult. & Nels. New Man. Bot. Rocky Mts. 444. 1909. <u>P</u>. <u>attenuatus</u> ssp. <u>pseudoprocerus</u> Keck, Am. Midl. Nat. 33:169. 1945. (<u>Rydberg</u> & <u>Bessey</u> <u>4919</u>, Bridger Mts., Mont.)

<u>P</u>. <u>pseudohumilis</u> Rydb. Mem. N. Y. Bot. Gard. 1:347. 1900. (<u>Rydberg</u> & <u>Bessey</u> <u>4915</u>, Mt. Chauvet, Fremont Co., Ida.) = var. pseudoprocerus.

PENSTEMON ATTENUATUS var. MILITARIS (Greene) Cronq. hoc loc. <u>P</u>. <u>militaris</u> Greene, Leafl. 1:166. 1906. <u>P</u>. <u>attenuatus</u> ssp. <u>militaris</u> Keck, Am. Midl. Nat. 33:171. 1945. (<u>Henderson</u> <u>3395</u>, Soldier Mts., Ida.)

<u>P</u>. <u>propinquus</u> Greene, Leafl. 1:166. 1906. (<u>Coville</u> <u>549</u>, Strawberry Butte, Grant Co., Oreg.) = var. palustris.

<u>P</u>. <u>veronicaefolius</u> Greene, Leafl. 1:167. 1906. (<u>Sandberg</u> <u>245</u>, Lake Waha, Ida.) A form of var. attenuatus with some of the leaves obscurely toothed.

<u>P</u>. <u>nelsonae</u> Keck & Thomps. Rhodora 37:419. 1935. (<u>Thompson</u> <u>10617</u>, base of Mt. Angeles, Clallam Co., Wash.) A form of var. attenuatus with obscurely toothed leaves, now believed to be merely introduced at the type station.

<u>P</u>. <u>assurgens</u> Keck, Am. Midl. Nat. 23:609. 1940. (<u>Davis</u> <u>441</u>, near the Craters of the Moon, Butte Co., Ida.) = var. militaris.

PENSTEMON ATTENUATUS var. PALUSTRIS (Pennell) Cronq. hoc loc. <u>P</u>. <u>palustris</u> Pennell, Not. Nat. Acad. Phila. 71:8. 1941. <u>P</u>. <u>attenuatus</u> ssp. <u>palustris</u> Keck, Am. Midl. Nat. 33:171. 1945. (<u>Ferris</u> & <u>Duthie</u> <u>703</u>, Prairie City, Grant Co., Oreg.)

<u>P</u>. <u>attenuatus</u> ssp. <u>hyacinthinus</u> Pennell, Not. Nat. Acad. Phila. 71:8. 1941. (<u>Pennell</u> <u>20879</u>, n. e. of Elk City, Idaho Co., Ida.) = var. attenuatus.

<u>P</u>. <u>attenuatus</u> ssp. <u>hyacinthinus</u> f. <u>multicolor</u> Pennell, Not. Nat. Acad. Phila. 71:8. 1941. (<u>Pennell</u> <u>20926</u>, Brown Creek, near Lolo Creek, Clearwater Co., Ida.) = var. attenuatus.

<u>P</u>. <u>cephalanthus</u> Pennell, Not. Nat. Acad. Phila. 95:3. 1942. (<u>Pennell</u> et al. <u>23679</u>, s.w. of Anaconda, Deerlodge Co., Mont.) = var. pseudoprocerus.

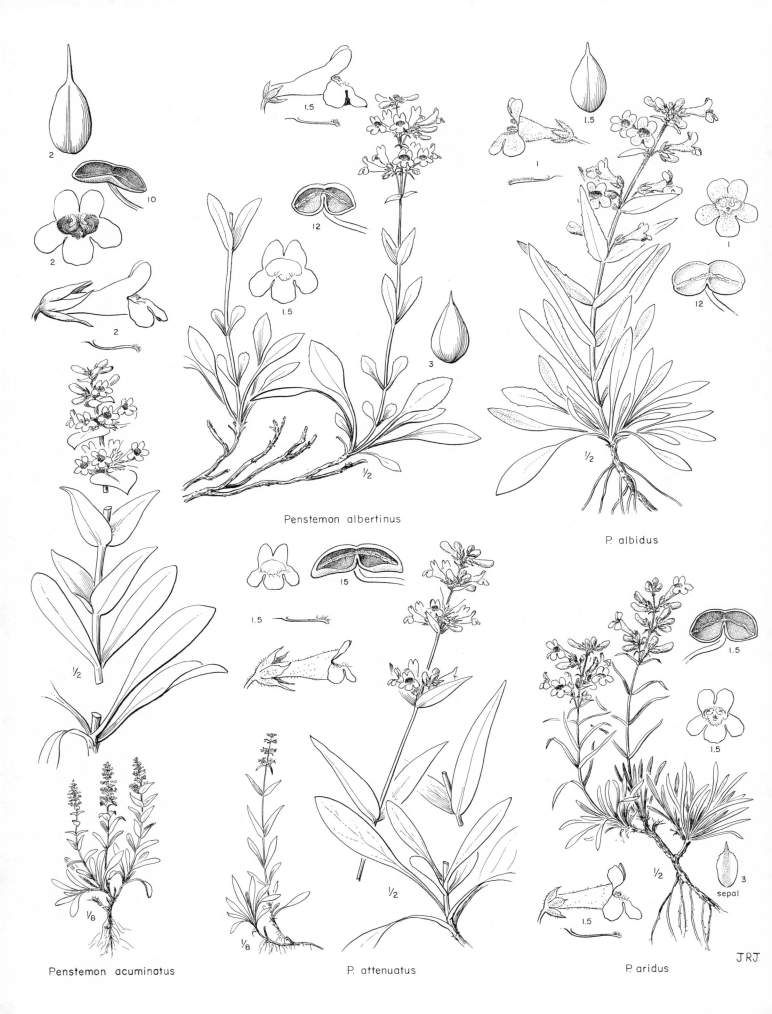

Penstemon albertinus

P. albidus

Penstemon acuminatus

P. attenuatus

P. aridus

JRJ

Plants tufted from a loose or compact, surficial, woody rhizome-caudex, 1-7(9) dm. tall, wholly glabrous below the glandular-hairy inflorescence, or the stem finely hirtellous-puberulent; leaves deep green, entire or nearly so, the basal ones generally well developed, petiolate, up to 17 cm. long and 4 cm. wide, the cauline ones mostly sessile, gradually or abruptly reduced; inflorescence of (1)2-several rather dense verticillasters; calyx 4-7 mm. long, the segments lanceolate to ovate or obovate, with narrow or broad and erose scarious margin; corolla glandular-hairy externally, blue or purple to pale yellow or nearly white, mostly 12-20(22) mm. long and with the tube distinctly expanded distally, sometimes shorter and more narrowly tubular in var. palustris; palate bearded, as also the expanded tip of the staminode; pollen sacs glabrous (except in var. militaris), ovate, 0.7-1.2 mm. long, becoming opposite, but scarcely explanate; capsules 6-8 mm. and seeds about 1 mm. long.

Drier meadows and moist, open or wooded slopes in the mountains and foothills, the var. palustris sometimes in wet meadows; w. Mont. and n.c. Wyo. to c. Wash., e. Oreg., and s. Ida. June-Aug.

Penstemon attenuatus var. attenuatus and var. pseudoprocerus are known from several counts to be hexaploids on a base of x=8, and it is assumed that the other varieties are likewise hexaploid. The species is believed to be of alloploid origin, and although the several varieties are now wholly confluent and evidently interfertile, it seems likely that they are not all of identical ancestry. Thus, although P. albertinus or something similar to it may well be a common ancestor to all of the varieties, it seems likely that P. confertus is involved in the origin of var. attenuatus and var. palustris, but not var. militaris and var. pseudoprocerus, that P. globosus is involved in the origin of var. militaris only, that P. procerus and/or P. rydbergii is involved in the origin of var. pseudoprocerus, and that P. procerus may possibly be involved also in the origin of var. palustris. The several varieties may be distinguished, with some difficulty, as follows:

1 Corollas relatively small, mostly (7)10-14 mm. long, the tube sometimes not much expanded; habit, anthers, and corolla-color of var. attenuatus; Blue Mt. region from s. Morrow Co. to Baker and Grant cos., Oreg., often in wet meadows, sometimes on drier slopes, not at high elevations                                    var. palustris (Pennell) Cronq.
1 Corollas larger, mostly 14-20 mm. long (or sometimes less than 14 mm. long in var. pseudoprocerus), the tube evidently expanded distally
    2 Pollen sacs incompletely dehiscent, their distal (free) ends shortly pouched, the partition between their proximal ends remaining intact, the sutures after dehiscence mostly setulose-ciliate and not becoming very widely separated; corolla deep blue; habit mostly of var. attenuatus, varying eastward to often nearly as in var. pseudoprocerus; moderate to rather high elevations in the mountains of c. Ida., from the Salmon R. axis on the n. to the Snake R. plains on the s., barely extending into adj. Mont.; also in the mountains of Cassia Co., Ida.                                    var. militaris (Greene) Cronq.
    2 Pollen sacs dehiscent throughout and eventually confluent, the glabrous sutures becoming widely separated
        3 Plants typically robust, 3-9 dm. tall; calyx lobes mostly lanceolate and narrowly scarious; corolla 14-20 mm. long, various shades of blue, purple, or lavender to pale yellow or nearly white, the several colors often occurring in the same colony; drier meadows and moist, open or wooded slopes in the foothills and at moderate elevations in the mountains, from n.w. Mont. (w. of the continental divide) across n. Ida. (chiefly n. of the Salmon R. axis, but also extending into w. Valley Co.) into the e. edge of Wash. (s. of the Spokane R.) and the Blue Mt. region of n.e. Oreg.; also along the e. side of the Cascade Range in Wash.                var. attenuatus
        3 Plants smaller, 1-4 dm. tall; calyx lobes mostly broader and more prominently scarious; corolla 12-18 mm. long, deep blue; drier meadows and open, often rocky slopes at moderate to rather high elevations in the mountains, e. (or barely w.) of the continental divide in Mont., w. to the Lost River Mts. in Ida., s. to w.c. Wyo. and adj. Ida.                                    var. pseudoprocerus (Rydb.) Cronq.

Penstemon barrettiae Gray, Syn. Fl. 2nd ed. 2¹:440. 1886. (Barrett, mountains of Hood R. Oreg., near its confluence with the Columbia)

Plants 2-4 dm. tall, the lower part shrubby and much branched; herbage and inflorescence glabrous and glaucous, or the sepals finely and inconspicuously glandular; leaves firm, irregularly serrulate or entire, the larger ones borne toward the base of the season's growth, on short, sterile shoots as well as on the fertile ones, up to 8 cm. long and 2.5 cm. wide, tapering to a narrow, sessile or shortly subpetiolar base; leaves of the flowering shoots (above the enlarged clusters) broad, sessile and clasping, mostly 1.5-3.5 cm. long and 0.8-2 cm. wide; inflorescence racemose or nearly so, the axillary peduncles simple and uniflorous, or some of them branched and 2-flowered; calyx 5-7 mm. long, the segments more or less ovate, rather thin and sometimes a little erose; corolla lilac or rose-purple, 33-38 mm. long, about 1 cm. wide at the mouth, glabrous outside, long-hairy near the base of the lower lip within, evidently keeled on the back and prominently 2-ridged on the lower side within; anthers densely long-woolly, the pollen sacs wholly dehiscent, becoming opposite and explanate; capsules narrow, nearly 1 cm. long; seeds irregularly compressed-prismatic, very narrowly wing-margined, about 1.5 mm. long. N=8.

Dry, rocky places at low altitudes near the e. end of the Columbia gorge, sometimes with sagebrush; Klickitat Co., Wash., and Hood River and Wasco cos., Oreg. Apr.-May.

A sharply marked local species.

Penstemon cardwellii Howell, Fl. N. W. Am. 510. 1901.

   P. fruticosus ssp. cardwellii Piper, Contr. U.S. Nat. Herb. 11:499. 1906. P. fruticosus
   var. cardwellii Krautter, Contr. Bot. Lab. U. Pa. 3:100. 1908. (Howell, Cascade Mts.
   near the base of Mt. Hood, Oreg.)

Low shrub 1-3 dm. tall, the stems often ascending rather than erect, and sometimes rooting near the base; stems of the season finely puberulent or subglabrous; leaves glabrous, the larger ones tending to be crowded toward the base of the season's growth or on short, sterile shoots, evidently serrulate to subentire, short-petiolate, with more or less elliptic blade 1.5-3.5 cm. long and 6-14 mm. wide, mostly (1.8)2-3.5 times as long as wide, rounded or obtuse at the tip; leaves of the flowering shoots mostly sessile, otherwise similar to the others or more often smaller and entire, seldom crowded; inflorescence sparsely to strongly glandular-hairy, rather crowded and few-flowered, essentially racemose, the simple pedicels opposite and axillary; calyx 8-12 mm. long, the segments thin, lanceolate or lance-ovate; corolla bright purple to deep blue-violet, (25)30-38 mm. long, about 1 cm. wide at the mouth, keeled on the back, glabrous outside, prominently long-white-hairy near the base of the lower lip within, especially along the 2 prominent, ventral ridges; anthers conspicuously long-woolly, the pollen sacs becoming opposite and peltately explanate; staminode slender, shorter than the fertile filaments, long-bearded at least toward the tip. N=8.

Open or wooded summits and slopes at moderate elevations in the mountains; Klamath region of s.w. Oreg., extending n. in the Cascade Range (w. of the summit) to Skamania Co., Wash., and in the Coast Ranges at least to Tillamook Co., Oreg. Late May-July.

The characters of P. cardwellii are such as to suggest that it may have arisen through hybridization between P. fruticosus and the more southern P. newberryi Gray.

Penstemon cinicola Keck, Carn. Inst. Wash. Pub. 520:294. 1940. (Keck & Clausen 3690, La-
   pine, Deschutes Co., Oreg.)

   P. truncatus Pennell, Not. Nat. Acad. Phila. 71:5. 1941. (Pennell 15550, Davis Lake,
   Klamath Co., Oreg.)

   P. truncatus f. puberulus Pennell, Not. Nat. Acad. Phila. 71:5. 1941. (Pennell 15562, Lon-
   roth, Klamath Co., Oreg.)

Plants tufted from a surficial, woody rhizome-caudex, 1.5-4 dm. tall, slender-stemmed, wholly glabrous, or sometimes finely puberulent throughout; leaves entire, relatively narrow, mostly linear or nearly so, up to 6 cm. long and 7 mm. wide, often arcuate-recurved, frequently crowded near the base, but scarcely forming a distinct rosette; inflorescence of several remote or approximate verticillasters, these mostly looser and fewer-flowered than in P. procerus; calyx 1.5-2.5 mm. long, the segments scarious-margined and tending to be erose, shortly mucronate to obtuse or subtruncate; corolla blue-purple, 6-10 mm. long, with spreading, obscurely bilabiate limb, the palate and the somewhat expanded tip of the staminode bearded; pollen sacs glabrous, subrotund, about 0.5 mm. long or less, becoming opposite and more or less explanate; capsules 3-4 mm. long and seeds nearly 1 mm. long. N=8, 16.

Mostly in dry, open places, often in volcanic sands, in the foothills and lowlands near the e. base of the Cascades; Deschutes Co., Oreg., to Lassen Co., Calif. June-July.

Penstemon cinicola sometimes approaches P. procerus var. brachyanthus in habitat and morphological characters, but seems more distinct than the varieties of P. procerus are from each other.

Penstemon confertus Dougl. in Lindl. Bot. Reg. 15: pl. 1260. 1829.
  P. confertus var. ochroleucus Trautv. Bull. Acad. St. Pétersb. 5:344. 1839. (Douglas, "between Salmon River and the Kettle Falls in the Columbia, lat. 48 degrees north")
Plants more or less tufted from a loose or compact, surficial, woody rhizome-caudex, 2-5 (7) dm. tall, glabrous throughout, or the stem finely hirtellous-puberulent, especially above; leaves entire, not blackening in drying, the lower ones petiolate, sometimes poorly developed, or often up to 15 cm. long and 2.5 cm. wide and forming a rosette; principal cauline leaves sessile, mostly lanceolate to lance-ovate or lance-oblong, up to 10 cm. long and 2.5 cm. wide; inflorescence of 2-10 fairly compact verticillasters, the lower generally remote, the upper commonly more or less confluent; bracts of the inflorescence, or many of them, with broadly scarious, erose margins, often resembling the calyx segments; calyx mostly 3-5 mm. long, the segments with broad, scarious, erose margins and generally with an abruptly slender tip; corolla ochroleucous or light yellow, often declined, mostly 8-12 mm. long, the tube narrow, mostly 2-3.5 mm. wide at the mouth, the short limb not very strongly bilabiate, the palate and staminode bearded; pollen sacs purple, glabrous, 0.4-0.7 mm. long, wholly dehiscent, becoming opposite and more or less explanate; capsules 4-5 mm. long and seeds scarcely 1 mm. long. N=16.

Mostly in fairly moist, open or wooded places, often in meadows or along streams, from the foothills and adj. lowlands to moderate elevations in the mountains, w. Mont. and s.w. Alta. to s.e. B.C. and n.e. Oreg. (Blue Mts. of Union and Umatilla cos.; along the Columbia R. in w. Sherman Co.) and to the foot of the Cascade Range in Wash. (extending well into the Wenatchee Mts.). May-Aug.

Penstemon confertus is obviously related to P. procerus, but the two appear to be sharply distinct.

Penstemon cusickii Gray, Proc. Am. Acad. 16:106. 1880. (Cusick, Powder R. or Eagle Creek, Baker Co., Oreg., June, 1880)
  P. macbridei A. Nels. Bot. Gaz. 52:272. 1911. (Macbride 105, Big Willow, Canyon Co., Ida.)
Perennial from a short, quickly deliquescent taproot, 1.5-4.5 dm. tall, tending to be shrubby at the base, the stems more or less numerous, rather slender and brittle; herbage and inflorescence finely hirtellous-puberulent, not glandular; leaves all cauline, the lowermost ones reduced, the others rather numerous and narrow, entire, sessile or nearly so, mostly 2-8 cm. long and 2-8(14) mm. wide; inflorescence a narrow, mixed panicle or nearly a raceme, leafy-bracteate below; calyx 4-6 mm. long, the segments ovate, tending to be narrowly scarious-margined and often a little erose; corolla glabrous inside and out, lavender to violet-purple, 15-24 mm. long, 5-9 mm. wide at the mouth; anthers permanently horseshoe-shaped, 1.3-1.9 mm. long, the pollen sacs dehiscent across their confluent apices, the lower portion remaining saccate and indehiscent, dentate-ciliolate along the sutures and tending to be obscurely hispidulous near the connective; staminode glabrous, the somewhat expanded, flattened tip often shortly exserted; capsules about 7 mm. long. N=8.

Sagebrush slopes and plains, especially in basaltic regions; w. Blaine Co. (Camas Prairie), Ida., to Malheur, Baker, and Wheeler cos., Oreg. May-early July.

There is sometimes a strong superficial resemblance between P. cusickii and P. seorsus, but these species of similar habit and habitat are sharply distinguished by the characters of the anthers and staminodes.

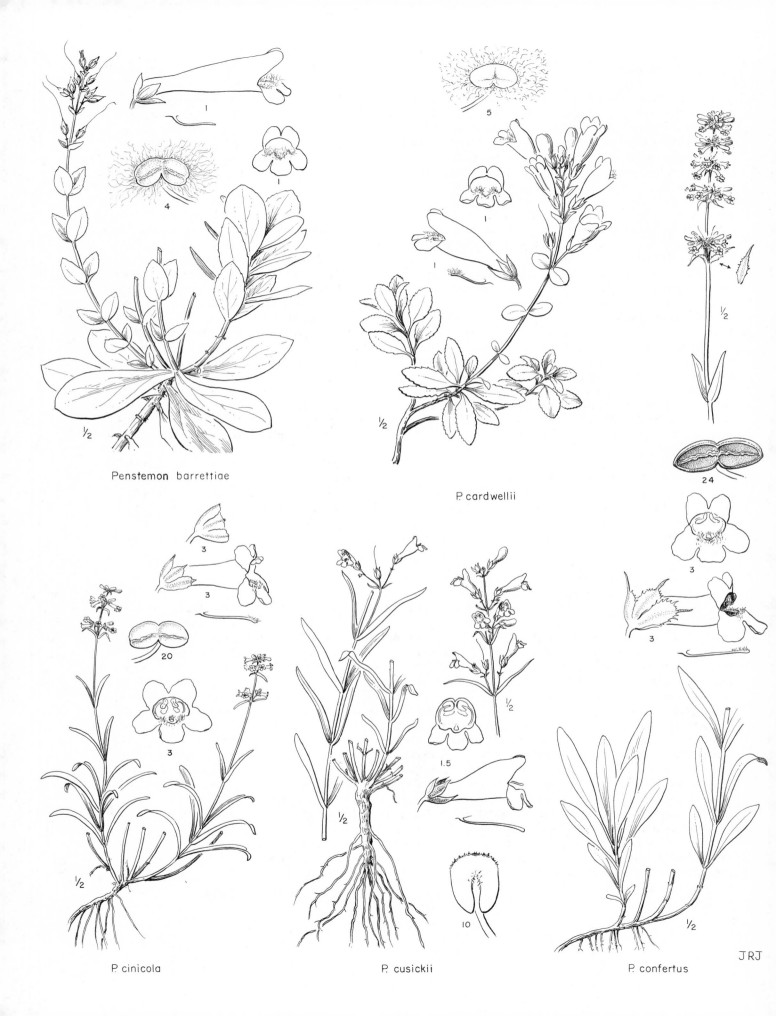

Penstemon barrettiae

P. cardwellii

P. cinicola

P. cusickii

P. confertus

JRJ

Penstemon cyaneus Pennell, Contr. U.S. Nat. Herb. 20:351. 1920. (Pennell 6046, n. of Ashton, Fremont Co., Ida.)

Perennial, 3-7 dm. tall, generally with several stout stems from a compact, branched caudex which tends to surmount a short, quickly deliquescent taproot; herbage and inflorescence essentially glabrous and sometimes a little glaucous; leaves entire, thick and firm, tending to darken in drying, the basal ones clustered, up to 16 cm. long and 2.5 cm. wide, with petiolate, oblanceolate to narrowly elliptic blade; cauline leaves, except the lowermost, mostly sessile, the middle and upper mostly lanceolate and often clasping, sometimes lance-ovate, up to 11 cm. long and 3 cm. wide; inflorescence of several, or rather numerous, loose verticillasters, more or less strongly secund in life; calyx 4-7 mm. long, the segments very broad and with prominently erose-scarious margins, inconspicuously or scarcely pointed; corolla bright blue, 25-35 mm. long, about 1 cm. wide at the mouth; pollen sacs 1.8-3.0 mm. long, divaricate or rarely becoming opposite, tending to be sigmoidally twisted, evidently dentate-ciliolate along the sutures and hispidulous on the surface, a short, proximal (apical) portion remaining indehiscent; staminode sparsely to moderately short-bearded toward the scarcely to evidently expanded tip; capsule 10-13 mm. and seeds 2-3 mm. long.

Open, rather dry places from the high plains to moderate elevations in the mountains, often with sagebrush; upper Snake R. plains from Bingham and Butte cos. (and also along the n. margin of the plains to Elmore Co.), Ida., e. to Park Co., Wyo., and Madison Co., Mont., and n. to Custer and s. Lemhi cos., Ida. June-Aug.

Penstemon davidsonii Greene, Pitt. 2:241. 1892.

P. menziesii davidsonii Piper, Mazama 2:99. 1901. P. menziesii ssp. davidsonii Piper, Contr. U.S. Nat. Herb. 11:499. 1906. P. menziesii var. davidsonii Jeps. Man. Fl. Pl. Calif. 908. 1925 (where attributed to Piper). P. menziesii f. davidsonii G. N. Jones, U. Wash. Pub. Biol. 5:226. 1936. (Davidson, Mt. Conness, Calif.)

PENSTEMON DAVIDSONII var. MENZIESII (Keck) Cronq. hoc loc. P. menziesii Hook. Fl. Bor. Am. 2:98. 1838, in greater part, excl. synonym. P. davidsonii ssp. menziesii Keck, Britt. 8:247. 1957. (Menzies, "Nutka," Vancouver I.) The name P. menziesii Hook. is illegitimate under the Rules of Nomenclature, since, as defined by Hooker, it included the type of the earlier Gerardia fruticosa Pursh and he should therefore have adopted the epithet fruticosus for his species.

P. menziesii ssp. thompsonii Pennell & Keck in Abrams, Ill. Fl. Pac. St. 3:768. 1951. (Thompson 6614, head of Beverly Creek, Chelan Co., Wash.) A robust form of var. davidsonii. Similar robust forms, chiefly from relatively low elevations, are found at least as far s. as s. Oreg.

Suffruticose perennial, with or without an evident taproot, forming dense mats (with creeping woody stems) on the surface of the substrate, and with scattered, erect flowering stems mostly 0.5-1(1.5) dm. tall, these finely spreading-hirtellous to strigulose-puberulent, sometimes in lines; mat leaves thick and firm, glabrous but scarcely glaucous, entire or serrulate, short-petiolate, the blade mostly 0.5-1.5(2) cm. long and 1-2.5 times as long as wide, the tip rounded to obtuse or somewhat acutish; leaves of the flowering shoots (above the mat) few and small, often bractlike, seldom as much as 1 cm. long; inflorescence a compact, few-flowered raceme, the bracts, pedicels, and main axis somewhat glandular-hairy, though mostly less strongly so than in P. ellipticus, the calyx inconspicuously so or glabrous; calyx mostly 7-10 mm. long, the thin segments lanceolate to narrowly ovate; corolla blue-lavender to purple-violet, 2-3.5 cm. long, keeled along the back, glabrous outside, sparsely or moderately hairy near the base of the lower lip within, especially along the two, prominent, ventral ridges; anthers densely long-woolly, the pollen sacs wholly dehiscent and eventually explanate; staminode slender, shorter than the fertile filaments, long-bearded toward the tip; capsule 8-10 mm. long; seeds irregularly angled, very narrowly winged, about 1.5 mm. long. N=8.

Ledges and among rocks, sometimes on talus, from moderate to high elevations in the mountains; Cascade-Sierra region from s. B.C. to Calif., and in the Olympic Mts. of Wash. and on s. Vancouver I. June-Aug.

Two varieties may be distinguished:

1 Leaves obscurely to evidently serrulate, tending to be broadest near, or even below, the

middle, and sometimes acutish; plants averaging smaller in all respects than var. davidsonii; s.B.C. and n. Wash., s. in the Cascades to Lewis Co., Wash., and occasionally to
Marion Co., Oreg.                                                                var. menziesii (Keck) Cronq.
1 Leaves entire, tending to be broadest above the middle, and more consistently rounded or
   obtuse; the only form of the species in Calif. and most of Oreg.; Wash. extending n. in the
   Cascade region to Okanogan Co. and passing freely into var. menziesii (as on Mt. Rainier)
                                                                                 var. davidsonii

Penstemon deustus Dougl. ex Lindl. Bot. Reg. 16: pl. 1318. 1830. (Douglas, scorched, rocky
   plains in the interior of n.w. Am.; one collection from "Lewis and Clark's River between the
   forks and the confluence with the Columbia")
PENSTEMON DEUSTUS var. HETERANDER (T. & G.) Cronq. hoc loc. P. heterander T. &
   G. Pac. R.R. Rep. 2²:123. 1857. P. deustus ssp. heterander Pennell & Keck, Am. Midl.
   Nat. 23:603. 1940.(Snyder, Sierra Nevada, Calif., June 30, 1854)
PENSTEMON DEUSTUS var. VARIABILIS (Suksd.) Cronq. hoc loc. P. variabilis Suksd.
   Deuts. Bot. Monats. 18:153. 1900. P. deustus ssp. variabilis Pennell & Keck, Am. Midl.
   Nat. 23:602. 1940. (Suksdorf 999, e. of the Klickitat R., Wash.)
   P. paniculatus Howell, Fl. N.W. Am. 513. 1901. (Howell, between the Klickitat Valley and
   the Columbia R., opposite The Dalles, Oreg.) = var. variabilis.
   Plants 2-6 dm. tall, more or less woody toward the much-branched base, the stems of the
season mostly simple and erect; plants usually more or less glandular in the inflorescence, at
least on the corollas, otherwise glabrous to finely puberulent or somewhat glandular; leaves
opposite to sometimes ternate, quaternate, or irregularly scattered, more or less sharply
toothed to sometimes entire, those on the short, sterile shoots commonly oblanceolate to obovate or elliptic and short-petiolate, those on the flowering stems more often sessile or nearly
so and often broad-based, up to 6 cm. long and 2.5 cm. wide; inflorescence of several or rather numerous, remote or subapproximate verticillasters, often rather loose; calyx 2.5-6 mm.
long, the segments lanceolate to narrowly ovate, sometimes narrowly scarious-margined, but
not erose; corolla commonly dull whitish with some purplish guide lines within, sometimes
faintly ochroleucous or washed with lavender, 8-20 mm. long, the tube narrow or ventricosely
expanded; pollen sacs 0.5-0.9 mm. long, glabrous, dehiscent throughout, becoming more or
less divaricate and often explanate; staminode glabrous or bearded, scarcely, or not at all,
expanded toward the tip; capsules 3-5 mm. and seeds about 1 mm. long. N=8, 16.
   Dry, open, often rocky places, from the lowlands to moderate or occasionally high elevations in the mountains; c. Wash. to w. Mont. and n. Wyo., s. to Calif., Nev., and Utah. May-
July.
   The species consists of about 5 wholly confluent, more or less geographically segregated
varieties, three of which occur within or near our range. Of these, var. variabilis was treated
as a subspecies of P. deustus by Pennell & Keck in 1940, with the comment (by Keck) that "intergrades with typicus are frequent in Grant County, Oregon." In later papers Keck has restored P. variabilis to specific rank, without comment, on the basis (personal communication)
that several chromosome counts indicate P. deustus to be a diploid, while a single count indicates P. variabilis to be a tetraploid. In view of the known variability in ploidy level of other
species of Penstemon (e.g., P. wilcoxii, P. cinicola) and of many other genera this evidence
seems insufficient to controvert the more abundant evidence, based on traditional criteria,
that P. variabilis constitutes only an ecologic-geographic race of P. deustus. It may be noted
that the only feature of P. variabilis that is not commonly found in one or another of the remaining varieties of P. deustus is the tendency to whorled or scattered leaves, but even this
character is not fully diagnostic, since many of the specimens of var. variabilis have wholly
opposite leaves.
   The varieties of P. deustus occurring within or near our range may be characterized as
follows:
1 Corolla essentially glabrous within, and only sparsely or scarcely glandular externally;
   plants tending to be compact and small-flowered, with small and often relatively narrow
   leaves that are usually sharply toothed; c. Deschutes Co., Oreg., to Nev. and n.e. Calif.
                                                                    var. heterander (T. & G.) Cronq.

1 Corolla glandular-hairy inside and out
   2 Leaves all opposite, relatively broad and evidently toothed, those of the flowering stems
      mostly sessile, 2-5 times as long as wide, rarely less than 6 mm. wide; staminode
      more often glabrous than bearded; fairly widespread e. of the Cascade summits (and
      near Eugene) in our area, but absent from most of the range of the other 2 varieties
                                                                              var. deustus
   2 Leaves, or some of them, frequently ternate, or quaternate, or scattered, all narrow
      and irregularly few-toothed to entire, often subpetiolate, more than 5 times as long as
      wide and not more than 6 mm. wide; inflorescence averaging looser and fewer-flowered
      than in var. deustus; staminode more often bearded than glabrous; dry foothills and low-
      lands from Klickitat Co., Wash., to Deschutes, Grant, and Umatilla cos., Oreg.
                                                                  var. variabilis (Suksd.) Cronq.

Penstemon diphyllus Rydb. Mem. N. Y. Bot. Gard. 1:349. 1900.
   P. triphyllus ssp. diphyllus Keck, U. Calif. Pub. Bot. 16:422. 1932. (Cooper, Mullan Pass,
      Mont., in 1860)
   Rather shortly taprooted perennial, tending to be shrubby at the base, (1)2-6 dm. tall, with
more or less numerous, rather slender and brittle stems, evidently glandular-hairy in the inflorescence, the stems otherwise merely pruinose-puberulent and the leaves essentially glabrous; leaves opposite, or often some of them a little offset, irregularly dentate or serrate-
dentate, all cauline, the lowermost ones reduced, the others sessile or nearly so but often
narrowed to the base, up to 6 cm. long and 1.5 cm. wide, averaging wider than in P. triphyllus; inflorescence a branching, often leafy-bracteate mixed panicle; calyx 4-6 mm. long,
the segments green or anthocyanic, entire, often distinctly unequal; corolla glandular-hairy
outside, glabrous within, or with a few hairs near the base of the lower lip, blue-lavender to
light purple-violet, 13-19 mm. long, not very strongly bilabiate, the upper lip cleft more than
half its length; anthers 0.9-1.3 mm. long, permanently horseshoe-shaped, the pollen sacs dehiscent across their confluent apices, saccate and indehiscent below, not becoming divaricate,
the sutures evidently dentate-ciliolate; staminode exserted, not much, if at all, expanded upward, the distal portion (often nearly half) moderately to very sparsely long-bearded; capsule
4-6 mm. and seeds 0.5-0.8 mm. long.
   Talus slopes, cliffs, and loose or rocky banks in the mountain regions, but not at high elevations; Missoula, Granite, and Beaverhead cos., Mont., to c. Idaho Co. (Whitebird) and c.
Valley Co., Ida.; apparently also at Palouse Falls, Whitman Co., Wash. Late June-early
Aug., a few flowers sometimes produced until Oct.
   Penstemon diphyllus is obviously allied to P. triphyllus, but populational intergradation be-
tween them has not yet been demonstrated.

Penstemon elegantulus Pennell, Not. Nat. Acad. Phila. 71:14. 1941. (Sheldon 8387, head of
   Horse Creek, Wallowa Co., Oreg.)
   Plants tufted from a compact, branching, surficial, woody caudex, 1-3 dm. tall, evidently
to obscurely glandular in the inflorescence, otherwise finely hirtellous-puberulent throughout,
or more often the basal and lower cauline leaves glabrous or nearly so; leaves entire or more
often with a few small, scattered, callous teeth, the basal ones up to about 10 cm. long and
1.6 cm. wide, the elliptic to lanceolate or ovate blade longer or shorter than the petiole; cauline leaves narrowly lanceolate and sessile to oblanceolate and subpetiolate, up to 5 cm. long
and 8 mm. wide; inflorescence of several few-flowered, not very dense verticillasters; calyx
3-6 mm. long, the segments ovate, more or less scarious-margined and sometimes erose;
corolla blue, generally somewhat glandular-hairy externally (sometimes sparsely or obscurely so), 15-22 mm. long, 5-6 mm. wide at the mouth; palate bearded; pollen sacs about 1 mm.
long, opening nearly throughout, essentially glabrous, becoming opposite; staminode bearded
toward the recurved, scarcely dilated tip.
   Open, grassy or rocky places overlooking the Snake River Canyon in Wallowa Co., Oreg.,
and Nez Perce and Idaho cos., Ida. May-June.
   Penstemon elegantulus differs from P. humilis in being mostly less pubescent than the bulk
of that species, in tending to have a few teeth on the leaves, and in the larger average size of

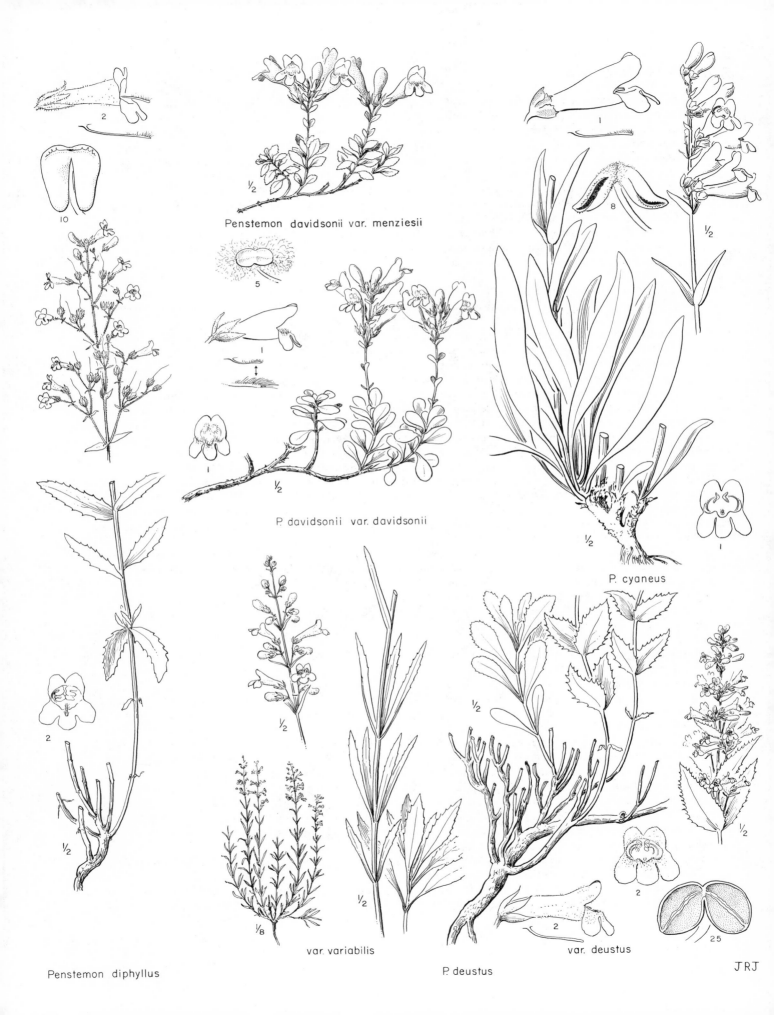

Penstemon davidsonii var. menziesii

P. davidsonii var. davidsonii

P. cyaneus

Penstemon diphyllus

var. variabilis

P. deustus

var. deustus

JRJ

the corolla. In all of these features it (P. elegantulus) approaches P. albertinus, but the distinctly hirtellous-puberulent condition of some (or all) of the leaves precludes its reference to that species. If the several collections of P. elegantulus were not concentrated in an area in which both of its allies are rare or wanting, they would probably be considered to represent hybrids or aberrant specimens, but under the circumstances it seems necessary to retain P. elegantulus in specific status at least until more is known about it.

Penstemon ellipticus Coult. & Fish. Bot. Gaz. 18:302. 1893. (MacDougal, near Lake Pend Oreille, Ida., Aug., 1892)

Suffruticose perennial, with or without an evident taproot, forming dense to rather loose mats (with creeping, woody stems) on the surface of the substrate, and with scattered, erect flowering stems mostly 0.5-1.5 dm. tall, these shortly spreading-hairy; leaves glabrous, those of the mat short-petiolate, the blade broadly elliptic to occasionally more ovate or obovate, rounded or obtuse, mostly (0.7)1-2.5 cm. long and 5-15 mm. wide, 1-2.5 times as long as wide, inconspicuously serrulate or subentire; leaves of the flowering shoots generally sessile or nearly so, often more ovate, generally fairly well developed, often fully as large as, and sometimes quite like, the mat leaves (but less crowded), seldom less than 1 cm. long; inflorescence a compact, few-flowered raceme, with simple axillary pedicels, more or less strongly glandular-hairy; calyx 8-15 mm. long, the segments lance-linear to linear-oblong; corolla deep lavender, 27-40 mm. long, about 1 cm. wide at the mouth, keeled along the back, glabrous outside, long-white-hairy near the base of the lower lip within, especially near the prominent ventral ridges; anthers densely long-woolly, the pollen sacs wholly dehiscent and becoming peltately explanate; staminode slender, shorter than the fertile filaments, sparsely to rather densely long-bearded for much of its length; capsule 8-11 mm. long; seeds 2-2.5 mm. long, irregularly compressed-prismatic and very narrowly winged.

Rocky places at moderate to high elevations in the mountains, often on cliffs, ledges, or in rock crevices, sometimes on talus; s.w. Alta. and s.e. B.C. to n. Ida. and n.w. Mont. and rarely to Custer Co., Ida., and n. Beaverhead Co., Mont. Late June-Sept.

Some small specimens with reduced cauline leaves, which evidently represent merely the extreme of variation in P. ellipticus, closely resemble the more western P. davidsonii var. menziesii. It is not unlikely that P. ellipticus will eventually have to be reduced to a variety of P. davidsonii, but there is a gap of over a hundred miles between the presently known ranges of the two, and I am reluctant to make the new combination until the necessity for it is more firmly established.

Penstemon eriantherus Pursh, Fl. Am. Sept. 737. 1814. (Bradbury, along the Missouri R. in the present state of S.D.) Pursh's publication of the name P. eriantherus for this species, which has essentially glabrous anthers, was the result of an unfortunate error in his attempt to identify P. eriantherus Nutt. Frasers' Cat. 1813, nom. nud. Under the Rules of Nomenclature, Pursh's application of the name must stand.

P. cristatus Nutt. Gen. Pl. 2:52. 1818. Chelone cristata Spreng. Syst. Veg. 2:813. 1825. (Nuttall, "from the confluence of Teeton River and the Missouri to the Mountain") The original publication includes Penstemon eriantherus Pursh as a synonym.

P. whitedii Piper, Bot. Gaz. 22:490. 1896. P. eriantherus var. whitedii A. Nels. Bot. Gaz. 54:148. 1912. (Whited 131, near Wenatchee, Wash.)

P. saliens Rydb. Mem. N.Y. Bot. Gard. 1:344. 1900. P. eriantherus ssp. saliens Pennell, Contr. U.S. Nat. Herb. 20:343. 1920. (Kennedy 53, Columbia Falls, Flathead Co., Mont.) = var. eriantherus.

P. dayanus Howell, Fl. N.W. Am. 511. 1901. P. whitedii ssp. dayanus Keck, Bull. Torrey Club 65:254. 1938. (Howell, Muddy Station, John Day Valley, Oreg.) = var. argillosus.

P. eriantherus var. argillosus M. E. Jones, Contr. West. Bot. 12:62. 1908. (Cusick 2803, John Day Valley, Oreg.)

P. eriantherus var. redactus Pennell & Keck, Bull. Torrey Club 65:252. 1938. (Pennell 20605, 16 miles e. of Monida, Beaverhead Co., Mont.)

P. eriantherus var. grandis Pennell & Keck, Bull. Torrey Club 65:251. 1938. (Pennell & Cotner 20511, e. of Bozeman, Mont.) = var. eriantherus.

P. whitedii ssp. tristis Pennell & Keck, Bull. Torrey Club 65:254. 1938. (Pennell 15200,
 Antelope Creek, Custer Co., Ida.) = var. redactus.

Perennial, 1-4 dm. tall, with several (occasionally solitary) stems arising from a usually
branched, largely subterranean caudex; plants evidently glandular-villous or glandular-hirsute
in the inflorescence, otherwise finely cinereous-puberulent to villous-hirsute (and sometimes
also glandular) throughout, or partly glabrate; leaves entire or more or less toothed, the basal ones often poorly developed, up to 13 cm. long and 2 cm. wide, the blade usually narrower
than in P. whippleanus; inflorescence of several distinct or approximate verticillasters; calyx
elongate, 7-13 mm. long, the segments lanceolate or narrower and wholly, or almost wholly,
herbaceous; corolla glandular-hairy externally, pale lavender to orchid, red-purple, or deep
blue-purple, 2-4 cm. long, strongly inflated distally, mostly 6-14 mm. wide at the mouth, the
lower lip moderately longer than the upper; palate strongly bearded; pollen sacs glabrous except for the finely setulose-dentate sutures, 1.1-1.8 mm. long, wholly dehiscent, becoming
opposite and sometimes eventually explanate; staminode more or less exserted from the orifice of the corolla, slightly expanded toward the tip, usually prominently long-bearded for
most of its length; ovary and capsule glandular-puberulent near the tip; capsule 7-12 mm. long
and seeds 2-3 mm. long. N=8.

Dry, open places in the plains, valleys, and foothills, sometimes ascending to moderate elevations in the mountains; N.D. to Neb. and n. Colo., w. to s.e. B.C., c. Wash., and c.
Oreg. May-July.

Four varieties of the species occur in our range. The varieties whitedii and argillosus are
sharply distinguished from var. eriantherus by the anthers, inflorescence, and geography,
and if only these three taxa had to be considered, the maintenance of two species would be well
justified. The var. redactus, however, so completely bridges the differences that all four taxa
must be considered as geographic races of a single species. These may be characterized as
follows:

1 Pollen sacs relatively broad, becoming explanate, 0.9-1.4 times as long as wide when fully
    mature, with a relatively long line of contact at the proximal end; corollas lavender, pale
    lilac, or pale to medium purple-violet, tending to be ascending and the verticillasters not
    appearing to be well separated; habit otherwise variable; Great Plains and adj. foothills
    and intermontane valleys, extending w. through n.w. Mont. and s.e. B.C. (and n. Ida?) to
    the vicinity of Spokane, Wash.                                            var. eriantherus
1 Pollen sacs mostly narrower
  2 Pollen sacs mostly 1.4-2.0 times as long as wide when fully mature, scarcely to evident-
      ly explanate, usually with a fairly long line of contact at the proximal end; inflorescence
      typically as in var. eriantherus, varying (especially toward the west) to sometimes as
      in the 2 following varieties; plants typically small (1-2 dm. tall) and narrow-leaved, but
      sometimes (especially toward the west) exactly matching var. whitedii in habit; s.w.
      Mont. (Deer Lodge, Silverbow, and Beaverhead cos.) to Clark and Butte cos., Ida.,
      and w. across c. Ida. to the Imnaha R. of e. Wallowa Co., Oreg., commonest in the
      Salmon R. drainage                                        var. redactus Pennell & Keck
  2 Pollen sacs mostly 1.8-2.5 times as long as wide when fully mature, not explanate, often
      constricted at the proximal end and thus with a short line of contact; corollas mostly
      spreading and the verticillasters appearing to be well separated
    3 Staminode densely bearded for most of its length (as in var. eriantherus and var. re-
        dactus); plants mostly 2-4 dm. tall, the cauline leaves mostly broadest near the base
        and often clasping, up to 2 cm. wide; corollas light blue to orchid; c. Wash. (Chelan
        and Douglas cos.)                                        var. whitedii (Piper) A. Nels.
    3 Staminode sparsely bearded or nearly glabrous; habit of var. whitedii, or with narrow-
        er leaves; corolla deep red-purple to blue-purple; drainage of the John Day and Des-
        chutes rivers in c. and n.c. Oreg.                            var. argillosus M. E. Jones

Penstemon miser Gray, a species closely related to P. eriantherus, but with the ovary and
capsule glabrous, and often with smaller flowers, is known from Nev. and n.e. Calif. n. to
Harper Ranch, Malheur Co., Oreg., and may eventually be found to occur in our range.

Penstemon euglaucus English, Proc. Biol. Soc. Wash. 41:197. 1928. (English 816, Mt. Hood, Oreg.)

Similar to P. rydbergii, but consistently glabrous and glaucous throughout; rosettes of basal leaves generally well developed; calyx 3.5-5 mm. long, the segments narrowly to broadly and conspicuously scarious-margined, often erose, abruptly or more gradually contracted to the narrow, pointed tip; pollen sacs often not opening quite to the tip, approaching but scarcely matching the condition of P. globosus, the sutures glabrous. N=24.

Dry, sandy, open or sparsely wooded slopes at moderate elevations in the Cascade Range; Mt. Adams, Wash., to the Three Sisters, Oreg. July-Aug.

Penstemon flavescens Pennell, Not. Nat. Acad. Phila. 95:4. 1942. (Pennell & Constance 20890, Coolwater Mt., near Lowell, Idaho Co., Ida.)

Plants tufted from a woody, surficial rhizome-caudex, forming clones sometimes 1 m. wide, glabrous throughout, or the stem very shortly hairy, 1.5-4 dm. tall; leaves entire, relatively thick and firm, tending to blacken in drying, the lowermost ones commonly well developed and forming rosettes, petiolate, up to 10 cm. long and 2 cm. wide, the others mostly sessile and rather few, as large as, or often distinctly smaller than, the lower; inflorescence of 1-4 compact verticillasters, the bracts, or many of them, with prominently scarious and erose margins; calyx mostly 5-9 mm. long, the segments with conspicuous, broad, erose or lacerate scarious margins and generally with an abruptly slender tip; corolla light yellow or slightly brownish, not declined, mostly 12-16 mm. long, the tube expanded distally, mostly 3.5-6 mm. wide at the mouth, the limb evidently bilabiate, the palate bearded; pollen sacs ovate, 0.7-0.9 mm. long, wholly dehiscent and becoming opposite, but scarcely explanate; staminode bearded at the slightly or scarcely dilated tip; capsule about 6 mm. long; seeds about 1.5 mm. long, evidently wing-margined.

Open or wooded, often rocky slopes well up in the mountains; Bitter Root Mts. of Idaho Co., Ida., and Ravalli Co., Mont., extending w. to Coolwater Mt., near Lowell, Idaho Co., Ida. July-Aug.

Penstemon fruticosus (Pursh) Greene, Pitt. 2:239. 1892.

Gerardia fruticosa Pursh, Fl. Am. Sept. 423. 1814. Dasanthera fruticosa Raf. Am. Monthly Mag. 2:267. 1818. Penstemon lewisii Benth. in DC. Prodr. 10:321. 1846. P. menziesii var. lewisii Gray, Proc. Am. Acad. 6:56. 1862. (Lewis, "pine forests of the Rocky Mts."; = along Lolo Creek, a tributary of the Clearwater R. in n. Ida., fide Pennell)

PENSTEMON FRUTICOSUS var. SCOULERI (Lindl.) Cronq. hoc loc. P. scouleri Lindl. Bot. Reg. 15: pl. 1277. 1829. P. menziesii var. scouleri Gray, Proc. Am. Acad. 6:57. 1862. P. fruticosus ssp. scouleri Pennell & Keck in Abrams, Ill. Fl. Pac. St. 3:765. 1951. (Douglas, Kettle Falls of the Columbia)

P. crassifolius Lindl. Bot. Reg. 24: pl. 16. 1838. P. menziesii var. crassifolius Schelle in Beissn. et al. Handb. Laubh. Bene 432. 1903. P. fruticosus var. crassifolius Krautter, Contr. Bot. Lab. Univ. Pa. 3:100. 1908. (Cultivated plants, from seeds collected by Douglas in "North West America") = var. fruticosus.

P. douglasii Hook. Fl. Bor. Am. 2:98. 1838. P. menziesii var. douglasii Gray, Proc. Am. Acad. 6:57. 1862. P. fruticosus var. douglasii Schneid. Ill. Handb. Laubh. 2:615. 1911. (Douglas, "Blue Mountains of N.W. America") = var. fruticosus.

P. adamsianus Howell, Fl. N.W. Am. 511. 1901. (Howell, Mt. Adams, Wash.) = var. fruticosus.

PENSTEMON FRUTICOSUS var. SERRATUS (Keck) Cronq. hoc loc. P. fruticosus ssp. serratus Keck in Abrams, Ill. Fl. Pac. St. 3:765. 1951. (Ownbey & Meyer 2068, Sheep Creek, Seven Devils Mts., Idaho Co., Ida.)

Ascending or erect, more or less bushy-branched shrubs or subshrubs mostly 1.5-4 dm. tall, evidently glandular-hairy in the inflorescence, otherwise essentially glabrous, or the stems of the season often finely puberulent or strigose-puberulent; larger leaves tending to be crowded toward the base of the season's growth and on short, sterile shoots, short-petiolate, the entire or toothed blade up to 6 cm. long and 1.5 cm. wide, 2-10 times as long as wide; flowering shoots with more or less reduced and less crowded leaves; inflorescence a short,

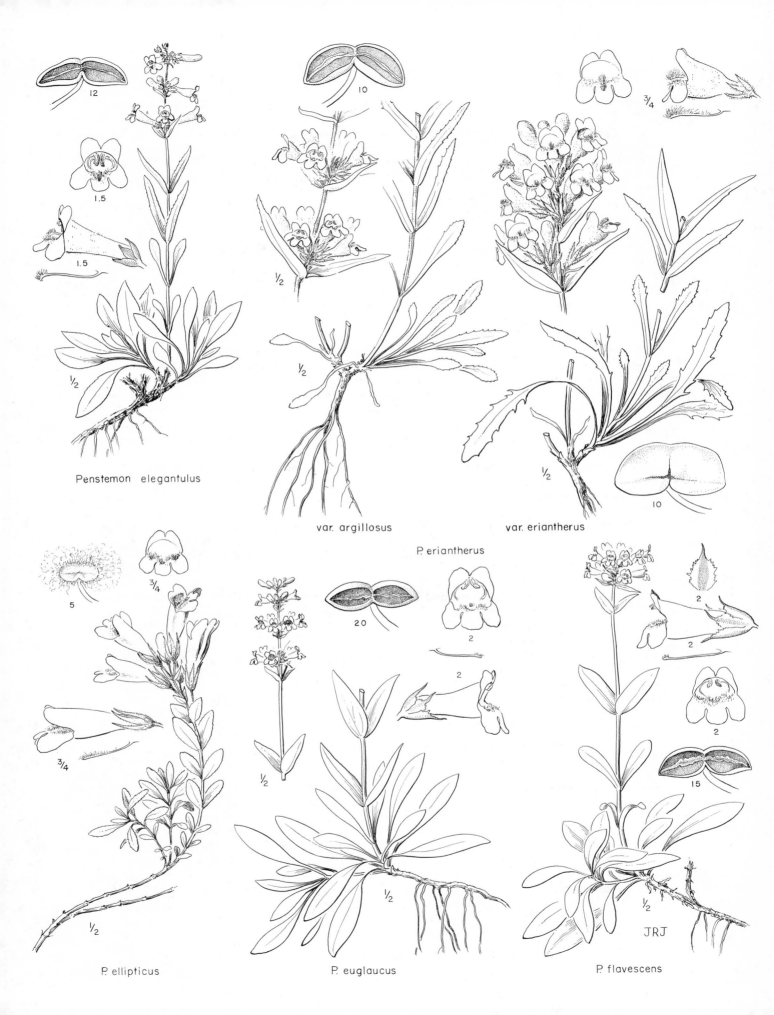

12

1.5

1.5

Penstemon elegantulus

10

½

½

var. argillosus

P. eriantherus

10

¾

½

var. eriantherus

5

¾

¾

P. ellipticus

20

2

2

2

½

P. euglaucus

2

2

2

15

½

JRJ

P. flavescens

few-flowered, bracteate raceme, with simple axillary pedicels; calyx 7-15 mm. long, with lance-linear to lance-ovate, long-acuminate segments; corolla blue-lavender to light purplish, (25)30-50 mm. long, about 1 cm. wide at the mouth, keeled on the back, glabrous outside, long-white-hairy near the base of the lower lip within, especially along the two prominent ventral ridges; anthers densely long-woolly, the pollen sacs wholly dehiscent and becoming explanate; staminode slender, shorter than the fertile filaments, long-bearded toward the tip; capsules 8-12 mm. long; seeds irregularly compressed-prismatic, very narrowly wing-margined, 1-2 mm. long. N=8.

Rocky, open or wooded places from the foothills to rather high elevations in the mountains; e. of the Cascade summits from s. B. C. to c. Oreg., e. to Mont. and Wyo. May-Aug.

This is a highly variable species with numerous slightly differing forms or local races. Two fairly well-marked varieties may be distinguished from a more heterogeneous third one, as follows:

1 Leaves prominently serrate or dentate, relatively small, the blade mostly 1-2.5 cm. long and 2-3.5 times as long as wide; corolla 3-4 cm. long; mountains near the Snake R., from Adams Co. to Latah Co., Ida., and e. of the Imnaha R. in Wallowa Co., Oreg., and in the Blue Mt. region of Asotin and Garfield cos., Wash. (plants from along the lower Salmon R. are referable here, but approach var. fruticosus)    var. serratus (Keck) Cronq.
1 Leaves otherwise, usually either longer, or subentire to entire, or both
  2 Leaves more or less toothed to entire, relatively very narrow, linear-oblanceolate or linear-elliptic, the larger ones with the blade mostly 2-5 cm. long and 3-5(7) mm. wide, mostly 6-10 times as long as wide; corolla 3.5-5 cm. long; Ferry Co., Wash., to Kootenai Co., Ida., n. to s. B. C.                              var. scouleri (Lindl.) Cronq.
  2 Leaves entire or slightly serrulate, wider in shape and often also in measurement, up to 6 cm. long and 1.5 cm. wide, mostly 2-7 times as long as wide; corolla (2.5)3-4 cm. long; range of the species, except for the areas occupied by varieties serratus and scouleri, and much more variable than either of these         var. fruticosus

Penstemon gairdneri Hook. Fl. Bor. Am. 2:99. 1838. (Douglas, Blue Mts.)

P. gairdneri var. oreganus Gray, Syn. Fl. 2nd ed. 2[1]:441. 1886. P. oreganus Howell, Fl. N. W. Am. 515. 1901. P. gairdneri ssp. oreganus Keck, Am. Midl. Nat. 23:596. 1940. (Cusick, mountains of e. Oreg.)

P. gairdneri var. hians Piper, Bull. Torrey Club 27:396. 1900. P. gairdneri ssp. hians Keck, Am. Midl. Nat. 23:597. 1940. (Vasey 432, e. Wash.) A form of var. gairdneri with rose-purple corolla and large limb.

P. puberulentus Rydb. Fl. Rocky Mts. 774, 1066. 1917. (Henderson 3169, Long Valley, Ida.) = var. oreganus.

Plants with several or many erect flowering stems 1-4 dm. tall from a branched, distinctly woody base, also producing short, densely leafy, sterile stems and tending to form loose mats; herbage finely and densely cinereous-puberulent, the inflorescence tending to become more loosely glandular-hairy; leaves numerous, alternate or irregularly scattered to essentially opposite, but their bases remaining well separated, not joined around the stem, all entire and linear or nearly so, seldom over 3(6) mm. wide; inflorescence of several or rather numerous few-flowered verticillasters, sometimes approaching the status of a simple raceme; calyx 3.5-8 mm. long, the segments lanceolate or ovate, obscurely, or not at all, scarious-margined, entire; corolla blue-purple or lavender to sometimes bright rose-purple, with or without guidelines, somewhat glandular-hairy externally, and also internally near the mouth, 14-22 mm. long, the tube more or less flared, the limb scarcely to very strongly spreading; pollen sacs essentially glabrous, more or less ovate, 0.8-1.2 mm. long, dehiscent throughout and becoming opposite but not explanate; staminode evidently bearded toward the scarcely expanded tip; capsule 6-8 mm. and seeds about 2 mm. long. N=8.

Dry, open, mostly rocky places, often on scab rock, frequently with sagebrush, or on soil too thin and sterile for sagebrush, from the valleys and plains to moderate elevations in the mountains; May-June.

The species consists of two well-marked but intergradient geographical varieties. The name P. gairdneri var. hians Piper was based on a plant from c. Wash. with bright rose-purple flowers and relatively large, spreading limb, otherwise like var. gairdneri. It is not yet established, however, that the plants with rose-purple corollas occupy a coherent geographic area separate from the more typical plants with darker purple flowers, nor is the correlation of the color of the flowers with the size of the limb established. Until these questions are resolved it seems better to submerge the var. hians in var. gairdneri. The varieties of the species here recognized may be characterized as follows:

1 Leaves mostly opposite or nearly so; plants averaging relatively robust, 1.5-4 dm. tall, the leaves mostly 2-7 cm. long and 2-3(5) mm. wide; Union and Baker cos., Oreg., to Adams, Washington, and w. Valley cos., Ida.                     var. oreganus Gray
1 Leaves mostly alternate or scattered; plants averaging smaller, 1-3 dm. tall, the leaves mostly 1-4 cm. long and 1-3 mm. wide; Ferry, Douglas, and Chelan cos., Wash., to Crook Co. and n. Harney and Malheur cos., Oreg.              var. gairdneri

Penstemon glandulosus Dougl. in Lindl. Bot. Reg. 15: pl. 1262. 1829. (Douglas, "Rocky Mts., lat. 47 degrees north, and at the base of the Blue Mountains on the banks of the Kooskoosky River"; the Kooskoosky is now known as the Clearwater R.)
  P. staticifolius Lindl. Bot. Reg. 21: pl. 1770. 1835. (Garden plants, from seeds collected by Douglas in "California")
  PENSTEMON GLANDULOSUS var. CHELANENSIS (Keck) Cronq. hoc loc. P. glandulosus ssp. chelanensis Keck in Abrams, Ill. Fl. Pac. St. 3:764. 1951. (Ward 315, Colockum Creek, Chelan Co., Wash.)
  Stout perennial herb from a branched caudex which surmounts a short, quickly deliquescent taproot, 4-10 dm. tall, viscidly glandular-hairy nearly or quite throughout, the several stems often more densely so than the leaves, of which the lower are sometimes subglabrous; leaves sharply toothed (entire in var. chelanensis), the basal ones mostly 10-35 cm. long and 2.5-9 cm. wide, petiolate, with lanceolate or lance-ovate to elliptic blade; lowermost cauline leaves generally reduced, the others well developed, sessile and often clasping, broadly lanceolate or lance-oblong to rotund-ovate, mostly 4-12 cm. long and 2-5 cm. wide; inflorescence of several distinct or occasionally subconfluent verticillasters; calyx 9-15 mm. long, the segments narrow and herbaceous; corolla blue-lavender, strongly glandular-hairy outside, glabrous inside or with a few long hairs near the base of the lower lip, 28-40 mm. long, often more than 1 cm. wide at the mouth; anthers permanently horseshoe-shaped, 1.7-2.3 mm. long; pollen sacs dehiscent across their confluent apices, the lower part remaining saccate and indehiscent, glabrous except for the dentate-ciliolate sutures; staminode glabrous, the tip flattened and somewhat expanded; capsules 10-14 mm. long, surpassed by the calyx; seeds 2-3 mm. long.

  Open, often rocky hillsides and banks in the foothills, valleys, and lower parts of the mountains; near the Snake R. from Payette (and n. Owyhee?) Co., Ida., and Baker Co., Oreg., to Nez Perce Co., Ida., and Garfield Co., Wash.; near the base of the Cascades in Klickitat Co., Wash., and Wasco and Hood River cos., Oreg.; near Wenatchee in Chelan Co., Wash. May-July.

  The plants from Chelan Co., Wash., have entire or subentire leaves and may be segregated as var. chelanensis (Keck) Cronq.

Penstemon globosus (Piper) Pennell & Keck, Carn. Inst. Wash. Pub. 520:294. 1940.
  P. confertus var. globosus Piper, Bull. Torrey Club 27:397. 1900. (Cusick 2328, Eagle Creek, Wallowa Mts., Baker Co., Oreg.)
  P. attenuatus var. glabratus G. N. Jones, Res. Stud. State Coll. Wash. 2:127. 1931. (Jones 679, Alder Creek, near Benewah, Benewah Co., Ida.)
  Plants more or less tufted from a loose or compact surficial, woody rhizome-caudex, mostly (1)2-6 dm. tall, glabrous throughout, or the stem hirtellous in lines above; leaves entire, the basal ones petiolate, oblanceolate to narrowly elliptic, up to 15 cm. long and 3.5 cm. wide; cauline leaves mostly sessile, well developed, lanceolate to ovate, up to 13 cm. long and 3.5 cm. wide; inflorescence of 1-several dense, often globose, remote or approximate verticillasters; calyx 5-8(10) mm. long, segments more or less strongly scarious-margined and often erose, usually abruptly contracted to the narrow, pointed tip; corolla blue or blue-purple,

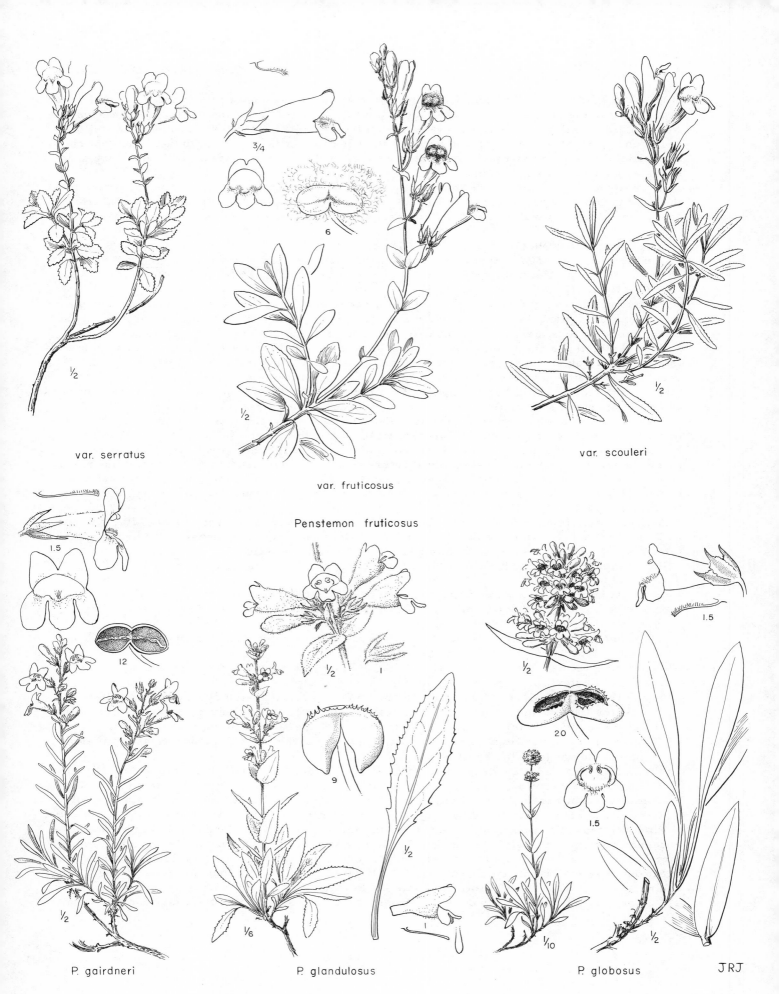

var. serratus

3/4

6

var. scouleri

var. fruticosus

Penstemon fruticosus

1.5

12

P. gairdneri

1

1/2

9

1/2

1

P. glandulosus

1/2

20

1.5

1/10

1/2

P. globosus

JRJ

(13)15-20(22) mm. long, evidently bilabiate, the gradually expanded tube mostly 4-7 mm. wide at the mouth, the palate bearded; staminode densely bearded above the middle; pollen sacs 0.7-1.2 mm. long, elliptic or elliptic-ovate, becoming opposite, incompletely dehiscent, the distal ends shortly pouched, the partition between the proximal ends remaining intact, the denticulate-ciliolate sutures not widely separated even after dehiscence; capsule 6-7 mm. long and seeds about 1 mm. long. N=16.

In wet or dry meadows and on moist open slopes, less often in timber or rocky places, from the base of the mountains to timber line; Benewah Co., Ida., to Adams, Valley, and Custer cos., extending into Wallowa and Baker cos., Oreg. June-Aug.

**Penstemon gracilis** Nutt. Gen. Pl. 2:52. 1818.

Chelone gracilis Spreng. Syst. Veg. 2:813. 1825. Penstemon digitalis var. gracilis Trautv. Bull. Acad. St. Pétersb. 5:345. 1839. P. pubescens var. gracilis Gray, Proc. Am. Acad. 6:69. 1862. (Nuttall, "from the Arikarees to Fort Mandan")

Stems arising singly or few together from a very short, simple, subterranean caudex; plants 1.5-5 dm. tall, glandular-hirsute in the inflorescence, otherwise glabrous or, especially the stem, more or less hirtellous-puberulent; leaves, or many of them, with scattered, short, callous teeth, these sometimes few and obscure; basal and lowermost cauline leaves short-petiolate, mostly oblanceolate or elliptic, up to 12 cm. long and 2 cm. wide, often poorly developed or soon withering; cauline leaves mostly sessile, lanceolate or lance-oblong to occasionally linear, mostly 2.5-10 cm. long and 2-15 mm. wide; inflorescence of several few-flowered and often approximate verticillasters, not very dense, sometimes secund; calyx 4-7 mm. long, the lanceolate to ovate segments with or without scarious margins; corolla glandular-hairy externally, pale lilac or pale violet, whitish within and with evident guidelines, 15-23 mm. long, 5-8 mm. wide at the mouth; palate bearded; pollen sacs 1-1.5 mm. long, becoming opposite, glabrous except for the setulose-dentate sutures, which do not become very widely separated even after dehiscence; staminode gradually expanded upward, straight, prominently yellow-bearded for much of its length, generally shortly exserted; capsule 4-8 mm. long; seeds less than 1 mm. long.

A characteristic species of the n. Great Plains, often in sandy or rocky soil; Alta. and n.e. B.C. to n. N.M., e. to n. Iowa, Wis., and s.w. Ont.; in our range occasionally found in the intermontane valleys as far w. as Sanders Co., Mont. June-July.

Penstemon gracilis is sharply distinct from all of our other species.

**Penstemon humilis** Nutt. ex Gray, Proc. Am. Acad. 6:69. 1862. (Nuttall, Rocky Mts., "a very depauperate doubtless alpine specimen . . . doubtless collected either in southwestern Wyoming or southeastern Idaho," fide Keck)

P. collinus A. Nels. Bull. Torrey Club 25:279. 1898. (Nelson 2960, Evanston, Wyo.) The well-developed form from lower elevations.

P. brevis A. Nels. Bot. Gaz. 54:417. 1912. (Nelson & Macbride 1457, Bear Canyon, near Mackay, Custer Co., Ida., at 10,000 ft.) The dwarf alpine phase.

P. cinereus Piper, Contr. U.S. Nat. Herb. 16:209. 1913. (Whited 3055a, near Laidlaw, Deschutes Co., Oreg.) A compact (but scarcely dwarf), small-leaved, evidently cinereous phase, which dominates that portion of the range of the species lying in n.e. Calif., s.e. Oreg., and the Deschutes R. valley, but which is matched by many specimens from elsewhere in the range of the species.

P. cinereus ssp. foliatus Keck, Am. Midl. Nat. 33:188. 1945. (Keck & Clausen 3662, near Dixie Pass, Grant Co., Oreg.) A robust, thinly cinereous or partly glabrate phase which dominates that portion of the range of the species lying in Wash. and the n.e. quarter of Oreg., but which is matched by many specimens from elsewhere in the range of the species.

Plants tufted from a compact, branching caudex, 1-6 dm. tall, glandular-hairy in the inflorescence, otherwise finely cinereous-puberulent throughout, or the puberulence sometimes more or less restricted to some of the cauline leaves and part of the stem; leaves entire, the basal ones generally tufted and persistent, petiolate, up to 12 cm. long and 2 cm. wide, with oblanceolate to more often elliptic or ovate blade, the cauline ones mostly sessile or nearly so and commonly more or less reduced, or sometimes fully as large as the basal ones; inflorescence of several, rather loose, discrete or sometimes confluent, few-flowered verticillasters; calyx 2.5-6 mm. long, the segments lanceolate to ovate, narrowly or obscurely scarious-margined, acute or acuminate, often anthocyanic; corolla glandular-hairy, blue-purple, 10-17 mm. long, the tube mostly 3-6 mm. wide at the mouth, the palate bearded; pollen sacs glabrous, 0.4-0.8 mm. long, becoming opposite, wholly dehiscent and sometimes explanate; staminode densely bearded toward the slightly, if at all, expanded tip; capsule 4-6 mm. long and seeds 1-1.5 mm. long. N=8.

Dry, open, often rocky places, frequently with sagebrush, from the plains and foothills to high elevations in the mountains, wholly e. of the Cascade-Sierra crest, and w. of the continental divide; c. Wash., e. Oreg., and c. Ida. (chiefly s. of the Salmon R. axis) to Mono Co., Calif., Nev., Utah, w. Wyo., and w. Colo. May-July.

The variation within P. humilis in stature, leaf size, and amount and distribution of pubescence is sufficiently correlated with geographic and ecologic factors to suggest that several incipient ecotypes are involved, but not sufficiently so to permit the ready recognition of infraspecific taxa (except for the well-marked and perhaps specifically distinct P. humilis var. brevifolius Gray, occurring wholly to the s. of our range, in the mountains of n. Utah, s.c. Ida., and n.e. Nev.). Were only the plants from Wash., Oreg., and Calif. to be considered, it might be possible, with some difficulty, to segregate the more northern, more robust, and less hairy plants (as done by Keck under the name P. cinereus ssp. foliatus) from the more southern, less robust, more hairy P. cinereus proper, but material of P. humilis from Ida., Nev., Wyo., and Utah encompasses the variation shown by the more western plants, without any clear or consistent geographic pattern. Plants from higher altitudes, throughout the range of the species as here defined, tend to be smaller and more compact than those from lower elevations, but any attempt to segregate a subalpine ecotype (to which, incidentally, the nomenclatural type of the species would seem to belong) is beset with so much difficulty as to be scarcely worth while. Another dwarf form, with consistently narrow leaves, occurs in the Lost River Mt. area of Ida., but is not confined to high elevations.

Penstemon laetus Gray, Proc. Boston Soc. Nat. Hist. 7:147. 1859. (Xantus, Ft. Tejon, Calif.)
   P. roezlii Regel, Gartenfl. 21:239. 1872. P. laetus var. roezlii Jeps. Man. Fl. Pl. Calif. 916. 1925. P. laetus ssp. roezlii Keck, U. Calif. Pub. Bot. 16:390. 1932. (Cultivated plants, from seeds collected by Roezl in the Sierra Nevada of Calif.)

Perennial from a short, quickly deliquescent taproot, tending to be shrubby at the base, 2-5 dm. tall, with several or many slender flowering stems and generally with short, sterile, leafy shoots from the base as well, evidently glandular-hairy in the inflorescence only, otherwise more or less hirtellous-puberulent throughout; leaves numerous, entire, up to 8 cm. long and 8 mm. wide, the lower oblanceolate and more or less petiolate, the others mostly sessile and linear to lance-linear or linear-oblong; verticillasters very loose and rather few-flowered, the inflorescence commonly appearing as a terminal mixed panicle or almost a raceme; calyx mostly 4-7 mm. long, the segments green or anthocyanic, entire, tending to be dissimilar; corolla deep blue-violet, 15-25 mm. long, up to about 1 cm. wide at the mouth, glandular-hairy outside, glabrous inside; anthers 1.5-2.0 mm. long, permanently horseshoe-shaped, the pollen sacs dehiscent across their contiguous apices and for about one-half of their length, or a little more, the basal portion remaining saccate and indehiscent, prominently dentate-ciliolate along the sutures and hispidulous on the surface near the connective, not becoming divaricate; staminode glabrous, slightly expanded at the tip; capsule 7-9 mm. long and seeds 1.5-2 mm. long. N=8.

Dry, open, often rocky or gravelly slopes and flats, from the sagebrush to the ponderosa pine zone; c. Oreg. (s. Wheeler Co.) to s.w. Oreg., Calif., and w. Nev. June-July.

Our plants, as described above, belong to the well-marked var. roezlii (Regel) Jeps., which might with some reason be treated as a distinct species. Several other varieties, all with larger flowers, occur to the s. of our range.

Penstemon laxus A. Nels. Bot. Gaz. 54:147. 1912.
   P. watsonii ssp. laxus Keck, Am. Midl. Nat. 23:610. 1940. (Nelson & Macbride 1196, Ketchum, Ida.)

Plants tufted from a surficial, somewhat woody rhizome caudex, 3-7 dm. tall, rather slender-stemmed; herbage finely hirtellous-puberulent throughout, or the leaves occasionally glabrous; leaves entire, all cauline, the lowermost 2-several pairs distinctly reduced and generally soon withering, the other mostly 5-8 pairs well distributed along the stem, sessile, more
or less lanceolate, mostly 4-9 cm. long and 5-17 mm. wide; flowers in 1-several compact, often approximate verticillasters, generally ascending, often strongly so; calyx 2-3.5 mm. long,
the segments scarious-margined and sometimes erose, acute or cuspidate to subtruncate; corolla blue-purple, mostly 11-15 mm. long, the limb rather obscurely bilabiate, the tube narrow, mostly 2-3 mm. wide at the mouth, the palate strongly bearded; staminode strongly to
weakly bearded, or even glabrous; pollen sacs glabrous, broadly ovate or subrotund, 0.5-0.9
mm. long, wholly dehiscent, becoming opposite and sometimes explanate; capsules 5-6 mm.
long and seeds 0.6-1.3 mm. long.

Dry meadows and open or sparsely wooded slopes in the foothills and at moderate elevations
in the mountains; Blaine, Elmore, and Boise cos., Ida., to the n. of the Snake R. plains.
June-July.

Penstemon laxus is closely allied to the more southern, geographically disjunct P. watsonii
Gray, which latter occurs from s.e. Owyhee Co., Ida., to Nev., Utah, n.w. Ariz., c. Colo., and
s.w. Wyo. It may well be that P. laxus should be subordinated to P. watsonii, as indicated by Keck,
but the proper combination in varietal status has not yet been made, and in view of the several differences from P. laxus exhibited by P. watsonii (spreading, much more flaring corolla with more
sparsely bearded palate, slightly larger and relatively narrower anthers with ciliolate-dentic-
ulate sutures, often broader leaves) and the lack of demonstrated intergradation or geographical contact between the two taxa, I am reluctant to disturb the nomenclature at this time.

Penstemon lemhiensis (Keck) Keck & Cronq. Britt. 8:248. 1957.
    P. speciosus ssp. lemhiensis Keck, Am. Midl. Nat. 23:612. 1940. (Blair s. n., Granite
    Mt., Lemhi Co., Ida.)

Similar in most respects to P. cyaneus, averaging a little narrower-leaved; herbage (?al-
ways) finely hirtellous-puberulent at least in part; calyx 7-11 mm. long, the segments lanceolate to narrowly ovate, evidently but not strongly scarious-margined below, tapering to a long-
acuminate or subcaudate tip; staminode glabrous.

In grassland and open ponderosa pine forests, from the valleys to moderate elevations in the
mountains; n. Lemhi Co., Ida. (from about Salmon northward), to s. Ravalli Co. and n. Beaverhead Co. (Big Hole Valley), Mont. June-July.

This appears to be a well-marked local species, fully as valid as the other species of the
group. A single specimen has been reported to have glabrous anthers.

Penstemon lyallii Gray, Syn. Fl. 2nd ed. 2$^1$:440. 1886.
    P. menziesii var. lyallii Gray, Proc. Am. Acad. 6:76. 1862. (Lyall, between Ft. Colville
    and the Rocky Mts.)
    P. linearifolius Coult. & Fish. Bot. Gaz. 18:302. 1893. P. lyallii var. linearifolius Kraut-
    ter, Contr. Bot. Lab. U. Pa. 3:104. 1908. (MacDougal, "Lake Pend d'Oreille," Ida., in
    1892)

Perennial from a stout taproot and branched caudex, 3-8 dm. tall, the more or less numerous stems glabrous below the evidently glandular-hairy inflorescence, or puberulent in lines,
wholly herbaceous, or often woody at the base; leaves all cauline, the lower reduced, the others narrow and elongate, sessile or tapering to an obscurely subpetiolar base, glabrous or occasionally rough-puberulent, rather remotely serrulate to subentire, mostly 3-13 cm. long
and 3-10(15) mm. wide; inflorescence more or less paniculate, some or all of the axillary peduncles generally branched and 2- to several-flowered; calyx 7-15 mm. long, the segments
lanceolate or narrower, entire and subherbaceous; corolla lavender, 3-4.5 cm. long, about
1 cm. wide at the mouth, glabrous outside, conspicuously woolly-villous along the prominent
ventral ridges within, the lower lip not much longer than the upper; anthers densely long-woolly, the pollen sacs dehiscent throughout and becoming explanate; filaments and staminode glabrous, the slender staminode mostly shorter than the fertile filaments; capsules 10-14 mm. long,
the seeds more or less prismatic, scarcely or very narrowly wing-margined, 1-2 mm. long.

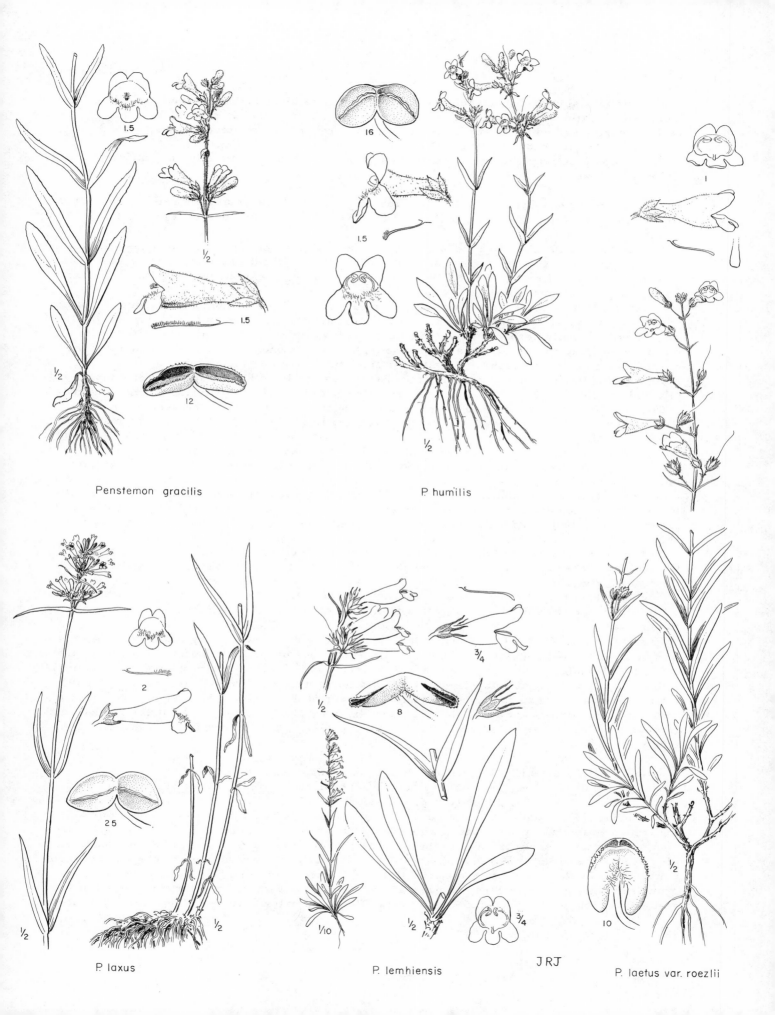

Penstemon gracilis

P. humilis

P. laxus

P. lemhiensis

JRJ

P. laetus var. roezlii

Steep, rocky slopes, cliffs, and banks, and gravel bars along streams, from the foothills to moderate elevations in the mountains; n. w. Mont. and s. w. Alta. to s. e. B. C. and n. Ida., perhaps barely extending into Spokane Co., Wash.; records from Stevens Pass, Wash., and Crook Co., Oreg., need confirmation. June-Aug.

This sharply marked species evidently hybridizes with the superficially very different P. ellipticus in Mont.

Penstemon montanus Greene, Pitt. 2:240. 1892. (Tweedy 866, mountains of Yellowstone Nat. Pk.)
   P. woodsii A. Nels. Bot. Gaz. 52:274. 1917. (Woods 265, Sawtooth Nat. Forest, Ida.) = var. montanus.
   PENSTEMON MONTANUS var. IDAHOENSIS (Keck) Cronq. hoc loc. P. montanus ssp. idahoensis Pennell & Keck in Davis, Fl. Idaho, 632. 1952, without Latin or type. P. montanus ssp. idahoensis Keck, Britt. 8:247. 1957. (Meyer 2276, rocky slopes, Big Roaring Lake, 20 miles n. of Pine, Elmore Co., Ida.)
   Perennial from a stout taproot and branched caudex, the latter commonly slender and often resembling a system of branching rhizomes; stems several or numerous, 1-3 dm. long, often lax, slender, wholly herbaceous, or often distinctly woody toward the base; sterile shoots not much if at all developed, or few and resembling the fertile ones; inflorescence and often the whole herbage evidently glandular-hairy and frequently clammy; leaves wholly cauline, the lower tending to be reduced, not forming a basal cluster, the others (especially the upper) sessile and often clasping, or narrowed to a short subpetiolar base, mostly 1.5-5 cm. long and 7-25 mm. wide, sharply toothed to entire; inflorescence essentially racemose, ordinarily with two pedicellate flowers at a node, each axillary to a bracteal leaf, and the main axis often also terminating in a flower; calyx 8-14 mm. long, with narrow, herbaceous, entire segments; corolla blue-lavender or lavender-orchid to light violet, 26-39 mm. long, about 1 cm. wide at the mouth, keeled along the back, glabrous outside, long-hairy within on the lower side, at least near the orifice, especially along the two prominent ventral ridges; anthers densely long-woolly, the pollen sacs wholly dehiscent and eventually explanate; staminode slender, slightly to very much shorter than the fertile stamens, glabrous to sometimes evidently long-bearded for much of its length; capsule 9-14 mm. long; seeds more or less flattened and tending to be somewhat wing-margined, 1.5-2 mm. long.
   Shifting talus slopes, or less often in rock crevices, well up in the mountains, often near timber line; Park Co., Mont., to the mountains of c. Ida. (as far w. as w. Valley Co.), s. through w. Wyo. and e. Ida. to c. Utah. July-Aug.
   The slenderly branched caudex of this species is directly correlated with the unstable habitat in which it usually occurs, and plants from cliff crevices are much more compact at the base.
   Plants from w. Custer Co. (near Bonanza) to n. Elmore and w. Valley cos., Ida., have entire or subentire, often glaucous leaves that are glabrous or rough-puberulent but ordinarily not at all glandular, and tend to have a very compact inflorescence, with the pedicels commonly well under 1 cm. long at anthesis; these constitute the well-marked var. idahoensis (Keck) Cronq. The var. montanus, occupying the remainder of the range of the species, has more or less strongly toothed, not at all glaucous, usually rough-hairy and often glandular leaves, and frequently has a looser inflorescence with longer pedicels, these sometimes as much as 2 cm. long.

Penstemon nemorosus (Dougl.) Trautv. Bull. Acad. St. Pétersb. 5:345. 1839.
   Chelone nemorosa Dougl. in Lindl. Bot. Reg. 14: pl. 1211. 1829. Apentostera triflora Raf. New Fl. N. Am. 2:73. 1836. (Douglas, "mountains of N. W. Am.")
   Perennial, (3)4-8 dm. tall, with several or many glabrous or finely puberulent stems arising from a branched, woody caudex which tends to surmount a short, quickly deliquescent taproot, herbaceous nearly or quite to the base; leaves glabrous, or somewhat hirsute beneath, all cauline, the lower ones reduced, the others distinctly short-petiolate, rather thin, conspicuously serrate, lanceolate to ovate, mostly (2.5)4-11 cm. long and (1)1.5-4 cm. wide; inflorescence evidently glandular-hirsute, generally more or less paniculate, with some or all of the axillary peduncles branched and 2- to several-flowered; calyx 6-10 mm. long, the segments

lanceolate or lance-ovate, entire; corolla pink-purple, glandular-hairy outside, glabrous inside, 25-33 mm. long, about 1 cm. wide at the mouth, strongly 2-ridged on the lower side within, the lower lip much longer than the upper; anthers densely long-woolly, the pollen sacs dehiscent throughout and eventually becoming explanate; filaments retrorsely scabrous-puberulent or granular, and conspicuously long-hairy at the base; staminode much shorter than the fertile filaments, bearded throughout its length; capsule 11-17 mm. long; seeds flattened, broadly wing-margined all around, 2-3 mm. long. N=15.

In woodlands and on moist, open, rocky slopes, from near sea level to 8000 ft. in the mountains; Wash. and s. Vancouver I. to n.w. Calif., from the Cascade Range to the coast, chiefly in the Cascade and Olympic mts. July-Aug. (Oct.)

A sharply marked species, and the only one in the genus known to have a chromosome number not based on x=8.


Penstemon nitidus Dougl. ex Benth. in DC. Prodr. 10:323. 1846.
    P. acuminatus var. minor Hook. Fl. Bor. Am. 2:97. 1838. (Douglas, "Red Deer and Eagle
    Hills" of the Saskatchewan R.)
    PENSTEMON NITIDUS var. POLYPHYLLUS (Pennell) Cronq. hoc loc. P. nitidus ssp. poly-
    phyllus Pennell, Not. Nat. Acad. Phila. 95:6. 1942. (Kirkwood 1274, Missoula, Mont.)

Perennial, commonly with several rather stout stems arising from a short, branched caudex which may surmount a quickly deliquescent taproot, mostly 1-3 dm. tall, wholly glabrous, the herbage glaucous; leaves thick and firm, entire, the basal ones tufted, oblanceolate or broader, petiolate, up to about 10 cm. long and 2.5 cm. wide; cauline leaves mostly sessile or nearly so, narrow or broad; inflorescence of several distinct or often approximate verticillasters, the subtending bracts broad as in P. acuminatus, or narrow and elongate; corolla bright blue, mostly 13-18 mm. long, the tube expanded distally; pollen sacs essentially glabrous, dehiscent throughout and becoming opposite, but not explanate, 0.8-1.2 mm. long; staminode tending to be shortly exserted, prominently yellow-bearded for 2-4 mm. toward the evidently expanded tip, the hairs mostly 0.5-1.0 mm. long, sometimes longer; capsules 9-12 mm. long and seeds about 3 mm. long.

Open, often grassy hillsides and plains, occasionally extending to moderate elevations in the mountains, where sometimes occurring on talus; w. Mont. and adj. Lemhi Co., Ida., to s.w. Alta., e. to n. Wyo., w. N.D., and s. Man. May-July.

Penstemon nitidus is closely allied to both P. angustifolius Pursh and P. acuminatus Dougl. The weak morphological cleavage between P. nitidus and P. acuminatus is bolstered by their wholly discrete ranges and by a difference in habitat as well, so that it seems advisable to maintain these two as distinct. P. angustifolius is consistently narrow-leaved, and has somewhat larger corollas and anthers than P. nitidus, but the distinction is not always very sharp, especially when P. nitidus var. polyphyllus is taken into account. On geographic grounds as well as by the abundance of intermediates, the var. polyphyllus is necessarily associated with P. nitidus rather than with P. angustifolius, which is not known to occur in our range. The ranges of P. nitidus and P. angustifolius come together in e. Mont. and w. N.D., and it is possible that monographic study will necessitate the treatment of these several taxa (except P. acuminatus) as geographic races of a single species, under the binomial P. angustifolius Pursh.

Two poorly defined varieties of P. nitidus may be recognized:
1 Leaves and bracts relatively narrow, the bracts mostly lanceolate or lance-ovate; calyx
    mostly 5-8 mm. long at anthesis; corolla generally with a few long hairs near the base of
    the lower lip within; a relatively uniform population, common near Missoula, Mont., and
    known also from Powell, Granite, and Silver Bow cos., Mont., and Lemhi Co., Ida.
                                                        var. polyphyllus (Pennell) Cronq.
1 Leaves and bracts mostly broader, the bracts mostly clasping and ovate to subrotund; calyx
    mostly 3-6 mm. long at anthesis; corolla generally glabrous; range of the species, except
    for the area occupied by var. polyphyllus, variable, often approaching var. polyphyllus in
    one or another respect                                        var. nitidus

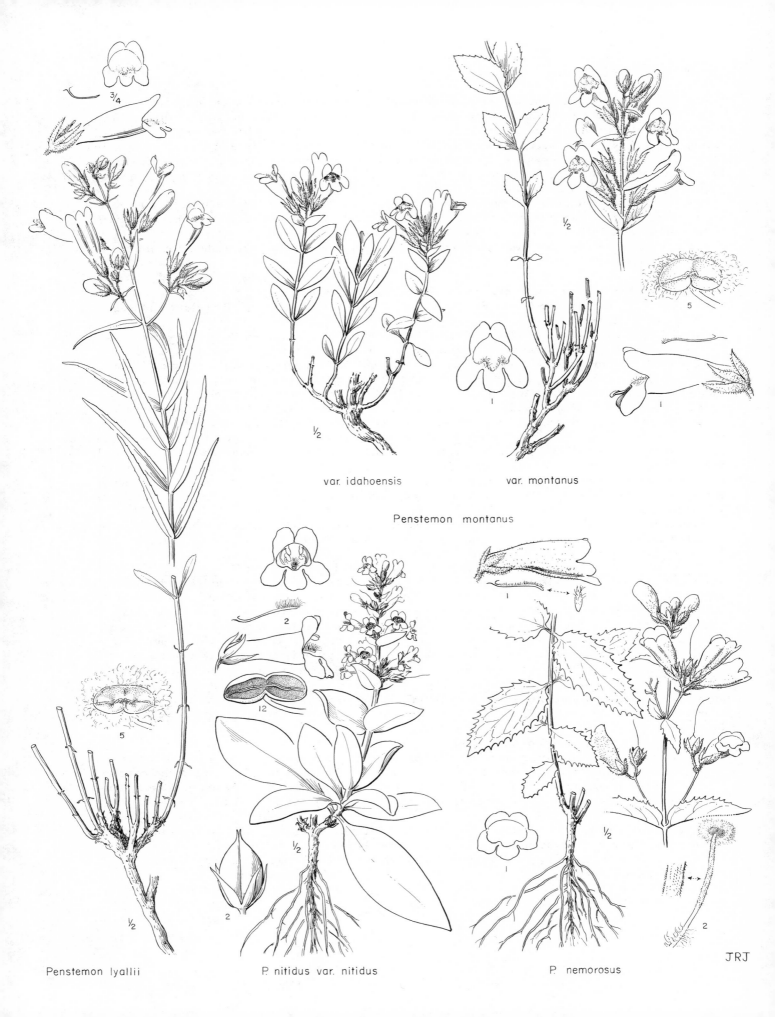

var. idahoensis

var. montanus

Penstemon montanus

Penstemon lyallii

P. nitidus var. nitidus

P. nemorosus

JRJ

Penstemon ovatus Dougl. in Hook. Curtis' Bot. Mag. 56: pl. 2903. 1829. (Douglas, near "the Grand Rapids of the Columbia River, at the distance of one hundred and forty miles from the ocean")

Robust plants, 3-10 dm. tall; stems clustered from a surficial woody caudex, spreading-hirsute below the inflorescence, sometimes rather shortly so; leaves hairy like the stem, especially (or at least) along the midrib beneath, or reputedly sometimes glabrous, all conspicuously serrate, often with slender teeth; basal leaves clustered, petiolate, with ovate or deltoid-ovate to subcordate blade up to 10 cm. long and half as wide, about equaling the petiole, often less sharply toothed than the well-developed, sessile and clasping cauline leaves, these often fully as large as the basal ones, the lower tending to be broadly oblong or ovate-oblong, the upper more ovate or triangular-ovate; inflorescence large and open, strongly glandular-hirsute; calyx 3-5 mm. long, the segments lanceolate or lance-ovate, often with evident sub-parallel veins, only narrowly or scarcely scarious-margined; corolla externally glandular-hairy, blue with paler, hairy palate and more or less evident guidelines, 15-22 mm. long, strongly bilabiate, the lower lip much longer than the upper; pollen sacs ovate or elliptic, 0.8-1.1 mm. long, wholly dehiscent and becoming opposite, sometimes eventually explanate, glabrous, or the sutures obscurely setulose-dentate; staminode bearded toward the more or less recurved, scarcely expanded tip; capsule 4-6 mm. and seeds 1-1.5 mm. long. N=8.

Open woods w. of the Cascade summits, at elevations of less than 3000 ft.; Wash., extending only a short distance into s. B.C. and n. Oreg. May-June (Aug.)

Penstemon payettensis Nels. & Macbr. Bot. Gaz. 62:147. 1916. (Payette Nat. Forest, Ida., sent to Aven Nelson by G. B. Mains under the herbarium number D-73)

Plants (1)2-7 dm. tall, with several (solitary) stout stems from a compact branched caudex (or a nearly simple crown) which commonly surmounts a short, quickly deliquescent taproot, the herbage and inflorescence essentially glabrous throughout; leaves thick and firm, often darkening in drying, the basal ones clustered, up to 15 cm. long, petiolate, with oblanceolate to broadly ovate blade up to 5 cm. wide; cauline leaves, except the lowermost, mostly sessile, the middle and upper mostly lance-ovate to ovate, broad-based and commonly clasping, (1)1.5-4 cm. wide; inflorescence of several or rather numerous distinct or approximate verticillasters; calyx 5-8 mm. long, the segments scarcely to evidently scarious-margined below, with a more or less elongate, acuminate to subcaudate tip; corolla bright blue, 18-28 mm. long, expanded distally, often 1 cm. wide at the mouth; pollen sacs 1.1-1.9 mm. long, straight, opening nearly or quite to the proximal end, but generally not confluent, more or less strongly setulose-dentate along the sutures, otherwise glabrous or often obscurely hispidulous toward the connective, becoming opposite or upwardly divaricate; staminode short-bearded toward the slightly or scarcely expanded tip, varying to sometimes essentially glabrous; capsule 8-10 mm. long and seeds about 2 mm. N=8.

Open places, in well-drained, often rather sandy or gravelly soil, from the valleys and foothills to moderate elevations in the mountains; c. Ida., from the Salmon R. axis on the n. to the margin of the Snake R. plains (and n. Owyhee Co.) on the s., more common toward the west than the east, and in the Wallowa Mts. of n. e. Oreg. May-Aug.

Penstemon peckii Pennell, Not. Nat. Acad. Phila. 71:12. 1941. (Pennell 15528, 9 miles n.w. of Sisters, Deschutes Co., Oreg.)

Plants tufted from a slender, superficial, woody rhizome-caudex, mostly 2.5-7 dm. tall, slender-stemmed, glabrous below the glandular-hairy inflorescence, or the stem slightly hirtellous-puberulent; leaves numerous, entire, linear or nearly so, up to about 7 cm. long and 5 mm. wide, all or nearly all cauline, the lower petiolate and rather crowded, but not forming rosettes, the others sessile and more distant; inflorescence of several fairly dense and often approximate verticillasters; calyx 2-3.5 mm. long, the segments abruptly pointed, with broad and often erose scarious margins; corolla tending to be declined, glandular-hairy, pale purplish blue to white, 8-10 mm. long, the tube narrow, only 2-3 mm. wide at the mouth, the palate bearded; staminode bearded toward the expanded tip; pollen sacs glabrous, subrotund or broadly ovate, 0.4-0.5 mm. long, wholly dehiscent, becoming opposite and more or less explanate; capsule about 4 mm. long and seeds less than 1 mm. long.

Dry soil, commonly with ponderosa pine, on the e. slope and along the e. base of the Cascade Range, from Mt. Hood to the Three Sisters, Oreg. June-Aug.

Penstemon pennellianus Keck, Am. Midl. Nat. 23:614. 1940. (Sprague s.n., Joseph Creek Canyon below Flora, Wallowa Co., Oreg., June 20, 1932)

Perennial, 2-6 dm. tall, generally with several stout stems from a compact branched caudex which tends to surmount a short, quickly deliquescent taproot; herbage and inflorescence glabrous; leaves entire, thick and firm, tending to darken in drying, the basal ones clustered, up to 27 cm. long and 4 cm. wide, petiolate, with oblanceolate or rather narrowly elliptic blade; cauline leaves, except the lowermost, mostly sessile, the middle and upper ones relatively broad and clasping, the better developed ones mostly 6-9 cm. long, 2.5-4 cm. wide, 2-3 times as long as wide; inflorescence of several distinct or approximate verticillasters, not secund; calyx 6-9 mm. long, the segments fairly narrow and evidently acuminate, their scarious margins not very broad; corolla bright blue, 26-33 mm. long, about 1 cm. wide at the mouth, glabrous; pollen sacs 1.9-2.5 mm. long, divaricate, tending to be sigmoidally twisted, evidently setose-dentate along the sutures and sparsely hispidulous on the surface toward the proximal end, which remains indehiscent; staminode short-bearded toward the slightly or scarcely expanded tip; capsules and seeds presumably as in P. speciosus.

Open, rocky or gravelly ridges and slopes at moderate elevations in the Blue Mts., from n. Wallowa Co., Oreg., to s. Asotin, Columbia, and Garfield cos., Wash. May-June.

Penstemon procerus Dougl. ex R. Grah. Edinb. New Phil. Journ. 7:348. 1829.

P. confertus var. violaceus Trautv. Bull. Acad. St. Pétersb. 5:344. 1839. P. confertus var. caeruleopurpureus Gray, Proc. Am. Acad. 6:72. 1862. P. confertus var. procerus Coville, Contr. U.S. Nat. Herb. 4:169. 1893. (Garden plants, from seeds collected by Drummond, probably near the Saskatchewan R. in Alta.) The plant which Douglas had in mind as P. procerus was the vigorous northwestern phase of P. rydbergii which has more recently been segregated as P. hesperius Peck, but the application of the name (P. procerus) under the Rules of Nomenclature is fixed by the specimens which Graham cited in its first formal publication.

P. micranthus Nutt. Journ. Acad. Phila. 7:45. 1834. Lepteiris parviflora Raf. New Fl. N. Am. 2:73. 1836. Penstemon procerus var. micranthus M. E. Jones, Bull. U. Mont. Biol. 15:45. 1910. (Wyeth, valleys of the Rocky Mts., near the sources of the Columbia) = var. procerus.

PENSTEMON PROCERUS var. TOLMIEI (Hook.) Cronq. hoc loc. P. tolmiei Hook. Fl. Bor. Am. 2:98. 1838. P. procerus ssp. tolmiei Keck, Britt. 8:249. 1957. (Tolmie, Mt. Rainier, Wash.)

PENSTEMON PROCERUS var. FORMOSUS (A. Nels.) Cronq. hoc loc. P. pulchellus Greene, Pitt. 3:310. 1898, not of Lindl. in 1828. P. formosus A. Nels. Proc. Biol. Soc. Wash. 17:100. 1904. P. cacuminis Pennell, Not. Nat. Acad. Phila. 71:2. 1941. P. tolmiei ssp. formosus Keck, Am. Midl. Nat. 33:147. 1945. P. procerus ssp. formosus Keck, Britt. 8:249. 1957. (Cusick 1720, alpine summits of the Blue Mts., Oreg., in 1897)

P. procerus ssp. pulvereus Pennell, Contr. U.S. Nat. Herb. 20:366. 1920. (Pennell 6036, n. of Swan Lake, Yellowstone Nat. Pk.) A form of var. procerus with puberulent sepals.

PENSTEMON PROCERUS var. BRACHYANTHUS (Pennell) Cronq. hoc loc. P. brachyanthus Pennell, Not. Nat. Acad. Phila. 71:3. 1941. P. tolmiei ssp. brachyanthus Keck, Am. Midl. Nat. 33:148. 1945. P. procerus ssp. brachyanthus Keck, Britt. 8:249. 1957. (Pennell 15710, Mt. Hood, Oreg.)

Plants more or less tufted from a loose or compact, surficial woody rhizome-caudex, 0.5-4(7) dm. tall, slender-stemmed, essentially glabrous throughout, or with puberulent sepals; leaves entire, the basal ones petiolate, oblanceolate to elliptic or ovate, up to 10 cm. long and nearly 1.5 cm. wide, often poorly developed or wanting; cauline leaves few, mostly sessile, up to 8 cm. long and 2 cm. wide, often much smaller, especially when the basal leaves are well developed; inflorescence of 1-several very dense verticillasters, the flowers more or less declined; calyx 1.5-6 mm. long, the segments truncate to caudate or gradually acute, the margins tending to be scarious and often erose; corolla deep blue-purple (sometimes ochroleucous

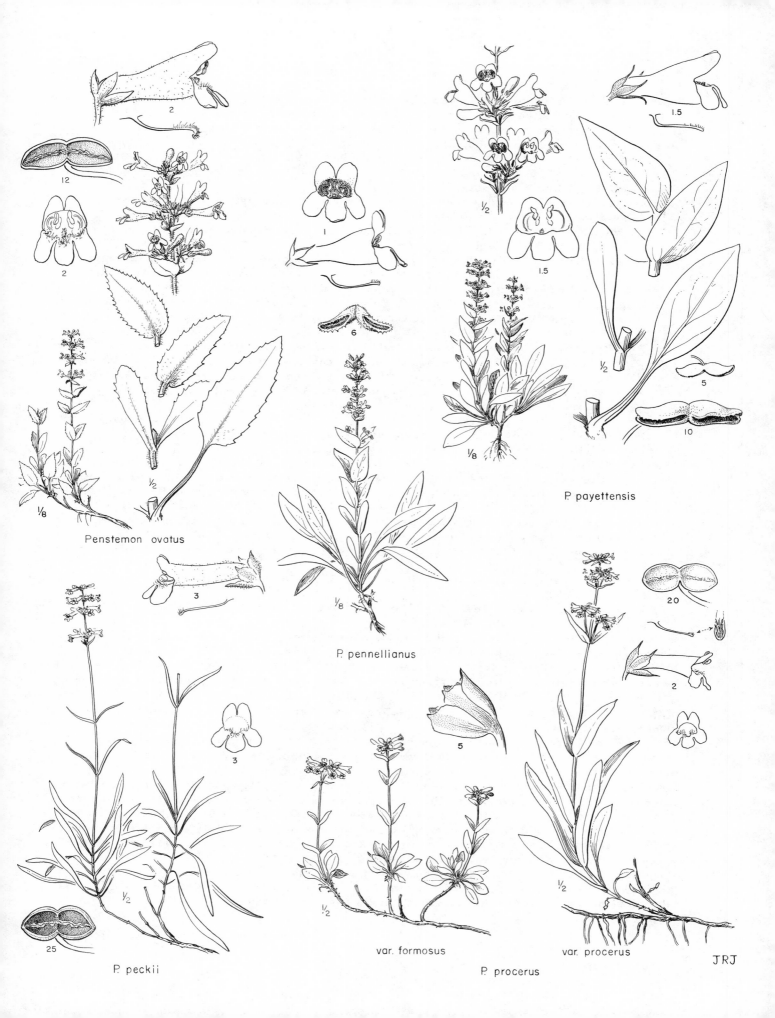

2

12

2

1

6

1.5

1.5

5

10

P. payettensis

Penstemon ovatus

3

3

P. pennellianus

20

2

5

var. formosus

var. procerus

P. peckii

25

P. procerus

JRJ

in var. tolmiei), 6-11 mm. long, the tube narrow, 2-3 mm. wide at the mouth, the limb more or less spreading and not very strongly bilabiate, the palate and the expanded tip of the sterile filament more or less bearded, or the sterile filament sometimes glabrous in var. formosus; pollen sacs glabrous, subrotund, becoming opposite and explanate, 0.3-0.7 mm. long; capsules 4-5 mm. long and seeds about 1 mm. long. N=8, 16.

Dry meadows and open or timbered slopes from the foothills to above timber line; Alas. and Yukon to Calif., Colo., and Sask. June-Aug.

The varieties of P. procerus occurring in our range may be distinguished, with some difficulty, as follows:

1 Calyx mostly 3-6 mm. long, the segments more or less strongly caudate-tipped, or narrower and more gradually tapering
   2 Basal rosette generally only poorly or scarcely developed, the enlarged basal leaves typically few or none; plants mostly (1)2-4(7) dm. tall, usually with more than 1 verticillaster of flowers per stem; corolla blue; foothills and moderate to less often high elevations in the Rocky Mts. region from Yukon and Alas. to Colo., extending w. (but not at high elevations) to the base of the Cascade Range in Wash., and to the Wallowa-Blue Mt. region in n.e. Oreg.                                          var. procerus
   2 Basal rosette well developed; plants mostly 0.5-1.5(3) dm. tall, with only 1 or less often 2-3 verticillasters of flowers per stem, the leaves averaging shorter and relatively broader than in var. procerus; corolla blue or sometimes ochroleucous; moderate to high elevations in the Cascade and Olympic mts. of Wash., extending well into s. B.C., often in more exposed and rocky habitats than var. procerus
                                                   var. tolmiei (Hook.) Cronq.
1 Calyx mostly 1.5-3 mm. long, the segments obtuse or subtruncate to often shortly cuspidate; basal rosette generally well developed; flowers blue
   3 Dwarf, alpine and subalpine plants, mostly 0.5-1.5 dm. tall, with relatively short, broad basal leaves and with the inflorescence commonly reduced to a single verticillaster; Wallowa and Strawberry mts. of e. Oreg., and in the mountains of Calif. and Nev.
                                      var. formosus (A. Nels.) Cronq.
   3 Taller, mostly montane to subalpine plants, mostly (1)1.5-3 dm. tall, the inflorescence usually of more than 1 verticillaster; Cascade Range of Oreg. and n. Calif., and apparently also in the Wallowa Mts. of n.e. Oreg.          var. brachyanthus (Pennell) Cronq.

**Penstemon pruinosus** Dougl. ex Lindl. Bot. Reg. 15: pl. 1280. 1828. (Douglas, "near the Priest's rapid of the Columbia")
  P. amabilis G. N. Jones, Res. Stud. State Coll. Wash. 2:126. 1930. (E. Nelson 1682, Blewett Pass, Kittitas Co., Wash.)

Plants tufted from a branched, often very stout and woody caudex, mostly 1-4(5) dm. tall, always glandular-hairy in the inflorescence, otherwise varying from glandular-hirsute throughout to merely hirtellous-puberulent or even essentially glabrous; leaves more or less strongly and sharply serrate, but often irregularly so, sometimes most of them entire; basal leaves well developed, up to about 10(15) cm. long and 2(3) cm. wide, the blade narrowly or broadly elliptic to ovate or lance-ovate, about equaling or shorter than the petiole; cauline leaves mostly sessile, triangular-ovate to narrowly lanceolate, mostly 1.5-6 cm. long and 0.5-2(3) cm. wide; inflorescence of several discrete or approximate verticillasters, these rather compact but seldom with very numerous flowers; calyx 3-6 mm. long, the segments lanceolate or narrowly ovate, inconspicuously or scarcely scarious-margined; corolla glandular-hairy externally, deep blue or less often lavender, the throat mostly paler purplish and marked with guide lines, 11-16 mm. long, the tube slightly or moderately expanded distally, mostly 2.5-5 mm. wide at the mouth; palate bearded; pollen sacs glabrous, 0.5-0.8 mm. long, dehiscent nearly or quite throughout and becoming opposite, sometimes eventually explanate; staminode only very slightly, if at all, expanded at the bearded tip; capsule 5-7 mm. long, glabrous or obscurely glandular-puberulent; seeds about 1 mm. long. N=8.

Open, mostly rocky places, from the valleys and plains to moderate elevations in the mountains; along the e. slope of the Cascades in Wash. from Kittitas Co. n. into adj. B.C., and e. into the scablands as far as Adams and Franklin cos. May-early July.

Penstemon pumilus Nutt. Journ. Acad. Phila. 7:46. 1834. (Wyeth, "Little Goddin River," now
the Little Lost R., Ida.)

Perennial from a compactly branched woody caudex which sometimes surmounts a short,
quickly deliquescent taproot, densely cinereous-puberulent throughout, also becoming glandular in the inflorescence; stems clustered, less than 2 dm. tall, prostrate to suberect; leaves
entire, narrow, the basal and lower cauline ones mostly oblanceolate and more or less petiolate, 1.5-5 cm. long and 2-6 mm. wide, the others becoming sessile and more linear-oblong or
lance-linear, scarcely to evidently reduced; inflorescence of 1-several small, often approximate verticillasters; calyx 5-8 mm. long, the segments lanceolate, herbaceous, entire; corolla stipitate-glandular outside, glabrous inside, 15-23 mm. long, 5-8 mm. wide at the mouth,
commonly with blue limb and more purplish tube; pollen sacs glabrous, dehiscent throughout,
becoming opposite but not explanate, 0.8-1.2 mm. long; sterile filament included or barely
exserted from the mouth of the corolla, prominently long-bearded for most of its length; capsule 5-8 mm. long and seeds 2-2.5 mm. long.

Dry, open places at lower elevations, often with sagebrush; valley of the Salmon R. and
nearby tributaries from near Salmon to near Challis, Ida., and s. through the valleys of the
Lost rivers and Birch Creek to the margin of the Snake R. plains. May-July.

Penstemon radicosus A. Nels. Bull. Torrey Club 25:280. 1898. (Nelson 2962, Evanston, Wyo.)

Plants tufted from a surficial, woody caudex, 1.5-4 dm. tall, rather slender-stemmed,
glandular-hairy in the inflorescence, otherwise finely hirtellous-puberulent throughout, or the
upper surfaces of the leaves sometimes nearly glabrous; leaves entire, all cauline, the lowermost ones distinctly reduced, the others well developed and rather uniform, mostly sessile,
lanceolate to lance-linear, 2-6 cm. long, 2-11(14) mm. wide, the stem appearing rather leafy;
inflorescence of several rather loose and few-flowered, often approximate verticillasters; calyx 5-9 mm. long, the segments lanceolate or ovate, acuminate, with narrow (or obscure) entire or subentire, scarious margins; corolla glandular-hairy, blue-purple, ventrally whitish
with darker guidelines, 16-23 mm. long, the lower lip projecting forward and evidently longer
than the reflexed upper one, the tube 4-7 mm. wide at the mouth, the palate moderately or
sparsely bearded; pollen sacs 0.8-1.1 mm. long, dehiscent nearly or quite throughout, becoming opposite, but not explanate, evidently to obscurely denticulate-ciliolate along the sutures,
otherwise glabrous; staminode densely bearded toward the slightly expanded tip; capsule 6-8
mm. long and seeds 1-1.5 mm. long.

Dry, open places, often with sagebrush, from the plains and foothills to moderate elevations
in the mountains; Wyo., extreme s.w. Mont., e. Ida. (w. to Custer, Butte, and Cassia cos.), n.
Nev., n. Utah, and extreme n. Colo. May-July.

A well-marked species, not to be confused with anything else.

Penstemon richardsonii Dougl. ex Lindl. Bot. Reg. 13: pl. 1121. 1828. (Douglas, bare dry
rocks in the vicinity of the Columbia and its branches; perhaps taken near Celilo Falls)
P. pickettianus St. John, Proc. Biol. Soc. Wash. 44:33. 1931. (Pickett 1390, Ribbon Cliff,
Columbia R. n. of Wenatchee, Wash.) = var. richardsonii.
PENSTEMON RICHARDSONII var. DENTATUS (Keck) Cronq. hoc loc. P. richardsonii ssp.
dentatus Keck, Britt. 8:250. 1957. (Cronquist 7260, 10 miles n.w. of Mitchell, Wheeler
Co., Oreg.)
PENSTEMON RICHARDSONII var. CURTIFLORUS (Keck) Cronq. hoc loc. P. richardsonii
ssp. curtiflorus Keck, Britt. 8:249. 1957. (Constance & Beetle 2705, 5 miles s. of Condon,
Gilliam Co., Oreg.)

Taprooted perennial, distinctly shrubby at the base, 2-8 dm. tall, with more or less numerous, rather brittle and often slender stems, evidently glandular-hairy in the inflorescence only, otherwise merely puberulent, or the leaves often glabrous; leaves opposite, all cauline,
the lowermost reduced, the others sessile or (especially the lower) short-petiolate, sharply
serrate-dentate to irregularly laciniate-pinnatifid, up to 7 cm. long and 3 cm. wide; inflorescence a terminal mixed panicle (sometimes rather simple and few-flowered), generally more
or less leafy-bracteate below; calyx 4-8 mm. long, the segments green or anthocyanic, entire
or with an occasional tooth, often distinctly unequal; corolla glandular-hairy outside, glabrous

or with some long, white hairs near the mouth within, bright lavender (the lower lip striped within), 22-32 mm. long (smaller in var. curtiflorus), ventricosely inflated distally and often more than 1 cm. wide at the mouth, strongly bilabiate, the upper lip cleft less than half its length; anthers 1.3-1.7 mm. long (essentially similar in all 3 varieties), permanently horse-shoe-shaped, the pollen sacs dehiscent across their confluent apices, saccate and indehiscent below, not becoming divaricate, the sutures more or less evidently dentate-ciliolate; staminode shortly exserted, sparsely or moderately long-bearded toward the scarcely expanded tip, or glabrous and more expanded in var. curtiflorus; capsules 5-8 mm. long and seeds about 1 mm. long or a little less. N=8.

   Cliff crevices and other dry, open, mostly rocky places at lower elevations; e. of the Cascade summits (and in the Columbia gorge) in Wash. and Oreg., and extending up the Okanogan Valley into s. B. C.; rarely as far e. as Wawawai, Wash., otherwise not closely approaching the Ida. boundary. May-Aug.

   The species consists of three geographic varieties, which may be characterized as follows:
1 Corolla 15-20 mm. long; staminode glabrous or nearly so, evidently expanded at the tip; leaves about as in var. dentatus, varying to almost as in var. richardsonii; s. Gilliam and n. Wheeler cos., to s. Wasco Co., Oreg.                          var. curtiflorus (Keck) Cronq.
1 Corolla 22-32 mm. long; staminode generally bearded, often conspicuously so
   2 Leaves relatively narrow and deeply laciniate-toothed to irregularly laciniate-pinnatifid; upper surfaces of the leaves glabrous to somewhat pruinose-puberulent; Wash., extending into s. B. C., and barely across the Columbia R. into Oreg.          var. richardsonii
   2 Leaves broader, ovate or lance-ovate, merely toothed, usually more regularly so than in var. richardsonii; upper surfaces of the leaves more or less strongly puberulent; Wasco Co., Oreg., to Union, Grant, and Crook cos.                          var. dentatus (Keck) Cronq.

Penstemon rupicola (Piper) Howell, Fl. N. W. Am. 510. 1901.
   P. newberryi var. rupicola Piper, Bull. Torrey Club 27:397. 1900. (Piper 2086, Mt. Rainier, Wash.)
Suffruticose perennial, with or without an evident taproot, forming dense mats on the surface of the substrate, and with scattered, fertile, erect or ascending stems about 1 dm. high or less, these shortly spreading-hairy; mat leaves thick and firm, strongly glaucous, glabrous or often spreading-hispidulous, irregularly and often inconspicuously serrulate, the subrotund to broadly elliptic or obovate blade 8-18 mm. long, 1-2 times as long as wide; leaves of the flowering shoots (above the mat) few and generally more or less reduced, less than 1 cm. long; inflorescence a compact, few-flowered raceme, the bracts, pedicels, and main axis more or less strongly glandular-hairy, the calyx often less so; calyx 6-11 mm. long, the segments narrowly lance-elliptic to ovate-oblong; corolla pink or nearly red to pink-lavender or rose-purple, 25-36 mm. long, keeled along the back, glabrous inside and out, or bearing a few long hairs near the base of the lower lip within; anthers evidently long-woolly, sometimes not very densely so; pollen sacs wholly dehiscent and becoming explanate; staminode slender, sparsely bearded or glabrous, often much shorter than the fertile filaments. N=8.

   On cliffs, ledges, and rocky slopes, usually fairly well up in the mountains, but descending to lower elevations in the Columbia gorge; Cascade region from Kittitas Co., Wash., to n. Calif., and extending w. nearly to the coast in the Klamath region. Late May-Aug.

Penstemon rydbergii A. Nels. Bull. Torrey Club 25:281. 1898. (Nelson 3214, Laramie Hills, Albany Co., Wyo.)
   ? P. ellipticus Greene, Leafl. 1:167. 1906, not of Coult. & Fish. in 1893. P. vaseyanus Greene, Leafl. 1:200. 1906. (Vasey 446, Wash.)
   P. oreocharis Greene, Leafl. 1:163. 1906. (Hall & Chandler 301, Pineridge, Fresno Co., Calif.)
   PENSTEMON RYDBERGII var. VARIANS (A. Nels.) Cronq. hoc loc. P. attenuatus var. varians A. Nels. Bot. Gaz. 54:146. 1912. (Macbride 974, Twilight Gulch, Owyhee Co., Ida.; lectotype by Keck)
   P. aggregatus Pennell, Contr. U.S. Nat. Herb. 20:367. 1920. P. rydbergii ssp. aggregatus Keck, Am. Midl. Nat. 33:158. 1945. (Pennell 5918, Evanston, Uinta Co., Wyo.)

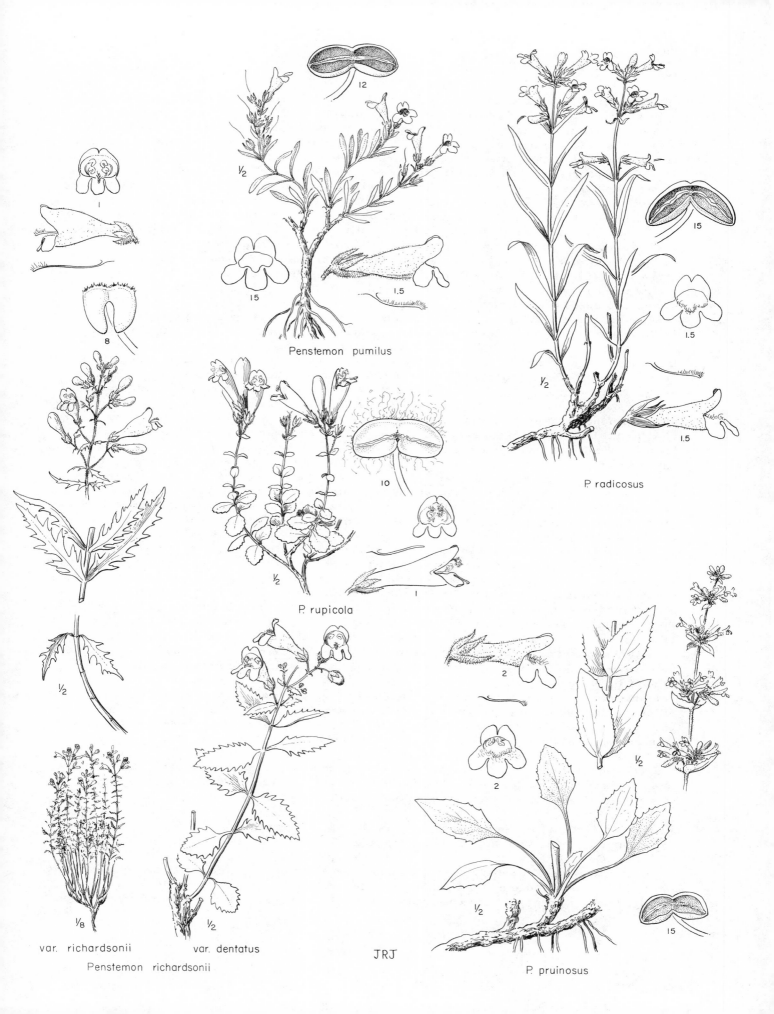

Penstemon pumilus

P. radicosus

P. rupicola

var. richardsonii    var. dentatus

Penstemon richardsonii

JRJ

P. pruinosus

P. hesperius Peck, Torreya 32:152. 1932. (Peck 16187, Gaston, Washington Co., Oreg.)

Plants more or less tufted from a loose or compact, surficial, woody rhizome-caudex, mostly 2-7(12) dm. tall, rather slender-stemmed, glabrous throughout, or sometimes puberulent in the inflorescence and along the stem; leaves entire, the basal ones petiolate, oblanceolate or narrowly elliptic, often forming distinct rosettes, up to 15 cm. long and 2 cm. wide, or sometimes poorly developed; cauline leaves few and mostly well developed, commonly sessile or nearly so, seldom 10 cm. long and 2 cm. wide; inflorescence of 1-several rather dense verticillasters, the flowers spreading at right angles to the stem; calyx 3-7(9) mm. long, the segments obscurely to fairly evidently scarious-margined and sometimes erose, tapering or more abruptly narrowed to the pointed tip; corolla blue-purple, mostly 11-15(18) mm. long, the limb less spreading and more distinctly bilabiate than in P. procerus, and the tube more expanded distally, mostly 3-5 mm. wide at the mouth ; palate bearded; staminode generally bearded at least at the expanded tip, rarely glabrous; pollen sacs glabrous, ovate, mostly 0.6-1.0 mm. long, dehiscent throughout and becoming opposite, but seldom explanate; capsule 5-6 mm. and seeds about 1 mm. long. N=8, 16.

Meadows and moist, open slopes, occasionally on drier slopes with sagebrush, chiefly in the foothills and at moderate elevations in the mountains; c. Wash. to s. w. Mont. (but apparently absent from c. Ida.) and n. Wyo., s. to the s. end of the Sierra Nevada in Calif., and to s. Utah and n. N. M., chiefly e. of the Cascades in our range, but also found occasionally in and near the n. end of the Willamette Valley in Oreg. and adj. Wash. Late May-July.

Typical P. rydbergii is a plant of the s. Rocky Mts., characterized by the very broad and conspicuously erose, white, scarious margins of the calyx segments. Our plants, as described above, belong to the mostly more northern or western var. varians (A. Nels.) Cronq. There is a partial bifurcation in the range of var. varians, and it has become customary to refer the plants of Wash., Oreg., w. Ida., n. Nev., and Calif. to one species (P. oreocharis Greene), and the plants of e. Ida., s. w. Mont., Wyo., Utah, and Colo. to another (P. aggregatus Pennell, or P. rydbergii ssp. aggregatus (Pennell) Keck). P. oreocharis, as thus treated, is known on the basis of 9 counts from widely separated stations to be a diploid, while the single recorded count of P. rydbergii ssp. aggregatus (Sevier Co., Utah) indicates it to be a tetraploid. Unfortunately, I have been able to find no morphologic difference to support the geographical and cytological ones. P. rydbergii ssp. aggregatus has been considered to have a larger corolla (15-18 mm.) than either P. rydbergii sens. strict. or P. oreocharis (10-15 mm.), but a large proportion of the specimens which, on the basis of their calyx, geographic origin, and authoritative annotation, must represent the ssp. aggregatus, have the corolla distinctly less than 15 mm. long, just as in P. oreocharis and typical P. rydbergii. There is at best merely a minor difference in average length of the corolla between ssp. aggregatus and the two related taxa under consideration. Therefore, keeping in mind the fact that diploid and tetraploid races are admitted without taxonomic distinction in several allied species, I am constrained to retain all of these plants (other than P. rydbergii proper) under the oldest (and only) varietal epithet, the var. varians. The type of var. varians comes from the geographical area of P. oreocharis.

Plants occurring w. of the Cascades have been segregated as P. hesperius Peck on the basis of their larger size and mostly puberulent inflorescence, but these characters are inconstant both e. and w. of the Cascades, and do not mark a clearly distinguishable population. The differences being at best statistical, it is not here considered necessary to give P. hesperius any formal taxonomic status.

Several collections from c. Wash. diverge from the norm in their relatively stout stems and in sometimes having a few obscure teeth on the leaves. These plants have been segregated as P. vaseyanus Greene, but their true biological status remains to be determined.

Penstemon seorsus (A. Nels.) Keck, Am. Midl. Nat. 23:595. 1940.

P. linarioides var. seorsus A. Nels. Bot. Gaz. 54:147. 1912. (Macbride 970, Twilight Gulch, Owyhee Co., Ida.)

Plants with several or many erect flowering stems 1-4 dm. tall from a branched, distinctly woody base, also producing short, densely leafy, sterile stems and tending to form loose mats; herbage finely and densely cinereous-puberulent, the inflorescence tending to become more loosely glandular-hairy; leaves numerous, all or nearly all opposite, the bases of a pair tending to meet or to be joined by a raised line around the stem, all linear or nearly so, 1-5.5 cm. long and 1-3(6) mm. wide; inflorescence of several or rather numerous loose and few-flowered verticillasters, sometimes approaching the status of a simple raceme; calyx 3-5.5 mm. long, the segments lanceolate or ovate, narrowly if at all scarious-margined; corolla blue or pink, glandular-hairy externally, glabrous internally, 15-23 mm. long, 4-6 mm. wide at the mouth, the limb not much if at all spreading, not over about 1 cm. wide; pollen sacs essentially glabrous, 0.8-1.2 mm. long, dehiscent throughout and becoming opposite, but not explanate; staminode slightly expanded distally, usually shortly exserted, evidently yellow-bearded for much of its length; capsules and seeds about as in P. gairdneri.

Dry, open, rocky places in the plains and foothills, often with sagebrush; Owyhee Co., Ida., to Harney and Jefferson cos., Oreg. June-July.

Penstemon seorsus is evidently allied to P. gairdneri, but intergradation between them has not yet been demonstrated.

Penstemon serrulatus Menzies ex Smith in Rees, Cycl. 26, Penstemon no. 5. 1813. (Menzies, w. coast of N. Am.)

P. diffusus Dougl. ex Lindl. Bot. Reg. 14: pl. 1132. 1828. (Douglas, "open grounds and banks of streams in the district around the mouth of the Columbia River")

P. diffusus f. albiflorus Hardin, Mazama 11:89. 1929. (Hardin 997, Mt. Hermann, Whatcom Co., Wash.)

Perennial from a branching, woody base, scarcely or not at all taprooted, with several or many glabrous or puberulent stems 2-7 dm. tall, not at all glandular; leaves essentially glabrous, all cauline, the lower ones reduced, the others mostly sessile or nearly so (or petiolate on vigorous young shoots), lanceolate to ovate-oblong, sharply serrate, mostly 3-8 cm. long and (0.7)1-3.5 cm. wide; inflorescence often of a single rather compact terminal verticillaster, sometimes of several verticillasters, these occasionally open and branched to form a more paniculate inflorescence; calyx 5-9 mm. long, the segments ciliolate and often also irregularly toothed, sometimes short-hairy on the back; corolla deep blue to dark purple, 17-25(28) mm. long, glabrous inside and out; fertile filaments glabrous; anthers permanently horseshoe-shaped, 1.1-1.6 mm. long; pollen sacs dehiscent across their confluent apices, the lower part remaining saccate and indehiscent, dentate-ciliolate along the sutures, otherwise glabrous; staminode shortly or scarcely exserted, with a long yellowish beard on about the upper half, the tip flattened and somewhat expanded; capsules 5-8 mm. and seeds 1-1.5 mm. long. N=8.

In a variety of moist to wet habitats from near sea level to 6000 ft. (or higher?) in the mountains; Cascade Range (chiefly w. of the summit) to the coast, from s. B.C. to s. Lane Co., Oreg. June-Aug.

Penstemon spatulatus Pennell, Not. Nat. Acad. Phila. 71:10. 1941. (Pennell 21091, about Ice Lake, Wallowa Mts., Oreg.)

Stems clustered from a surficial, woody caudex, 1-2.5 dm. tall, wholly glabrous below the glandular-hairy inflorescence, or the stem finely hirtellous-puberulent; leaves entire, the basal ones well developed, petiolate, up to 6 cm. long (petiole included) and 18 mm. wide, the blade more or less elliptic; cauline leaves few, mostly sessile, up to about 3.5 cm. long and 8 mm. wide; inflorescence of 1-4 dense, often crowded verticillasters; calyx 2.5-5 mm. long, the segments with more or less evidently scarious, often slightly erose margins; corolla glandular-hairy, blue-violet, marked with guide lines within, 10-13 mm. long, the tube 4-5 mm. wide at the mouth; palate bearded, as also the scarcely dilated tip of the staminode; pollen sacs glabrous, more or less ovate, 0.6-0.8 mm. long, dehiscent nearly or quite throughout, eventually becoming opposite, but scarcely explanate; capsules and seeds unknown.

Open slopes at fairly high elevations in the Wallowa Mts., Oreg.; known from only a few collections. July-Aug.

Penstemon spatulatus is so similar to some forms of P. attenuatus var. pseudoprocerus from Mont. and e. Ida. that one is tempted to consider it (P. spatulatus) as merely another variety of P. attenuatus, but there appear to be no specimens from the immediate geographical area of P. spatulatus that are transitional to P. attenuatus. Since P. spatulatus is also suggestive of P. humilis in habit and in the marking of the corolla, no taxonomic change seems to be clearly warranted at this time.

Penstemon speciosus Dougl. ex Lindl. Bot. Reg. 15: pl. 1270. 1829.
   P. glaber var. occidentalis Gray, Proc. Am. Acad. 6:60. 1862. P. glaber var. speciosus
   Rydb. Mem. N. Y. Bot. Gard. 1:344. 1900. (Douglas, "Spokan R.")
   Perennial, 2-9 dm. tall, generally with several stout stems arising from a compact, branched caudex which tends to surmount a short, quickly deliquescent taproot, glabrous throughout, or often finely cinereous-puberulent; leaves entire, thick and firm, tending to darken in drying, narrow, the basal ones clustered, up to 15 cm. long and 2 cm. wide, petiolate, with oblanceolate to narrowly elliptic or even lance-ovate blade; cauline leaves (except the lower) mostly sessile and occasionally somewhat clasping, up to 2 or occasionally 2.5 cm. wide, mostly 3.5-10 times as long as wide; inflorescence of several or usually numerous rather loose verticillasters, sometimes secund; calyx 4-8 mm. long, the segments mostly ovate and with more or less prominently erose-scarious margins, acutish to acuminate; corolla bright blue, 26-38 mm. long, about 1 cm. wide at the mouth; pollen sacs 1.9-3.0 mm. long, divaricate, tending to be sigmoidally twisted, glabrous except for the evidently dentate-ciliolate sutures, a short proximal portion remaining indehiscent; staminode glabrous or occasionally with a sparse, short beard toward the slightly or scarcely expanded tip; capsule 9-12 mm. and seeds 2-3 mm. long. N=8.
   Dry, open or sparsely wooded slopes, often with sagebrush, juniper, or ponderosa pine, mostly in the foothills and lowlands, sometimes at moderate elevations in the mountains; e. of the Cascade summits in Wash., from Lincoln, Douglas, and Chelan cos., s. through e. Oreg. to Calif., and e. through Nev. to n. e. Utah, entering Ida. only s. of the Snake R. plains. May-July.
   Only the typical variety, as described above, occurs in our range.

Penstemon subserratus Pennell, Not. Nat. Acad. Phila. 71:13. 1941. (Pennell 15732, Gotchen
   Creek Ranger Station, Yakima Co., Wash.)
   Perennial from a surficial. somewhat woody, branched caudex, evidently glandular-hirsute in the inflorescence, otherwise essentially glabrous, or the stem very finely and inconspicuously hirtellous-puberulent; leaves, or some of them, generally with a few small, irregularly distributed, and often inconspicuous teeth, sometimes all entire; basal leaves well developed, up to 12 cm. long and 2.5 cm. wide, the blade narrowly to broadly elliptic to ovate, longer or shorter than the petiole; cauline leaves well developed, mostly sessile, broadly lance-triangular or oblong to lance-linear, mostly 2-6 cm. long and 0.5-2 cm. wide; inflorescence of several rather loose and few-flowered verticillasters; calyx 3-5 mm. long, the segments lanceolate to ovate, evidently though narrowly scarious-margined and erose; corolla glandular-hairy externally, blue or purplish, mostly 11-16(18) mm. long, the tube mostly 3-5 mm. wide at the mouth; palate bearded; pollen sacs 0.7-1.1 mm. long, dehiscent essentially throughout and becoming opposite, sometimes explanate, finely (or scarcely) dentate-ciliolate along the sutures, otherwise glabrous; staminode barely or scarcely dilated toward the bearded, recurved tip; capsule 4-5 mm. and seeds about 1 mm. long. N=16.
   Open woods and clearings at lower elevations along and near the e. side of the Cascade Range in Yakima, Klickitat, and e. Skamania cos., Wash., and n. Hood River Co., Oreg. May-June (July).

Penstemon triphyllus Dougl. in Lindl. Bot. Reg. 15: pl. 1245. 1829. (Douglas, Blue Mts.)
   Taprooted perennial, distinctly shrubby at the base, 2.5-8 dm. tall, with more or less numerous, rather slender and brittle stems, evidently glandular-hairy in the inflorescence only, otherwise merely pruinose-puberulent, or the leaves often glabrous; leaves irregularly arranged, many of them ternate or even quaternate, others opposite or single, all cauline, the lowermost reduced, the others rather numerous, sessile or subsessile from a narrow base, nearly linear to narrowly lance-elliptic, up to 5 cm. long and 1 cm. wide, irregularly and rather sharply toothed, or some entire; inflorescence a loose, mixed panicle, often leafy-bracteate below; calyx 4-6 mm. long, the segments green or anthocyanic, entire, often distinctly

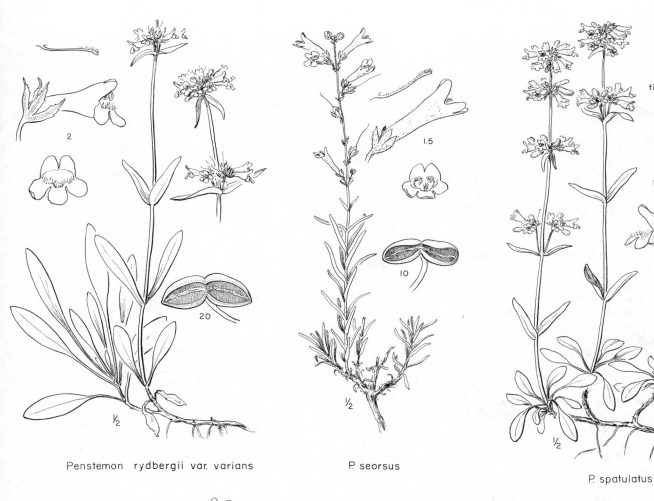

2

1.5

10

tip of staminodium

18

2

20

Penstemon  rydbergii var. varians

P. seorsus

P. spatulatus

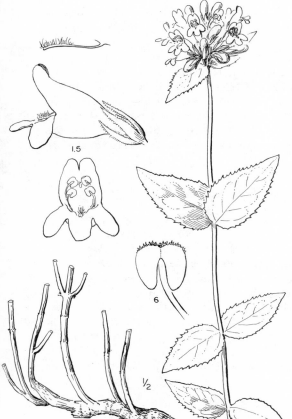

1.5

6

P. serrulatus

6

1

1

sepal

1/2

1/8

P. speciosus

JRJ

unequal; corolla glandular-hairy outside, glabrous within, or with a few hairs near the base of the lower lip, blue-lavender to light purple-violet, 13-19 mm. long, not very strongly bilabiate, the upper lip cleft more than half its length; anthers 0.9-1.3 mm. long, permanently horseshoe-shaped, the pollen sacs dehiscent across their confluent apices, saccate and indehiscent below, not becoming divaricate, the sutures evidently dentate-ciliolate; staminode exserted, scarcely expanded upward, rather densely long-bearded for about half its length; capsule 4-6 mm. long and seeds 0.7-1.1 mm. long. N=8.

Cliff crevices and dry, rocky banks and slopes at low elevations; near the Snake R. and its tributaries in n. e. Oreg. (Baker and Wallowa cos.), s. e. Wash. (to e. Franklin Co.), and adj. Ida. (Nez Perce Co. to Adams Co.), extending up the Salmon R. to at least 18 miles above Riggins, and once collected as a waif along the Columbia in Klickitat Co., Wash. May-July.

Penstemon venustus Dougl. ex Lindl. Bot. Reg. 16: pl. 1309. 1830. (Douglas, "dry channels of rivers among the mountains of North-west America")

Perennial from a stout taproot, shrubby at the base, with numerous stems 3-8 dm. long, forming a rounded clump; herbage and inflorescence not at all glandular, essentially glabrous (and sometimes glaucous) throughout, or the stem often puberulent in lines; leaves all cauline, the lowermost reduced, the others numerous, sessile, lanceolate or lance-ovate to broadly oblong, mostly (2)4-10(12) cm. long and (0.7)1-3(3.5) cm. wide, sharply serrate, varying to rarely subentire; inflorescence a narrow terminal thyrsoid panicle, or nearly a raceme; calyx 2.5-6.5 mm. long, the segments often scarious-margined and erose-toothed, sometimes ciliolate, otherwise glabrous; corolla bright lavender to purple or purple-violet, 25-38 mm. long, often more than 1 cm. wide at the mouth, glabrous inside and out except for the evidently ciliate-margined lobes; fertile filaments more or less pubescent (often conspicuously long-white-hairy) toward the tip, the lower pair evidently exserted, the upper scarcely so; anthers permanently horseshoe-shaped, 1.6-2.0 mm. long; pollen sacs dehiscent across their confluent apices, the lower half remaining saccate and indehiscent, evidently dentate-ciliolate along the sutures, otherwise glabrous or inconspicuously hispidulous toward the connective; staminode shortly exserted, with a prominent long white beard near the flattened and slightly expanded tip; capsules 6-9 mm. and seeds 1.2-2.5 mm. long. N=32.

Open, rocky slopes, from the foothills and valleys to moderate elevations in the mountains; Shoshone Co. to Washington Co., Ida., w. to Walla Walla Co., Wash., and Umatilla and Baker cos., Oreg. May-Aug.

The species is ordinarily sharply marked and not likely to be confused with any other. A single collection (Hitchcock & Muhlick 13866) from 20 miles s. of New Meadows, Adams Co., Ida., is so aberrant in its elongate, narrow, essentially entire leaves (the larger ones about 8 cm. long and scarcely 1 cm. wide) as to suggest hybridization with some other species, although in other respects it would pass readily as one of the smaller-flowered plants of P. venustus. A few other specimens from the same general area diverge from typical P. venustus only in their irregularly and inconspicuously serrulate to subentire leaves; the taxonomic significance of this variation is not yet understood.

Penstemon washingtonensis Keck, Am. Midl. Nat. 33:150. 1945. (Kelly 21, n. of Lake Chelan, Chelan Co., Wash.)

Plants more or less tufted from a rather loose, surficial, woody rhizome-caudex, mostly 1-2.5 dm. tall, glabrous below the glandular-hairy inflorescence, or the stem finely hirtellous-puberulent; leaves entire, the basal ones well developed and rosette-forming, 2.5-6 cm. long, 5-18 mm. wide; cauline leaves few, mostly sessile, often reduced; inflorescence of 1-3 dense verticillasters; calyx 4-6 mm. long, the segments lance-attenuate, with scarious and often erose margins; corolla more or less declined, glandular-hairy, deep blue or occasionally ochroleucous, 9-12 mm. long, the tube narrow, only 2-3 mm. wide at the mouth, the palate bearded; staminode bearded toward the expanded tip; pollen sacs glabrous, subrotund, 0.5-0.6 mm. long, wholly dehiscent, becoming opposite and more or less explanate; capsules and seeds presumably as in P. procerus.

Open slopes and flats at moderate elevations in the mountains of Chelan and Okanogan cos., Wash. July-Aug.

Except for its glandular-hairy inflorescence, P. washingtonensis has much the appearance of P. procerus var. tolmiei.

Penstemon whippleanus Gray, Proc. Am. Acad. 6:73. 1862. (Bigelow, Sandia Mts., N.M.)
   P. glaucus var. stenosepalus Gray, Proc. Am. Acad. 6:70. 1862. P. stenosepalus Howell, Fl. N.W. Am. 514. 1901. (First collection cited is James, Pikes Peak, Colo.)

Plants tufted from a surficial, branched caudex, 2-6 dm. tall, essentially glabrous below, becoming strongly glandular-hairy in the inflorescence; leaves all entire or nearly so, the basal with elliptic to ovate blade up to 6 cm. long and 3.5 cm. wide, longer or shorter than the petiole, the cauline mostly sessile and oblong or lanceolate, up to about 6 cm. long and 1.5 cm. wide; inflorescence of 2-7 verticillasters, these not very dense; calyx elongate, 7-11 mm. long, the segments lanceolate or narrower and wholly or almost wholly herbaceous; corolla glandular-hairy externally, blue or violet to dull purple, lavender, or cream, sometimes varicolored, 18-28 mm. long, strongly inflated distally, mostly 7-11 mm. wide at the mouth, strongly bilabiate, the lower lip much the longer; palate bearded; pollen sacs broadly ovate, glabrous, 1-1.4 mm. long, wholly dehiscent, becoming opposite and eventually explanate; staminode evidently exserted from the orifice of the corolla, usually bearded toward the scarcely expanded tip; ovary and capsule ordinarily glandular-puberulent near the tip; capsule 6-9 mm. and seeds about 1-1.5 mm. long.

Dry meadows and open or lightly wooded, often rocky slopes well up in the mountains, often near timber line; Taylor Mts., Madison Co., Mont., s. through Wyo. and e. Ida. to Utah, n. Ariz., Colo., and N.M. July-Aug.

The somewhat similar species P. rattanii Gray, with mostly toothed, often larger leaves, and with glabrous capsule, approaches our range in s.w. Oreg., and may eventually be found within our borders.

Penstemon wilcoxii Rydb. Bull. Torrey Club 28:28. 1901. (Wilcox 370, Kalispell, Mont.)
   P. ovatus var. pinetorum Piper in Piper & Beattie, Fl. Palouse Region 158. 1901. P. pinetorum Piper, Contr. U.S. Nat. Herb. 11:500. 1906. (Piper 1662, Cedar Mt., Latah Co., Ida.)
   P. leptophyllus Rydb. Fl. Rocky Mts. 773, 1066. 1917. (Jones 6493, Rush Creek, Washington Co., Ida.) The robust extreme.

Stems clustered from a surficial, branched, woody caudex, 3-10 dm. tall; plants more or less evidently glandular-hairy in the inflorescence, otherwise commonly glabrous, or the stem (occasionally) and the leaves (rarely) rather shortly spreading-hirsute; leaves more or less strongly serrate with mostly slender teeth, the basal ones with more or less well developed, elliptic or lance-ovate to deltoid-ovate or subcordate blade up to 9 cm. long and 5 cm. wide, longer or shorter than the petiole; cauline leaves well developed, often larger than the basal ones, mostly sessile, the larger ones mostly 4-10 cm. long and (0.8)1.5-4 cm. wide; inflorescence of several relatively large, loose, and confluent verticillasters, typically much more open than in all allied species except P. ovatus, becoming more congested in forms approaching P. albertinus; calyx 2.5-5 mm. long, the segments broadly lanceolate to ovate, narrowly or obscurely scarious-margined; corolla externally glandular-hairy, blue, the paler throat provided with guide lines, 15-23 mm. long, strongly bilabiate, the lower lip much longer than the upper, the tube 4-8 mm. wide at the mouth; palate bearded; pollen sacs 0.7-1.0 mm. long, wholly dehiscent, becoming opposite and often explanate, glabrous except for the often somewhat dentate-ciliolate sutures; staminode scarcely dilated toward the recurved, bearded tip; capsule 4-6 mm. and seeds about 1 mm. long. N=8, 16.

In open or often wooded, sometimes rocky places, from the foothills to moderate elevations in the mountains; c. and n. Ida., extending into n.e. Oreg. (Wallowa and Baker cos.), extreme e. Wash. (Whitman Co.), and n.w. Mont. (w. of the continental divide). May-July.

Penstemon wilcoxii hybridizes freely with P. albertinus, and herbarium specimens are frequently difficult to identify. Although no nomenclatural change is here proposed, the type collection of P. wilcoxii includes specimens with relatively very narrow leaves which may actually represent such hybrids.

The more robust extreme of P. wilcoxii approaches the more western P. ovatus in appear-

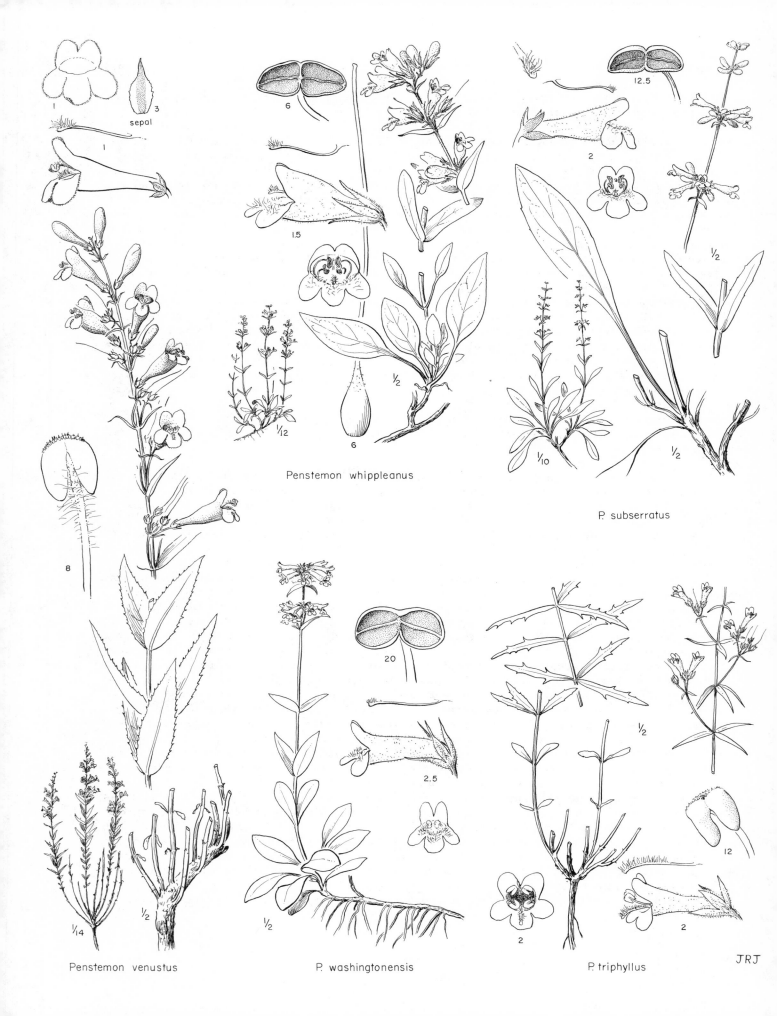

1

3
sepal

1

6

1.5

Penstemon whippleanus

8

1/12

6

12.5

2

1/2

1/10

1/2

P. subserratus

1/14

1/2

Penstemon venustus

20

2.5

1/2

P. washingtonensis

1/2

2

12

2

P. triphyllus

JRJ

ance, and when such specimens also have more or less hirsute herbage, as is sometimes the case, they might easily pass for P. ovatus in the absence of geographic data. On and near the Grangeville prairie of Ida., P. wilcoxii appears to pass into P. attenuatus, although the latter species is supposed to be hexaploid, while P. wilcoxii contains only diploid and tetraploid. races so far as known.

In spite of the specimens which tend to connect P. wilcoxii with each of several quite different species, there remains a considerable body of specimens which clearly indicates the existence of a self-perpetuating natural population not properly referable to any of the species with which P. wilcoxii intergrades.

## Rhinanthus L.

Flowers in terminal, leafy-bracteate spikes; calyx somewhat inflated at anthesis, accrescent and very conspicuously inflated in fruit, the 4 teeth relatively short; corolla yellow, bilabiate, galeate, the upper lip hooded and enclosing the anthers, its teeth short or obsolete; lower lip 3-lobed, shorter than the galea, external in bud; stamens 4, didynamous; pollen sacs equal; stigma entire; capsule flattened, orbicular, loculicidal; seeds numerous, flattened, orbicular, evidently winged; annuals with opposite, wholly cauline, sessile, toothed leaves.

The genus, in the restricted sense, contains probably only two species, both of which were included in R. crista-galli by Linnaeus in 1753. R. major L. (1756) differs from R. crista-galli as now restricted (=R. minor L.) chiefly in its larger, coarser corollas (17-24 mm. long) with more closed throat. Most or all of the numerous later segregates from R. crista-galli as here treated are ill-defined and scarcely significant. (From the Greek, rhinos, snout, and anthos, flower, in reference to the irregular corolla.)

Rhinanthus crista-galli L. Sp. Pl. 603. 1753.
  R. minor L. Amoen. Acad. 3:54. 1756. Alectorolophus minor Dum. Fl. Belg. 33. 1827. (Europe)
  Alectorolophus borealis Stern. Ann. Cons. Jard. Bot. Genèv. 3:25. 1899. Rhinanthus borealis Chab. Bull. Herb. Boiss. 7:429. 1899. Rhinanthus minor ssp. borealis A. Löve, Bot. Not. 1950:52. 1950. (Fischer, Unalaska)
  Rhinanthus kyrollae Chab. Bull. Herb. Boiss. 7:511. 1899. Alectorolophus kyrollae Stern. Abh. Zool. Ges. Wien $1^2$:119. 1901. Rhinanthus borealis ssp. kyrollae Pennell in Abrams, Ill. Fl. Pac. St. 3:802. 1951. (Several N. American specimens, from Wash. to Me., are cited)
  Rhinanthus rigidus Chab. Bull. Herb. Boiss. 7:516. 1899. Alectorolophus rigidus Stern. Abh. Zool. Bot. Ges. Wien $1^2$:121. 1901. (Several Rocky Mountain specimens are cited)
  Alectorolophus pacificus Stern. Abh. Zool. Bot. Ges. Wien $1^2$:120. 1901. (Palliser s.n., Brit. N. Am.)

Stem 1.5-8 dm. tall, simple or few-branched, thinly villous-puberulent on two of the four sides; leaves triangular-lanceolate to oblong, firm, serrate or crenate-serrate, 2-6 cm. long, 4-15 mm. wide, scabrous above, scabrous-hispidulous beneath; bracts gradually reduced upward, resembling the leaves, but mostly shorter, relatively broader, and with longer and sharper teeth; fruiting calyx 12-17 mm. long and nearly or quite as wide, reticulate-veiny; corolla 9-14 mm. long; galea generally with a pair of short and broad subapical teeth. N=7.

Meadows, fields, and moist slopes, at various elevations; circumboreal, extending s., in America, to N.Y., Colo., and n.w. Oreg. Rather infrequently collected. June-Aug.

## Scrophularia L. Figwort

Flowers numerous in elongate, terminal, branched, open, nearly naked inflorescences; calyx deeply 5-cleft, the segments almost free; corolla greenish-yellow or greenish-purple to dark maroon, firm, bilabiate, the upper lip 2-lobed, flat, projecting forward, external in bud, the lower shorter and with the central lobe deflexed; fertile stamens 4, the filaments expanded upward, the pollen sacs divergent; fifth stamen represented by a knob or scale less than 2 mm.

long on the upper lip; stigma capitate; capsule septicidal; seeds numerous, turgid; odorous perennial herbs with quadrangular stems and opposite, petiolate, toothed leaves.

More than 100 species of the Northern Hemisphere, especially Eurasia. (Named for its reputed value in treating scrofula. )

1 Sterile stamen purple or brown, mostly 0. 5-1 mm. wide and a little longer than wide; w. of the Cascades                                                                   S. CALIFORNICA
1 Sterile stamen yellow-green, mostly 1-1. 8 mm. wide and a little wider than long; widespread                                                                         S. LANCEOLATA

Scrophularia californica Cham. & Schlecht. Linnaea 2:585. 1827.
S. nodosa var. californica M. E. Jones, Contr. W. Bot. 12:67. 1908. (Chamisso, San Francisco)
S. oregana Pennell, Bull. Torrey Club 55:316. 1928. (Harvey 57, Newport, Lincoln Co., Oreg. )

Very similar to S. lanceolata; petioles averaging longer (frequently 4-5 cm. long); sepals often more pointed, but variable; corolla more prominently maroon, at least as to the upper lip; sterile stamen purple or brown, spatulate to obovate, mostly 0. 5-1 mm. wide and a little longer than wide.

In moist low ground; s. w. B. C. to Calif. and adj. Nev. , not extending e. of the Cascades in our range. June-Aug.

Scrophularia lanceolata Pursh, Fl. Am. Sept. 419. 1814.
S. nodosa var. lanceolata M. E. Jones, Contr. W. Bot. 12:67. 1908. (Pa. )
S. nodosa var. occidentalis Rydb. Contr. U.S. Nat. Herb. 3:517. 1896. S. occidentalis Bickn. Bull. Torrey Club 23:315. 1896. S. lanceolata var. occidentalis Pennell, Torreya 22:82. 1922. (Rydberg 914, Rapid City, S. D. )
S. serrata Rydb. Bull. Torrey Club 36:688. 1909. (Sandberg s.n., in 1887, Granite, n. Ida.)
S. lanceolata f. velutina Pennell, Monog. Acad. Phila. 1:276. 1935. (St. John 7696, s. of Omak Lake, Okanogan Co. , Wash. ) The commoner phase of the species in our range, with the herbage more or less hairy.

Stems clustered from thickened roots, stout, 5-15 dm. tall; herbage finely spreading-hairy or subglabrous, more or less stipitate-glandular in the inflorescence; leaves all cauline, only gradually reduced upward; petioles short, mostly 1-3(5) cm. long; blades triangular-ovate or rather broadly triangular-lanceolate, with acute or acuminate tip and truncate or subcordate to sometimes merely rounded base, mostly 5-15 cm. long and 2-7 cm. wide, conspicuously and often sharply (sometimes doubly) toothed, occasionally subincised; inflorescence 1-5 dm. long, 3-10 cm. wide, its branches mostly opposite, the bracts narrow and inconspicuous; calyx lobes 2-3 mm. long, thin, somewhat scarious-margined, broad and blunt or rounded; corolla 9-14 mm. long, yellowish green, with light maroon overcast especially above; sterile filament flabellate to subreniform, mostly wider than long, 1-1. 8 mm. wide, yellow-green; capsule ovoid, 6-8 mm. long.

In moist low ground, from the valleys to moderate elevations in the mountains; s. B. C. to Nova Scotia, s. to n. Calif. , n. N. M. , Okla. , and Va. June-Aug.

## Synthyris Benth.

Flowers borne in terminal racemes; calyx of 4 essentially distinct sepals; corolla blue, occasionally pink or lavender to white, campanulate to subrotate, unequally 4-lobed, the upper lobe the largest and internal to the others in bud; stamens 2; stigma capitate; capsule somewhat compressed, sometimes notched at the summit, loculicidal; seeds 2-many per locule; fibrous-rooted perennial herbs with well-developed, petiolate, basal leaves and erect or lax, naked or sparsely bracteate scapes or basal peduncles, the bracts (except in S. schizantha) alternate when present.

Nine species, native to the cordilleran region of w. N. Am. (Name from the Greek syn, together or joined, and thyris, a little door or valve, referring to the valves of the capsule. )

Synthyris missurica, S. reniformis, and S. schizantha are of proven or potential value for moist areas in gardens w. of the Cascades.
Reference:
Pennell, F. W. A revision of Synthyris and Besseya. Proc. Acad. Phila. 85:77-106. 1933.

1 Leaves merely toothed or shallowly lobulate; fruits, except in S. missurica, about twice as wide as high, or even wider; corollas diverse
   2 Corolla campanulate, the lobes a little shorter than the tube; peduncles weak and lax, more or less reclining in fruit; seeds with thick, strongly incurved margins; w. of the Cascade summits, and in the Columbia gorge                    S. RENIFORMIS
   2 Corolla subrotate, the lobes much longer than the tube, more or less spreading; peduncles or scapes more or less erect; seeds flat, thin-margined
      3 Corolla lobes entire or slightly erose; capsule scarcely wider than high, scarcely notched; e. of the Cascades, and in the Columbia gorge          S. MISSURICA
      3 Corolla lobes laciniately incised; capsules nearly or fully twice as wide as high, strongly notched; w. of the Cascade summits, and in n. Ida.
         4 Corolla 4-6 mm. long; leaves withering at the end of the first season; Cascade, Olympic, and Coast ranges of Wash. and n. Oreg.          S. SCHIZANTHA
         4 Corolla 2.5-3.5 mm. long; leaves tending to persist until flowering time of the second season; mountains of n. c. Idaho Co., Ida.          S. PLATYCARPA
1 Leaves deeply cleft to pinnately dissected; fruits about as wide as high; corolla lobes about equaling, or a little shorter than, the tube
   5 Leaves pinnipalmately or close-pinnately cleft to below the middle, the segments broadly confluent; Mission Range, Mont.          S. CANBYI
   5 Leaves pinnately dissected, with narrow rachis; more southern or western
          S. PINNATIFIDA

Synthyris canbyi Pennell, Proc. Acad. Phila. 85:93. 1933. (MacDougal 345, Sin-yale-a-min Peak, Mission Range, Mont.)
   Perennial from a stout rhizome, somewhat hirsute in the inflorescence, otherwise subglabrous; leaves surpassing the inflorescence at anthesis, the petiole 5-9 cm. long, the blade 2-3 cm. long and wide, pinnipalmately or close-pinnately cleft to beyond the middle, the segments broadly confluent, sharply toothed or again cleft, often overlapping distally; corolla 5-7 mm. long, the lobes about equaling the tube; immature capsule 5 mm. long, pubescent.
   Known only from the Mission Range, Mont., at an altitude of about 8500 ft. June-July.
   The related S. laciniata (Gray) Rydb., of Utah, differs in its distinctly palmately, more shallowly cleft leaves.

Synthyris missurica (Raf.) Pennell, Proc. Acad. Phila. 85:89. 1933.
   Veronica reniformis Pursh, Fl. Am. Sept. 10. 1814. V. missurica Raf. Am. Monthly Mag. 3:175. 1818. V. purshii G. Don, Gen. Syst. 4:573. 1837. (Lewis & Clark, Hungry Creek, near the Lochsa Fork of the Clearwater R., Ida., according to Pennell)
   Synthyris reniformis var. major Hook. Journ. Bot. & Kew Misc. 5:257. 1853. Wulfenia major Heller, Cat. N. Am. Pl. 7. 1898. Synthyris major Heller, Muhl. 1:5. 1900. S. missurica ssp. major Pennell, Proc. Acad. Phila. 85:91. 1933. S. missurica var. major Davis, Fl. Ida. 637. 1952. (Geyer, Snowy Mts., highlands of the Nez Perces) The larger form from lower elevations, perhaps properly to be recognized as a distinct variety.
   S. stellata Pennell, Proc. Acad. Phila. 85:89. 1933. (Thompson 4030, near Oneonta tunnel, Columbia gorge, Oreg.)
   S. missurica f. rosea St. John, Fl. S. E. Wash. 382. 1937. (St. John 6069, Kamiak Butte, Whitman Co., Wash.)
   Perennial from a short, stout rhizome, 1-6 dm. tall, often puberulent or villosulous in the inflorescence, otherwise glabrous; leaves long-petiolate, with suborbicular (basally deeply cordate) to reniform-cordate blade 2.5-8 cm. wide and nearly or quite as long, palmately veined, shallowly lobulate or coarsely dentate all around, the lobules or primary teeth again few-dentate, the teeth blunt or sharp; corolla 4-7 mm. long, the lobes somewhat spreading,

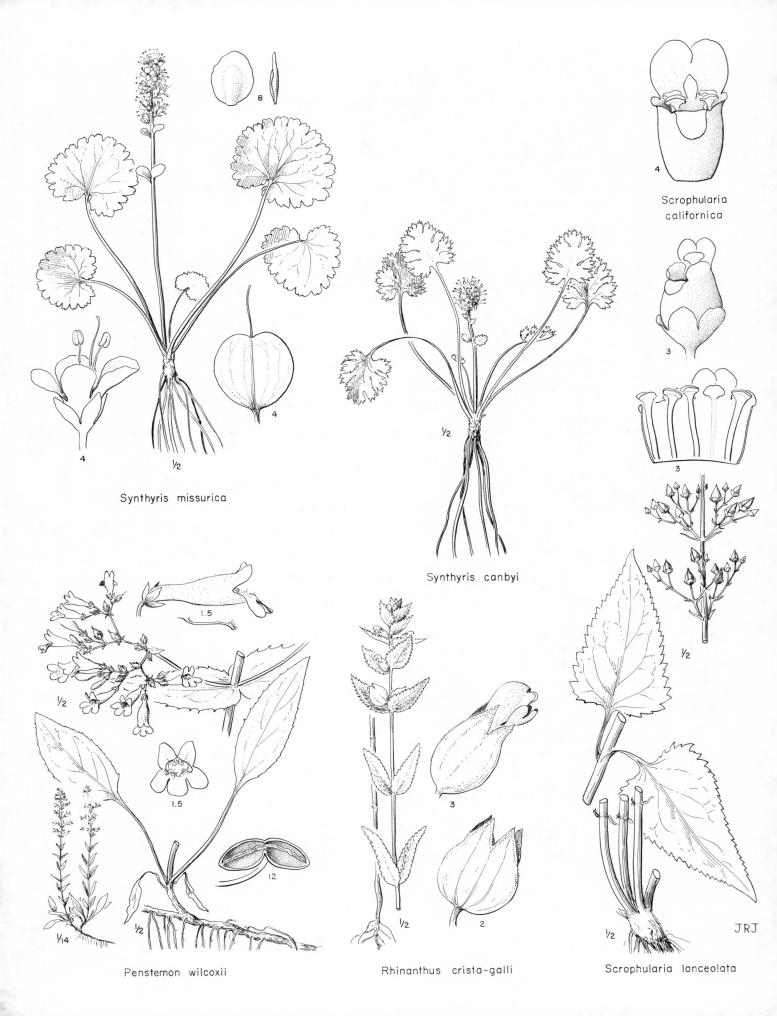

**Synthyris missurica**

**Synthyris canbyi**

Scrophularia
californica

Penstemon wilcoxii

Rhinanthus crista-galli

Scrophularia lanceolata

JRJ

much longer than the tube, entire or slightly erose; capsule 5-7 mm. high, only slightly, if at all, wider, shallowly or obscurely notched; seeds flat, thin-margined.

Moist, open or shaded slopes, from the foothills to moderate elevations in the mountains; n. and w. c. Ida., s. e. Wash., and the Blue and Wallowa mts. region of n. e. Oreg.; Columbia gorge region in Oreg. and Wash.; Lake Co., Oreg., and Modoc Co., Calif. Apr.-July.

The form of the species which occurs in and about the Columbia gorge tends to have somewhat better developed bracts beneath the inflorescence, and more sharply toothed leaves, than do more characteristic forms of the species, but the differences are slight and scarcely beyond the range of variation of the principal population. The Columbia gorge plants might perhaps properly be considered to represent a rather weak variety, but no new combination is here proposed.

Synthyris pinnatifida Wats. Bot. King Exp. 227. 1871
  Wulfenia pinnatifida Greene, Erythea 2:83. 1894. (Watson 802, head of American Fork Canyon, Wasatch Mts., Utah)
  SYNTHYRIS PINNATIFIDA var. LANUGINOSA (Piper) Cronq. hoc loc. S. pinnatifida ssp. lanuginosa Piper, Contr. U. S. Nat. Herb. 11:504. 1906. S. lanuginosa Pennell & Thomps. Proc. Acad. Phila. 85:93. 1933. (Flett 815, Olympic Mts., Wash.)
  S. dissecta Rydb. Bull. Torrey Club 36:691. 1909. (Chesnut & Jones 199, Bozeman, Mont.) = var. canescens.
  S. hendersoni Pennell, Proc. Acad. Phila. 85:94. 1933. (Kirtley, Salmon City, Ida.) = var. canescens.
  S. cymopteroides Pennell, Proc. Acad. Phila. 85:94. 1933. (Macbride & Payson 3120, Clyde, Blaine Co., Ida.) = var. canescens.
  SYNTHYRIS PINNATIFIDA var. CANESCENS (Pennell) Cronq. hoc loc. S. cymopteroides ssp. canescens Pennell, Proc. Acad. Phila. 85:95. 1933. (Locke 470, head of the Salmon R., Custer Co., Ida.)
  S. paysoni Pennell & Williams, Proc. Acad. Phila. 85:96. 1933. (Payson & Armstrong 3453, Sheep Mt., Teton Co., Wyo.) = var. pinnatifida.

Perennial from a short, stout rhizome, 0.5-2(2.5) dm. tall; leaves pinnately dissected, with narrow rachis and small ultimate segments, mostly surpassed by the inflorescence; corolla 4-7 mm. long, the lobes about equaling, or a little shorter than, the tube; capsule 4-8 mm. high and nearly as wide; seeds flat, thin-margined.

Open, often rocky slopes at high altitudes in the mountains; Utah and w. Wyo. to s. w. Mont. and c. Ida.; Olympic Mts. of Wash. June-Aug.

Three varieties may be recognized, as indicated in the following key. The var. canescens is intermediate between the other two varieties in pubescence, the extremes being scarcely distinguishable in that regard from the var. lanuginosa, on the one hand, and var. pinnatifida on the other. Since var. lanuginosa appears to differ from var. canescens only in pubescence, a specific distinction between the two can scarcely be upheld, even with full regard for the impressive geographic hiatus. The var. canescens is a little more sharply distinguished from the geographically adjacent var. pinnatifida by the shape of the floral bracts, but there is some intergradation in this character as well, and the plants are so similar in general that it seems better to treat them as conspecific. The selection of a name for the intermediate variety provides a difficult choice. The somewhat inappropriate name var. canescens, based on the pubescent extreme of the population, is here selected in order to have the same epithet as that which must, according to the Rules, be used by those who treat the three infraspecific taxa as subspecies instead of varieties.

1 Bracts mostly obovate and rounded-obtuse; plants scantily villous or glabrous; capsules glabrous; petioles often conspicuously woolly on the margins at the base; Wasatch and Bear River mts. of Utah, and the mountains of w. Wyo., perhaps extending to the Bitter Root Mts. of Mont.                                                                var. pinnatifida
1 Bracts mostly rhombic or ovate and acute to caudate-acuminate; plants often more hairy; petioles seldom conspicuously more hairy at the base than above
  2 Herbage subglabrous to more or less villous or tomentulose, usually glabrate in age; capsules glabrous or hairy; s. w. Mont. and c. Ida.          var. canescens (Pennell) Cronq.

2 Herbage permanently white-tomentulose, the tomentum mostly a little finer than in the pubescent extreme of var. canescens; capsules tomentulose; Olympic Mts. of Wash.

var. lanuginosa (Piper) Cronq.

Synthyris platycarpa Gail & Pennell, Am. Journ. Bot. 24:40. 1937. (Gail, Coolwater Mt., Selway Nat. Forest, Idaho Co., Ida.)

Similar to S. schizantha, differing as indicated in the key; leaves eventually glabrous beneath, but with some persistent long hairs above, crenate to lobulate, the sinuses 1.5-10 mm. deep, the lobules or teeth often entire or subentire; capsules 3.5-5 mm. wide.

Open woods at 5000-6000 ft. in the mountains of n. c. Idaho Co., Ida. June.

Although obviously allied to S. schizantha, S. platycarpa retains its distinguishing features in cultivation.

Synthyris reniformis (Dougl.) Benth. in DC. Prodr. 10:454. 1846.

Wulfenia reniformis Dougl. ex Benth. Scroph. Ind. 46. 1835. (Douglas, mountains near the Columbia R.; perhaps taken in the Columbia gorge)

Synthyris rotundifolia Gray, Syn. Fl. 2$^1$:285. 1878. (Several collections from w. Oreg. and Wash. are cited)

Perennial from a short rhizome or caudex; leaves long-petiolate, the blade cordate-ovate to reniform-cordate, palmately veined, mostly 2-8 cm. long and wide, sparsely hirsute-strigose above, similarly hairy or glabrous beneath, shallowly lobulate all around, with the lobules again toothed, the teeth rounded to acutish; peduncles several, naked, weak, curved, less than 1.5 dm. long, more or less reclining in fruit; flowers few in a short raceme; corolla 5-7 mm. long, campanulate, the lobes a little shorter than the tube; capsule 2-4 mm. high, 6-8 mm. wide, conspicuously long-ciliate on the margins; seeds 2 per locule, with thick, strongly incurved margins. N=12.

Coniferous woods, less commonly in open places, at low elevations; s. w. Wash. to San Francisco Bay, chiefly or probably wholly w. of the Cascades, except for the Columbia gorge. March-May.

Synthyris schizantha Piper, Bull. Torrey Club 29:223. 1902. (Lamb 1343, Baldy Peak, Olympic Mts., Wash.)

Perennial from a short stout rhizome, 1-3 dm. tall; scape and petioles villous-puberulent; leaves long-petiolate, with reniform-cordate to cordate-rotund blade 4-12 cm. wide and nearly or quite as long, palmately veined, somewhat villous-hirsute along the veins beneath, glabrous or somewhat glandular above, shallowly lobulate all around, the sinuses mostly 5-15 mm. deep, the lobules again few-toothed with rounded to subacute teeth; corolla 4-6 mm. long, the lobes much longer than the tube, more or less spreading, deeply and irregularly laciniate; fruit 3 mm. high, 6 mm. wide, evidently notched; seeds somewhat flattened and thin-margined.

Moist, often shaded cliffs and ledges in the mountains; Olympic Mts. near Lake Quinault; Saddle Mt., Clatsop Co., Oreg.; Cascade Range near Mt. Rainier. May-Aug.

### Tonella Nutt. ex Gray

Flowers 1-several in the axils of normal or more or less reduced upper leaves (bracts); calyx with 5 subequal lobes exceeding the tube; corolla blue and white, subrotate, with short tube and spreading, somewhat bilabiate limb, the upper lobes external in bud; stamens 4, equal, exserted, the filaments pubescent; pollen sacs confluent at the tip; stigmas wholly united; capsule dehiscing along 4 sutures; seeds 1-2 in each cell, turgid; annual with opposite, mostly tripartite or trifoliate leaves.

The genus consists of only the following species. (Derivation unknown)

1 Corolla showy, 6-12 mm. wide; pedicels and upper part of the stem evidently stipitate-glandular; s. e. Wash., n. e. Oreg., and adj. Ida.                    T. FLORIBUNDA
1 Corolla inconspicuous, 2-4 mm. wide; pedicels and stem glabrous or nearly so; w. of the Cascades (and in the Columbia gorge)                    T. TENELLA

Tonella floribunda Gray, Bot. Calif. 1:556. 1876.

  Collinsia floribunda Greene, Pitt. 1:55. 1887. (Presumably the Spalding collection from the
    Kooskooskie [Clearwater] R.)

  Plants 1-3.5 dm. tall, often branched, the stem glabrous below, becoming stipitate-glandular above; leaves glabrous or nearly so, the principal ones short-petiolate or sessile, deeply tripartite, with lance-elliptic or narrower, entire or few-toothed segments mostly 1.5-4
cm. long; lower leaves more petiolate, often more distinctly trifoliate, and with small but relatively broad leaflets, or the lowermost leaves simple; leaves of the inflorescence reduced
and sometimes entire; flowers numerous, on elongate, slender, stipitate-glandular pedicels;
corolla showy, 6-12 mm. wide.

  Open, often rocky places in the canyon of the Snake R. and adjacent tributaries (extending up
the Salmon R. at least to Riggins); s.e. Wash., n.e. Oreg., and adj. Ida. Apr.-June.


Tonella tenella (Benth.) Heller, Muhl. 1:5. 1900.

  Collinsia tenella Benth. in DC. Prodr. 10:593. 1846. Tonella collinsioides Nutt. ex Gray,
    Proc. Am. Acad. 7:378. 1868. (Nuttall, Columbia R.)

  Plants 0.5-2.5 dm. tall, slender and weak, often branched, glabrous throughout, or the
pedicels sparsely stipitate-glandular distally; leaves all small, the blade not over 2 cm. long;
lowermost leaves petiolate, with relatively broad, often merely toothed blade, those next above
trifoliate or tripartite, with fairly broad segments; middle and upper leaves with progressively narrower segments and shorter petioles, the uppermost reduced and often entire; flowers
fewer than in T. floribunda, small and relatively inconspicuous, the corolla 2-4 mm. wide.

  Fairly moist, open or partly shaded places; along the Columbia from the e. end of the gorge
to near the coast, s. in the Puget trough to c. Calif. March-Apr.


Verbascum L. Mullein

  Flowers borne in terminal racemes, spikes, or spikelike thyrses; calyx of 5 essentially distinct sepals; corolla yellow or occasionally white, rotate, slightly irregular, 5-lobed, the upper lobes external in bud and slightly shorter than the lower; stamens 5, all anther-bearing,
some or all of the filaments densely hairy; stigma capitate; capsule ellipsoid to subglobose,
septicidal; seeds numerous; biennial or perennial herbs with well-developed basal and alternate
cauline leaves.

  Perhaps 200 species, native to Eurasia. (The classical Latin name for some of the species.)

1 Plants more or less densely stipitate-glandular upward, essentially glabrous below, the
    leaves green; inflorescence open, the pedicels surpassing the bracts, 8-15 mm. long at
    anthesis; filaments covered with purple-knobbed hairs               V. BLATTARIA
1 Plants copiously tomentose throughout with branched, glandless hairs; inflorescence very
    dense, the flowers sessile or nearly so; filaments provided with slender yellow hairs, or
    some of them glabrous                                                 V. THAPSUS

Verbascum blattaria L. Sp. Pl. 178. 1753. (s. Europe)

  Taprooted biennial, producing in the first year a rosette of basal leaves which may persist
through the second, and in the second a single erect stem 4-15 dm. tall; plants more or less
densely stipitate-glandular upward, especially in the inflorescence, otherwise essentially glabrous; basal leaves oblanceolate or a little broader, 5-15 cm. long, 1-3 cm. wide, tapering to
a short petiolar base, evidently toothed, and often also shallowly lobed; cauline leaves numerous, progressively reduced upward, soon becoming sessile and clasping, toothed but seldom
lobed; raceme open and elongate, the pedicels 8-15 mm. long at anthesis, much surpassing
the lanceolate bracts; corolla yellow or white, 2-3 cm. wide; filaments covered with purple-
knobbed hairs; capsule ovoid-globose, 6-8 mm. high. N=15, 16. Moth mullein

  Roadsides, fields, and waste places; native of Eurasia, now widely naturalized in temperate
N. Am. May-Sept.


Verbascum thapsus L. Sp. Pl. 177. 1753. (Europe)

  Coarse, taprooted biennial, producing in the first year a rosette of basal leaves which may

Synthyris pinnatifida var. canescens

Synthyris reniformis

Tonella floribunda

Tonella tenella

Synthyris schizantha

JRJ

Synthyris platycarpa

persist through the second, and in the second a single erect stem 0.4-2 m. tall; stem, leaves, and inflorescence copiously and persistently tomentose with stellately or dendritically branched hairs; basal leaves broadly oblanceolate, tapering to a petiolar base, 1-4 dm. long, 3-12 cm. wide, obscurely round-toothed or entire; cauline leaves numerous, progressively reduced upward, becoming sessile and clasping the stem, which is evidently winged by their decurrent bases; inflorescence very dense in flower and fruit, the flowers sessile or nearly so; corolla yellow or rarely white, 1-2 cm. wide; upper three filaments densely yellow-hairy, lower two longer and glabrous; capsule broadly ovoid, 7-10 mm. long. N=17, 18. Common mullein.

A common weed of roadsides and waste places, native to Eurasia, now established throughout most of temperate N. Am. June-Aug.

Verbascum speciosum Schrad., a related, less densely hairy species with somewhat more open inflorescence, with the filaments all hairy, and with the leaves not decurrent at the base, has been collected at Portland, Oreg.


Veronica L. Speedwell


Flowers borne in terminal or axillary racemes or spikes, or solitary in the upper axils; calyx of (5)4 essentially distinct sepals; corolla blue or violet to pink or white, nearly rotate, irregularly 4-lobed, the upper lobe the largest and internal to the other 3 in bud, the lower lobe the smallest; stamens 2; stigma capitate; capsule somewhat compressed, often notched or lobed at the tip, loculicidal; seeds few-many; annual or perennial herbs; leaves all cauline (at least in our species), entire or toothed, opposite, or those of the inflorescence alternate.

A large genus, of more than 200 species, widely distributed in the N. Temp. Zone, especially in the Old World. (Name of doubtful origin; perhaps commemorating St. Veronica.)

Some of our native species, notably V. cusickii, are desirable garden subjects, but difficult to grow at low altitudes.

Reference:

Pennell, F. W. "Veronica" in North and South America. Rhodora 23: 1-22; 29-41. 1921.

In addition to the following species, V. filiformis Sm., a native of Asia Minor sometimes grown in rock gardens, has become a lawn weed in n.e. U.S., and has recently been reported to be established at Seattle and Bellingham, Wash. It is a slender, trailing perennial from slender rhizomes, also rooting at the nodes, with cordate-orbicular to reniform, slightly toothed, short-petiolate leaf blades mostly less than 1 cm. long, otherwise much like V. persica.

1 Main stem terminating in an inflorescence, its flowers either densely crowded or more remote and axillary, the upper bract-leaves usually alternate
  2 Rhizomatous perennial; seeds numerous
    3 Style elongate, (5)6-10 mm. long; filaments 4-8 mm. long; leaves entire, glabrous; capsules and habit nearly of V. wormskjoldii; high altitudes in the mountains
                                                                     V. CUSICKII
    3 Style shorter, 1-3.5 mm. long; filaments 1-4 mm. long; leaves glabrous or hairy, slightly toothed or sometimes entire
      4 Capsules higher than wide; stem (and often also the otherwise glabrous leaves) sparsely to densely villous-hirsute with loosely spreading hairs; stems simple, erect or merely decumbent at the base; filaments 1-1.5 mm. long; moderate to high altitudes in the mountains            V. WORMSKJOLDII
      4 Capsules wider than high; stem finely and closely puberulent; leaves glabrous or nearly so; stems tending to creep at the base, or to produce prostrate lower branches; filaments (1)2-4 mm. long; lowlands to rather high elevations in the mountains                       V. SERPYLLIFOLIA
  2 Annual
    5 Pedicels very short, only 1-2 mm. long even in fruit; corolla only 2-3 mm. wide; seeds 5 or more per locule
      6 Principal leaves linear-oblong to oblong or oblanceolate, 3-10 times as long as wide; corolla white or whitish; style nearly obsolete, only 0.1-0.3 mm. long; seeds numerous                          V. PEREGRINA

  6   Principal leaves ovate or broadly elliptic, 1-2 times as long as wide; corolla blue-
      violet; style 0.4-1.0 mm. long; seeds (5)8-11 per locule          V. ARVENSIS
 5   Pedicels elongate, mostly 5-35 mm. long in fruit; plants either with larger corollas or
     with fewer seeds.
      7   Leaves, or many of them, palmately 3- to 5-lobulate; capsule slightly or scarcely
          notched; seeds 1-2 per locule; fruiting pedicels 0.5-1.5 cm. long; corolla about 3
          mm. wide                                                       V. HEDERAEFOLIA
      7   Leaves mostly more or less toothed, not palmately lobulate; capsule evidently notched;
          corollas, seeds, and pedicels various
           8   Fruiting pedicels mostly 0.5-1.5 cm. long; corolla 2-4 mm. wide; seeds 2-4 per
               locule; capsule barely wider than high, narrowly notched to the middle or usually
               beyond; style evidently shorter than the not very divergent lobes of the capsule
                                                                         V. BILOBA
           8   Fruiting pedicels mostly 1.5-4 cm. long; corolla 5-11 mm. wide; seeds 5-10 per
               locule; capsule conspicuously wider than high, broadly notched to not beyond the
               middle, the style equaling or surpassing the strongly divergent lobes
                                                                         V. PERSICA
 1  Main stem never terminating in an inflorescence, the leaves opposite throughout and the
    flowers all in axillary racemes
     9   Plants evidently pubescent; leaves relatively broad, 1-3 times as long as wide; seeds 6-
         12 per locule, about 1 mm. long; mesophytes
          10   Leaves coarsely toothed, with mostly 5-11 teeth on each side, more or less broad-
               based, and sessile or nearly so; mature pedicels 5-9 mm. long, surpassing the
               subtending bracts; capsules (seldom produced) shorter than the sepals; corolla 9-
               12 mm. wide                                               V. CHAMAEDRYS
          10   Leaves finely toothed, the larger ones with mostly 12-20 teeth on each side, more or
               less elliptic or elliptic-obovate, narrowed to a short petiole or subpetiolar base;
               fruiting pedicels 1-2 mm. long, shorter than the subtending bracts; capsules sur-
               passing the sepals; corolla 4-8 mm. wide                  V. OFFICINALIS
     9   Plants essentially glabrous, or merely finely glandular in the inflorescence (occasionally
         evidently hairy in V. scutellata, which has linear or lanceolate leaves 3-20 times as
         long as wide); seeds various; plants growing in water or in wet places
          11   Leaves all short-petiolate; capsules and seeds about as in the 2 following species
                                                                         V. AMERICANA
          11   Leaves (at least the middle and upper ones of the flowering shoots) sessile
               12   Capsule turgid, slightly or scarcely notched, slightly, if at all, wider than high;
                    seeds numerous, 0.5 mm. long or less; leaves 1.5-5 times as long as wide
                     13   Leaves 1.5-3 times as long as wide; fruiting pedicels mostly strongly ascend-
                          ing, or upcurved; capsules about as high as wide, or a little higher; flowers
                          blue or violet                                 V. ANAGALLIS-AQUATICA
                     13   Leaves (2.5)3-5 times as long as wide; fruiting pedicels divaricately spread-
                          ing; capsules mostly a little wider than high; flowers white to pink or pale
                          bluish                                         V. CATENATA
               12   Capsule flattened, conspicuously notched, evidently wider than high; seeds 5-9
                    per locule, 1.2-1.8 mm. long; leaves (3)4-20 times as long as wide
                                                                         V. SCUTELLATA

Veronica americana Schwein. ex Benth. in DC. Prodr. 10:468. 1846. (N. Am., presumably
from Bethlehem, Pa.)
  V. americana var. crassula Rydb. Mem. N.Y. Bot. Gard. 1:353. 1900. (Flodman 778, Little Belt Pass, Mont.) A reduced, montane form.
  V. americana f. rosea Henry, Ott. Nat. 31:56. 1917. (Pt. Alberni, B.C.)
  Perennial from shallow creeping rhizomes, glabrous throughout, with erect or ascending
simple stems 1-10 dm. long; leaves opposite, all shortly petiolate, evidently serrate to subentire, lanceolate to lance-ovate or narrowly subtriangular, or the lower more elliptic, mostly 1.5-8 cm. long, 0.6-3 cm. wide, generally 2-4 times as long as wide, or the lower a lit-

tle wider; racemes axillary, pedunculate, open, mostly 10- to 25-flowered; corolla 5-10 mm. wide, blue; style 2.5-3.5 mm. long; fruiting pedicels divaricate, 5-14 mm. long; capsule turgid, 3 mm. high and about as wide or slightly wider, scarcely notched; seeds numerous, 0.5 mm. long or less. N=18.

Wet places, from the lowlands to moderate elevations in the mountains; widespread in temperate N. Am., and common in our range. May-July.

The original publication of V. americana contained no reference to V. beccabunga var. americana Raf. 1830, or Torrey 1847, and is therefore not considered to be a transfer.

Veronica anagallis-aquatica L. Sp. Pl. 12. 1753.

Fibrous-rooted, probably biennial or short-lived perennial plants, more or less erect, 2-10 dm. tall, glabrous throughout, or slightly glandular-puberulent in the inflorescence; leaves all opposite, mostly elliptic or elliptic-ovate to elliptic-oblong, sessile and mostly clasping (or the lower narrowed to a subpetiolar base), mostly 2-10 cm. long and 0.7-5 cm. wide, 1.5-3 times as long as wide, sharply serrate to entire (sterile, autumnal shoots with more rounded and petiolate leaves); racemes axillary, pedunculate, many-flowered; sepals highly variable in form and size; corolla blue, about 5 mm. wide; fruiting pedicels generally strongly ascending, or upcurved, 3-8 mm. long; capsule turgid, 2.5-4 mm. high, scarcely notched, about as high as wide, or a little higher; style 1.5-2.5 mm. long; seeds numerous, 0.5 mm. long or less. N=18.

Along ditches and slowly moving streams, or in other wet places, frequently in shallow water, but largely emersed; native of Europe, now widely established in the U.S.; rarely found in our range. June-Sept.

Veronica arvensis L. Sp. Pl. 13. 1753. (Europe)

Taprooted annual, somewhat villous-hirsute below, generally more puberulent above, erect or subprostrate, the stems 0.5-3 dm. long, simple or branched especially below; principal leaves opposite, ovate or broadly elliptic, sometimes subcordate at the base, 0.5-1.5 cm. long, 1-2 times as long as wide, crenate-serrate, the lower generally short-petiolate; inflorescence terminal, condensed or elongate; bracts alternate, slightly or conspicuously narrower than the leaves, each subtending a single subsessile flower; corolla blue-violet, 2-2.5 mm. wide; fruiting pedicels 1-2 mm. long; capsule 3 mm. high, exactly obcordate; style 0.4-1.0 mm. long; seeds mostly (5)8-11 per locule, 0.8-1.2 mm. long. N=7, 8.

An occasional weed in fields and gardens, less commonly in open, relatively undisturbed places; native of Eurasia, now widely naturalized in N. Am. Apr.-Sept.

Veronica biloba L. Mant. 2:172. 1771. (Cappadocia in Asia Minor)
V. campylopoda Boiss. Diagn. Pl. Nov. 4:80. 1844. (Several collections from the Near East are cited)

Taprooted annual, villous-hirsute, and somewhat viscid upward, the stems simple or branched especially below, 5-15 cm. long, erect or loosely ascending; principal leaves opposite, the lower short-petiolate; blade lance-ovate to broadly deltoid-ovate or elliptic, 0.5-2.5 cm. long, 1-3 times as long as wide, sharply serrate; inflorescence terminal, elongate; bracts alternate, more or less reduced, each subtending a single rather short-pedicellate flower, the pedicel elongating to 5-15 mm. in fruit; sepals prominent, often veiny; corolla blue, 2-4 mm. wide; style 0.5-1.2 mm. long, evidently surpassed by the not very spreading lobes of the capsule; capsule evidently hairy, 3-4 mm. high, about 4 mm. wide, lobed to the middle or generally beyond, the notch rather narrow; seeds 2-4 in each locule, 1.7-2 mm. long, minutely to strongly rugulose. N=7, 14, 21.

Waste places; native of Eurasia, now known from several collections in the interior arid w. U.S. (Wash., Ida., Utah). Apr.-June.

Veronica catenata Pennell, Rhodora 23:37. 1921.
V. connata ssp. glaberrima Pennell, Monog. Acad. Phila. 1:368. 1935. V. connata var. glaberrima Fassett, Rhodora 41:525. 1939. V. comosa var. glaberrima Boiv. Nat. Can. 79:174. 1952. (Rydberg 926, Hot Spgs., S.D.) These names all based on V. catenata Pennell.

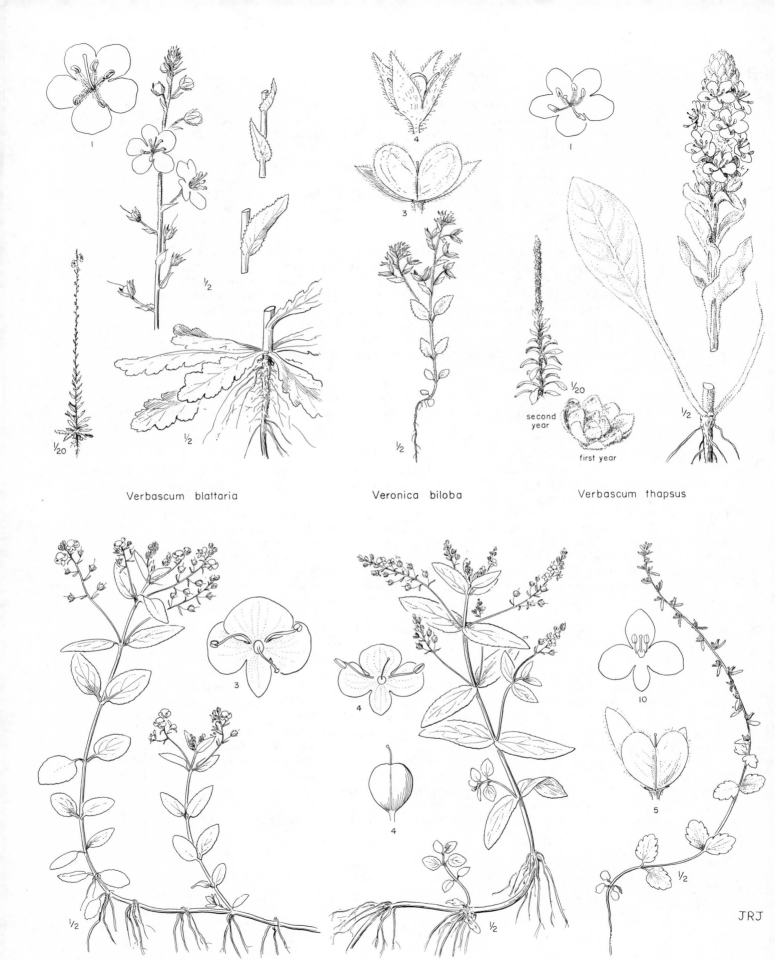

Verbascum blattaria

Veronica biloba

Verbascum thapsus

Veronica americana

Veronica anagallis-aquatica

Veronica arvensis

Closely related to V. anagallis-aquatica, differing in the characters given in the key, and in its entire or subentire leaves, often fewer-flowered racemes, less variable, mostly broader and blunter sepals, and sometimes more evidently notched capsules; petiolate-leaved autumnal shoots not produced. N=18.

In slow-flowing streams and ditches; widespread in the U.S. and adj. Can., and occasionally found at lower elevations in our range; also in Europe. June-July.

Several older names (V. aquatica Bernh., V. comosa Richt., V. connata Raf., V. salina Schur.) have been applied to this species, but V. aquatica Bernh. (1834) is a later homonym of V. aquatica S.F. Gray (1821), and the other names probably do not properly apply. See Watsonia 1:349-53. 1950, for discussion of these names by J.H. Burnett. European specimens tend to have broader, slightly more toothed leaves than do the American plants, but do not appear to be taxonomically separable. A sterile hybrid with V. anagallis-aquatica is reported to be produced frequently in Europe. In America, eastern plants have the pedicels, rachis of the raceme, and distal part of the stem finely stipitate-glandular, while western plants are wholly glabrous. Pennell supposed the name V. connata Raf. to be based on the glandular form, and applied the name V. connata ssp. glaberrima (=V. catenata) to the glabrous form. No such geographical correlation is evident in Europe, where glabrous plants have been named V. aquatica f. laevipes Beck, Fl. Nied.-Österr. 2²:1051. 1893 (V. salina f. laevipes Fern. Rhodora 41:568. 1939).

Veronica chamaedrys L. Sp. Pl. 13. 1753. (Europe)

Perennial from slender rhizomes, the stems ascending or loosely erect, 1-3 dm. tall; stem evidently spreading-hairy, the leaves often more sparsely so; leaves all opposite, ovate, broad-based and sessile or nearly so, mostly 1.5-3 cm. long and 0.8-2 cm. wide, 1-2 times as long as wide, coarsely crenate-serrate, with mostly 5-11 teeth on each side; racemes axillary, pedunculate, loose, the flowers evidently pedicellate; pedicels 5-9 mm. long at maturity, surpassing the subtending bracts, these alternate; corolla blue, 9-12 mm. wide; style 3-5 mm. long; capsules (rarely produced) 3-3.5 mm. high, broadly triangular-obcordate, with apparently about 6 seeds per locule. N=16.

A weed in lawns and open, mostly disturbed sites; native of Europe, introduced in e. U.S., and occasionally found in our range, chiefly w. of the Cascades. May-June.

Veronica cusickii Gray, Syn. Fl. 2¹:288. 1878. (Cusick, alpine region of the Blue Mts., Oreg.)
    V. allenii Greenm. Bot. Gaz. 25:263. 1898. V. cusickii var. allenii Macbr. & Pays. Contr.
    Gray Herb. n.s. 49:67. 1917. V. cusickii f. allenii Pennell, Rhodora 23:13. 1921. (Allen
    95a, Mt. Rainier, Wash.) A white-flowered form.

Perennial from a loose or compact system of shallow rhizomes; stems simple, erect, or curved at the base, 0.6-2 dm. tall, rather thinly villous-puberulent and often glandular, the inflorescence more strongly glandular and hairy; leaves opposite, elliptic to ovate or obovate, 1-2.5 cm. long, 5-14 mm. wide, rounded to acute at the tip, glabrous, entire; flowers pedicellate in well-defined, terminal racemes, usually at least the upper bracts alternate; sepals unequal; corolla deep blue-violet, 8-13 mm. wide when expanded; filaments 4-8 mm. long; style conspicuously exserted, (5)6-10 mm. long, longer than the glandular-pubescent, evidently to scarcely notched capsule, this 4-6 mm. high and tending to be a little higher than wide; seeds rather numerous, about 1 mm. long.

Moist, open rocky slopes, or sometimes along streams or in meadows, at rather high elevations in the mountains, sometimes above timber line; Olympic and Cascade mts. of Wash., e. to the mountains of n. and c. Ida. and w. Mont., and in the Wallowa and higher Blue mts. of n.e Oreg.; a smaller-leaved, perhaps varietally separable plant occurs in Yosemite Nat. Pk., Calif. July-Aug.

Veronica hederaefolia L. Sp. Pl. 13. 1753. (Europe)

Taprooted annual, branched at the base, with prostrate or weakly ascending stems 0.5-4 dm. long, the herbage sparsely to moderately spreading-hirsute; leaves 0.5-2 cm. wide, mostly wider than long, palmately veined and most of them palmately 3- to 5-lobulate, the petiole equaling or shorter than the blade; lowest leaves opposite; floriferous portion of the stem elongate, with alternate, fully developed leaves, each bearing in the axil a single flower, the pedicel rather short at anthesis, elongating to 0.5-1.5 cm. in fruit; sepals accrescent, becoming 5-7 mm. long, then deltoid-ovate and conspicuously ciliate; corolla pale bluish, about 3 mm. wide; style 0.6-1.0 mm. long; capsule 2.5-3.5 mm. high, slightly or scarcely notched, globose or broader than high; seeds 1, or less often 2, per locule, transversely rugose. N=28.

A weed in waste places and more or less disturbed habitats; native of Europe, now widely naturalized in n. e. U.S., and known from a single collection in our range (Almota, Snake River Canyon, Whitman Co., Wash.) Apr.-June.

Veronica officinalis L. Sp. Pl. 11. 1753. (Europe)
    V. tournefourtii Vill. Prosp. Hist. Pl. Dauph. 20. 1779. V. officinalis var. tournefourtii
    Reichb. Icon. Fl. Germ. 20:49. 1862. (France) A small-leaved form.

Fibrous-rooted perennial, the lower portion of the stem creeping, rooting at the nodes, more or less elongate and often with reduced leaves, the tips and branches assurgent and 0.5-2.5 dm. long; herbage and inflorescence evidently spreading-hairy; leaves all opposite, more or less elliptic or elliptic-obovate, narrowed to a short petiole or subpetiolar base, mostly 1.5-5 cm. long and 0.6-3 cm. wide, 1.5-3 times as long as wide, rather finely serrate, the larger ones with mostly 12-20 teeth on each side; racemes axillary, pedunculate, spiciform, the pedicels only 1-2 mm. long even in fruit and surpassed by the small, alternate, subtending bracts; corolla light blue, sometimes with lavender stripes, 4-8 mm. wide; capsule 4 mm. high, broadly triangular-subcordate, shallowly retuse, evidently surpassing the sepals; style 2.5-4.5 mm. long; seeds mostly 6-12 per locule, about 1 mm. long. N=9, 18.

Roadsides, fields, and other disturbed sites; native to Europe, now widely distributed in e. U.S., and found w. of the Cascades in our range. Apr.-July.

Veronica peregrina L. Sp. Pl. 14. 1753. (Europe)
    V. xalapensis H.B.K. Nov. Gen. & Sp. 2:389. 1818. V. peregrina (ssp.) xalapensis Pennell, Tor-
    reya 19:167. 1919. V. peregrina var. xalapensis St. John & Warren, N.W. Sci. 2:90. 1928.
    (Humboldt & Bonpland, near Xalapa, Mex.)
    V. sherwoodii Peck, Torreya 28:56. 1928. (Sherwood 439, Wallowa Lake, Oreg.) An over-
    wintering form.

Fibrous-rooted annual, erect, or curved at the base, 0.5-3 dm. tall, simple, or branched especially below; principal leaves opposite, oblong or linear-oblong to oblanceolate, 0.5-3 cm. long, 1-9 mm. wide, irregularly toothed or entire; inflorescence terminal, elongate, lax; bracts alternate, gradually reduced upward, often not much different from the leaves, each subtending a single subsessile flower; corolla white or whitish, inconspicuous, only about 2 mm. wide; fruiting pedicels only 1-2 mm. long; capsule 3-4 mm. high, more or less obcordate, the notch broad and shallow or a little deeper and very narrow; style nearly obsolete, only 0.1-0.3 mm. long; seeds numerous, 0.4-0.8 mm. long. N=26.

Swales, wet meadows, stream banks, and other moist places, from the lowlands to moderate elevations in the mountains; widespread in temperate parts of N. Am. and S. Am., and introduced in Eur. Apr.-Sept.

Our common, native phase of the species, the var. xalapensis (H.B.K.) St. John & Warren, has the stem (and commonly also the sepals and capsules) more or less pubescent with short, gland-tipped hairs. The typical variety of the species, a native of e. U.S., differs in being wholly glabrous. It is occasionally found w. of the Cascades as a supposed introduction.

Veronica persica Poir. in Lam. Encyc. Meth. 8:542. 1808. (Persia)
Lax, more or less taprooted, somewhat villous-hirsute annual, the stems 1-4 dm. long, simple or branched, especially below, loosely ascending, often rooting at the lower nodes; principal leaves opposite, short-petiolate, the blade broadly elliptic or ovate, crenate-serrate, 1-2 cm. long, 1-1.5 times as long as wide; floriferous portion of the stem mostly elongate, the bracts alternate, resembling the leaves, gradually reduced upward, each subtending a single long-pedicellate flower, the pedicel becoming 1.5-4 cm. long in fruit; sepals prominent, often veiny; corolla blue, 5-11 mm. wide; style 1.5-2.5 mm. long, equaling or surpassing the lobes of the capsule; capsule 5-9 mm. wide, 3-5 mm. high, the tips of the broad lobes near the mar-

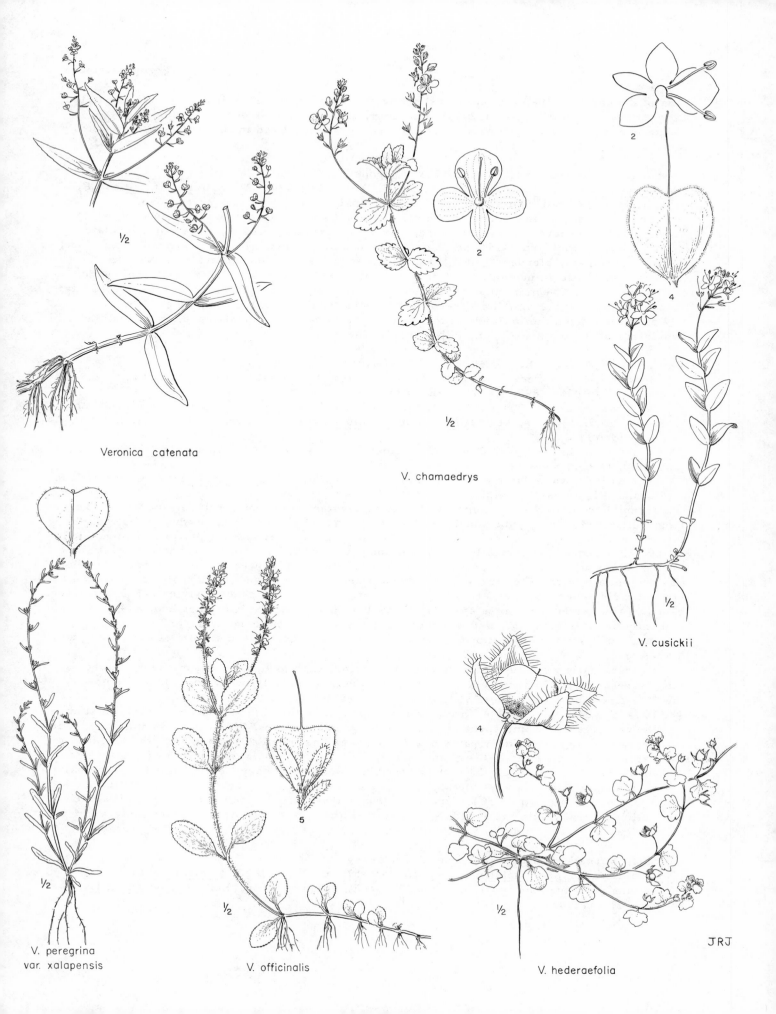

Veronica catenata

V. chamaedrys

V. cusickii

V. peregrina
var. xalapensis

V. officinalis

V. hederaefolia

JRJ

gins of the capsule, the broad notch extending scarcely (or evidently less than) halfway to the base; seeds 5-10 per locule, 1.2-1.8 mm. long, transversely strongly rugose. N=14.

Lawns and waste places; native of Eurasia, widely naturalized in N. Am., and occasionally found in our range. March-May.

**Veronica scutellata** L. Sp. Pl. 12. 1753. (Europe)
  V. scutellata var. villosa Schum. Enum. Pl. Saell. 1:7. 1801. V. scutellata f. villosa Pennell, Monog. Acad. Phila. 1:370. 1935. (Denmark) A more or less hairy form.

Perennial from shallow, creeping rhizomes, glabrous throughout, or occasionally hairy; stems erect or ascending, 1-4 dm. long; leaves all opposite, sessile, linear to lanceolate, mostly 2-8 cm. long and 2-15 mm. wide, (3)4-20 times as long as wide, entire or with a few, remote, divergent, slender, small teeth; racemes axillary, pedunculate, mostly 5- to 20-flowered, the filiform pedicels 6-17 mm. long at maturity; corolla bluish, 6-10 mm. wide; fruit flattened, 2.5-4 mm. high, evidently wider than high, conspicuously and rather broadly notched, the style 2-4 mm. long; seeds 5-9 in each locule, 1.2-1.8 mm. long. N=9.

Wet places, from the lowlands to moderate elevations in the mountains; Eurasia and the n. two-thirds of the temperate zone in N. Am. May-Sept.

**Veronica serpyllifolia** L. Sp. Pl. 12. 1753. (Europe and N. Am.)
  V. tenella All. Fl. Pedemont. 1:75. 1785. V. serpyllifolia var. tenella Beck, Fl. Nied. -Österr. 2²:1056. 1893. (Italian Alps) = var. humifusa.
  V. humifusa Dickson, Trans. Linn. Soc. 2:288. 1794. V. serpyllifolia var. humifusa Vahl, Enum. 1:65. 1806. V. serpyllifolia ssp. humifusa Syme ex Claph. et al. Fl. Brit. Isl. 882. 1952. (Scotland)
  V. funesta Macbr. & Pays. Contr. Gray Herb. n. s. 49:68. 1917. (Applegate 424, Swan Lake Valley, Oreg.) = var. humifusa.
  V. serpyllifolia var. decipiens Boiv. Nat. Can. 79:176. 1952. (Breitung 4357, Cypress Hills Park, Sask.) = var. humifusa.

Perennial from a loose or compact, branching system of creeping rhizomes; stems 1-3 dm. long, finely and closely puberulent, tending to creep at the base, or to produce prostrate, lower branches, otherwise simple; leaves opposite (or some reduced upper ones alternate), elliptic to broadly ovate, mostly 1-2.5 cm. long and 0.5-1.5 cm. wide, rounded to acutish at the tip, glabrous or nearly so, slightly toothed or entire, the lower often short-petiolate; flowers pedicellate in definite, terminal racemes which become loose and elongate, at least the upper bracts generally alternate; corolla 4-8 mm. wide when expanded; styles 2-3.5 mm. long, capsules finely and sometimes sparsely glandular-pubescent, evidently notched, 3-4 mm. high, evidently broader than high; seeds numerous, less than 1 mm. long. N=7.

Moist meadows and other moist places, often in disturbed sites; more or less cosmopolitan. May-Aug.

Two technically well-marked varieties, in N. Am. approaching the distinctness of species, may be recognized. In some other parts of the world, especially where both taxa are native, the cleavage appears to be less sharp.

1  Flowers usually bright blue; inflorescence more hairy than the vegetative part of the stem, the pedicels with some spreading, viscid or glandular hairs; filaments 2-4 mm. long; widespread in the mountainous parts of the world, and native in our range from the lowlands to rather high elevations in the mountains                                    var. humifusa (Dickson) Vahl
1  Flowers white or pale bluish, with darker blue lines; inflorescence not evidently more hairy than the vegetative part of the stem, the pedicels finely and closely puberulent; filaments 1-2.5 mm. long; native of Europe and perhaps elsewhere, widely introduced at lower elevations elsewhere in the world, common in e. U.S., and occasionally found (only w. of Cascades) in our range                                                      var. serpyllifolia

**Veronica wormskjoldii** Roem. & Schult. Syst. Veg. 1:101. 1817.
  V. alpina var. wormskjoldii Hook. Bot. Mag. 57: pl. 2975. 1830. (Wormskjold, Greenl.)
  V. alpina var. unalaschcensis Cham. & Schlecht. Linnaea 2:556. 1827. (Chamisso & Escholtz, Unalaska)

V. alpina var. geminiflora Fern. Rhodora 41:454. 1939. (Baker 607, near Pagosa Peak, Colo.)

V. alpina var. alterniflora Fern. Rhodora 41:455. 1939. V. wormskjoldii ssp. alterniflora Pennell in Abrams, Ill. Fl. Pac. St. 3:793. 1951. (Macbride & Payson 3632, Cape Horn, Custer Co., Ida.)

V. alpina var. cascadensis Fern. Rhodora 41:456. 1939. (Allen 277, Mt. Rainier, Wash.)

Perennial from a loose or compact system of shallow rhizomes; stems simple, erect, or curved at the base, 0.7-3 dm. tall, sparsely to densely villous-hirsute with loosely spreading hairs, the inflorescence more densely so and somewhat viscid or glandular; leaves all cauline, opposite (or some reduced upper ones alternate), elliptic to lanceolate or ovate, mostly 1-4 cm. long and 0.5-2 cm. wide, rounded to acute at the tip, villous-hirsute like the stem, or often glabrous, slightly toothed or entire; flowers pedicellate in well-defined, terminal racemes, these at first compact, later often elongate; at least the upper bracts usually alternate; corolla deep blue-violet, 6-10 mm. wide when expanded; filaments 1-1.5 mm. long; style short, 1-3 mm. long; capsule glandular-pubescent, broadly notched, 4-7 mm. high, a little higher than wide; seeds numerous, about 1 mm. long. N=18.

Moist meadows, stream banks, bogs, and moist open slopes, at moderate to high elevations in the mountains; Alas. to Greenl., s. to Calif., N.M., and N.H. July-Aug.

Plants from the less extreme habitats may have the mature raceme more elongate, with the fruits less crowded, than more strictly alpine or boreal specimens, but the variation is contin- uous and probably not worthy of taxonomic notice. These looser forms were considered by Fernald to represent 3 separate varieties, and by Pennell to be a single subspecies.

Veronica wormskjoldii is closely related to the European and high-boreal n. e. American V. alpina, differing chiefly in some details of pubescence. Fernald has disposed of V. wormskjol- dii as a series of American varieties of V. alpina, but these varieties are scarcely distinguish- able among themselves, while on the other hand, true populational intergradation between any of them and the admittedly similar V. alpina remains to be demonstrated. It therefore seems advisable to follow Pennell in recognizing V. wormskjoldii as distinct, at least until more evi- dence is available.


OROBANCHACEAE. Broomrape Family

Flowers gamopetalous, perfect, or rarely some of them unisexual; calyx regular or irreg- ular, 4- to 5-lobed, or the segments sometimes more or less connate; corolla tubular, bila- biate or rarely regular, 5-lobed (or 2 of the lobes connate); stamens 4, didynamous, epipetal- ous, the filaments often reverse-bent beneath the anthers; ovary superior, 2- to 3-carpellary, ordinarily 1-locular; placentae parietal, (1)2 per carpel, intruded and often bifurcate, rarely confluent and the capsules thus 2-locular; style solitary, with a capitate or disciform, often 2- to 4-lobed stigma; fruit a 2- to 3-valved, longitudinally dehiscent capsule; seeds ordinarily numerous and small, with well-developed endosperm and very small, few-celled embryo; her- baceous, often fleshy, root parasites, lacking chlorophyll, commonly yellowish to brownish or purplish in color, with bract-like, alternate leaves and spicate to corymbose or more or less paniculate inflorescence, or the flowers solitary.

A rather small family, of cosmopolitan distribution. Beck (Pflanzenr. IV. 261. 1-348. 1930) recognized 14 genera and 156 species.

1 Pollen sacs blunt, barely or not at all mucronate; bracts of the densely spicate inflorescence very broad, blunt, and conspicuous; our species with a tuft of long hairs at the base of each filament; coastal and subcoastal (Puget Sound included)            BOSCHNIAKIA
1 Pollen sacs pointed and more or less mucronate at the base; bracts narrower, less con- spicuous, and mostly more pointed; inflorescence variously shaped; lower lip of the corolla, at least in our species, well developed and not markedly shorter than the upper; filaments glabrous, or hairy in part, but not with a tuft of long hairs at the base; throughout our range            OROBANCHE


Boschniakia C. A. Mey. Ground Cone

Flowers borne in a dense, spicate inflorescence, subtended by conspicuous, broad, blunt bracts; calyx irregularly lobed or toothed, or truncate and lobeless; corolla curved, bilabiate, the upper lip often somewhat boat-shaped, shallowly if at all cleft, the lower lip often more or less reduced; filaments more or less hairy below, often densely so; pollen sacs blunt, seldom at all mucronate; ovary 2- to 3(4)-carpellary, 1-locular, with 2-4 bifurcate, parietal placentae; stigma entire or 3- to 4-lobed; coarse and fleshy root-parasites.

About half a dozen habitally very similar species, native to Asia and w. N. Am. The genus is technically poorly defined and not well separated from Orobanche; it is sometimes segregated into 3 genera of only 2 species each. Our species belongs to the segregate genus Kopsiopsis, Beck. (Named for Boschniaki, Russian botanist. )

Boschniakia hookeri Walpers Rep. 3:479. 1844-45.
  Orobanche tuberosa Hook. Fl. Bor. Am. 2:92. 1838, not of Vell. in 1825. O. hookeri Beck, Monog. Orob. 85. 1890. Boschniakia tuberosa Jeps. Man. Fl. Pl. Calif. 954. 1925. Kopsiopsis tuberosa Beck, Pflanzenr. IV. 261:305. 1930. (Menzies, n. w. coast of America)
  Plants yellow to dark red or purple, 8-12 cm. tall, 3 cm. or less thick in the inflorescence, often thicker at the cormlike base; stems solitary or clustered, copiously scaly-bracteate, the bracts, especially those of the inflorescence, conspicuous, broad, and mostly blunt; flowers with or without a pair of basal bractlets in addition to the subtending bract; calyx commonly with 2-3 short lobes; corolla 1-1.5 cm. long, firm, constricted near or below the middle, the lower lip shorter than the upper; filaments with a dense tuft of hairs at the base; stigma obscurely to evidently 2- to 3-lobed; placentae 2-3(4); capsule 1-1.5 cm. long.

Parasitic on Gaultheria shallon, along and near the coast (including the Puget Sound area), from n. B. C. to n. Calif. June-July.

### Orobanche L. Broomrape; Cancerroot

Flowers borne in spicate to paniculate or corymbose inflorescences, or solitary; calyx irregularly or nearly regularly 4- to 5-cleft, or divided to the base above and below; corolla bilabiate, both lips developed; filaments glabrous or sometimes hairy, especially below; pollen sacs ordinarily with well-separated, pointed, more or less mucronate bases; ovary 2-carpellary, 1-locular, with 4 parietal placentae; stigma entire or 2- to 4-lobed; root parasites, often more or less fleshy.

By far the largest genus of the family, cosmopolitan in distribution, consisting of 4 well-defined sections that have sometimes been treated as genera. (Name from the Greek orobos, vetch, and anchein, to choke. )

Beck recognized 99 species in 1930. In addition to Beck's treatment in Das Pflanzenreich, the following papers treat the North American species:
  Achey, Daisy M. A revision of the section Gymnocaulis of the genus Orobanche. Bull. Torrey Club 60:441-51. 1933.
  Munz, P. A. The North American species of Orobanche, section Myzorrhiza. Bull. Torrey Club 57:611-24. 1930.
1 Calyx more or less deeply cleft into 5 subequal lobes; native species
  2 Flowers sessile or on pedicels up to about 3 cm. long, provided with a pair of bractlets just beneath the calyx, in addition to the subtending bract (Myzorrhiza)
    3 Calyx mostly 5-8 mm. long, the lobes about equaling the tube, or a little shorter; stem ordinarily with many short, slender branches, so that the yellowish flowers are borne in a loose, paniculiform inflorescence; mostly in coniferous woods, nearly throughout our range                                                                O. PINORUM
    3 Calyx mostly 10-20 mm. long, the lobes much longer than the tube; stem simple or few-branched, but not forming an open, paniculiform inflorescence; corolla mostly purplish or pink; mostly on open slopes or in meadows, often in sagebrush areas
      4 Flowers, especially the lower ones, generally more or less pedicellate, the pedicels up to about 3 cm. long; inflorescence mostly short and stout, often corymbose; ours with woolly anthers and with the corollas mostly 22-30 mm. long
        5 Lower lip of the corolla strongly spreading, 7-15 mm. long; both sides of the Cascades in Oreg. and Wash. and adj. B. C.                                          O. GRAYANA

5 Lower lip of the corolla erect (i. e., continuous with the line of the tube), or slightly arcuate-spreading toward the tip, 4-6 mm. long; e. of the Cascades
O. CALIFORNICA
4 Flowers sessile or nearly so; inflorescence spicate, tending to be elongate; ours with glabrous anthers and with the corolla mostly 15-22 mm. long; e. of the Cascades
O. LUDOVICIANA
2 Flowers conspicuously long-pedicellate, without bractlets; throughout our range (Gymnocaulis, Thalesia)
6 Pedicels mostly 4-10, up to about as long as the more or less elongate stem; calyx lobes equaling or shorter than the tube          O. FASCICULATA
6 Pedicels 1-3, much longer than the mostly very short stem; calyx lobes evidently longer than the tube          O. UNIFLORA
1 Calyx divided to the base into 2 lateral segments, these generally cleft into two, mostly unequal lobes; flowers sessile, without bractlets; Mediterranean species, introduced about Portland, Oreg., and adj. Wash. (Osproleon)          O. MINOR

Orobanche californica Cham. & Schlecht. Linnaea 3:134. 1829.
  Phelipaea californica G. Don, Gen. Syst. 4:632. 1837. Aphyllon californicum Gray, Bot. Calif. 1:584. 1876. Myzorrhiza californica Rydb. Bull. Torrey Club 36:695. 1909. (Chamisso, vicinity of San Francisco)
  Myzorrhiza corymbosa Rydb. Bull. Torrey Club 36:696. 1909. Orobanche californica var. corymbosa Munz, Bull. Torrey Club 57:618. 1930. (Mulford, Reynold's Cr., Ida., in 1892)
  Plants 5-12(18) cm. tall, glandular-puberulent throughout; inflorescence short and dense, often more or less corymbose; flowers borne on pedicels up to about 3 cm. long, or the upper sometimes sessile, subtended by a pair of narrow bractlets just beneath the calyx, as well as by a bract at the base of the pedicel; calyx 10-20 mm. long, deeply cleft, the 5 narrow, subequal lobes elongate, caudate-attenuate, much longer than the tube; corolla purplish, mostly 22-28 mm. long, the lower lip only slightly or scarcely arcuate-spreading, 4-6 mm. long; anthers long-woolly.
  Open slopes in the foothills and valleys, commonly parasitic on Artemisia tridentata; Wash. (e. of the Cascades) and adj. s. B. C. to s. Calif., e. to w. Mont. and Utah. June-Aug.
  Our plants, as described above, represent the var. corymbosa (Rydb.) Munz. The nomenclaturally typical phase, occurring in cismontane Calif., is mostly taller and has glabrate anthers; a third variety occurs in s. Calif.
  Orobanche californica, O. grayana, and O. ludoviciana are closely allied species, not well separated when the total geographic range and morphologic variability of each is considered, and there appears to be some intergradation. In our area, however, the separation is seldom difficult.

Orobanche fasciculata Nutt. Gen. Pl. 2:59. 1818.
  Phelipaea fasciculata Spreng. Syst. Veg. 2:818. 1825. Anoplon fasciculatum G. Don, Gen. Syst. 4:633. 1837. Anoplanthus fasciculatus Walpers Rep. 3:480. 1844-45. Aphyllon fasciculatum Gray, Syn. Fl. 2[1]:312. 1878. Thalesia fasciculata Britt. Mem. Torrey Club 5:298. 1894. (Nuttall, Ft. Mandan, N. D.)
  Phelipaea lutea Parry, Am. Nat. 8:214. 1874. Aphyllon fasciculatum var. luteum Gray, Syn. Fl. 2[1]:312. 1878. Thalesia fasciculata var. lutea Britt. Mem. Torrey Club 5:298. 1894. Thalesia lutea Rydb. Bull. Torrey Club 36:693. 1909. Orobanche fasciculata f. lutea Beck, Pflanzenr. IV. 261:51. 1930. O. fasciculata var. lutea Achey, Bull. Torrey Club 60:449. 1933. (Parry 202, Owl Creek, Wyo.) The form with pale stems and flowers.
  Stems solitary or clustered, slender or moderately stout, 3-15 cm. long, yellowish to more often purplish, the bracts mostly few and remote; plants glandular-puberulent or glandular-villous throughout; pedicels mostly 4-10, 2-15 cm. long, not much, if at all, longer than the stem, the lower often longer than the upper, or all more or less approximate in origin, so that a loose, flat-topped corymb that much surpasses the stem is formed; flowers without bractlets; calyx about 1 cm. long, the 5 narrow, acute lobes equaling, or shorter than, the

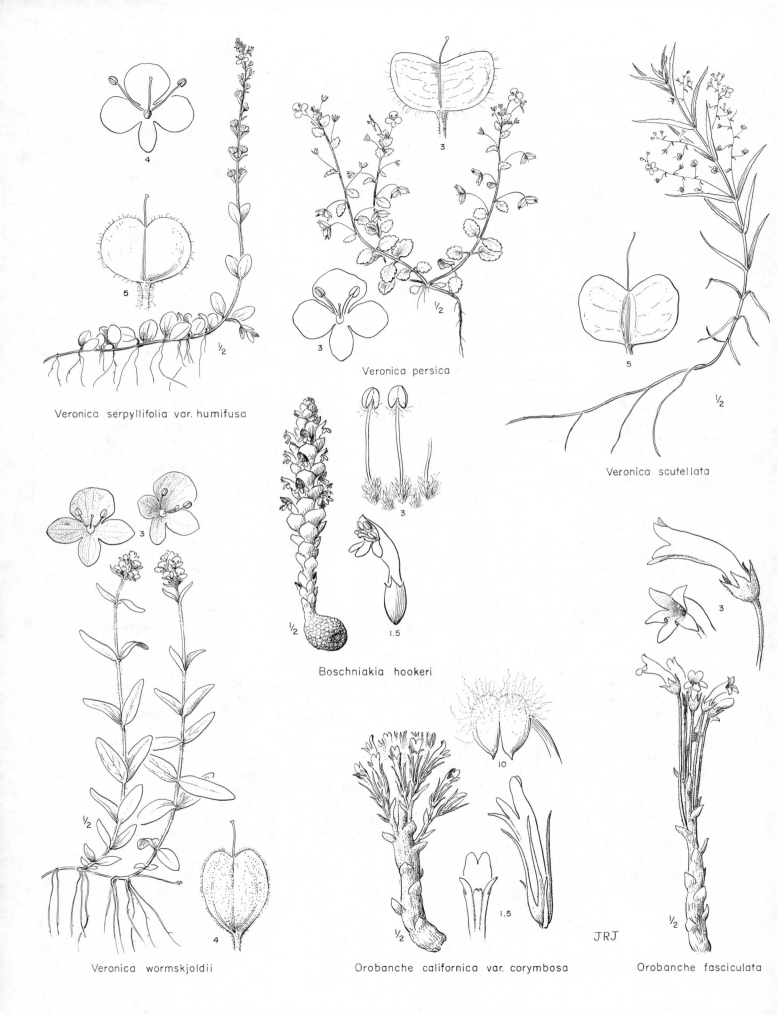

4

5

½

Veronica serpyllifolia var. humifusa

3

3

½

Veronica persica

5

½

Veronica scutellata

3

3

1.5

Boschniakia hookeri

3

½

4

Veronica wormskjoldii

10

½

1.5

Orobanche californica var. corymbosa

3

½

Orobanche fasciculata

JRJ

tube; corolla purple or yellowish, 15-30 mm. long, the lobes rounded to acute; anthers glabrous or hairy.

Open slopes in the foothills and valleys, parasitic on a wide variety of hosts, but particularly on Artemisia tridentata and other species of Artemisia; B. C. to Alta., Mich., and Ind., s. to Calif. and n. Mex. May-July.

Orobanche grayana Beck, Bibl. Bot. 4:79. 1890.
  Orobanche comosa Hook. Fl. Bor. Am. 2:92. 1838, not of Wallr. in 1822. Anoplanthus
    comosus Walpers Rep. 3:480. 1844-45. Phelipaea comosa Gray, Pac. R. R. Rep. 4:118.
    1857. Aphyllon comosum Gray, Bot. Calif. 1:584. 1876. Myzorrhiza grayana Rydb. Bull.
    Torrey Club 36:695. 1909. (Douglas, banks of the Columbia)
  O. grayana var. nelsonii Munz, Bull. Torrey Club 57:616. 1930. (Nelson 2479, near Salem,
    Oreg.)
  Similar to O. californica var. corymbosa, differing chiefly in the characters given in the key.

Meadows and open slopes, parasitic chiefly on various species of Compositae; both sides of the Cascades in Oreg. and Wash., extending into s. B. C., and southward through Calif. and adj. Nev. to Baja Calif. June-Sept.

Our plants may be referred, in the broad sense, to var. grayana. A much larger-flowered variety occurs on the coast of Calif.

Orobanche ludoviciana Nutt. Gen. Pl. 2:58. 1818.
  Phelipaea ludoviciana G. Don, Gen. Syst. 4:632. 1838. Aphyllon ludovicianum Gray, Bot.
    Calif. 1:585. 1876. Myzorrhiza ludoviciana Rydb. in Small, Fl. S. E. U. S. 1093. 1903.
    (Nuttall, Ft. Mandan, N. D.)
  OROBANCHE LUDOVICIANA var. ARENOSA (Suksd.) Cronq. hoc loc. Aphyllon arenosum
    Suksd. Allg. Bot. Zeit. 12:27. 1906. Orobanche multiflora var. arenosa Munz, Bull. Tor-
    rey Club 57:623. 1930. (Suksdorf 2781, Bingen, Klickitat Co., Wash.)
  Plants 7-20 cm. tall, glandular-puberulent throughout; inflorescence dense, spicate, often elongate; flowers sessile, or the lower inconspicuously short-pedicellate, subtended by a pair (or only 1) of narrow bractlets, as well as by a bract; calyx 10-20 mm. long, deeply cleft, the 5 narrow, subequal lobes long-pointed, much longer than the tube; corolla purplish, or pink striped with white, mostly 15-22 mm. long, the lips short and not much spreading, with blunt or somewhat pointed lobes; anthers essentially glabrous.

Open places in the foothills and valleys; parasitic on a variety of hosts, particularly species of Artemisia and other Compositae; Wash. (e. of the Cascades) to s. Calif. and n. Mex., e. to Sask., Ind., and Tex. July-Sept.

Our plants, as described above, belong to the var. arenosa (Suksd.) Cronq., which ranges through the drier parts of the cordilleran area n. of the s. w. deserts. Typical O. ludoviciana, a plant of the Great Plains, with mostly more pointed corolla-lobes and with the anthers generally somewhat woolly along the sutures after dehiscence, may be expected along the e. edge of our range in Mont. At least two other varieties occur to the s. of our range. See comment under O. californica.

Orobanche minor J. E. Smith, Engl. Bot. 6:422. 1797. (Sutton, near Sheringham, Eng.)
  O. columbiana St. John & English, Proc. Biol. Soc. Wash. 44:34. 1931. (English 1069, near
    Fisher, Clark Co., Wash.)
  Plants 1.5-5 dm. tall, strict and simple from a more or less bulbous-thickened base, more or less glandular-villous throughout; inflorescence spicate, more or less elongate, not very dense, especially below, the axis exposed; flowers sessile, subtended by a bract but without bractlets; calyx split to the base above and below, the 2 lateral segments thus formed slender and elongate, mostly bifurcate, commonly unequally so, the lower fork the smaller; corolla white or yellowish, marked with violet, 10-15 mm. long; anthers glabrous or nearly so. N=19.

Parasitic on a variety of hosts, particularly Leguminosae and Compositae, but most commonly on cultivated clover, becoming a serious weed in clover fields in Europe. Native of the Mediterranean region, widely introduced in temperate regions elsewhere in the world, especially in association with clover; established in America in the Atlantic seaboard states, and in the vicinity of Portland, Oreg. and adjacent Wash.; to be expected to spread in our range at least on the w. side of the Cascades. May-June.

Orobanche pinorum Geyer, Journ. Bot. & Kew Misc. 3:297. 1851.
   Phelipaea pinetorum Gray, Proc. Am. Acad. 7:371. 1867. Aphyllon pinetorum Gray, Bot.
    Calif. 1:585. 1876. Myzorrhiza pinorum Rydb. Bull. Torrey Club 36:695. 1909. (Geyer,
    "top of the high mountains near St. Joseph, Coeur d'Aleine country," Ida.)

   Plants 1-3 dm. tall, viscid-puberulent throughout; stems 1 or several from an evidently thickened, somewhat tuberous base, soon tapering and becoming more slender, with more or less numerous, short, slender branches in the upper half or two thirds, so that a loose, panic-uliform inflorescence is formed; flowers short-pedicellate or sessile, subtended by a pair of bractlets at the base of the calyx as well as by a bract at the base of the pedicel; calyx 5-8 mm. long, the slender lobes about equaling the tube, or a little shorter; corolla yellowish, marked with purplish brown, 13-20 mm. long, the lips short, with acute lobes; anthers glabrous or nearly so.

   Mostly in coniferous woods, where parasitic on various conifers; Wash. and n. Ida. to n. w. Calif. July-Aug.

Orobanche uniflora L. Sp. Pl. 633. 1753.
   O. biflora Nutt. Gen. Pl. 2:59. 1818. Phelipaea biflora Spreng. Syst. Veg. 2:818. 1825.
    Anoplon biflorum G. Don, Gen. Syst. 4:634. 1837. Aphyllon uniflorum Gray, Man. 290.
    1848. Thalesia uniflora Britt. Mem. Torrey 5:298. 1894. (Va.)
   Thalesia purpurea Heller, Bull. Torrey Club 24:313. 1897. Orobanche porphyrantha Beck,
    Pflanzenr. IV. 261:49. 1930. O. uniflora var. purpurea Achey, Bull. Torrey Club 60:445.
    1933. (Heller 3099, mouth of the Potlatch R., near Lewiston, Ida.)
   Aphyllon minutum Suksd. Deuts. Bot. Monats. 18:155. 1900. Thalesia minuta Rydb. Bull.
    Torrey Club 36:692. 1909. Orobanche uniflora var. minuta Beck, Pflanzenr. IV. 261:49.
    1930. (Suksdorf, Bingen, Klickitat Co., Wash., in 1892)
   Aphyllon sedi Suksd. Deuts. Bot. Monats. 18:155. 1900. Thalesia sedi Rydb. Bull. Torrey
    Club 40:485. 1913. Orobanche sedi Fern. Rhodora 28:236. 1926. O. uniflora f. sedi Beck,
    Pflanzenr. IV. 261:49. 1930. (Suksdorf 2130, Chenowith, Skamania Co., Wash.) = var.
    minuta.
   Aphyllon inundatum Suksd. Allg. Bot. Zeit. 12:27. 1906. Orobanche uniflora f. inundatum
    Beck, Pflanzenr. IV. 261:49. 1930. (Suksdorf 205, near Bingen, Klickitat Co., Wash.) =
    var. minuta.

   Stems short, 1-5 cm. long, not stout, much shorter than the 1-3 elongate pedicels, which are 3-10(15) cm. long; plants finely glandular-villous, especially upward; flowers without bractlets; calyx 4-12 mm. long, the 5 narrow, subequal lobes evidently longer than the tube; corolla ochroleucous to purple, 15-35 mm. long, the lobes finely fringed-ciliate; anthers gla-brous or hairy. N=18, 36.

   Mostly in open, moist or dry places, or in open woods, from the lowlands to moderate ele-vations in the mountains, parasitic on many species, particularly Sedum and various Saxifra-gaceae and Compositae; Yukon to Newf., s. to s. Calif. and Fla. Apr.-Aug.

   A variable but sharply defined species. Four varieties may be recognized, two occurring throughout the w. cordilleran region, one (the var. uniflora) from the Great Plains to the At-lantic, and a fourth in Newf. Our two varieties, which have not yet been clearly demonstrated to show any consistent differences in host or habitat, are morphologically fairly well marked, and may be distinguished as follows:

1 Anthers glabrous; corolla ochroleucous to purple, 1.5-2.5 cm. long, the throat not much
    expanded (mostly 3-5 mm. wide as pressed), and the relatively small limb not much
     spreading                                                                    var. minuta (Suksd.) Beck
1 Anthers usually more or less woolly; corolla purple, 2-3.5 cm. long, with relatively broad
    throat (mostly 5-9 mm. wide as pressed) and expanded, spreading limb
                                    var. purpurea (Heller) Achey

## LENTIBULARIACEAE. Bladderwort Family

Flowers gamopetalous, perfect, irregular; calyx laterally 2-parted, often to the base or nearly so, the segments entire, or sometimes the upper lip 3-lobed and the lower lip 2-lobed; corolla strongly or rarely obscurely bilabiate, the lips evidently lobed (upper 2, lower 3) or entire, the proper tube sometimes very short, the lower lip commonly prolonged into a basal spur; stamens 2, attached to the corolla tube near its base, each with a single pollen sac, but this, except in Pinguicula, transversely generally more or less constricted; filaments often contorted; ovary superior, 1-locular, with free-central placenta and usually numerous ovules; style short or obsolete; stigma 2-lobed, with one lobe smaller or obsolete; fruit a capsule, dehiscing by 2 or 4 valves, or bursting irregularly; seeds ordinarily numerous and small, without endosperm, the embryo not much differentiated; annual or perennial herbs, aquatic or of wet soil, with alternate (or all basal, or in part or wholly whorled), well-developed to minute leaves, these when submersed usually dissected, and producing numerous small bladders which entrap crustaceans and other small aquatic animals; flowers borne singly or more often in few-flowered (rarely branched) racemes, at the ends of erect scapes or peduncles.

A rather small family, of cosmopolitan distribution, best developed in the tropics. There are perhaps 300 species, usually divided into 5 genera. Four of these genera are small and fairly homogeneous. The fifth, Utricularia, is so complex and variable that Barnhart (Mem. N.Y. Bot. Gard. 6:39-64. 1916) divided it into 12 genera. The dissected leaves have sometimes (perhaps properly) been considered to represent in reality systems of branching stems; the slender, rootlike organs of genera of the Lentibulariaceae other than Pinguicula may also be modified stems.

1 Calyx 5-lobed, the lobes partly connate into an upper lip of 3 and a lower lip of 2; flowers solitary on bractless scapes; palate wanting; terrestrial herbs of wet places, with well-developed, entire, rosulate basal leaves        PINGUICULA
1 Calyx deeply 2-lobed; flowers solitary or more commonly racemose, each subtended by a bract; corolla tube generally closed by a well-developed palate at the base of the lower lip; aquatic plants with submersed, dissected leaves        UTRICULARIA

### Pinguicula L. Butterwort

Flowers solitary at the ends of erect, bractless scapes; calyx 5-lobed, the lobes partly connate into an upper lip of 3 and a lower lip of 2; corolla bilabiate, the upper lip 2-lobed, the lower 3-lobed and prolonged into a basal spur, without a palate; capsule 2-valved; fibrous-rooted, scapose, terrestrial herbs of wet places, with well-developed, entire basal leaves.

About 30 species, according to Barnhart (see reference above), widely distributed in the Northern Hemisphere and in the South American Andes. (Name a diminutive of the Latin pinguis, fat, referring to the greasy appearance of the leaves of P. vulgaris.)

Pinguicula vulgaris L. Sp. Pl. 17. 1753. (Europe)
  P. macroceras Link. Jahrb. Gewächsk. 1³:54. 1820. P. vulgaris var. macroceras Herder, Acta Hort. Petrop. 1:380. 1873. (Pallas, Unalaschka)
  Fibrous-rooted perennial; leaves succulent, broadly oblanceolate or subelliptic, short-petiolate, 2-5 cm. long, 7-18 mm. wide, slimy on the upper surface, digesting the small insects that are caught; scapes 5-15 cm. high, obscurely glandular; calyx about 3 mm. long; corolla lavender-purple or rarely nearly white, 1.5-2.5 cm. long including the slender, 5-9 mm. spur, with flaring throat and broad, rounded lobes. N=32.
  In bogs and wet soil in the mountains; circumboreal, in America extending s. to s.w. Oreg., Mont., Mich., and N.Y. July-Aug.

### Utricularia L. Bladderwort

Flowers solitary or more commonly several in a raceme (this rarely branched) at the end of an erect peduncle, each subtended by a bract; calyx deeply 2-lobed, the lobes entire or nearly so; corolla bilabiate, the upper lip entire or shallowly 2-lobed, the lower entire or 3-

lobed, in most species elevated at the base into a prominent palate, the tube prolonged at the base into a spur or sac; rootless or apparently fibrous-rooted herbs, typically (including all our species) submersed and with much-dissected leaves, less commonly terrestrial and with much-reduced leaves, the leaves alternate, or partly or wholly whorled; submersed leaves sparingly to copiously provided with small, buoyant, valve-lidded bladders which serve as traps for small crustaceans and other aquatic animals, or the traps borne on separate branches.

A large, cosmopolitan, variable genus of perhaps nearly 300 species, best developed in the tropics. (Name from the Latin utriculus, a little bag, referring to the bladders or traps.)

1 Flowers large, the lower lip mostly 10-20 mm. long; leaves mostly 2-parted at the base, then unequally and quasipinnately several times dichotomous, the segments more or less terete, the ultimate ones filiform, strongly acuminate; bladders abundant, borne on ordinary leaves                                                                                          U. VULGARIS
1 Flowers smaller, the lower lip mostly 4-12 mm. long; leaves mostly 3-parted at the base, then 1-3 times dichotomous, often unequally, but not quasipinnately so, the segments flat
  2 Bladders borne on specialized branches distinct from the dissected leaves; corolla with the palate well developed and with the spur nearly as long as the lower lip, which is mostly 8-12 mm. long; ultimate leaf segments more or less obtuse; fruiting pedicels erect                                                                                                   U. INTERMEDIA
  2 Bladders mostly borne on ordinary leaves; corolla with the palate small or obsolete and with the spur reduced and much shorter than the lower lip, which is about 4-8 mm. long; ultimate leaf segments sharply acuminate; fruiting pedicels arcuate-recurved
                                                                                                                            U. MINOR

Utricularia intermedia Hayne in Schrad. Journ. Bot. 1$^1$:18. 1800. (1801).
  Lentibularia intermedia Nieuwl. & Lunell, Am. Midl. Nat. 5:9. 1917. (Germany)
  Submersed plants with very slender stems, commonly creeping along the bottom; leaves numerous, alternate, mostly 0.5-2 cm. long, commonly 3-parted at the base and then 1-3 times dichotomous, the segments often unequal, slender, flat, not much narrower in successive dichotomies, the ultimate ones rather blunt; bladders borne on specialized branches distinct from the leaves, 2-4 mm. wide; winter buds ovoid or ellipsoid, 5-7 mm. long, flowers mostly 2-4 in lax racemes at the end of an emergent peduncle 6-20 cm. long; corolla yellow, the proper tube very short, the lower lip commonly 8-12 mm. long, with a well-developed palate; upper lip not much more than half as long as the lower; spur nearly as long as the broad, slightly lobed lower lip; fruiting pedicels suberect.
  In shallow, standing or slowly moving water; circumboreal, in America extending s. to Calif., Ind., and Del. Mostly July-Aug.

Utricularia minor L. Sp. Pl. 18. 1753. (Europe)
  U. occidentalis Gray, Proc. Am. Acad. 19:95. 1883. (Suksdorf, Falcon Valley, Klickitat Co., Wash.) A form with both the spur and the palate a little better developed than usual.
  Submersed plants with very slender stems, often creeping along the bottom; leaves numerous, alternate, mostly 0.3-1 cm. long, commonly 3-parted at the base and then dichotomously or irregularly 1-3 times divided, the segments slender, flat, the ultimate ones strongly acuminate; bladders borne on the leaves, often rather few, 1-2 mm. wide; winter buds subglobose, 1.5-5 mm. long; flowers mostly 2-9 in a lax raceme at the end of an emergent peduncle 4-15 cm. long; corolla yellow, the proper tube very short, the lower lip commonly 4-8 mm. long, its palate not much, if at all, developed; upper lip evidently shorter, not more than about half as long as the lower; spur poorly developed, up to about half as long as the broad, shallowly or scarcely lobed lower lip; fruiting pedicels arcuate-recurved. N=18-20.
  In shallow, standing or slowly moving water; circumboreal, in America extending s. to Calif., Colo., Ind., and N.J. June-Sept.

Utricularia vulgaris L. Sp. Pl. 18. 1753.
  Lentibularia vulgaris Moench. Meth. 521. 1794. (Europe)
  Utricularia macrorhiza LeConte, Ann. Lyc. N.Y. 1:73. 1824. U. vulgaris var. americana

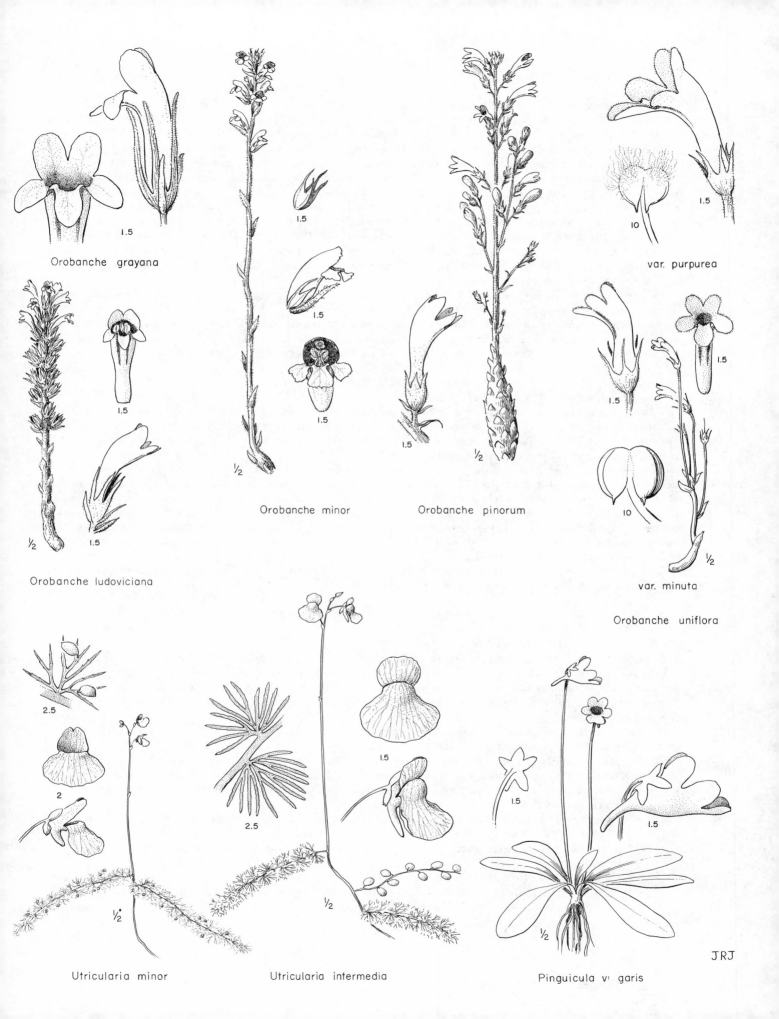

Orobanche grayana

Orobanche minor

Orobanche pinorum

var. purpurea

Orobanche ludoviciana

var. minuta

Orobanche uniflora

Utricularia minor

Utricularia intermedia

Pinguicula vulgaris

JRJ

Gray, Man. 5th ed. 318. 1867. Lentibularia vulgaris var. americana Nieuwl. & Lunell, Am. Midl. Nat. 5:9. 1917. Utricularia vulgaris ssp. macrorhiza R. T. Clausen, Cornell U. Agr. Exp. Sta. Mem. 291:9. 1949. (Can. to Carolina)

Submersed, free-floating plants, the stems coarser than in our other two species, often 1 mm. or more thick; leaves numerous, alternate, 1-5 cm. long, mostly 2-parted at the base and then repeatedly and unequally (quasipinnately) dichotomous, the segments terete, progressively more slender after each division, the ultimate ones filiform and strongly acuminate; bladders numerous, borne on the leaves, 1-3 mm. wide; winter buds ovoid or ellipsoid, 1-2 cm. long; flowers mostly 6-20 in a lax raceme at the end of a stout, emergent peduncle 6-20 cm. long, the axis becoming elongate; corolla yellow, the proper tube very short, the lower lip mostly 1-2 cm. long, broad, slightly lobed, with well-developed palate; upper lip not much, if at all, shorter than the lower; spur well developed, falcate, directed forward; fruiting pedicels arcuate-recurved. N=18-20.

In ponds, lakes, marshes, and slowly moving streams; circumboreal, in America extending s. to Calif., Ariz., Tex., and Fla. June-Aug.

American plants tend to have the spur somewhat more slender and pointed than do the European ones, and on this basis they have been segregated by some as a separate variety, subspecies, or species, but the difference is minor and scarcely worth taxonomic notice.

## PLANTAGINACEAE. Plantain Family

Flowers regular (or the calyx sometimes somewhat irregular), perfect or rarely unisexual, gamopetalous, ordinarily 4-merous as to the calyx, corolla, and androecium, or the stamens rarely only 1-3; corolla scarious, persistent; stamens epipetalous, alternate with the corolla lobes; ovary superior, 2-celled, with 1-many ovules in each cell, the cells sometimes each with a median, partitionlike outgrowth from the septum, or in Littorella and Bougeria the ovary 1-celled; placentation axile (basal in Littorella and Bougeria); style solitary, with a simple, elongate stigma; fruit a circumscissile capsule, the top deciduous (2-seeded and indehiscent in one species of Plantago, an achene in Littorella and Bougeria); endosperm present; annual or perennial herbs, rarely suffrutescent, with entire or toothed, simple leaves, these usually basal in a close spiral (morphologically alternate), or in a few species cauline and opposite; inflorescence a bracteate spike.

The family consists of three genera, Plantago, Littorella, and Bougeria, the last two genera composed of only three and one species, respectively.

### Plantago L. Plantain; Ribgrass

Characters essentially of the family.

A large, cosmopolitan genus; Pilger (Pflanzenr. IV. 269:1-466. 1937) recognized 257 species, but this is doubtless too many. With the exception of the species-pairs P. patagonica-aristata and P. eriopoda-tweedyi, our species are all very sharply limited, at least in our region, and not very closely allied among themselves. (The classical Latin name, from planta, the sole of the foot.)

1 Leaves cauline, opposite; peduncles axillary                                    P. PSYLLIUM
1 Leaves all basal, the plants scapose
  2 Plants evidently perennial (some species may bloom during the first year)
    3 Corolla tube glabrous; leaves broader than linear, generally at least some of them over 1 cm. wide, more or less evidently petiolate; plants (except P. hirtella) not maritime in our range
      4 Ovules and seeds 6-30; leaves broad, the well-defined blade broadly elliptic to cordate-ovate and mostly 1.3-2.3(3) times as long as wide; plants not woolly at the base; seeds reticulate, 1 mm. long; native and introduced, mostly weedy plants
                                         P. MAJOR
      4 Ovules 2-4; leaves narrower, the elliptic or narrower blade mostly (2)2.5-10 times as long as wide, or even longer; plants often woolly at the base; seeds not reticulate, well over 1 mm. long

    5  Sepals free; bracts obtuse to acute; native species, not weedy
      6  Capsule circumscissile at or below the middle, 2-4.5 mm. long (2)3- to 4-seeded,
          the seeds 1.7-3 mm. long
        7  Corolla lobes 2-4 mm. long, erect after anthesis; coastal in our area
                                                              P. HIRTELLA
        7  Corolla lobes 1.5 mm. long or less, spreading or reflexed; not coastal in our
            area
          8  Plants densely and generally conspicuously brown-woolly at the base; spike
              elongate, mostly 5-20 cm. long at maturity; alkaline places in the plains and
              intermontane valleys, in our range known only in Mont. and c. Ida.
                                                              P. ERIOPODA
          8  Plants sparsely and inconspicuously or scarcely brown-woolly at the base;
              spike short, about 2-7 cm. long at maturity; nonalkaline meadows at moderate
              elevations in the mountains; in our range known from w. Mont. to c. Ida.
                                                              P. TWEEDYI
      6  Capsule indehiscent (unique in the genus in this respect), 6-7 mm. long; seeds 2, 4-
          5 mm. long; w. Wash. and northward            P. MACROCARPA
    5  Outer sepals (the two next to the bract) connate, entire or with a mere notch at the summit
        (unique among our species in this respect); bracts mostly acuminate or caudate-acu-
        minate; ovules 2; introduced weed                P. LANCEOLATA
  3  Corolla tube pubescent (unique among our species in this respect); leaves linear or near-
      ly so, scarcely or not at all petiolate, seldom as much as 1 cm. wide; seeds 2-4; mari-
      time plants                                        P. MARITIMA
2  Plants annual (occasionally short-lived perennial in P. aristata, which is characterized by
    the conspicuously long-exserted bracts of the inflorescence)
  9  Capsules 4(6)-seeded; stamens mostly 2; leaves puberulent or glabrous, rarely as much
      as 3 mm. wide; inflorescence wholly glabrous; corolla very small, the lobes less than
      1 mm. long; bracts ovate, succulent, 2 mm. long; plants of saline situations
                                                          P. ELONGATA
  9  Capsules 2-seeded; stamens 4; leaves more or less woolly-villous to glabrate, often over
      3 mm. wide; inflorescence sparsely to densely long-hairy; corolla larger, the lobes
      mostly 1.5-2 mm. long; bracts linear or nearly so, firm, mostly longer; plants of non-
      saline situations
    10  Bracts mostly inconspicuous and not much elongate, occasionally (especially the lower
        ones) moderately surpassing the more or less densely woolly spike, not blackening
        in drying; leaves more or less conspicuously woolly-villous; common native species
        e. of the Cascades                               P. PATAGONICA
    10  Bracts much elongate, conspicuously exserted from the mostly not very woolly spike,
        tending to blacken in drying; leaves mostly not very hairy, often glabrate; occasion-
        al weed w. of the Cascades                       P. ARISTATA

Plantago aristata Michx. Fl. Bor. Am. 1:95. 1803.
    P. patagonica var. aristata Gray, Man. 2nd ed. 269. 1856. P. gnaphaloides var. aristata
        Hook. Fl. Bor. Am. 2:124. 1838. P. purshii var. aristata M. E. Jones, Bull. U. Mont.
        Biol. 15:46. 1910. (Ill.)
    Plants taprooted, annual or occasionally short-lived perennial, similar to P. patagonica,
but averaging more robust and less pubescent, the petioles evidently dilated and papery at the
sheathing, striate base; differing sharply from P. patagonica in the conspicuously long-exserted bracts of the inflorescence, which generally blacken in drying, the lower ones exserted 5-
25 mm., the upper ones often shorter, but still conspicuously exserted. N=10.
    Dry, open places, especially in disturbed sites; native from Ill. to La. and Tex., now natu-
ralized over most of e. U.S. and adj. s.e. Can., and occasionally found as a weed w. of the
Cascades in our range. June-Aug.

Plantago elongata Pursh, Fl. Am. Sept. 729. 1814. (Bradbury, upper La.)
  P. bigelovii Gray, Pac. R.R. Rep. 4:117. 1857. (Bigelow, Benicia, Calif.)
  P. myosuroides Rydb. Mem. N.Y. Bot. Gard. 1:369. 1900. (Watson 749, Salt Lake Valley,
    near the mouth of the Jordan R., Utah)
    Slender annual from a well-developed taproot, 3-20 cm. tall; leaves and scapes rough-puber-
ulent to occasionally glabrous, often becoming somewhat woolly at the base; leaves linear,
succulent, 2-10 cm. long, 0.3-1.5(3) mm. wide; spike wholly glabrous, 1-10 cm. long, 5 mm.
wide or less, loosely to rather closely flowered, the rachis partly exposed; bracts fleshy,
broadly ovate, spurred-carinate, 2 mm. long; plants subdioecious, some with small and perhaps nonfunctional stamens, others with reduced and mostly nonfunctional pistils; corolla
small, the lobes less than 1 mm. long, erect to reflexed; stamens mostly 2; capsule 2-3 mm.
long; seeds 4, or reputedly sometimes 5-6, 1.3-2 mm. long, fuscous, more or less pitted or
rugulose at maturity.
    Moist, somewhat saline places; s. B.C. to s. Calif., e. to Sask., w. Minn., and Tex.
Apr.-June.
    Plants of the Puget trough and the Pacific Coast have usually been segregated as P. bigelo-
vii Gray, but these do not appear to differ from typical, inland P. elongata in any respect save
the often slightly more densely flowered spikes, and there is no definite cleavage into two morphologically recognizable populations.
    Plantago pusilla Nutt., an eastern American species of nonsaline soils, has been collected
as a waif at Albina, Portland, and has been reported from other stations w. of the Cascades
in our area. It differs from P. elongata, with which it was long confused, in its smaller capsules and seeds, mostly more slender spikes, and particularly in its very short and quickly
deliquescent taproot, the plants appearing fibrous-rooted.

Plantago eriopoda Torr. Ann. Lyc. N.Y. 2:237. 1827. (James, depressed and moist situations
  along the Platte)
    Similar to P. tweedyi, sometimes larger and coarser, copiously and generally conspicuously
brown-woolly at the crown; leaves brittle and somewhat fleshy; spike elongate, mostly 5-20
cm. long at maturity; corolla lobes 1-1.5 mm. long; capsules 3-4 mm. long; seeds 2.0-2.7
mm. long.
    Alkaline meadows in the plains and intermontane valleys; plains states, chiefly from Colo.
and Neb. northward, extending n. to s. Mack. and Yukon, w. to c. Ida. and through the Great
Basin region to Calif. and apparently the s. Oreg. coast; apparently in n. Mex.; also in salt
marshes along the lower St. Lawrence R. in Que. June-July.

Plantago hirtella H.B.K. Nov. Gen. & Sp. 2:187. 1818.
  P. virginica var. hirtella Kuntze, Rev. Gen. 2:532. 1891. (Peru)
  P. subnuda Pilger, Notizb. Bot. Gart. Berl. 5:260. 1912. P. durvillei var. subnuda Pilger,
    Pflanzenr. IV. 269:234. 1937. (Heller 6764, Pacific Grove, Monterey Co., Calif.)
    Fibrous-rooted perennial from a short, stout, erect apparent caudex, somewhat brown-
woolly at the crown; leaves firm-succulent, hispid-hirsute to glabrous, several-nerved, elliptic or elliptic-oblanceolate, 5-20(35) cm. long (including the mostly short petiole) and 1.5-
6 cm. wide; scapes stout, 0.5-4 dm. tall, hairy; spikes dense, 5-25 cm. long and nearly or
fully 1 cm. thick; bracts firm, keeled, 3 mm. long, tending to be acute; corolla lobes 2-4 mm.
long, narrow, acute, spreading only at anthesis, forming a persistent closed beak over the
capsule; stamens 4; capsule 2.5-4.5 mm. long; seeds 2-4, dull black, 2-3 mm. long.
    Widespread in tropical America, extending northward, on bluffs and tidal flats along the
coast, to Grays Harbor Co., Wash. May-Sept.

Plantago lanceolata L. Sp. Pl. 113. 1753. (Europe)
    Fibrous-rooted perennial with a short, stout, erect caudex, often with an imperfectly de-
veloped taproot as well, more or less tan-woolly at the crown; leaves villous to glabrous, 3-
to several-nerved, narrowly elliptic or lance-elliptic, long-acute, often remotely denticulate,
commonly (5)10-40 cm. long and (0.5)1-4 cm. wide; scapes several, striate-sulcate, mostly
strigose, 1.5-6 dm. tall; spike very dense, ovoid-conic at first, at maturity cylindric, 1.5-

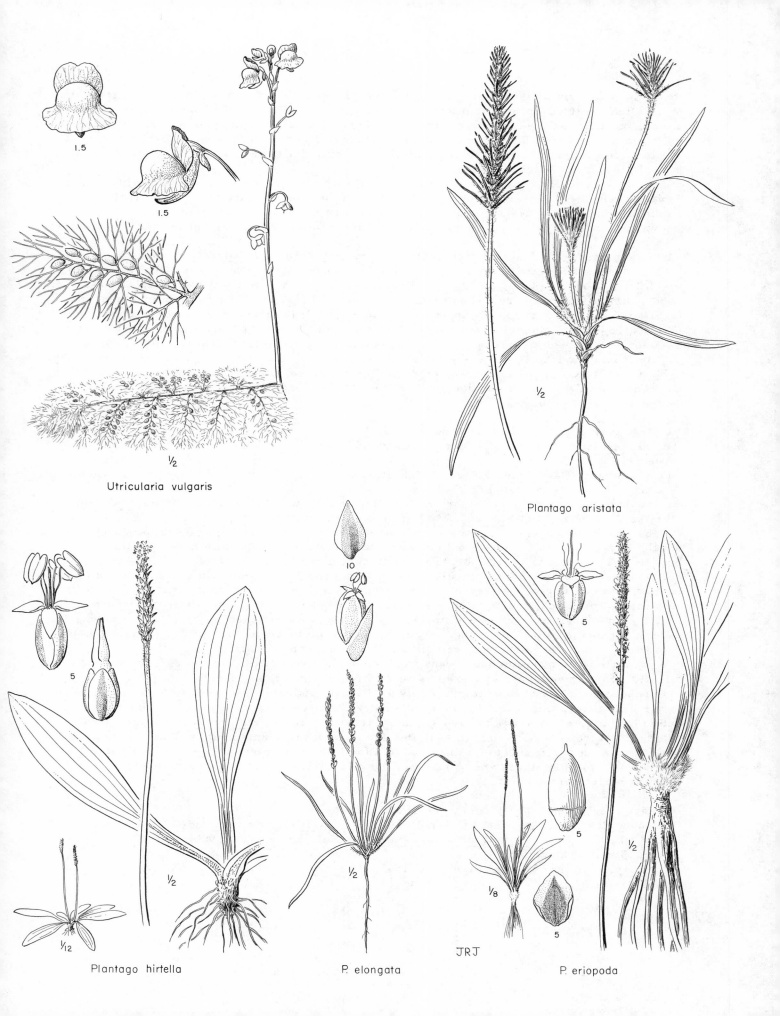

1.5

1.5

1/2

Utricularia vulgaris

1/2

Plantago aristata

5

10

5

5

1/2

1/12

Plantago hirtella

1/2

P. elongata

1/8

5

5

1/2

JRJ

P. eriopoda

8 cm. long, nearly or fully 1 cm. thick; bracts thin, ovate, with acuminate or caudate-acuminate, often exserted tip; sepals often villous-ciliate toward the tip of the strong midnerve, the 2 outer ones connate, with separate midveins, merely notched at the summit, or entire; corolla lobes 2-2.5 mm. long, spreading or reflexed; stamens 4, exserted, conspicuous; capsule 3-4 mm. long, the (1)2 seeds shining, blackish, 2 mm. long, deeply concave on the adaxial face. N=6, 12, 48. English plantain.

A weed of roadsides, pastures, and other disturbed sites where it is not too dry; native of Eurasia, now a cosmopolitan weed, particularly in the moister parts of the temperate zones; common w. of the Cascades, less frequent elsewhere in our area. Apr.-Aug.

European botanists recognize numerous infraspecific taxa on inconstant technical characters.

Plantago macrocarpa Cham. & Schlecht. Linnaea 1:166. 1826. (Chamisso, Unalaschka)

Stout perennial from a heavy root, essentially glabrous throughout, except that the rachis of the inflorescence and the upper part of the peduncle are commonly woolly-villous at least when young; leaves elongate, several-nerved, commonly (5)10-45 cm. long (petiole included) and (0.5)1-3.5 cm. wide; scapes 1- to several, 1-6 dm. tall; inflorescence short and dense at anthesis, 1.5-4 cm. long, elongating to 3-10 cm. in fruit; bracts broadly elliptic or ovate, thin, 3-4 mm. long; corolla lobes spreading, 1.5-2 mm. long; anthers 4, long-exserted; capsule 6-7 mm. long, indehiscent, falling entire; seeds 2, black, minutely roughened, 4-5 mm. long.

Sphagnum bogs, lake shores, and other cold, wet places; subcoastal (but not maritime) from Grays Harbor Co., Wash., to the Aleutian Islands; also recently reported from Lincoln Co., Oreg. Mostly May-June.

Plantago major L. Sp. Pl. 112. 1753. (Europe)
P. nitrophila A. Nels. Bull. Torrey Club 29:405. 1902. (Nelson 8417, near Manville, Converse Co., Wyo.)
P. major var. flavovirens Pilger, Fedde Rep. Sp. Nov. 18:275. 1922. (Specimens cited from Oreg., Calif., and Colo.)
P. major var. pachyphylla Pilger, Fedde, Rep. Sp. Nov. 18:277. 1922. (Jones 1030, Salt Lake City, Utah, and Nelson 7673, Laramie, Wyo; syntypes)

Fibrous-rooted perennial from a short, stout, erect caudex, glabrous or rather inconspicuously hairy, especially below, but not woolly at the crown; leaves with broadly elliptic to broadly ovate or cordate-ovate blade abruptly contracted to the well-defined petiole, the blade entire or irregularly toothed, mostly 4-18 cm. long and 2.5-11 cm. wide, 1.3-2.3(3) times as long as wide, strongly 3- to several-nerved; scapes 5-25 cm. long; spike dense but narrow, less than 1 cm. thick, mostly elongate, commonly 5-30 cm. long, essentially glabrous; bracts broad, thin-margined, mostly 2-4 mm. long; corolla lobes reflexed, 1 mm. long; stamens 4, exserted; capsule 2.5-4 mm. long, circumscissile near or below the middle; seeds 6-30, black or brown, 1 mm. long, strongly reticulate. N=6. Common plantain.

A common weed of roadsides, lawns, and other disturbed sites, consisting of numerous slightly differing biotypes, and now cosmopolitan in distribution. May-Aug.

European botanists recognize many only arbitrarily definable infraspecific taxa. In addition to our introduced, weedy forms, we have a native, scarcely weedy type, occurring chiefly in saline habitats, which tends to have succulent leaves. The name P. major var. pachyphylla Pilger (P. nitrophila A. Nels.) is available for these latter plants, but, at least in the herbarium, they are very difficult to distinguish from the introduced forms. The closely related P. asiatica L., to which these and some other American plants have often been referred, is apparently strictly Asiatic.

Plantago maritima L. Sp. Pl. 114. 1753. (N. Europe)
P. juncoides Lam. Tab. Encyc. 1:342. 1791. P. maritima var. juncoides Gray, Man. 2nd ed. 268. 1856. P. maritima ssp. juncoides Hultén, Fl. Alas. 9:1431. 1949. (Commerson, Strait of Magellan)
P. decipiens Barn. Monog. Plant. 16. 1845. P. juncoides var. decipiens Fern. Rhodora 27: 100. 1925. (Labr.)

P. juncoides var. californica Fern. Rhodora 27:100. 1925. P. maritima var. californica
Pilger, Pflanzenr. IV. 269:187. 1937. (Copeland 3331, Montara Point, Calif.)

Perennial with a stout and often elongate taproot or erect caudex, often slightly woolly at the
crown; leaves linear or nearly so, scarcely or obscurely petiolate, fleshy, glabrous or nearly
so, but often white-pustulate, evidently to scarcely several-nerved; scapes hirsute-strigose,
5-25 cm. long; spikes dense, 2-10 cm. long, less than 1 cm. thick; bracts fleshy, broadly
triangular-ovate, obtuse to acuminate, 1.5-4 mm. long; corolla tube densely and persistently
short-hairy; corolla lobes 1-1.5 mm. long; anthers 4, exserted; ovules in ours mostly 3 to 4,
but the seeds more often 2 or 3, brown or black, 2-2.5 mm. long. N=6, 12.

A widespread and variable maritime species, very rarely occurring in salty places in the
continental interior (e.g., reported from Great Salt Lake, Utah). June-Aug.

The American plants, as described above, belong to the ssp. juncoides (Lam.) Hultén, which
differs from the Old World ssp. maritima in its broader bracts and often 4-ovulate, mostly 2-
to 4-seeded (as opposed to mostly 3-ovulate and 1- to 2-seeded) capsules. Within the ssp. juncoides, two varieties may be distinguished in our range. The var. californica (Fern.) Pilger,
occurring mostly on bluffs and cliffs from Tillamook Co., Oreg., to s. Calif., has short, relatively broad and blunt, often toothed leaves mostly 3-7 cm. long. The remainder of our
plants, occurring more often on beaches and in salt marshes, have more elongate leaves (up
to 25 cm. long), and have been referred to var. juncoides. Patagonian material of var. juncoides, however, is dwarf and has very narrow leaves, mostly only 1-2 mm. wide; it is matched
in those respects by many specimens from the higher latitudes on the N. Am. Atlantic Coast,
but only by the extreme individuals from the N. Am. Pacific Coast. Our plants, with the leaves
up to as much as 1 cm. wide, closely resemble eastern American specimens which have been
treated as P. juncoides var. decipiens (Barn.) Fern. However, in view of some experimental
work by Rousseau (Contr. Inst. Bot. U. Montreal 44:59-64. 1942) which strongly suggests that
the dwarf, narrow-leaved habit is merely a direct response of the individual to a rigorous habitat, no new combination is here proposed.

Plantago patagonica Jacq. Icon. Pl. Rar. 2:9, pl. 306. 1786-93. (Patagonia)
P. lagopus Pursh, Fl. Am. Sept. 99. 1814, not of L. P. purshii Roem. & Schult. Syst. Veg.
3:120. 1818. P. gnaphaloides Nutt. Gen. Pl. 1:100. 1818. P. patagonica var. gnaphaloides
Gray, Man. 2nd ed. 269. 1856. (Nuttall, near the confluence of the "Jauke" R., probably
the present day Dakota R., and the Missouri)
P. spinulosa Dcne. in DC. Prodr. 13:713. 1852. P. patagonica var. spinulosa Gray, Syn. Fl.
2¹:391. 1878. P. purshii var. spinulosa Shinners, Field & Lab. 18:117. 1950. (Wright,
Texas in 1848-49)
P. picta Morris, Bull. Torrey Club 28:118. 1901, not of Colenso in 1890. P. xerodea Morris,
Bull. Torrey Club 36:515. 1909. P. purshii var. picta Pilger, Pflanzenr. IV. 269:369.
1937. (Parish 2643, mouth of Santa Ana Canyon, Calif.) = var. spinulosa.
P. oblonga Morris, Bull. Torrey Club 28:119. 1901. P. spinulosa var. oblonga Poe, Bull.
Torrey Club 55:411. 1928. P. purshii var. oblonga Shinners, Field & Lab. 18:117. 1950.
(Orcutt, Colorado desert, San Diego Co., Calif.) = var. spinulosa.

Annual, 5-20(30) cm. tall; herbage more or less strongly woolly-villous; leaves 3-13 cm.
long and 1-7 mm. wide, oblanceolate or nearly linear, tapering gradually to the petiolar base
and to the acute tip, often 3-nerved; spike dense, cylindric, more or less strongly woolly,
mostly 1.5-10 cm. long and well under 1 cm. wide; bracts firm, linear or nearly so, sometimes with narrow and inconspicuous dilated scarious margins at the base, commonly about 5
mm. long or less and scarcely or not at all exserted from the spike, or in var. spinulosa the
lower (less commonly all) more elongate and evidently short-exserted; corolla lobes spreading, 1.5-2 mm. long; stamens 4; seeds 2, brown, minutely roughened, 2 mm. long. N report-
edly=10, 12.

Dry, open places in the valleys, foothills, and plains; s. B.C. to Sask., s. to Calif. and
Tex., and introduced farther east; in our range not extending w. of the Cascade summits; also
widespread in Argentina and Chile, where several varieties and forms have been described.
Mostly May-June.

Our plants, as principally described above, are scarcely if at all to be distinguished from

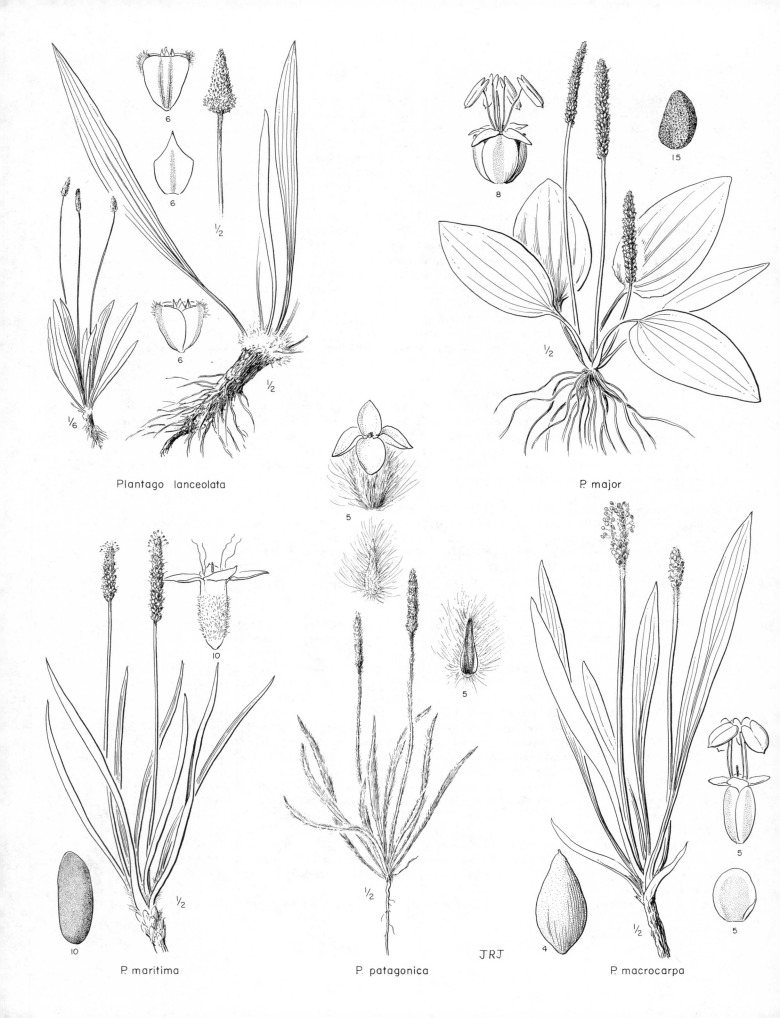

Plantago lanceolata

P. major

P. maritima

P. patagonica

JRJ

P. macrocarpa

the South American ones, but the name P. patagonica var. gnaphaloides (Nutt.) Gray is available should a segregation prove possible. In the eastern and southern portions of its native North American range, P. patagonica often has somewhat elongate and exserted bracts, thus approaching the mostly more eastern P. aristata, and some specimens seem truly intermediate to that closely allied species. Toward the s., especially in s. Ariz. and adj. areas, long-bracted plants may be more common than the short-bracted ones, with which they intergrade freely. Farther n., as along the e. edge of our range, the long-bracted plants tend to occur as occasional colonies, or to grow intermingled with the short-bracted ones. The name P. patagonica var. spinulosa (Dcne.) Gray is available for the long-bracted plants, which are probably the result of introgression from P. aristata.

Plantago psyllium L. Sp. Pl. 115. 1753. (S. Europe)

    P. indica L. Syst. Nat. 10th ed. 2:896. 1759. Psyllium indicum DuMont, Bot. Cult. 2nd ed. 2:492. 1811. (Egypt)

    Plantago arenaria Waldst. & Kit. Descr. Icon. Pl. Rar. Hung. 1:51. 1802. Psyllium arenarium Mirbel, Hist. Pl. 14:333. 1804-5. (Hungary)

    Taprooted annual or occasionally short-lived perennial, 1-6 dm. tall, evidently hirsute, the hairs often of 2 distinct lengths, and tending to be a little viscid; stems solitary or several, branched when well developed; leaves numerous, cauline, opposite, linear or nearly so, entire, 2-8 cm. long, 1-3 mm. wide; peduncles numerous in well-developed plants, 2-8 cm. long, axillary, or some seemingly subterminal because of a foreshortening of the upper internodes; spikes 0.5-1.5 cm. long, dense, nearly 1 cm. thick; bracts broad, rounded, conspicuously scarious-margined, the lowermost ones abruptly and firmly foliaceous-caudate; corolla lobes soon reflexed, 1.5-2 mm. long; ovules 2, often only 1 maturing; seeds shining, brown, 2.0-2.8 mm. long.

    A weed in waste places, especially along railroad tracks; native to s.e. Europe and adj. Asia and n. Afr., now well established in e. U.S., and collected once at Seattle. June-Oct.

    It has been customary to use the name P. indica L. for this species, and to attach the name P. psyllium to a related species which lacks the caudate tips to the lower bracts. Linnaeus' original characterization of P. psyllium, which included the phrase, "spicis foliosis," clearly applies to the present plant, however.

Plantago tweedyi Gray, Syn. Fl. 2nd ed. 2[1]:390. 1886. (Tweedy, e. fork of the Yellowstone R. in 1885)

    Perennial from a stout but rather short taproot (or erect caudex?) which gives rise to slender, fibrous roots, inconspicuously or scarcely woolly at the crown; leaves elliptic, several-nerved, glabrous or with scattered hairs, not thickened, commonly 8-22 cm. long (petiole included) and 1-4 cm. wide; scapes not much, if at all, surpassing the leaves, villous-hirsute or glabrate; spikes short and rather dense, 2-7 cm. long at maturity, 5-8 mm. thick, glabrous except for the often hairy rachis; bracts broad, hyaline-margined, 2 mm. long, blunt or acutish; corolla lobes spreading or reflexed, barely or scarcely 1 mm. long; stamens 4; capsule 2-3 mm. long, circumscissile below the middle; seeds (3)4, olive to brown or blackish, 1.7-2.3 mm. long.

    Nonalkaline meadows at moderate to high elevations in the mountains; Alta. and Sask. (?) to Ida., Utah, and Colo. June-Aug.

    The high-northern species Plantago canescens Adams (P. septata Morris) has been attributed to Mont. on the basis of a specimen at the N.Y. Bot. Gard. which bears on the label the data: Plantago - L.M. Umbach 377 - Plains - Midvale, Mont. - July 14, 1903. This is so far from the otherwise known range of the species (Jasper Nat. Pk., Alta., to Mack., Yukon, Alas., and n. Asia) that the record needs confirmation. In the key here presented, P. canescens would run, with some difficulty, to P. tweedyi, from which it may be distinguished by its longer corolla lobes (nearly or fully 2 mm. long) and mostly narrower leaves (0.5-2 cm. wide) that are often conspicuously long-hirsute.

## RUBIACEAE. Madder Family

Flowers regular, perfect or rarely unisexual, gamopetalous, epigynous; calyx entire or toothed, or obsolete; corolla (3)4- to 5-lobed; stamens epipetalous, alternating with the corolla lobes; pistil (1)2- to 10-locular, with 1-many ovules in each cell, the placentation axile, basal, or apical; style simple or divided, or the styles separate; fruit dry or fleshy, dehiscent or indehiscent; endosperm copious to wanting; trees or shrubs, less often (but including all our species) herbs, with opposite (or whorled), simple and mostly entire leaves; stipules mostly interpetiolar, sometimes reduced to a mere interpetiolar line, or sometimes enlarged and leaflike, the leaves thus appearing whorled; flowers in various sorts of inflorescences of cymose origin.

One of the larger families of angiosperms, with more than 300 genera and perhaps 7000 species, the great majority tropical. Cinchona, Coffea, and Gardenia are among the economically more important genera.

1 Leaves opposite, with small stipules; ovules basally attached; corolla salverform-funnel-
    form                                                                              KELLOGGIA
1 Leaves whorled (or the uppermost sometimes merely opposite); ovules laterally attached
    (to the partition); corolla diverse
    2 Corolla funnel-shaped, the tube not much, if at all, shorter than the lobes
        3 Calyx teeth evident; flowers in small heads with a basal involucre of leaflike bracts
          (unique among our genera in this regard)                          SHERARDIA
        3 Calyx teeth obsolete; flowers in an open inflorescence          ASPERULA
    2 Corolla rotate or nearly so, the tube much shorter than the lobes          GALIUM

### Asperula L.

Similar to Galium, differing in its funnelform corollas, with the tube not much, if at all, shorter than the lobes.

About 80 species of the Old World, most abundant in the Mediterranean region. (Name a diminutive of the Latin asper, rough, presumably referring to the leaves.)

Asperula odorata L. Sp. Pl. 103. 1753. (N. Europe)

Rhizomatous perennial, vanilla-scented, somewhat resembling Galium triflorum; stems 1.5-5 dm. tall, retrorsely hispid at the nodes, otherwise glabrous or nearly so; leaves mostly in whorls of 6-10, typically 8, oblanceolate or narrowly elliptic, cuspidate, 1.5-5 cm. long, 4-12 mm. wide, antrorsely scabrous-ciliate on the margins and sometimes on the midrib beneath, otherwise glabrous or nearly so; inflorescence terminal, branched, nearly naked; corolla white, 3-5 mm. long, the lobes mostly somewhat longer than the tube; fruit about 3 mm. high, covered with hooked bristles. N=22. Sweet woodruff, Waldmeister.

Woodlands; native of Europe, sparsely introduced in e. U.S., and occasionally found w. of the Cascades in our range. May-June.

### Galium L. Bedstraw

Flowers perfect or rarely unisexual, borne in small or large, basically cymose inflorescences, or in axillary peduncles; calyx lobes obsolete; corolla rotate or nearly so, the 3-4 lobes much longer than the tube, valvate; ovary 2-celled, with a solitary ovule attached near the middle of the septum in each cell; styles 2, short; stigmas capitate; fruit dry (in all ours) or in some species fleshy, the carpels approximate or divergent, indehiscent; endosperm well developed; annual or perennial herbs with square stems and whorled, entire leaves (upper leaves merely opposite in Galium bifolium), exstipulate except insofar as some of the leaves of each whorl may be considered to be enlarged stipules.

A large genus, of perhaps 300 species, widely distributed, especially in temperate regions. (Name from the Greek gala, milk, from the use of G. verum for curdling.)

1 Plants perennial from creeping rhizomes
    2 Fruits without hooked hairs, not at all scabrous-muricate

  3 Flowers perfect; fruits and young ovaries glabrous or in G. boreale rather inconspicuously short-hairy
    4 Flowers more or less numerous in a terminal, compound and much-branched, rather
showy inflorescence; stems prostrate or ascending to more often erect, glabrous
or pubescent, but not retrorsely scabrous on the angles; corollas 4-lobed
      5 Leaves in whorls of 4, 3-nerved, not cuspidate; fruits short-hairy or occasionally
glabrous; common and widespread native species      G. BOREALE
      5 Leaves mostly in whorls of 6-8(12), 1-nerved, cuspidate; fruits glabrous; occa-
sional introduced weeds w. of the Cascades
        6 Flowers white; stem glabrous in the inflorescence; leaves broadly linear to
linear-oblong or oblanceolate      G. MOLLUGO
        6 Flowers bright yellow (unique among our species in this respect); stem general-
ly puberulent or hirtellous at least in the inflorescence; leaves narrowly linear
          G. VERUM
    4 Flowers solitary or few in small, rather inconspicuous inflorescences; stems weak,
tending to recline or scramble on other vegetation, commonly retrorse-scabrous
on the angles; corollas 3-lobed or less often 4-lobed
      7 Flowers solitary or 2-3 at the ends of axillary or terminal peduncles which may
themselves be borne in threes; corolla small, mostly 1-1.5 mm. wide; widespread species      G. TRIFIDUM
      7 Flowers several in small, irregularly branched, basically cymose inflorescences;
corolla larger, mostly 2-3 mm. wide; chiefly coastal species    G. CYMOSUM
  3 Flowers unisexual, the plants dioecious; fruits and young ovaries conspicuously pubescent with long, spreading, straight or merely flexuous hairs      G. MULTIFLORUM
 2 Fruits provided with hairs that are hooked at the tip (uncinate), or in G. asperrimum the
fruit sometimes merely muricate-scabrous
  8 Leaves 4 in a whorl, evidently trinerved; stem glabrous, erect; Cascade summits and
westward
    9 Stem with mostly 2-4(5) whorls of leaves; leaf bases evidently narrow, more or less
cuneate, the margins tending to be concavely rounded; inflorescence few-flowered,
the flowers commonly 2-3(6) on each of the 1-3 terminal peduncles; boreal species,
extending s., rarely, to the Cascades of n. Wash.      G. KAMTSCHATICUM
    9 Stem with mostly 5-8(9) whorls of leaves; leaf bases broader, the margins tending
to be convexly rounded; inflorescence with more numerous flowers, each primary
peduncle tending to be cymosely branched and several-flowered; Cascades to the
coast      G. OREGANUM
  8 Leaves mostly 5-6 in a whorl, 1-nerved; stem usually retrorsely scabrous on the angles, mostly either prostrate, ascending, or scrambling on other vegetation; widespread
    10 Bristles of the fruit very short, only 0.15-0.3 mm. long, or the fruit sometimes
merely muricate-scabrous, with the short bristles very stout and scarcely
hooked; flowers commonly borne in loose, irregularly branched, somewhat leafy-
bracteate inflorescences at the ends of the main stem and the branches
      G. ASPERRIMUM
    10 Bristles of the fruit of moderate length, mostly 0.5-1.0 mm. long; flowers com-
monly borne in threes at the ends of axillary peduncles, or the peduncles sometimes cymosely branched and several-flowered      G. TRIFLORUM
1 Plants annual from a short taproot
 11 Leaves 2-4 in a whorl, not cuspidate; stem glabrous; fruit with hooked hairs
      G. BIFOLIUM
 11 Leaves mostly 5-8 in a whorl, cuspidate; stem retrorsely scabrous or rarely glabrous;
fruits diverse
    12 Fruits very small, only about 1 mm. long or less, granular-roughened or with hooked
hairs; plants tending to be freely branched and loosely many-flowered
      G. PARISIENSE

12  Fruits larger, (1.5)2-4(5) mm. long; plants generally not much branched and with relative-
      ly few flowers
      13  Fruits scabrous-muricate; pedicels strongly recurved in fruit; peduncles equaling or
            shorter than the subtending whorl of leaves; leaves mostly glabrous on the upper sur-
            face (margins excepted); rare species                          G. TRICORNE
      13  Fruits with hooked hairs, or very rarely glabrous; pedicels divaricate or ascending,
            straight; peduncles equaling or mostly exceeding the subtending whorl of leaves;
            leaves mostly uncinate-hispid above; common species             G. APARINE

Galium aparine L. Sp. Pl. 108. 1753. (Europe)
      G. spurium L. Sp. Pl. 106. 1753. G. aparine var. spurium Wimm. & Grab. Fl. Siles. 1:119.
      1827. G. aparine ssp. spurium Dusen, Öfvers. Svensk. Vet. Akad. Förhandl. 58:238. 1901.
      (Europe) The glabrous-fruited form of var. echinospermum.
      G. vaillantii DC. in Lam. & DC. Fl. Fr. 3rd ed. 4:263. 1805. G. agreste var. echinosper-
      mum Wallr. Sched. Crit. 59. 1822. G. aparine var. vaillantii Koch, Syn. Fl. Germ. 330.
      1837. G. spurium var. vaillantii Beck, Fl. Nied. Österr. 2²:1122. 1893. G. spurium var.
      echinospermum Hayek, Fl. Steierm. 2:393. 1912. G. aparine var. echinospermum Farw.
      Rep. Mich. Acad. Sci. 19:260. 1917. (Vicinity of Paris)
      G. aparine var. minor Hook. Fl. Bor. Am. 1:290. 1833. (Douglas, near Ft. Vancouver)
      Annual, 1-10 dm. tall, the stems weak but not especially slender, retrorsely hooked-sca-
brous on the angles, and tending to scramble on other vegetation, seldom much branched;
leaves mostly in whorls of (6)8, narrow, 1-nerved, cuspidate, mostly 1-4 cm. long, retrorsely scabrous-ciliate on the margins and often also on the midrib beneath, uncinate-hispid above;
inflorescences small, mostly 3- to 5-flowered, borne on axillary peduncles which mostly sur-
pass the subtending whorl of leaves and commonly bear a partial or complete whorl of reduced
leaves at the summit, or the peduncles borne in threes at the ends of the short, axillary
branches; pedicels well developed, essentially straight, divaricate or ascending; corolla greenish-white, 4-parted, 1-2 mm. wide; fruit (1.5)2-4(5) mm. long, the segments approximate,
covered with hooked bristles, or very rarely glabrous.
      A common, rather weedy species occurring in a variety of habitats; circumpolar, probably
native to both hemispheres, and now found over most of temperate N. Am. Apr.-June.
      Most of our plants have small fruits, up to about 3 mm. long, and may be referred to the
var. echinospermum (Wallr.) Farw., which has been reported to have a somatic chromosome
number of 20. Occasional specimens in our range represent the nomenclaturally typical variety, which has larger fruits, mostly 3-4(5) mm. long, and in which somatic chromosome numbers of 22, 44, 64, and about 88 have been reported. Among the American specimens, at least,
there is no clear morphological break between the two groups, in spite of the supposed differences in chromosome number.

Galium asperrimum Gray, Mem. Am. Acad. II. 4:60. 1849. (Fendler 289, Santa Fe, N.M.)
      G. asperrimum var. asperulum Gray, Bot. Calif. 1:284. 1876. G. asperulum Rydb. Fl.
      Rocky Mts. 809. 1917. (Specimens cited from Calif. and Nev.)
      Perennial from creeping rhizomes; stems 2-8 dm. long, retrorsely scabrous on the angles,
tending to scramble on other vegetation; leaves mostly (5)6 in a whorl, or only 4 on the smaller branches, narrowly elliptic, cuspidate, commonly 1-3.5 cm. long, 1-nerved, retrorsely
scabrous-ciliate on the margins and on the midrib beneath, or antrorsely so on the margins,
rarely wholly glabrous; flowers commonly borne in loose, irregularly branched, somewhat
leafy-bracteate inflorescences at the ends of the main stem and branches; corollas 4-parted,
2-3 mm. wide, whitish; fruit 1.5-2 mm. high, the segments approximate, covered with very
short, hooked bristles only 0.15-0.3 mm. long, or sometimes merely muricate-scabrous,
with the short bristles very stout and scarcely hooked.
      Thickets, from the foothills to moderate elevations in the mountains; Wash. and Oreg., e.
of the Cascades, to w. Mont., s. to Calif., N.M., and n. Mex. June-Aug.

Galium bifolium Wats. Bot. King Exp. 134. 1871. (Watson 480, "in the Trinity, Battle, and
      East Humboldt mountains, Nev., and in the Wahsatch")

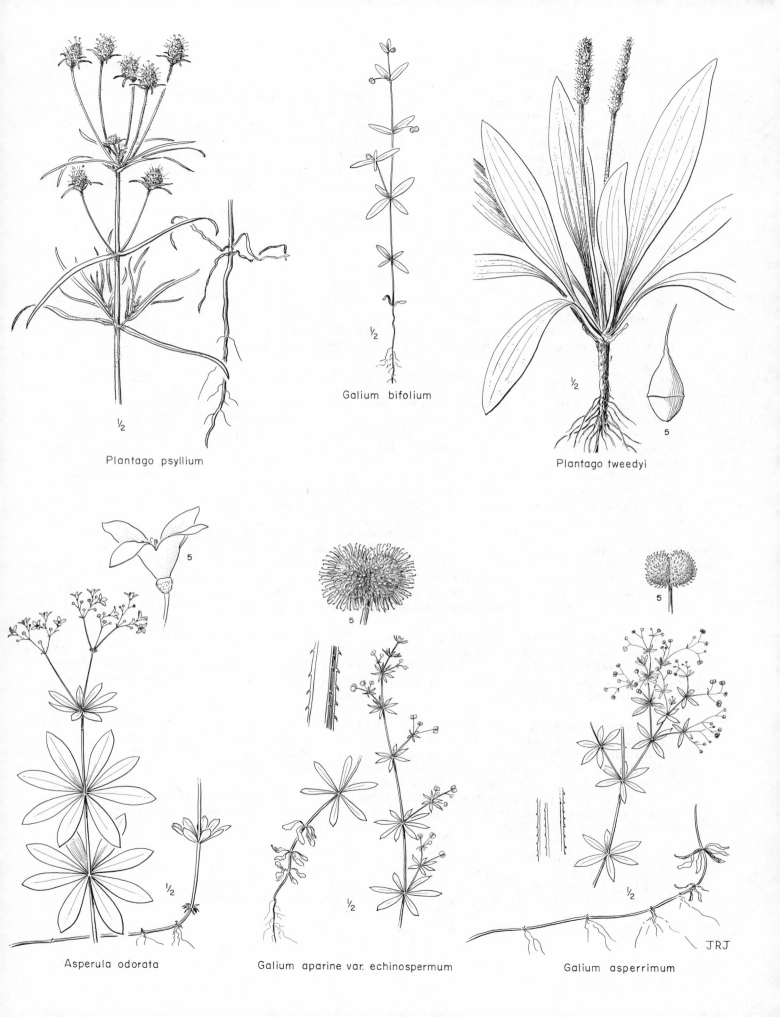

Plantago psyllium

Galium bifolium

Plantago tweedyi

Asperula odorata

Galium aparine var. echinospermum

Galium asperrimum

JRJ

Annual, slender and mostly erect, 5-20 cm. tall, simple or moderately branched, the herbage glabrous; leaves linear-elliptic or a little broader, mostly 1-2(2.5) cm. long, up to 5 mm. wide, obtuse or merely acute, the lower in whorls of 4, often with one pair smaller, generally some of the upper merely opposite; flowers solitary in the axils, at first short-pedunculate and erect, the peduncles later elongate (to as much as 3 cm.), divaricate, and with nodding fruit; corolla white, 3-lobed, minute and generally soon deciduous; fruit 2.5-3.5 mm. high, pubescent with short, uncinate hairs, the segments approximate; endosperm in cross section shaped like a thickened horseshoe.

In moist or sometimes rather dry places, from the foothills to high elevations in the mountains; s. B. C. and Mont. to s. Calif. and Colo., scarcely, if at all, extending w. of the Cascade summits in our range. May-Aug.

Galium boreale L. Sp. Pl. 108. 1753. (Europe)
  G. hyssopifolium Hoffm. Deutschl. Fl. 2nd ed. 1:71. 1800. G. boreale var. hyssopifolium
    DC. Prodr. 4:600. 1830. (Germany) The form with glabrous or glabrate fruits.
  G. septentrionale Roem. & Schult. Syst. Veg. 3:254. 1818. G. boreale ssp. septentrionale
    Iltis, Rhodora 59:40. 1957. (Lakes of Can. and N.Y., and near Wilkesbarre, Pa.)
  G. boreale var. intermedium DC. Prodr. 4:601. 1830. (Europe). The form with short,
    curled hairs on the fruit.
  G. boreale var. linearifolium Rydb. Mem. N.Y. Bot. Gard. 1:375. 1900. (Rydberg 134,
    Pumpkin Seed Valley, Neb.)
  Perennial, mostly 2-8 dm. tall, with numerous erect stems from well-developed, creeping rhizomes, the stems commonly short-bearded just beneath the nodes, otherwise glabrous or scaberulous; leaves borne in fours, sessile, glabrous or scabrous, lanceolate or nearly linear, obtuse to acute with minutely rounded tip, not at all mucronate, commonly 1.5-4.5 cm. long, more or less strongly 3-nerved, often bearing axillary, mostly sterile branches with somewhat smaller leaves and the whole plant thus appearing very leafy; flowers numerous in terminal, rather showy, cymose panicles, the corollas 4-lobed, white or slightly creamy, 3.5-7 mm. wide; fruits about 2 mm. long, pubescent with short, straight or curled (but not hooked) hairs that are inconspicuous to the naked eye, or glabrous.

In a wide variety of not too dry habitats, from sea level to timber line; circumpolar, in America extending s. to Calif., Ariz., Tex., Mo., and W. Va. June-Aug.

Löve and Löve have noted (Am. Midl. Nat. 52:88-105. 1954) that European and w. Asian material of G. boreale is tetraploid (N=22), while American and east-Asian material is hexaploid (N=33), but the other differences which they point out between the two types do not withstand re-examination, so that taxonomic segregation does not seem to be warranted at this time.

The hairs of the fruit may be either straight, curled, or wanting in both American and European plants. Although the three types are not uniformly distributed, no one of them has a coherent total geographic area in which it is the sole or even the principal representative of the species, nor is any edaphic or other ecologic correlation evident; they are here considered to be without taxonomic significance.

Galium cymosum Wieg. Bull. Torrey Club 24:401. 1897. (Flett s.n., Tacoma, Wash., in 1896)
  Similar to G. trifidum, and perhaps no more than a variety of it, often larger, the stems up to 8 dm. long and the leaves up to 2.5 cm. long; leaves 4-6 in a whorl; inflorescences irregularly branched and several-flowered, the corollas 4-lobed or less often 3-lobed, a little larger than in G. trifidum, commonly 2-3 mm. wide.

Moist places, chiefly along the coast (including the Puget Sound area), from s. B. C. to Oreg., extending inland, less commonly, to the w. slope of the Cascades and apparently to Missoula, Mont. June-Aug.

Galium kamtschaticum Steller ex Schult. & Schult. Mantissa 3:186. 1827.
  G. rotundifolium var. kamtschaticum Kuntze, Rev. Gen. 1:282. 1891. (Steller, Kamtchatka)
  G. littellii Oakes, Mag. Hort. 7:179. 1841. (Oakes, White Mts., N.H.)
  Perennial from slender creeping rhizomes, the stems arising singly, erect, 1-2 dm. tall,

glabrous; leaves in 2-4(5) whorls of 4, antrorsely scabrous-ciliate on the margins, and commonly with scattered similar hairs on at least the upper surface, thin and often somewhat veiny, evidently trinerved, broadly obovate to broadly ovate-elliptic, mostly 1-3 cm. long and 1/2 or 2/3 as broad, mucronate at the otherwise obtuse or rounded tip, notably narrowed below, the base cuneate and tending to have more or less concave margins; peduncles commonly 1-3 from the uppermost whorl of leaves, rather elongate and commonly 2- to 3-flowered, or sometimes branched and 5- to 6-flowered, the inflorescence very open, especially in fruit; corolla 4-lobed, 3 mm. wide, the lobes acutish to obtuse; fruit 1.5 mm. high, the segments approximate, beset with numerous well-developed, hooked bristles.

Moist, cold, coniferous woods; Kamchatka, Korea, and through the Aleutian Islands and the Alaska panhandle to the Cascades of n. Wash.; also in s.e. Can. and adj. New England and N.Y., and on the n.e. side of Lake Superior. Rare and local, only one station known in our area (Stevens Pass). July-Aug.

Galium mollugo L. Sp. Pl. 107. 1753. (Mediterranean region of Europe)

Rhizomatous perennial, mostly 3-12 dm. tall, with numerous weak but not especially slender, reclining to usually more or less erect stems from a decumbent base, the stems glabrous or sometimes shortly spreading-hairy below; short, slender, ascending, leafy, perennial offshoots produced from the base in summer or fall, some of these probably becoming the principal stems of the following year; leaves borne in whorls of 6-8 (or on the smaller branches only 4), sessile, antrorsely scabrous-ciliate on the margins, otherwise glabrous, broadly linear to linear-oblong or somewhat oblanceolate, commonly 1-3 cm. long (or on the branches smaller), 1-nerved, cuspidate; flowers numerous in terminal, often divaricately branched, rather showy cymose panicles, the corollas 4-lobed, white or nearly so, 2-4 mm. wide; fruits glabrous, small, 1-1.5 mm. high, with appressed segments. N=11, 22, 33. Also 2N=55.

A Eurasian weed, now well established in n.e. U.S. and adj. Can., and reported to be sparingly established in lawns in the Willamette Valley of Oreg. Mostly May-July.

Galium multiflorum Kell. Proc. Calif. Acad. Sci. 2:97. 1863. (Veatch, Washoe, Nev.)

G. bloomeri Gray, Proc. Am. Acad. 6:538. 1865. (Bloomer 250, Virginia City, Nev.)

G. multiflorum var. watsoni Gray, Syn. Fl. 1$^2$:40. 1878. G. watsoni Heller, Bull. Torrey Club 25:627. 1898. G. multiflorum ssp. watsoni Piper, Contr. U.S. Nat. Herb. 11:527. 1906. (Watson 484 in part, "Wahsatch Mts.," Utah, August, 1869, at 8000 ft.; lectotype by Heller)

G. multiflorum ssp. puberulum Piper, Contr. U.S. Nat. Herb. 11:527. 1906. G. multiflorum var. puberulum St. John, N.W. Sci. 2:90. 1928. G. watsonii ssp. puberulum Ehrend. Contr. Dudley Herb. 5:16. 1956. (Elmer 414, Ellensburg, Wash.)

G. watsonii f. scabridum Ehrend. Contr. Dudley Herb. 5:16. 1956. (Whited 338, Deschutes R. near Redmond, Oreg.)

Plants perennial, spreading freely from creeping, often eventually lignescent rhizomes; stems several from the base, up to about 4 dm. tall, freely branched when well developed, and often more or less woody below; herbage glabrous to densely spreading-hirtellous; leaves mostly in whorls of 4, sessile, narrowly lanceolate to broadly ovate, acute and often more or less cuspidate, (3)5-15(22) mm. long, with the midrib prominent beneath, and often with a more or less evident pair of lateral veins from the base; inflorescences individually small and few-flowered, numerous at the ends of the branches, or in reduced forms borne on short, axillary, peduncular branches; corolla pale greenish-yellow, inconspicuous, 2-4 mm. wide; plants dioecious, the staminate flowers with no, or abortive, ovaries; fruit densely clothed with long, spreading, straight or flexuous, flattened, whitish bristles.

Dry, open, often rocky slopes; Wash. to s. Calif., e. of the Cascade summits (and mostly e. of the Sierra Nevada summits as well), e. to c. Ida., the Wasatch Range of Utah, and n. Ariz. May-Aug., depending partly on the altitude.

A complex and highly variable species of uncertain limits, but well separated from all of our other species. The variations in leaf shape and vesture are not well correlated with each other, and only very imperfectly correlated with geography. In our area most of the plants from Oreg. and all of those from Ida. and extreme s.e. Wash. are glabrous or nearly so, while plants

from elsewhere in Wash. are mostly pubescent, but when the range of the species as a whole
is considerd, the ranges of the pubescent and glabrate forms are neither coherent nor even
approximately mutually exclusive. There is some correlation between geography and leaf shape,
our plants tending to have narrower leaves than those from Calif., but there is no clear differentiation of the species into narrow-leaved and broad-leaved populations, whether geography
is considered or not. Taxonomic segregation on these bases is therefore considered unwarranted, at least until a more thorough study is made.

A recent study by F. Ehrendorfer (Survey of the Galium multiflorum complex in western
North America, Contr. Dudley Herb. 5:1-36. 1956) has not wholly clarified our problems.
Ehrendorfer refers all of our plants to one or another form or subspecies of G. watsonii, but
he incorrectly typifies G. watsonii on a Cusick specimen from Oreg. The earlier and proper
lectotype of G. watsonii by Heller (see Leafl. West. Bot. 8:145-47. 1957) is from the Wasatch
Range of Utah, and is referred by Ehrendorfer to G. hypotrichium ssp. utahense Ehrend.

Galium oreganum Britt. Bull. Torrey Club 21:31. 1894.
  G. kamtschaticum ssp. oreganum Piper, Contr. U.S. Nat. Herb. 11:526. 1906. (Howell,
    Oreg. )
    Perennial from slender, creeping rhizomes, the stems arising singly, erect, 1-4 dm. tall,
glabrous; leaves mostly in 5-8(9) whorls of 4, antrorsely hispid-ciliate on the margins, often
with similar hairs along the 3 main veins, and sometimes with hairs scattered over the surface
as well, elliptic to elliptic-ovate, mostly 1.5-5 cm. long and 1/3-2/3 as wide, tending to be
acutish at the slightly mucronate tip, the margins convexly rounded toward the not especially
narrow base; inflorescence more branched and with more numerous flowers than in G. kamtschaticum, more often with peduncles from the upper axils as well as from the summit, the
peduncles tending to be cymosely branched and several-flowered; corolla 4-lobed, 3 mm. wide,
the lobes acute or acuminate; fruit 2 mm. high, the segments approximate, beset with numerous, well-developed, hooked bristles.

Moist woods and meadows, from near sea level to 5000 ft. in the mountains; Oreg. and
Wash., from the Cascades to near the coast. June-Aug.

Although obviously allied to G. kamtschaticum, G. oreganum appears to be wholly distinct.

Galium parisiense L. Sp. Pl. 108. 1753. (France)
  G. anglicum Huds. Fl. Angl. 2nd ed. 69. 1778. G. parisiense ssp. anglicum Gaud. Fl. Hel-
    vet. 1:439. 1828. G. parisiense var. leiocarpum Tausch, Flora 18: 354. 1835. G. parisien-
    se var. anglicum G. Beck, Fl. Nied.-Österr. 2²:1122. 1893. (England)
  G. divaricatum Lam. Encyc. Meth. 2:580. 1788. G. parisiense var. divaricatum DeVis. Fl.
    Dalmat. 3:8. 1852. (France) The diffuse extreme of var. leiocarpum.
  G. parvifolium Roem. & Schult. Syst. Veg. 3:246. 1818. G. parisiense ssp. parvifolium
    Gaud. Fl. Helvet. 1:439. 1828. G. anglicum var. parvifolium DC. Prodr. 4:607. 1830.
    (Thoiry, region of Gex, France) This would seem to provide an earlier varietal epithet for
    var. leiocarpum, to which it evidently applies, but the proper nomenclatural combination
    appears not to have been made (?), and in view of the fact that a thorough search of the
    European literature might provide a still older varietal epithet, no new combination is here
    proposed.
    Slender annual 1-4 dm. tall, the stem evidently to obscurely or scarcely retrorse-scabrous
on the angles; leaves mostly in whorls of 5-8, often 6, linear or linear-oblanceolate, mucronate, 1-nerved, 4-10 mm. long, antrorsely scabrous on the margins; flowers borne in small
cymes at the ends of the mostly numerous branches, the better developed plants appearing
abundantly floriferous; corolla minute; fruit only about 1 mm. long or a little less, the segments approximate, granular-roughened or provided with short hooked hairs. N=11, 22, 33.

A weed in fallow fields, roadsides, lawns, and other disturbed sites; native of Europe, introduced into e. U.S. and Calif., and reported from the upper Willamette Valley in Oreg.
June-July.

The nomenclaturally typical phase of the species, with hooked hairs on the fruits, is established in Calif., but is apparently not otherwise known in America; our plants, with slightly
granular-roughened, otherwise glabrous fruits, belong to the var. leiocarpum Tausch.

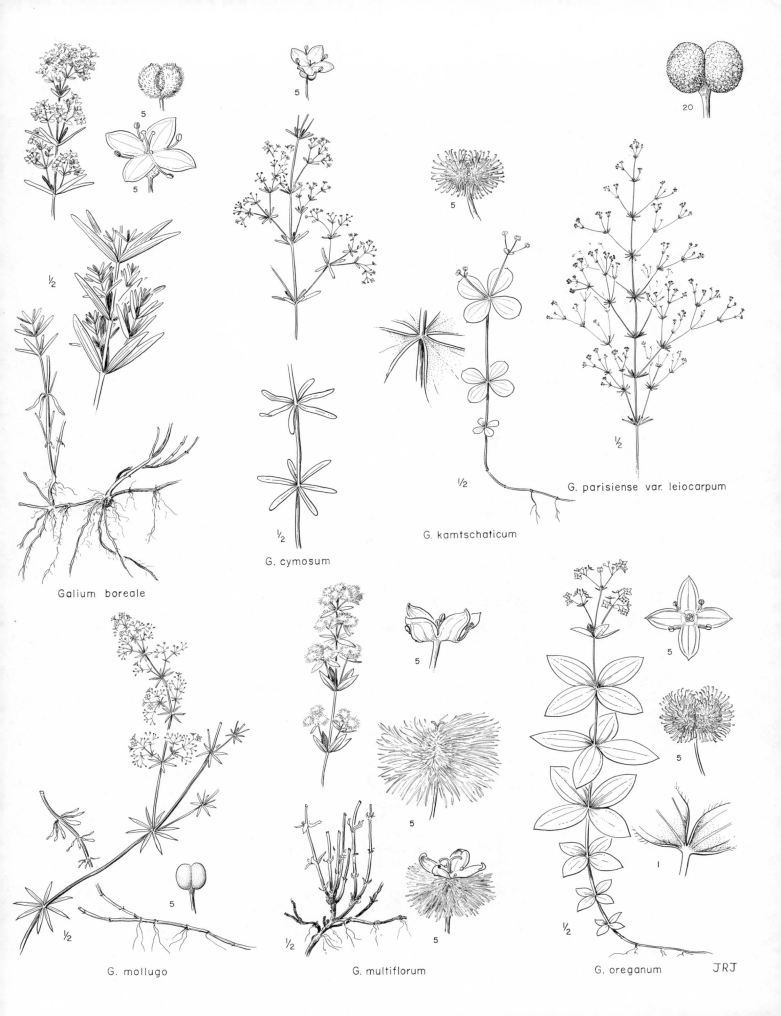

Galium boreale

G. cymosum

G. kamtschaticum

G. parisiense var. leiocarpum

G. mollugo

G. multiflorum

G. oreganum

JRJ

Galium tricorne Stokes in With. Bot. Arr. Brit. Pl. 2nd ed. 1:153. 1787. (England)
   Similar to G. aparine, differing chiefly in the characters given in the key, the stout, recurved fruiting pedicels very characteristic. N=22.
   An Old World weed, adventive along the American Atlantic Coast, established in Calif., and
reported from western Oreg.

Galium trifidum L. Sp. Pl. 105. 1753. (Kalm, Can.)
   G. tinctorium L. Sp. Pl. 106. 1753. G. trifidum var. tinctorium T. & G. Fl. N. Am. 2:22.
      1841. G. trifidum ssp. tinctorium Hara, Rhodora 41:388. 1939. (Kalm, N. Am., probably
      near Crown Point, N.Y.)
   G. claytoni Michx. Fl. Bor. Am. 1:78. 1803. (Can. and N.J.) = var. tinctorium.
   G. brandegei Gray, Proc. Am. Acad. 12:58. 1876. (Brandegee, valley of the Rio Grande, in
      N.M., at 9000 ft.) A depauperate form of var. pacificum, but the name often misapplied to
      similar depauperate forms of var. trifidum.
   G. trifidum var. pacificum Wieg. Bull. Torrey Club 24:400. 1897. G. trifidum ssp. paci-
      ficum Piper, Contr. U.S. Nat. Herb. 11:526. 1906. (Carpenter s. n., Placer Co., Calif.,
      in 1892)
   G. trifidum var. subbiflorum Wieg. Bull. Torrey Club 24:400. 1897. G. subbiflorum Rydb.
      Bull. Torrey Club 33:152. 1906. G. trifidum ssp. subbiflorum Piper, Contr. U.S. Nat.
      Herb. 11:526. 1906. G. claytoni var. subbiflorum Wieg. Rhodora 12:229. 1910. G. tinc-
      torium var. subbiflorum Fern. Rhodora 39:320. 1937. (Hall & Harbour 230, Colo.) = var.
      pacificum, the two names of the same date having been combined under var. pacificum by
      Hara in 1939.
   G. tinctorium var. submontanum Wright, Zoë 5:54. 1900. (Several specimens from the Pacif-
      ic states are cited) = var. pacificum.
   G. tinctorium var. diversifolium Wright, Zoë 5:54. 1900. (Specimens are cited from Ariz.,
      Calif., and Mex.) = var. pacificum.
   G. columbianum Rydb. Fl. Rocky Mts. 808. 1917. G. trifidum ssp. columbianum Hultén, Fl.
      Aleut. Isl. 307. 1937. (Sandberg, MacDougal, & Heller 600, St. Joe R., "Kootenai" Co.,
      Ida.) = var. pacificum.
   Stems numerous from very slender, creeping rhizomes, 0.5-6 dm. long, often much branched,
slender, weak, and lax, tending to scramble on other vegetation, commonly retrorse-scabrous on the angles, sometimes essentially glabrous; leaves borne in fours, or often some of
them in fives and sixes, sessile, linear to narrowly elliptic, blunt, 1-nerved, 5-20 mm. long,
often retrorsely scabrous along the margins and sometimes also along the midrib beneath,
otherwise glabrous; peduncles 1- to 3-flowered, numerous, axillary or terminal, commonly 1-
3 together at the ends of short, axillary branches; corolla small, about 1-1.5 mm. wide, whitish, 3-parted or occasionally 4-parted; fruits glabrous, the segments divergent and nearly distinct at maturity, each globular and 1-1.75 mm. thick; endosperm spherical and hollow, in
cross-section annular. N=12.
   Moist places, from sea level to high elevations in the mountains; circumpolar, in America
extending s. to Calif., n. Mex., Tex., Ala., and Ga. June-Sept.
   The American phases of the species as here defined may be divided, with some difficulty, into 3 varieties as follows:
1 Leaves 4 in a whorl; flowers mostly solitary (sometimes 2) at the ends of flexuous, more or
      less elongate peduncles which are often over 1 cm. long (these commonly borne 3 together),
      or the peduncles shorter and firmer in depauperate plants; circumpolar, chiefly boreal,
      but in e. U.S. extending southward to Neb., Ohio, and N.Y.                    var. trifidum
1 Leaves 4-6 in a whorl, usually some of the whorls with at least 5 leaves; peduncles 1- to 3-
      flowered and often shorter than in var. trifidum.
   2 Flowers mostly solitary or paired on each peduncle, these often elongate as in var. tri-
         fidum, the extremes of variation closely approaching characteristic var. trifidum, on
         the one hand, and var. tinctorium on the other; cordilleran, extending also into e. Asia
                                                                            var. pacificum Wieg.
   2 Flowers mostly 2-3 on each peduncle, the pedicels and peduncles relatively short and

straight, the pedicels rarely more than 5 mm. long; U.S. and adj. Can., e. of the cordillera                                                                                var. tinctorium (L.) T. & G.

Galium triflorum Michx. Fl. Bor. Am. 1:80. 1803. (Can.)
  G. triflorum var. asprelliforme Fern. Rhodora 37:445. 1935. (Fernald & Long 4205, near
  Great Bridge, Norfolk Co., Va.)
  G. triflorum f. hispidum Leyendecker, Iowa St. Coll. Journ. Sci. 15:180. 1941. (Pammel et
  al. 3934, Ledges State Pk., Boone Co., Iowa)
  G. triflorum f. glabrum Leyendecker, Iowa St. Coll. Journ. Sci. 15:180. 1941. (Bush 6029,
  Webb City, Jasper Co., Mo.) The form with essentially glabrous stem.
    Perennial from creeping rhizomes; stems 2-8 dm. long, usually retrorsely hooked-scabrous
on the angles at least below, prostrate or sometimes ascending or scrambling on other vegetation, often forming a loose rosette; leaves vanilla-scented, mostly (5)6 in a whorl, or only 4
on the smaller branches, narrowly elliptic to somewhat oblanceolate, cuspidate, commonly
1.5-4.5 cm. long, 1-nerved, generally antrorsely scabrous-ciliate on the margins and retrorsely hooked-scabrous on the midrib beneath, otherwise mostly glabrous; peduncles axillary, elongate, generally 3-flowered at the end, with divergent pedicels, sometimes branched
and several-flowered, occasionally the uppermost several whorls of leaves reduced so that
something of a terminal inflorescence is formed; corolla 4-parted, 2-3 mm. wide, whitish;
fruit 1.5-2 mm. high, the segments approximate, covered with hooked bristles mostly 0.5-
1.0 mm. long.
    Moist woods, from near sea level to moderate elevations in the mountains; circumboreal,
extending s., in America, to Calif., Mex., and Fla. Mostly June-Aug.
    Occasional western American specimens with branching peduncles and reduced upper leaves re-
semble the otherwise southeastern plants which have been segregated as var. asprelliforme
Fern., but in the West, at least, such specimens do not seem taxonomically separable from
the more usual forms of the species.

Galium verum L. Sp. Pl. 107. 1753. (Europe)
  Similar to G. mollugo, generally erect; stem generally densely puberulent or hirtellous in
the inflorescence, or even throughout; leaves narrowly linear, with revolute margins, often
finely hairy, sometimes up to 12 in a whorl; flowers bright yellow. N=11, 22.
  A native of Europe, escaped from cultivation and found as a weed of roadsides and fields in
n.e. U.S. and adj. Can., and sparingly established in lawns w. of the Cascades in our range.


                                         Kelloggia Torr.


    Flowers perfect, borne in open, terminal, cymose inflorescences, mostly 4- to 5-merous;
calyx teeth short; corolla funnelform-salverform, valvate; ovary 2-celled, with a solitary,
erect, basally attached ovule in each cell; style slender, bifid at the tip; fruit small, dry, indehiscent, covered with hooked bristles after the manner of Galium; endosperm well developed;
perennial herbs with opposite, entire, sessile leaves and small interpetiolar stipules.
    Two species, the following, and a similar one from Yunnan, China. (Named for Albert Kel-
log, 1813-1887, California botanist.)

Kelloggia galioides Torr. Bot. Wilkes Exp. 332. 1874. (Pickering & Brackenridge, Walla
  Walla R., Wash.)
    Perennial from well-developed creeping rhizomes. often with a taproot as well; herbage gla-
brous; stems more or less clustered, 1-6 dm. tall; leaves lanceolate or lance-linear, mostly
1.5-5 cm. long and 2-15 mm. wide; flowers long-pedicellate, (3)4(5)-merous; corolla pink or
white, 4-8 mm. long, the shortly hispid, ascending-spreading, narrow lobes nearly as long as
the slender tube; stamens and style shortly exserted; fruit 3-4 mm. long, more or less oblong,
the 2 halves readily separable.
    Wooded or open slopes in the mountains, sometimes among rocks or along streams, from
moderate to high elevations; c. Wash. to n. Baja Calif., e. to Ida., w. Wyo., Utah, and Ariz.,
in our area not extending w. of the Cascades summits. June-Aug.

## Sherardia L.

Flowers perfect, borne in small heads with a basal involucre of leaflike bracts; calyx teeth 4-6, lanceolate, well developed; corolla funnelform, the slender tube evidently longer than the 4 valvate lobes; ovary 2-celled, with a solitary ovule attached near the base of the septum in each cell; style bifid at the tip; stigmas capitate; fruit dry, crowned by the persistent sepals, dicoccous, the carpels indehiscent; endosperm well developed; annual with the habit of Galium.

The genus consists of only the following species. (Named for William Sherard, 1659-1728, a patron of botany.)

Sherardia arvensis L. Sp. Pl. 102. 1753. (N. Europe)

Slender annual 0.5-3 dm. tall, simple or branched especially at the base; stem spreading-hairy or retrorsely scabrous; leaves in whorls of about 6, 0.5-2 cm. long, stiffly hirsute above, the antrorsely scabrous, somewhat cartilaginous margins confluent distally into a firm point; heads on axillary and quasi-terminal, naked peduncles; involucre 4-9 mm. high, the bracts (7)8(10), shortly connate below; corolla 3 mm. long, pinkish; fruit scabrous-strigose, 2 mm. high exclusive of the prominent, pointed sepals. N=11.

Orchards, waste places, and other open, mostly disturbed sites; native from the Mediterranean region n. to Scandinavia, now widely distributed as a weed elsewhere in the world, and well established w. of the Cascades in our range. Apr.-July.

## CAPRIFOLIACEAE. Honeysuckle Family

Flowers regular or irregular, perfect (or the marginal ones sometimes neutral), gamopetalous, epigynous; calyx more or less evidently 3- to 5-lobed; corolla mostly 5-lobed, sometimes bilabiate, the tube sometimes spurred or gibbous; stamens epipetalous, ordinarily 5 and alternating with the corolla lobes, only 4 in Linnaea; pistil 2- to 5-locular, with 1-several pendulous ovules in each locule, sometimes only 1 locule fertile; stigma capitate or 2- to 5-parted, the style elongate to obsolete; fruit indehiscent (except in Diervilla), usually fleshy; endosperm copious; shrubs or woody vines, less often herbaceous or arborescent plants, with opposite (rarely alternate or whorled), mostly exstipulate leaves, the stipules when present usually small and adnate to the petiole; flowers in various sorts of inflorescences of mostly cymose origin.

About a dozen genera and perhaps 400 species, of wide distribution, chiefly in the N. Temp. Zone, or of mountainous areas in the tropics, one genus in New Zealand. All of our genera contain species of proven horticultural value, but most of them are too large for the home garden.

1 Style very short or none, the stigmas sessile or nearly so; inflorescence branched and mostly with more or less numerous flowers, umbelliform to corymbiform or paniculiform; corolla rotate to shortly open-campanulate; fruit drupaceous

  2 Leaves pinnately or bipinnately compound; fruits with 3-5 small, seedlike stones

                                                                         SAMBUCUS

  2 Leaves simple, sometimes lobed; fruit with 1 large stone      VIBURNUM

1 Style well developed, more or less elongate; inflorescence various, of paired flowers at the end of a peduncle, or of short racemes or spikes, or of sessile verticels on an elongate or foreshortened axis, but not umbelliform, corymbiform, nor paniculiform; corolla short-campanulate to elongate and tubular, not at all rotate; leaves simple; fruits diverse

  3 Stamens as many as the corolla lobes, mostly 5; fruits fleshy; plants not at once trailing and with the flowers paired on terminal peduncles

    4 Corolla regular or merely ventricose (the tube unequally expanded, somewhat bulged on one side); ovary with 2 fertile uniovulate locules and 2 sterile locules with several abortive ovules each; fruit white, drupaceous, with 2 seedlike stones

                                                   SYMPHORICARPOS

    4 Corolla evidently irregular, either 2-lipped, or spurred at the base, or both; ovary mostly 2- to 3-locular, with several ovules in each locule; fruit a several-seeded berry, red to blue or black           LONICERA

3  Stamens 4, corolla lobes 5; fruits dry; trailing plants with the flowers mostly paired on terminal peduncles                                                                            LINNAEA

## Linnaea L.  Twinflower

Flowers regular or nearly so, 5-merous as to the calyx and corolla, pedicellate, nodding, paired at the end of a long naked peduncle; corolla funnelform to campanulate, pink or pinkish; stamens 4, inserted toward the base of the corolla, 2 of them shorter than the others; ovary 3-locular, two of the cells containing several abortive ovules each, the other with a solitary pendulous normal ovule; style elongate; stigma capitate or obscurely lobed; fruit small, dry, indehiscent, unequally 3-locular, one-seeded; creeping evergreen herbs with small, exstipulate, few-toothed to entire leaves.

Genus probably monotypic, although some additional names have been proposed for Chinese plants. (Named for Carolus Linnaeus, 1707-78, originally by Gronovius.)

Linnaea borealis L. Sp. Pl. 631. 1753.

Obolaria borealis Kuntze, Rev. Gen. 1:275. 1891. (Sweden)

Linnaea americana Forbes, Hort. Woburn. 135. 1833. L. borealis var. americana Rehd. Rhodora 6:56. 1904. L. borealis ssp. americana Hultén, Fl. Aleut. Isl. 310. 1937. (America)

L. borealis var. longiflora Torr. Bot. Wilkes Exp. 327. 1874. L. longiflora Howell, Fl. N. W. Am. 280. 1900. L. borealis ssp. longiflora Hultén, Fl. Aleut. Isl. 310. 1937. (Oreg. to B.C., Sitka, and the Rocky Mts.; no type cited, the first collector mentioned is Lyall)

L. borealis var. longiflora f. angustissima Wittr. Acta. Hort. Berg. 4[7]:173. 1907. (Fidalgo I., Wash.; collector not directly cited, perhaps Lyall)

L. borealis var. longiflora f. insularis Wittr. Acta. Hort. Berg. 4[7]:173. 1907. (Rosendahl & Brand 20, Renfrew Dist., Vancouver I., B.C.)

Stems slender but woody, elongate, trailing and creeping, more or less hairy, at least when young, and often also glandular, producing numerous short, suberect, leafy stems mostly less than 10 cm. long, these each bearing a slender terminal peduncle 3.5-8 cm. long; leaves short-petiolate, firm, rather broadly elliptic or obovate to subrotund, with a few shallow teeth or sometimes entire, mostly 7-25 mm. long and 5-15 mm. wide, glabrous, or long-hairy especially along the margins and veins; peduncle bearing a pair of minute bracts at the summit, usually forking into a pair of pedicels 1-2.5 cm. long, occasionally the axis continuing and producing one or two extra pairs of flowers; calyx 2-5(6) mm. long; corolla hairy within, mostly 9-16 mm. long, with a definite slender tube at the base, flaring from the summit of the calyx, or often from above it. N=16.

In open or dense woods and brush at various elevations throughout our range; circumpolar, extending s. in America to Calif., Ariz., N.M., S.D., Ind., and W. Va. June-Sept.

One of our most desirable native ground-covering shrubs, easily introduced from the wild and tending to spread rapidly but not agressively. Two varieties of the species may be recognized. Most of the American plants, as described above, belong to the var. longiflora Torr. The Eurasian plants, and also those from the Aleutian Islands and some parts of continental Alas., belong to the var. borealis, which differs from the var. longiflora in its more consistently subrotund leaves and in its somewhat shorter corollas (mostly 6-10 or 11 mm. long) which flare from within the calyx, the tube proper (as opposed to the flaring throat) being very short or even wanting. The material from w. N. Am. often shows the differences from typical var. borealis in more pronounced fashion than does the material from e. N. Am., but no further taxonomic segregation seems reasonable.

Linnaea borealis ssp. americana (Forbes) Hultén and L. borealis ssp. longiflora (Torr.) Hultén are here considered synonymous. For the taxon in this rank, L. borealis ssp. longiflora is here selected as the correct name, thus fixing the choice according to Article 57 of the International Code of Botanical Nomenclature, 1956 edition. The validity of publication of Hultén's combinations has sometimes been questioned, but his indirect reference (through Fernald, Rhodora 24:210-12. 1922) to the respective basonyms is adequate under Article 32 of the 1956 Code. Although the two epithets (longiflora and americana) are of the same date in subspecific rank, the epithet longiflora has many years priority in varietal status.

## Lonicera L. Honeysuckle

Flowers regular or more often (including all our species) irregular, 5-merous as to the calyx, corolla, and androecium (or the calyx lobes sometimes obsolete), borne on 2-flowered, axillary peduncles or in verticels (opposite, 3-flowered, sessile cymules) on terminal or terminal and axillary raches; corolla regularly or nearly regularly 5-lobed, or often evidently bilabiate, with 4-lobed upper lip, the tube often gibbous or spurred near the base; ovary mostly 2- to 3-locular, each cell with 3-8 pendulous ovules, the placentation axile; style elongate, with a capitate stigma; fruit a small, several-seeded berry; shrubs or woody vines with exstipulate or rarely stipulate, mostly entire leaves.

About 150 species, occurring mostly in temperate and subtropical regions of the Northern Hemisphere, about 20 species in N. Am. (Named for Adam Lonitzer, 1528-86, German botanist. )

Reference:

Rehder, Alfred. Synopsis of the genus Lonicera. Ann. Rep. Mo. Bot. Gard. 14:27-232. 1903.

1 Flowers paired on axillary peduncles; leaves all distinct; more or less erect shrubs
  2 Bracts at the summit of the peduncle enlarged, broad and foliaceous, often anthocyanic, forming an involucre, the outer pair 8-15 mm. long or more; ovaries and fruits wholly distinct (Distegia)    L. INVOLUCRATA
  2 Bracts at the summit of the peduncle narrow and mostly small, relatively inconspicuous, less than 5 mm. long except in forms of L. caerulea (Xylosteon)
    3 Bractlets (inner bracts) wholly connate into a narrow-mouthed cup which tightly encloses the ovaries and grows with them into a fruit, the ovaries thus appearing wholly united, but actually distinct within the cup; corolla yellow, obscurely or scarcely bilabiate, the lobes about as long as the tube    L. CAERULEA
    3 Bractlets small and inconspicuous, or obsolete, not enclosing the ovaries; ovaries and fruit united at least at the base, but obviously paired
      4 Flowers ochroleucous or light yellow, obscurely or scarcely bilabiate, the slightly unequal lobes much shorter than the tube; ovaries and fruits divergent, united only at the base; widespread species    L. UTAHENSIS
      4 Flowers dark reddish purple, strongly and deeply bilabiate, the lips longer than the tube, the upper lip shallowly 4-lobed; ovaries and fruits united to the middle or usually above; Cascades from Mt. Adams southward    L. CONJUGIALIS
1 Flowers in terminal or terminal and axillary inflorescences of several or many flowers; uppermost leaves ordinarily connate-perfoliate; vines or climbing shrubs
  5 Corolla slightly bilabiate, 2.5-4 cm. long, the tube densely hairy within, about 3 or 4 times as long as the lips; filaments attached well down in the tube; widespread native species    L. CILIOSA
  5 Corolla strongly bilabiate, the tube from shorter than to about twice as long as the lips; filaments attached nearly at the orifice of the corolla
    6 Corolla 1.2-3 cm. long, the tube densely hairy within; native species
      7 Some of the leaves with well-developed, connate stipules (unique among our species in this regard); corolla tube from shorter than to about equaling the lips; w. of the Cascades    L. HISPIDULA
      7 None of the leaves stipulate; corolla tube from slightly longer than to about twice as long as the lips; eastern species, extending westward to s. e. B. C.    L. DIOICA
    6 Corolla 3-5 cm. long, the tube glabrous within; introduced along the coast from Lane Co. , Oreg. , southward    L. ETRUSCA

Lonicera caerulea L. Sp. Pl. 174. 1753.
  Caprifolium caeruleum Lam. Fl. Fr. 3:336. 1778. Chamaecerasus caerulea Delarbre, Fl. d'Auvergne, 2nd ed. 131. 1800. Xylosteon caeruleum Dumont, Bot. Cult. 2:575. 1802. Isika caerulea

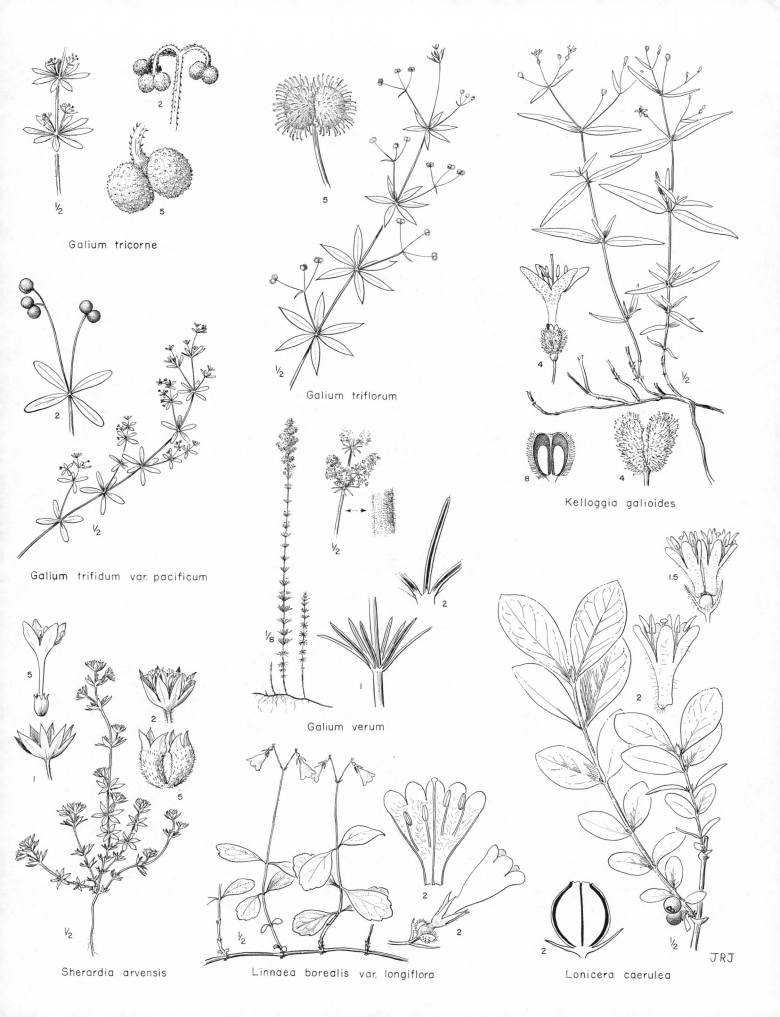

Galium tricorne

Galium trifidum var. pacificum

Galium triflorum

Galium verum

Kelloggia galioides

Sherardia arvensis

Linnaea borealis var. longiflora

Lonicera caerulea

JRJ

Borkh. Handb. Forstbot. 2:1682. 1803. Euchylia caerulea Dulac, Fl. Haut-Pyren. 463.
1867. (Switzerland)

Lonicera cauriana Fern. Rhodora 27:10. 1925. (Suksdorf 559, Mt. Paddo, Wash.)

Shrub 0.2-2 m. tall; leaves short-petiolate, elliptic to somewhat oblong or obovate, entire,
2-6 cm. long and 1-3 cm. wide, rounded or obtuse at the tip, more or less hirsute at least beneath; peduncles axillary, borne toward the base of the twigs of the season, short, only 2-
5 mm. long, 2-flowered, bearing at the summit a pair of narrow bracts 3-6 mm. long, or
these occasionally up to 1 cm. long and somewhat foliaceous in texture, but still relatively narrow and inconspicuous; ovaries 2-locular, distinct, but appearing united because they are enclosed by a narrow-mouthed cup derived from the connate bractlets; corolla yellow, 10-13 mm.
long, obscurely or scarcely bilabiate, the subequal lobes equaling or a little shorter than the
tube, which is hairy inside and usually also outside, and has a short, thick spur or gibbosity at
the base; style and stamens glabrous; fruit nearly 1 cm. long, typically bluish-glaucous, but in
ours apparently red, the cup fleshy but thin. N=9, 18.

Stream banks and other moist or sometimes rather dry places, mostly at moderate elevations
in the mountains, seldom collected in our range; Eurasia; Can. s. of about 60 degrees lat.,
s. to Calif., Nev., Wyo., Minn., and Pa. June-July.

Lonicera caerulea is a highly variable species which is sharply defined if taken in the broad
sense. In recent years it has become customary to restrict the name L. caerulea to Eurasian
plants, and to segregate the American plants under two other binomials, L. cauriana Fern.
for cordilleran specimens, and L. villosa (Michx.) R. & S. for more eastern ones. Asiatic,
e. American, and w. American plants all look very much alike, however, in spite of the minor
technical differences on which the segregation has been based. It is not improbable that L. cauriana Fern. should be maintained, in infraspecific status, for the cordilleran, apparently red-
fruited phase of L. caerulea, but no new combination is here proposed.

Lonicera ciliosa (Pursh) DC. Prodr. 4:333. 1830.

Caprifolium ciliosum Pursh, Fl. Am. Sept. 160. 1814. (Lewis, "on the banks of the Kooskoosky") It has been customary to attribute the combination L. ciliosa to Poir. in Lam.
Encyc. Meth. Suppl. 5:612. 1817, but the name was not there clearly published.

C. occidentale Lindl. Bot. Reg. 17: pl. 1457. 1831. Lonicera occidentalis Hook. Fl. Bor.
Am. 1:282. 1833. L. ciliosa var. occidentalis Rehd. Ann. Rep. Mo. Bot. Gard. 14:171.
1903 (where incorrectly attributed to Nicholson in 1901). (Garden specimens from seeds
brought by Douglas from Ft. Vancouver)

Lonicera suksdorfii Gand. Bull. Soc. Bot. France 65:33. 1918. (Suksdorf 5678, Bingen,
Wash.)

Twining vine, sometimes climbing to a height of 6 m.; twigs hollow, glaucous when young;
leaves glabrous except for the more or less ciliate margins, strongly glaucous beneath, the principal ones short-petiolate or subsessile, broadly elliptic, rounded to acute at the summit, mostly
4-10 cm. long and 2-5.5 cm. wide, one or more of the uppermost pairs on each twig a little
smaller, sessile, and connate-perfoliate; flowers in a dense, short, terminal inflorescence
consisting of several verticels on a more or less foreshortened axis, blooming in centripetal
sequence; ovaries distinct, 3-locular; corolla orange-yellow to orange-red, sometimes taking
a purplish cast in drying, 2.5-4 cm. long, shallowly and not very strongly bilabiate, the tube
about 3 or 4 times as long as the lips, swollen or gibbous on one side above the base; fruits
red, nearly 1 cm. thick.

Woods and thickets, from sea level to moderate elevations in the mountains; s. B.C. to n.
Calif., especially w. of the Cascade summits, extending e. to w. Mont. May-July.

An attractive vine that deserves to be included in the native garden more frequently.

Lonicera conjugialis Kell. Proc. Calif. Acad. Sci. 2:67. 1863.

Caprifolium conjugiale Kuntze, Rev. Gen. 1:274. 1891. Xylosteon conjugialis Howell, Fl.
N.W. Am. 282. 1900. (Veatch, Washoe, Nev.)

Shrub 6-15 dm. tall; leaves short-petiolate, elliptic or rhombic to sometimes broadly ovate
or subrotund, acute to broadly rounded at the summit, mostly 2.5-7.5 cm. long and 1.5-4.5
cm. wide, more or less hirsute beneath and often also above; peduncles axillary, 1-4.5 cm.

long, their bracts minute and commonly more or less adnate to the base of the ovary, or obsolete; flowers paired; corolla dark reddish-purple, 8-11 mm. long, deeply bilabiate, the upper lip shallowly 4-lobed, the lower narrower and entire; corolla tube shorter than the lips, strongly gibbous-inflated on the lower side, but scarcely spurred, densely and conspicuously long-hairy within, as also the lower parts of the style and filaments; ovaries and fruits 3-locular, united at least to the middle, often only the tips separate; fruits nearly 1 cm. long, reputedly red or reddish-black, appearing bluish to black in the herbarium.

In woods, meadows, and moist open slopes; Mt. Adams, Wash., s. through the Cascades into the Sierra Nevada of Calif.; common southward, but rarely collected in our range. June-July.

Lonicera dioica L. Syst. Veg. 12th ed. 2:165. 1767.
 Caprifolium dioicum Roem. & Schult. Syst. Veg. 5:260. 1819. (No type given)
 Lonicera hirsuta var. glaucescens Rydb. Contr. U.S. Nat. Herb. 3:503. 1896. L. glauces-
  cens Rydb. Bull. Torrey Club 24:90. 1897. L. dioica var. glaucescens Butters in Clements,
  Rosendahl, and Butters, Minn. Trees & Shrubs 289. 1912. (Typification obscure; several
  specimens are cited)
 Resembling L. ciliosa; differing, in addition to the characters given in the key, in its eciliate leaves that may be short-hairy beneath, and in its yellowish to purplish or reddish corollas.
 Woods, often in wet places; Que. to N.C., w. to Mack., n.e. B.C., S.D., and Okla., and extending into our range in s.e. B.C. (Field). June-July.
 Several varieties have been recognized. Ours is var. glaucescens (Rydb.) Butters, occupying the more western part of the range of the species, and characterized by having the leaves more or less persistently short-hairy beneath.

Lonicera etrusca Santi, Viagg. Mont. 113. 1795.
 Caprifolium etruscum Roem. & Schult. Syst. Veg. 5:261. 1819. (Region of Tuscany, Italy)
 Similar to L. ciliosa, but less glaucous, and often evidently pubescent, our form becoming glandular in the inflorescence; leaves less strongly connate, sometimes all free; corolla 3-5 cm. long, pale yellowish, often tinged with purple, glabrous within, more deeply and conspicuously bilabiate than in L. ciliosa, the lips sometimes equaling the tube; filaments attached nearly at the orifice of the corolla. N=9.
 Native of the Mediterranean region, now established in thickets along the coast of Oreg., from Florence, Lane Co., s. to Curry Co., and on s. Vancouver I. July.

Lonicera hispidula (Lindl.) Dougl. ex T. & G. Fl. N. Am. 2:8. 1841.
 Caprifolium hispidulum Lindl. Bot. Reg. 21: pl. 1761. 1836. Lonicera hispidula var. doug-
  lasii Gray, Proc. Am. Acad. 8:628. 1873. (Garden plants, from seeds collected by Doug-
  las in N.W. Am.)
 Lonicera microphylla Hook. Fl. Bor. Am. 1:283. 1833, not of Willd. in 1819. (Douglas,
  Cascade Mts.)
 L. californica T. & G. Fl. N. Am. 2:7. 1841. Caprifolium californicum K. Koch, Hort.
  Dendr. 294. 1853. Lonicera hispidula var. vacillans Gray, Proc. Am. Acad. 8:628. 1873.
  Caprifolium hispidulum var. californicum Greene, Fl. Fran. 347. 1892. Lonicera hispidula
  var. californica Rehd. in Bail. Cycl. Am. Hort. 3:943. 1900 (where incorrectly attributed
  to Greene). (Douglas, Calif.)
 Vine or climbing shrub with hollow twigs that are glaucous when young; leaves firm, glaucous beneath, variously hirsute or puberulent to glabrous, 2-7 cm. long, 1.5-5 cm. wide, often subcordate at the base, the uppermost ordinarily connate-perfoliate as in L. ciliosa, the others except the lowermost short-petiolate and with well-developed, connate, interpetiolar stipules; flowers in terminal or terminal and axillary inflorescences, in several compact verticels on a well-developed axis that varies from glabrous to hirsute or even glandular, the lower verticels blooming first; ovaries distinct, 3-locular; corolla pink, or yellow tinged with pink, 12-18 (20) mm. long, deeply bilabiate, the lips about as long as the tube, or longer, the tube swollen or gibbous on one side above the base; fruits red, nearly 1 cm. thick.
 Woods and thickets w. of the Cascade summits; s. B.C. to s. Calif. June-Aug.

Lonicera involucrata (Rich.) Banks ex Spreng. Sys⁺ Veg. 1:759. 1825.

Xylosteum involucratum Richards. App. Frankl. Journ. 733. 1823. Caprifolium involucratum
Kuntze, Rev. Gen. 1:274. 1891. Distegia involucrata Cockerell, U. Colo. Stud. 3:50. 1905.
(Richardson, "wooded country between 54 and 64 degrees latitude")

Lonicera ledbourii Esch. Mém. Acad. St. Pétersb. 10:284. 1826. Caprifolium ledebourii
Greene, Fl. Fran. 346. 1892. Distegia ledebourii Greene, Man. Bay Reg. 164. 1894. Xy-
losteon ledebourii Howell, Fl. N. W. Am. 282. 1900. Lonicera involucrata var. ledebourii
Jeps. Man. Fl. Pl. Calif. 968. 1925. (Eschscholtz, Calif., probably near San Francisco)

L. flavescens Dippel, Gartenfl. 37:7. 1888. Caprifolium flavescens Kuntze, Rev. Gen. 1:274.
1891. Lonicera involucrata var. flavescens Rehd. Ann. Rep. Mo. Bot. Gard. 14:100. 1903.
Distegia flavescens Cockerell, U. Colo. Stud. 3:50. 1905. (Lyall, Fraser R., B. C.)

Shrub 0.5-4 m. tall; young twigs quadrangular; leaves short-petiolate, elliptic or elliptic-
oblong to elliptic-ovate or elliptic-obovate, acuminate or sometimes merely acute, mostly 5-14
cm. long and 2-8 cm. wide, glabrous or inconspicuously glandular above, glabrous or more commonly somewhat hirsute beneath, even if only along the main veins; peduncles axillary, 0.5-5 cm.
long, bearing at the summit a pair of conspicuous, broad, green or more or less purple-tinged
bracts about (0.8)1-1.5(2) cm. long, these enclosing another shorter, but very broad, pair of
bracts which eventually enlarge to nearly the same size, both pairs spreading or reflexed in
fruit and then usually more or less purplish-red; flowers paired, closely subtended by the
bracts; corolla yellow, sometimes tinged with red, glandular-pubescent, mostly 1-2 cm. long,
shortly and subequally lobed, with a short thick spur at the base; fruits 3-locular, wholly distinct, black, globose, nearly or fully 1 cm. thick. N=9.

Woodlands and thickets, from sea level to rather high elevations in the mountains, mostly in
fairly moist or wet soil; Alaska panhandle to s. Calif. and Chihuahua, e. to Mont. and N. M.,
and less commonly to Mich. and Que. Apr.-Aug., depending partly on the elevation.

The fruit is said to be somewhat poisonous, but it is so bitter and nauseous that there is little danger of its being eaten in sufficient quantities to cause concern. Our plants belong to the
widespread var. involucrata, a shrub (0.5)1-2(3) m. tall, with the anthers about equaling the
corolla or often slightly exserted, the corolla seldom over 1.5 cm. long. The rather poorly
defined var. ledebourii (Esch.) Jeps., a taller plant (mostly 1.5-4 m.), with the anthers shortly included, sometimes reaching only to the sinuses, and with the corollas up to 2 cm. long,
replaces the typical variety in the Calif. Coast Ranges, and extends northward along the coast,
rarely as far as s. Lane Co., Oreg.


Lonicera utahensis Wats. Bot. King Exp. 133. 1871.

Caprifolium utahense Kuntze, Rev. Gen. 1:274. 1891. Xylosteon utahensis Howell, Fl. N. W.
Am. 282. 1900. (Watson 477, Cottonwood Canyon, Wasatch Mts., Utah)

Lonicera ebractulata Rydb. Mem. N. Y. Bot. Gard. 1:372. 1900. Xylosteon ebractulatum
Rydb. Fl. Rocky Mts. 814. 1917. Lonicera utahensis f. ebractulata St. John, Fl. S. E.
Wash. 393. 1937. (Rydberg & Bessey 5010, Spanish Basin, Mont.)

Shrub mostly 1-2 m. tall; leaves short-petiolate, elliptic to somewhat ovate or oblong, sometimes subcordate at the base, obtuse to broadly rounded at the summit, mostly 2-8 cm. long
and 1-4 cm. wide, or some of them smaller, glabrous above, glabrous or often hirsute beneath; peduncles axillary, 5-15 mm. long (or to 35 mm. in fruit), bearing at the summit a pair
of minute bracts 1-3 mm. long; bractlets minute, 1 mm. long or less, or obsolete; flowers
paired; corolla ochroleucous or light yellow, 1-2 cm. long, obscurely or scarcely bilabiate,
the slightly unequal lobes much shorter than the tube, which is hairy within and bears a short,
thick spur at the base; style and stamens glabrous; ovaries 2-locular, divergent, weakly or
scarcely united at the base when young, firmly so when mature, the individual berries about
1 cm. thick or a little less, bright red.

Moist, wooded or open slopes at moderate to rather high elevations in the mountains; s. B. C.
to n. Calif., e. to Alta., Mont., Wyo., and Utah; not known in the Oreg. Coast Ranges. May-July.


### Sambucus L. Elderberry

Flowers regular, typically 5-merous as to the calyx, corolla, and androecium, borne in

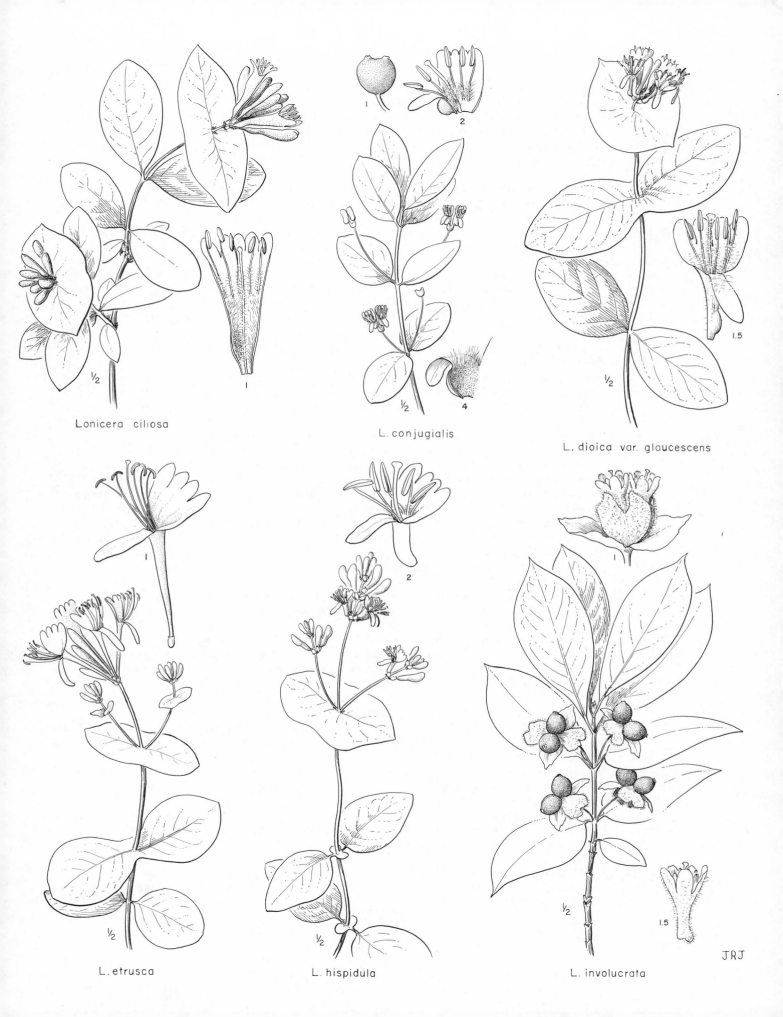

Lonicera ciliosa

L. conjugialis

L. dioica var. glaucescens

L. etrusca

L. hispidula

L. involucrata

JRJ

compound, umbelliform or paniculiform, basically cymose inflorescences; calyx inconspicuous; corolla rotate or nearly so; ovary 3- to 5-celled, with 1 pendulous ovule in each cell; style very short, almost obsolete, 3- to 5-lobed; fruit berrylike, juicy, with 3-5 small, seedlike stones (each enclosing a seed); shrubs, or sometimes coarse herbs or small trees, with pithy stems and large, stipulate or exstipulate, pinnately or even bipinnately compound leaves and serrate leaflets.

Nearly 20 species, chiefly of temperate regions and of mountainous parts of the tropics, in both the Old and the New Worlds. (The classical Latin name for the plant now called Sambucus nigra L. )

Our species are both worthy of cultivation in the wild garden because of their fruits which attract countless birds, especially robins and band-tailed pigeons.

1 Inflorescence pyramidal or strongly convex, paniclelike, with the main axis extending well beyond the lowest branches; fruits not glaucous                                                     S. RACEMOSA
1 Inflorescence flat-topped, with the axis scarcely, or not at all, produced beyond the mostly 4 or 5 subumbellately clustered principal branches; fruits strongly glaucous
                                                                                                            S. CERULEA

Sambucus cerulea Raf. Alsogr. Am. 48. 1838. (Woodlands on the w. side of the Rocky Mts., in Oreg. territory)
  S. glauca Nutt. ex T. & G. Fl. N. Am. 2:13. 1841. S. cerulea var. glauca Schwerin, Mitt. Deuts. Dendrol. Ges. 1909:37. 1909. (Nuttall, plains of the Oreg., near the Blue Mts. )
  S. decipiens M. E. Jones, Bull. U. Mont. Biol. 15:46. 1910. (Jones, St. Ignatius Mission, Mont. )
  S. ferax A. Nels. Bot. Gaz. 53:225. 1912. (Macbride 631, Trinity, Ida. )
  Coarse shrub, sometimes arborescent, normally with several stems from the base, (1)2-4(6) m. tall, with soft, pithy, glaucous twigs; stipules linear, small and commonly soon deciduous, or apparently wanting; leaves petiolate, mostly 5- to 9(11)-foliolate, the leaflets lanceolate or lance-ovate to elliptic, strongly acuminate, sharply serrate, often strongly inequilateral at the base, commonly 5-15 cm. long and 2-6 cm. wide, glabrous or rarely a little hairy beneath; inflorescence pedunculate or sessile, flat-topped in life, commonly 4-15 or 20 cm. wide at anthesis, umbelliform, with several (commonly 4-5) rays from the base, these often again subumbellately branched, though eventually cymose; occasionally there may be a pair of smaller inflorescence-branches arising from the leaf axils at the base of the peduncle of the main inflorescence; flowers white or creamy, 4-6 or 7 mm. across, the corolla lobes evidently longer than the short flat tube; fruit edible, globose, 4-6 mm. thick, bluish-black beneath the dense, waxy bloom, thus appearing pale powdery blue; nutlets rugulose.

Valley bottoms and open slopes where not too dry, from near sea level to moderate elevations in the mountains; s. B.C. to w. Mont., s. to Calif., Ariz., and N.M. May-July.

Only the nomenclaturally typical variety, as described above, occurs in our range; two or three other varieties are found well to the southward.

Sambucus racemosa L. Sp. Pl. 270. 1753. (Mountains of s. Europe)
  S. pubens Michx. Fl. Bor. Am. 1:181. 1803. S. racemosa var. pubens Koehne, Deuts. Dendrol. 532. 1893. S. racemosa ssp. pubens House, Woody Pl. West. N.C. 32. 1913. (Mountains of Pa., Can., and Carolina)
  S. pubens var. leucocarpa T. & G. Fl. N. Am. 2:13. 1841. S. racemosa f. leucocarpa House, Bull. N.Y. State Mus. 243-44:69. 1923. S. pubens f. leucocarpa Fern. Rhodora 35:310. 1933. (Hogg, Catskill Mts., N.Y.) = var. pubens.
  S. pubens var. arborescens T. & G. Fl. N. Am. 2:13. 1841. S. racemosa var. arborescens Gray, Syn. Fl. 1²:8. 1884. S. arborescens Howell, Fl. N.W. Am. 279. 1900. (Typification obscure, but the name doubtless intended to apply to the phase of the species for which it is here used)
  S. melanocarpa Gray, Proc. Am. Acad. 19:76. 1883. S. racemosa var. melanocarpa McMinn, Ill. Man. Calif. Shrubs. 529. 1939. (Watson, mountains of Mont.; lectotype)
  S. callicarpa Greene, Fl. Fran. 342. 1892. S. racemosa var. callicarpa Jeps. Fl. W. Mid-

dle Calif. 471. 1901. (Coast Ranges of Calif.; see discussion given by Greene under his S. maritima, Pitt. 2:297. 1892.) = var. arborescens.

S. leiosperma Leib. Proc. Biol. Soc. Wash. 11:40. 1897. (Coville & Leiberg 370, Crater Lake, Oreg.) = var. arborescens, approaching var. microbotrys.

S. microbotrys Rydb. Bull. Torrey Club 28:503. 1897. S. racemosa var. microbotrys Kearney & Peebles, Journ. Wash. Acad. Sci. 29: 492. 1939. (Bessey, Pike's Peak, Colo.)

S. melanocarpa (form?) fuerstenbergii Schwerin, Mitt. Deuts. Dendrol. Ges. 1909:43. 1909. (Fuerstenberg, Glacier, Selkirk Mts., B.C.) A form of var. arborescens or var. pubens with chestnut-brown fruits.

Shrub, sometimes arborescent, with soft, pithy, commonly somewhat glaucous twigs; stipules consisting of a pair of thickened, glandulous appendages, in ours small and often soon deciduous; leaves petiolate, mostly 5- to 7-foliolate, the leaflets lanceolate or lance-ovate to elliptic, strongly acuminate, sharply serrate, about 4.5-17 cm. long and 2-6 cm. wide, often strongly inequilateral at the base; inflorescence paniculiform, with a definite central axis projecting well beyond the lowermost pair of branches, relatively small, only 4-10 cm. long in fruit, and of nearly the same size at anthesis; flowers white or creamy, 3-6 mm. across, the corolla lobes evidently longer than the short, flat tube; fruit globose, 5-6 mm. thick, red or purple-black, or less often yellow, brown, or even white, not glaucous. N=18.

In a wide variety of reasonably mesic habitats; circumboreal. Mar.-July.

Typical European S. racemosa is glabrous, and differs from all our forms of the species in its better developed and often more persistent stipules, but this latter character apparently does not hold true for all of the Old World forms of the species. In the absence of a thorough modern monograph of the group it is convenient to follow Hultén in recognizing all of the American forms as a single subspecies, ssp. pubens (Michx.) House, with several varieties, especially since this course permits the retention of the well-known epithet for the common e. American variety, and avoids the necessity which would otherwise arise under the Rules to adopt instead an unfamiliar and misleading varietal name. The American varieties of the species may then be treated as follows:

1 Fruit black or purplish-black; nutlets with slightly rugulose or pebbly surface; leaflets somewhat pubescent beneath (especially when young) to not infrequently glabrous; shrub commonly 1-2 m. tall; plants chiefly of the n. Rocky Mts. (Mont., Wyo., Ida., and adj. Alta. and B.C.), but extending also, less commonly, to e. Wash. (even as far as Mt. Adams) and Oreg., the Sierra Nevada of Calif., Nev., n. Ariz., and apparently n. N.M., mostly occurring fairly well up in the mountains, and flowering from May to July
                                        var. melanocarpa (Gray) McMinn

1 Fruit bright red, varying to sometimes yellow, chestnut, or even (rarely) white

    2 Large, sometimes arborescent shrubs, mostly 2-6 m. tall; nutlets mostly smooth, rarely roughened; leaves generally more or less pubescent beneath, the hairs sometimes confined chiefly to the midrib and main veins; plants of coastal and subcoastal Alas. and the Aleutian Islands, southward through w. B.C. to w. Wash. and Oreg. (from the Cascade Range to the coast), and extending in the Calif. Coast Ranges to the vicinity of San Francisco Bay, commonly occurring at moderate or low elevations and blooming from March to July      var. arborescens (T. & G.) Gray (S. callicarpa of some floras)

    2 Smaller shrubs, mostly 0.5-3 m. tall; nutlets mostly with slightly rugulose or pebbly surface, rarely smooth

        3 Leaves essentially glabrous; shrub mostly 0.5-2 m. tall, occurring chiefly from the s. Rocky Mts. (s. Wyo. to N.M.) westward to the Sierra Nevada of Calif., extending n., rarely, to Fremont Co., Ida., found from the foothills to rather high elevations in the mountains, and blooming mostly in June and July
                                        var. microbotrys (Rydb.) Kearney & Peebles

        3 Leaves generally evidently pubescent beneath, rarely glabrous; shrub mostly 1-3 m. tall, widespread in n. e. U.S. and adj. Can., extending westward to the Black Hills, and across s. Can. to Alta. and apparently e. B.C., blooming commonly in May and June                                var. pubens (Michx.) Koehne

## Symphoricarpos Duhamel. Snowberry

Flowers regular or the corolla somewhat ventricose, 5-merous or occasionally 4-merous as to the calyx, corolla, and androecium, borne in terminal or axillary short racemes or spikes, these sometimes reduced to single flowers; corolla short-campanulate to elongate-campanulate or subsalverform, pink to white; ovary 4-locular, two of the cells containing several abortive ovules each, the other two each with a solitary, pendulous, normal ovule; style well developed, with a capitate or slightly 2-lobed stigma; fruit white, or in some extralimital species red or black, berrylike, with 2 seedlike stones each enclosing a seed; erect or trailing shrubs with exstipulate, simple, entire to sometimes coarsely toothed or somewhat lobed leaves.

About 10 species, one native to China, the rest to N. Am. (Name from the Greek syn, together, phorein, to bear, and karpos, fruit, referring to the closely clustered fruits.)

Reference:

Jones, George N. A monograph of the genus Symphoricarpos. Journ. Arn. Arb. 21: 201-52. 1940.

1 Corolla relatively short and broad, short-campanulate, not much, if at all, longer than wide (pressed), often bulged on one side (ventricose), the lobes varying from half as long to longer than the tube
  2 Style elongate, (3)4-7(8) mm. long, more or less exserted, long-hairy near the middle or occasionally glabrous; anthers about 1.5-2 mm. long, evidently shorter than the filaments; corolla lobes mostly equaling or a little longer than the tube
                                                                               S. OCCIDENTALIS
  2 Style short, 2-3 mm. long, glabrous; anthers about 1-1.5 mm. long, not much, if at all, shorter than the filaments; corolla-lobes often evidently shorter than the tube
    3 Erect shrubs; corolla mostly 5-7 mm. long; fruits 6-15 mm. long; nutlets 4-5 mm. long                                                                          S. ALBUS
    3 Trailing shrubs; corolla 3-5 mm. long; fruits 4-6 mm. long; nutlets 2.5-3 mm. long
                                                                                   S. MOLLIS
1 Corolla relatively long and narrow, elongate-campanulate, evidently longer than wide, not evidently ventricose, the lobes mostly 1/4-1/2 as long as the tube; erect shrubs with short styles                                                                   S. OREOPHILUS

Symphoricarpos albus (L.) Blake, Rhodora 16:118. 1914.
  Vaccinum album L. Sp. Pl. 350. 1753. Xylosteum album Moldenke, Rev. Sudam. Bot. 5:3. 1937. (Kalm, hills near the St. Lawrence R., Can.)
  Symphoricarpos racemosus Michx. Fl. Bor. Am. 1:107. 1803. Lonicera racemosa Pers. Syn. 1:214. 1805. Symphoria racemosa Pursh, Fl. Am. Sept. 162. 1814. (Michaux, hills near Mistassini, Que.)
  Symphoricarpos racemosus var. laevigatus Fern. Rhodora 7:167. 1905. S. albus var. laevigatus Blake, Rhodora 16:119. 1914. S. albus f. laevigatus G. N. Jones, U. Wash. Pub. Biol. 5:236. 1936. (Garden specimens, from material collected by Lewis & Clark, "beyond the Rocky Mountains")
  S. rivularis Suksd. Werdenda 1:41. 1927. (Suksdorf 7557, Falcon Valley, Klickitat Co., Wash.)

Erect, branching shrubs, (0.5)1-2(3) m. tall, less conspicuously rhizomatous than S. occidentalis; twigs glabrous or obscurely puberulent; petioles 2-4 mm. long; leaves elliptic or elliptic-ovate, sometimes broadly so, entire or with a few coarse, irregular teeth, mostly 1.5-5 cm. long and 1-3.5 cm. wide, or larger on sterile shoots and then often irregularly lobed, essentially glabrous above, glabrous or often sparsely hirsute-puberulent beneath; racemes short, dense, subsessile, few-flowered, produced at the ends of the twigs and often also in the upper axils; corolla 5-7 mm. long, densely hairy within, relatively short and broad, nearly as wide as long, commonly somewhat ventricose, the lobes equaling or merely half as long as the tube; anthers mostly 1-1.5 mm. long, not much, if at all, shorter than the filaments; style glabrous, 2-3 mm. long; fruits subglobose or ellipsoid, the larger ones mostly 1-1.5 cm. long; nutlets mostly 4-6 mm. long and 2.5-3.5 mm. wide. N= about 27.

Thickets, woodlands, and open slopes, from the lowlands to moderate elevations in the mountains; Alaska panhandle to Quebec, s. to Calif., c. Ida., Colo., Neb., and Va. May-Aug.

Most of our plants belong to the Pacific slope phase of the species, the var. laevigatus (Fern.) Blake, as described above, which also commonly extends onto the Atlantic slope in Mont. At the e. edge of our range the var. laevigatus passes into the characteristic Atlantic slope phase of the species, the var. albus, which differs in its smaller size (less than 1 m. tall), mostly smaller fruits (up to 1 cm. long), and mostly more hairy (but occasionally glabrous) twigs and lower surfaces of the leaves. The var. laevigatus is commonly cultivated, and often escapes in e. U.S.

Symphoricarpos mollis Nutt. in T. & G. Fl. N. Am. 2:4. 1841.
  S. albus var. mollis Keck, Bull. So. Calif. Acad. Sci. 25:72. 1926. (Nuttall, Santa Barbara, Calif.)
  S. mollis var. acutus Gray, Syn. Fl. 1²: 14. 1884. S. acutus Dieck, Hamb. Gart. Blumenzeit. 44:562. 1888. S. rotundifolius var. acutus Frye & Rigg, N.W. Fl. 366. 1912. (Austin, Lassen Peak, Calif.)
  SYMPHORICARPOS MOLLIS var. HESPERIUS (G. N. Jones) Cronq. hoc loc. S. hesperius G. N. Jones, Journ. Arn. Arb. 21:220. 1940. (Allen 105, upper valley of the Nisqually R., Pierce Co., Wash.)

Trailing shrub 1-3 m. long, often rooting at the nodes, the branches rising less than 0.5 m.; young twigs more or less puberulent to subglabrous; petioles 1-3 mm. long; leaves elliptic or sometimes ovate, mostly 1-3 cm. long and 5-20 mm. wide, entire or sometimes with a few coarse teeth or shallow lobes (those on vigorous shoots sometimes larger and more frequently toothed or lobed), glabrous or inconspicuously puberulent above, sparsely to moderately hirsute-puberulent beneath; flowers in short, dense, few-flowered terminal racemes, sometimes also in the upper axils; corolla short-campanulate, not much if at all longer than wide, sometimes somewhat ventricose, 3-5 mm. long, the lobes equaling or often a little shorter than the tube, sparsely to rather conspicuously long-hairy within at the base of the lobes and top of the tube; anthers 1 mm. long, a little shorter than the filaments; style glabrous, 2-3 mm. long; fruit subglobose, 5-6 mm. long; nutlets 2.5-3 mm. long, 1.5-2 mm. wide.

Woodlands and open slopes at moderate or low elevations; s. B.C. to s. Calif., chiefly in and w. of the Cascades and Sierra Nevada, extending inland, in our area, to n. Ida. and s.e. Wash. June-July.

Our plants, as described above, belong to the var. hesperius (G. N. Jones) Cronq., which extends s. to n.w. Calif. In cismontane Calif. the var. hesperius is replaced by the more densely hairy var. mollis. A third variety (var. acutus Gray), occurring chiefly in the Sierra Nevada and s. Cascades, differs from var. mollis in having some longer spreading hairs intermingled with the numerous short, curled hairs of the young twigs, the var. mollis having the hairs of the twigs more uniform in character. There are no sharp lines between these varieties, and some specimens might be misidentified in the absence of geographic data.

Symphoricarpos occidentalis Hook. Fl. Bor. Am. 1:285. 1833. (Richardson, "wooded country between latitude 54 and 64")

Erect, more or less branching shrubs 3-10 dm. tall, less bushy than S. albus, spreading freely from rhizomes and often forming dense colonies; young twigs puberulent or rarely glabrous; petioles 3-10 mm. long; leaves elliptic or ovate, often broadly so, entire or often with a few coarse, blunt, irregular teeth, mostly 2.5-8 cm. long and 1.5-5 cm. wide, or larger on sterile shoots and then often more lobed, glabrous above, usually hirsute-puberulent beneath, at least along the main veins; short, dense, pedunculate or subsessile, few to rather many-flowered racemes produced at the ends of the twigs and in the upper axils; corolla 5-8 mm. long, often wider than long, densely hairy within, the lobes more spreading than in S. albus, equaling or commonly a little longer than the tube; anthers mostly 1.5-2 mm. long and evidently shorter than the filaments; style elongate, (3)4-7(8) mm. long, evidently long-hairy near the middle, varying to occasionally glabrous; fruit subglobose, 6-9 mm. long; nutlets about 3.5 mm. long and 2-2.5 mm. wide.

Open prairies, and moist low ground along streams or lakes; B.C. to Man., s. to n. Wash., Utah, N.M., Mich., and Mo.; a species characteristically of the Rocky Mts. and northern

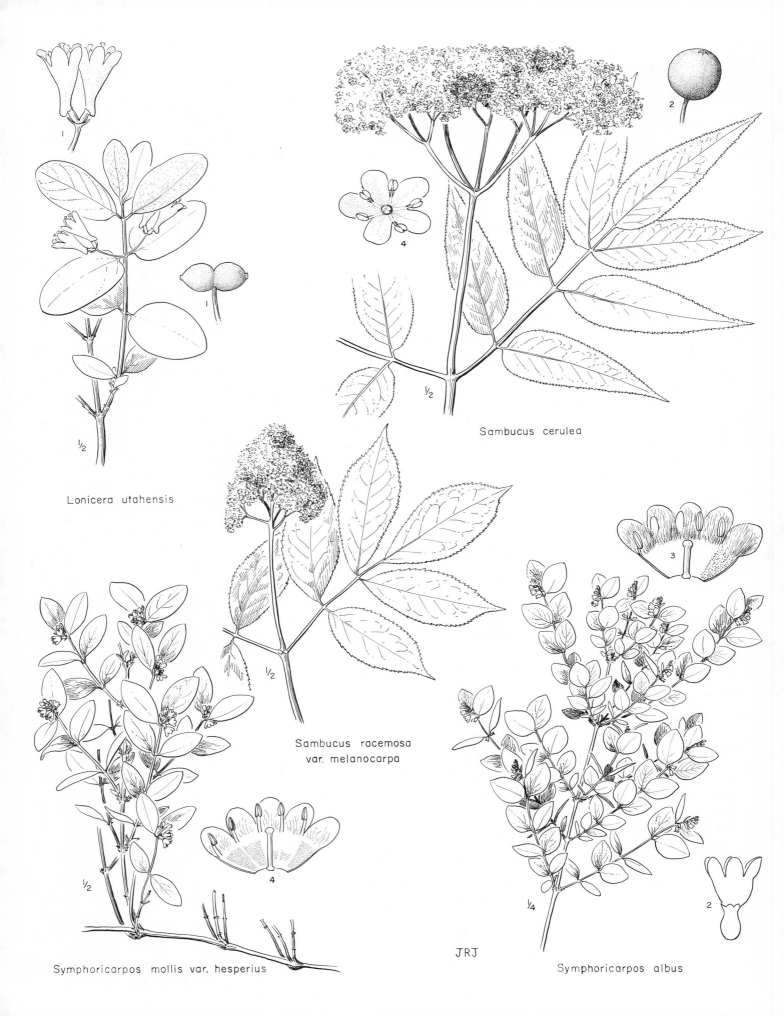

Lonicera utahensis

Sambucus cerulea

Sambucus racemosa
var. melanocarpa

Symphoricarpos mollis var. hesperius

JRJ

Symphoricarpos albus

plains, common in Mont., rarely collected in Ida. and n. Wash. (Okanogan Co.). June-Aug.

Symphoricarpos oreophilus Gray, Journ. Linn. Soc. 14:12. 1873.
  S. rotundifolius var. oreophilus M. E. Jones, Proc. Calif. Acad. Sci. II. 5:690. 1895. (Par-
    ry 223, headwaters of South Clear Creek, and alpine ridges e. of Middle Park, Rocky Mts.
    of Colo.)
  S. utahensis Rydb. Bull. Torrey Club 26:544. 1899. S. oreophilus var. utahensis Nels. in
    Coult. & Nels. New Man. Bot. Rocky Mts. 470. 1909. (Rydberg s.n., Logan, Utah)
  S. vaccinioides Rydb. Mem. N.Y. Bot. Gard. 1:371. 1900. S. rotundifolius var. vaccinioi-
    des Nels. in Coult. & Nels. New Man. Bot. Rocky Mts. 471. 1909. (Rydberg & Bessey 5017,
    forks of the Madison R., Mont.)
  S. tetonensis A. Nels. Bull. Torrey Club 31:246. 1904. (Merrill & Wilcox 205, Teton Mts.,
    Wyo.)

Erect, branching shrubs 0.5-1.5 m. tall; leaves and young twigs densely puberulent to glabrous; leaves more or less elliptic or elliptic-ovate, mostly 1-3.5(5) cm. long and 5-25 mm.
wide; petioles 1-4 mm. long; flowers borne on short, drooping pedicels in the upper axils, and
often also in short, few-flowered, terminal racemes; corolla elongate-campanulate, evidently
longer than wide, mostly (6)7-10(11) mm. long (dry), the lobes 1/4-1/2 as long as the tube, the
tube commonly hairy within below the level of insertion of the filaments, or sometimes glabrous; anthers mostly 1-2 mm. long, about as long as the filaments; style short, 2-4 mm. long,
glabrous; fruits broadly ellipsoid, mostly 7-10 mm. long; nutlets 4-6.5 mm. long, obtuse or
acutish at the base.

Open slopes and dry meadows, from the foothills to high elevations in the mountains, southward confined to the higher elevations; s. B.C. to Mont., s. to Calif., N.M., and n. Mex.,
e. of the Cascade summits. June-Aug.

Our plants, as described above, belong to the var. utahensis (Rydb.) A. Nels., which occupies the northern and western part of the range of the species, and extends s. occasionally
into s.e. Utah and s. Colo. The var. oreophilus, differing in its longer, (9)10-13 mm., corollas, and often sharper-based nutlets, occupies the more southern and eastern part of the range
of the species, chiefly from the Rocky Mts. of Colo. and the mountains of s. Utah southward,
extending n.w. occasionally as far as the Wasatch Range of Utah and the Ruby and Toiyabe mts.
of Nev.

Plants with the leaves and young twigs densely puberulent pass by easy stages into the less
common, wholly glabrous phase, without any other differences being apparent, and the two
forms sometimes occur together, so that it does not seem wise to base any taxonomic segregation on this character.

The type of S. utahensis, from Logan, Utah, has the corollas a scant 9 mm. long when dry,
within the normal size range for the variety to which the epithet is here attached, and about at
the lower limit for var. oreophilus. Geographically, the specimen is better associated with the
variety for which the name is here used, rather than with var. oreophilus. Many of the species
known in the Wasatch region and southward do not extend n. to Logan.

What appears to be S. longiflorus Gray, a Great Basin species, has been collected (Peck
18902) without flowers near Millican, Deschutes Co., Oreg., nearly within the s. border of
our range. It is distinguished from S. oreophilus by its subsalverform corollas (mostly 10-
13 mm. long) with somewhat spreading lobes, subsessile anthers, and smaller leaves (up to
1.5 cm. long).


Viburnum L.


Flowers perfect, regular (or the marginal ones sometimes neutral and irregular), 5-merous
as to the calyx, corolla, and androecium, borne in umbelliform (all our species) or sometimes
paniculiform, basically cymose inflorescences; corolla open-campanulate to subrotate, mostly
white; ovary 3-locular, two locules reduced, sterile, and more or less vestigial, the third with
a single pendulous ovule; stigmas minute, subsessile, 3, or apparently only 1; fruit a 1-locu-
lar, 1-seeded drupe with soft pulp; shrubs or small trees, exstipulate or with small petiolar
stipules, with opposite (rarely whorled), simple, entire or toothed to lobed leaves.

About 150 species, of wide distribution, mostly in temperate regions, or in mountainous parts of the tropics. (The classical Latin name of the plant now called V. lantana L.)

1 Inflorescences with the marginal flowers neutral and enlarged, the corollas mostly 1.5-2.5 cm. broad; leaves trilobed                                       V. OPULUS

1 Inflorescences with the flowers all perfect and alike, their corollas less than 1 cm. broad; leaves lobed or lobeless

  2 Inflorescences small (mostly 1-2.5 cm. wide at anthesis) and relatively few-flowered (less than 50); stamens inconspicuous, the filaments 1 mm. long or less; leaves, or many of them, tending to be trilobed, as well as sharply toothed, the unlobed leaves mostly acuminate                                       V. EDULE

  2 Inflorescences larger (mostly 2.5-5 cm. wide at anthesis) and with more numerous flowers; stamens exserted, the filaments 3-5 mm. long; leaves coarsely and often rather bluntly toothed, not at all trilobed, acutish to rounded at the tip       V. ELLIPTICUM

Viburnum edule (Michx.) Raf. Med. Repos. N.Y. II. 5:254. 1808.

  V. opulus var. edule Michx. Fl. Bor. Am. 1:180. 1803. V. edule Pursh, Fl. Am. Sept. 203. 1814. (Rafinesque's publication of the name, cited above, is only doubtfully valid under Article 33 of the International Rules of Botanical Nomenclature, 1956 edition) (Can.)

  V. pauciflorum Pylaie ex T. & G. Fl. N. Am. 2:17. 1841. (Pylaie, Newf.)

Straggling to suberect shrubs 0.5-2.5 m. tall; winter buds with 2 connate outer scales; stipules wanting; petioles, peduncles, and commonly also the leaf blades (especially along the veins) with small and inconspicuous, scattered, sessile or stipitate glands; leaf blades more or less hirsute beneath, especially along the main veins and in their forks, mostly palmately veined at the base and shallowly 3-lobed, sometimes pinnately veined and lobeless, sharply toothed, commonly 3-10 cm. long and about as wide, or narrower and mostly acuminate when lobeless, commonly bearing a pair of glandular teeth or projections near the junction with the petiole; inflorescences borne on short axillary shoots which bear a single pair of leaves, relatively small and few-flowered, commonly 1-2.5 cm. wide at anthesis and with less than 50 flowers; flowers all alike, the whitish corolla 4-7 mm. across; stamens inconspicuous, their filaments 1 mm. long or mostly less; fruits acid, edible, red or orange, 1-1.5 cm. long, with a large flattened stone. Squashberry.

Moist woods and swamps; Alas. to Newf., s. to n. Oreg. (Cascade Range), n. Ida., Colo., Minn., and Pa. May-July.

This species, like V. opulus, is worthy of cultivation for its brilliant autumnal foliage.

Viburnum ellipticum Hook. Fl. Bor. Am. 1:280. 1833. (Douglas, "on the branches of the Columbia near its confluence with the Pacific")

  V. ellipticum var. macrocarpum Suksd. Deuts. Bot. Monats. 18:97. 1900. (Suksdorf 1213, near Bingen, Klickitat Co., Wash.)

Shrub 1-3 m. tall; winter buds with two pairs of outer scales; stipules petiolar, narrowly linear, 3-10 mm. long, or wanting; petioles 6-20 mm. long, spreading-hirsute and finely glandular to subglabrous; leaf blades more or less hirsute beneath, with 3 or sometimes 5 main veins from the base, broadly elliptic to subrotund, sometimes subcordate at the base, mostly 3-8 cm. long, coarsely and often rather bluntly toothed; inflorescences borne at the ends of leafy branches, mostly 2.5-5 cm. wide at anthesis, many-flowered; flowers all alike, the whitish corolla 5-9 mm. wide; stamens exserted, the filaments 3-5 mm. long; fruits red, ellipsoid, 1-1.6 cm. long, with a large, flattened stone.

Thickets, bottom lands, and open woods, w. of the Cascade Range, from s. Wash. to n. Calif. May-June.

Viburnum opulus L. Sp. Pl. 268. 1753. (Europe)

  V. opulus var. americanum Ait. Hort. Kew. 1:373. 1789. (N. Am.) Not V. americanum Mill., which may prove to be an older name for the e. Am. species now called V. alnifolium Marsh.

  V. trilobum Marsh. Arb. Am. 162. 1785. V. opulus ssp. trilobum R. T. Clausen, Cornell U. Agr. Exp. Sta. Mem. 291:10. 1949. (Mountains in the interior parts of Pa.)

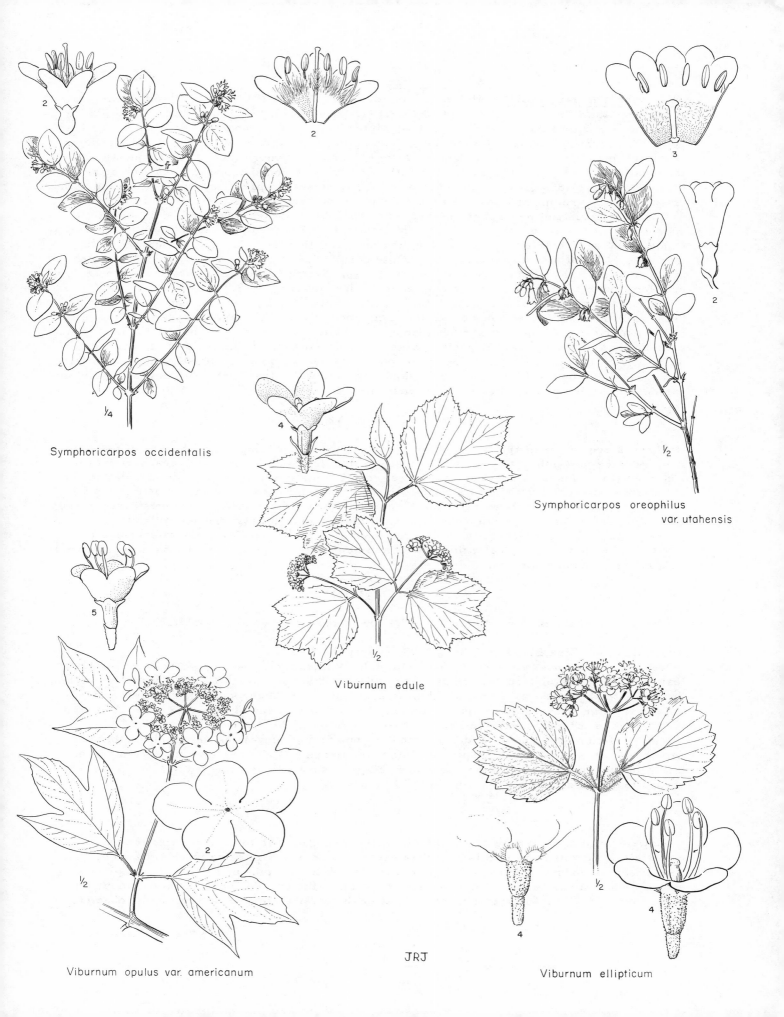

Symphoricarpos occidentalis

Symphoricarpos oreophilus
var. utahensis

Viburnum edule

Viburnum opulus var. americanum

JRJ

Viburnum ellipticum

V. oxycoccos var. subintegrifolium Hook. Fl. Bor. Am. 1:281. 1833. V. opulus var. subintegrifolium T. & G. Fl. N. Am. 2:18. 1841. ("Columbia," by Scouler and by Douglas)
Coarse shrub or small tree, 1-4 m. tall; winter buds with 2 connate outer scales; stipules
petiolar, narrowly linear, 2-6 mm. long, tending to be blunt or even thickened and glandular
at the tip; petioles 1-3 cm. long, glabrous except for one or more thickened, glandular projections (or stalked glands) near the summit, these mostly round-topped and as high or higher
than wide; leaf blades glabrous or more often stiffly hirsute in part, especially along the main
veins beneath, prominently 3-veined from the base and ordinarily evidently trilobed, mostly
4-12 cm. long and not much if at all narrower, the lobes coarsely few-toothed or sometimes
subentire, mostly acuminate, the lateral often falcately divergent, the sinuses entire; inflorescence short-pedunculate, 5-15 cm. wide, the marginal flowers enlarged and neutral, with rotate, deeply lobed, somewhat irregular, white corolla 1.5-2.5 cm. across; perfect flowers
much smaller, the corolla mostly 3-4 mm. across; stamens exserted; fruits acid, red, 1-1.5
cm. long, with a large, flattened stone. High-bush cranberry. N=9.
Moist woods; Newf. to s. B.C., s. to Pa., Ill., S.D., Wyo., n. Ida., and Wash. (Columbia gorge); Eurasia; rare in our area, more common in n.e. U.S. May-July.
The American plants, as described above, constitute the var. americanum Ait., from which
the nomenclaturally typical, Eurasian phase of the species may be distinguished, with some
difficulty, by its more attenuate, often somewhat longer stipules, and mostly shorter, sessile
or subsessile, concave-topped petiolar glands that are wider than high. The Eurasian plant occasionally escapes from cultivation, at least in e. U.S.

## ADOXACEAE

Flowers regular or nearly so, perfect, gamopetalous, semi-epigynous, the tips of the carpels
free from the perianth; lateral flowers mostly with 3 sepals and 5 corolla lobes; terminal flowers mostly with 2 sepals and 4 corolla lobes; corolla rotate; stamens twice as many as the corolla lobes, paired at the sinuses, each with only a single pollen sac; ovary mostly 4- to 5-
celled, the cells commonly as many as the corolla lobes, with a single pendulous ovule in each
cell; style short, distinct; fruit a small, dry drupe with 4-5 nutlets; endosperm copious; exstipulate herbs with several basal leaves and a single pair of opposite cauline leaves, the
scales of the rhizome alternate; flowers borne in compact, headlike, mostly 5-flowered cymes.
A single genus and species.

### Adoxa L. Moschatel

Characters of the family. (Name from the Greek adoxos, obscure, ignoble.)

Adoxa moschatellina L. Sp. Pl. 367. 1753. (Europe)
Delicate herb with a musky odor, 5-20 cm. tall, arising from a short, scaly rhizome; basal
leaves long-petiolate, ternate, the primary divisions discrete, again 1- to 2-ternate or parted,
the ultimate segments rather broad, thin, round-toothed and mucronate; cauline leaves similar but smaller and less dissected, commonly borne a little above the middle of the flowering
shoot; corolla inconspicuous, yellowish green, 5-8 mm. wide. N=18, 27.
Moist, often mossy places in the forested regions at high elevations in the mountains; circumboreal, in America extending s. to Colo., Iowa, and N.Y. I have seen no specimens of
this seldom-collected plant from our range, but it is surely to be expected in Mont., since it
occurs in Yellowstone Nat. Pk. and in Colo. June-Aug.

## VALERIANACEAE

Flowers regular or irregular, perfect or unisexual, gamopetalous, epigynous; calyx segments either inrolled at anthesis and later expanded and pappuslike, or much reduced or obsolete; corolla mostly 5-lobed, often somewhat bilabiate, the tube often spurred or gibbous;
stamens epipetalous, alternating with, but fewer than, the corolla lobes, 1-4, commonly 3;
pistil basically 3-carpellary, one carpel fertile, the other 2 sterile and sometimes obsolete;

style with a simple, bilobed, or more often trilobed stigma; ovule solitary, pendulous; fruit dry, indehiscent; endosperm wanting; opposite-leaved, exstipulate herbs with the flowers borne in various sorts of basically determinate inflorescences.

About 11 genera and 300 or more species, of wide distribution, but chiefly in the N. Temp. Zone. Both Plectritis and Valeriana include native species of some horticultural promise.

1 Calyx segments about 9-20, inrolled at anthesis, expanded on the mature fruit; our species
　　perennial                                                                          VALERIANA
1 Calyx segments obsolete; annuals
　2 Stem dichotomously branched above, the flowers borne in cymose glomerules at the ends
　　　of the branches; ovary 3-celled (only 1 cell fertile); stigma 3-lobed    VALERIANELLA
　2 Stem simple or with opposite, axillary branches, the inflorescence of terminal, subcap-
　　　itate or interrupted-spicate cymose glomerules; ovary 1-celled, the sterile cells obso-
　　　lete; stigma 2-lobed or occasionally 3-lobed                              PLECTRITIS

### Plectritis DC.

Inflorescence of subcapitate or interrupted-spicate clusters of basically determinate nature, the stem simple or with few, paired axillary branches; calyx obsolete; corolla subequally 5-lobed or evidently bilabiate, the tube spurred or rarely merely gibbous; stamens 3; ovary 1-locular, the 2 sterile carpels apparently obsolete; stigma 2-lobed or rarely 3-lobed; fruit dry, indehiscent , commonly with a pair of lateral wings, or these sometimes reduced or obsolete; annuals with opposite, entire or obscurely toothed, sessile or short-petiolate leaves.

Probably only 3 species, as described below, although a number of others have been proposed on inconstant variations in technical features of the flowers and fruit. All of the species are very similar in general appearance. (Name from the Greek, plektos, plaited, referring presumably to the complex inflorescence. )

Reference:

Nielsen, Sarah Dyal. Systematic studies in the Valerianaceae. Am. Midl. Nat. 42:480-501. 1949.

1 Convex side of the fruit keeled, not grooved; cotyledons transverse to the ventral face of the
　　fruit; wings of the fruit, when developed, tending to be connivent toward the base and di-
　　vergent above; hairs of the fruit, when present, more or less pointed    P. CONGESTA
1 Convex side of the fruit broader, scarcely keeled, bearing a narrow groove down the center;
　　cotyledons parallel to the ventral face of the fruit; wings of the fruit, when present, tend-
　　ing to be about equally divergent above and below; hairs of the fruit, when present, clavate
　　or long-cylindrical and blunt
　2 Corolla mostly white or pinkish, about equally 5-lobed, with a thick spur; hairs on the
　　　fruit, when present, without any definite arrangement at least on the convex side; inflo-
　　　rescence appearing terete in life                                    P. MACROCERA
　2 Corolla mostly pink or deep pink, evidently bilabiate, with a slender spur; hairs of the
　　　fruit, when present, usually unequally distributed, being more numerous in a band on
　　　each side of the groove on the convex face; inflorescence appearing somewhat quadran-
　　　gular in life                                                          P. CILIOSA

Plectritis ciliosa (Greene) Jeps. Man. Fl. Pl. Calif. 971. 1925.
　Valerianella ciliosa Greene, Proc. Acad. Phila. 1895:548. 1896. Aligera ciliosa Suksd.
　　Deuts. Bot. Monats. 15:146. 1897. (Greene, Oakville, Napa Co. , Calif., in 1895)
　Aligera rubens Suksd. Deuts. Bot. Monats. 15:146. 1897. Plectritis californica var. rubens
　　Dyal,* Am. Midl. Nat. 42:495. 1949. (Suksdorf, Klickitat Co., Wash. ) A small-flowered,
　　glabrous-fruited form.
　Aligera macroptera Suksd. Deuts. Bot. Monats. 15:146. 1897. Plectritis macroptera Rydb.
　　Fl. Rocky Mts. 818. 1917. (Suksdorf, Klickitat Co., Wash. ) A large-flowered form.
　Aligera patelliformis Suksd. W. Am. Sci. 12:53. 1901. Plectritis macroptera var. patelli-

---

*For purposes of botanical nomenclature, Mrs. Nielsen retains her maiden name, Dyal.

formis Dyal, Am. Midl. Nat. 42:497. 1949. (Parish 4539, Cuyamaca Mts., Calif.) A small-
flowered form with the hairs of the fruit evenly distributed.

Aligera macroptera var. obtusa Suksd. Allg. Bot. Zeit. 12:6. 1906. (Suksdorf 2678, near
Bingen, Wash.)

A. californica Suksd. Werdenda 1:44. 1927. Plectritis californica Dyal, Am. Midl. Nat. 42:
495. 1949. (Curran, Sweetwater, Calif., May, 1884.) A large-flowered form.

Aligera barbata Suksd. Werdenda 1:44. 1927. (Suksdorf 7006, Bingen Mt., Klickitat Co.,
Wash.) A small-flowered form.

A. glabrior Suksd. Werdenda 1:44. 1927. (Suksdorf 6959, Bingen, Klickitat Co., Wash.) A
small-flowered, glabrous-fruited form.

A. intermedia Suksd. Werdenda 1:44. 1927. (Suksdorf 7007, Bingen Mt., Klickitat Co.,
Wash.) A small-flowered form.

Plants slender, 1-5 dm. tall, the herbage glabrous or nearly so; leaves mostly 1-3 cm.
long and 3-10 mm. wide, distant, the lowermost ones more or less obovate and short-petiolate,
the others more oblong and sessile; inflorescence appearing somewhat quadrangular in life;
corolla 2-8 mm. long, evidently bilabiate, commonly deep pink with one or two red dots at the
base of the middle lobe of the lower lip, the spur relatively slender and elongate; fruit 2-4 mm.
long, the convex side grooved, scarcely keeled, the wings about equally divergent above and
below, or wanting; hairs of the fruit clavate or long-cylindrical and blunt, commonly forming
a conspicuous band on each side of the groove, and more scattered over the rest of the surface,
or the fruit occasionally glabrous; cotyledons parallel to the ventral face of the fruit.

Vernally moist, open slopes and meadows; Klickitat Co., Wash., s. through w. Oreg. to
Calif. Apr.-May.

Apparently not common in our area.


Plectritis congesta (Lindl.) DC. Prodr. 4:631. 1830.

Valerianella congesta Lindl. Bot. Reg. 13: pl. 1094. 1827. (Horticultural specimens, from
seeds collected "on the north-west coast of America" by Douglas) The large-flowered,
wing-fruited form.

Betckea samolifolia DC. Prodr. 4:642. 1830. Plectritis samolifolia Hoeck. Engl. Bot. Jahrb.
3:37. 1882. Valerianella samolifolia Gray, Proc. Am. Acad. 19:83. 1883. (Bertero, near
la Punta de Corets, Chile) A small-flowered, wingless-fruited form.

Betckea major Fisch. & Mey. Ann. Sci. Nat. Bot. II. 5:189. 1836. Plectritis major Hoeck.
Engl. Bot. Jahrb. 3:37. 1882. P. congesta var. major Dyal, Am. Midl. Nat. 42: 486.
1949. (Calif.) A large-flowered, wingless-fruited form.

Valerianella aphanoptera Gray, Proc. Am. Acad. 19:83. 1883. Plectritis aphanoptera Suksd.
Deuts. Bot. Monats. 15:144. 1897. (Suksdorf, w. Klickitat Co., Wash., June 9, 1882) A
small-flowered, narrow-winged form.

Valerianella anomala Gray, Proc. Am. Acad. 19:83. 1883. Plectritis anomala Suksd. Deuts.
Bot. Monats. 15:144. 1897. (Suksdorf 26, Klickitat Co., Wash.; lectotype by Nielsen) An
extreme form with the spur reduced to a mere gibbosity at the base of the corolla throat.

Plectritis microptera Suksd. Deuts. Bot. Monats. 15:119. 1897. (Lower Columbia R.) A
large-flowered, wing-fruited form.

P. involuta Suksd. Deuts. Bot. Monats. 15:144. 1897. P. samolifolia var. involuta Dyal, Am.
Midl. Nat. 42:489. 1949. (Suksdorf 666, Klickitat Co., Wash.) A small-flowered, wing-
fruited form.

P. congesta var. alba Suksd. Allg. Bot. Zeit. 12:6. 1906. (Suksdorf 2552, Bingen, Wash.)
A white-flowered form.

P. racemulosa Gand. Bull. Soc. Bot. France 65:35. 1918. (Suksdorf 973, Whatcom Co.,
Wash.) A large-flowered, wingless-fruited form.

P. suksdorfii Gand. Bull. Soc. Bot. France 65:36. 1918. (Suksdorf 110, Bingen, Wash.) A
large-flowered, wing-fruited form.

P. anomala var. lactiflora Suksd. Werdenda 1:43. 1927. (Suksdorf 11988, Bingen, Wash.)
A white-flowered form with reduced spur.

P. gibbosa Suksd. Werdenda 1:43. 1927. P. anomala var. gibbosa Dyal, Am. Midl. Nat.

42:488. 1949. (Suksdorf 9921, Falcon Valley, Klickitat Co., Wash.) A form with wingless fruit and reduced corolla spur.

Plants slender or rather stout, 1-6 dm. tall, the herbage glabrous or nearly so; leaves mostly 1-6 cm. long and 3-22 mm. wide, distant, the lowermost ones spatulate or obovate and short-petiolate, the others more oblong or elliptic and sessile; corolla 2-8 mm. long, pink or white, evidently bilabiate, and bearing a well-developed, thick spur; fruit 2-4 mm. long, the convex side more or less sharply keeled, not grooved, the wings tending to be connivent below and spreading above, or wanting; hairs of the fruit when present more or less pointed; cotyledons transverse to the ventral face of the fruit.

Open, vernally moist slopes and meadows; s. Vancouver I. to s. Calif., scarcely extending e. of the Cascade Range. Apr.-June.

The form from the Puget Sound area is especially valuable in the wild garden, maintaining itself readily by seed. In this and other species of Plectritis, as noted by Mrs. Nielsen, wing-fruited and wingless-fruited plants often occur in the same colony, in such a fashion as to suggest that they may represent Mendelian segregates.

Plectritis macrocera T. & G. Fl. N. Am. 2:50. 1841.

P. congesta var. minor Hook. Fl. Bor. Am. 1:291. 1833. Valerianella macrocera Gray, Proc. Am. Acad. 19:83. 1883. Aligera macrocera Suksd. Deuts. Bot. Monats. 15:147. 1897. A. minor Heller, Cat. N. Am. Pl. 7. 1898. Plectritis minor Abrams, Fl. Los Angeles 382. 1904. (Douglas, Calif.)

Aligera grayi Suksd. Deuts. Bot. Monats. 15:147. 1897. Plectritis macrocera var. grayi Dyal, Am. Midl. Nat. 42:491. 1949. (Presumably a Suksdorf collection from Klickitat Co.)

Aligera mamillata Suksd. Deuts. Bot. Monats. 15:147. 1897. Plectritis macrocera var. mamillata Dyal, Am. Midl. Nat. 42:491. 1949. (Suksdorf, Simcoe Mts., Wash., in 1884)

Aligera ostiolatata Suksd. Deuts. Bot. Monats. 15:147. 1897. (Curran, Antioch, Calif., is the first of two specimens cited)

Plants slender or rather stout, 1-6 dm. tall, often finely glandular in the inflorescence, the herbage otherwise glabrous or nearly so; leaves 1-4.5 cm. long and 3-18 mm. wide, distant, the lowermost ones more or less obovate and short-petiolate, the others more oblong or elliptic and sessile; inflorescence appearing terete in life; corolla 2-6 mm. long, mostly white or pinkish, about equally 5-lobed, with a rather short, thick spur; fruit 2-4 mm. long, the convex side grooved, scarcely keeled, the wings about equally divergent above and below, or wanting; hairs of the fruit when present clavate or long-cylindrical and blunt, not forming bands on the dorsal face; cotyledons parallel to the ventral face of the fruit.

Stream banks and vernally moist, open places; s. B.C., to s. Calif., e. to Mont. and Utah. Mar.-June.

## Valeriana L.

Flowers borne in corymbiform to paniculiform or thyrsoid inflorescences of basically determinate nature, perfect or unisexual; calyx initially involute and inconspicuous, later enlarged and spreading, usually (including all our species) with several or numerous long, setaceous, plumose, pappus-like segments; corolla gamopetalous, the tube sometimes gibbous at the base, the 5 lobes equal or subequal; stamens 3; ovary basically 3-carpellary, the 2 abaxial carpels vestigial; stigma 3-lobed; fruit a nerved achene; annual or perennial herbs with opposite, entire to bipinnatifid leaves.

Perhaps as many as 200 species, occurring on all the continents except Australia. Meyer (Ann. Mo. Bot. Gard. 38:377-503. 1951) has recently recognized 31 species in N. Am. In addition to the following species, V. officinalis L., the garden heliotrope, has been reported in our range as an occasional escape from cultivation. It is somewhat similar to V. scouleri or V. occidentalis, but differs in its narrower and more numerous leaf segments. (Name of Latin origin, possibly from valere, to be strong.)

1 Plants with a stout taproot and short, branched caudex; basal leaves tapering more or less gradually to the petiolar base; inflorescence more or less paniculiform even at anthesis
V. EDULIS

1 Plants with a stout rhizome or caudex and numerous fibrous roots; lower leaves with mostly
    sharply differentiated blade and petiole; inflorescence corymbiform at anthesis, though of-
    ten more expanded in fruit
  2 Corolla small, mostly 2-4 mm. long, the lobes not much, if at all, shorter than the tube;
      plants gyno-dioecious, some with chiefly perfect, others with chiefly pistillate flowers
    3 Relatively small plants, 1-4(6) dm. tall, not very leafy, the lateral lobes of the cau-
       line leaves mostly well under 1 cm. wide; achenes lanceolate, glabrous
                                                      **V. DIOICA**
    3 Relatively robust plants, 3-9 dm. tall, tending to be amply leafy, the lateral lobes of
       some of the cauline leaves often over 1 cm. wide; achenes a little broader, mostly
       lance-ovate, short-hairy or occasionally glabrous        **V. OCCIDENTALIS**
  2 Corolla larger, mostly 4-18 mm. long, the lobes not more than about half as long as the
      tube; flowers usually all perfect
    4 Stamens much surpassing the corolla lobes; corolla 4-9 mm. long; widespread
      5 Robust plants, mostly 3-12 dm. tall, with ample cauline leaves, the basal leaves
         when present varying from smaller to a little larger than the cauline ones, the
         leaf segments commonly but not always coarsely crenate or wavy; fruits mostly
         ovate to oblong-ovate                       **V. SITCHENSIS**
      5 Smaller, less leafy plants, commonly 1-7 dm. tall, the cauline leaves equaling or
         smaller than the well-developed and persistent basal ones, the leaf segments com-
         monly entire or nearly so, sometimes more or less toothed; fruits narrower, most-
         ly lance-oblong or lance-linear
        6 Basal leaves, or some of them, usually pinnatifid; chiefly w. of the Cascade sum-
           mits                                          **V. SCOULERI**
        6 Basal leaves mostly lobeless; more eastern species       **V. ACUTILOBA**
    4 Stamens equaling or mostly shorter than the corolla lobes; corolla 11-18 mm. long;
      Wenatchee Mts.                            **V. COLUMBIANA**

Valeriana acutiloba Rydb. Bull. Torrey Club 28:24. 1901.
  V. capitata ssp. acutiloba F. G. Mey. Ann. Mo. Bot. Gard. 38:407. 1951. (Rydberg & Vree-
  land 5576, Sangre de Cristo Range, Colo.)
  VALERIANA ACUTILOBA var. PUBICARPA (Rydb.) Cronq. hoc loc. V. pubicarpa Rydb.
  Bull. Torrey Club 36:697. 1909. V. capitata ssp. pubicarpa F. G. Mey. Ann. Mo. Bot.
  Gard. 38:406. 1951. (Rydberg & Carlton 7717, Mt. Nebo, Utah)
  Fibrous-rooted perennial from a stout branched rhizome or caudex, mostly 1-6 dm. tall,
the stem usually minutely spreading-hirtellous; leaves glabrous, the basal ones (mostly on
separate short shoots) well developed, petiolate, with well-marked, mostly undivided, obovate-spatulate to oblong or ovate blade that is up to 8 cm. long and 3.5 cm. wide; cauline
leaves 1-3 pairs, the lowest often petiolate and undivided, the others mostly pinnatifid, with
few and reduced lateral lobes which are rarely as much as 1 cm. wide; inflorescence mostly
1.5-5 cm. wide at anthesis, often not much enlarged in fruit; flowers ordinarily all perfect;
corolla white, 4-7 mm. long, commonly somewhat hairy outside, the lobes scarcely half as
long as the somewhat gibbous tube; stamens well exserted; plumose calyx segments mostly
10-17; fruits 3.5-5.5 mm. long, mostly lance-oblong or lance-linear, short-hairy or sometimes glabrous.
  Open, often rocky slopes at moderate to high elevations in the mountains, often near snow
banks; s.w. Mont. to s. Oreg., s. to Calif., Ariz., and N.M. June-July.
  Only the var. pubicarpa (Rydb.) Cronq., as described above, occurs in our range; it is typically a Great Basin plant, extending northward into c. Ida. and s.w. Mont. Typical V. acutiloba, differing in having narrower, more acute leaf segments and more consistently expanded
fruiting inflorescence, and in being more glabrous, occurs chiefly in the s. Rocky Mts. A
third phase occurs in Calif. and adj. Oreg. The morphological differences (as given by Meyer)
between V. acutiloba as here defined and the Alaskan V. capitata Pall. appear to be as significant and as stable as those found between other related taxa recognized as species in Valeriana, and the 1500-mile range gap indicates that no practical difficulty in identification is to
be expected.

Valeriana columbiana Piper, Bot. Gaz. 22:489. 1896. (Whited 140, Wenatchee, Wash.)

Fibrous-rooted perennial from a stout branched rhizome or caudex, mostly 1-5 dm. tall, the stem short-hairy below, the leaves glabrous or sometimes short-hairy along the veins; basal leaves petiolate, with well-defined, broad, lobeless, usually somewhat sinuate or toothed blade up to 5.5 cm. long and 3.5 cm. wide; cauline leaves 1-4 pairs, the lower evidently petiolate, the upper subsessile, all generally pinnatifid, with the terminal lobe up to 6 cm. long and 2.5 cm. wide, the 1-2 pairs of lateral lobes smaller or sometimes nearly as large; inflorescence compact and 3-6 cm. wide at anthesis, expanded and becoming diffuse in fruit; flowers all perfect; corolla white, 11-18 mm. long, the lobes about half as long as the somewhat gibbous tube; stamens equaling or usually a little surpassed by the corolla lobes; plumose calyx segments 11-16; fruits glabrous, lanceolate or narrower, 5-7 mm. long.

Open slopes in the Wenatchee Mts. of Wash., from 1500 to 6500 ft. May-July.

This, the most showy of our species, merits a place in any rock garden, but it is less easily grown than the taller, more moisture-loving, less beautiful V. sitchensis.

Valeriana dioica L. Sp. Pl. 31. 1753. (Europe)

V. sylvatica Richards. App. Frankl. Journ. 730. 1823, not of F. W. Schmidt in 1795. V. dioica var. sylvatica Wats. Bot. King Exp. 136. 1871. V. septentrionalis Rydb. Mem. N.Y. Bot Gard. 1:376. 1900. V. dioica ssp. sylvatica F. G. Mey. Ann. Mo. Bot. Gard. 38:417. 1951. (Richardson, "on the Clearwater River" in n. Alta.)

V. wyomingensis E. Nels. Erythea 7:167. 1899. V. micrantha var. wyomingensis Nels. in Coult. & Nels. New Man. Bot. Rocky Mts. 476. 1909. (Nelson & Nelson 5686, Undine Falls, Yellowstone Nat. Pk.)

V. psilodes Gand. Bull. Soc. Bot. France 65:37. 1918. (Rydberg & Bessey 5001, Bridger Mts., Gallatin Co., Mont.)

Fibrous-rooted perennial from a stout branched rhizome or caudex, mostly 1-4(6) dm. tall, glabrous or nearly so; basal leaves well developed, petiolate, the well-defined blade rarely as much as 8 cm. long and 3 cm. wide, mostly undivided; cauline leaves 2-4 pairs, mostly short-petiolate or subsessile, pinnatifid, the terminal lobe ovate-oblong or narrower, up to 5 cm. long and 2.5 cm. wide, the 1-7 pairs of lateral lobes smaller, rarely as much as 1 cm. wide; inflorescence compact and 1.5-3 cm. wide at anthesis, elongating and becoming diffuse in fruit; plants gynodioecious, some with perfect, others with pistillate flowers; corolla white, 2-4 mm. long (or the pistillate shorter), the lobes not much, if at all, shorter than the tube; stamens well exserted; plumose calyx segments mostly 9-15; fruit lanceolate, 3-5 mm. long, less than 2 mm. wide, glabrous. N=8 has been reported for the European phase of the species.

Moist places in the mountains, often in wet meadows; transcontinental in Can. s. of 60 degrees lat., extending s. to n.w. Wash., c. Ida., and n.w. Wyo.; also in Europe. May-July.

The American plant has been distinguished on minor, perhaps insignificant characters as var. sylvatica (Rich.) Wats. or ssp. sylvatica (Rich.) F. G. Mey.

Valeriana edulis Nutt. ex T. & G. Fl. N. Am. 2:48. 1841. (Nuttall, Walla Walla, Wash.)

Patrinia ceratophylla Hook. Fl. Bor. Am. 1:290. 1833. Valeriana ceratophylla Piper, Contr. U.S. Nat. Herb. 11:532. 1906, not of H.B.K. in 1818. (Douglas, "between the Kettle Falls and Spokan")

V. trachycarpa Rydb. Bull. Torrey Club 31:645. 1904. (Underwood & Selby 352, Red Mt., Colo.)

V. edulis f. glabra St. John, Fl. S.E. Wash. 397. 1937. (Elmer 822, Pullman, Wash.)

Perennial from a long stout taproot and short branched caudex, 1-12 dm. tall, generally glabrous or nearly so; basal leaves numerous, firm, linear to obovate, the blade tapering rather gradually to the petiole or petiolar base, the whole leaf entire and mostly 7-40 cm. long by 7-55 mm. wide, or often some of them with a few narrow, forward-pointing, lateral lobes; cauline leaves 2-6 pairs, smaller, nearly always pinnatifid with narrow segments, and becoming sessile or subsessile upward; inflorescence paniculiform, often rather compact at anthesis, expanded and more open in fruit; plants polygamodioecious; corollas of perfect and staminate flowers mostly 2.5-3.5 mm. long, those of pistillate flowers scarcely 1 mm. long; corolla lobes not much shorter than the tube; stamens not much exserted; plumose calyx segments

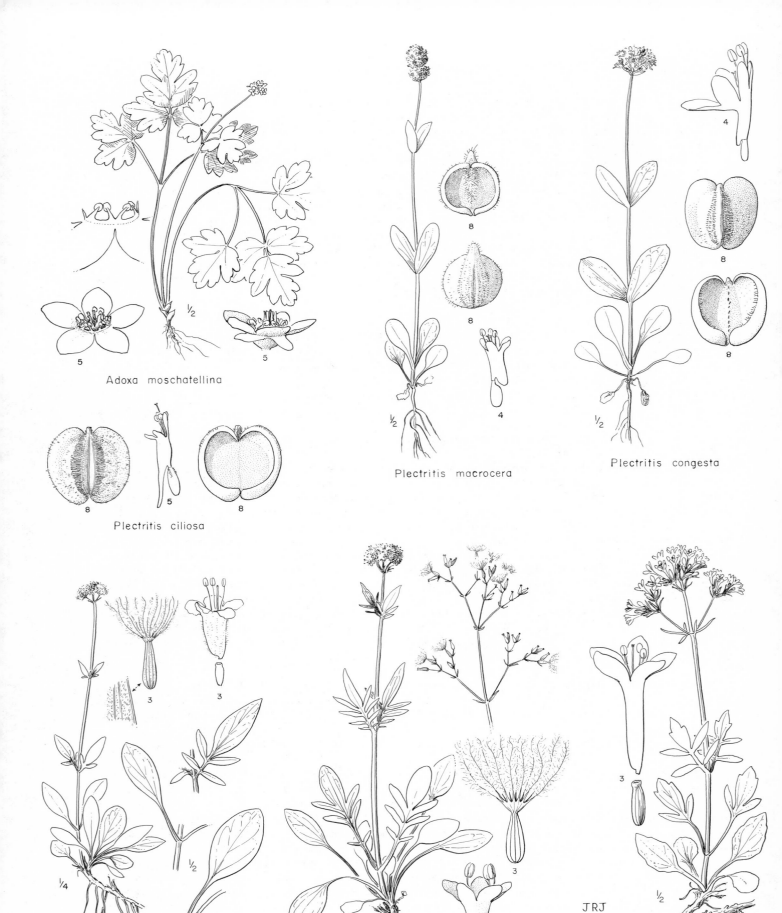

Adoxa moschatellina

Plectritis ciliosa

Plectritis macrocera

Plectritis congesta

Valeriana acutiloba var. pubicarpa

Valeriana dioica

Valeriana columbiana

JRJ

mostly 9-13; fruits more or less ovate or ovate-oblong, 2.5-4.5 mm. long, short-hairy or glabrous.

Moist, sometimes saline meadows, and in a wide variety of other mostly open and not too dry habitats, from the foothills to rather high elevations in the mountains; s. B.C., e. Wash. (e. of the Cascades), and Mont., s. to Mex., and also from Minn. and Iowa to Ont. and Ohio. June-Aug.

Only the nomenclaturally typical phase, as described above, occurs in our range. Forms tending to differ in minor features are found in Mex. and n.c. U.S., but are separable only with difficulty.

Valeriana occidentalis Heller, Bull. Torrey Club 25:269. 1898. (Heller 3353, incorrectly cited by Heller as 2353, Lake Waha, Nez Perce Co., Ida.)
   V. micrantha E. Nels. Erythea 7:166. 1899. (Nelson 793, Wind R., Fremont Co., Wyo.)
   Similar to V. dioica, but more robust, commonly 3-9 dm. tall, and tending to be more amply leafy, the lateral lobes of some of the cauline leaves often over 1 cm. wide, the terminal lobe up to 9 cm. long and 4 cm. wide; often some of the basal leaves with some lateral lobes; fruits slightly broader, mostly lance-ovate, short-hairy or occasionally glabrous.

Moist, open or shaded places from the foothills to rather high elevations in the mountains; n. Ida. and adj. Mont., s. through e. Oreg. to n.w. Calif., and to w. S.D., Colo., and n. Ariz. May-July.

In c. Ida., w. Mont., and n. Wyo., the range of V. occidentalis overlaps that of the closely allied V. dioica, and in that region the distinctions between the two often break down, so that it might eventually be considered desirable to reduce V. occidentalis to varietal status under V. dioica.

Valeriana scouleri Rydb. Mem. N.Y. Bot. Gard. 1:377. 1900.
   V. sitchensis ssp. scouleri Piper, Contr. U.S. Nat. Herb. 11:533. 1906. V. sitchensis var. scouleri G. N. Jones, U. Wash. Pub. Biol. 5:237. 1936. (Scouler, Columbia R. near Oak Point)
   Fibrous-rooted perennial from a stout branched rhizome or caudex, mostly 1.5-7 dm. tall, glabrous or nearly so; leaves or their segments all entire or slightly (rarely evidently) crenate; basal leaves well developed and persistent, petiolate, mostly pinnatifid with enlarged terminal segment up to 8 cm. long and 4 cm. wide, and one or two (4) pairs of much smaller lateral segments up to 3 cm. long and 1.5 cm. wide, or some of them often lobeless, the blade than resembling the terminal segment of the lobed leaves; cauline leaves commonly 2-4 pairs, shorter-petiolate or the upper subsessile, pinnatifid like the basal ones, but often smaller; inflorescence compact and about 2-7 cm. wide at anthesis, expanded and becoming diffuse in fruit; flowers ordinarily all perfect; corolla white or pinkish, 5-9 mm. long, glabrous outside, the lobes mostly less than half as long as the somewhat gibbous tube; stamens well exserted; plumose calyx segments 12-18; fruit glabrous, lance-oblong or lance-linear, 5-6 mm. long, 1-2 mm. wide.

Moist woods, wet cliffs, and wet meadows, from sea level to 4000 ft. in the mountains; s. Vancouver I. and the adj. mainland of B.C., s. to n.w. Calif., mostly w. of the Cascade summits. Apr.-July.

This species appears to be reasonably distinct from V. sitchensis, with which it has often been confused. From V. acutiloba, however, it is separable only with some difficulty.

Valeriana sitchensis Bong. Mém. Acad. St. Pétersb. VI. 2:145. 1833. (Mertens s.n., Sitka)
   V. hookeri Shuttlew. Flora 20:450. 1837. V. capitata var. hookeri T. & G. Fl. N. Am. 2:48. 1841. V. sitchensis var. hookeri G. N. Jones, U. Wash. Pub. Biol. 7:153, 175. 1939. (Drummond s.n., woods in the Rocky Mts. about lat. 56 degrees)
   V. frigidorum Gand. Bull. Soc. Bot. France 65:37. 1918. (Cusick 1715, e. Oreg.)
   V. suksdorfii Gand. Bull. Soc. Bot. France 65:36. 1918. (Suksdorf 6060, Mt. Paddo, Wash.)
   Fibrous-rooted perennial from a stout, branched rhizome or caudex, mostly 3-12 dm. tall, glabrous or occasionally short-hairy; leaves chiefly cauline, mostly 2-5 pairs, with the lowermost one or two pairs more or less reduced, those next above well developed, petiolate, pin-

natifid, with enlarged terminal segment up to 10 cm. long and 7 cm. wide, the 1-4 pairs of lateral segments less reduced than in related species, up to 7.5 cm. long and 4.5 cm. wide, rarely less than 1 cm. wide; upper leaves more or less reduced and becoming sessile or subsessile; basal leaves when present similar to and sometimes larger than the larger cauline ones, or smaller, or sometimes undivided; leaves or their segments all coarsely crenate to occasionally entire; inflorescence compact and 2.5-8 cm. wide at anthesis, expanded and becoming diffuse in fruit; flowers ordinarily all perfect; corolla white, 4.5-7 mm. long, glabrous or short-hairy outside, the lobes mostly less than half as long as the somewhat gibbous tube; stamens well exserted; plumose calyx segments mostly 12-20; fruits glabrous, mostly ovate or oblong-ovate, 3-6 mm. long, about 2-2.5 mm. wide.

Moist, open or wooded places at middle and upper altitudes in the mountains, often in wet meadows; s. Yukon and s. Alas. (n. of the panhandle) to w. Mont., c. Ida., and n. Calif., apparently absent from the coastal ranges of Oregon and the lowlands of the Puget trough in Oreg. and Wash., the range thus contiguous with, but only slightly overlapping, that of V. scouleri. June-Aug.

## Valerianella Mill.

Inflorescence dichotomously branched, the flowers borne in cymose glomerules at the ends of the branches; calyx lobes 3-6 or (in our species) obsolete; corolla gamopetalous, the tube often gibbous or minutely spurred, the 5 lobes about equal; stamens 3; ovary 3-carpellary; stigma 3-lobed; fruit dry, 3-locular, the 2 abaxial locules sterile and ordinarily with an evident groove between them; annual or biennial herbs with opposite, entire or toothed, commonly sessile leaves.

About 60 species, native of the Northern Hemisphere, most numerous in s. Europe. Both of our species are introduced. (Name a diminutive of Valeriana.)

Reference:

Dyal, Sarah C. Valerianella in North America. Rhodora 40:185-212. 1938.

1 Fertile cell of the fruit with an enlarged corky mass attached to the back; groove between
    the sterile cells narrow, shallow, and relatively inconspicuous     V. LOCUSTA
1 Fertile cell of the fruit without a corky mass; groove between the sterile cells relatively
    wide, deep and conspicuous                                          V. CARINATA

Valerianella carinata Loisel. Not. Pl. Fl. Fr. 149. 1810. (Fields about Paris)

Similar to V. locusta, except for the fruit, which averages slightly smaller, lacks the corky mass on the fertile cell, and has a wide deep groove, with thin lateral walls and an evident midvein, between the two sterile cells. N=9.

Moist, open, often disturbed places; native of Europe, reported by Dr. Dyal from the Willamette Valley of Oreg., and more recently (in Davis' Fl. Idaho) from Calif. and Ida. as well. Apr.-May.

Valerianella locusta (L.) Betcke, Anim. Val. 10. 1826.

   Valeriana locusta L. Sp. Pl. 33. 1753. Valeriana locusta var. olitoria L. Sp. Pl. 33. 1753.
   (The alpha variety of Linnaeus). Valerianella olitoria Poll. Hist. Pl. Palat. 1:30. 1776.
   Fedia olitoria Vahl, Enum. 1:19. 1804. (Europe)

Annual, 1-4 dm. tall, the weak stem sparsely retrorse-hispidulous, simple below the inflorescence, or dichotomously few-branched; leaves coarsely ciliolate and sometimes short-hairy on the upper surface as well, the lower broadly oblanceolate or broader, and more or less petiolate, the others sessile and more oblong, entire or the upper irregularly few-toothed near the base, 1-7 cm. long, 3-18 mm. wide; glomerules of flowers several, commonly 5-13 mm. wide; corolla 1.5-2 mm. long, the lobes white or pale bluish, about equaling the funnelform tube, which bears a small, saccate, ventral gibbosity; fruit about 2 mm. or rarely 4 mm. long, glabrous or minutely hairy, the fertile cell bearing a thick, corky mass on the back; groove between the sterile cells narrow, shallow, and inconspicuous. N=7.

Moist, open places, often in disturbed soil; native of Europe, now widely established in the

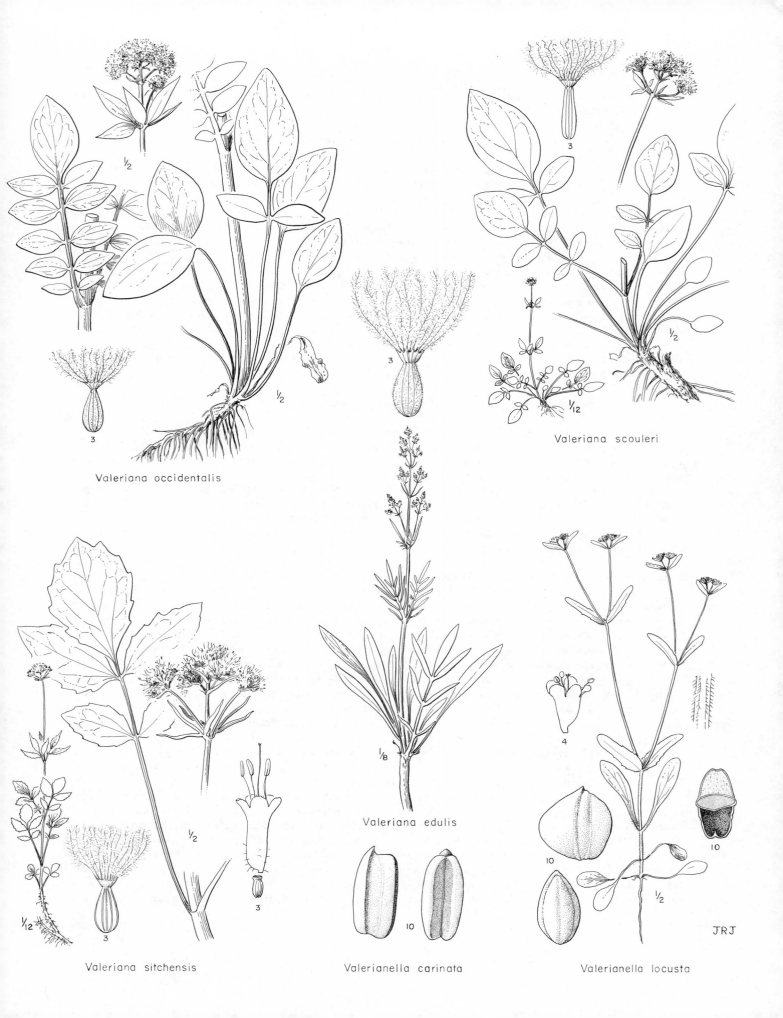

Valeriana occidentalis

Valeriana scouleri

Valeriana edulis

Valeriana sitchensis

Valerianella carinata

Valerianella locusta

JRJ

U.S. and occasionally found in our range; established near Lewiston, Ida. at least as early as 1892. Apr. -May.

## DIPSACACEAE. Teasel Family

Flowers more or less irregular, perfect, gamopetalous, epigynous; calyx small, cupulate or more or less deeply cut into 4 or 5 segments or into more numerous teeth or hairs; corolla 4- to 5-lobed, often more or less 2-lipped; stamens 4 or 2, epipetalous, alternate with the corolla lobes, exserted, not connate; style with a simple or 2-lobed stigma; ovary 1-celled, with a solitary pendulous ovule; fruit dry, indehiscent, enclosed (except at the tip) by a gamophyllous, apically cupulate-toothed or subentire involucel which may be adnate to the ovary below; seeds with thin or copious endosperm; herbs with opposite or whorled, exstipulate leaves, the flowers ordinarily borne in dense, involucrate heads.

About 10 genera and 150-200 species, all native to the Old World, most numerous in the Mediterranean region. The heads of the inflorescence are probably secondary, each representing an aggregation of uniflorous heads, after the manner of <u>Echinops</u> and some other genera of the Compositae.

1  Receptacle chaffy; calyx cupulate, 4-angled or 4-lobed; stem prickly               DIPSACUS
1  Receptacle merely hairy; calyx of at least 8 elongate teeth or bristles, united below; stem
    glabrous or hairy, not prickly                                                                KNAUTIA

### Dipsacus L. Teasel

Heads dense, ovoid or cylindric; involucral bracts linear, spine-tipped, usually elongate; receptacular bracts ovate or lanceolate, acuminate into an awn that surpasses the flowers; involucel 4-angled, truncate or 4-toothed at the summit; calyx short, cupulate, 4-angled or 4-lobed, often persistent at the summit of the achene; corollas all about alike, with a long tube and 4 more or less unequal lobes; stamens 4; stigma oblique, entire; coarse biennial or perennial herbs with prickly stems and large, opposite, sessile, often connate leaves.

About a dozen species of Europe, w. Asia, and n. Africa. (The Greek and Latin name, presumably from the Greek dipsa, thirst, referring to the accumulation of water in the cuplike base of the connate leaves.)

Dipsacus sylvestris Huds. Fl. Angl. 49. 1762.
  <u>D</u>. <u>fullonum</u> ssp. <u>sylvestris</u> Claph. in Claph. et al. Fl. Brit. Isl. 1013. 1952. (Europe)
  Stout, taprooted biennial, 0.5-2 m. tall, the stem striate-angled, increasingly prickly upward; leaves more or less prickly, especially on the midrib beneath, otherwise glabrous or nearly so, the basal ones oblanceolate, crenate, generally dying early in the second season, the cauline ones lanceolate, up to 3 dm. long, becoming entire upward, commonly connate into a water-collecting cup at the base; heads erect, ovoid or subcylindric, 3-10 cm. long, terminating the long, naked peduncles; involucral bracts more or less prickly, curved upward, unequal, the longer ones surpassing the heads; receptacular bracts ending in a conspicuous, stout, straight awn; calyx silky, 1 mm. long; corolla slender, pubescent, 10-15 mm. long, the tube whitish, the short (1 mm.) lobes generally pale purple; fruit about 5 mm. long. N reportedly=8, 9.

Generally in moist, low places (or in better drained soil in more humid regions, as w. of the Cascades), especially in disturbed sites; native of Europe, now widespread as a weed in N. Am., and occurring throughout our range. July-Sept.

The closely related cultivated species, <u>D</u>. <u>fullonum</u> L., fuller's teasel, differing principally in the stouter, strongly recurved spine tips of the receptacular bracts, is introduced in Calif. Linnaeus considered the two to be conspecific, under the name <u>D</u>. <u>fullonum</u>, his unnamed var. β being the true fuller's teasel (the heads of which are still used in England for raising the nap on some kinds of cloth). It might be argued that <u>D</u>. <u>fullonum</u> L. should be typified by the wild plant (i.e., the form later segregated as <u>D</u>. <u>sylvestris</u> Huds.), with the cultivated plant taking the name <u>D</u>. <u>sativus</u> Honck. (1792). The weight of historical practice, however, has been to accept the more logical, if perhaps less legally proper typification of Hudson, who in 1762 considered the two phases to represent different species and restricted the name D. fullonum L. to the cultivated plant with recurved receptacular bracts.

## Knautia L.

Heads dense, hemispheric; involucral bracts lanceolate, herbaceous; receptacle densely hairy, without bracts; involucel compressed but more or less 4-angled, the shortly cupulate apex commonly obscurely 2-toothed; calyx of at least 8 elongate, usually suberect teeth or bristles that are shortly connate below; corollas more or less unequally 4- to 5-lobed, those of the marginal flowers often the largest; stamens 4; stigma emarginate; annual or perennial herbs with opposite leaves.

About 40 species of Europe, w. Asia, and n. Africa. (Named for Christian Knaut, 1654-1716, German physician and botanist.)

Knautia arvensis (L.) Coult. Mem. Dipsac. 29. 1823.
  Scabiosa arvensis L. Sp. Pl. 99. 1753. (Europe)
  Perennial from a taproot and branching, sometimes elongate caudex, 3-10 dm. tall, more or less hirsute; lowermost leaves usually merely coarsely toothed, the others more or less deeply pinnatifid with narrow lateral and broader terminal segments, up to 2.5 dm. long, reduced upward; heads terminating the long, naked peduncles, 1.5-4 cm. wide; involucral bracts 8-12 mm. long; calyx mostly 3-4 mm. long, the 8-12 teeth bristle-like; corollas lilac-purple, 4-lobed, the marginal ones larger than the central; fruit 5-6 mm. long, densely hairy. N=10, 20.
  A European weed, naturalized in e. U.S., and becoming established in Mont. and s. B.C. June-Aug.

## CUCURBITACEAE. Cucumber or Gourd Family

Plants monoecious or dioecious, the flowers regular, gamopetalous or seemingly polypetalous, the pistillate ones epigynous; limb of the calyx and corolla commonly more or less combined; stamens 1-5, typically 3, with two 2-celled and one 1-celled anther, free or variously monadelphous; style solitary, with a thickened, entire or lobed stigma; placentae thickened, parietal, or commonly confluent and partitioning the 1- to 4-locular ovary; fruit fleshy or sometimes dry, usually relatively large; seeds large, commonly more or less flattened, without endosperm; annual or perennial vines, trailing or climbing by tendrils, with small to large, mostly white or yellow to greenish flowers, and simple, alternate, often lobed leaves.

A large family of about 90 genera and probably 900 species, chiefly tropical and subtropical. Watermelons, cantaloupes, cucumbers, and gourds, all of Old World origin, and squashes and pumpkins, of New World origin, belong to this family. Our two genera, although sharply limited and distinguished by abundant characters, are closely allied, and have frequently been combined, along with the more southern Echinopepon, into a single genus Echinocystis.
1 Annual; seeds flattened, roughened; germination epigaeous; flowers ordinarily 6-merous
ECHINOCYSTIS
1 Perennial from a much enlarged, woody root; seeds turgid, smooth; germination hypogaeous;
  flowers 5- to 8-merous																MARAH

## Echinocystis T. & G. Nom. Conserv. Balsam Apple; Wild Cucumber

Plants monoecious, the staminate flowers in axillary, pedunculate, narrow panicles, the pistillate ones from the same axils, solitary or sometimes in pairs; flowers ordinarily 6-merous, the calyx lobes small and bristlelike, inserted at the sinuses of the corolloid perianth; stamens 3, the filaments and anthers more or less connate; ovary 2-celled; ovules 2 in each cell, attached to the wall or to the partitions; style short, with a broad, lobed stigma; fruit weakly spiny, more or less bladdery-inflated, at length dry, bursting irregularly at the apex, fibrous-netted within; seeds flattened; germination epigaeous; annual climbing vines with branched tendrils and alternate, lobed leaves.

A single species. (Name from the Greek echinos, hedgehog, and kystis, bladder, referring to the fruits.)

Echinocystis lobata (Michx.) T. & G. Fl. N. Am. 1:542. 1840.
  Sicyos lobata Michx. Fl. Bor. Am. 2:217. 1803. Micrampelis lobata Greene, Pitt. 2:128.
    1890. (W. Pa., along the Ohio R.)
  High-climbing annual; leaves petiolate, 5-15 cm. long, scaberulous, cordate at the base, palmately 5-lobed, the mostly triangular-acute lobes commonly remotely callous-toothed; perianth greenish-white, the tube short, the corolla lobes spreading, narrowly lanceolate, commonly 3-6 mm. long and about 1 mm. wide; fruit ellipsoid or subglobose, 3.5-5 cm. long; seeds flattened but rather thick, more or less ovate, 1.5 cm. long.
  Moist bottom lands and thickets; N.B. to Fla., w. to Sask., Mont., s. Ida., Ariz., and Tex. July-Sept.


### Marah Kell. Bigroot; Manroot

  Plants monoecious, the staminate flowers in axillary, pedunculate racemes or narrow panicles, the pistillate ones from the same axils, mostly solitary; flowers 5-merous or sometimes 6- to 8-merous, the calyx lobes inconspicuous, inserted at the sinuses of the corolloid perianth; stamens 3, the filaments and anthers more or less connate; ovary 2- to 4-celled; ovules 1-several in each cell, attached to the wall or to the partitions; style short, with a broad, lobed stigma; fruit spiny or sometimes smooth, more or less bladdery-inflated, at length dry, bursting irregularly at the apex, fibrous-netted within; seeds turgid, the cotyledons thickened; germination hypogaeous; perennial from a much enlarged, woody root, the annual stems climbing or trailing, provided with branched tendrils; leaves alternate, mostly lobed.
  About 6 species, native to w. U.S. and adj. Can. and Mex., entirely on the Pacific slope. (Name from the Hebrew marah, bitter, in allusion to the intensely bitter root.)
  Reference:
  Stocking, K. M. Some taxonomic and ecological considerations of the genus Marah (Cucurbitaceae). Madroño 13:113-37. 1955.

Marah oreganus (T. & G.) Howell, Fl. N.W. Am. 239. 1898.
  Sicyos oregana T. & G. Fl. N. Am. 1:542. 1840. Megarrhiza oregana Torr. ex Wats. Proc.
    Am. Acad. 11:138. 1876. Echinocystis oregana Cogn. Mem. Cour. Acad. Belg. 28:87.
    1878. Micrampelis oregana Greene, Pitt. 2:129. 1890. ("On the Oregon from near its
    mouth to Kettle Falls"; specimens collected by Scouler, Douglas, and Tolmie are cited)
  High-climbing perennial; leaves petiolate, scabrous-hispid above, sparsely hairy or subglabrous beneath, seldom over 2 dm. long, cordate at the base, rather irregularly (and sometimes very shallowly) palmately lobed; staminate flowers in racemes, the perianth campanulate, 5- to 8-merous, whitish, the tube 3-6 mm. long, the corolla lobes deltoid to ovate or ovate-oblong, commonly 3-8 mm. long and 2-5 mm. wide; fruit ellipsoid to ovoid-oblong, commonly 3-8 cm. long, 2- to 4-locular, weakly spiny or almost smooth; seeds 1 or 2 in each locule, turgid, broad, rounded at both ends, 2 cm. long, smooth. N=16.
  Bottom lands, fields, thickets, and open hillsides; s. B.C. to n. Calif., mostly w. of the Cascades, extending e., rarely, as far as the Snake R. along the Oregon-Idaho boundary. Apr.-June.


### CAMPANULACEAE. Harebell Family

  Flowers regular or irregular, perfect, gamopetalous, wholly epigynous, or with the apex of the ovary sometimes free, mostly 5-merous as to the calyx, corolla, and androecium; stamens generally (including all ours) free from the corolla or nearly so, alternate with the lobes; style 1; stigma lobes usually as many as the carpels; ovary 1- to 5-celled, with axile or parietal placentation and few to numerous ovules, 2- to 5-carpellary; fruit (in all ours) a capsule, opening by valves or pores (or irregularly), or sometimes a berry; annual, biennial, or perennial herbs (all ours) or sometimes shrubs or trees, with mostly alternate (or all basal), simple, exstipulate leaves, the flowers borne in diverse types of determinate or indeterminate inflorescences.

About 60 genera and 1300 species, of cosmopolitan distribution. The family is divided into two well-marked subfamilies (often considered as distinct, closely related families) of almost equal size, the Campanuloideae (or Campanulaceae proper), of chiefly n. temperate and boreal distribution, with regular corolla, distinct stamens, and mostly determinate or mixed inflorescence; and the Lobelioideae (or Lobeliaceae), of chiefly tropical and subtropical distribution, with irregular corolla, filaments and anthers connate into a tube, and with indeterminate inflorescence.

Reference:

McVaugh, Rogers. Campanulaceae (Lobelioideae), in North American Flora, vol. 32A, 1943.

1 Filaments and anthers distinct; corolla regular (or wanting in some flowers)
  2 Perennials; capsules opening laterally (outside the sepals); cleistogamous flowers wanting, the flowers all with well-developed corollas; filaments expanded and ciliate at the base; our species all with evidently pedicellate or pedunculate flowers     CAMPANULA
  2 Annuals; capsules, flowers, and filaments diverse; flowers sessile or subsessile
    3 Capsule opening only at the apex, within the calyx; flowers all with well-developed corollas, technically solitary and terminal, but appearing irregularly scattered because of the sympodial branching of the stem; filaments only slightly (if at all) expanded at the base, not ciliate     GITHOPSIS
    3 Capsule opening laterally, sometimes near the summit, but outside the calyx; lower flowers cleistogamous, with reduced or no corolla; filaments dilated and ciliate at the base; inflorescence more or less spiciform or interrupted-spiciform
      4 Corollas shallowly lobed; inflorescence a sympodial false spike, the bracts opposite the flowers; capsule opening near the base     HETEROCODON
      4 Corollas deeply lobed (to below the middle); inflorescence a monopodial false spike, the flowers 1-several in the axil of each bract; capsule opening near the summit, or near the middle     TRIODANIS
1 Filaments and anthers united into a tube, two of the anthers shorter than the others, the orifice of the tube thus oblique or appearing lateral; corolla distinctly irregular (or wanting in some flowers)
  5 Flowers pedicellate (sometimes subsessile in Howellia); hypanthium and fruit fusiform to ellipsoid or globose; capsule subapically or irregularly dehiscent; duration and habit various
    6 Corolla well developed, in ours mostly 7-20 mm. long; ovary and fruit 2-locular, with axile placentation, the fruit dehiscent by subapical valves; seeds numerous, minute; habit and habitat various, but not at once aquatic and with linear-filiform leaves
      7 Corolla tube cleft at least halfway to the base on the dorsal side (between the two lobes of the upper lip), the sinus thus formed much deeper than the next adjacent lateral sinuses; our species perennial     LOBELIA
      7 Corolla tube not deeply cleft dorsally, the depth of the dorsal sinus about equaling that of the lateral ones; annual     PORTERELLA
    6 Corolla minute, scarcely 3 mm. long, or wanting; ovary and fruit 1-locular, with parietal placentation, the fruit irregularly dehiscent by rupture of the thin lateral walls; seeds few and relatively large; aquatic annuals, immersed or with floating branches, the leaves linear or linear-filiform     HOWELLIA
  5 Flowers sessile in the axils of leaves or foliaceous bracts, but appearing long-stalked because of the elongate, narrow hypanthium; capsule elastically dehiscent by long slits on the sides; annuals, commonly of moist places, but not truly aquatic     DOWNINGIA

### Campanula L. Harebell; Bellflower; Bluebells-of-Scotland

Flowers typically pedicellate in a racemiform inflorescence, in which the terminal flower blooms first and the subsequent progression is centripetal, or the terminal flower abortive and the inflorescence thus apparently indeterminate, or not infrequently the flowers solitary; corolla regular, more or less campanulate, the 5 lobes short or elongate; stamens free from the corolla and from each other, the filaments short, expanded at the usually ciliate base; ovary 3-locular (in all our species) or sometimes 5-locular, with axile placentation; stigma lobes as many as the locules; capsules short or elongate, opening by lateral pores which vary in position from near the base to near the apex; seeds numerous, compressed; perennial (in all our species) or rarely annual or biennial herbs, with alternate leaves.

About 300 species, chiefly of the n. temp. and arctic regions. The cultivated Canterbury bell is C. medium L. (Name diminutive of Latin campana, bell, referring to the shape of the flowers.)

Several of our species, especially C. piperi, are highly prized garden plants, especially for rock gardens.

In addition to the following species, C. rapunculoides L., a European plant now well established as a roadside weed in n. e. U.S., has been collected at Portland, Oreg. It is erect and fairly robust, with well-developed, lanceolate to subcordate, crenate, mostly short-petiolate cauline leaves and rather numerous corollas 1.5-3.5 cm. long in an elongate inflorescence.

1 Style well exserted from the corolla; corolla lobes more or less spreading or recurved, longer than the tube; capsules subglobose, 3-6 mm. long; leafy-stemmed plants with toothed leaves and usually several or many flowers, occurring at low elevations mostly w. of the Cascade summits

  2 Leaves all sessile or subsessile, mostly 12-35 below the inflorescence; plants mostly 3-8 dm. tall; corolla lobes narrowly lanceolate; Polk Co., Oreg., and southward

                                                              C. PRENANTHOIDES

  2 Leaves all petiolate, except the reduced upper ones, mostly 4-10 below the inflorescence; plants mostly 1-4 dm. tall; corolla lobes broader, mostly ovate-oblong; Alas. to Calif.

                                                              C. SCOULERI

1 Style equaling or shorter than the corolla; corolla lobes slightly or scarcely spreading, longer or shorter than the tube; capsules diverse, often narrower and much longer; habit and habitat diverse, but not combined as above

  3 Capsules erect, opening by pores near the summit; plants mostly dwarf, up to 1.5 dm. (or in C. parryi rarely to 3.5 dm.) tall, with few or more often solitary flowers, occurring in alpine and subalpine habitats

    4 Anthers 1-2.5 mm. long; capsule 12-20 mm. long; corolla 6-12 mm. long, the lobes equaling or longer than the tube; leaves entire or nearly so; Rocky Mts.

                                                      C. UNIFLORA

    4 Anthers 3-5.5 mm. long; capsule 3-12 mm. long; corolla and leaves diverse

      5 Hypanthium moderately to densely woolly-villous with long, loosely spreading hairs; corolla mostly 18-30 mm. long, the lobes much shorter than the tube; leaves slightly to strongly toothed; Cascades of n. Wash., Selkirk Range of B.C., and northward

                                                C. LASIOCARPA

      5 Hypanthium glabrous to merely scabrous or minutely hirtellous; corolla mostly 6-16 mm. long, the lobes longer or shorter than the tube

        6 Leaves sharply toothed; capsule subglobose, 3-5 mm. long; Olympic Mts., Wash.

                                        C. PIPERI

        6 Leaves entire or nearly so; capsule narrower; not of the Olympic Mts.

          7 Plants minutely spreading-hirtellous throughout; capsule 5-7 mm. long; Cascade Range of Wash., s. to Calif., and rarely in c. Ida. and w. Mont.

                                    C. SCABRELLA

          7 Plants glabrous or nearly so except for the ciliate-margined bases of the lower leaves; capsule 7-11 mm. long; chiefly of n.c. Ida. and adj. Mont., also in the Wenatchee Mts. of Wash.             C. PARRYI

  3 Capsules nodding, opening by pores near the base; plants when well developed taller (up to 8 dm.) and with several or many flowers, subalpine dwarf specimens habitally similar to C. parryi, but with smooth to merely scabrous leaf margins; widespread plants, commonly occurring at moderate and low elevations      C. ROTUNDIFOLIA

Campanula lasiocarpa Cham. Linnaea 4:39. 1829. (Unalaschka)

Perennial, creeping below ground, with 1 or several lax stems up to 1.5 dm. tall; plants glabrous or nearly so except for the moderately to densely woolly-villous hypanthium and the ciliate petioles or proximal margins of the leaves; leaves evidently to sometimes obscurely serrate with callous teeth, the basal ones petiolate, with oblanceolate to elliptic or subrhombic blade 6-25 mm. long, the cauline ones few and more or less reduced; flowers solitary, or rarely 2; calyx lobes foliaceous but narrow, 5-18 mm. long, commonly some or all of them with a few slender teeth; corolla blue, 18-30 mm. long, much surpassing the style, the broad lobes much shorter than the tube; anthers 3-5 mm. long; capsule subcylindric, not much enlarged upward, about 1 cm. long, opening near the summit.

Alpine situations; Cascade Range of Snohomish Co., Wash., and Selkirk Range of B.C., n. to Mack. and Alas., thence through the Aleutian Islands to Kamchatka and s. to Hokkaido, Japan. July-Aug.

Campanula parryi Gray, Syn. Fl. 2nd ed. 2$^1$:395. 1886. (Parry, Mts. of Colo.)
   C. parryi var. idahoensis McVaugh, Bull. Torrey Club 69:240. 1942. (Kirkwood & Severy 1638, Swamp Lake, Selway Nat. Forest, Ida.)
   C. rentonae Senior, Rhodora 51:302. 1949. (Renton s.n., Mt. Stuart, Chelan Co., Wash.)
   Perennial from branching, slender rhizomes, up to 2.5 or rarely 3.5 dm. tall, glabrous or nearly so except for the prominently ciliate, proximal portions of the margins of the lower leaves; leaves entire or nearly so, the basal ones mostly elliptic to oblanceolate and subpetiolate, 7-30 mm. long and 3-12 mm. wide, the cauline ones well developed but narrower, mostly 2-5 cm. long and 2-5 mm. wide; flowers typically solitary, sometimes 1-3 additional lateral ones present; calyx lobes mostly 2-6(9) mm. long; corolla blue, 9-15 mm. long, the broad lobes equaling or merely half as long as the tube; style shorter than the corolla; anthers 4-5.5 mm. long; capsule erect, broadest above the middle, 7-11 mm. long and less than half as wide, opening near the apex.

Moist subalpine meadows and other open places in the mountains of c. and n.c. Ida. and adj. Mont., and also in the Wenatchee Mts. of Wash. July-Aug.

Only the var. idahoensis McVaugh, as described above, occurs in our range. The var. parryi, a plant of the s. Rocky Mts., has larger flowers with notably longer, commonly toothed calyx lobes and remotely toothed leaves. There is some intergradation between C. parryi var. idahoensis and C. scabrella, and further study may warrant a taxonomic realignment.

Campanula piperi Howell, Fl. N. W. Am. 409. 1901. (Piper 2217, Mt. Steele, Olympic Mts., Wash.)
   C. piperi f. sovereigniana English, Little Gardens 12$^3$:13. 1940. (E. H. English 1085, Hurricane Ridge, Clallam Co., Wash.) A white-flowered form.
   Perennial, creeping below ground, with 1 or several lax stems up to 1 dm. tall; plants glabrous, or finely scabrous-hirtellous especially above; leaves sharply serrate-dentate with slender, firm teeth, narrow-based, mostly broadest above the middle, 1-3 cm. long and a third as wide, the cauline ones sometimes as large as the basal; flowers mostly 1-3; calyx lobes foliaceous but narrow, 5-10 mm. long, commonly some or all of them with a few slender teeth; corolla blue, 12-16 mm. long, the broad lobes twice as long as the tube; anthers 3.5-5 mm. long; style not much shorter than the corolla; capsule subglobose, 3-5 mm. long and nearly or quite as wide, opening at or above the middle. N=17.

Rock crevices at high altitudes in the Olympic Mts. of Wash. July-Aug.

Campanula prenanthoides Dur. Journ. Acad. Phila. II. 2:93. 1855.
   Asyneuma prenanthoides McVaugh, Bartonia 23:36. 1945. (Pratt, Calif.)
   Perennial from an apparent taproot, the clustered, erect stems often rhizomatous at the base, mostly 3-8 dm. tall; herbage glabrous or shortly spreading-hairy; leaves more or less sharply serrate with callous teeth, mostly 12-35 below the inflorescence, excluding the reduced and soon deciduous lowermost ones, sessile or subsessile, narrowly to broadly ovate or elliptic-ovate, mostly 1.5-3.5(6) cm. long, the upper reduced; inflorescence usually more or less pedunculate, falsely racemose or thyrsoid-racemose, often with several flowers at each node, and sometimes branched below; flowers similar to those of C. scouleri, but more numerous and with narrower, mostly narrowly lanceolate lobes, and the tube averaging a little short-

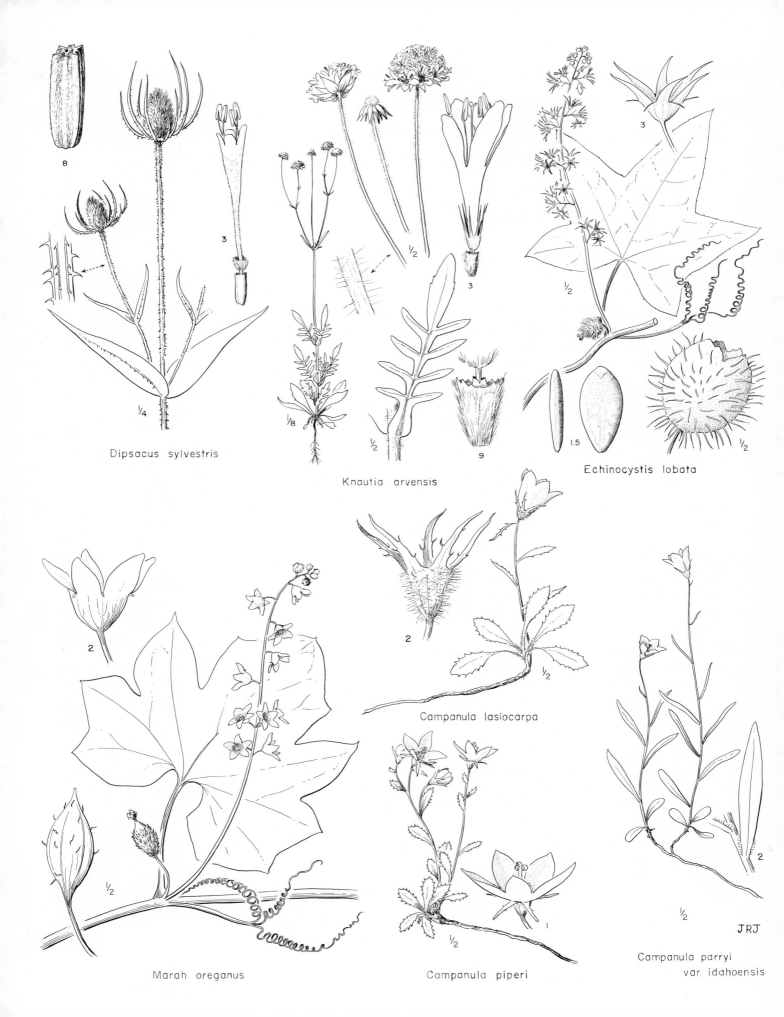

Dipsacus sylvestris

Knautia arvensis

Echinocystis lobata

Campanula lasiocarpa

Marah oreganus

Campanula piperi

Campanula parryi
var. idahoensis

JRJ

er (1-3 mm. long); capsules 3-5 mm. long, subglobose, tending to be retuse at the base, dehiscing at or below the middle.

Open woods; Polk Co., Oreg., s. to Calif. June-Aug.

## Campanula rotundifolia L. Sp. Pl. 163. 1753. (Europe)

C. petiolata A. DC. Monog. Campan. 278. 1830. C. rotundifolia var. petiolata Henry, Fl. So. B. C. 283. 1915. (Richardson, Slave Lake)

C. macdougalii Rydb. Bull. Torrey Club 28:25. 1901. (MacDougal 66, Priest Lake, Ida.) A shade form with relatively broad cauline leaves.

C. sacajaweana Peck, Proc. Biol. Soc. Wash. 50:123. 1937. (Peck 16549, Matterhorn, Wallowa Mts., Oreg.) A widely distributed dwarf alpine form which may prove to be a distinct ecotype.

Perennial with a slenderly branched caudex or system of rhizomes arising from an eventual taproot, 1-8 dm. tall, glabrous or inconspicuously hispidulous, the leaf margins smooth or with some short, stiff hairs up to 0.2 mm. long; basal leaves petiolate, sometimes with broadly ovate to subrotund or cordate-rotund, angular-toothed blade up to 2 cm. long, sometimes merely oblanceolate, often deciduous; cauline leaves more or less numerous, narrower and more elongate, commonly linear or nearly so, mostly 1.5-8 cm. long and seldom as much as 1 cm. wide; flowers typically several or rather numerous in a lax, racemiform or elongate-paniculiform, determinate inflorescence, solitary in depauperate or subalpine specimens; flowers erect or nodding; calyx lobes 4-12 mm. long, usually entire; corolla blue, 1.5-3 cm. long, the lobes much shorter than the tube; style equaling or generally shorter than the corolla; anthers 4.5-6.5 mm. long; capsule nodding, broadly obconic to narrowly cyathiform, 5-8 mm. long, dehiscing near the base. N=17, 34.

In a wide variety of habitats, commonly at lower elevations, sometimes subalpine; circumboreal, but not at high latitudes, extending s. in the mountains to Tex., n. Mex., and n. Calif.

Very easily cultivated, but apt to become somewhat weedy. June-Aug. or Sept.

## Campanula scabrella Engelm. Bot. Gaz. 6:237. 1881. (Mt. Scott, n. Calif.)

Perennial from a taproot and more or less slenderly branched caudex, minutely spreading-hirtellous throughout, the several stems up to 1 dm. tall; leaves entire, the basal ones oblanceolate, 0.5-4 cm. long, the cauline ones narrower and slightly to strongly reduced; flowers solitary, or sometimes 2-5, erect; calyx lobes 2-6 mm. long; corolla blue, 6-12 mm. long, the lobes from a little shorter to a little longer than the tube; anthers 3.5-5 mm. long; style about equaling the corolla; capsule cylindric-obconic, 5-7 mm. long, opening near the summit.

Talus slopes and other rocky places at high altitudes in the mountains; Cascade region, from c. Wash. to Mt. Scott, Calif.; also in c. Ida. and w. Mont. June-Aug.

## Campanula scouleri Hook. ex A. DC. Monog. Campan. 312. 1830.

C. scouleri var. hirsutula Hook. Fl. Bor. Am. 2:28. 1834. (Scouler, Ft. Vancouver on the Columbia) The α variety of Hooker.

C. scouleri var. glabra Hook. Fl. Bor. Am. 2:28. 1834. (Specimens from the "Northwest Coast" and "Fort Vancouver," collected by Douglas and by Scouler, are cited)

Perennial from branching, slender rhizomes, lax, often curved at the base, mostly 1-4 dm. tall, glabrous or inconspicuously short-hairy; leaves sharply serrate with callous teeth, the lowermost mostly with ovate or rotund-ovate (sometimes narrower) blade 1-4 cm. long borne on a petiole of nearly equal or greater length, the others progressively narrower, more elongate, and less petiolate, passing more or less abruptly into the reduced, sessile, linear bracts of the false lax racemes; usually 4-10 leaves below the inflorescence; flowers ordinarily several, on slender, often elongate pedicels, relatively small, the pale blue corolla mostly 8-12 mm. long, with ovate-oblong, more or less recurved-spreading lobes that are longer than (or rarely merely equal to) the tube; style well exserted; capsule 3-6 mm. long, subglobose, broadly rounded or subtruncate at the base, dehiscing near the middle.

Open or dense woods, less often on rock outcrops or talus; Alaska panhandle to n. Calif., mostly w. of the summits of the Cascade Range, rarely extending onto their e. slope, at elevations up to about 4000 ft. June-Aug.

Campanula uniflora L. Sp. Pl. 163. 1753. (Lapland)

Perennial with a taproot and slenderly branched caudex, the several stems lax, seldom over 1 dm. tall; plants glabrous except sometimes for a few long, loose hairs especially on the hypanthium; leaves entire or obscurely callous-toothed, the basal ones oblanceolate or broader, 1-3 cm. long including the petiole, the cauline ones sometimes much reduced, sometimes fully as large, becoming sessile; flowers solitary, erect; calyx lobes narrow, 2-5 mm. long; corolla blue, 6-12 mm. long, evidently surpassing the style, the lobes equaling or longer than the tube; anthers 1-2.5 mm. long; capsule elongate, tapering to the base, 12-20 mm. long, opening near the summit.

Rocky or grassy places at high altitudes in the mountains; circumboreal, extending s. in the Rocky Mts. to Colo.; in our area known only from Mont. July-Aug.

## Downingia Torr. Nom. Conserv.

Flowers sessile, solitary in the axils of the middle and upper leaves (or forming a leafy-bracted spike); corolla bilabiate, inverted so that the 3-lobed, morphological upper lip is on the lower or abaxial side; filaments and anthers connate, two of the anthers shorter than the others; ovary 1-locular or 2-locular, with correspondingly parietal or axile placentation; stigma 2-lobed; fruit elongate, slender, subcylindric, opening elastically by 3-5 longitudinal slits which extend nearly the length of the capsule; seeds numerous; alternate-leaved, somewhat succulent annuals.

About a dozen species, native to w. N. Am., especially Calif., one Calif. species occurring also in Chile. (Named for A. J. Downing, 1815-52, American horticulturist.)

The plants are beautiful when blooming en masse, and they are worth a trial in wet spots in the native garden.

Reference:

McVaugh, Rogers. A monograph on the genus Downingia. Mem. Torrey Club 19[4]:1-57. 1941.

1 Corollas inconspicuous, scarcely, if at all, surpassing the calyx, mostly 4-7 mm. long; capsule 2-locular, with axile placentation                                          D. LAETA

1 Corollas showy, mostly well surpassing the calyx, seldom less than 7 mm. long; capsule 1-locular, with parietal placentation

  2 Anther tube scarcely exserted from the corolla, not at all or only slightly incurved
                                                       D. YINA

  2 Anther tube evidently exserted from the corolla, more or less strongly incurved, commonly standing almost at right angles to the filament tube          D. ELEGANS

Downingia elegans (Dougl.) Torr. Bot. Wilkes Exp. 375. 1874.

  Clintonia elegans Dougl. ex Lindl. Bot. Reg. 15: pl. 1241. 1829. Bolelia elegans Greene, Pitt. 2:126. 1890. (Garden specimens from seeds collected by Douglas, "plains of the Columbia, near Wallawallah River, and near the head springs of the Multnomah")

  Clintonia corymbosa A. DC. Prodr. 7:347. 1838. Downingia corymbosa Nels. & Macbr. Bot. Gaz. 55:382. 1913. (Douglas, Columbia R.)

  Downingia brachypetala Gand. Bull. Soc. Bot. France 65:55. 1918. D. elegans var. brachypetala McVaugh, Mem. Torrey Club 19[4]:55. 1941. (Suksdorf 2762, Falcon Valley, Klickitat Co., Wash.)

  D. elegans f. rosea St. John, Res. Stud. State Coll. Wash. 1:105. 1929. (St. John 9627, Princeton, Latah Co., Ida.)

Erect or curved-ascending, simple or branched, fibrous-rooted annual commonly 1-5 dm. tall, glabrous except for the scabrous hypanthium; leaves sessile, mostly 0.5-2.5 cm. long, the lower narrow and often soon deciduous, the others generally broader, commonly lanceolate or lance-ovate, up to about 9 mm. wide; calyx lobes linear or narrowly elliptic, commonly 4-10(14) mm. long; corolla showy, blue to pink or white, with a yellow-ridged white eye, mostly 8-18 mm. long, the tube short and widely flaring so that the large (apparent) lower lip stands at a broad angle to the hypanthium without being evidently reflexed from the tube; anther tube evidently exserted, more or less strongly incurved, commonly standing almost at right angles to the 4.5-10.5 mm. filament tube; fruit 2-5 cm. long, 1-2 mm. thick, broadest near the base, 1-locular, with parietal placentae.

Vernal pools, wet meadows, and edges of ponds; Oreg. and n. Calif. to c. and e. Wash., n. Ida., and n. Nev. June-Aug.

Most of our plants belong to the var. elegans, as described above. Occasional specimens in our range, and many from farther south, have smaller flowers (corolla 5-9 mm. long; filament tube 3-4.5 mm. long), and in some of these the corolla tube is relatively better developed, as in D. yina. These small-flowered plants have been segregated as var. brachypetala (Gand.) McVaugh; as indicated by McVaugh, their origin and proper taxonomic status remain uncertain.

Downingia laeta Greene, Leafl. 2:45. 1910.
  Bolelia laeta Greene, Erythea 1:238. 1893. (Greene s.n., Humboldt Wells, Nev., July 6, 1893)
  B. brachyantha Rydb. Mem. N.Y. Bot. Gard. 1:483. 1900. Downingia brachyantha Nels. & Macbr. Bot. Gaz. 55:382. 1913. (Williams 712, Augusta, Mont.)
  Erect or recurved-ascending, simple or branched, fibrous-rooted, wholly glabrous annuals commonly 0.5-2 dm. tall; leaves sessile, mostly 0.5-2 cm. long, the lower narrow and often soon deciduous, the others generally broader, commonly lanceolate or lance-ovate, up to about 4 mm. wide; calyx lobes linear or linear-elliptic, mostly 3-8 mm. long; corolla inconspicuous, light blue or purplish, marked with white or yellow, mostly 4-7 mm. long, the tube relatively well developed, the lips not widely divergent; anther tube about 1.5-2 mm. long, slightly if at all incurved, its long axis nearly parallel with that of the 2-2.5 mm. filament tube; fruit 2-4.5 cm. long and 1-2 mm. thick, 2-locular, with axile placentae.

Marshes, wet (often alkaline) meadows, and edges of ponds; s.w. Sask. and w. Mont. to s.e. Oreg., n.e. Calif., c. Nev., c. Utah, and w. Wyo. June-July.

Downingia yina Appleg. Contr. Dudley Herb. 1:97. 1929. (Applegate 4479, Four Mile Lake, Klamath Co., Oreg.)
  D. willamettensis Peck, Proc. Biol. Soc. Wash. 47:187. 1934. D. yina var. major McVaugh, N. Am. Fl. 32A:24. 1943. (Peck 16291, Aumsville, Marion Co., Oreg.)
  Plants mostly 0.5-3 dm. tall, resembling the less robust forms of D. elegans in most respects; corolla mostly 7-12 mm. long, the well-developed tube 3.5-6 mm. long, the (apparent) lower lip rather sharply spreading or reflexed; anther tube usually partly included, only slightly or scarcely curved, its long axis nearly parallel with that of the 2-4.5 mm. filament tube.

Marshes, wet meadows, and edges of ponds; w. Wash., s. mostly in the Puget trough to the Klamath region of s.w. Oreg. and adj. Calif.; also in s.e. Oreg. Apr.-Aug.

Typical D. yina is restricted by McVaugh to a few stations in the Klamath region; the more common and widespread phase of the species is distinguished on minor technical characters as var. major McVaugh.

### Githopsis Nutt.

Flowers technically solitary and terminal, but appearing irregularly scattered because of the more or less sympodial branching of the stems; corolla regular, tubular-campanulate, the 5 lobes equaling or shorter than the tube; stamens free from the corolla and from each other, the filaments short, smooth, and only slightly (if at all) dilated at the base; ovary 3-locular, with axile placentation; stigma 3-lobed; capsule elongate-obconic, opening within the calyx by terminal pores; seeds numerous, angular; low, usually much-branched annuals with alternate, narrow, toothed leaves.

The genus probably consists of only the following species, although several segregates have been described. ("Named in allusion to the resemblance of the flowers with those of Githago segetum.")

Githopsis specularioides Nutt. Trans. Am. Phil. Soc. II. 8:258. 1842. (Nuttall, "Plains of the Oregon, near the outlet of the Wahlamet")

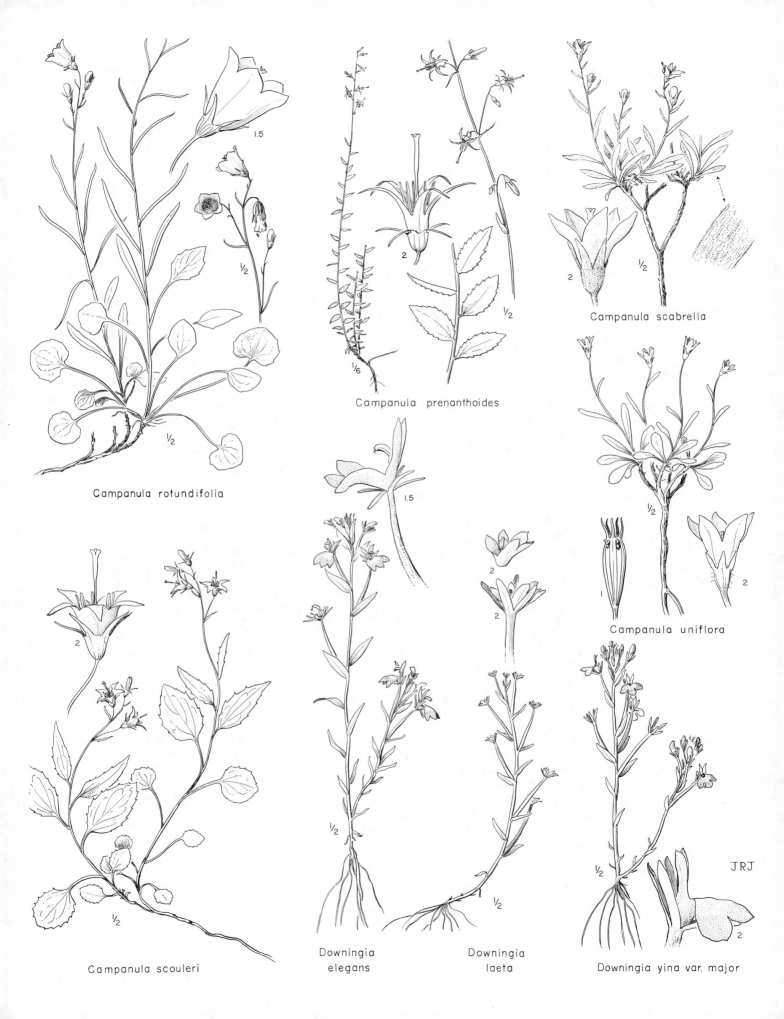

Campanula rotundifolia

Campanula prenanthoides

Campanula scabrella

Campanula uniflora

Campanula scouleri

Downingia elegans

Downingia laeta

Downingia yina var. major

JRJ

G. specularioides var. hirsuta Nutt. Trans. Am. Phil. Soc. II. 8:258. 1842. (Nuttall,
    "Plains of the Oregon, near the outlet of the Wahlamet")
    Plants spreading-hirtellous or glabrous, up to about 3 dm. tall; leaves oblong or narrower,
up to about 15 mm. long and 3 mm. wide, remotely serrate; calyx divided to the hypanthium,
the lobes elongate, linear or nearly so, commonly 5-15(20) mm. long; corollas blue with whitish throat, dimorphic, sometimes less than 1 cm. long and shorter than the calyx lobes, sometimes (especially to the s. of our range) much larger, up to 2 cm. long; capsules prominently
ribbed, 6-15 mm. long.
    Dry, open places in the valleys and foothills; both sides of the Cascades, from s. Wash. to
s. Calif. May-June.

Heterocodon Nutt.

    Flowers subsessile in a lax, sympodial false spike, borne opposite the bracts; lower flowers
cleistogamous, with reduced or abortive corolla; upper flowers with normal, regular corolla,
the 5 lobes shorter than the tube; stamens free from the corolla and from each other, the filaments short, ciliate at the expanded base; ovary inferior, 3-locular, with axile placentation;
stigma 3-lobed; capsule short and broad, opening tardily by inconspicuous irregular pores
near the base; seeds numerous, angular; slender annuals with alternate, sessile, toothed,
short and broad leaves.
    The genus consists of only the following species. (Name from the Greek heteros, different,
and kodon, bell, referring to the two kinds of flowers.)

Heterocodon rariflorum Nutt. Trans. Am. Phil. Soc. II. 8:255. 1842.
    Specularia rariflora McVaugh, Leafl. West. Bot. 3:48. 1941. (Nuttall, "grassy plains of the
    Wahlamet and Oregon")
    Lax, very slender, simple or sparingly branched annual, commonly 0.5-3 dm. tall; herbage
glabrous, or not infrequently hispid on the margins of the leaves and angles of the stem; leaves
somewhat clasping, distant, rotund or rotund-ovate, sharply toothed, small, seldom as much
as 1 cm. long; calyx divided to the hypanthium, the lobes foliaceous, veiny, ovate or broader,
2-4 mm. long; corollas of the upper flowers blue, 3-6 mm. long, the others abortive; hypanthium commonly spreading-hispid.
    Moist, open places in the foothills and valleys; s. B.C. and Ida. to Calif., Nev., and re-
putedly Wyo. June-Aug.

Howellia Gray

    Flowers axillary, pedicellate or subsessile, both petaliferous and apetalous, the corolla
when present bilabiate, with the tube deeply cleft dorsally; filaments and anthers connate, two
of the anthers shorter than the others; ovary 1-locular, with parietal placentation; stigma 2-
lobed; fruit irregularly dehiscent by the rupture of the very thin lateral walls; ovules few;
seeds large, up to 4 mm. long; annual aquatic plants, immersed or with floating branches.
    The genus as defined by McVaugh consists of only the following species. (Named for the
brothers Thomas and Joseph Howell, 1842-1912 and 1830-1912, respectively, who were among
the earliest resident botanists of the Pacific Northwest.)

Howellia aquatilis Gray, Proc. Am. Acad. 15:43. 1879. (Thomas & Joseph Howell 137, Sau-
    vies I., Multnomah Co., Oreg.)
    Plants rooted, naked below, branched above, the branches spreading or floating; whole plant
glabrous, green, 1-6 dm. long; leaves numerous, alternate, or some of them seemingly opposite or whorled in threes, flaccid, linear or linear-filiform, entire or nearly so, 1-4.5 cm.
long, up to 1.5 mm. wide; flowers mostly 3-10, axillary, often scattered, the stout pedicels
1-4(8) mm. long, merging gradually with the base of the capsule; calyx lobes 1.5-7 mm. long;
corolla about 2-2.7 mm. long, or wanting from the earlier flowers; fruit 5-13 mm. long, 1-
2 mm. thick; seeds about 5 or fewer, 2-4 mm. long, shiny brown.
    In ponds and lakes; n.w. Oreg., w. Wash., and n. Ida.; rarely collected. May-July.

## Lobelia L.

Flowers pedicellate, borne in terminal racemes, or solitary in the upper axils; corolla irregular, inverted so that the 3-lobed, morphological upper lip is on the lower or abaxial side; corolla tube dorsally split (between the two lobes of the actual upper lip) to below the middle, commonly nearly to the base, often fenestrate as well; filaments and anthers connate, two of the anthers shorter than the others; ovary 2-locular, with axile placentation; stigma 2-lobed; fruit fusiform to ellipsoid or globose, dehiscent near the apex; annual or perennial herbs, or shrubs, with alternate (or all basal), toothed or subentire leaves.

More than 300 species, of wide geographic distribution. (Named for Matthias de L'Obel, 1538-1616, Belgian botanist.)

1 Leaves flat, cauline as well as basal, only the basal ones sometimes immersed
                                                                 L. KALMII
1 Leaves terete, hollow, in a basal rosette, the stem essentially naked; aquatics with only the inflorescence ordinarily emergent                  L. DORTMANNA

**Lobelia dortmanna** L. Sp. Pl. 929. 1753.

   Lobelia lacustris Salisb. Prodr. 128. 1796. Dortmanna lacustris G. Don, Gen. Hist. Pl. 3: 715. 1834. Rapuntium dortmanna Presl, Prodr. Monog. Lob. 18. 1836. (Europe)

Fibrous-rooted aquatic perennial, glabrous throughout, the hollow, upright, mostly unbranched stem up to 1 m. tall; leaves in a basal rosette, linear, fleshy, hollow, somewhat falcate, 2-8 cm. long; cauline leaves few and inconspicuous, reduced to mere filiform bracts; raceme emergent, minutely bracteate; flowers few and well spaced; pedicels without bractlets; calyx lobes deltoid or narrower, blunt, 1.5-2.5 mm. long; corolla pale blue or white, 1-2 cm. long, the lower lip pubescent at the base and nearly equaling the tube, which is entire except for the deep dorsal fissure; capsule 5-10 mm. long and 3-5 mm. wide, the apex free from the hypanthium; seeds less than 1 mm. long, roughened, with a prominent square base at one end.

In shallow water at the margins of lakes and ponds; Newf. to Minn.; Vancouver I. and the adj. mainland in Wash. and B.C., s. to the Cascades of n. Oreg.; n.w. Europe. June-Aug.

**Lobelia kalmii** L. Sp. Pl. 930. 1753.

   Rapuntium kalmii Presl, Prodr. Monog. Lob. 23. 1836. Dortmanna kalmii Kuntze, Rev. Gen. 2:380. 1891. (Kalm, Can.)

   Lobelia kalmii var. strictiflora Rydb. Mem. N.Y. Bot. Gard. 1:378. 1900. L. strictiflora Lunell, Bull. Leeds Herb. 2:8. 1908. Petromarula strictiflora Nieuwl. & Lunell, Am. Midl. Nat. 5:13. 1917. (Scribner 130, Teton R., Mont.)

Slender, fibrous-rooted, simple or sometimes branched perennial, commonly 1-4 dm. tall, glabrous or inconspicuously pubescent; basal leaves spatulate, commonly 1-3 cm. long, often deciduous; cauline leaves 4-15, narrower, commonly linear or the lower oblanceolate, mostly 1-5(7) cm. long and 1-5(8) mm. wide; racemes loose and rather few-flowered; pedicels commonly bibracteolate near the middle; calyx lobes narrowly triangular, acute, mostly 2-4 mm. long; corolla mostly blue with a white or white and yellow eye, or sometimes wholly white, 7-13 mm. long, the lower lip glabrous and mostly longer than the tube, which is entire except for the dorsal fissure; capsule 4-8 mm. long and 3-4 mm. wide, the apex free from the hypanthium; seeds less than 1 mm. long, roughened, acute at both ends.

In marl or peat bogs, along shores, and in other wet places; Newf. to Great Slave Lake, Can., s. to Pa., Minn., Mont., and n.e. Wash. July-Aug.

## Porterella Torr.

Flowers axillary, pedicellate; corolla bilabiate, inverted, so that the 3-lobed, morphological upper lip is on the lower or abaxial side; filaments and anthers connate, two of the anthers shorter than the others; ovary 2-locular, with axile placentation; stigma 2-lobed; fruit narrow, more or less obconic, dehiscent near the apex; seeds numerous, smooth; alternate-leaved annuals with the aspect of Downingia.

A single species, named for Prof. T.C. Porter, American botanist, 1822-1901.

Porterella carnosula (H. & A.) Torr. Rep. U.S. Geol. Surv. Terr. 4:488. 1872.
  Lobelia carnosula H. & A. Bot. Beech. Voy. 362. 1838. Laurentia carnosula Benth. ex Gray,
    Bot. Calif. 1:444. 1876. (Tolmie, Blackfoot R., Snake country)
  Porterella eximia A. Nels. Bull. Torrey Club 27:270. 1900. Laurentia eximia Nels. in Coult.
    & Nels. New Man. Bot. Rocky Mts. 475. 1909. (Nelson 6544, Jackson Lake, Wyo.)
    Erect or curved-ascending, simple or branched, glabrous, somewhat succulent, fibrous-
rooted annual commonly 0.5-3 dm. tall; leaves sessile, mostly 0.5-2(3) cm. long, the lower
linear or nearly so and often soon deciduous, the others broader, commonly lanceolate, up to
about 4 mm. wide; flowers solitary in the axils of the middle and upper leaves, the pedicels
well developed, expanded gradually into the base of the capsule, becoming 0.5-2(3.5) cm. long
in fruit; calyx lobes linear or nearly so, commonly 3-8(11) mm. long; corolla very irregular,
blue with a yellow or whitish eye, 8-15 mm. long, the deeply 3-lobed lower lip equaling or
slightly longer than the tube; fruit commonly 6-13 mm. long and 2-3 mm. thick, narrowly obconic, the apex sometimes free from the hypanthium; seeds numerous, about 1 mm. long.
N reported by different authors as 11, 12.
    Wet meadows and edges of ponds; Yellowstone Nat. Pk., westward along the s. border of
our range to s.e. Oreg., and s. at increasing elevations to n. Ariz. and Tulare Co., Calif.
June-Aug.


                              Triodanis Raf.


    Flowers sessile or subsessile in the axils of the middle and upper leaves, 1-several in each
axil, forming a dense or interrupted false spike on a monopodial axis, blooming from the bottom upward, except that the terminal flower develops before those immediately beneath it; lower flowers cleistogamous, with reduced or abortive corolla; calyx lobes 5, or only 3 or 4 in
the cleistogamous flowers; upper flowers with normal, regular corolla, the 5 lobes longer than
the tube; stamens free from the corolla and from each other, the filaments short, expanded and
ciliate at the base; ovary mostly 3-locular, with axile placentation, varying to sometimes 1-
locular and with parietal placentation; stigma generally 3-lobed, sometimes 2-lobed; capsules
linear to ellipsoid or clavate, opening by rather regular pores at or usually above the middle;
seeds numerous, more or less lenticular; annuals with alternate, petiolate or more often sessile, short or elongate, toothed leaves.
    Eight species, native chiefly to N. Am., one species native to the Mediterranean region, and
two North American species occurring also in S. Am. (Name said by Rafinesque to refer to the
three unequal teeth of the calyx.)
  Reference:
  McVaugh, Rogers. The genus Triodanis Rafinesque, and its relationships to Specularia and
    Campanula. Wrightia 1:13-52. 1945.
1 Middle and upper leaves (floral bracts) lanceolate or narrower, mostly 5-10 times as long
    as wide; capsules of the cleistogamous flowers mostly 8-15 mm. long, 1-locular; capsules
    of the open flowers 1.5-2.5 cm. long, 1-locular or 2-locular; Mont. and eastward
                                                              T. LEPTOCARPA
1 Leaves (and floral bracts) broader, mostly rotund-ovate, seldom more than twice as long as
    wide; capsules of the cleistogamous flowers mostly 4-7 mm. long, those of the open flow-
    ers up to 1 cm. long, all 2- or 3-locular; throughout our range     T. PERFOLIATA

Triodanis leptocarpa (Nutt.) Nieuwl. Am. Midl. Nat. 3:192. 1914.
  Campylocera leptocarpa Nutt. Trans. Am. Phil. Soc. II. 8:257. 1842. Specularia leptocarpa
    Gray, Proc. Am. Acad. 11:82. 1876. Pentagonia leptocarpa Kuntze, Rev. Gen. 2:381.
    1891. Legouzia leptocarpa Britt. Mem. Torrey Club 5:309. 1894. (Ark.)
    Simple or basally branched annual mostly 1-5 dm. tall, often floriferous to near the base,
scabrous or shortly spreading-hairy below, less evidently so or becoming glabrous above;
leaves sessile, lanceolate or oblanceolate to lance-elliptic, or the upper (floral) ones linear,
inconspicuously crenate or subentire, mostly 1.5-3.5 cm. long and 2-7 mm. wide, mostly 5-
10 times as long as wide, or the lowermost ones sometimes short-petiolate and a little broader; calyx divided to the hypanthium, the lobes elongate and narrow, 6-15 mm. long, or smaller

in the cleistogamous flowers; corollas of the upper flowers blue-violet, 7-10 mm. long, the tube scarcely 2 mm. long; capsules linear, those of the cleistogamous flowers mostly 8-15 mm. long and 1-locular, those of the open flowers 15-25 mm. long and 1-locular or 2-locular; seeds 0.7-1.0 mm. long.

Open, often barren sites in the plains and foothills; Mont. to Minn., s. to Tex. and Ark. June-Aug.

Triodanis perfoliata (L.) Nieuwl. Am. Midl. Nat. 3:192. 1914.

Campanula perfoliata L. Sp. Pl. 164. 1753. C. amplexicaulis Michx. Fl. Bor. Am. 1:108. 1803. Prismatocarpus perfoliatus Sweet, Hort. Brit. 1st ed. 251. 1826. Specularia perfoliata A. DC. Monog. Campan. 351. 1830. Dysmicodon perfoliatum Nutt. Trans. Am. Phil. Soc. II. 8:256. 1842. Pentagonia perfoliata Kuntze, Rev. Gen. 2:381. 1891. Legouzia perfoliata Britt. Mem. Torrey Club 5:309. 1894. (N. Europe)

Erect annual commonly 1-6 dm. tall, simple or occasionally with a few long branches, often floriferous to near the base, more or less scabrous or spreading-hispid at least below; leaves (and floral bracts) sessile and cordate-clasping, rotund-ovate or broader, palmately veined, 0.5-3 cm. long and nearly as wide, or wider, or a few of the lowermost ones relatively narrower, more obovate, and sometimes short-petiolate; calyx divided to the hypanthium, the lobes narrowly triangular-acuminate, 5-8 mm. long, or smaller in cleistogamous flowers; corollas of the upper flowers deep purple to pale lavender, 8-13 mm. long, the tube about 2-4 mm.; capsules oblong or narrowly obovoid, those of the cleistogamous flowers mostly 4-7 mm. long, those of the open flowers up to 1 cm. long, all 2-locular or 3-locular; seeds 0.5-0.6 mm. long, smooth or roughened.

Various habitats, from the valleys and plains to moderate elevations in the mountains, often in disturbed soil; throughout the U.S. and adj. parts of Can. and Mex.; also in tropical Am., where perhaps only introduced, and introduced in Europe. May-Aug.

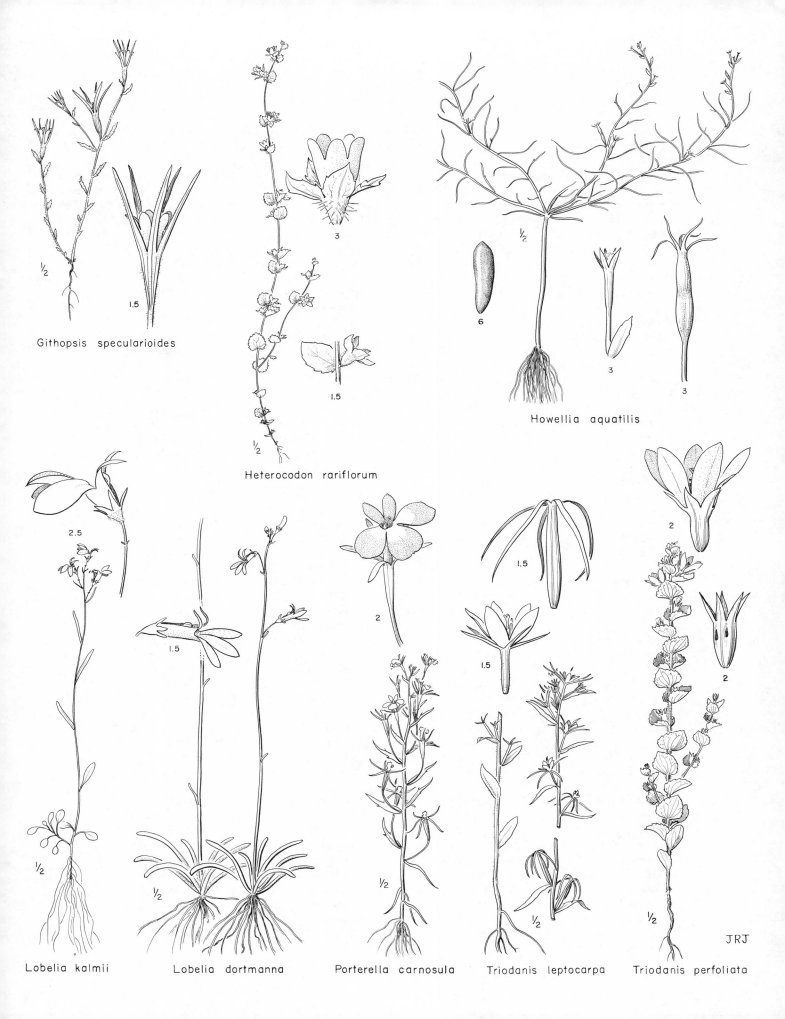

Githopsis specularioides

Heterocodon rariflorum

Howellia aquatilis

Lobelia kalmii

Lobelia dortmanna

Porterella carnosula

Triodanis leptocarpa

Triodanis perfoliata

JRJ

# Index and Partial List of Synonyms

Those names listed in the left-hand column which are followed by page reference in the right-hand column include well-known common names, accepted names of families and genera, and certain species incidentally mentioned but not formally described. Those names not followed by page citation are nonaccepted names (synonyms) and are followed by reference to the genus (and usually the species) under which they may be found in the text. It has not in general been considered necessary to include in this second category the synonymy for genera that have fewer than half a dozen species in our area unless it happens to be particularly involved. Accepted species will be found in alphabetical sequence under the genus.